CARBONATE CEMENTS

**BASED ON A SYMPOSIUM
SPONSORED BY THE SOCIETY OF
ECONOMIC PALEONTOLOGISTS AND
MINERALOGISTS**

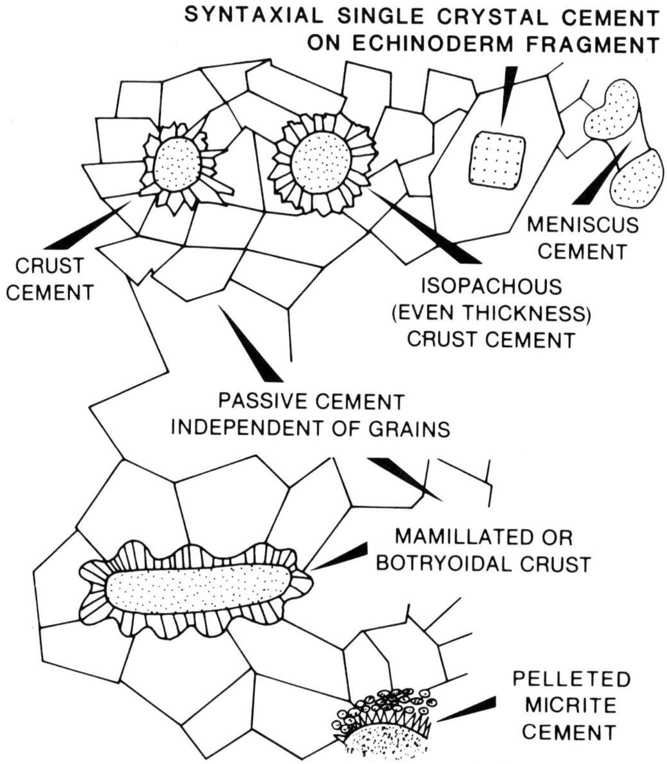

Edited by

Nahum Schneidermann and Paul M. Harris

Gulf Oil Exploration and Production Company, and

Gulf Research and Development Company

Houston, Texas

SOCIETY OF ECONOMIC PALEONTOLOGISTS AND MINERALOGISTS

R.L. Ethington, Editor of Special Publications

Tulsa, Oklahoma, U.S.A. ISBN 0-918985-37-4 *April 1985*

A PUBLICATION OF

THE SOCIETY OF ECONOMIC PALEONTOLOGISTS AND MINERALOGISTS

a division of

THE AMERICAN ASSOCIATION OF PETROLEUM GEOLOGISTS

PREFACE

In February 1983, a proposal was submitted to the SEPM Research Committee to organize a SEPM Research Symposium entitled "Carbonate Cements Revisited." Over twenty-seven tentative titles of papers had already been obtained and this initial response indicated the timeliness of the proposed symposium. Indeed, such a symposium was held during the 1983 AAPG/SEPM/EMD/DPA Annual Convention in Dallas. The program included, in addition to the standard morning and afternoon sessions, also a poster session and a special evening session held in conjunction with the SEPM Carbonate Research Group. This late-night session, attended by more than 150 people, attested to the success of the symposium; it was held in competition with a performance by the legendary Roy Clark.

The symposium program, chaired by the co-editors, included the following papers:

L.M. Walter: New Data on Relative Stability of Carbonate Minerals: Implications for Diagenesis and Cementation.

P.A. Sandberg: Evaluation of Ancient Aragonite Cements and Their Temporal Distribution.

C.T. Feazel, R.A. Schatzinger: Prevention of Carbonate Cementation in Petroleum Reservoirs.

I.G. Macintyre: Submarine Cements—The Peloidal Question.

B.J. Pierson, E.A. Shinn: Distribution and Preservation of Carbonate Cements in the Pleistocene Limestones of Hogsty Reef Atoll, Southeast Bahamas.

R.G. Lighty: Submarine Cementation Patterns of Holocene Reefs Provide Models for Porosity Development in Ancient Reef Reservoirs.

M. Esteban, F. Calvet: Cementation of Upper Miocene Reefs in Western Mediterranean.

R.A. Walls, G. Burrowes: The Role of Cementation in the Diagenetic History of Devonian Reefs, Western Canada.

B.H. Wilkinson, A.L. Smith, S.U. Janecke: Low Magnesium Sparry Calcite Marine Cements in Phanerozoic Hardground Grainstones.

R.L. Folk, H.S. Chafetz: Diagenetic Calcite Varieties from Travertines of Central Italy.

R.E. Woronick, L.S. Land: Burial Cements in the Lower Cretaceous Pearsall Formation and Lower Glen Rose Formations, South Texas.

R.G. Loucks: Distinguishing Diagenetic Environments of Equant Calcite Cementation: Example from Lower Cretaceous Pearsall Formation in South Texas.

D. Prezbindowski: Burial Cementation—Is it Important? A Case Study, Stuart City Trend, South Central Texas.

C.H. Moore, Y. Druckman, J.A.D. Dickson, A.H. Stueber: Jurassic Subsurface Calcite Cementation, Central Gulf of Mexico.

R.C. Burruss, K.R. Cercone, P.M. Harris: Regional Distribution of Hydrocarbon Fluid Inclusions in Carbonate Fracture Filling Cements: Geohistory Analysis and Timing of Oil Migration, Oman Foredeep.

K.C. Lohmann: Role of Fluid Inclusions in Diagenesis of Metastable Marine Cements.

A.C. Kendall: New Cements for Old Radiaxial Fibrous Calcites—A Reassessment.

P.A. Scholle, R.B. Halley: Burial Diagenesis in Carbonate Rocks.

H.S. Chafetz, B.H. Wilkinson: Freshwater Carbonate Cements.

G.M. Friedman: The Problem of Cements in Classifying Carbonate Rocks, Especially Reefs.

R.M. Mitterer, R. Cunningham, Jr.: The Interaction of Natural Organic Matter with Grain Surfaces: Implications for Calcium Carbonate Precipitation.

I. Lerche, C.G. Kendall, P.M. Harris: Controls on Carbonate Cementation and Solution.

C.H. Moore: Intertidal Cementation: Some Geochemical, Mineralogical and Petrographic Considerations.

W.J. Meyers, K.C. Lohmann: Geochemistry of Regionally Extensive Calcite Cement Zones, Mississippian of New Mexico.

J.E. Warme, N. Schneidermann: Patch-Reef Cementation: Holocene of Enewetak Atoll and Jurassic of Morocco.

As a result of the large audience participation and the generated interest in the topic, the SEPM Publications Committee and the Executive Council authorized publication of the proceedings of our research symposium as a SEPM Special Publication. For various reasons, some of the original speakers were unable to contribute to this volume; conversely, a few papers were added by contributors not on the original program.

The editors express their profound thanks to all contributors to this volume for their cooperation. Special thanks are due to the SEPM Carbonate Research Group and the Publication Committee. Gulf Oil Corporation permitted our participation in this endeavor, and allocated necessary secretarial help. Last, but not least, we thank our spouses for their understanding and encouragement.

N. SCHNEIDERMANN
P.M. HARRIS

INTRODUCTION

In the annals of carbonate sedimentology, few fields have undergone more discussion than the area of carbonate cements. From early discussions by Sorby and other European geologists, through such descriptions as provided by Cullis (1904), into the explosion of cement papers in the late fifties and sixties, carbonate cements have received disproportionate attention. This interest was further expanded with the discovery of a diversity of modern submarine cements (Friedman, 1964; Milliman, 1966; Cifelli *et al.*, 1966; and many others) culminating in the appearance of two landmark publications edited by Füchtbauer (1969) and Bricker (1971).

Why this interest? Studies of carbonate cements provide visual gratification to carbonate petrographers and mineralogists. Economic geologists exploring for minerals and hydrocarbons worry about their effect on porosity occlusion or preservation. Inorganic geochemists are now able to precipitate carbonate cements under controlled conditions, and organic geochemists can observe their interactions with living or fossil organic matter. Students of modern carbonate environments of deposition can observe almost instantaneous cementation processes in a diverse group of environments ranging from fresh water to the deep sea floor.

In the last two decades, substantial progress was achieved in recognizing and documenting modern and ancient carbonate cements. There is now abundant petrographic evidence permitting recognition and classification of modern cements and their ancient analogs, as well as their diagenetic alterations. Descriptions and discussions presented in textbooks by Carozzi (1960), Chilingar and others (1967), Cayeux (1970), Bathurst (1971), Milliman (1974), Wilson (1975), Friedman and Sanders (1978), and many others introduced new interpretations about carbonate cements to everyday use by students, researchers and other professionals. For example, Flügel (1982) lists nine categories of carbonate cement fabrics and about 20 cement type synonyms. With this extensive knowledge, a precise definition of environments of precipitation and diagenesis of carbonate cements is possible.

The SEPM Special Symposium was designed to bring together specialists working on modern marine and fresh-water cements, on their ancient analogs and their proposed relationship to burial conditions. Special emphasis was put on discussions of deep burial cementation environments or regional cementation trends as related to near-surface or subsurface conditions and on recognition of controls on carbonate cementation. This special publication is divided into four sections. In the first part, authors review geochemical and geological constraints on cement mineralogy, morphology and distribution through geologic time. In the second segment, early stable and metastable marine cements from a diverse group of environments and geologic ages are discussed. Examples of integrated geochemical and petrographic studies on regional distribution of burial cements and the transition from very early realms of cementation into deeper burial environments are presented in the third part. Examples of fresh-water cements, representing either the beginning or end of the carbonate diagenetic cycle are shown in the final part. Individual introductions to each of the sections will follow.

We began planning this symposium on the previously stated assumption that petrographic analysis of carbonate cements can provide us with a clear, concise interpretation of their genetic and diagenetic history. Indeed, several of the papers in this symposium indicate the necessity to continue and expand such optical observations. However, considerable evidence is also presented to indicate the complexity and pitfalls in such interpretations. New techniques, discussed in detail by several authors, are now able to provide researchers with a much better understanding of the regional distribution of the cements, of the nature of the brines from which they crystallized and the relative timing of their precipitation. Isotope and trace element analyses and liquid inclusion studies on carefully selected microsamples from cement zones defined by staining and cathodoluminescence seem to provide precise answers to the riddle of timing and sequence in cement precipitation, especially when such studies are combined with integrated regional analyses and burial histories. We hope that the future direction of carbonate cement studies will follow the lead presented by the authors participating in this volume.

N. SCHNEIDERMANN P.M. HARRIS

REFERENCES

BATHURST, R.G.C., 1971, Carbonate sediments and their diagenesis: Amsterdam, Elsevier, Dev. Sediment., v. 12, 620 p.

BRICKER, O.P. (ed.), 1971, Carbonate Cements: Johns Hopkins Univ. — Stud. Geol. no. 19, 376 p.

CAROZZI, A.V., 1960, Microscopic Sedimentary Petrography: New York-London, Wiley, 485 p.

CAYEUX, L., 1970, Sedimentary Rocks of France, Carbonate Rocks (Limestone and Dolomites) (translated and updated by A. V. Carozzi): Darien Hafner, 506 p.

CHILINGAR, G.V., BISSEL, H.J., FAIRBRIDGE, R.W. (eds.) 1967, Carbonate Rocks: Amsterdam, Elsevier, Dev. Sediment. v. 9 (A and B), 804 p.

CIFELLI, R., BOWEN, T.V., AND SIEVER, R., 1966, Cemented foraminiferal oozes from the mid-Atlantic ridge: Nature, v. 209, p. 32–34.

CULLIS, C.G., 1904, The mineralogical changes observed in the cores of the Funafuti borings, *in* T.G., Bonney (ed.), The Atoll of Funafuti: Rept. of the Coral Reef Committee of the Royal Soc. (London), p. 392-420.

FLÜGEL, E., 1982, Microfacies Analysis of Limestones (translated by K. Christenson): Berlin-Heidelberg-New York, Springer-Verlag, 633 p.

FRIEDMAN, G.M., 1964, Early diagenesis and lithification in carbonate sediments, Jour. Sed. Petrol., v. 34, p. 777–813.

______, AND SANDERS, J.E., 1978, Principles of Sedimentology: New York-London, Wiley, 792 p.

FÜCHTBAUER, H. (ed.), 1969, Lithification of Carbonate Sediments, Sedimentology, v. 12, 327 p.

MILLIMAN, J.D., 1966, Submarine lithification of carbonate sediments, Science, v. 153, p. 994–997.

______, 1974, Marine carbonates, Part 1: Berlin-Heidelberg-New York, Springer-Verlag, 375 p.

WILSON, J.L., 1975, Carbonate Facies in Geologic History: Berlin-Heidelberg-New York, Springer-Verlag, 471 p.

CONTENTS

SECTION ONE: INTRODUCTION

In this section, the authors review various geochemical and geological constraints on cement mineralogy, morphology and distribution through geologic time. Walters conducted experiments of dissolution rates of various biogenic carbonates and the effects of mineralogy, grain microstructure and solution saturation state on their stability. She found that in certain diagenetic settings microstructural detail may exert a more important control on the diagenetic sequence than mineral stability. The effect of organic molecules on solution and precipitation of carbonate cement, still poorly understood, is discussed in length by Mitterer and Cunningham. Their laboratory studies show that aspartic-rich organic matter affects the kinetics of calcium carbonate reactions by complexing calcium cations or adsorbing to mineral surfaces.

Sandberg discusses the distribution of aragonite, which is a major cement type in modern carbonates, in ancient limestones. His controversial conclusions on the validity of morphological observations on interpretation of ancient cements are well taken. An example of such often debated topic is presented by Kendall who chose various interpretations of cavity-filling radiaxial fibrous calcites as his topic of discussion. His new observations indicate that most of the features of this cement type, originally considered of neomorphic origin are primary and therefore, the interpretation that it replaces acicular carbonate cement is incorrect. The next paper in this section by Harris, Kendall and Lerche briefly summarizes the styles of carbonate cementation occurring in various diagenetic environments. Finally, Feazel and Schatzinger discuss a topic of utmost importance to economic geologists—the preservation of porosity in the subsurface and resultant reservoir quality. Their summary of the important mechanism includes minimal burial, reduced burial stress, increased framework rigidity, petroleum emplacement, stable mineralogy, permeability barriers and reopening of cemented porosity.

N.S., P.M.H

RELATIVE REACTIVITY OF SKELETAL CARBONATES DURING DISSOLUTION: IMPLICATIONS FOR DIAGENESIS

LYNN M. WALTER[1]
Rosenstiel School of Marine and Atmospheric Science, Division of Marine Geology and Geophysics, University of Miami, 4600 Rickenbacker Causeway, Miami, Florida 33149

ABSTRACT

An accurate knowledge of the potential reaction paths and sequences followed during early diagenesis is central to models of limestone textural development and the chemical evolution of rock-water systems. The lithification of chemically metastable carbonate sediments involves extensive dissolution of aragonite and magnesian calcite and the ultimate rock texture and chemistry is controlled to a large extent by the relative rates of these dissolution reactions. This paper summarizes results from a laboratory investigation of the dissolution rates of various biogenic carbonates and evaluates the roles mineralogy, grain microstructure, and solution saturation state play in determining the relative stability of aragonite and magnesian calcite during the dissolution phase of diagenesis.

Dissolution experiments were performed in seawater and meteoric type solutions using crushed samples of different skeletal grain types composed of calcite, aragonite, and magnesian calcite at a common grain size (37–125 microns). The results indicate that microstructural complexity can control the relative reactivities of carbonate grains and override differences in mineralogic stability. Aragonite grains having more complex microstructures can dissolve more rapidly than the reportedly less stable magnesian calcites inferring that the difference of their reactive surface areas apparently is greater than that of their thermodynamic stabilities.

The degree of microstructural control on dissolution rate is related to the saturation state of the solution. The dissolution rate data at various saturation states have been used to delineate regions of microstructural versus mineralogic control. In solutions undersaturated with respect to calcite, aragonites having complex microstructures can dissolve more rapidly than magnesian calcites. In solutions supersaturated with respect to calcite but still undersaturated with respect to aragonite, mineralogic control becomes more significant and magnesian calcites containing > 12 mole% $MgCO_3$ dissolve at rates similar to or greater than aragonites. In solutions supersaturated with respect to both calcite and aragonite, only magnesian calcites containing > 12 mole% $MgCO_3$ actively dissolve. Dissolution experiments using untreated natural carbonate sediments in solutions undersaturated with respect to calcite reveal that the amount of Sr^{2+} released during dissolution can only be accounted for by preferential dissolution of aragonites. Thus, the commonly accepted sequence of diagenetic alteration (magnesian calcites > aragonite > calcite) is only strictly observed in experiments conducted at or near aragonite saturation.

The composition of natural diagenetic fluids may sometimes fall below calcite and aragonite equilibrium, the region where grain microstructure has the greatest impact on relative dissolution rates. Because dissolution rates increase exponentially as saturation state decreases, more diagenetic "work" may be accomplished by short times spent in undersaturated solutions than long times spent closer to equilibrium. Therefore, the microstructural detail of component sediment grains may be a more important control on the sequence of diagenetic alteration than mineralogic stability in certain diagenetic settings.

INTRODUCTION

Shallow water carbonate sediments are very reactive chemically, and much of their diagenetic alteration and porosity development can occur prior to deep burial (Friedman, 1975; Longman, 1980). As a result, the modeling of textural and compositional trends in limestones relies on the accurate understanding of the rate and sequence of early diagenetic processes occurring in both marine and meteoric environments. The results presented in this paper will demonstrate that some commonly accepted assumptions concerning controls on early carbonate diagenetic reactions may be oversimplifications.

Most shallow marine carbonate sediments are derived from the skeletal breakdown of carbonate-secreting organisms. Sediment mineralogy is variable, but can be generally characterized as a mixture of the chemically metastable minerals aragonite and magnesian calcite. Magnesian calcite commonly comprises 15 to 50% by weight and has an average composition of about 14 mole% $MgCO_3$ (Bathurst, 1971). Diagenetic alteration can begin in the marine depositional environment. Although aragonite and magnesian calcite cementation is locally common, dissolution may also occur in undersaturated temperate ocean waters (Alexandersson, 1976, 1978) or where CO_2 generation

[1]Present address: Department of Earth and Planetary Sciences, Washington University, St. Louis, Missouri 63130

from decay of organic detritus decreases the pH of sediment pore waters (Aller, 1982).

Upon exposure of the sediment to meteoric water circulation, intense diagenetic alteration occurs. The overall diagenetic process in meteoric systems involves dissolution of magnesian calcites and aragonites and eventual replacement by stable low-Mg calcite cement.

Three fundamental chemical controls determine the path and extent of early diagenetic reactions; (1) the thermodynamic stability of the carbonate minerals being dissolved or precipitated, (2) the saturation state of the diagenetic fluid, and (3) the available surface area for reaction which is related to differences in microstructure among skeletal carbonate grain types. It has been generally assumed that the large difference in mineralogic stability between aragonite and magnesian calcites is the prime control of a sequence of diagenetic alteration which invariably leads to selective loss of the most unstable phases, the magnesian calcites.

Mineralogic stability doubtless plays a significant role in the relative rates of diagenetic alteration. However, the results of this study indicate that in some situations, solution saturation state and grain microstructure can also play important modifying roles on the relative reactivity of aragonite and magnesian calcites during diagenesis.

Previous Work on Carbonate Mineral Stabilities

Before presenting the experimental results and discussing their implications, it is important to review the current state of our knowledge on the most basic chemical control on diagenesis, the relative thermodynamic stabilities of calcite, aragonite and magnesian calcites. The thermodynamic stabilities of low-Mg calcite and aragonite are well known and generally agreed upon (Plummer and Busenberg, 1982, Sass *et al.*, 1983). The thermodynamic state of the magnesian calcites, however, is a topic of considerable dispute. Unlike aragonite and calcite, the stabilities of magnesian calcites cannot be measured directly because the magnesian calcite dissolution reaction becomes incongruent due to the precipitation of low-Mg calcite before equilibrium can be attained (Plummer and Mackenzie, 1974). Estimations of the thermodynamic stability of a given magnesian calcite composition vary by nearly a factor of 4 (Mackenzie *et al.*, 1983). The most current and complete set of magnesian calcite stability data is that of Plummer and Mackenzie (1974). Their results indicate greater instabilities for magnesian calcites than any other set of measurements found in the literature. Sample preparation may be responsible for some of the discrepancy among the results obtained in different studies (Walter and Morse, 1984a). Samples are generally crushed prior to use in most studies which can induce significant crystal strain and generate submicron particles. Both of these influences can lead to significantly greater chemical instability (Chave and Schmalz, 1966). Determinations of stabilities of 12 and 18 mole% magnesian calcites which were ultrasonically cleaned and annealed reveal that these solids may be nearly three times more stable than the results of Plummer and Mackenzie (1974) indicate (Walter and Morse, 1984a).

Thermodynamic stability constants currently used for low-Mg calcite, aragonite, and several magnesian calcite compositions are listed in Table 1, along with expressions for the ion activity product at equilibrium. The data of Plummer and Mackenzie (1974) are used for the magnesian calcites because their data set has received the widest degree of acceptance. However, it should be kept in mind that these numbers probably represent maximum estimates of the instability of magnesian calcites. The solubility difference between aragonite and calcite is calculated from the ratio of the numerical values of their equilibrium constants. Aragonite is about 1.5 times more soluble than pure calcite, regardless of the molar Mg:Ca ratio of the solution. However, magnesian calcite stability is re-

TABLE 1.—COMMONLY ACCEPTED EQUILIBRIUM CONSTANTS AND RELATIVE SOLUBILITIES FOR CARBONATE MINERALS

Mineralogy	Equilibrium constant*	Solubility Relative to Calcite (Mg:Ca = 5)	(Mg:Ca = 1)	(Mg:Ca = 0.2)
low-Mg calcite	$10^{-8.481} = a_{Ca} \times a_{CO3}$	1	1	1
aragonite	$10^{-8.301} = a_{Ca} \times a_{CO3}$	1.5	1.5	1.5
Mg-calcites:				
8 mole% $MgCO_3$	$10^{-8.21} = a_{Ca}^{.92}\, a_{Mg}^{.08}\, a_{CO3}$	1.6	1.9	2.2
12 mole% $MgCO_3$	$10^{-7.87} = a_{Ca}^{.88}\, a_{Mg}^{.12}\, a_{CO3}$	3.4	4.1	5.0
18 mole% $MgCO_3$	$10^{-7.43} = a_{Ca}^{.82}\, a_{Mg}^{.18}\, a_{CO3}$	8.4	11.2	15.0

*Equilibrium constant values for low-Mg calcite and aragonite are taken from Plummer and Busenberg (1982) and Sass *et al.* (1983). Values for magnesian calcite equilibrium constants are derived from data of Plummer and Mackenzie (1974) using the ion activity product at "stoichiometric saturation" proposed by Thorstensen and Plummer (1977).

lated to the molar Mg:Ca ratio in the solution. The following equation can be used to calculate the solubilities of the various magnesian calcites relative to pure calcite;

$$\frac{K_{\text{Mg-calcite (x)}}}{K_{\text{calcite}}} = \frac{a_{Ca}^{(1-x)}\, a_{Mg}^{x}\, a_{CO_3}}{a_{Ca}\, a_{CO_3}} \quad (1)$$

where x = mole % $MgCO_3$ and a_i = activity of ion. The desired relation for the solubility ratio of a given magnesian calcite to pure calcite follows from rearrangement of Eqn. 1;

$$\frac{a_{CO_3\ \text{equil. Mg-calcite}}}{a_{CO_3\ \text{equil. calcite}}} = \frac{K_{\text{Mg-calcite (x)}}}{K_{\text{calcite}}} \left(\frac{a_{Mg}}{a_{Ca}}\right)^{-x} \quad (2)$$

This equation expresses the difference in solubility as the ratio of the activity of carbonate ions in two different solutions having the same concentrations of dissolved calcium and magnesium, one solution in equilibrium with magnesian calcite, the other pure calcite. According to the data of Plummer and Mackenzie (1974) given in Table 1, magnesian calcites containing 7.5 mole% $MgCO_3$ are nearly equal in stability to aragonite. The solubility of magnesian calcites increases sharply with increasing mole% $MgCO_3$ and also increases as the solution Mg:Ca molar ratio decreases, as calculated by Eqn. 2 (Table 1).

Assuming the solubility data listed in Table 1 are correct, it is possible to estimate the relative dissolution rates of the three carbonate minerals as a function of solution pH. Dissolution rates are known to increase exponentially with increasing undersaturation with respect to the dissolving carbonate mineral (Berner and Morse, 1974; Walter and Hanor, 1979a; Kier, 1980). Figure 1 illustrates the predicted trends for carbonate mineral dissolution rates as a function of pH in a seawater solution assuming the different mineral types have equivalent reactive surface areas. The solubility differences listed in Table 1 predict that magnesian calcites should dissolve most rapidly at any given pH because they are always much farther from equilibrium than are aragonite or calcite. Therefore, the selective loss of magnesian calcites would be expected from sediments undergoing diagenesis.

Results from the field study of diagenetic alteration of metastable carbonates provide another measure of the relative chemical stabilities of different carbonate minerals. The selective loss of magnesian calcite has been documented from many field studies of meteoric Pleistocene limestone diagenesis (Friedman, 1964; Land *et al.*, 1967; Matthews, 1968; Gavish and Friedman, 1969; Plummer *et al.*, 1976). However, there are also meteoric diagenetic systems where selective dissolution of aragonite is observed (Müller and Tietz, 1975; Buddemeier, 1976; Walter and Hanor, 1979b; Have *et al.*, 1982), implying a different trend in the relative reactivities of magnesian calcite and aragonite. Neumann (1965) observed preferential dissolution of fine-grained magnesian calcites from marine sediments in Harrington Sound (Bermuda), but preferential dissolution of aragonite has also been reported in other marine environments. Preferential dissolution of aragonite occurred in experiments where permeable sacks containing magnesian calcite and aragonite ooids were suspended at various depths in the open ocean (Milliman, 1977). The selective loss of aragonite over magnesian calcites has also occurred in Red Sea sediments (Friedman, 1965). Berner *et al.* (1976) report equal dissolution rates of Mg-calcite and aragonite from Bermuda Pedastal sediments. Purdy (1968) noted a lack of mineralogic control over grain recrystallization and micritization in the shallow marine environment.

Are there any mechanisms which may be invoked to resolve the discordance between the relative reactivities of aragonite and magnesian calcites in nature and those predicted from thermodynamic data? Carbonate grains having more finely crystalline microstructures present a greater amount of surface area to the aqueous solution and may dissolve more rapidly than grains with smoother surfaces. It is possible that differences in the microstructural detail (available surface area) of carbonate grain types may reverse the relative reactivity predicted from existing thermodynamic data for the different carbonate minerals. For example, if a *Halimeda* grain (aragonite) had sufficiently greater available surface area for reaction than an echinoid grain (12 mole% $MgCO_3$), it could potentially dissolve more rapidly and override the factor of two difference in their solubilities predicted by the data

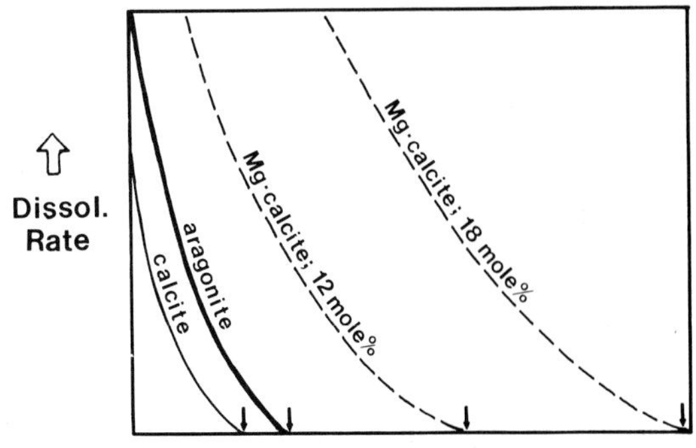

FIG. 1.—Schematic model for dissolution rates (moles per unit time) of carbonate minerals as a function of solution pH at equivalent reactive surface areas. Arrows denote the relative equilibrium pH values for each mineral based on the solubility data given in Table 1.

of Plummer and Mackenzie (1974). A second possibility is that the estimated Mg-calcite solubilities of Plummer and Mackenzie (1974) are too large and that the difference in thermodynamic stabilities between aragonite and magnesian calcites is smaller than their data would indicate (Walter and Morse, 1984a).

The influences of grain microstructure and saturation state on the relative dissolution rates of aragonite and magnesian calcites have not been evaluated in previous laboratory experiments. The purpose of the experiments described in this paper was to determine the relative dissolution rates of a number of biogenic carbonates characterized by different mineralogies and crystalline microstructures at known saturation states in solution. Although previous studies have been made of the dissolution behavior of carbonates (Chave *et al.*, 1962; Land, 1967; Weyl, 1967; Schmalz, 1967; Plummer and Mackenzie, 1974; Walter and Hanor, 1979a) most of these have used crushed materials containing fine particles and did not control the saturation state of the fluid. In the present study, care was taken to remove more reactive fine particles by ultrasonic cleaning. Dissolution experiments were also performed using natural uncrushed sediments as a cross check on the results obtained from the prepared skeletal carbonates. The amount of dissolved strontium released during natural sediment dissolution was used to trace dissolving phase mineralogy and to evaluate the potential for microstructural control over relative dissolution rates of aragonite and magnesian calcites.

METHODS

Determination of Dissolution Rates of Biogenic Carbonates

A group of common biogenic carbonates (magnesian calcites, aragonites, and a low-Mg calcite) were investigated in the dissolution experiments. Table 2 lists the type and composition of the carbonates, as determined by X-ray diffraction analysis. The $MgCO_3$ content of the magnesian calcites was determined by both X-ray diffraction and bulk chemical analysis. The mole% $MgCO_3$ of each magnesian calcite was determined by X-ray diffraction by the peak displacement from that of pure calcite using the data of Goldsmith *et al.* (1961). Bulk chemical analysis of mole% $MgCO_3$ from atomic absorption spectroscopy agreed well with the X-ray diffraction results for the echinoids and foraminifera (±0.5%). However, the red algae both had significantly higher mole% $MgCO_3$ values from bulk chemical analysis (close to 23 mole% $MgCO_3$) than that indicated from X-ray diffraction (18 mole% $MgCO_3$). This discrepancy may be related to the possible presence of brucite ($Mg(OH)_2$) in the red algae (Schmalz, 1965, Weber and Kaufman, 1965).

The solids were crushed to a 37–125 micron size range and ultrasonically cleaned to remove the abundant, adhering submicron particles and were annealed at 170°C for 6 hours to reduce the crystal strain induced by crushing. No discernable change in mineralogy occurred as a result of the preparation procedures. The removal of submicron particles is imperative in kinetic studies because their small size and great surface areas can lead to anomalous dissolution rates (Holdren and Berner, 1979; Holdren and Adams, 1982). Microstructural detail of each skeletal grain type was examined with a scanning electron microscope.

Experimental procedure is described in detail in Walter (1983) and will be briefly summarized here. The dissolution experiments were conducted in seawater (35‰ salinity) and a low salinity solution having a Mg:Ca molar ratio of 1, similar to many meteoric diagenetic fluids. The individual biogenic grain types were added to the solution and kept in suspension by an overhead stirrer. The dissolution rates were measured at different pH values (saturation states) in the solutions at 25°C and fixed P_{CO_2} ($10^{-2.5}$ atm.) using the pH-stat method (Morse, 1974). Acid of known normality (0.01 N HCl) was injected into the solution to maintain the pH at a constant value during the dissolution reaction. The calcium released during dissolution changed the dissolved calcium concentration by less than 2%, an insignificant increase. Because pH, calcium concentration, and P_{CO_2} were held constant during each experiment, the dissolution rate for each carbonate was measured at a constant degree of undersaturation. The dissolution rate was calculated from the rate of acid addition and its normality. The saturation state was calculated from the run pH value, P_{CO_2}, and-dissolved calcium concentration using appropriate equilibrium constants for the carbonic acid system and equilibrium constants for each carbonate mineral. Calculation procedures are described given in Walter (1983).

TABLE 2.—SOURCE AND MINERALOGY OF BIOGENIC CARBONATES USED IN THE DISSOLUTION EXPERIMENTS

Mineralogy	Source
Low-Mg calcite (<2.5 mole% $MgCO_3$)	barnacle (*Balanus*)
Aragonite	coral (*Fungia*)
	coral (*Acropora*)
	pelecypod (*Hippopus*)
	gastropod (*Strombus*)
	green algae (*Halimeda*)
Mg-calcite:	
11 mole% $MgCO_3$	echinoid (*Echinus*)
12 mole% $MgCO_3$	echinoid (*Clypeaster*)
12 mole% $MgCO_3$	echinoid (*Tripneustes*)
15 mole% $MgCO_3$	forams (*Peneroplis*)
18 mole% $MgCO_3$	red algae (*Goniolithon*)
18 mole% $MgCO_3$	red algae (*Neogoniolithon*)

The dissolution rates reported are those obtained after about 30 minutes of reaction. Initial rates are usually faster presumably because of selective dissolution of fine particles remaining after ultrasonic cleaning. The rates become linear after this period indicating that the bulk solid, and not more reactive fine particles is controlling the dissolution reaction. As mentioned, brucite ($Mg(OH)_2$) may exist as a discrete non-carbonate phase in the red algae (Schmalz, 1965; Weber and Kaufman, 1965). Because brucite is more undersaturated than the bulk magnesian calcite, it may be preferentially removed during the early stages of dissolution (Walter and Hanor, 1979a). In the present study, red algal dissolution rates dropped sharply during the first 45 minutes, but were linear therafter, even up to 6 hours duration, the longest experiment. As with the other solids, the reported rate is derived from this linear region. The rates as a function of saturation state are reproducible to ±10%.

The effect of sample treatment on dissolution rate was investigated. Ultrasonic cleaning reduced dissolution rates sharply, in some cases by as much as a factor of 100. Heating the solids to 170°C reduced dissolution rates by as much as 80%. In some cases, however, heating the material had no discernable effect on dissolution rate. This was generally observed for grain types such as *Halimeda* which were more easily crushed and so presumably suffered less crystal strain.

Dissolution Experiments on Natural Carbonate Sediments

Additonal pH-stat dissolution experiments were conducted in a strontium-free solution having a Mg:Ca molar ratio of 5 using three natural carbonate sediments. The dissolution experiments were conducted below calcite saturation (about 0.3 times calcite saturation) using the procedures described earlier. Less than 1% of the total sediment was dissolved in each experiment.

The sediments were sieved to a restricted size range to minimize potential grain size effects. Sediment mineralogy after sieving was determined by X-ray diffraction analysis using a standard curve for (calcite + magnesian calcite)/aragonite peak area ratios (weight %) as described by Milliman (1974). The modal mole% $MgCO_3$ in the magnesian calcite fraction was determined from the modal peak d-spacing using the data of Goldsmith *et al.* (1961). The sediments were examined petrographically to identify dominant grain types and estimate their relative abundance. The Sr/(Ca + Mg) ratio of each sediment was determined by atomic adsorption spectroscopy with a precision of ±5%. The results of the sediment analyses are given in Table 3.

The purpose of the sediment dissolution experiments was to identify dissolving phase composition by analysis of the amount of strontium released to the strontium-free solution during dissolution. Aragonites incorporate more strontium than do the magnesian calcites, with repective averages of 8500 ppm and 2000 ppm (Milliman, 1974). Therefore, the amount of strontium released to the solution during the dissolution of a known amount of sediment can be used to estimate the mineralogic composition of the dissolving phase. Strontium was chosen as a chemical tracer for dissolving phase composition because, unlike magnesium, it is generally released congruently (Amiel *et al.*, 1973).

The amount of standardized acid added and its normality were used to calculate the amount of sediment dissolved during the experiment. Because strontium is a trace constituent, the amount of acid added was used to calculate the total moles of calcium and magnesium released during dissolution. The solutions were filtered after the experiment and analyzed for dissolved Sr^{+2} using atomic absorption spectroscopy and matched matrix standards. A blank of the solution used in the experiment was analyzed to correct for possible strontium contamination. The Sr/(Ca + Mg) ratio of the

TABLE 3.—MINERALOGY, COMPOSITION, AND CHEMISTRY OF THREE SEDIMENT TYPES USED IN DISSOLUTION EXPERIMENTS

Sediment Source	Size range (microns)	*Mineralogy** (wt. .%) C	A	MC	mode mole% $MgCO_3$	Sr^{+2} (ppm)	dominant grain types
beach[1]	180-300	<1	55.0	45.0	15.5	4560	red algae = coral + *Halimeda*
lagoonal[2]	125-180	9	64.5	26.5	13.5	6075	*Halimeda* > peneroplid forams
reef[3]	125-180	<1	71.0	29.0	13.5	7660	coral > forams (*Homotrema*) = *Halimeda*

*C = low-Mg calcite, A = aragonite, MC = Mg - calcite
[1]Boiler Bay, St. Croix, U.S.V.I.
[2]Bahia Honda State Park, Big Pine Key, Florida
[3]Tague Bay, St. Croix, U.S.V.I.

dissolving phase was calculated from the atomic absorption analyses for Sr^{2+} and the total moles of carbonate sediment dissolved (Ca + Mg).

EFFECTS OF SKELETAL MICROSTRUCTURE AND SOLUTION SATURATION STATE ON DISSOLUTION RATES OF BIOGENIC CARBONATES

This section presents laboratory dissolution rate data for a variety of common skeletal carbonate grain types. The results of these experiments clarify the roles grain microstructure and solution saturation state play in determining the relative reactivity of natural carbonate minerals undergoing diagenetic alteration.

A complete listing of dissolution rates for each grain type as a function of saturation state is given in Walter (1983). The main question addressed here is how the relative reactivities of the different skeletal grain types is related to their microstructural complexity and to the saturation state in the solution. The microstructural detail of representative grain types is shown in SEM photomicrographs in Figure 2. Only one representative is shown for the corals (*Fungia*), echinoids (*Clypeaster*), and red algae (*Goniolithon*) because microstructural detail within each of these groups is similar. The gastropod and pelecypod grains also have similar microstructural detail, and only the pelecypod (*Hippopus*) is shown. The photomicrographs reveal that there are large differences in crystal size, arrangement and microporosity among the various grain types. Because of the severe restriction of fluid transport in and out of small pore throats and between crystallites of the grains, however, much of the internal surface area of these grains does not participate actively during dissolution (Walter and Morse, 1984b). The relative rankings of external surface roughness (available surface area) of the grain types depicted in Figure 2 estimated from visual examination would be : *Halimeda* > pelecypod = gastropod > peneroplid forams > red algae >= barnacle > echinoid.

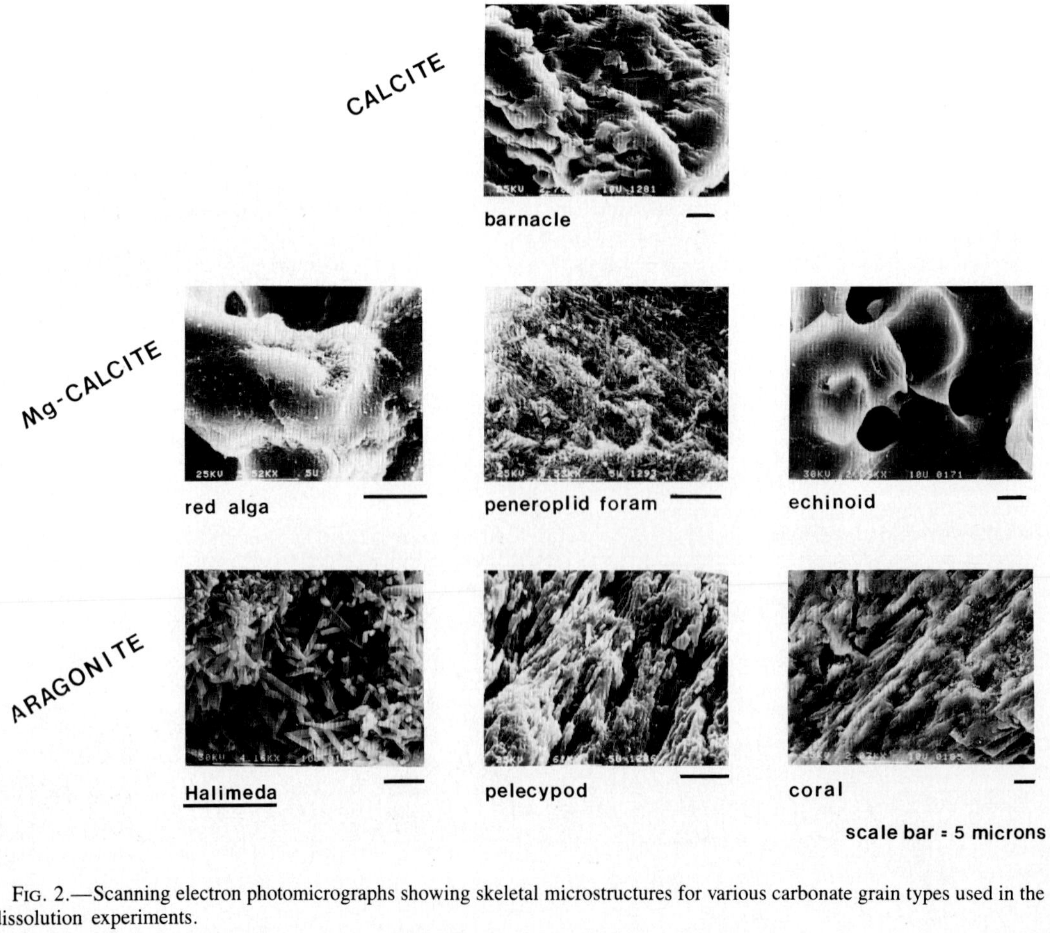

FIG. 2.—Scanning electron photomicrographs showing skeletal microstructures for various carbonate grain types used in the dissolution experiments.

The relative dissolution rates for the various grain types in seawater undersaturated with respect to calcite are shown in histogram format in Figure 3. The gastropod and pelecypod, red algae, corals, and echinoids all dissolved at about the same rate within each group. The ranking of the relative dissolution rates in seawater below calcite saturation is: *Halimeda* > pelecypod = gastropod > red algae > peneroplid forams > corals > barnacle > echinoids. This ranking does not correspond with mineralogic stability but, instead, is similar to that of microstructural detail. This indicates that differences in available surface area for reaction can be more significant in controlling relative dissolution rates than mineralogic stability. Thus, selective dissolution of many aragonitic skeletal grain types can occur in solutions undersaturated with respect to calcite.

A histogram of the relative reactivities observed in seawater supersaturated with respect to calcite but still below aragonite saturation (0.8 times undersaturated) is shown in Figure 4. Note that the absolute dissolution rates are roughly five times slower than those shown in Figure 3, which were obtained farther from equilibrium. The ranking of relative grain reactivities in Figure 4 is: red algae > peneroplid forams ≅ *Halimeda* > pelecypod = gastropod > corals > echinoids. Differences in mineralogic stability as well as microstructure are important controls on dissolution rates in solutions between aragonite and calcite saturation. Here, magnesian calcites containing 18 mole% $MgCO_3$ may dissolve more rapidly than aragonites. In general, however, the net difference in aragonite and magnesian calcite dissolution rates is not extreme.

In seawater supersaturated with respect to aragonite, only magnesian calcites containing > 12 mole% $MgCO_3$ actively dissolve. Echinoids (12 mole% $MgCO_3$) do not dissolve detectably in solutions supersaturated with respect to aragonite. This implies that the solubility in seawater of magnesian calcite containing 12 mole% $MgCO_3$ is similar to aragonite as shown by Walter and Morse (1984a).

The ranking of grain susceptibility to marine recrystallization observed in numerous shallow water carbonate environments by Purdy (1968) is: red algae > *Halimeda* > peneroplid forams > gastropod = pelecypod > corals > echinoids. This is very similar to the ranking observed in the present laboratory dissolu-

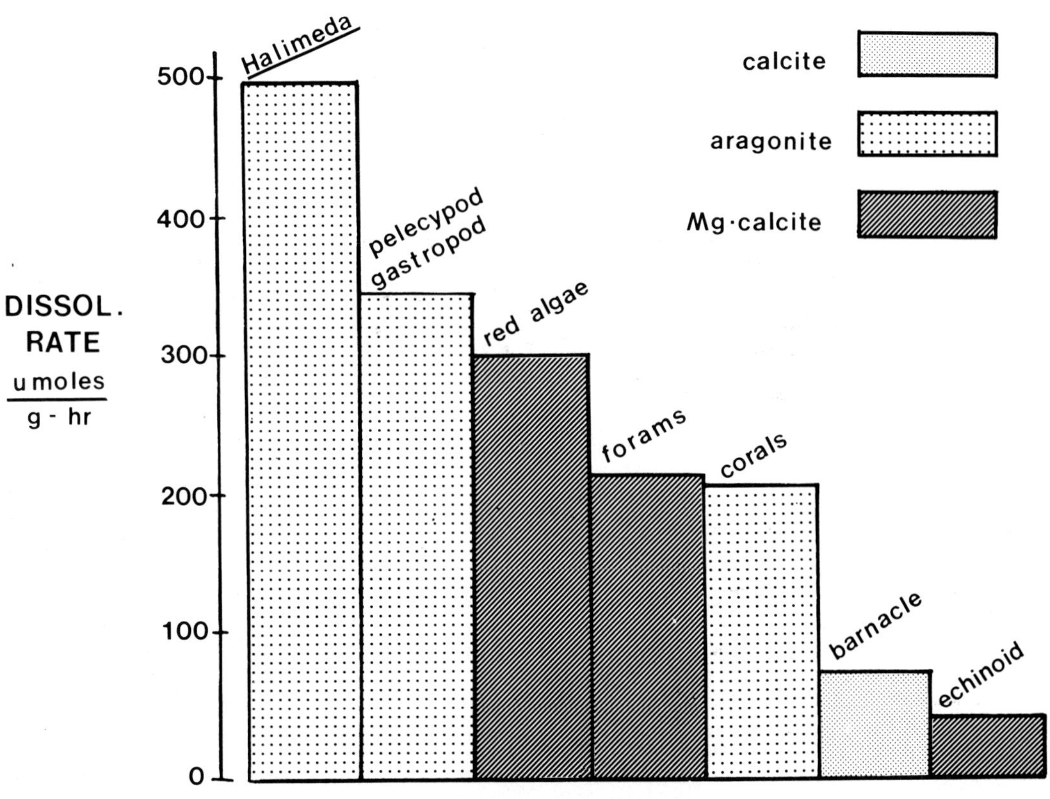

FIG. 3.—Histogram of relative dissolution rates of biogenic carbonates in seawater below calcite saturation (solution 0.5 times saturated with respect to calcite).

tion experiments in solutions supersaturated with respect to calcite but undersaturated with respect to aragonite. The "recrystallization" Purdy reports may result from physical processes (i.e. micritization by algal boring) rather than dissolution/precipitation reactions. Hence, the degree of recrystallization may be as much a function of the physical durability (i.e. low surface areas and less delicate microstructures) of skeletal grains as of their chemical stability. However, results of the present study demonstrate that grains which have greater apparent physical durability also have lower chemical reactivity. Therefore the distinction between chemical and physical processes occurring during recrystallization in marine environments (Purdy, 1968) is perhaps not possible.

The relative reactivity sequence below calcite saturation determined in a low salinity meteoric type solution having a Mg:Ca molar ratio of 1 is compared with the sequence observed in seawater in Table 4. No significant alteration in the reactivity sequence is observed in this meteoric solution although the magnesian calcites dissolve somewhat more rapidly relative to the aragonites as a result of their increased solubility in solutions of lower Mg:Ca molar ratio as predicted by Eqn. 2.

The potential for aragonites to dissolve at rates similar to or greater than magnesian calcites may not be

TABLE 4.—RELATIVE DISSOLUTION RATES IN SEAWATER VERSUS A LOW SALINITY SOLUTION HAVING A MG:CA MOLAR RATIO OF 1.0

grain type	relative rate* (seawater)	relative rate* (Mg:Ca = 1)
Halimeda	1.00	1.00
red algae	0.60	0.73
coral	0.42	0.46
echinoid	0.08	0.31

*Rate of *Halimeda* dissolution defined as 1.00 for both solutions.

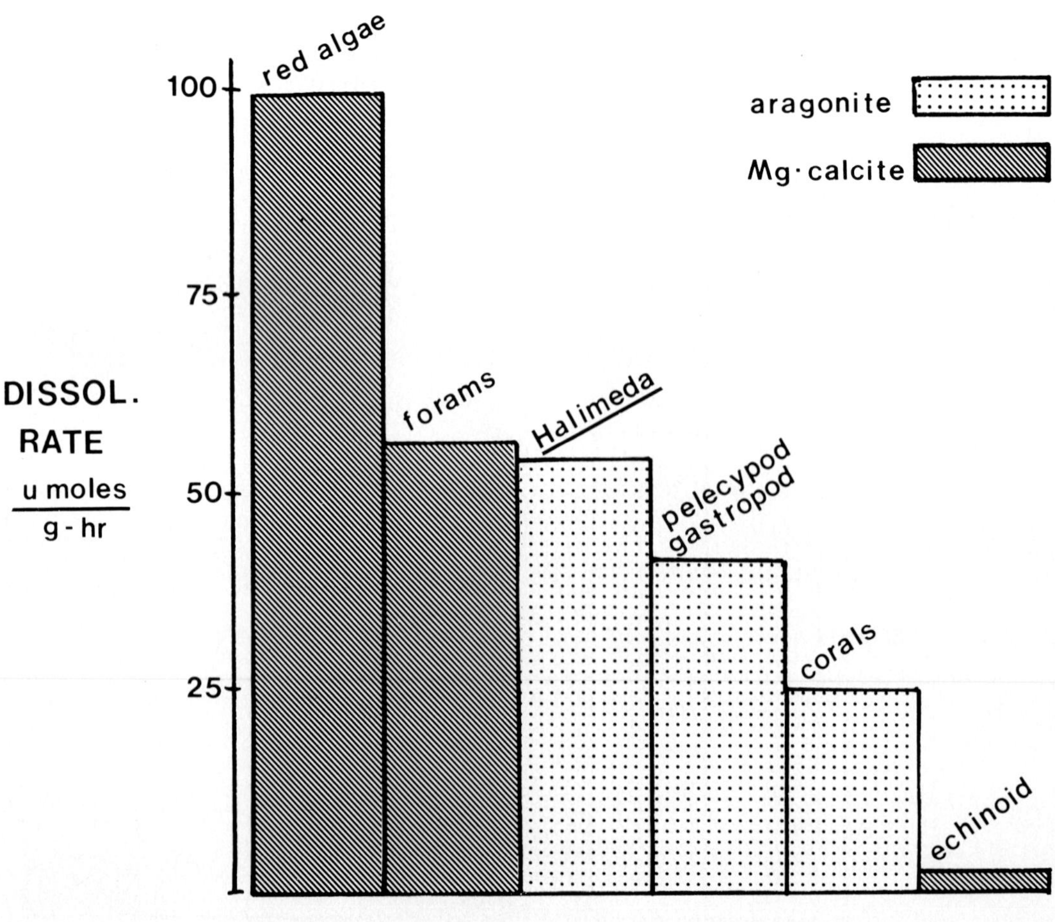

FIG. 4.—Histogram of relative dissolution rates of biogenic carbonates in seawater between calcite and aragonite saturation (solution 0.8 times saturated with respect to aragonite).

strictly a result of their greater surface areas available for reaction. As mentioned, the thermodynamic stability of magnesian calcites is not well known. Although the data of Plummer and Mackenzie (1974) predict that aragonite is equivalent in stability to a 7.5 mole% magnesian calcite, other results indicate that aragonite is equivalent in stability to 12 mole% magnesian calcite (Walter and Morse, 1984a). This paper addresses itself primarily to the relative reactivity of different carbonate phases, but the results support the strong possibility that the magnesian calcites may not be as thermodynamically unstable as was previously supposed. Thus, the more rapid rates of aragonite dissolution relative to magnesian calcite may be related not only to microstructural differences, but also to a smaller difference in thermodynamic stability between aragonite and magnesian calcite.

A model delineating different regions of dissolution behavior can be generated which is applicable to both marine and meteoric systems where metastable carbonate deposits are undergoing diagenetic alteration. This model delineates three regions of dissolution behavior based on the saturation state of the solution (Fig. 5). In solutions undersaturated with respect to calcite (Region 1), microstructural detail, rather than mineralogic stability, controls the relative dissolution rates of the skeletal grain types. Aragonites having complex microstructures may dissolve more rapidly than thermodynamically less stable magnesian calcites. Between calcite and aragonite saturation (Region 2), mineralogic stability becomes more important in controlling relative dissolution rates, although microstructure still exerts some influence. Here, magnesian calcites containing >12 mole% $MgCO_3$ may dissolve at similar or more rapid rates than aragonite. Above aragonite saturation (Region 3), only magnesian calcites containing > 12 mole% $MgCO_3$ actively dissolve. Thus, the relative reactivity of aragonite and magne-

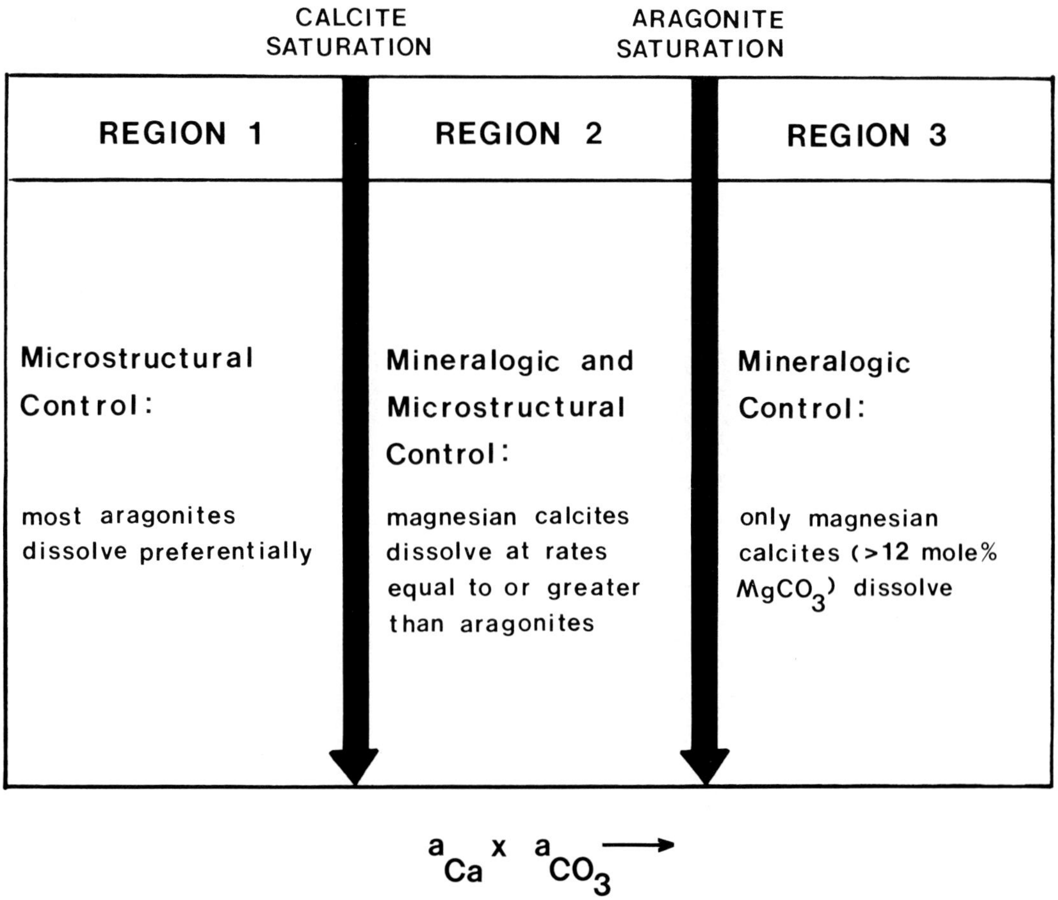

FIG. 5.—Regions of microstructural versus minerologic control during dissolution as a function of solution saturation state.

sian calcite appears to be related to of both grain microstructure and mineralogy, and the relative importance of these parameters is controlled by saturation state of the diagenetic fluid. The possible mineralogic trends predicted during diagenetic alteration to stable low-Mg calcite limestone in the three regions are shown schematically in Figure 6. Thus, the commonly accepted sequence of diagenetic alteration (magnesian calcite > aragonite > calcite) is only strictly observed in Region 3.

RESULTS FROM SEDIMENT DISSOLUTION EXPERIMENTS

The results of sediment dissolution experiments in strontium-free, calcite undersaturated solutions (Region 1) agree well with those from experiments using prepared individual biogenic carbonates. The strontium/(calcium + magnesium) ratios in the fluid after dissolution are compared with those of the initial bulk sediment in Table 5. In all cases, the dissolving phase is more strontium-rich than the bulk sediments, indicating preferential dissolution of aragonite. The weight% aragonite in the dissolving phase can be estimated using the relation:

$$\frac{\text{bulk sediment } Sr^{+2} \text{ (ppm)}}{\text{dissolving phase } Sr^{+2} \text{ (ppm)}} = \frac{\text{bulk sediment wt. \% aragonite}}{\text{dissolving phase wt. \% aragonite}} \quad (3)$$

The calculated aragonite weight % in the dissolving phase for each sediment is always significantly greater than that of the bulk sediment (Table 5). These experi-

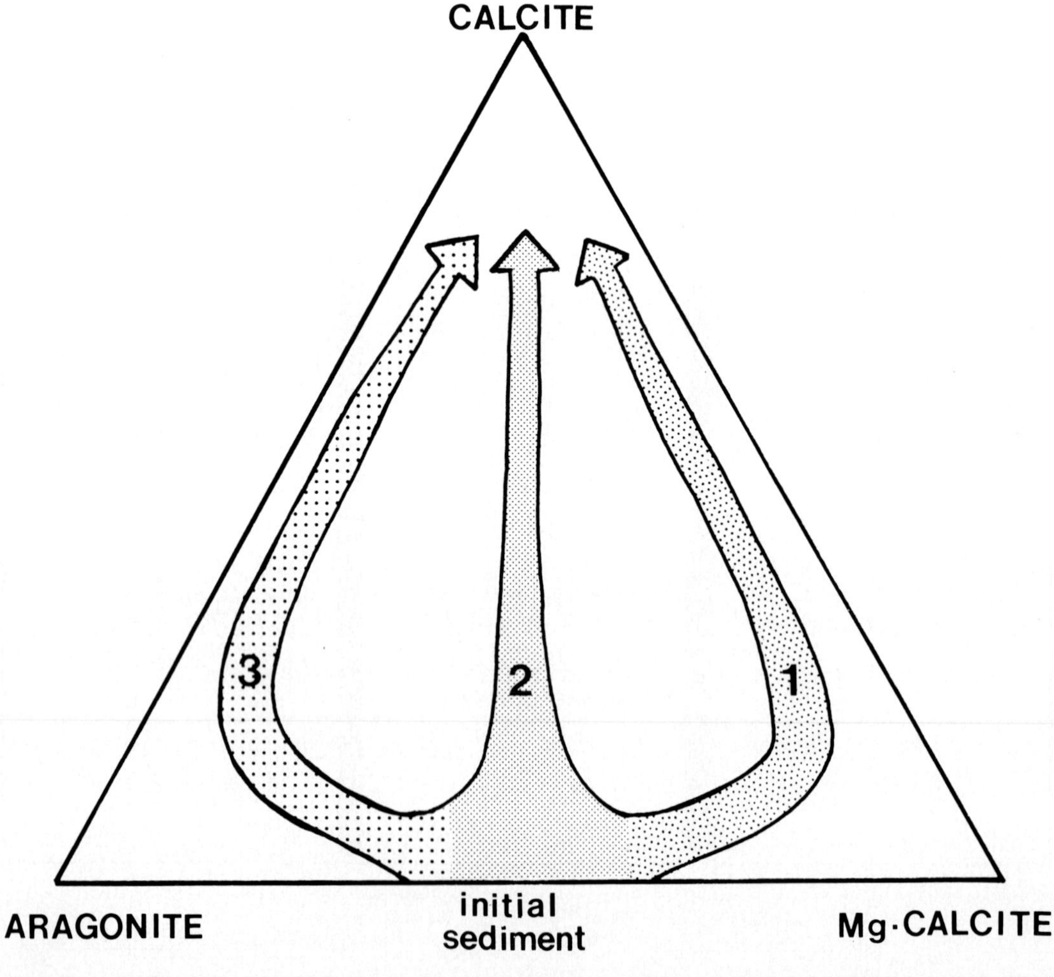

FIG. 6.—Possible compositional trends occurring during diagenesis in Region 1 (preferential dissolution of aragonite), Region 2 (approximately equal dissolution rates of aragonite and magnesian calcites), and Region 3 (preferential dissolution of magnesian calcites containing >12 mole% $MgCO_3$).

ments demonstrate that natural sediments behave in the same manner as the prepared carbonate grains and that selective dissolution of aragonite can occur in diagenetic systems having pore water chemistries below calcite saturation (Region 1).

Thorstensen *et al.* (1972) performed sediment dissolution experiments in distilled water using a natural carbonate sediment (50 wt.% magnesian calcite, 36 wt.% aragonite, 14 wt.% low-Mg calcite). The final Mg:Ca molar ratio in the experimental solution was used to determine to dissolving phase composition. The observed ratio was 1:8, indicating that magnesian calcites in the sediment accounted for 75% of the total carbonate dissolved. However, magnesium may not provide an unambiguous measure of dissolving phase composition. The laboratory study of Amiel *et al.* (1973) demonstrates that magnesium was preferentially released during coral dissolution. Solution Mg:Ca molar ratios of 1:4 to 1:10 were observed by Amiel *et al.* (1973) during the first weeks of their experiments. They attribute their results to incongruent release of trapped sea salts or adsorbed magnesium in nonlattice sites. Therefore, the Mg:Ca molar ratio of 1:8 observed by Thorstensen *et al.* (1972) could potentially reflect preferential dissolution of aragonites.

DISCUSSION

Marine and Meteoric Diagenesis

When metastable sediments are exposed to meteoric fluid circulation, extensive dissolution of aragonites and magnesian calcites occurs and is usually accompanied by precipitation of low-Mg calcite cement. The textural and compositional evolution of the limestone will be strongly influenced by the relative reactivity of the carbonate grains during dissolution and the relative rates of calcite nucleation and growth on grain surfaces. From the evidence presented in previous sections it is expected that grains with more complex microstructures will dissolve first and will probably be preferentially cemented by calcite because of the larger surface areas available for calcite nucleation. For example, Land (1966) observed that grains having complex microstructures such as *Halimeda*, red algae, and forams were usually the first grains to become cemented with calcite during the diagenesis of Bermuda limestones.

The saturation state of the diagenetic fluids also plays a significant role in determining the style of the diagenetic alteration. If the waters are near saturation with respect to aragonite (Region 3), magnesian calcites may be preferentially dissolved. However, the dissolution rate data for magnesian calcites indicates that only magnesian calcites containing >12 mole% $MgCO_3$ will dissolve appreciably in such a diagenetic system. In waters below aragonite saturation (Regions 1 and 2), grain microstructure can significantly affect the relative stability of aragonites and magnesian calcites. As long as the waters are supersaturated with respect to calcite (Region 2), it is likely that microstructural complexity will also control the relative growth rates of cement on different grain types.

Few data exist on the saturation state of pore waters in meteoric diagenetic systems to aid in the chemical classification of diagenetic environments. Plummer *et al.* (1976) report that in Bermudan ground waters, about half of the analyses indicate undersaturation with respect to calcite, placing many of these waters in Region 1. There is a variable, but usually low amount of magnesian calcite and aragonite remaining in these rocks, however. Analyses of saturation state are also available from pore waters in a carbonate beach sequence in St. Croix, U.S.V.I., containing aragonite and magnesian calcite and infiltrated by meteoric fluids. Here, saturation states are variable but about half of the analyses indicated undersaturation for aragonite, and even occasionally calcite undersaturation (Walter, unpublished data). Walter and Hanor (1979b) report preferential dissolution of aragonites from this beach sequence based on mineralogic trends and strontium/(magnesium + calcium) ratios in the beach pore waters. These trends were originally explained by selective inhibition of magnesian calcite dissolution by orthophosphate (Walter and Hanor, 1979b) but it has since been reported that aragonite dissolution rates are inhibited to the same degree as those of magnesian calcite dissolution (Walter and Morse, 1981). Thus, the compositional trends in the St. Croix diagenetic system may be more readily explained by the influence of grain microstructure and solution saturation state. Although much more work is required to define the saturation state of fluids in natural diagenetic systems and test the model proposed here, it is reasonable to expect that a substantial number may periodically fall in Regions 1 and 2.

The dynamic nature of diagenetic systems must also be considered. A pore water analysis from a given point in time may reflect the aftermath of diagenetic

TABLE 5.—COMPARISON OF BULK SEDIMENT COMPOSITION WITH COMPOSITION OF THE DISSOLVING PHASE IN DISSOLUTION EXPERIMENTS CONDUCTED BELOW CALCITE SATURATION (REGION 1)

Sediment type	Bulk Sr^{2+} (ppm) (sediment)	wt. % Aragonite (sediment)	Sr^{2+} (ppm) (dissolving phase)	Wt. % Aragonite (dissolving phase)
beach	4560	55.0	6412	77
lagoon	6075	64.5	8400	87
reef	5410	71.0	7660	100

reactions, and not necessarily provide information about the pore water composition that has had the largest diagenetic impact on the rock-water system. For example, seasonal rains may introduce large amounts of organic detritus into the system, leading to a rapid increase in pore water P_{CO_2} levels from bacterial respiration. This would create an initially large degree of undersaturation for the carbonate minerals in the sediment. It is reasonable to expect an initially rapid dissolution response from the sediment because dissolution rates increase exponentially with increasing undersaturation. Rapid dissolution would eventually drive the pore water system towards equilibrium. Therefore, although a pore water analysis might indicate the equilibrium point of the system, it may not necessarily reveal much about the saturation state where a significant portion of the diagenetic alteration took place. It is possible, then, that considerable diagenetic alteration occurs in Regions 1 and 2 in nature, regions where microstructural complexity can override the differences in mineralogic stability of the grains.

Given the dynamic nature of diagenetic systems, it is possible for a given ground water initially to dissolve aragonites preferentially (Region 1), then dissolve aragonites and magnesian calcites at approximately equal rates (Region 2) and, perhaps, become supersaturated with respect to aragonite (Region 3) and dissolve magnesian calcites only, although at very slow dissolution rates. Therefore, the sequence of diagenetic alteration in a given sediment may vary with time. As the water moves into Regions 2 and 3, calcite may precipitate on grain surfaces and its growth rate will probably be controlled by grain microstructure.

Diagenesis may also occur in buried shallow marine sediments where aerobic respiration increases P_{CO_2} values sufficiently to drive the pore water below saturation. Berner (1966) analysed the saturation state of pore waters from shallow water carbonate sediments and reports significant undersaturation with respect to aragonite and sometimes calcite. The same phenomenon has been observed in shell-bearing terrigeneous muds (Aller, 1982). Thus, as in meteoric environments, the potential exists for diagenetic alteration in Regions 1 and 2. Most mineralogic studies of shallow water carbonate cores have not found evidence for selective dissolution of magnesian calcites in marine systems (Pilkey, 1964; Berner, 1966; Rosenfeld, 1979). However, even extensive dissolution in marine sediments may not produce a notable change in sediment mineralogy because of the potentially similar rates of aragonite and magnesian calcite dissolution given the complex microstructures of many aragonites.

In addition to microstructure, grain size can also affect the surface area available for chemical reactions. Investigation of the laboratory dissolution rates of different biogenic carbonate grain size fractions shows that their dissolution rates follow a simple inverse radius (1/R) dependence (Walter and Morse, 1984b). For example, a carbonate grain having a 100 micron radius disolves at twice the rate as one having a 200 micron radius. However, the role of grain size may be reduced somewhat in natural diagenetic systems. The laboratory experiments were performed at low solid:solution ratios (<5 g l^{-1}) whereas solid:solution ratios in most diagenetic systems are much greater. In a grain supported framework, there is a strong feedback from grain size to fluid flow rates, which in turn affects the saturation state of the diagenetic fluid. Although smaller grains could potentially dissolve more rapidly, the reduced permeability in their vicinity may tend to reduce their effective dissolution rates. Thus, the effect of grain size on dissolution rates in natural systems may not be as significant as the effect of grain microstructure.

CONCLUSIONS

1. Laboratory dissolution experiments using skeletal carbonate grain types and natural sediments demonstrate that dissolution rates can be more strongly controlled by the microstructural detail (available surface area) of a grain than by its mineralogic stability. Aragonites having complex microstructures such as *Halimeda*, can dissolve more rapidly than thermodynamically less stable magnesian calcites (red algae and peneroplid forams). Microstructural complexity should also have significant effects on relative rates of cementation because the rate of precipitation reactions are also generally controlled by available surface area.
2. The degree of microstructural control over skeletal carbonate dissolution rates is related to the saturation state of the solution. In solutions undersaturated with respect to calcite (Region 1), most aragonites dissolve more rapidly than magnesian calcites because of their complex microstructures. In solutions between aragonite and calcite saturation (Region 2), mineralogic control becomes more significant and magnesian calcites dissolve at rates equal to or greater than aragonite rates. In solutions supersaturated with respect to aragonite (Region 3), only magnesian calcites containing >12 mole% $MgCO_3$ dissolve. Therefore, the commonly accepted sequence of diagenetic alteration (magnesian calcites $>$ aragonite $>$ low-magnesian calcite) is only strictly observed in solutions near or above saturation for aragonite. These data indicate that 12 mole% magnesian calcite appears to be similar in stability to aragonite, rather than the 7.5 mole% equivalence point predicted by Plummer and Mackenzie (1974).
3. Existing data on saturation states of pore waters from marine and meteoric environments indicate that many diagenetic systems periodically fall in Regions 1 and 2 where microstructural differences exert the greatest control on dissolution rate. Because natural diagenetic systems are dynamic, fluids may initially fall in Region 1 and gradually

evolve in composition towards Region 3. Dissolution rates increase exponentially with increasing undersaturation so it is likely that a considerable portion of diagenetic alteration may occur in Regions 1 and 2. In addition, calcite nucleation and growth in meteoric systems would be expected in water compositions falling in Region 3 because of the high degree of calcite supersaturation.

ACKNOWLEDGMENTS

This research was supported by the Marine Chemistry Section National Science Foundation Grant OCE 79-19242 (John W. Morse) as a portion of Walter's dissertation research.

Jeffrey S. Hanor, Mark W. Longman, and John W. Morse are thanked for critically reviewing this manuscript. Harold R. Wanless, Elizabeth Burton, and Jack Meeder also provided helpful comments and criticisms on an early draft of this work. Helpful discussions with L. Neil Plummer, Jane Schoonmaker and Fred T. MacKenzie are gratefully acknowledged. Charles Bukowski is thanked for his philosophical guidance.

REFERENCES

ALEXANDERSSON, E.T., 1976, Actual and anticipated petrographic effects of carbonate undersaturation in shallow seawater: Nature, v. 262, p. 653–657.

______, 1978, Destructive diagenesis of carbonate sediments in the eastern Skagerak, North Sea: Geology, v. 6, p. 324–327.

ALLER, R.C., 1982, Carbonate dissolution in nearshore terrigenous muds: the role of physical and biological reworking: Jour. Geol., v. 90, p. 79–95.

AMIEL, A.J., FRIEDMAN, G.M., AND MILLER, D.S., 1973, Distribution and nature of incorporation of trace elements in modern aragonitic corals: Sedimentology, v. 20, p. 47–64.

BATHURST, R.G.C., 1971, Carbonate Sediments and Their Diagenesis: Elsevier, New York, 658 p.

BERNER, R.A., 1966, Chemical diagenesis of some modern carbonate sediments: Amer. Jour. Sci., v. 264, p. 1–36.

______, AND MORSE, J.W., 1974, Dissolution kinetics of calcium carbonate in seawater. IV. Theory of calcite dissolution: Amer. Jour. Sci., v. 274, p. 108–134.

______, BERNER, E.K., AND KIER, R.S., 1976, Aragonite dissolution on the Bermuda Pedestal: its depth and geochemical significance: Earth and Planet. Sci. Lett., v. 30, p. 169–178.

BUDDEMEIER, R.W., 1976, The hydrogeochemistry of Enewetak Atoll: Annual Report, 1975-1976, for E.R.D.A. Contract E-(26-1)-641, Hawaii Institute of Geophysics.

CHAVE, K.E., DEFFEYES, K.S., WEYL, P.K., GARRELS, R.M., AND THOMPSON, M.E., 1962, Observations on the solubility of skeletal carbonates in aqueous solutions: Science, v. 137, p. 33–34.

______, AND SCHMALZ, R.F., 1966, Carbonate-seawater reactions: Geochim. Cosmochim. Acta, v. 30, p. 1037–1048.

FRIEDMAN, G.M., 1964, Early diagenesis and lithification in carbonate sediments: Jour. Sed. Petrology, v. 34, p. 777–813.

______, 1965, Occurrence and stability relationships of aragonite, high-magnesian calcite, and low-magnesian calcite under deep-sea conditions: Geol. Soc. America Bull., v. 76, p. 1191–1196.

______, 1975, The making and unmaking of limestones or the downs and ups of porosity: Jour. Sed. Petrology., v. 34, p. 777–813.

GAVISH, E., AND FRIEDMAN, G.M., 1969, Progressive diagenesis in Quarternary to Late Tertiary carbonate sediments: Sequence and time scale: Jour. Sed. Petrology, v. 39, p. 980–1006.

GOLDSMITH, J.R., GRAF, D.L., AND HEARD, H.C., 1961, Lattice constants of the calcium magnesium carbonates: Amer. Mineralogist, v. 46, p. 453–457.

HAVE, T.T., HEIJNEN, N., AND NICKEL, E., 1982, Alterations in guanophosphates and Mio-Pliocene carbonates of Table Mountain Santa Barbara, Curacao: Sed. Geol., v. 31, p. 141–165.

HOLDREN, G.R., AND BERNER, R.A., 1979, Mechanism of feldspar weathering-I. Experimental studies: Geochim. Cosmochim. Acta, v. 43, p. 1161–1171.

______, AND ADAMS, J.E., 1982, Parabolic dissolution kinetics of silicate minerals: An artifact of nonequilibrium precipitation processes?: Geology, v. 10, p. 186–190.

KIER, R.S., 1980, The dissolution kinetics of biogenic calcium carbonates in seawater: Geochim. Cosmochim. Acta, v. 44, p. 241–252.

LAND, L.S., 1966, Diagenesis of metastable skeletal carbonates: Ph.D. thesis, Marine Sci. Center, Lehigh Univ., Bethlehem, Pa., 141 p.

______, 1967, Diagenesis of skeletal carbonates: Jour. Sed. Petrology, v. 37, p. 914–930.

______, MACKENZIE, F.T., AND GOULD, S.J., 1967, Pleistocene history of Bermuda: Geol. Soc. America Bull., v. 78, p. 993–1006.

LONGMAN, M.W., 1980, Carbonate diagenetic textures from near surface diagenetic environments: Amer. Assoc. Petroleum Geologists Bull., v. 64, p. 461–487.

MACKENZIE, F.T., BISCHOFF, W.D., BISHOP, F.C., LOIJENS, J., SCHOONMAKER, J., AND WOLLAST, R., 1983, Magnesian calcites: low-temperature occurrence, solubility and solid solution behavior: Mineral. Soc. Amer. Short Course, "Carbonates, Mineralogy and Chemistry," R.J. Reeder, Organizer, *Reviews in Mineralogy*, v. 11, p. 97–144.

MATTHEWS, R.K., 1968, Carbonate diagenesis: Equilibration of sedimentary mineralogy to the subaerial environment; coral cap of Barbados, West Indies: Jour. Sed. Petrology, v. 38, p. 1110–1119.
MILLIMAN, J.D., 1974, Marine Carbonates: Recent Sedimentary Carbonates, Part 1, Springer-Verlag, New York, 375 p.
______, 1977, Dissolution of calcium carbonate in the Sargasso Sea (Northwest Atlantic), N.R. Anderson and A. Malahoff (eds.), *in* The Fate of Fossil Fuel CO_2 in the Oceans: Plenum Press, New York., pp. 641–654.
MORSE, J.W., 1974, Dissolution kinetics of calcium carbonate in seawater. III. A new method for the study of carbonate reaction kinetics: Amer. Jour. Sci., v. 274, p. 97–107.
MÜLLER, G., AND TIETZ, G., 1975, Transformation of carbonate sands into limestone and dolostone, Fuertaventura, Canary Islands, Spain: IXth International Congress of Sedimentology, Nice, France, p. 143–150.
NEUMANN, A.C., 1965, Processes of recent carbonate sedimentation in Harrington Sound, Bermuda: Bull. Marine Sci., v. 15, p. 987–1035.
PILKEY, O.H., 1964, Mineralogy of the fine fraction of certain carbonate cores: Bull. Marine Sci. Gulf and Caribbean, v. 14, p. 125–139.
PLUMMER, L.N., AND MACKENZIE, F.T., 1974, Predicting meneral solubility from rate data: Application to the magnesian calcites: Amer. Jour. Sci., v. 274, p. 61–83.
______, AND BUSENBERG, E., 1982, The solubilities of calcite, aragonite and vaterite in CO_2-H_2O solutions between 0 and 90°C, and an evaluation of the aqueous model for the system $CaCO_3$-CO_2-H_2O: Geochim. Cosmochim. Acta, v. 46, p. 1011–1040.
______,VACHER, H.L., MACKENZIE, F.R., BRICKER, O.P., AND LAND, L.S., 1976, Hydrogeochemistry of Bermuda: A case history of groundwater diagenesis of biocalcarenites: Geol. Soc. America Bull., v. 87, p. 1301–1316.
PURDY, E.D., 1968, Carbonate diagenesis: An environmental survey: Geol. Romana, v. 7, p. 183–228.
ROSENFELD, J.K.,. 1979, Interstitial water and sediment chemistry of two cores from Florida Bay: Jour. Sed. Petrology, v. 49, p. 989–994.
SASS, E., MORSE, J.W., AND MILLERO, F.J., 1983, Dependence of the values of calcite and aragonite thermodynamic solubility products on ionic models: Amer. Jour. Sci., v. 283, p. 218–229.
SCHMALZ, R.F., 1965, Brucite in carbonate secreted by the red alga, *Goniolithon* sp: Science, v. 149, p. 993–996.
______, 1967, Kinetics and diagenesis of carbonate sediments: Jour. Sed. Petrology, v. 37, p. 60–67.
THORSTENSEN, D.C., MACKENZIE, F.T., AND RISTVET, B.L., 1972, Experimental vadose and phreatic cementation of skeletal carbonate sand: Jour. Sed. Petrology, v. 42, p. 162–167.
______, AND PLUMMER, L.N., 1977, Equilibrium criteria for two-component solids reacting with fixed composition in an aqueous phase-example: The magnesian calcites: Amer. Jour. Sci., v. 277, p. 1202–1223.
WALTER, L.M., 1983, The dissolution kinetics of shallow water carbonate grain types: Effects of mineralogy, microstructure, and solution chemistry: Ph.D. thesis, Univ. Miami, Miami, Florida, 320 pp.
______, AND HANOR, J.S., 1979a, Effects of orthophosphate on the dissolution kinetics of biogenic magnesian calcites: Geochim. Cosmochim. Acta, v. 43, p. 1377–1385.
______, AND ______, 1979b, Orthophosphate: effect on the relative stability of aragonite and magnesian calcite during early diagenesis: Jour. Sed. Petrology, v. 49, p. 937-944.
______, AND MORSE, J.W. 1981, Inhibition of aragonite dissolution by orthophosphate [abs]: Geol. Soc. America Abstracts with Programs, v. 13, p. 380.
______, AND ______, 1984a, A re-evaluation of magnesian calcite stabilities: Geochim. Cosmochim. Acta, v. 48, p. 1059–1069.
______, AND ______, 1984b, Reactive surface area of skeletal carbonates during dissolution: Effect of grain size: Jour. Sed. Petrology, v. 54, (4), p. 1081–1090.
WEBER, J.N. AND KAUFMAN, J.W., 1965, Brucite in the calcareous algae *Goniolithon*: Science, v. 149, p. 996-997.
WEYL, P.K., 1967, The solution behavior of carbonate minerals in seawater: Proc. Int. Conf. on Tropical Oceanogr., Miami, Fla. (Univ. of Miami), p. 178-228.

THE INTERACTION OF NATURAL ORGANIC MATTER WITH GRAIN SURFACES: IMPLICATIONS FOR CALCIUM CARBONATE PRECIPITATION

RICHARD M. MITTERER AND ROBERT CUNNINGHAM, JR.
Program in Geosciences, The University of Texas at Dallas, Richardson, Texas 75080, and Exxon Production Research Company, Houston, Texas 75252

ABSTRACT

Seawater is a complex solution of inorganic ions and organic molecules in contact with solid phases. Because of its reactivity and sorptive properties, some of the organic matter (OM) will directly affect the kinetics of inorganic reactions by modifying the rates and perhaps the types of reactions occurring between the inorganic ions in solution and the solids. The interaction of natural OM with calcium carbonate systems occurs in two ways: (1) adsorption of the OM to calcium carbonate surfaces and (2) complexation or chelation of free cations by dissolved or adsorbed OM. Both processes involve polar functional groups on the organic molecules, with the carboxylate anion ($RCOO^-$) being the most likely interacting species, although other functional groups may also be important.

OM associated with a variety of skeletal and non-skeletal calcium carbonate, including skeletal organic matrix, OM within ooids, and OM extracted from carbonate grain surfaces, was studied for chemical characterization, adsorption phenomena, and cation-binding ability. Skeletal OM is largely protein while ooid and adsorbed OM are humic substances with proteinaceous components comprising about one-third or more of the composition. Aspartic acid is the most abundant amino acid in both skeletal protein and humic substances. Conversely, OM associated with non-carbonate sediments is poor in proteinaceous constituents and relatively depleted in aspartic acid.

Aspartic acid-rich protein and humic substances bind or complex with metal ions in proportion to the concentration of carboxyl groups present. Blockage of carboxyl groups to make them inactive destroys the ability of the OM to bind metal ions. Many, if not most, of the carboxyl groups available for metal ion complexation in both calcified protein and aspartic acid-rich humic substances are on aspartic acid. Thus, this amino acid provides a significant portion of the metal-binding ability of the different types of OM.

Aspartic acid-rich OM is preferentially adsorbed by calcite compared to quartz. Again, the carboxyl group is the likely function to be involved in this adsorption. Blockage of carboxyl groups significantly reduces the ability of humic substances to adsorb to mineral surfaces. The similarity in geometry, charge, and composition enables the carboxylate anion to substitute for the carbonate anion in complexing calcium ions or in adsorbing to calcite surfaces.

Competition between organic and inorganic ions for dissolved species and surface adsorption sites is driven by the requirement of the system to remain electroneutral. Concentration variations in dissolved organic and inorganic ions in the pore waters brought about by bioturbation and organic and inorganic diagenesis result in variations in the tendency of OM to affect the chemistry of the system. Most of the calcium carbonate formed in the marine environment consists of skeletal material. This calcium carbonate contains protenaceous OM that is thought to be involved in formation of the mineral phase (biological calcification). By analogy, naturally-occurring OM of somewhat similar composition and properties and with identical functional groups may also be involved in the precipitation of calcium carbonate in the sedimentary environment (geological calcification)

INTRODUCTION

An understanding of the association of organic matter (OM) with calcium carbonate is important in two different contexts. One of these involves the role of OM in nucleating and precipitating calcium carbonate, both skeletal and non-skeletal. The other context concerns the significance of carbonates as source rocks for hydrocarbons. Each of these aspects requires a thorough understanding of the organic geochemistry of carbonates beginning with modern sediments.

Compared to siliciclastic sediments, carbonates differ in their association with OM by (1) containing an internal organic matrix in skeletal and non-skeletal systems, and (2) by showing the effect of both the internal organic matrix and surface-adsorbed OM on carbonate reactions. Figure 1 summarizes the different types of OM associated with carbonate sediments.

Besides the usual complement of dissolved inorganic species and reactive sedimentary particles, seawater contains organic matter in the dissolved (DOM), particulate (POM), and living forms, with the DOM reservoir the largest. Concentrations of DOM are about 1 mg/l in surface seawater but are usually greater in shallow, highly productive coastal environments (Ogura, 1970; Fredericks and Sackett, 1970). Within the sediment, DOM in pore waters may be greater than 100 mg/1 (Krom and Sholkovitz, 1977). The DOM is very incompletely characterized and our knowledge and understanding of this fraction is much less in comparison to the inorganic system. A large variety of organic molecules are present in seawater, many in trace concentrations, ranging from simple molecules that can readily be identified to complex macromolecules that are only sketchily known. Most of the DOM in seawater and in pore water is a high molecular weight polymer variously called "gelbstoffe", yellow organic acids or humic substances. Because of the presence of abundant DOM in natural waters and because of the potential importance of the DOM in affecting calcium carbonate reactivity, an increasing number of analytical and experimental studies have sought to elucidate the role of dissolved and adsorbed OM on the calcium carbonate system in the marine environment.

Other substances besides OM are known to affect the rate of precipitation of calcium carbonate. The role of Mg^{2+} in inhibiting calcite precipitation (Pytkowicz, 1965; Berner, 1975) and the role of PO_4^{3-} in inhibiting both calcite and aragonite precipitation are well established (Berner *et al.*, 1978). The importance of these inorganic ions, individually or combined, on calcium carbonate precipitation may be greater or lesser than the effect of OM.

The interaction of DOM with calcium carbonate systems occurs in two ways: (1) adsorption of DOM to calcium carbonate surfaces, and (2) complexation or chelation of free cations by the dissolved or adsorbed OM. Both processes involve polar functional groups on the organic molecules, with the carboxylate anion ($RCOO^-$) being the most likely interacting species because of its charge and its similarity to the carbonate anion, although other functional groups may also be important. However, a detailed understanding of the DOM-calcium carbonate system is lacking due to a number of factors that are only poorly known, including: (1) the complex nature of the DOM, (2) the factors affecting or controlling the adsorption of DOM to mineral surfaces, (3) the composition of the adsorbed phase, and (4) the role of OM in nucleation of calcium carbonate after adsorption. This paper is a synthesis of studies on the chemical composition, adsorption, and cation-binding ability of natural organic matter from a variety of calcium carbonate, both skeletal and non-skeletal.

TYPES OF ORGANIC MATTER IN CARBONATE SEDIMENTS

WITHIN GRAINS
- **SKELETAL PROTEIN**
- **NON-SKELETAL OM** (MAY INCLUDE CELLULAR OM)

ON GRAIN SURFACES
- **ADSORBED** (MAY INCLUDE CELLULAR OM)

EXTRAGRANULAR
- **DISSOLVED**
- **PARTICULATE**
- **CELLULAR** (BACTERIAL, ALGAL, FUNGAL, ETC.)

FIG. 1.—The three major types of organic matter associated with calcium carbonate sediments. The first category is unique to calcium carbonate while the other two categories describe organic matter associated with any sediment.

CARBONATE-ORGANIC INTERACTION: EARLIER WORK

The importance of calcium carbonate-organic interaction was recognized in the 1960s through pioneering efforts proceeding along three lines. First, through the study of mineral-organic matrix relationships in mollusc shells, came the recognition that a proteinaceous matrix is intimately associated with skeletal carbonates. Organisms apparently utilize the protein membranes to nucleate the mineral phase and, probably, to establish the mineralogy and ultrastructural arrangement of the crystals. Second, as an outgrowth of the previous studies, the effect of some pure organic substances on calcium carbonate precipitation in the laboratory was investigated. Kitano and Hood (1965) and Kitano and Kanamori (1966) found that some pure organic compounds induced calcite (or high Mg-calcite) precipitation while other compounds caused aragonite to precipitate. In addition, they found that the rate of precipitation was influenced by the nature of the organic compound present. A third major work was the discovery that suspended carbonate minerals are coated with an organic membrane hindering the equilibration of the grain with seawater (Chave, 1965). Subsequent investigations by Chave and Suess (1967) and Suess (1970, 1973) demonstrated that carbonates adsorb organic coatings from seawater and that all carbonate grains in the shallow sedimentary environment are apparently coated with organic substances.

Since these pioneering efforts, a number of experimental and analytical investigations have further explored the calcium carbonate-organic system. DOM was found to adsorb onto calcium carbonate nuclei during laboratory precipitation (Chave and Suess, 1970; Otsuki and Wetzel, 1973). In addition, the DOM decreased the rate of precipitation of the mineral. In cases of high DOM concentration, precipitation did not occur in short experimental times of a few hours. Suess (1970) and Meyers and Quinn (1971, 1973) studied the adsorption of pure protein and fatty acid molecules onto calcite, which was found to adsorb an amount of OM equivalent to a monomolecular layer. Berner *et al.*, (1978) investigated the kinetics of inhibition of aragonite precipitation by a number of pure organic compounds and natural OM (humic substances). The polymeric humic substances strongly inhibited aragonite precipitation compared to many organic monomers, such as amino acids and fatty acids, which had little or no effect on the rate of precipitation. Comparable results were found for the organic-rich pore waters of marine sediments where high degrees of supersaturation persisted with no evidence of calcium carbonate precipitation (Berner *et al.*, 1970).

In addition to inhibiting equilibration, organic compounds may also play a more direct role in the precipitation of non-skeletal calcium carbonate. Ooids and marine cements are examples of non-skeletal calcium carbonate minerals forming in the marine environment, presumably in the presence of dissolved and adsorbed OM. Ooids, for example, are known to contain concentric laminations of OM alternating with layers of mineral. On the basis of the similar amino acid compositions between skeletal protein and the organic matrix within ooids, Mitterer (1968, 1972) suggested that the OM may also play an important role in the precipitation of the non-skeleton mineral phase. In support of this suggestion are the experimental results on the formation of radial ooids in the presence of humic substances (Suess and Fütterer, 1972; Davies *et al.*, 1978; and Ferguson *et al.*, 1978). In each study, ooids were formed containing organic membranes interspersed within the mineral layers. In the experiments by Davies *et al.* (1978) and Ferguson *et al.* (1978), the presence of compounds such as free amino acids, amino sugars and pure proteins did not result in the formation of radial ooids. Control experiments with strictly inorganic solutions also produced no radial ooids.

Based on these studies, it is evident that organic substances are ubiquitously associated with skeletal and non-skeletal calcium carbonate and that the OM probably exerts a strong influence on the rate of precipitation as well as on the mineralogy and fabric of calcium carbonate in organisms and in the laboratory.

COMPOSITION OF ORANIC MATTER ASSOCIATED WITH CALCIUM CARBONATE

Skeletal Organic Matter

Skeletal carbonates are characterized by the presence of an organic matrix that is largely protein in composition. Although the exact details are uncertain, calcification is thought to proceed through the interaction of the organic matrix with the dissolved inorganic constituents in the fluids of the organisms. (See Weiner *et al.*, 1983, for a recent review.) Considerable research has focused on the composition of the organic matrix. A possible difficulty with this approach is that the active calcifying substance may comprise only a portion of the organic matrix, and bulk analyses do not distinguish between calcifying and non-calcifying organic matrix. Nevertheless, bulk compositions may give clues to the calcification mechanism from patterns of protein compositions.

Analytical investigations of the nature and composition of the calcified OM in a variety of organisms, including molluscs, corals, serpulid worms, and calcareous algae, have shown that the protein is characterized by a predominance of aspartic acid (Fig. 2). The skeletal protein of corals, for example, contains from 25 per cent (scleractinians) to 75 per cent (alcyonarians) aspartic acid, while in serpulid worms and calcareous algae aspartic acid comprises about 30 per cent of the organic matrix (Mitterer, 1971, 1972, 1978). Figure 2 also illustrates the relative amino acid composition of "average" skeletal OM from a mixed skeletal sand. The aspartic acid content is about 35 per cent. The skeletal protein composition of mollusks has been examined in greatest detail. Although a variety of compositions have been reported, the amino acid analyses of the protein matrix are generally characterized by a predominance of aspartic acid (Degens, 1976). Weiner and Hood (1975) have established the presence of a repeating sequence of alternating residues of aspartic acid in the soluble skeletal protein from a number of molluscs.

Despite the general predominance of aspartic acid in the organic matrix of mollusc shells, the relative concentration of this amino acid is not uniformly high, and a wide range of values exists. For example, the skeletal protein in some molluscs contains as little as 4 per cent aspartic acid. The amount of protein within mollusc shells also varies over a wide range, with protein contents in excess of 2 per cent and as low as 0.1 per cent (Hare and Abelson, 1965). The protein content of the

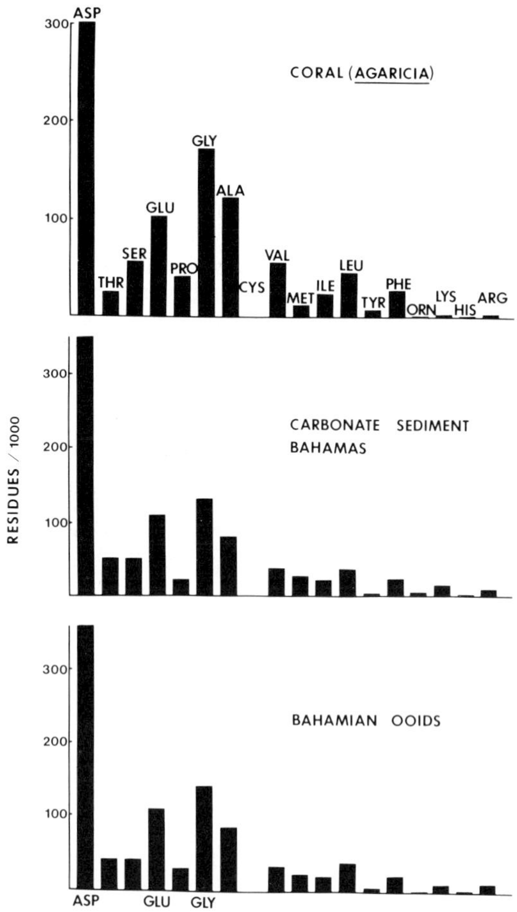

FIG. 2.—Amino acid composition of the organic matrix in various skeletal and non-skeletal calcium carbonate sediments. Relative concentration is in amino acid units (residues) per thousand units. Amino acid abbreviations are standard. Amino acids mentioned in text are: aspartic acid (asp), serine (ser), glutamic acid (glu), glycine (gly), and alanine (ala). Data from Mitterer (1972, 1976) and unpublished.

shell may be an indication of the efficiency of the calcification process. Organisms requiring a high protein content are relatively inefficient at calcification, while organisms with a low protein content have developed the ability to calcify more efficiently (Hare and Abelson, 1964). The protein content of the bivalve shell correlates inversely with the variations in aspartic acid. That is, organisms with a high skeletal protein content tend to have a relatively low aspartic acid concentration, while organisms with very little organic matrix tend to have a relatively high aspartic acid concentration (Fig. 3). Evidently, some organisms have evolved a more efficient calcifications ability by modification of the skeletal protein composition leading to an enhanced proportion of aspartic acid and a concomitant decrease in the amount of protein required for calcification. The change in composition could have occurred through evolutionary modification of the calcifying protein or by selected loss of non-calcifying proteins from the organic matrix. In either case, an increase in the relative concentration of aspartic acid results in an increase in the efficiency of calcification by bivalves.

Non-skeletal Organic Matter

Aspartic acid- rich OM is also associated with non-skeletal grains such as ooids (Fig. 2). Ooids from many localities exhibit the same aspartic acid predominance (Mitterer, 1968, 1972; Trichet, 1968; Mopper and Degens, 1972). Although many ooids contain algal and fungal colonies, not all ooids contain these "contaminants". In addition, microchemical techniques permit the analysis of only a few selected ooid grains, and microscopic observation of the dissolution process and the organic membranes released reveal only amorphous OM. Consequently, the amino acid analyses probably reflect the composition of the organic laminae and not cellular tissues.

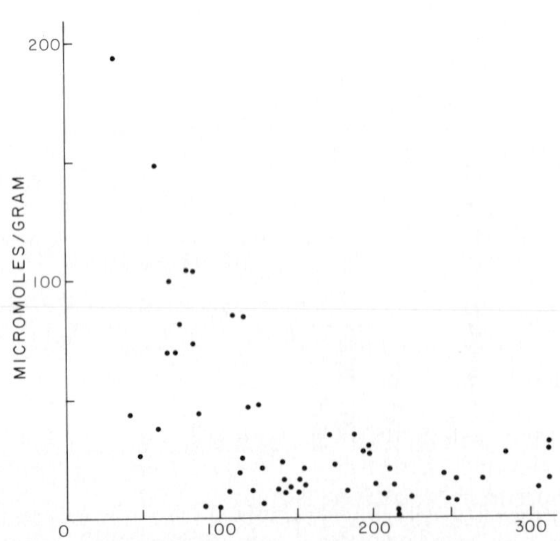

FIG. 3.—Relative aspartic acid content of the organic matrix in the shells of Bivalves as a function of the total amount of protein in micromoles of amino acid per gram of shell. Data from Degens and Spencer (1966).

OM associated with ooids must first be adsorbed to the grain before it can be incorporated within the laminae by subsequent precipitation of mineral. Therefore, the composition of OM adsorbed to carbonate surfaces in general should also be relatively high in aspartic acid if indeed the pattern established for ooid OM is meaningful. Aspartic acid-rich OM is present as an adsorbed phase on carbonate grains from Florida Bay (Mitterer and Carter, 1977; Carter and Mitterer, 1978; Cunningham, 1980). Müller and Suess (1977) also found evidence for aspartic acid enrichment in OM adsorbed to carbonate surfaces from a Pacific atoll. All of the analyses indicate that the adsorbed OM, just as with skeletal OM, has a somewhat variable composition and a range in relative aspartic acid concentration, although the adsorbed OM exhibits an enrichment in aspartic acid relative to the DOM and POM in the environment. Convincing evidence for this generalization has recently been obtained (Caughey, 1982). The amino acid composition of pore waters from carbonate sediments in Florida Bay exhibits a relative depletion in aspartic acid compared to the OM on the grain surfaces. Thus, there is a partitioning of the OM within the sediments with an aspartic acid-rich fraction preferentially adsorbed to grain surfaces and an aspartic acid-depleted fraction enriched in the pore waters.

Composition of Organic Matter in Non-carbonate Sediments and Seawater

Amino acid analyses have been conducted on non-carbonate sediments from a number of locations and a variety of marine environments ranging from coastal to deep sea. The OM associated with these sediments is characterized by a relatively low aspartic acid concentration, with other amino acids exhibiting a greater concentration. Figure 4 shows the amino acid composition for several examples of OM associated with non-carbonate sediments and seawater. Non-calcified sabellarid worm tubes provide an interesting comparison to calcified OM. Sabellarid worms live in agglutinated dwelling tubes formed by the cementation of local sediment with an organic mucus. Serpulid worms, conversely, live in dwelling tubes formed by the secretion of an organic mucus that undergoes calcification. The amino acid composition of the OM in serpulid worm tubes is characterized by the predominance of aspartic acid. The organic binder in the agglutinated tubes of sabellarid worms is depleted in aspartic acid and enriched in glycine and serine. These contrasting amino acid compositions between analogous structures of related organisms provide perhaps the best illustration of the differences between calcified and non-calcified OM.

Non-carbonate sediments and seawater also have high concentrations of glycine and other amino acids, with aspartic acid present in lesser abundance (Fig.4).

In addition, amino acids usually comprise a smaller proportion of the OM in non-carbonate sediments (Carter and Mitterer, 1978). Thus, carbonate sediments are not only enriched in aspartic acid but they are also enriched in amino acids compared to non-carbonate sediments.

Summary of the Analytical Results

The picture that emerges from the analytical investigations can be summarized as follows:

1) Carbonate sediments, both skeletal and non-skeletal, contain OM within and on the grains that is rich in amino acids, with aspartic acid the prevalent amino acid. Some of this OM is the calcified protein of the skeletal carbonates; part of the OM is adsorbed onto the grains from the seawater or porewater.

2) Non-carbonate sediments do not exhibit this amino acid enrichment or compositional pattern. Accordingly, the presence of aspartic acid-rich OM is a unique feature of calcium carbonate.

3) Neither the bulk composition of a variety of soft tissues nor the DOM in seawater is rich in aspartic acid. Thus, there is no uniform background of aspartic acid-rich OM in seawater.

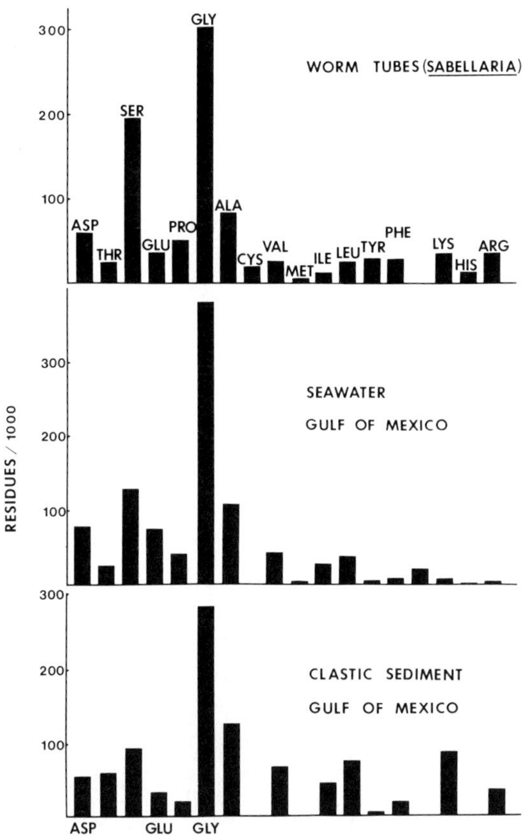

FIG. 4.—Relative amino acid composition of various non-carbonate sources. Abbreviations as in Figure 2. Data from Mitterer (1971), Carter and Mitterer (1978), and unpublished.

ADSORPTION EXPERIMENTS

Laboratory experiments can aid in understanding the factors affecting the association of an organic coating with carbonate grains in the marine environment. Experiments have been conducted using both pure organic compounds and natural OM obtained from sediments. Advantages and disadvantages exist for each type. Pure compounds have known compositions and functional groups and their adsorptive properties can be understood readily in terms of these characteristics. However, pure compounds generally are not very abundant in seawater and are not representative of the bulk of the DOM. The advantage of natural OM in adsorption experiments is that it is more realistic of the sedimentary environment; however, it is generally not well characterized either by composition or functional groups, and the processes of extraction and purification for experimental uses may result in changes that modify the adsorptive properties.

DOM consists largely of high molecular weight polymers (humic substances). Based on the isolation procedures used, the OM extractable from carbonate surfaces and that contained within ooids can also be classified as humic substances. The humic substances are further subdivided into humic acids and fulvic acids. Because of their abundance in the skeletal and non-skeletal organic matrix, our experimental studies have emphasized the amino acid portion of humic OM. Amino acids are present primarily in a polymerized, hydrolyzable form in natural OM rather than in the free state.

The amino acid composition of fulvic acids extracted from carbonate and non-carbonate sources is illustrated in Figure 5. The aspartic acid predominance typical of OM within carbonate grains is also present in OM adsorbed onto carbonate surfaces, whereas plant debris and non-carbonate sediment have a predominance of glycine.

Many studies have shown that the addition of powdered calcium carbonate to solutions containing either pure organic compounds or natural OM results in the adsorption of some of the organic phase to the mineral surface. Previous work in this laboratory examined the adsorption of fulvic acid to reagent calcite surfaces in deionized water (Carter, 1978). We have extended this work to saline (NaCl) and artificial seawater systems (Fig. 6). Neither the salinity nor the composition of the solution has a great effect on the adsorbed amount of OM, measured as amino acids. Roughly comparable, although not identical, amounts are adsorbed from each solution. The differences in the adsorption isotherms can be explained by electrostatic effects.

The adsorption isotherm shows least OM adsorption onto the calcite surface in the DI water system. This is probably due to preferential adsorption of negatively charged molecules via electrostatic attraction for surface calcium ions. Because many of the negatively charged groups on each adsorbed molecule will not be interacting with the surface and thereby neutralized, an increasingly negative electrical potential accumulates at the interface as more molecules are adsorbed. At some point the electrical potential becomes large enough to act as an effective barrier against further adsorption of OM.

The two systems with high ionic strength, artificial seawater and saline solution (with the same ionic strength and pH as seawater), show greater adsorption of OM than the DI water system and are comparable except at the highest OM/mineral values when the saline system shows approximately 40 per cent more adsorption than the artificial seawater system. Apparently, the charge effects controlling adsorption in the artificial seawater and DI water systems are diminished in the saline system and maximum adsorption can occur. Overall, we conclude from these experiments that the relatively minor variations in salinity or composition of seawater will have little effect on the amount of OM adsorbed to carbonate surfaces. Pore water compositions can exhibit greater variations, especially in anoxic sediments, but the consequence of these differences on the adsorbed phase is not known. Other factors such as microbial activity probably play a greater role than composition of the pore water in the adsorption process.

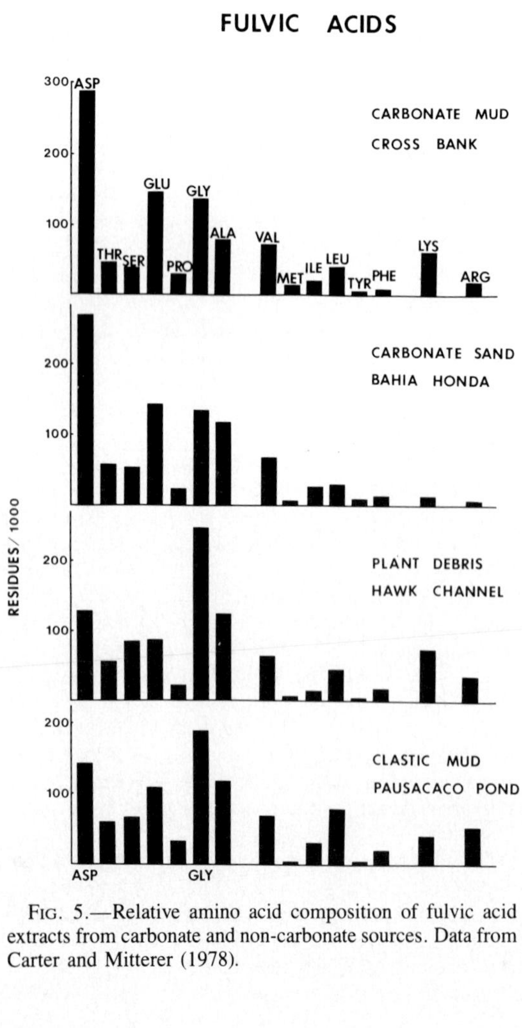

FIG. 5.—Relative amino acid composition of fulvic acid extracts from carbonate and non-carbonate sources. Data from Carter and Mitterer (1978).

The amino acid composition of the adsorbed OM has a greater contrast than the bulk adsorptive property. In preliminary adsorption experiments it was noted that calcite adsorbs a more aspartic acid-rich fraction of OM from solutions containing protein from *Mercenaria* and OM from ooids (Mitterer, 1972). More detailed experiments showed that the aspartic acid content of the OM adsorbed by calcite is a function of the ratio of OM to mineral in the experiment (Fig. 7). At high ratios, aspartic acid comprised almost 70 per cent of the amino acids in the OM adsorbed to calcite, whereas the stock solution contained OM in which aspartic acid comprised only about 29 per cent of the amino acids. The amino acids in the OM adsorbed by quartz powder, from solutions of the same composition, contained a predominance of glycine and serine, but not aspartic acid. Quartz preferentially adsorbs the aspartic acid-poor fraction. These results are qualitatively similar to the analyses of natural adsorbed OM (Fig. 5).

Such excessive aspartic acid enrichments are unrealistic and are not found normally in natural carbonate systems or in experiments using saline and artificial seawater systems. The only known natural example of such high aspartic acid enrichment is the calcified protein of alcyonarian coral spicules, which contain about 70 per cent aspartic acid. Instead, the amino acid component of the adsorbed OM on carbonate grains in the sedimentary environment contains about 20–30 per cent aspartic acid. A possible explanation for the excessively high aspartic acid compositions of the adsorbed OM in experimental systems with DI water is the lack of metal counter ions in solution. Experiments run in deionized water had no metal ions in solution other than low concentrations of the ions in the calcium carbonate system. The presence of high concentrations of metal ions in solution would be expected to modify the adsorptive properties of the OM through cation-OM interaction. In addition, fulvic acids contain molecular structures (e.g., amino acids, peptides, etc.) that are ampholytic in character; that is, they possess both acidic and basic functional groups on the same molecule. Due to their high carboxyl (COOH) group content fulvic acids are essentially weak acids at natural pH's.

In the DI water system at pH 8.5 (Carter, 1978) the acidic functional groups of the fulvic acid molecules are highly dissociated yet there are no metal counter ions in solution to counterbalance the negative charge. Consequently, a large electrostatic attraction exists between the fulvic acid molecules and the surface calciums of the calcite. The high positive enrichment

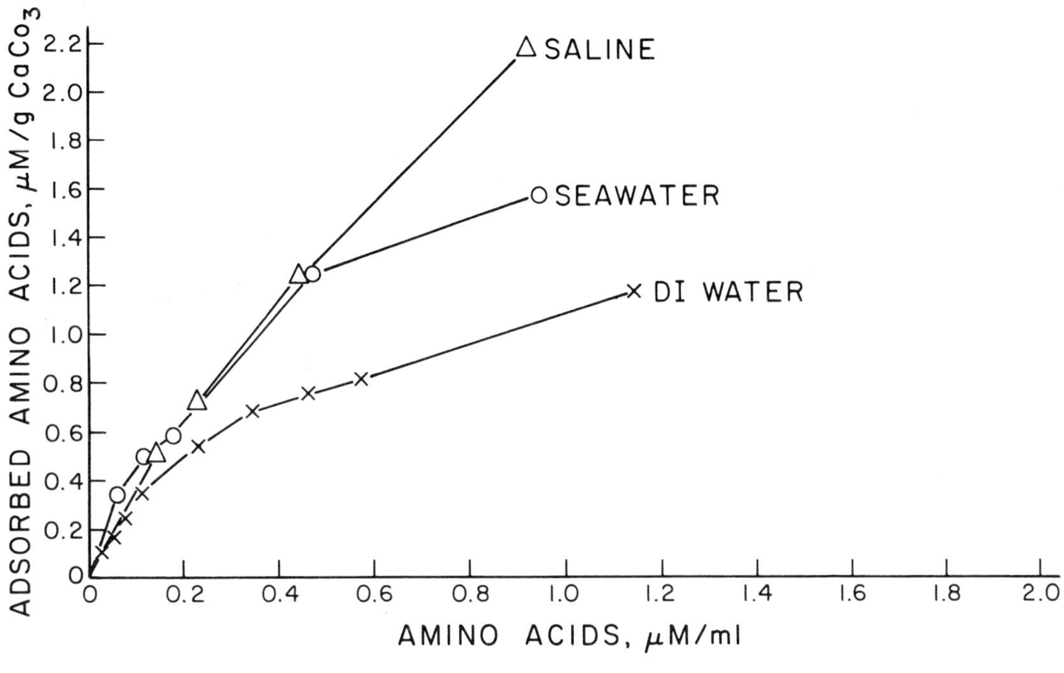

FIG. 6.—Adsorption isotherms of fulvic acid extracts onto calcium carbonate in deionized water, saline solution, and artificial seawater. Data from Cunningham (1980).

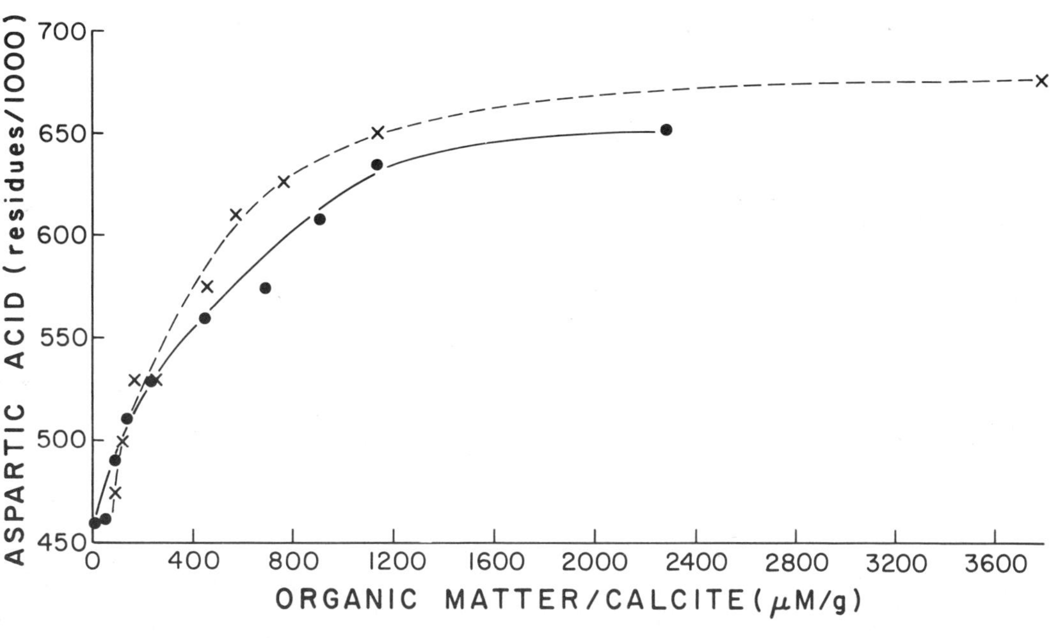

FIG. 7.—Relative aspartic acid content of fulvic acid adsorbed to calcium carbonate surfaces in deionized water as a function of the ratio of organic matter (OM) to mineral. Experiments were run with constant concentration of OM and variable amounts of calcite (X) as well as with constant amount of calcite and variable concentrations of OM (●). Data from Carter (1978).

factor (EF) of aspartic acid at all OM/mineral values provides evidence for this. As OM/mineral decreases, the EF also decreases presumably due to the concentration-related lower probability of a negatively charged molecule (aspartic acid-rich) striking and remaining at the surface (Fig. 8).

In the saline system the EF of aspartic acid at the various OM/mineral values is much reduced from those in the DI water system. Presumably, the electrostatic attraction of the surface calciums for the negative fulvic acid polymers is diminished by the formation of the Na-fulvic acid ion pair. There is, however, an indication that the relationship of a decreasing aspartic acid EF with decreasing OM/mineral values still exists. Interestingly, the basic amino acids show a large increase in their EF in the saline system compared to the DI water system (Fig. 8). Due to lower acidic functional group interaction with surface calciums the basic functional groups may be more able to interact with negatively charged surface sites on the calcite (carbonate anions).

The artificial seawater system shows the lowest EF values for aspartic acid of the three systems. It is possible that the high concentration of divalent cations in solution causes a large percentage of the acidic functional groups to form cation-fulvic acid ion pairs diminishing the electrostatic interaction with the mineral surface even more than for the saline case. In fact, dissolved calcium bound to the fulvic acid polymers may be electrostatically repulsed by surface calciums, further lowering the EF. Decreased adsorption relative to saline at the highest OM/mineral value provides supporting evidence. The basic amino acids show very high EFs in artificial seawater at the highest OM/mineral value. These positively-charged amino acids possibly may compete effectively with free and bound calcium in solution for carbonate sites on the mineral surface. Both acidic and basic amino acid EF values converge toward zero as the OM/mineral value decreases in the seawater system. Presumably, at lower OM concentrations competitive adsorption effects are diminished, and a phase resembling the stock solution

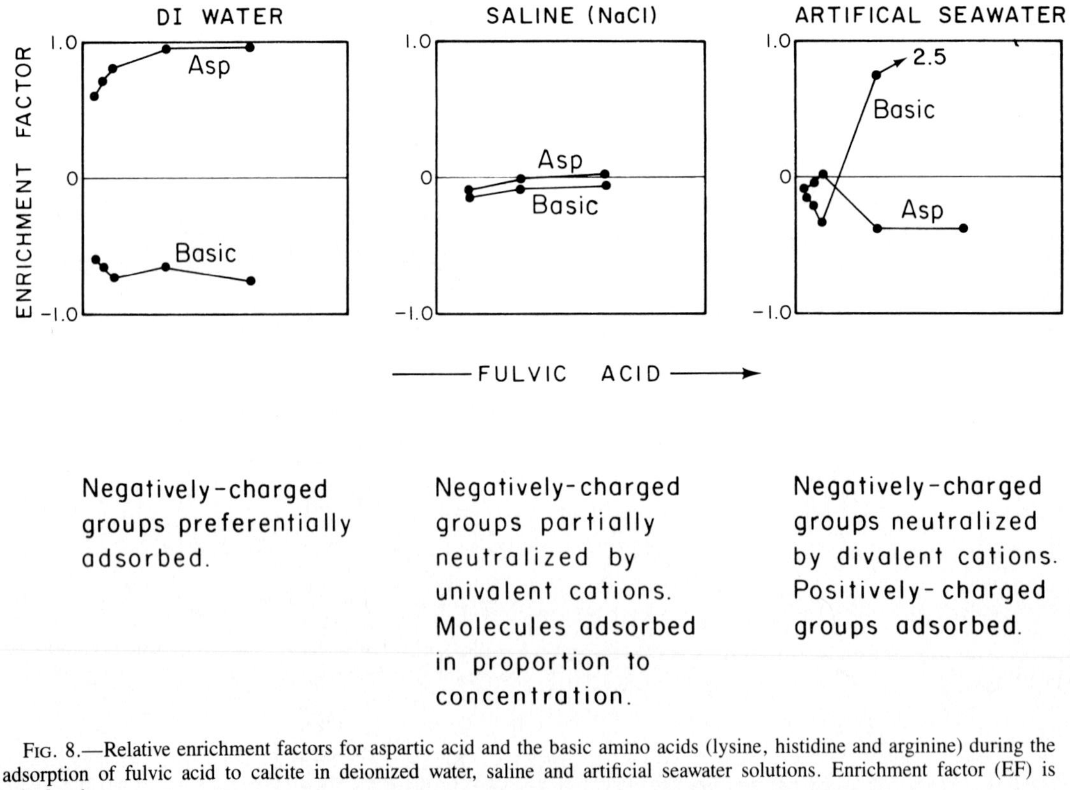

FIG. 8.—Relative enrichment factors for aspartic acid and the basic amino acids (lysine, histidine and arginine) during the adsorption of fulvic acid to calcite in deionized water, saline and artificial seawater solutions. Enrichment factor (EF) is calculated as:

$$EF = \frac{[\text{adsorbed amino acid}] - [\text{solution amino acid}]}{[\text{solution amino acid}]}$$

with all concentrations in residues/1000. An EF of 0 indicates equal relative concentrations in adsorbed and solution phases, an EF of +1 indicates a doubling of the relative concentration in the adsorbed compared to the solution phase, etc. Negative EF's indicate lower relative concentrations in the adsorbed compared to the solution phases.

in relative concentration of the amino acids is adsorbed.

The role of carboxyl groups, provided in part by aspartic acid, in adsorption of OM to mineral surfaces is evident from a comparison of the adsorption characteristics of the OM prior to and subsequent to modification of the carboxyl functions (Fig. 9). Chemical blocking of the carboxyl groups virtually eliminates the adsorptive property of the OM. Consequently, without the negatively charged carboxyl groups, adsorption of natural OM to mineral surfaces through electrostatic interaction is greatly reduced. Thus, the carboxyl group largely provides natural OM with its surface-active or adsorptive properties on carbonate surfaces.

The process of adsorption of a dissolved organic molecule onto a solid interface involves changes in the enthalpy and entropy of the three components in the system: surface, solvent, and adsorbate. For adsorption to occur the sum of these six values must be negative so that the free energy of the process will be minimized. Generally, adsorption begins with diffusion of the molecule to the interface. Once it arrives at the interface an interfacial pressure barrier must be overcome for the molecule to be admitted into the interface (MacRitchie, 1978). This involves compressing molecules already in the interface and clearing a space for the incoming molecule. With higher interfacial pressures it becomes increasingly difficult to adsorb. In addition to the interfacial pressure barrier an electrical potential barrier may also exist. This is especially significant for ampholytic molecules that have developed a net charge at pH conditions removed from their characteristic isoelectric point. Charge similarities or differences between molecules and the solid surface may inhibit or enhance adsorption. At the isoelectric point electrostatic interadsorbate repulsion is at a minimum favoring maximum attraction and packing in the interface (Eirich, 1977).

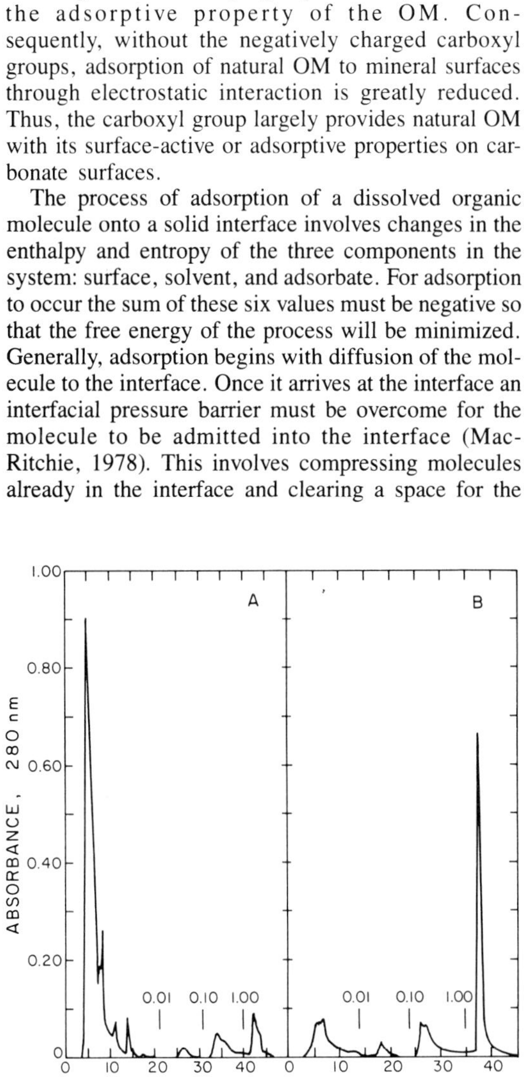

FIG. 9.—(A) Column chromatographic elution pattern of fulvic acid extract after treatment of the fulvic acid with an agent that blocks carboxyl groups. Most of the treated fulvic acid elutes rapidly indicating that it is not adsorbed to or retarded by the column substrate. (B) Column chromatographic elution pattern of untreated fulvic acid. Most of the fulvic acid elutes in the later fractions indicating adsorption to or retardation by the column substrate (hydroxyapatite).

The aspartic acid enrichments in the adsorbed layer that are observed in nature may occur via selective desorption, rather than adsorption, of organic molecules from the mineral-liquid interface. Desorption requires that all segments of the molecule leave the interface together, thus requiring a greater activation energy than adsorption (MacRitchie, 1978). Desorption has been observed to be a much slower process than adsorption (MacRitchie, 1972). In sediments, desorption may occur due to bioturbation, agitation, or chemical and physical changes affecting pH, ionic strength, and temperature of pore waters. Molecules that exhibit relatively rapid adsorption at the solid/liquid interface due to high affinity for the surface may, conversely, be desorbed much more slowly also due to their high affinity for the surface.

In summary, the amount and amino acid composition of OM adsorbed to calcite differ somewhat in DI, saline, and artificial seawater systems. These differences can largely be explained by the effects of the presence or absence of metal counter ions in solution. The carboxyl group is the likely function to be involved in the adsorption process.

METAL COMPLEXING

Extensive literature exists on metal complexes with terrestrial humic substances. Much less is known about metal complexes with marine humic substances and even less about metal complexes with humic substances from carbonate sediments. An investigation of the metal-binding ability of the OM from carbonate sediments is significant for two reasons. First, it adds to the scant body of data on metal-marine humic interactions, particularly from carbonate sediments. Second, it is a necessary step in understanding the influence of this OM on the reactivity of calcium carbonate in the marine environment.

Organic acids behave as weak acids, undergoing partial dissociation in accordance with the following equilibrium

$$HA \underset{K_a}{\overset{K_d}{\rightleftharpoons}} H^+ + A^- \qquad (1)$$

where HA is the undissociated weak acid, A is the dissociated anion of the weak acid and k_d and k_a are dissociation and association constants, respectively.

Carboxyl groups are a major source of hydrogen ions in natural OM. In systems where a relatively strong association is formed between a metal ion and an organic ligand, such as a carboxylate anion, the equilibrium expression is

$$2(RCOOH) + M^{2+} \rightleftharpoons (RCOO^{-})_2 M^{2+} + 2H^{+} \quad (2)$$

Because of differences in binding strengths, the equilibrium binding constant for various cations with the same ligand will differ. Various organic ligands will, of course, bind the same metal ion to different degrees. We studied the metal complexing ability of several different types of carbonate humic substances. Metals used included Mn^{2+} and Gd^{3+} as well as Ca^{2+}. The former are used in these studies because they have similar ionic radii and ionic charges to Ca^{2+}. From equations (1) and (2) it is apparent that a knowledge of free and (or) bound metal concentrations is necessary for calculation of association or dissociation constants. To determine free and bound metal concentrations in solution after complexing had occurred we used spectroscopic techniques including pulsed nuclear magnetic resonance (for bound Gd^{3+}), electron paramagnetic resonance (for free Mn^{2+}) and an ion-specific electrode (for free Ca^{2+}). Binding experiments were conducted at 25°C and 0.2 M ionic strengths (Cunningham and Mitterer, 1980). Bound (or free) metal is calculated as the difference between the known total metal and free (or bound) metal determined at several points in a metal-organic ligand titration. Using a suitable graphic technique (Scatchard plot) the intrinsic association constant and number of binding sites can be obtained (Scatchard, 1949).

A previous investigation of the metal-binding ability of OM extracted from carbonate sediments has demonstrated that non-skeletal (e.g., ooid) and adsorbed OM bind metal ions to a significant degree (Cunningham and Mitterer, 1980). High molecular weight (>10,000 MW) fulvic acids extracted from Florida Bay sediment samples have two classes of binding sites (for Mn^{2+}), a high affinity class and a low affinity class. Fulvic acid extracted from Bahamian ooids has an affinity for Mn^{2+} similar to that of the low affinity sites of the OM from the Florida Bay sediments, yet the ooid OM binds about twice as much total Mn^{2+}. The fulvic acids generally bind Mn^{2+} in proportion to their aspartic acid content.

Additional metal-binding studies using different metal ions and OM, including Ca^{2+} and skeletal OM, are illustrated graphically on a Scatchard plot (Figs. 10 and 11). Skeletal protein from *Mercenaria mercenaria* was titrated with Gd^{3+}. The protein exhibits one class of binding sites with a concentration of 390 micromoles/g of protein. The stability constant, K_a, is 5880 M^{-1}. The high association constant indicates that the skeletal protein of *M. mercenaria* has a very high affinity for Gd^{3+}.

Fulvic acid extract from carbonate sediment collected at White Bank, Florida Keys, was found to have two classes of Ca^{2+} binding sites, a high affinity class with a stability constant approximately equal to 4167 M^{-1} and a low affinity class with a stability constant approximately equal to 560 M^{-1} (Fig. 11). The number of Ca^{2+} binding sites in the high affinity class is 75 micromoles/g of fulvic acid and the number of low affinity binding sites is 250 micromoles/g of fulvic acid. The low affinity class of binding sites holds about three to four times as much Ca^{2+} as the high affinity class. The aspartic acid content of White Bank fulvic acid is 385 micromoles/g, of fulvic acid, which is more than adequate to account for the number of high and low affinity Ca^{2+} binding sites. Evidently, OM associated with carbonate sediments binds metal ions, especially Ca^{2+}, in proportion to the amount of aspartic acid or total acidic amino acids.

Further evidence that the carboxyl groups are involved in metal binding is obtained by comparing the complexing ability of the OM after chemically blocking the carboxyl groups. Blocked carboxyl groups result in almost complete loss of complexing ability. Thus, the results are identical in both the adsorption and metal-binding experiments. Almost total loss of both adsorptive property and metal-binding ability occurs when the carboxyl groups are chemically blocked.

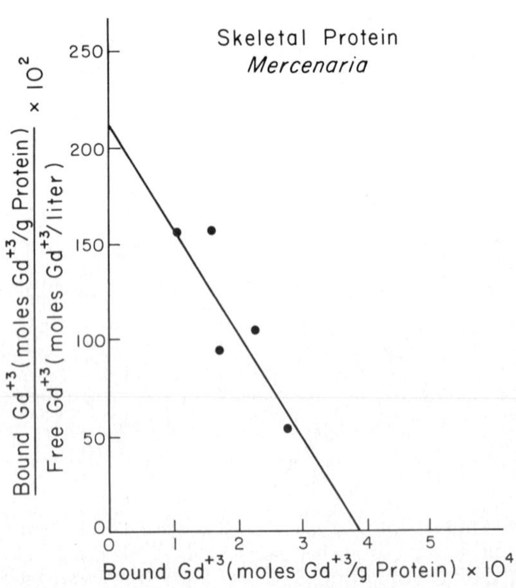

FIG. 10.—Scatchard Plot of gadolinium bound to soluble skeletal protein from *Mercenaria*. See discussion in text.

CARBONATE-ORGANIC INTERACTION—A MODEL

Invertebrates precipitate characteristic skeletal structures that are permeated with organic filaments. It is generally thought that the protein matrix is involved in the calcification process. The matrix may provide nucleating sites for the mineral phase by acting as a template for epitaxial nucleation or by attracting the ions electrostatically. The potential significance of aspartic acid-rich protein to invertebrate calcification has been emphasized by Hare (1963), Degens *et al.* (1967), Mitterer (1971), Weiner and Hood (1975), Degens (1976) and Weiner *et al.* (1983). This amino acid may provide binding sites for Ca^{2+} because of the presence of a negatively-charged carboxyl group for each aspartic acid residue. In essence, it is thought that the protein matrix of mineralized tissues has inherent properties that impart an ability to calcify.

OM associated with shallow marine carbonate sediments is similar in composition to skeletal OM but is distinctly different from that associated with non-carbonate marine sediments with respect to amino acid content and composition (Mitterer and Carter, 1977; Carter and Mitterer, 1978). Amino acids comprise a significantly greater proportion of the OM associated with carbonate sediments, with more than 50 per cent of the organic carbon consisting of amino acids. In non-carbonate sediments, amino acids comprise only about 10 per cent of the organic carbon. A striking feature of the amino acid composition of OM associated with carbonate sediments is the predominance of the acidic amino acids, especially aspartic acid. OM within skeletal and non-skeletal calcium carbonate and OM extracted from carbonate surfaces exhibit this pattern, whereas the OM of siliciclastic sediments is comprised primarily of the neutral amino acids glycine and alanine. Thus, the nature of the OM associated with carbonate and siliciclastic sediments is fundamentally different. These contrasting compositions may have an important bearing on the role of carbonates as source rocks for hydrocarbons. In addition, the unique nature of the OM may influence the precipitation of calcium carbonate.

The role of adsorbed OM is widely presumed to be that of an inhibitory agent in carbonate precipitation. There is no doubt that OM retards calcium carbonate precipitation based on inorganic kinetic experiments. However, there is also considerable data to suggest that OM adsorbed onto sediment grains does not entirely play the passive role of merely coating a carbonate grain with an unreactive sheath. The formation of natu-

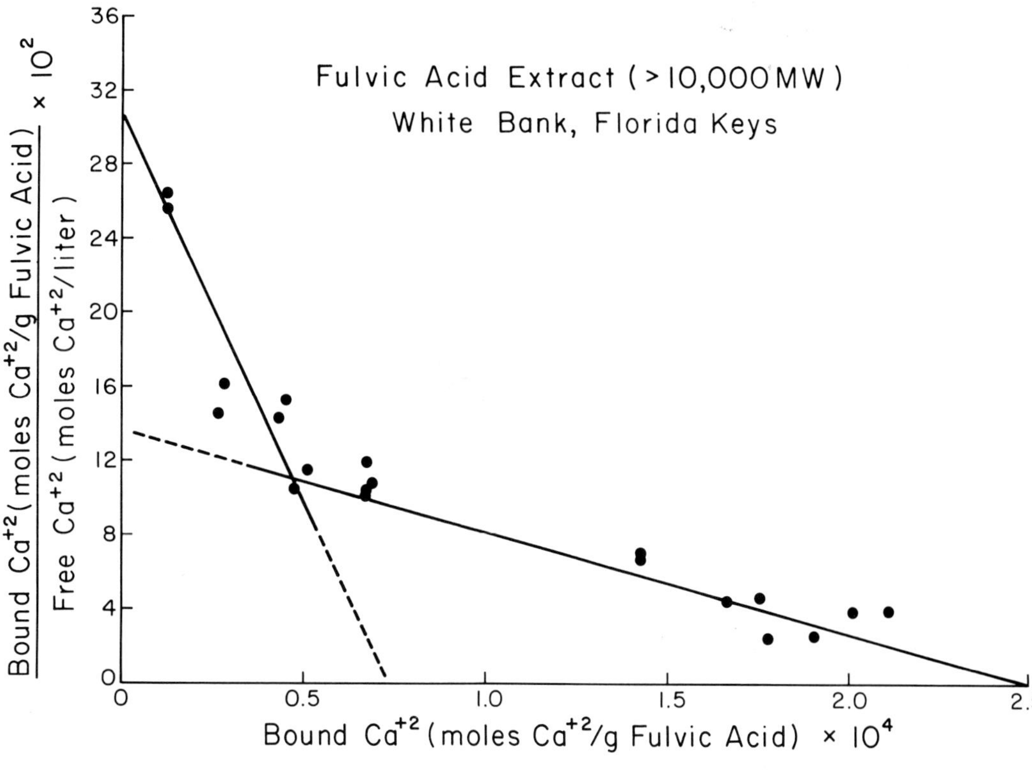

FIG. 11.—Scatchard Plot of calcium bound to fulvic acid extract. See discussion in text.

ral ooids in association with an organic matrix together with the observations that humic acids may induce (or at least not prevent) the precipitation of ooids in laboratory systems (Suess and Fütterer, 1972; Davies *et al.*, 1978; Ferguson *et al.*, 1978) suggest that some of the OM on sediment grains may not entirely inhibit calcium carbonate precipitation. The apparent conflict between these postulated roles for the OM is obvious. If the organic coating is strictly inhibitory, then precipitation of carbonate minerals should not occur in the shallow marine environment where the organic coatings are ubiquituous. Perhaps a partial solution to the conflict is that the kinetics of simple inorganic laboratory systems should not be used as the frame of reference. The seawater carbonate-organic system is much more complicated than simple laboratory systems that do not contain OM, and if the frame of reference were to be the natural seawater system itself, the influence of OM would not be viewed as exclusively inhibitory of carbonate precipitation.

The adsorption of aspartic acid-rich OM to carbonate surfaces both in organisms and in the sedimentary environment is highly specific and implies an ordered interaction between functional groups of the OM and the constituents of the carbonate crystal surface. The carboxyl function (present in biomonomers and biopolymers such as fatty acids, amino acids, and proteins, and in geopolymers such as fulvic and humic acid) is the most likely interactive group on the OM based on the selective association of aspartic acid with carbonate surfaces. The nature of this interaction could be simple electrostatic interaction between the negatively-charged carboxyl groups and unsatisfied positive charges from surface calcium ions in the calcite or aragonite lattice. A geometrical comparison between the carboxyl group and the carbonate anion, however, points to an even more specific interaction (Fig. 12). Both groups are planar structures. The O-C-O bond angle in both groups is about 120–125°. The C-O bond length is 12.8 nm in the carbonate anion and 12.7 nm in the carboxyl group. Finally, the negative charge on the oxygens is ⅔ for each carbonate oxygen and ½ for each carboxyl oxygen. Evidently, the similarities in geometry and ionic charge are sufficiently close that the carboxyl group can "fit" into the carbonate lattice. This would account for the adsorption of molecules containing carboxyl functions. Aspartic acid-rich proteins and humic materials, being polymeric and sometimes containing greater than 30 per cent aspartic acid, may be able to satisfy the positive charges on many surface calciums simultaneously. The crystal surface charges may be equally or better satisfied by adsorption of such polymeric material.

In summary, then, it appears that the interaction between aspartic acid-rich OM, such as proteins and humic materials, and carbonate surfaces is highly specific and is due to the similarity in geometry and charge between the carboxyl group and the carbonate anion permitting substitution of the organic group for the inorganic anion. Whether such substitution acts exclusively as an inhibitor, or permits or induces nucleation possibly by epitaxy, is uncertain. The evidence in favor of inducing nucleation is circumstantial and based primarily on the unique composition of the OM, its adsorptive properties, and its complexing ability. However, precipitation of non-skeletal calcium carbonate does occur in the marine environment in the presence of OM, as does precipitation of skeletal calcium carbonate, and laboratory experiments have demonstrated the precipitation of calcium carbonate in association with DOM. In view of these circumstances it is appropriate to propose a model for the role of natural OM in calcium carbonate precipitation in the marine environment. A range of interactions may occur depending on the composition and abundance of the OM (Fig. 13).

1. OM is very abundant.

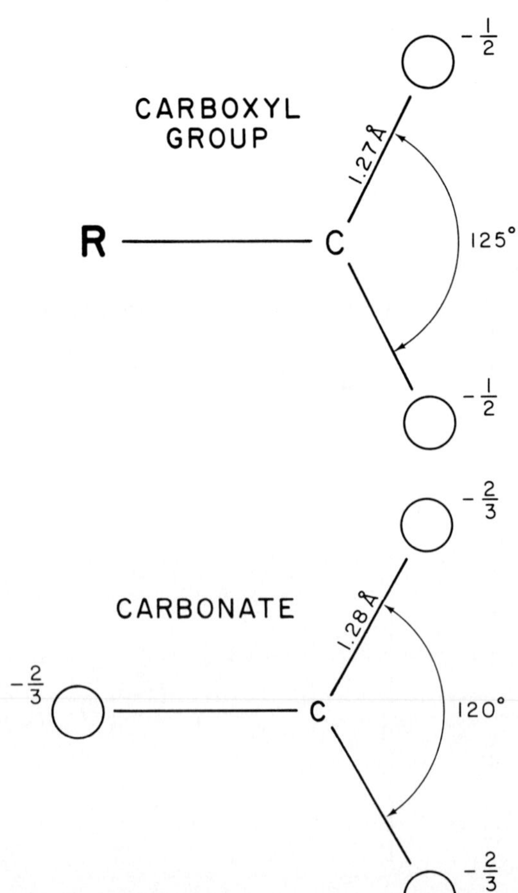

FIG. 12.—Geometric comparison of carboxyl group and carbonate anion. R represents the remainder of the molecule to which the carboxyl group is attached.

The high concentration of DOM will complex a major portion of the total calcium in solution and will reduce the ion activity product of Ca^{2+} and CO_3^{2-}. Precipitation will not occur because the concentration of free Ca^{2+} will be too low. This condition will occur in anoxic pore waters of fine-grained sediments rich in OM and in other low energy environments where the amount of OM is very high (e.g., Berner et al, 1970; Chave and Suess, 1970).

2. OM is moderately abundant and largely hydrophobic.

Any functional groups will interact with the mineral substrate, with the hydrophobic portion of the molecules directed outward toward the surrounding seawater. The organic coating, possibly lipid-rich, will inhibit further interaction between the substrate and the seawater. This condition will prevail in the open marine environment, dominated by lipid-rich algal OM, particularly in regions of low calcium carbonate productivity, and in the deep subsurface where grains have been coated with migrated hydrocarbons.

3. OM is moderately abundant and largely hydrophilic.

The surface-active OM will be adsorbed to the mineral substrate and will complex some of the dissolved Ca^{2+}. Either by epitaxial nucleation or by enhancement of the concentration of Ca^{2+} due to complexing in the microenvironment adjacent to the substrate, or a combination of both, the local concentration of Ca^{2+}, complexed and free, may be raised to a level allowing nucleation of aragonite or calcite (Fig. 14). This condition will exist in shallow marine environments of moderate to high energy where carbonate productivity is high and in the near surface part of the marine phreatic zone.

4. OM is sparsely present.

In the absence or very low concentration of OM, either due to highly oxidizing conditions or to removal by biological or diagenetic processes, calcium carbonate precipitation will follow inorganic kinetics. This condition will prevail in the highly oxidizing fresh water vadose zone and in the deeper portion of the marine phreatic ("stagnant") zone, far from any source of fresh OM, where any previously adsorbed or interstitial OM will have been oxidized or isolated by early cementation.

Competition between organic and inorganic species and surface adsorption sites is driven by the requirement of the system to remain electroneutral. Concentration variations in dissolved organic and inorganic

ORGANIC MATTER — $CaCO_3$ INTERACTION

TOC	O_2 LEVEL	OM TYPE	EFFECT OF OM	TYPICAL ENVIRONMENT
HIGH	ANOXIC	HYDROPHOBIC (LIPID-RICH)	COMPLEXING OF Ca BY DOM; NO PRECIPITATION	PORE WATERS IN FINE-GRAINED SEDIMENTS
MODERATE	↓ INCREASING LEVEL OF O_2	↓ INCREASING DEGREE OF OXIDATION	LIPID COATING; ISOLATION OF GRAIN; NO PRECIPITATION	OPEN MARINE BASIN
LOW		HYDROPHILIC (OXYGEN-RICH)	ADSORPTION OF OM AND COMPLEXING OF Ca; POSSIBLE PRECIPITATION OF $CaCO_3$ (EPITAXY?)	SHALLOW MARINE GRAINSTONE AND BOUNDSTONE
"ABSENT"	OXIC		NONE; PRECIPITATION FOLLOWS INORGANIC KINETICS	VADOSE

FIG. 13.—A model for the interaction of organic matter with calcium carbonate. Variables are total organic carbon (TOC), dissolved oxygen, and composition of organic matter.

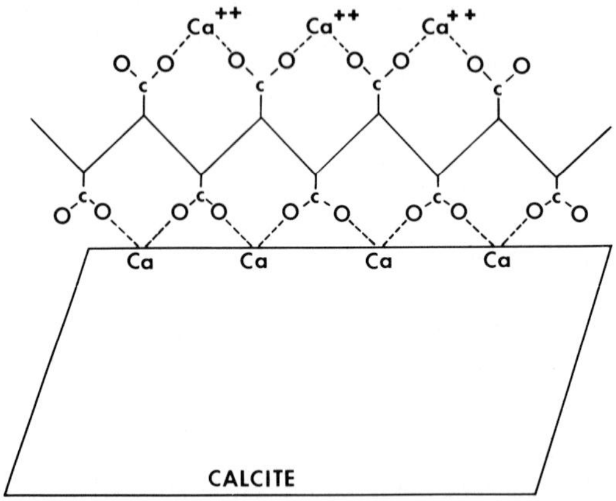

FIG. 14.—Highly idealized illustration of the interaction of carboxyl groups with calcite surface.

ions in the pore waters brought about by bioturbation and organic and inorganic diagenesis result in variations in the tendency of OM to affect the chemistry of the system. In shallow marine sediments, therefore, the microenvironment surrounding a carbonate grain is a mixture of inorganic ions and organic molecules; these interact to result either in isolation of grains from the pore fluid or in further precipitation of calcium carbonate, depending on the types of molecules and environmental parameters.

ACKNOWLEDGMENTS

We thank D.A. Jeffrey, Michele Thomas and D. Yurewicz for their helpful reviews. This research was supported by NSF Grant OCE 79-25389. Contribution No. 442 of the Geosciences Program, University of Texas at Dallas.

REFERENCES

BERNER, R.A., 1975, The role of magnesium in the crystal growth of calcite and aragonite from sea water: Geochim. Cosmochim. Acta, v. 39, p. 489–504.

______, SCOTT, M.R., AND THOMLINSON, C., 1970, Carbonate alkalinity in the pore waters of anoxic marine sediments: Limnol. Oceanogr., v. 15, p. 544–549.

______, WESTRICH, J.T., GRABER, R., SMITH, J., AND MARTENS, C.S., 1978, Inhibition of aragonite precipitation from supersaturated seawater: A laboratory and field study: Amer. Jour. Sci., v. 278, p. 816–837.

CARTER, P.W., 1978, Adsorption of amino acid-containing organic matter by calcite and quartz: Geochim. Cosmochim. Acta, v. 42, p. 1239–1242.

______, AND MITTERER, R.M., 1978, Amino acid composition of organic matter associated with carbonate and non-carbonate sediments: Geochim. Cosmochim. Acta, v. 42, p. 1231–1238.

CAUGHEY, M.E., 1982, A study of the dissolved organic matter in the pore waters of carbonate-rich sediment cores from Florida Bay: M.S. Thesis, Univ. of Texas at Dallas, 69 p.

CHAVE, K.E., 1965, Calcium carbonate: Association with organic matter in surface seawater: Science, v. 148, p. 1723–1724.

______, AND SUESS, E., 1967, Suspended minerals in seawater: Trans. New York Acad. Sci., v. 29, p. 991–1000.

______, AND ______, 1970, Calcium carbonate saturation in seawater: Effects of dissolved organic matter: Limnol. Oceanogr. 15, p. 633–637.

CUNNINGHAM, R., JR., 1980, Organic-inorganic interactions of marine humic substances from carbonate sediments: Metal binding and adsorption studies: Ph.D. dissertation, Univ. of Texas at Dallas, 127 p.

______, AND MITTERER, R.M., 1980, Metal binding study of fulvic acids from carbonate sediments using manganese as a magnetic resonance probe, *in* P.E. Hare, T.C. Hoering, and K. King, Jr. (eds.), Biogeochemistry of Amino Acids: New York, Wiley, p. 129–143.

DAVIES, P.J., BUBELA, B., AND FERGUSON, J., 1978, The formation of ooids: Sedimentology, v. 25, p. 703–730.

DEGENS, E.T., 1976, Molecular mechanisms on carbonate, phosphate and silica deposition in the living cell: Topics in Current Chemistry, v. 64, p. 1–112.

______, AND SPENCER, D.W., 1966, Data file on amino acid distribution in calcified and non-calcified tissues of shell-forming organisms: Tech. Rep. 66–27, Woods Hole Oceanographic Inst., unpublished manuscript.

______, ______, AND PARKER, R.H., 1967, Paleobiochemistry of molluscan shell proteins: Comp. Biochem. Physiol., v. 20, p. 553–579.

EIRICH, R.R., 1977, The conformational states of macromolecules adsorbed at solid-liquid interfaces: J. Colloid Interface Sci., v. 58, p. 423–436.

FERGUSON, J., BUBELA, B., AND DAVIES, P.J., 1978, Synthesis and possible mechanism of formation of radial carbonate ooids: Chem. Geol., v. 22, p. 285–308.

FREDERICKS, A.D., AND SACKETT, W.M., 1970, Organic carbon in the Gulf of Mexico: J. Geophy. Res., v. 75, p. 2199–2206.

HARE, P.E., 1963, Amino acids in the proteins from aragonite and calcite in the shells of *Mytilus californianus*: Science, v. 139, p. 216–217.

______, AND ABELSON, P.H., 1964, Proteins in mollusk shells: Carnegie Inst. Wash. Yearb., v. 63, p. 267–270.

______, AND ______, 1965, Amino acid composition of some calcified proteins: Carnegie Inst. Wash. Yearb., v. 64, p. 223–232.

KITANO, Y., AND HOOD, D.W., 1965, The influence of organic material on the polymorphic crystallization of calcium carbonate: Geochim. Cosmochim. Acta, v. 29, p. 29–41.
______, AND KANAMORI, N., 1966, Synthesis of magnesian calcite at low temperatures and pressures: Geochem. Jour., v. 1, p. 1–10.
KROM, M.D., AND SHOLKOVITZ, E.R., 1977, Nature and reactions of dissolved organic matter in the interstitial waters of marine sediments: Geochim. Cosmochim. Acta, v. 41, p. 1565–1573.
MACRITCHIE, F., 1972, The adsorption of proteins at the solid/liquid interface: J. Colloid Interface Sci., v. 38, p. 484–488.
______, 1978, Proteins at interfaces, *in* C.B. Anfinson, J.T. Edsall, and F.M. Richards (eds.), Advances in protein chemistry: New York, Academic Press, p. 283–326.
MEYERS, P.A., AND QUINN, J.G., 1971, Interaction between fatty acids and calcite in seawater: Limnol. Oceanogr., v. 16, p. 992–997.
______, AND______, 1973, Factors affecting the association of fatty acids with mineral particles in sea water: Geochim. Cosmochim. Acta, v. 37, p. 1745–1759.
MITTERER, R.M., 1968, Amino acid composition of organic matrix in calcareous oolites: Science, v. 162, p. 1498–1499.
______, 1971, Comparative amino acid composition of calcified and non-calcified polychaete worm tubes: Comp. Biochem. Physiol., v. 38b, p. 405–409.
______, 1972, Biogeochemistry of aragonite mud and oolites: Geochim. Cosmochim. Acta, v. 36, p. 1407–1422.
______, 1978, Amino acid composition and metal binding capability of the skeletal protein of corals: Bull. Mar. Sci., v. 28, p. 173–180.
______, AND CARTER, P.W., 1977, Some analytical and experimental data on organic-inorganic interaction: Proceedings, Third International Coral Reef Symposium, University of Miami, Florida, p. 541–547.
MOPPER, K., AND DEGENS, E.T., 1972, Aspects of the biogeochemistry of carbohydrates and proteins in aquatic environments: Tech. Rep. Woods Hole Oceanographic Inst., Ref. No. 72-68, unpublished manuscript.
MÜLLER, P.J., AND SUESS E., 1977, Interaction of organic compounds with calcium carbonate—III. Amino acid composition of sorbed layers: Geochim. Cosmochim. Acta, v. 41, p. 941–949.
OGURA, N., 1970, Dissolved organic carbon in the equatorial region of the central Pacific: Nature, v. 227, p. 1335–1336.
OTSUKI, A., AND WETZEL, R.G., 1973, Interaction of yellow organic acids with calcium carbonate in freshwater: Limnol. Oceanogr., v. 18. p. 490–493.
PYTKOWICZ, R.M., 1965, Rates of inorganic calcium carbonate nucleation: Jour. Geol., v. 73, p. 196–199.
SCATCHARD, G., 1949, The attraction of proteins for small molecules and ions: Ann. New York Acad. Sci., v. 51, p. 660–672.
SUESS, E., 1970, Interaction of organic compounds with calcium carbonate—I. Association phenomena and geochemical implications: Geochim. Cosmochim. Acta, v. 34, p. 157–168.
______, 1973, Interaction of organic compounds with calcium carbonate—II. Organic-carbonate association in recent sediments: Geochim. Cosmochim. Acta, v. 37, p. 2435–2448.
______, AND FÜTTERER, D., Aragonitic ooids: Experimental precipitation from seawater in the presence of humic acid: Sedimentology, v. 19, p. 129–139.
TRICHET, J., 1968, Etude de la composition de la fraction organique des oolites. Comparaison avec celle des membranes de bacerries et des cyanophycees: Compt. Rend., v. 267, p. 1492–1494.
WEINER, S., AND HOOD, L., 1975, Soluble protein of the organic matrix of mollusk shells: a potential template for shell formation: Science, v. 190, p. 987–989.
______,TRAUB, W., AND LOWENSTAM, H.A., 1983, Organic matrix in calcified exoskeletons, *in* P. Westbrook and E.W. de Jong (eds.), Biomineralization and Biological Metal Accumulation: Boston, D. Reidel, p. 205–224.

ARAGONITE CEMENTS AND THEIR OCCURRENCE IN ANCIENT LIMESTONES

PHILIP SANDBERG
Department of Geology, 245 NHB, University of Illinois, Urbana, Illinois 61801

ABSTRACT

Modern aragonite cements have a variety of morphologies, predominantly laths or needles. Needle fringes are commonly somewhat splayed (except where developed as epitaxial continuations of well-ordered substrate crystals, e.g., in molluscan shells). Similar needle fringes preserved as aragonite engulfed in second generation calcite spar have been found in Pennsylvanian to Triassic rocks. Many other isopachous fibrous calcite cements with regular optic orientation have been commonly inferred to be the product of pseudomorphic replacement (paramorphic calcitization) of an aragonite precursor cement. Alteration textures in calcitized skeletons, ooids, and cements known to have been originally aragonite do not support that inference. Even if pseudomorphism were supportable, morphologies of aragonite and calcite cements fringes are dissimilar. Botryoidal aragonites, now calcitized to various irregular mosaics, occur in Mississippian to possibly lower Jurassic rocks, often with an elevated Sr^{2+} content and/or rare oriented relics of the original aragonite fibers. Morphologically similar botryoids have been reported from the lower Cambrian and from the middle Precambrian. Comparable botryoids and engulfed aragonite needle fringes have not been encountered in rocks of intervening ages.

In some Pleistocene cement fringes from Japan, mixed radiaxial fibrous calcite (RFC) and fascicular-optic calcite (FOC) overlie both aragonitic skeletons and first-generation aragonite cement. This occurrence and other properties of RFC-FOC refute earlier interpretations of genesis of those cements by replacement of a metastable fibrous precursor (possibly aragonite) at a distally migrating alteration front. That RFC-FOC and other cements with undulose extinction overlying still-aragonitic cements suggest that undulose extinction is not a secure criterion for replacement origin. Similar reservations apply for non-planar crystal boundaries when the cements are composite crystals (as in the RFC-FOC fringes).

INTRODUCTION

Carbonate sedimentologists have long turned to the Florida-Bahamas region, Barbados, Bermuda, and the Persian Gulf for an understanding of carbonate deposition in ancient seas. One cannot deny the necessity of consulting active modern processes and their products for clues to observed ancient carbonates and the unobserved processes inferred to have formed them. However, ironically, the very successes in the study of modern carbonates which provided so many insights into ancient carbonates may also have fostered an overreliance on the modern correlatives. For many topics, such as the original mineralogy of ancient marine ooids and cements, there has persisted a sort of Lyellian substantive uniformitarian view (Gould, 1965; Austin, 1979) of carbonates. The strong hold which the so-called "Modern Analog" has had on views in carbonate sedimentology is reflected in the generally uncritical repetition of the inference that ancient radial-concentric calcite ooids originate from alteration of concentric aragonite ooids like those now forming in the Bahamas. That inference has persisted, even in quite recent journal articles and reference books, despite Sorby's (1879) cogent arguments to the contrary and numerous subsequent works on the textural and optic disruption which accompanies calcitization of aragonite constituents (including ooids, skeletons, and cements).

In recent years, there has been a growing awareness of the limitations of modern examples as correlatives of ancient carbonates. This more circumspect view of the "Modern Analog" is perhaps nowhere better captured than in the aphorism attributed to Lowenstam by Neumann & Land (1975): "The present is the key to perhaps the Pleistocene." Freed from the constraints of a substantive uniformitarian view, sedimentary geologists have been reevaluating ancient carbonates and their sedimentary precursors and diagenesis. For example, some ancient sparry calcite cements have been attributed to marine deposition (Folk, 1973; Wilkinson *et al.*, 1982). Echoing Sorby (1879), ancient radial-concentric calcite ooids have been taken to represent original deposition as calcite (Simone, 1981). In these reevaluations, ooids have received perhaps more than their fair share of attention, whereas cements and micrites have been slighted. The micrites are complex polymineralic systems beyond the scope of this paper. What I hope to do here is to apply some of the insights gained in the study of diagenesis of aragonitic ooids and skeletons to the evaluation of the original mineralogy of ancient cements and the temporal distribution of aragonite cements.

It is vital to make a distinction between constituents for which an original aragonite mineralogy has been assumed on the basis of modern correlatives, and those in which original aragonite mineralogy is supported by more secure evidence. An important factor here was an entrenched earlier view, noted by Alexandersson (1969, p. 47), that aragonite was the normal or even the only primary marine cement. The effects of that view are still with us, despite copious new evidence to the contrary. A correlated problem is the "case-hardening" effect of citation. For example, Newell (1955) has been cited as support for original aragonite mineralogy of the isopachous calcite cements in the Permian reefs of the S.W. USA. A reexamination of Newell's paper shows that he said (1955, p. 308) only that "characteristics of the fibrous calcite and its uniform relationships with other constituents of the reefs suggest that it was deposited from solution, perhaps as aragonite . . ." Numerous other examples exist of such speculative comments offered as "supporting evidence". However, more detailed petrography and various new approaches (e.g., Lohmann & Meyers, 1978; Wenk *et al.*, 1983) give reason for confidence in efforts to determine properties of ancient cements.

What then constitutes good evidence for original aragonite mineralogy of constituents of an ancient limestone? A variety of textural states, sometimes supplemented with compositional information, have been taken in the literature as criteria for recognizing calcites that were formerly aragonite. As noted earlier (Sandberg, 1983), those criteria vary in their reliability and can be ranked in a hierarchy of decreasing reliability (Table 1).

In the remainder of this paper, several abbreviations (some following the usage of Brand & Veizer, 1980) will be used for frequently encountered lengthy terms. They are: HMC (high- magnesium calcite), LMC (low-magnesium calcite; used in reference to original state of constituents, as in "LMC cements" or "LMC brachiopod shell"), dLMC (diagenetic low-magnesium calcite: this refers to the products of diagenetic alteration of aragonite, HMC, or LMC.), FOC (fascicular optic calcite; see Kendall, 1977), RFC (radiaxial fibrous calcite: see Kendall & Tucker, 1973).

MODERN ARAGONITE CEMENTS

Modern aragonite cements are quite diverse in morphology, including broad laths (Schroeder, 1972a, Fig. 4), apparently equant spar (Land & Moore, 1980, Figs. 9, 10; James & Ginsburg 1979, Fig. 6-6E), spatulas (Scherer, 1975, Fig. 2d), and "micrite" (Allen *et al.*, 1969). However, the most prevalent morphology (especially at the scale of the light microscope) is a fiber or needle of aragonite with a high length to width ratio. The needles vary in packing and ordering, from very loose and random (Jorgensen, 1976, Figs. 8, 10) through overlapping splayed and porous arrays radiating up from the substrate (Illing, 1954, pl. 3, Figs. 8, 9; Alexandersson, 1969, Figs. 9, 10, 11B; Schroeder, 1972a, Fig. 1e; Longman, 1980, Fig. 5C; Jindrich, 1983, Fig. 8f) to very solid masses of close packed, nearly parallel fibers, such as the spherulites (e.g., Schroeder, 1972a, Fig. 2) or botryoids (Ginsburg & James, 1976, Figs. 3, 4). Those botryoids and even the spherulites are unusual in their dense, highly ordered arrays with relatively smooth distal surfaces (i.e., adjacent crystals are essentially of equal length). Continuing growth in that configuration is reflected by the smooth curves of the accretionary banding throughout the botryoids and spherulites (James & Ginsburg 1979, Fig. 6-5B; Schroeder, 1972a, Fig. 2b).

Aragonite fringe cements may be crudely isopachous (certainly more so than freshwater calcite cements), especially when viewed in the light microscope at low magnifications (Fig. 1A; Moore, 1973, Fig. 4a; Dravis 1979, Figs. 7A, 8B; Longman, 1980, Fig. 5a; see however, James & Ginsburg, 1979, p. 111). However, in detail (Figs. 1, 2), the crystals are usually of quite variable lengths and diverge considerably from perpendicular to the substrate (Schroeder, 1972b, Figs. 17, 18; Moore, 1973, Fig. 4b; Schroeder, 1974, Fig. 1a; Alexandersson, 1969, pl. 2, Fig. C; Alexandersson, 1972, Figs. 9–11; Friedman *et al.*, 1974, Figs. 1, 3; Bathurst, 1974, Fig. 1; Schlager & James, 1978, Fig. 12; Füchtbauer, 1974; Fig. 5–60; Blatt *et al.*, 1980, Fig. 14-4). Buchbinder & Friedman (1980) suggested that the aragonite cement in their Pleistocene material "occurs as a perpendicular fringe around grains." However, their figures of such cements on non-molluscan substrates (Fig. 5g, right side; Fig. 5h, lower portion) show splayed aragonite needles similar to those illustrated by Schroeder (1972), James &

TABLE 1.—CRITERIA FOR RECOGNITION OF ANCIENT ARAGONITES[1]

1.	*Still aragonite*. Encountered in quite old rocks (Figs. 3A–D).
2.	*Mosaic of generally irregular calcite spar containing oriented aragonite relics*. Crystal boundaries commonly cross-cut original structure defined by organic or other inclusions. Replacement crystals normally 10 – 10^3 times larger than replaced aragonite crystals (Fig. 5).
3.	*Calcite mosaic as for Criterion 2, but without aragonite relics*. Original aragonite mineralogy supported by elevated Sr^{2+} content, relative to levels reasonably expected in dLMC resulting from alteration of originally calcitic constituents.
4.	*Calcite mosaic as for Criterion 3, but Sr^{2+} not elevated, or not measured*. Differing Sr^{2+} content in the resulting calcites can reflect differing degrees of openness of the diagenetic system.
5.	*Mold or subsequently calcite-filled mold*. Common for aragonitic consituents, but equivocal because observed for non-aragonitic constituents (e.g., miliolid foraminifera, Schroeder, 1979) in undolomitized limestone.
Note:	Constituents preserving regularity both in texture and in optical orientations are sometimes suggested as products of pseudomorphic or paramorphic calcitization of aragonite. Not supported by observations on diagenesis of known aragonites. Even the unusual hot spring alteration of some aragonite ooids reported by Richter and Besenecker (1983, fig. 4) produces a 10–50 fold increase in crystal size and leaves aragonite relics (Sandberg, unpublished).

[1]Arranged in order of decreasing reliability, from Sandberg (1983).

Ginsburg (1979) and herein (Figs. 1–3). If the isopachous intergranular cement composed of parallel crystals of uniform length illustrated by Flügel (1982, pl. 3, Figs. 1, 2) is aragonite, it is quite exceptional. The spherulitic aragonite which Schroeder (1972a, Fig. 3) found lining pelecypod borings is almost certainly biogenic. Calcified linings of tubes which the bivalve has excavated for the siphons are characteristic of several groups of endolithic bivalves.

Substrate control of epitaxial cement crystal orientations (most commonly in intraskeletal cavities) produces degrees of ordering intermediate between the botryoids and cement fringes just discussed. Cement overgrowths on scleractinian corals (Macintyre, 1977, Fig. 4D, d) often have an oblique orientation. That orientation reflects their development as epitaxial continuations of obliquely oriented crystals of the trabecular skeletal ultrastructure Figs. 7B–D; Cullis, 1904, Figs. 25, 29, 32; Friedman, 1973, Fig. 2). On many molluscan shells or shell fragments, aragonite cements occur as densely packed palisades of nearly parallel fibers oriented generally perpendicular to the substrate (Gevirtz & Friedman, 1966, Figs. 4–6, 8; Lucia, 1968, Fig. 18a, b; Shinn, 1969, Figs. 24, 27; Schroeder, 1972a, Figs. 1a, b; Alexandersson, 1972, Fig. 6; Jorgensen, 1976, Fig. 9; James & Ginsburg, 1979, Fig. 6-4D; Jindrich, 1983, Fig. 8d), and competetive growth interference is not generally evident. The strong control which the regularly oriented crystals of those molluscan substrates exert on the ce-

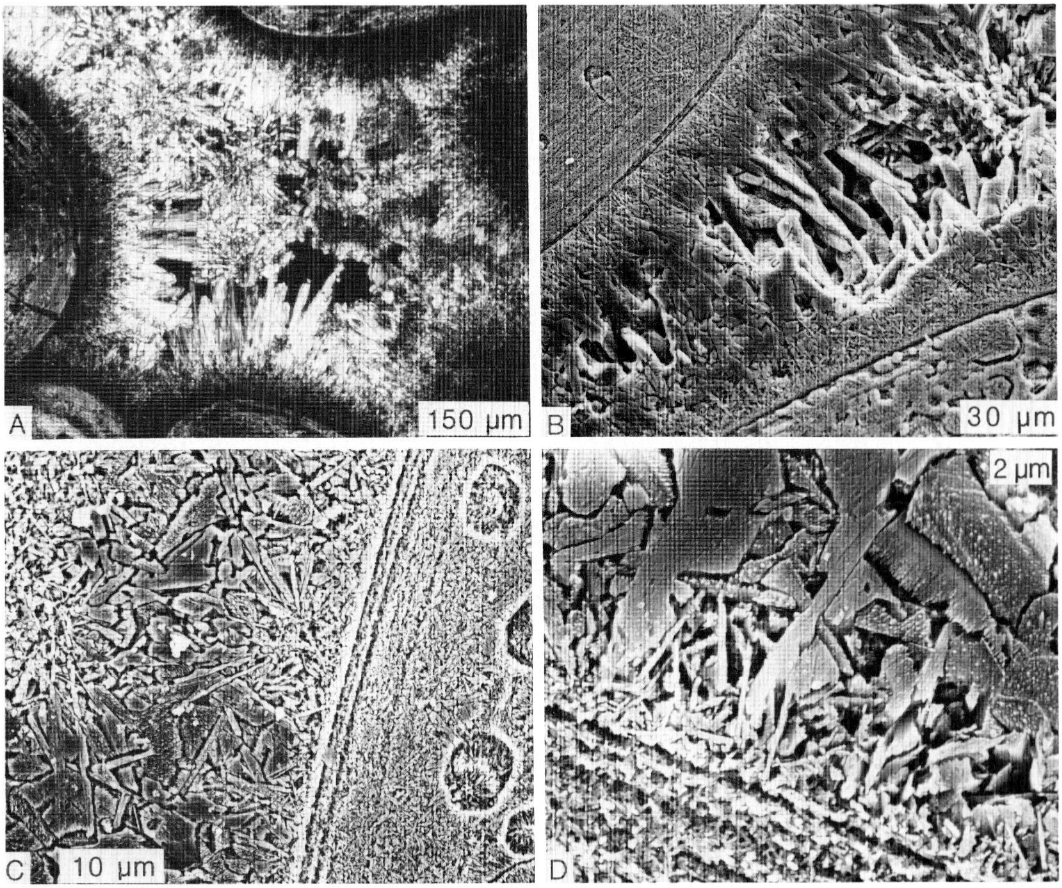

FIG. 1.—Aragonite cements in ooid hardground, Eleuthera Bank, Bahamas (see Dravis, 1979). *A*, Thin section, crossed nicols. Note the dark micritic proximal portion of the cement and the irregular lengths and slightly splayed orientations of the larger distal needles. *B*, Detail of aragonite cement fringe between two ooids. Note the irregular, micritic proximal part of the cement. The ooids are penetrated with several different types of aragonite filled endoliths, "micritization"), SEM. *C*, A very disordered cement developed on an ooid with aragonite filled endoliths, SEM. *D*, Detail view of cement-ooid junction, SEM. Note the very disordered proximal portion, the very rapid distal increase in size of some crystals, and the intricate intergrowth of the crystals. The needles of the ooid (lower left) show considerable departures from the statistically tangential orientation, SEM.

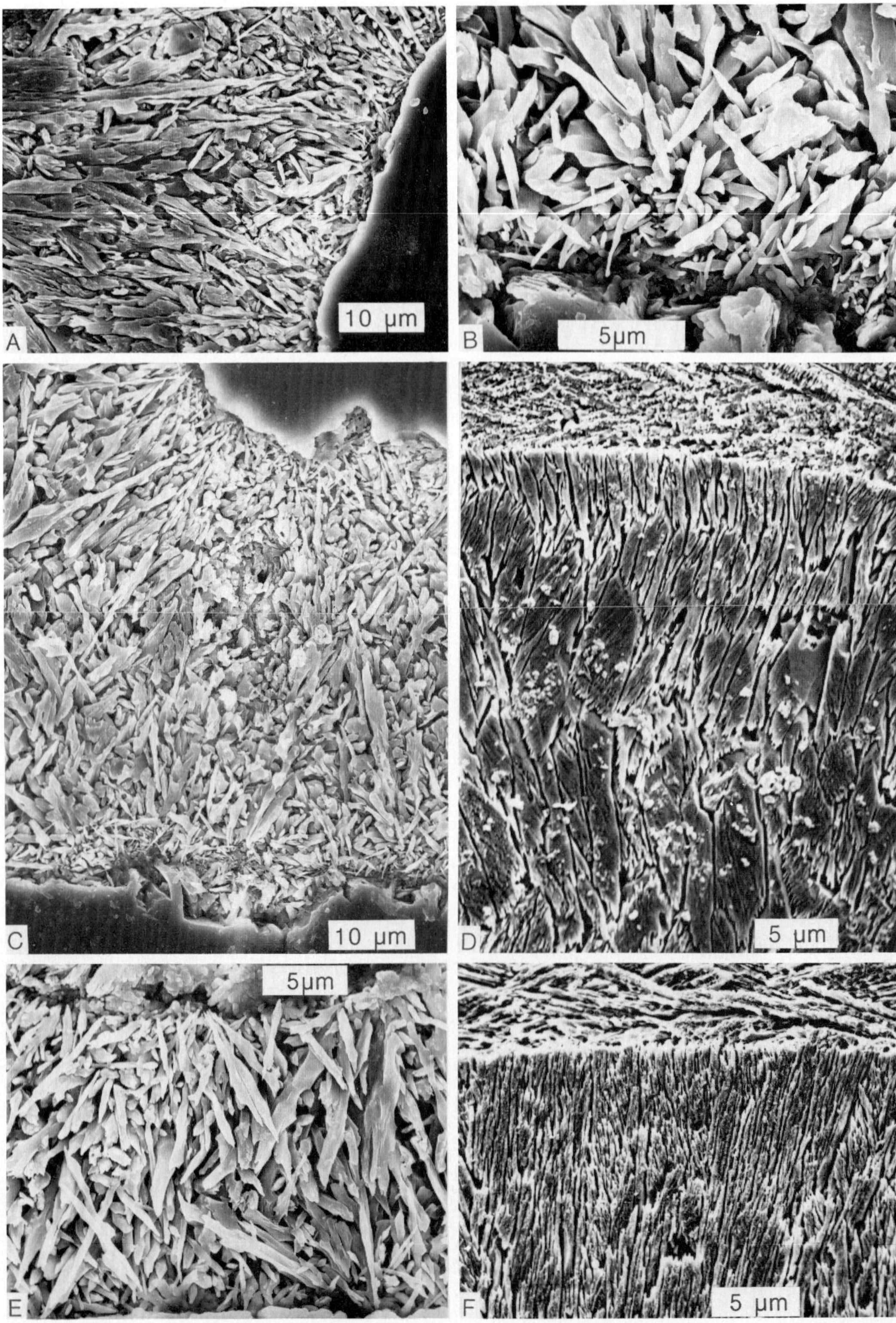
10 µm
A
5µm
B
10 µm
C
5 µm
D
5µm
E
5 µm
F

ment orientation is underscored by the occurrence of "crossed-lamellar cements" on crossed-lamellar molluscan ultrastructures (Schroeder, 1972a, Fig. 3a; Ginsburg *et al.*, 1971, Fig. 6). Even the well-ordered, mainly perpendicular cements on molluscs are distinguishable, by virtue of their variable lengths and somewhat disordered arrays, from calcite palisade cements which are typically equal length crystals oriented perpendicular to the substrate (e.g., Schroeder, 1973, Fig. 4b). If one compares calcite palisade cements to the more generally occuring intergranular aragonite cements (discussed above) found on diverse carbonate and non-carbonate grains, the differences are even more striking. Intergranular aragonite cement fringes may have little angular divergence from perpendicular to the substrate, usually on molluscan or other grains composed of oriented aragonite. However, they commonly are quite splayed and form an overlapping, porous meshwork (Schroeder, 1972a, Fig. 1e; James & Ginsburg, 1979, Fig. 6-4A-C), in the finer proximal portion (Figs. 1, 2A–C, E). That proximal portion appears micritic and dark (Fig. 1A), often brownish, in the LM, but the continuity into the coarser, "radial" portion is clear. The point to be made is that such "micritic & radial" cement "couplets" found in older rocks may commonly represent only growth stages of a single cement generation. James & Ginsburg (1979, p. 111) noted that, in their material from the Belize barrier reef, HMC cement was "present as isopachous rinds" but that "aragonite, in contrast, never forms isopachous rinds but instead the crystals either extend out from the grain surface in all directions producing a mesh-like intergranular cement or occur as spherulites . . ." Illustrations of isopachous calcite cements and aragonite cements by other authors generally corroborate that interpretation (e.g., Cullis, 1904, Figs. 26–28; James, *et al.*, 1976 Fig. 16; Taylor & Illing, 1969, pl. 5, Figs. c, d; Shinn, 1971, Fig. 4; Schroeder, 1972a, Fig. 7; 1973, Fig. 4b; Ward, 1975, Figs. 31a, b; Land & Moore, 1980; Fig. 11; Jindrich, 1983, Fig. 8e). Although such nearly equal lengths are most common in calcite cement fringes, James & Ginsburg (1979, Figs. 6-2b, c, d) illustrated a variety of cement fabrics, some with quite variable crystal lengths (see also Macintyre et. al., 1968, Fig. 4). Ancient examples of variable length calcite cement fringes were illustrated by Purser (1969, Fig. 11B) and Talbot (1971, Fig. 7). Note that Talbot's Figures 7 and 8 were reversed in publication.

Some studies suggest that, even in shallow shelf environments, HMC cements can be as common as aragonite or even more common (Land & Goreau, 1970; Ginsburg *et al.*, 1971; Schroeder, 1972a; Friedman *et al.*, 1974; James *et al.*, 1976; James & Ginsburg, 1979), although aragonite may predominate locally. The exact controls on cement mineralogy are not well known, but the occurrence of different mineralogies in adjacent intraskeletal cavities or even sequentially in the same cavity (Schroeder, 1972a; 1973) suggests that modern conditions facilitating one or the other polymorph cannot be terribly different. Although substrate control of cement mineralogy may occur, as in the molluscan examples just discussed, increasing evidence suggests a common absence of such control (e.g., Dravis, 1979; Alexandersson & Milliman, 1981). Sequencing of mineralogies in cement successions shows no consistent pattern nor is there optic continuity between the two mineralogies. HMC cements may be preceded (James, 1983; Halley *et al.*, 1983) or followed (Alexandersson, 1972) by aragonite cements. Alexandersson (1972, p. 454) noted that intervention of a HMC micrite cement inhibited the development of ordered aragonite cement otherwise seen on molluscan substrates.

A radiaxial fibrous form of aragonite, inferred (based on analogy with modern cements) to have been the original composition of RFC cements in some Mississippian grainstones (Cotter, 1966), is unknown among modern cements. Given the inferred mode of growth of RFC cements (see Kendall, this volume), equivalent growth of aragonite seems highly unlikely.

DIAGENESIS

Recognition of altered aragonites.—The general diagenetic behavior of aragonite has been covered in some detail in earlier papers (e.g., Bathurst, 1964; James, 1974; Sandberg, 1975a, b; Pingitore, 1970, 1976; Folk & Assereto, 1976; Assereto & Folk, 1980). Those studies, and others uncited, emphasized the texturally disruptive nature of the alteration, despite the fine scale of the alteration process, as well the frequent preservation of aragonite constituents as molds of LMC spar cement filled molds. A large number of other articles have suggested that skeletons, ooids, and diverse cements with fine uniform textures and regular optic behavior were the products of pseudomorphic transformation of aragonite. The topic of

FIG. 2.—*A–C, E,* Aragonite needle cements on volcanic rock fragment grains, St. John, Virgin Islands, SEM. In all cases, the cements are quite splayed in the initial portion and become only somewhat more ordered away from the substrate grains. Figs. C & E show cements between two adjacent grains. *D, F,* Epitaxial aragonite cements on molluscan shell substrates (at the top of each photo), Persian Gulf sample ES119 from the study of Shinn, 1969; SEM. The degree of ordering is considerably more than in the intergranular cements, either in Figs. 2A–C, E, or in Fig. 1. The substrate grains are located as follows: Fig. A (right side), Fig. B (bottom), Figs. C & E (top and bottom), Figs. D & F (top).

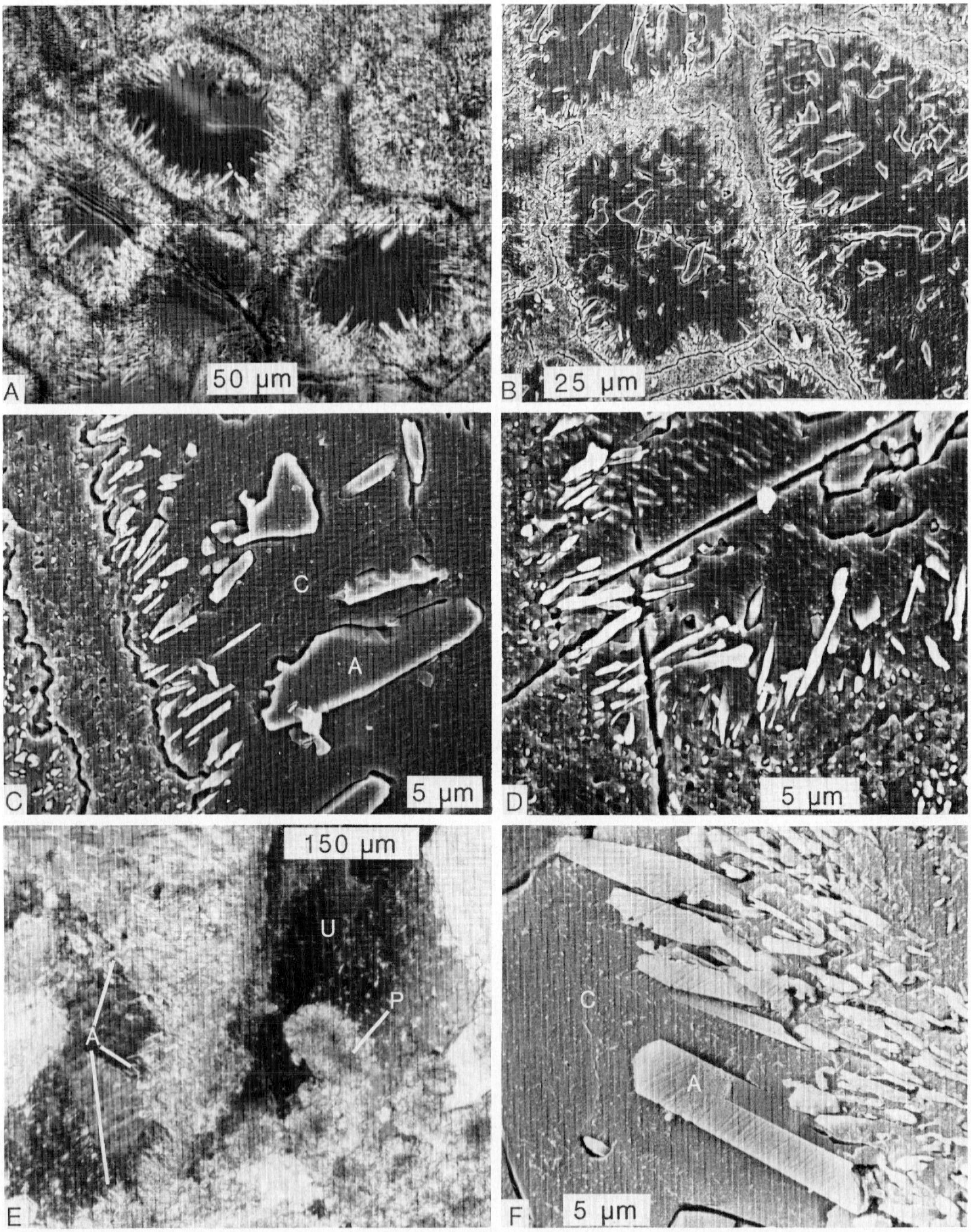

FIG. 3.—*A–D*, Aragonite needle fringe cements growing on the cell walls of the phylloid alga *Archaeolithophyllum* and engulfed distally in secondary sparry calcite cement, U. Pennsylvanian, E. Kansas. Fig. A is a light micrograph, crossed nicols; Figs. B–D are SEM. The aragonite needles look much less common in the SEM (compare Figs. A and B) because it images only a surface, compared to the 30 µm thickness of a thin section shown in a light micrograph. Note the drusy texture of the aragonite cement (Figs. B–D). *E*, Aragonite needle cements (A) and peloidal (P) calcite cements engulfed in undulose second-generation sparry calcite (U) in a Triassic aragonite coral (Cassian beds, sample of Scherer, 1977), light micrograph, crossed nicols. *F*, Aragonite needle cement engulfed between calcitized coral skeleton (*Montastrea annularis*) to the right and second generation sparry calcite cement (C) to the left, SEM. One quite large crystal (A) appears to have 110 twinning. Key Largo ls. (Pleistocene), Key Largo, Florida.

pseudomorphing deserves special attention and will be discussed later. For the evaluation of the temporal distribution of aragonite cements (and ooids) in the geologic record the crucial question is: "are there consistent properties which characterize products of calcitization of aragonites, whether skeletons, ooids, or cements?" My experience indicates that there are. I will attempt to summarize them here:

1. The prime petrographic trait, used since the time of Sorby (1879), is the presence, in calcitized aragonites, of an irregular, generally rather coarse calcite mosaic. The boundaries of the crystals of that mosaic irregularly cross-cut the relics of the original structure commonly noted in calcitized skeletons (Bathurst, 1964; Pingitore, 1976; Sandberg & Hudson, 1983). Those relics (which may include organic matter, non-carbonates, and aligned fluid inclusions) are probably the major feature differentiating a calcitization mosaic from a void-fill calcite (Bathurst, 1964). Those aragonite cements which have a relatively massive structure, e.g., botryoids (Otte & Parks, 1963; Mazzullo & Cys, 1979; Mazzullo, 1980), spherulites ("spar spots" of Ravn & Heckel, 1978) or giant rays (the "raggioni" of Assereto & Folk, 1976), also have such relics (Fig. 5). The morphology and size of the crystals in the calcitization mosaic may vary considerably among different materials and even from point to point within constituents with original structural diversity in the aragonite. However, that mosaic has two important properties: A. it encompasses the entire constituent, unless a part remains as unaltered aragonite, i.e., there are no residual regions of fine calcite with regular ultrastructure. B. the replacement calcite crystals are characteristically much larger than the aragonite crystals they are replacing and are disordered optically and texturally in comparison to that of aragonite.

2. The replacement calcite is often brownish and pseudopleochroic in calcitized skeletons (Hudson, 1962). Somewhat similar appearances occur in calcitized Pennsylvanian botryoids from Kansas and Missouri.

3. With increasing frequency and in progressively older materials, relics of the original aragonite are being found as solid inclusions in the replacement calcite (Fig. 5; Sandberg, 1975a, b, in press; Sandberg & Hudson, 1983; Sandberg *et al.*, 1973; Sandberg & Popp, 1981, in prep.). It should be noted, however, that relics do not occur in all calcitized aragonites (Sandberg & Hudson, 1983), even in the Pleistocene.

4. Elevated Sr^{2+} content is common in calcitized aragonites, although the amount present can be quite variable, depending on starting material and on the nature of the chemical system. Nevertheless, elevated Sr^{2+} values above levels reasonable for pirmary calcites (given Sr^{+} concentrations in seawater and the partition coefficient of Sr^{+} into calcite) corroborate the textural indications of former aragonite mineralogy (Assereto & Folk, 1976; Davies, 1977a; Mazzullo, 1980). Data of Sandberg & Hudson (1983, Table I) indicate that Sr^{2+} in aragonite relics does not account for the elevated Sr^{2+} content in the combined aragonite and neomorphic calcite. The calcite itself has a high Sr^{2+} content.

In contrast to calcitization of aragonite, the HMC or LMC to dLMC transformation typically preserves the optic regularity and may even lack changes in crystallite morphology that are readily detectable by light microscopy (Friedman, 1964; Banner & Wood, 1964; Schneidermann, 1970; Sandberg, 1975a) or even by means of SEM (Towe & Hemleben, 1976). The latter authors argued that the alteration process was from fine HMC to fine dLMC to coarse dLMC. As noted by Steinen (1978), once the magnesium is lost, little driving force remains for further alteration. Surely crystal size alone cannot be a major factor, or else why should we have any early Paleozoic micrites left? I believe it is more probable that, under different conditions (as yet not well understood), differing degrees of textural modification produce either fine dLMC or a dLMC that is coarse relative to original calcite crystal sizes. Extreme changes (e.g., Sandberg, 1983, Fig. 3) in crystal size appear rare. Later, secondary recrystallization is a possible, but probably infrequent fate for such initially stabilized diagenetic products (Veizer, 1977; Sandberg & Hudson, 1983). Elevated Sr^{2+} content, and even aragonite relics, in many ancient constituents argues against later multiple recrystallizations.

Wagner & Matthews (1982) have argued for a synergistic combination of petrography and geochemistry in the study of carbonates. That multiple technique approach has much to recommend it, especially if the different components of the limestone are segregated and analyzed separately (e.g., Brand & Veizer, 1980; Popp *et al.*, in press). However, I disagree, at least in part, with the conclusion of Wagner & Matthews that textural criteria do not exist that allow determination of initial mineralogies of grains. It is the obscuring influence of alleged pseudomorphic calcitization, that masks the fact that once-aragonitic grains are distinctive in their textural disruption. The unusually fine hot spring alteration reported by Richter and Besenecker (1983, Fig. 4), still represents a large size increase to irregular 5–50+ μm crystals which generally contain aragonite relics (Sandberg, unpublished).

Calcitization of aragonites in near surface meteoric environments or in deeper burial environments does not appear to produce any distinctive differences in the resulting calcitization mosaics (Sandberg & Hudson, 1983). Füchtbauer (1974, p. 295) suggested that calcitization of skeletal aragonites occurred mainly in "rocks which presumably had initial low permeability, e.g., shales, marls, or micritic limestones." My experience has been that many highly porous grainstones have well developed calcitization, e.g., the Pleistocene of south Florida, the Pennsylvanian of eastern Kansas, the Triassic of the Alps.

In units in which aragonitic skeletons and ooids occur together, their diagenetic behavior upon calcitization is essentially identical (Sandberg, 1975b, Sandberg & Popp, in prep.). The same is true for coexisting aragonite cements and skeletons. In the upper Pennsylvanian of Kansas & Missouri, phylloid algal blades (*Archaeolithophyllum*) and small botryoidal cement overgrowths are both preserved as coarse, brownish, pseudopleochroic calcite with crystal boundaries irregularly cross-cutting the original structure that is reflected by non-carbonate inclusions.

When non-botryoidal aragonite cement is a solid palisade, as it is in most epitaxial overgrowths on molluscan shells, then it is a starting material equivalent to the solid aragonite shells (or the botryoids). Its diagenesis should then produce the same sort of irregular replacement mosaic found in calcitized shells (Sandberg & Hudson, 1983, Fig. 1) or botryoids (Fig. 4; Mazzullo & Cys, 1979, Fig. 4; Otte & Parks, 1963, pl. 3, Figs. A, B). There is no evidence that aragonite cements undergo a different diagenesis, but there is abundant evidence that those cements may be less susceptible to alteration than the aragonite shells on which they are deposited and may persist long after the skeleton is calcitized (Fig. 3; Land, 1967, Fig. 12; Schroeder, 1973, Figs. 8, 10). The reasons for this are not clear, but may in part relate to the lesser development, in crystals of the cements, of twins and defect structures which increase solubility (Wenk *et al.*, 1983).

Micrites are not included in this discussion because the precursor muds are probably always polymineralic, hence different alteration behavior will result (Lasemi & Sandberg, 1983, and in prep.).

Füchtbauer (1974, p. 295) suggested that aragonite-calcite transformations could be either "homoaxial" or "heteroaxial", referring to retention or loss of optical and morphologic orientations. As examples of homoaxial transformation he cited the experimental work of Hathaway & Robertson (1961), the work by Fenninger & Flajs (1974) on hydrozoans, and Land (1967). In the high-temperature compression experiments of Hathaway & Robertson (1961), one run (at 1000 psi and 430°C) produced a carbonate indicated as "100% calcite." Crystal morphology in the illustration they gave for that sample (1961, Fig. 3) is not greatly different from that of the starting sediment (Fig. 1). All other runs produced (at lower T and/or P) strong departures in morphology, tending toward larger, equant crystals. Their one anomalous run remains as a possible example of morphologic homoaxial transformation (nothing is known of the optic properties of the run products). It would be useful to test that interpretation with a replication of the experimental run. The relevance of those extreme experimental conditions to natural diagenesis is uncertain.

Texture-preserving transformation of aragonite to calcite was interpreted to have occurred in Jurassic and Lower Cretaceous supposed hydrozoans (Fenninger & Flajs, 1974, Pl. 7) identified as stromatoporoids. In order to account for the presently calcitic spherulitic textures, Fenninger & Flajs invoked a "transformation" of aragonite to calcite. Although not defined or characterized, that process was regarded as distinct from alteration by solution-redeposition. Current understanding of aragonite diagenesis does not support such a distinction. In the face of the consistency of diagenetic behavior of aragonite in other taxa, in ooids and in cements, it is more parsimonious to infer that the calcitic Mesozoic stromatoporoids with preserved fine texture studied by Fenniger & Flajs were originally calcitic. Land (1967) found preferred orientation of the coarse replacement crystals in a calcitized mollusc that reflected the orientation of the crossed-lamellar aragonite of the original shell. However, the replacement mosaic, in that and other calcitized molluscans, is coarse and irregular relative to the replaced aragonite structures. There is not a smooth extinction sweep as the stage is rotated. Rather there is a scattering of crystals at extinction throughout the field of view at any position of stage rotation. Morphologically, those calcitization mosaics are examples of the "heteroaxial transformation" of Füchtbauer, regardless of what U-stage investigations might reveal about preferred optic orientation.

Preserved ancient aragonites.—Common interpretations of carbonate diagenesis would suggest that unaltered aragonite cements should not occur in ancient rocks, unless protected in impermeable lithologies like the well-known Buckhorn Asphalt. It was quite surprising then, when Wendt (1977) reported unaltered first-generation aragonite cements encased in second-generation sparry calcite in intraskeletal voids of Permian fossils. Some earlier data at least suggested that such preservation might be possible. Cullis (1904, Figs. 32, 35–37, 56), Shinn (1969, Fig. 25) and Schneidermann *et al.* (1972, Fig. 3) had reported such "engulfed" aragonites from the Pleistocene. Pingitore (1970) found that needle fringe aragonite cements were absent or rare in lithified Pleistocene reefal limestones of Barbados. My studies of the Key Largo Limestone of south Florida indicate that aragonite fringes are rather commonly preserved, engulfed in second generation LMC spar cement and abutting proximally against calcitized skeletal substrates (Fig. 3F). Schroeder (1973, Figs. 8, 10) illustrated similar aragonite cements from the Pleistocene of Bermuda, in which the skeletal substrate had been calcitized, leaving the cement surrounded by dLMC.

Such geologically young occurrences could be regarded as "partially altered" transitory states in rocks destined soon to be altered to the final state, LMC and

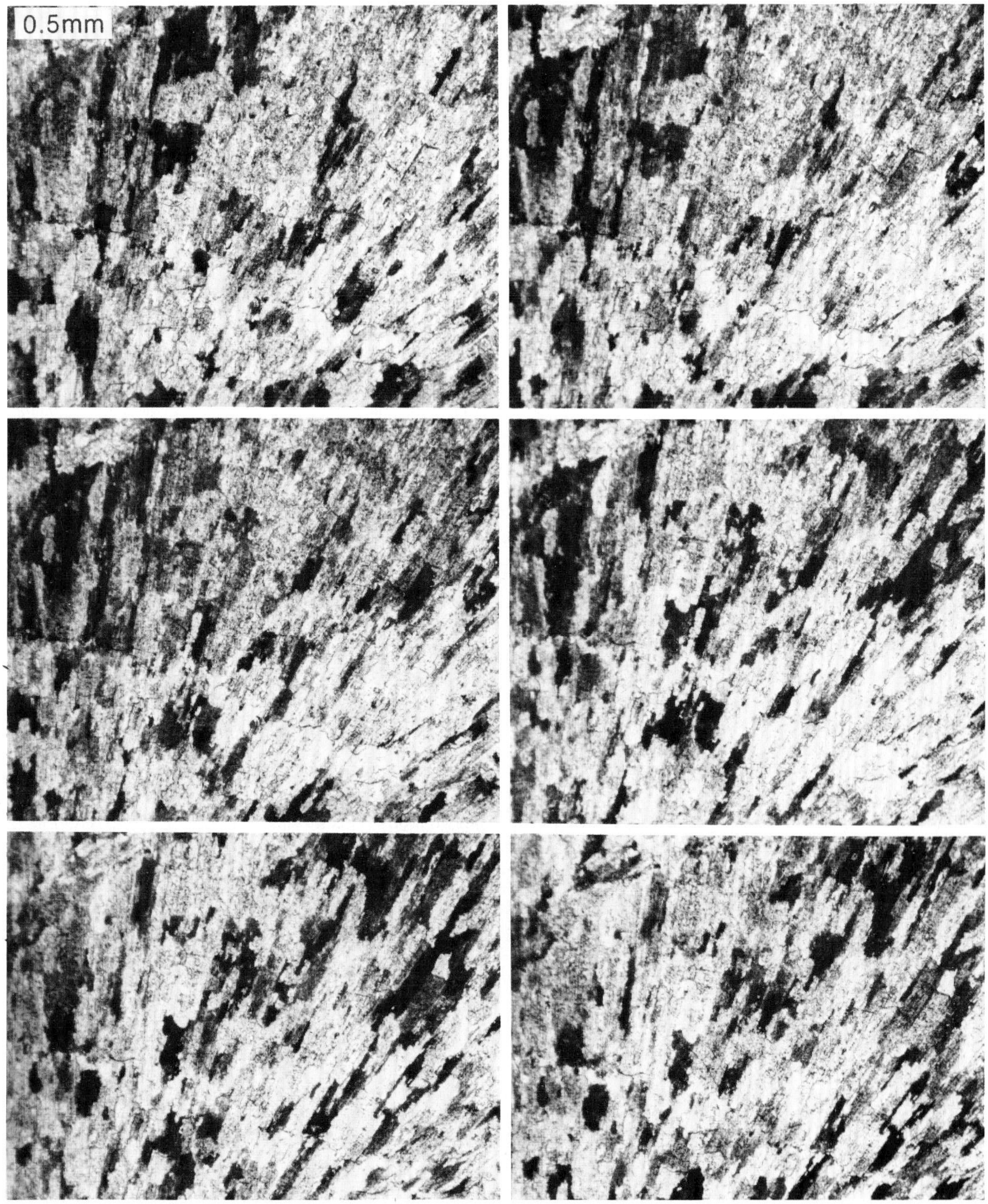

FIG. 4.—Calcitized botryoidal aragonite cement, Permian reef complex, Laborcita Fm., New Mexico (Sample of Mazzullo, 1980). The figures show the same portion of one botryoid at successive 12° rotations of the polarizer-analyzer orientation, relative to the specimen. In this field of approximately 60° angular divergence of original crystallite long axes, there are replacement crystals at extinction through the array at every rotation position. Note that different segments of the length of former fibers are replaced by calcite crystals of quite different extinction angles, i.e., there is no systematic relationship between optic orientations of original and replacement crystals.

dLMC with or without dolomite, regarded as characteristic of ancient limestones in general. It is, therefore, significant that comparable engulfed aragonite needle cements remain in ancient rocks (Wendt, 1977), not only in protective, relatively impermeable lithologies (Scherer, 1977; Scherer & Wendt, 1978), but also in normal grainstones and packstones (Fig. 3; Sandberg & Popp, 1981, in prep.).

For several reasons, these aragonite needles are taken as engulfed first-generation cements rather than relics of calcitization of a more dense aragonite cement. In transmitted light, the needles can be seen to be entire and continuous to the substrate (In contrast, aragonite relics in calcitized cements are typically irregular segments of the length of the precursor aragonite needles, Fig. 5). The calcite which surrounds the needles is the same clear calcite that occurs out in the pore center. That clear spar lacks the brownish color, pseudopleochroism, and abundant fluid inclusions that characterize adjacent calcitized aragonite skeletons. It also lacks the "ghosts" of precursor structure that characterize calcitization mosaics of other aragonite cements (Mazzullo, 1980; Otte & Parks, 1963). All these properties indicate origin as a porous, first-generation aragonite needle cement later engulfed in a LMC spar cement. The interpretation by Schneidermann *et al.* (1972) of what Pingitore (1976) later called "cross-cutting relationships" as indicative of former aragonite pore-fill cement was an error for the samples in our figures 1 and 2. The coarse needles we illustrate from another sample (1972, Fig. 3) are engulfed aragonite cements, not relics of a calcitization (compare Shinn, 1969, Fig. 25).

Although it is not yet clear how, the presence of the engulfing second-generation LMC spar cement obviously affords protection to needle fringes even as old as Pennsylvanian (Figs. 3A–D; Wendt, 1977, Fig. 1). The story is somewhat complicated by Cullis's interpretation (1904, p. 398–400) of aragonite cement in part as the product of recrystallization of calcareous muds. He noted (p. 399) that those aragonite cements inferred to have replaced mud are formed as continuation of skeletal aragonite crystals and inferred that mud within or between calcitic organisms was changed into calcite "often similarly disposed", *i.e.*, in optic continuity with the skeletal calcite. There is no compelling reason to regard those aragonite or calcite cements as a replacement of mud. The equivalent epitaxial aragonite cements in corals from the Pleistocene of Japan are morphologically identical in mud-filled and adjacent mud-free cavities (Fig. 7). They, like Cullis' aragonite needles, are more reasonably interpreted as prior cement precipitation, followed by mud emplacement in some of the cavities. Examples most likely to be replacement of carbonate sediment by aragonite involve quite massive aragonite spherulites irregularly encroaching on enclosing sediment (Warren, 1982, Fig. 6), apparently in hypersaline waters.

Alteration of botryoids and spherulites.—Because of their size and distinctive morphologies, the most obvious of ancient aragonite cements are the botryoids. Cements of this morphology, commonly with an inference of original aragonite mineralogy, have been reported from rocks of Late Mississippian to Early Jurassic age (see Sandberg, 1983, Fig. 6), and also from the Early Cambrian (James & Klappa, 1983) and the Middle Precambrian (1.8–1.9 Ga; Grotzinger & Read, 1983). The lower Jurassic botryoids of Burri *et al.* (1973) have not been studied petrographically for indications of original mineralogy, but may possibly be calcitized aragonite like the Mississippian to Triassic examples discussed in this section.

The range of alteration accompanying calcitization of ancient botryoids is exemplified by the botryoids (Figs. 4, 5) from the Permian of New Mexico (Otte & Parks, 1963; Mazzullo & Cys, 1979; Mazzullo, 1980) and the Pennsylvanian of S. E. Kansas (Heckel, 1978; Ravn & Heckel, 1978; Sandberg & Popp, in prep.). The Permian botryoids are now composed of numerous calcite crystals predominantly elongate in the same direction as the original aragonite fibers (Fig. 4). The Pennsylvanian botryoids, in contrast, are each replaced by relatively few or even only one calcite crystal. Despite the difference in size, replacement crystals in botryoids of both ages contain rare oriented relics of the original aragonite (Fig. 5). The Permian material

FIG. 5.—Aragonite relic crystals in calcitized aragonite botryoids and spherulites. Figs. A–C are SEM; D–F are LM, crossed nicols. *A*, Aragonite relics in a calcitized botryoid from material of Mazzullo (1980), Laborcita Fm., Permian, New Mexico. Note that the relics are discontinuous segments of the length of the original aragonite crystals. *B*, Aragonite relics in a calcitized spherulite ("spar spot" of Ravn & Heckel, 1978), Plattsburg Ls., U. Pennsylvanian, eastern Kansas. Note that the relics are considerably smaller than those in the Permian botryoid, although the Permian relics give some indication of being composite crystals made up of smaller fibers (see fig. 5A). *C*, Aragonite relics in another spherulite. Same sample as in Fig. 5B. the relics, oriented NE-SW, have a higher brightness in emissive mode SEM in this and other samples, probably in part because of the influence of the incorporated Sr^{+2} on the mean atomic number and hence on backscattered primary electron emission. *D*, Aragonite relics in irregular neomorphic calcite. Same sample as Fig. 5A. The relics are quite bright against the enclosing calcite crystal at extinction. *E*, Abundant aragonite fiber relics in a calcitized "spar spot". Same sample as Fig. 5B, C. The relics are clearly discernible against the extinguished crystal in the center, but can even be seen in the adjacent, non-extinguished calcite crystal to the lower right. *F*, Rare aragonite relics in a massive replacement calcite crystal of a botryoid, Captain Creek Ls., U. Pennsylvanian, eastern Kansas. Note the abundant dark, non-carbonate (non-birefringent) inclusions in the same NE-SW orientation as the aragonite relics.

often has elevated Sr^{2+} content (Mazzullo, 1980, Table 1); Sr^{2+} content of the Pennsylvanian material has not yet been determined.

Davies (1977a) reported botryoidal cements (illustrated only in plain light) which had abundant retention of relic fibrous texture (presumably in coarser replacement crystals). Davies (1977a, b) did not present crossed nicols views of botryoids, but his plain light figure (1977b, Fig. 8, 2nd photo on left) does suggest irregular crystal boundaries cross-cutting relics of the original fibrous texture. Strontium values were very high (7000–8000 ppm), suggesting not only an ara-

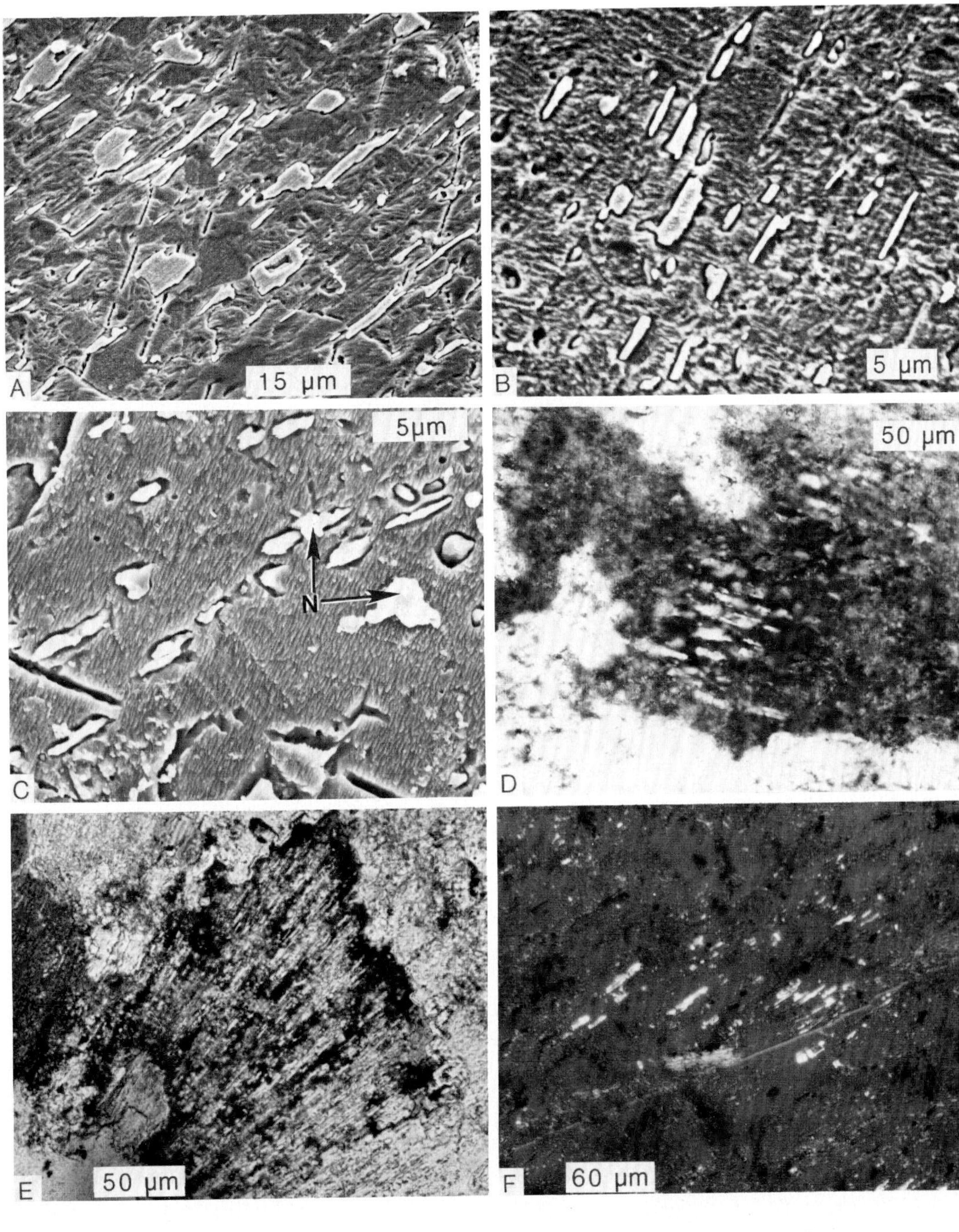

gonite precursor, but also a nearly closed system calcitization, similar to that inferred by Wardlaw *et al.* (1978) for strontium-rich calcite replacing aragonite molluscs. The generally heavy, "marine" oxygen isotopic values in the botryoids of Davies (1977a) support this interpretation.

Mazzullo (1980) hypothesized a complex diagenetic history for the Permian botryoids involving calcitization of aragonite fibers to fattened and enlarged calcite fibers (called "primary fibers"), aggregation of those "primary" fibers into elongate, parallel-sided "ray crystals", and finally amalgamation of "ray crystals" into more irregularly shaped "composite crystals". Cross & Klosterman (1981, Figs. 4A, B) illustrated relic outlines of the original fibers (shown in their plain light photographs). The enclosing, larger, calcite spar crystals, whose boundaries cross-cut those relics, were shown in their crossed nicols photographs of virtually identical areas. They interpreted those simultaneously present features as two of three "stages of alteration". Mazzullo and Cys (1979) suggested that "primary fibers" (i.e., inclusions outlining locations of former aragonite fibers) occurred in all but the largest replacement crystals. Those largest crystals were taken to be "the most neomorphosed". That interpretation is refuted by the presence of relics of the original aragonite (Fig. 5) in those largest replacement calcite crystals, including the immense (up to 2 cm) calcite crystals replacing some Pennsylvanian botryoids from Kansas.

The general presence of relic structure, (including aragonite relics), and/or elevated Sr^{2+} content in the Permian and Pennsylvanian calcitized aragonite botryoids argues against hypotheses of multiple recrystallization events. The same results can be obtained by a more parsimonious single-step calcitization process. In that interpretation, original structure of the aragonite botryoid is reflected in various ways in the products of that alteration: A. retention of non-carbonate or fluid inclusions which outline the locations of original aragonite fibers produced the "primary fibers" observed in plain light. B. strong influence of the original aragonite crystal boundaries on calcite mosaic boundaries (compare Sandberg & Hudson, 1983, Fig. 1) produced straight-sided "ray crystals". C. local absence of such inherited control resulted in the more irregular "composite crystals".

The "ray crystals" are often just local extensions of "composite crystals", whose boundaries otherwise transect the relics ("primary fibers") of the original structure. The "ray" and "composite" crystals are not temporally successive stages, but only spatial variants of the alteration process. The whole mosaic was formed in one event as the diagenetic front passed through the original aragonite botryoid. "Ray crystals" could be expected to be smaller than "composite crystals" simply because the control of replacement crystal boundaries by non-carbonate impurities along the original aragonite fiber boundaries would probably not extend uninterrupted over long distances. Extensive, well-organized organic sheets can produce widespread control of replacement crystal boundaries during calcitization of molluscan shells (e.g., Sandberg & Hudson, 1983, Fig. 1). However, such sheets are not essential or even likely parts of an aragonite botryoid.

The elongate fibrous habit of the precursor aragonite in the Permian botryoids is reflected in the preferred elongation of the replacement calcite crystals parallel to the relic structure. Those crystals have usually replaced parts of many laterally adjacent aragonite fibers (Fig. 4), are large relative to those fibers, and irregularly cross-cut the original structure reflected by non-carbonate relics and fluid inclusions. It is clearly not the "paramorphic inversion" suggested by Mazzullo (1980, p. 416); textural and optic disruption is considerable.

Mazzullo & Cys (1979, p. 923) suggested that "c-axial directions are parallel or subparallel to crystal lengths . . . or to the long axes of included primary fibers." However, the illustrations (Mazzullo & Cys, 1979, Fig. 4b; Mazzullo, 1980, Figs, 4a, 6c) show an array of "primary fibers" diverging over a range of 60–70°. Irregular replacement crystals at extinction are scattered across the entire range of angular positions in those arrays. If the same area of an equivalent sample is observed at successive 12° rotations of a polarizer-analyzer orientation (Fig. 4) the general pattern of scattered crystals at extinction remains. That observation is underscored by the fact that portions of the length of any given group of essentially parallel aragonite fibers go to extinction at quite different angular positions. Neither the published photos nor material of Mazzullo's which I have studied support the interpretation (Mazzullo, 1980, p. 411, Figs. 4B, 6C) that the c-axes of the replacement calcite are oriented in a "divergent-radial" manner, or that they are "consistently oriented at angles of approximately 45° to the elongation of the primary fibers . . ." (p. 413), or parallel to that elongation (Mazzullo & Cys, 1979, p. 923).

In the Pennsylvanian botryoids, replacement calcite crystals are very large. In thin section, there is no clear evidence of precursor control on either optic orientation or crystal boundaries. The absence of optical or textural regularity in both botryoid types is an important point. Despite the well-ordered aragonite precursor, the resulting calcite shows the same disordered, cross-cutting mosaic found in calcitized aragonite skeletons and ooids. Some aragonite cements (like the engulfed needle fringes) may outlast their aragonite substrates, but where aragonite cements are sufficiently massive (i.e. more comparable to aragonite skeletons and ooids), their calcitization products are also comparable. Mazzullo & Cys (1979, p. 919) seemed to suggest that the calcite pseudospar replacing their botryoids was syndepositional marine, but later (p. 933) suggested meteoric alteration. They suggested that the calcitization with preservation of details of the precursor aragonite is "abnormal" diagenesis. I have found that non-carbonate relics of original structure occur, often accompanied by relic aragonite inclusions (e.g., Sandberg & Hudson, 1983; Folk & Assereto, 1976), in some Paleozoic and many Mesozoic-Cenozoic rocks.

The view that neomorphic crystals cannot replace crystals with wide angular divergence of optic axes (Kendall & Tucker, 1973, p. 383) was influenced by the then-current interpretation of replacement origin of RFC and the observation of 360° cement fringes divided into numerous radial RFC crystals (Tucker & Kendall, 1973). However, single calcite crystals commonly replace numerous adjacent layers of crossed-lamellar ultrastructure (with angular divergence of nearly 90°).

The "primary fibers" in the Permian botryoids are elongate ghosts of original aragonite crystals, outlined by fluid and non-carbonate inclusions (perhaps largely organic) in the larger replacement calcite crystal of which they are a part. They consistently have the same optic orientation and birefringence as the surrounding parts of the crystal, regardless of the angle between the extinction direction and the long axis of the "primary fibers." In contrast, a relic aragonite inclusion (Figs. 5D–F) stands out brightly against the extinguished enclosing calcite crystal, particularly as the angle between their optic axes approaches 45°. All the massive calcitized aragonite cements which I have thus far investigated (Permian, Pennsylvanian), do contain some aragonite relics. Folk & Assereto (1976) reported aragonite relics from their large Holocene aragonite rays ("raggioni"). It is quite possible that most of the other massive cements whose inferred aragonite mineralogy is supported by calcitization textures and/or high strontium (e.g., those of Davies, 1977a) may be found to enclose some aragonite relics.

Longman (1980, p. 465) suggested that, for isopachous marine cements, "the fibrous texture may be preserved in spite of neomorphism to calcite." That statement could be easily misinterpreted. The square-end shapes which Longman mentioned may be preserved. However, when calcitized, such crystals (Assereto & Folk, 1976, pl. 6, Fig. B; Folk & Assereto, 1976, Fig. 2) are internally a mosaic of irregular crystals. The square ends are recognizable because prior to calcitization, they have been distally surrounded, usually by carbonate mud. The suggested retention of fibrous texture, particularly the implied radial optic regularity, is not supported by the alteration behavior of known aragonites, including cement.

The search for "square-end" terminations in ancient aragonite fringe cements is likely to be a generally unsuccessful one. In my experience and that of Torbjörn Alexandersson and of Johannes Schroeder (pers. comm.), such terminations are not common in inter- or intragranular aragonite fringe cements. The square ends may be more common in larger arrays (e.g., Assereto & Folk, 1976, 1980; Folk & Assereto, 1976, Loucks & Folk, 1976). It is noteworthy that what appears at low magnification to be single square-ended crystal, may appear, at higher magnifications, to be composed of numerous parallel fibrous subcrystals (compare Figs. 7D, 11).

The suggestion that botryoidal aragonite cements are the precursors of RFC (Cross & Klosterman, 1981) is not supported by observation on aragonite diagenesis, or current interpretations of RFC and its genesis (Kendall, this volume).

It appears that, in the presence of unusual water chemistries, dense spherulitic aragonite may grow as a replacement of surrounding matrix and even some allochems. In a study of salinas from S. Australia, Warren (1982) found spherulitic aragonite which, on the basis of abundant enfacial junctions, grew in place in the sediment. The unusual "spar spots" (Ravn & Heckel, 1978) from the Pennsylvanian of S.E. Kansas may also be replacing the surrounding matrix, engulfing some peloids and other allochems in the process. Even though they were later replaced by rather coarse, irregular calcite spar, those "spar spots" frequently contain relic aragonite fibers in their original radial orientations (Fig. 5E).

It is possible that the aragonite fan druses of Shinn (1969, Fig. 18) may represent replacive growth under rather extreme water compositions, although that aragonite could represent the filling of one of the endolith borings that are common in that Persian Gulf material. The other inferred replacive aragonites of Shinn (1969, Figs. 23, 24) appear to be only aragonite cement overgrowths and normal molluscan aragonite ultrastructures (Sandberg, in prep.).

Pseudomorphic alteration.—The concept of pseudomorphic calcitization of aragonite is a tenacious, but, in its common usage, unsupportable view which pervades the carbonate literature. If by the term "pseudomorph" one is referring to preservation of the external form of large scale features, such as the aragonite pseudomorphs after gypsums of Warren (1982), the giant aragonite rays of Assereto & Folk (1980) or of Kostecka (1972) or whole fossils pseudomorphed by sparry calcite, then it is in accord with common usage of the term in mineralogy and other areas of geology. However, that is not the general usage in most references to replacement of aragonite, in which "pseudomorphic" or "paramorphic" replacement typically implies duplication of fine crystal morphology at the ultrastructural level. There is, in my experience, no support in the geologic record for the latter view.

It is difficult to trace the precise origins of so widely applied a view, but I believe that the influence of what might be called the "ooid analogy" and the "coral analogy" (Sandberg, in press) has been great. Specifically, those are the interpretations that ancient radial calcitic ooids and trabecular to lamellar calcitic rugosan corals were originally aragonitic, like their modern correlatives, Bahamian ooids and scleractinian corals. Both those analogies were founded on assumption (whether explicit or implicit) of substantive equivalence between original states of the modern and ancient carbonates being compared. The validity of those assumptions was rarely treated further, and major effort was directed toward "explaining" the apparent texturally retentive calcitization.

Numerous references have been made to presumed original aragonite mineralogy for presently calcitic cements of quite regular fine texture and optic orientation (e.g., Wolf, 1965; Cotter, 1966; Fenninger, 1968). Noteworthy among such cements are the generally isopachous cements common in many grainstones, such as the Jurassic oolitic hardgrounds of northern Burgundy (Purser, 1969). It is evident, by means of both the light microscope and SEM, that those cements are composed of uniform crystals which are both morphologically and optically radial, densely packed, and oriented perpendicular to the ooid substrates (Fig. 6). Radial calcite ooids in those units were once regarded

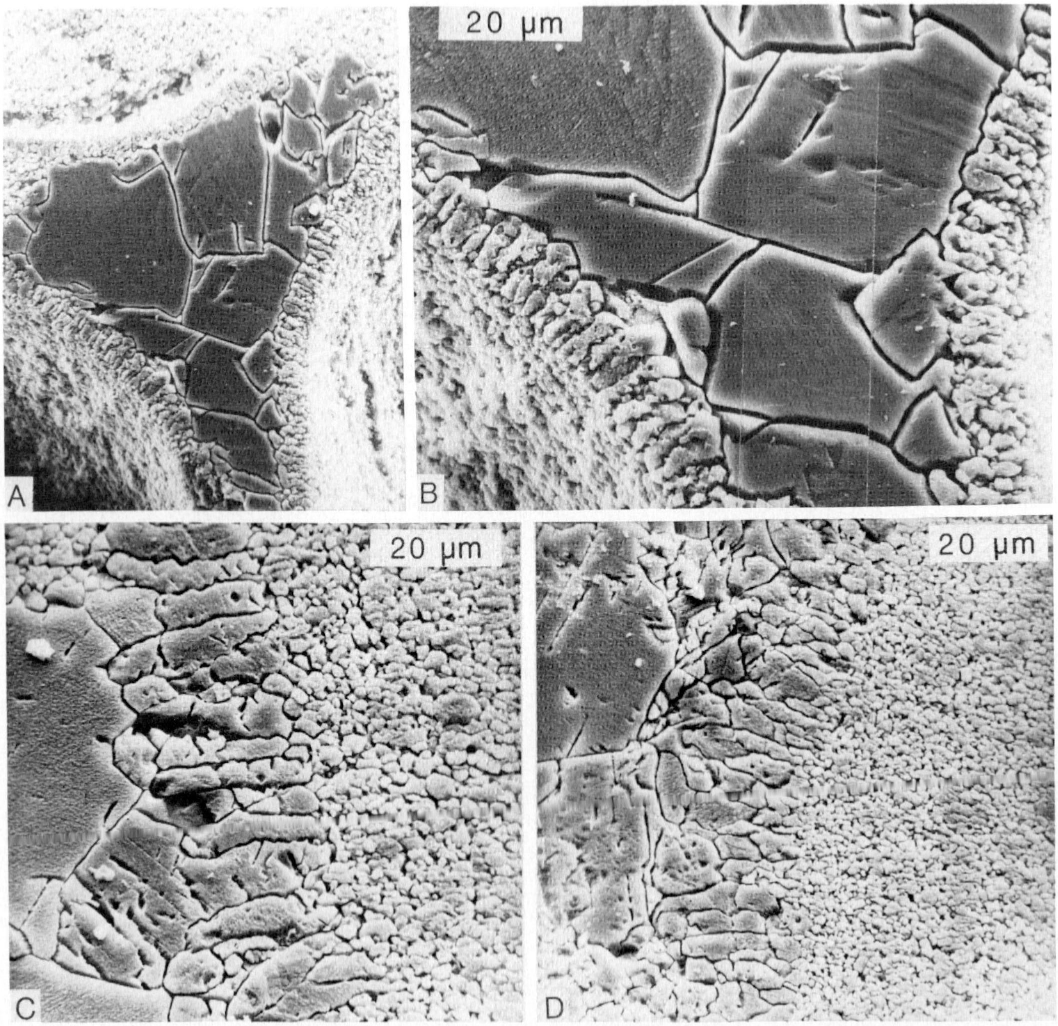

FIG. 6.—Isopachous calcite cement fringes on ooids, SEM. A–C are Jurassic (Callovian), E. Paris Basin (material of Purser, 1969); D is Upper Mississippian (Chester) S. W. Indiana. *A*, Isopachous cement engulfed in second generation calcite spar. The ooids have been weathered out. *B*, Detail of lower center of Fig. A. Note the uniform radial arrangement of crystals in the isopachous fringe, the general lack of a drusy texture, and the pointed crystal terminations. *C*, Note the somewhat longer crystals of the isopachous fringe. *D*, Note the similarity to the cements in the Jurassic sample in Fig. C and the fine, radial crystallites of the ooid.

as originally aragonitic, but are now generally acknowledged to have been originally radial calcite (Simone, 1981). Ironically, even quite recent literature suggests that the same reasoning does not seem to have extended to the similarly radial calcite cements which bind those ooids together. The relatively well-ordered radial fibrous fabrics of those Jurassic cements, and their Mississippian equivalents (Fig. 6) are probably determined by their development as overgrowths on radially oriented crystals of the calcitic ooids they encrust (a situation analogous to the aragonite molluscan overgrowths discussed earlier). Kostecka (1978) presented a careful appraisal of the possible origins of such fine radial cements from the Triassic of Poland, rejecting an inference of pseudomorphic calcitization and concluding that they must have been originally calcite.

Replacement of aragonite by calcite on a microscale with an intervening solution-redeposition phase, but without intervening void spaces and resulting in "perfect preservation of all details" (Friedman, 1964, p. 810), was envisioned in the process termed "paramorphic replacement". The inapplicability of that term, and its implied process, for the description of aragonite diagenesis (see Sandberg, 1975b, p. 519–523) is indicated by the observed textures of known calcitized aragonites and by the diagenetic behavior in materials in which diagenetic fronts are "caught in the act" (Purdy, 1968; Sandberg, *et al.*, 1973; James, 1974; Tebbutt, 1975; Pingitore, 1976; Sandberg & Hudson, 1983). In such material, one can see replacement on a microscale across very thin alteration fronts. Nevertheless, it is evident that extensive textural disruption has occurred. Wardlaw *et al.* (1978, Figs. 4, 5) illustrated (SEM) a calcitization front in a gastropod in which no void was discernible between the skeletal aragonite and the replacement calcite at 10,000X magnification. The extremely fine scale of the alteration (and the lack of significant void space along which fluids might move) is further attested to by the trace element data. The Sr^{2+} content of the calcite is essentially that of the original aragonite (Wardlaw *et al.*, 1978, Table 1). Despite that nearly closed-system alteration across an extremely thin front, the original texture of the aragonite shell is thoroughly obliterated.

Even if evidence from alteration of known aragonites supported an inference of pseudomorphic calcitization, the typical morphologies of intergranular aragonite cements are not like those of the intergranular calcite cements which some authors suggest are derivable by calcitization (compare Fig. 6 and Figs. 1, 2). The refutation is similar to arguments against inferred secondary development of the large aragonite rays in Great Salt lake ooids from the intervening fine, porous, relatively less-ordered aragonite (Sandberg, 1975b, p. 510, Figs. 3, 8–10). In the typically somewhat splayed aragonite cement fringes, growth increments are only in the crudest average sense parallel to the substrate and are discontinuous laterally (obviously not occurring in intercrystalline spaces). Even in relatively parallel dense arrays the increments could be expected to lack lateral continuity because of the variable crystal lengths. Therefore, aragonite cements still could not reasonably be regarded as precursors of isopachous, solid calcite cements with continuous growth increments essentially parallel with the substrate. It merits repeating here that calcitization of known aragonites does not produce the optic or morphologic regularity which such an inferred paramorphic transformation would require.

Mirsal & Zankl (1979, p. 944) inferred that their Devonian and Triassic cements with radiaxial optic behavior (their "Cement I") and overlying more bladed cement in optic continuity (their "Cement II") had originally been composed of "alternative layers of aragonite and calcite". Although cement couplets (calcite followed by aragonite, or vice versa) occur in modern marine carbonates rather frequently, the only multiple interlayerings of the two mineralogies of which I am aware are in the non-marine Pyramid Lake ooids (Popp & Wilkinson, 1983). In that material, there is nothing approximating the crystal continuity implied by Mirsal & Zankl (1979, Fig. 10). More compelling reasons exist, however, for rejecting the inference by Mirsal & Zankl (1979) of original aragonite mineralogy for parts of their cement succession. Their Cement I sequence is an optically continuous radiaxial cement. As discussed elsewhere in this paper (see also Kendall, this volume), derivation of radiaxial cements (or other calcite cements with consistent optic orientation relative to the substrate) from aragonite precursors is not supported by the observed optic behavior of calcites after known aragonites. The maximum range in Sr content throughout the entire cement sequence was 34 ppm at Adnet and 133 ppm at Wirbelau (Mirsal & Zankl, 1979, Table 1) and the highest Sr^{2+} value was 389 ppm, a value in the range common for calcite, but low for calcitized aragonites, especially in the "semi-closed system" they envisioned. Their inference (p. 948) that precipitation of aragonite from the "semi-closed" sea water system would lower the salinity enough to change $\delta^{18}0$ is inconsistent both with the insignificant contribution (relative to other ions) of $(Ca,Sr)CO_3$ to marine salinity and with the dependence of salinity-related diminishment of $\delta^{18}O$ on admixture of ^{18}O-depleted fresh water. As suggested by Dickson & Coleman (1980, p. 116) for other (Jurassic) materials "the preservation of extremely fine zones (both cation and isotope) . . . . indicates an absence of modification since precipitation."

Undulose extinction and replacement origin.— Bathurst (1983) has argued that some coarse intergranular spar in some Jurassic grainstones is neomorphic after a marine pore-fill cement (possibly aragonite). There is an intriguing correlation between his inferred neomorphic intergranular cements and petrographic indications of lesser compaction than in the grainstones with inferred primary intergranular spar. That relationship certainly suggests early cementation. Bathurst's inferred neomorphic cements are certainly not like "typical" modern early cements (see Figs. 1–3, 6). However, there are temporal differences in the nature of submarine cements (Sandberg, 1983). It has been proposed that sparry cements precipitated in submarine environments in the Ordovician (Wilkinson *et al.*, 1982) and Jurassic (Wilkinson *et al.*, 1983). My experience has been that, if the intergranular space had been filled with an aragonite cement, one would not expect calcitization to produce a clear neomorphic spar without relics (Bathurst, 1983, Fig. 1f). Aragonite cements which completely occlude intergranular porosity are not very common. When they occur as substrate-perpendicular palisades, a distinctive polygonal suture (Shinn, 1975) can result. Such pore-filling cements are solid masses of aragonite, and their calcitization products should resemble the calcitized botryoids and spherulites discussed herein, i.e., in possessing irregular calcite mosaics and some aragonite relics. If the pores were only partially filled with aragonite cements, then one could anticipate some engulfed aragonite cements (compare Fig. 3, and Schroeder, 1973, Figs. 8, 10), or at least that the replacement features mentioned above would be found only along the pore margins.

The undulose or patchy extinction of Bathurst's cements (1983, Fig. 2) was taken (like the undulose extinction of RFC, Kendall & Tucker, 1973) as an indication of a replacement origin. The presence of highly undulose cements (some of them with RFC-type reversed extinction sweep) overlying first-generation, metastable cements in both the Triassic and the Pleistocene (see Figs. 3E, 10, and discussion below) precludes use of such undulosity as a secure indicator of replacement. Similarly, in materials known by other criteria (2, 3, Table 1) to have been originally aragonitic, extinction in the replacement calcite may be only faintly undulose or not at all.

In some samples from the Alpine Triassic material studied by Scherer (1977), still-aragonitic corals are overgrown by a coarse interseptal cement with very undulose extinction. Close examination shows that to be a second generation cement engulfing fine acicular aragonite and peloidal calcite (HMC?) cements on the septal surfaces (Fig. 3E), a condition incompatible with an inference of replacement origin. Some of the corals in other subsamples of that material have been calcitized to a coarse, irregular, relic-bearing calcite and overgrown with bladed to equant, clear calcite crystals with unit extinction. The last central pore space, distal to that clear calcite, is filled by a cement with very undulose extinction. Neither its placement nor its distinct difference from the calcite replacing the known aragonite of the corals are supportive of an inference of either replacement origin of original aragonite mineralogy for that final cement.

I would interpret the undulose patchy extinction of the Jurassic cements (Bathurst, 1983) as a reflection of subcrystal domains of probable primary origin. That interpretation, I believe, approximates the current views of Kendall (this volume) on undulose extinction in RFC. The presence, in those Jurassic grainstones, of an early (pre-compaction) cement quite different from early cements in modern grainstones is not unreasonable, given apparent temporal differences in submarine cements discussed above.

Those Jurassic cements have possibly a wider significance. The general rule that intercrystalline contacts of crystals precipitating from a fluid should be planar compromise boundaries (Buckley, 1951) refers essentially to single crystals which are large relative to the zone of contact being considered. The undulose or patchy extinction found in the Pleistocene "RFC" cements (Figs. 7A, B) and the Jurassic cements of Bathurst (1983) suggests that those cements are composed of numerous subcrystals which meet, with small lattice misalignments, along low angle grain boundaries which do not show as petrographic features in plain light. The petrographically discernible, clearly non-planar intercrystalline boundaries in those cements may well be composite boundaries made up of numerous non-parallel, minute, local compromise boundaries. The presence of non-planar boundaries, especially in the Pleistocene "RFC", (Figs. 7A, B, 8–10), in which replacement origin is not a reasonable interpretation (see discussion below), suggests a need for reevaluation of earlier assumptions that "non-planar = replacement" in cement crystal boundaries (see similar conclusion in Kendall & Broughton, 1978). What about the extinction behavior of known calcitized aragonites? Undulose extinction is fairly common in calcitized skeletons, but, even in molluscan shells, there is a wide range in optic behavior from very undulose to essentially unit extinction. In the calcitized botryoids of the Kansas Pennsylvanian, the extinction is often somewhat undulose. The considerably smaller elongate replacement crystals in Mazzullo's Permian botryoids generally have close to unit extinction.

RFC, FOC, and aragonite diagenesis.—Both radiaxial fibrous calcite (RFC) (Kendall & Tucker, 1973) and fascicular optic calcite (FOC) (Kendall, 1977) have been inferred by various authors to be derivable by replacement of an acicular aragonite (or HMC) precursor cement. The envisioned tightly bundled acicular precursor of Kendall & Tucker (1973, Fig. 8) radiated, with considerable angular divergence, from spaced centers of nucleation on the substrate. Fibrous cements of that morphology are not common, particularly among recent cements. Modern submarine cements are generally either palisades or loose needle fringes, both quite unlike the inferred precursor of RFC. The only modern "cement" which closely resembles that supposed RFC precursor is the aragonite lining of some pelecypod borings (Schroeder, 1972a, Fig. 3). If those are biogenic deposits, as interpreted here, they have little if any relevance to intergranular cements.

In the model presented by Kendall & Tucker (1973), RFC crystals were inferred to have developed by nucleation of replacement crystals along boundaries between precursor bundles, followed by an upward migration of an alteration front generally parallel to the substrate. The boundaries between crystals were believed to be thus shifted to the centers of the inferred precursor bundles. Kendall & Tucker (1973, p. 379) suggested that retention of the c-axis orientation of the precursor by the replacement RFC "will produce the peculiar optic axis pattern that characterizes radiaxial crystals." However, if there were a retention of optic orientations, than only the boundaries of the crystal units should be shifted. There should not be the odd reversed extinction sweep, and the sweep should be continuous and uninterrupted across the RFC crystal boundaries. Adjacent margins of contiguous RFC crystals should extinguish at nearly the same angle because those margins correspond to the centers of the inferred precursor bundles, where fibers are nearly perpendicular to the substrate (Kendall & Tucker, 1973, Fig. 8).

In RFC cements of diverse ages, the actual extinction sweeps in the same direction as the image rotation and shows marked discontinuities (often 15–30°) between extinction for contiguous portions of adjacent RFC crystals. Not uncommonly there can be diagonal, or even logitudinal, sweep patterns also not "inheritable" from an inferred bundled precursor.

Although Kendall & Tucker (1973, p. 372) noted a number of characteristics of RFC that suggest a cement origin, they rejected that interpretation based in part on non-planar boundaries and undulose extinction (see discussion earlier in this paper). In addition (*loc. cit.*), cement origin was rejected because: "It is currently believed that coarsely-crystalline calcite does not precipitate from sea-water (Friedman, 1964, 1968)." The use of the coexistence of RFC and FOC in the same cement fringe as "support" for replacement origin of FOC (Kendall, 1977) was based on the assumption that RFC is also of replacement origin. The assumption of replacement origin than lead to inference of a two-stage alteration from an inferred acicular aragonite or HMC precursor, although "no direct evidence can be advanced to substantiate the two-stage replacement mechanism" (Kendall, 1977, p. 1061). Kendall concluded (1977, p. 1062) that "whether the original cement had a low-magnesian calcite, high-magnesian calcite, or aragonitic mineralogy is unknown and cannot be established from fabric evidence alone." Although I believe that those RFC and FOC cements are not the product of replacement, I would suggest that, had they originated in the inferred manner, presence of an aragonitic precursor would have produced distinctive, disruptive textures. Diagenetic fronts in aragonite constituents undergoing calcitization are characteristically non-planar and transect growth features of the precursor very irregularly. They are quite unlike the inferred, substrate-parallel alteration front of Kendall & Tucker (1973, Fig. 8). A non-replacement origin of RFC (and by extension, FOC as well) is supported by the "marine" oxygen isotopic values determined by Hudson & Colemen (1978) and by Walls *et al.* (1979). Marshall (1981) and Marshall & Ashton (1980) followed the Kendall & Tucker model in their interpretation of genesis of some Jurassic RFC (but from HMC). Their material showed numerous increments distinctive in both isotopic and trace elemental composition. For those zoned RFC cements, as for the cements of Mirsal & Zankl (1979), I believe the interpretation of Dickson & Coleman (1980) is most reasonable, i.e., "an absence of modification since precipitation."

In the model of Kendall & Tucker, if altering fluids were to enter a precursor fibrous cement from its distal surface, then those fluids must paradoxically alter only the last encountered portion of the precursor at the alteration front migrating up from the substrate. If the fluids entered along the substrate cement interface, the developing RFC crystals would progressively occlude fluid pathways up to the alteration front. The only alternative, fluid movement laterally through the cement sequence, should produce an alteration gradient in quite a different orientation from that which they hypothesized. Calcitized shells (Marshall, 1981), on which RFC cements may occur, not only block hypothetical fluid pathways, but manifest alteration textures which could reasonably be expected in the RFC had it also been originally aragonite.

Both RFC and FOC are regarded in the literature as features of ancient, altered limestones. It is therefore significant that cement fringes composed of both RFC and FOC occur in scarcely altered, aragonite-rich Pleistocene limestones from Japan (Figs. 7–10). Those cements occur in peloid-floored primary pores between calcitic and aragonitic skeletons and in primary pores within still aragonitic corals (Figs. 7A, B). The crystals are somewhat elongate perpendicular to the substrate, although not to the extent seen in many ancient RFC-FOC cement fringes. The intermixing of RFC and FOC in the same cement fringe suggest it is not reasonable to infer a difference in genesis. That intermixing has been noted in ancient rocks (Kendall, 1977; Chafetz, 1979) but has commonly been explained by complex diagenetic models.

In the Japanese material, corals and molluscan skeletons are still aragonitic (Figs. 7, 8, 10). On the corals, aragonite cements in optic continuity with the skeletal

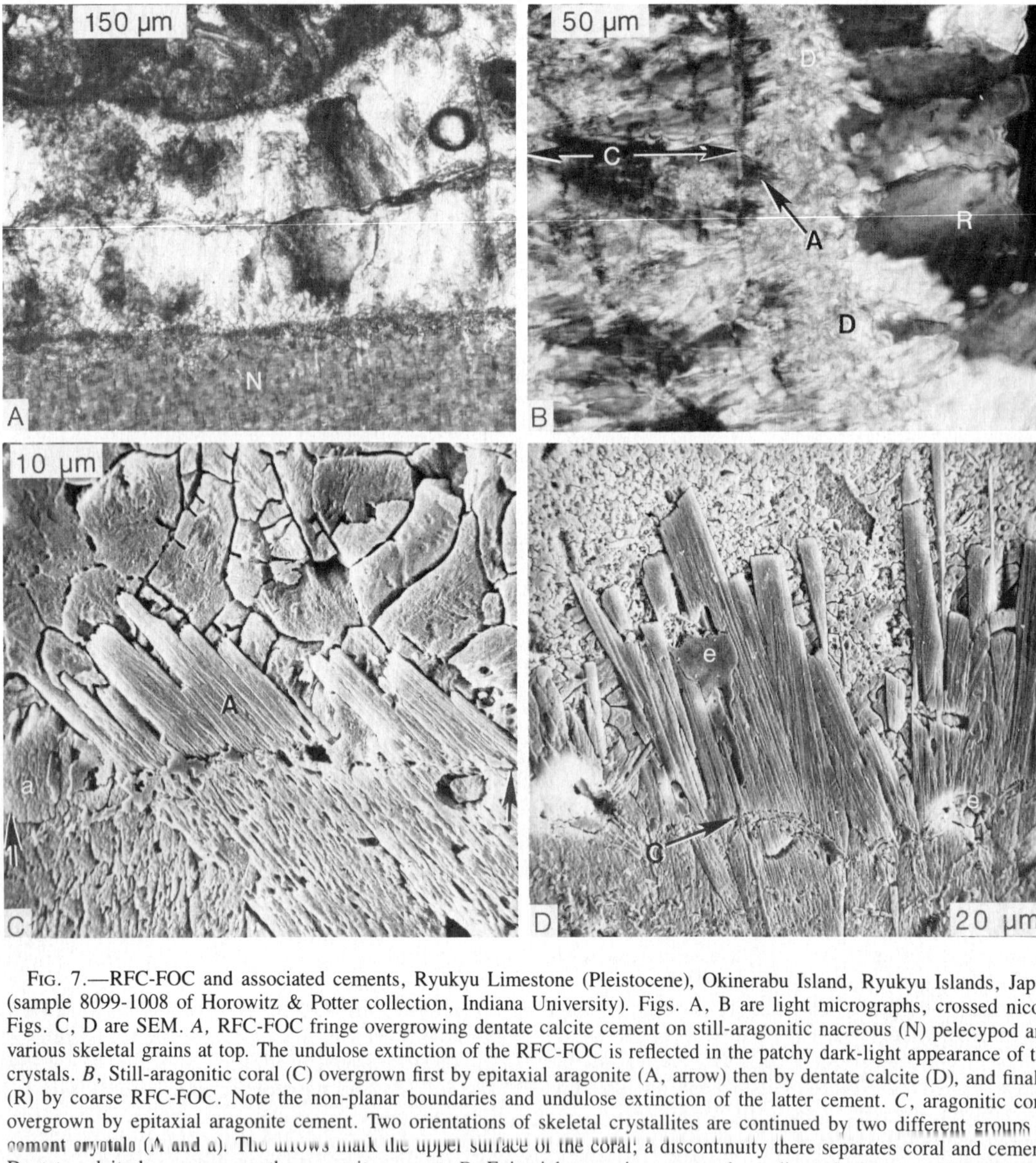

FIG. 7.—RFC-FOC and associated cements, Ryukyu Limestone (Pleistocene), Okinerabu Island, Ryukyu Islands, Japan (sample 8099-1008 of Horowitz & Potter collection, Indiana University). Figs. A, B are light micrographs, crossed nicols Figs. C, D are SEM. *A*, RFC-FOC fringe overgrowing dentate calcite cement on still-aragonitic nacreous (N) pelecypod and various skeletal grains at top. The undulose extinction of the RFC-FOC is reflected in the patchy dark-light appearance of the crystals. *B*, Still-aragonitic coral (C) overgrown first by epitaxial aragonite (A, arrow) then by dentate calcite (D), and finally (R) by coarse RFC-FOC. Note the non-planar boundaries and undulose extinction of the latter cement. *C*, aragonitic coral overgrown by epitaxial aragonite cement. Two orientations of skeletal crystallites are continued by two different groups of cement crystals (A and a). The arrows mark the upper surface of the coral; a discontinuity there separates coral and cement. Dentate calcite has overgrown the aragonite cement. *D*, Epitaxial aragonite cement above discontinuity at coral surface (C, arrow). Both coral and cements are penetrated by endolith borings (e) now filled by cement, micrite, and embedding plastic. Square end terminations occur, but the crystals appear on closer examination to be composed of numerous finer crystals. The cement is covered distally by micritic carbonate.

aragonite (Figs. 7B–D) are overlain by a RFC-FOC fringe, commonly with a dentate calcite cement generation intervening (Figs. 7B, 10). The RFC-FOC fringes are dominated by FOC, but both cement types occur in all examples studied. The fringe is much smaller than many ancient examples and some of the features noted in older examples (e.g., curved twin lamellae) have not yet been encountered. Nevertheless, the distinctive optic behavior, non-planar boundaries, and subcrystal units described earlier (Kendall & Tucker, 1973; Kendall, 1977) are well developed (Figs. 8–10). The exact crystals in those SEM photos have not been studied petrographically, so it is uncertain which are RFC or FOC (FOC comprises 60–90% of any given cement fringe in the Pleistocene material studied). It is unreasonable to advance an interpretation of an alteration to produce RFC-FOC that would simultaneously leave the adjacent first generation aragonite

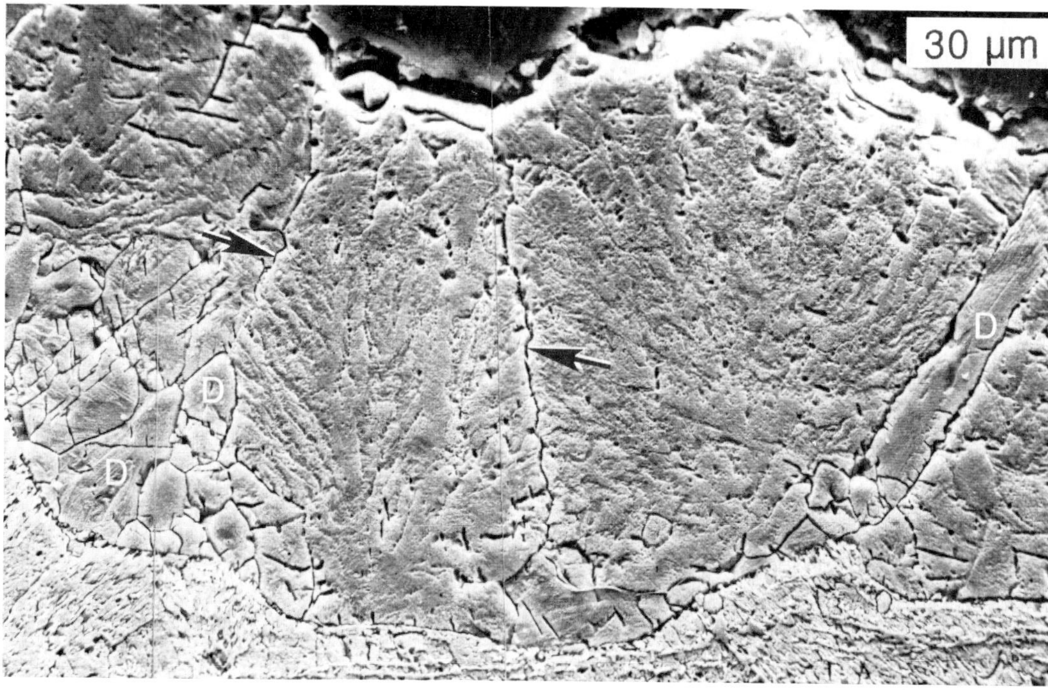

FIG. 8.—RFC-FOC and associated cements, Pleistocene, Japan (same sample as in Fig. 7), on still-aragonitic coral. First generation aragonite cements are not present in this view, but dentate (D) calcite does occur. Note the non-planar boundaries (arrows) and radiating subcrystals of the RFC-ROC. SEM mosaic.

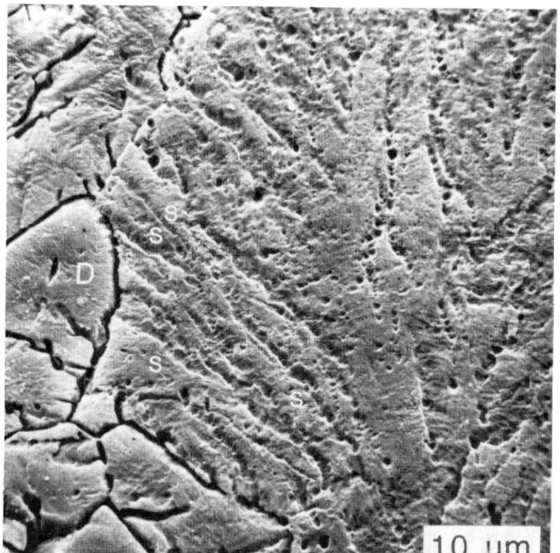

FIG. 9.—Detail of Fig. 8, SEM. The RFC-FOC has distinct subcrystals (s). Boundaries between subcrystals etch somewhat more strongly than the general crystal surface, but nonetheless are not open surfaces like the boundaries between the layer RFC-FOC crystals (Fig. 8) or between those crystals and the dentate calcite (D).

cement and the aragonite skeletons untouched. The most reasonable interpretation is that the RFC-FOC cement fringe is a primary precipitate and not the replacement envisioned by Kendall & Tucker originally. The environment of its deposition is uncertain. Kendall & Tucker (1973, p. 386) gave evidence for formation of RFC in the same environment as the inferred fibrous precursor (i.e., presumably marine). The Japanese material suggests that RFC-FOC may also form in conditions other than normal marine. The underlying dentate calcite (Figs. 7B, 10) has morphologies similar to those associated with freshwater deposition. Furthermore, the endolith borings that occur in the aragonite skeletons and overlying aragonite cements (Fig. 10) have thus far not been encountered in the dentate calcite or the RFC-FOC fringe. Oxygen and carbon isotope analyses on this material will be attempted to seek an answer to this question. It may be that RFC-FOC cements can form in a variety of water chemistries, but it is curious that they have not been encountered in modern environments.

TEMPORAL DISTRIBUTION

The acceptance, in earlier literature, of inferences of pseudomorphic calcitization of aragonite meant that inferred aragonite occurred throughout the Phanerozoic. That interpretation masked an apparent oscillatory Phanerozoic temporal trend in non-skeletal carbonate mineralogy (Sandberg, 1983). Earlier discussions (Wilkinson 1979; see Sandberg 1975a for additional references) tended to support changes in the oceanic Mg/Ca ratio as the prime control on mineralogy of inorganic marine carbonates. There is good evidence (beyond the influence of LMC plankton) for increase in oceanic Mg/Ca since the Jurassic, based on the marked decrease in dolomite formation during the last 100 million years, and the consequent rise in oceanic Mg/Ca needed in order for seawater cycling through mid-ocean ridges to be able to maintain a steady state composition (Holland, 1984). Although oceanic Mg/Ca ratio may have been influential since the Jurassic, the presence of aragonite cements in much older rocks precludes the general attribution of mineralogical change to Mg/Ca fluctuation. As discussed elsewhere (Mackenzie & Pigott, 1981; Fischer, 1981), oscillations in pCO_2 probably characterized the Phanerozoic. Those times of elevated pCO_2 (times of submerged continents) appear to coincide with times of aragonite absence in non-skeletal marine carbonates. That is, a plot of the temporal distribution of engulfed aragonite fringes (Criterion 1) and calcitized botryoids and spherulites (Criteria 2, 3, 4) suggests that aragonite precipitation appears to have been only intermittently facilitated (Sandberg, 1983). During aragonite-facilitating conditions (Late Precambrian–Early (?) Cambrian, Late Miss.–(?) Early Jur., Early or Mid Cenozoic–Recent), both aragonite and HMC precipitated inorganically in marine waters. During the intervening, aragonite-inhibiting conditions, HMC (and LMC ?) only precipitated. The inferred temporal trend is strengthened by the coincidence with the cement data, of the temporal distribution of ooids showing properties (Criteria 1–4) of calcitized aragonites. The inhibition is circumstantially linked to times of higher pCO_2. Given & Wilkinson (in press) have suggested that CO_3^{2-} concentration is responsible for mineralogy of marine inorganic carbonates, with high CO_3^{2-} concentration favoring aragonite. Lynn Walter (pers. comm.) has found experimentally that the inhibitory effect of SO_4^{2-} is many times stronger for calcite than for aragonite. The relationships those workers have observed may be of major importance in explaining the temporal trends. During the conditions of elevated pCO_2 which characterize the aragonite-inhibiting episodes, activity of CO_3^{2-} would be reduced, favoring calcite. Certainly calcite-favoring conditions can occur during aragonite-facilitating episodes, as they do today

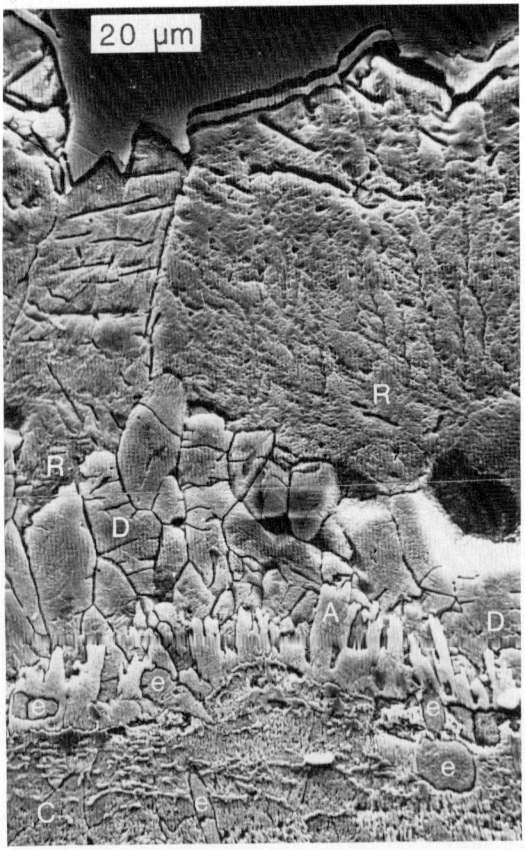

FIG. 10.—Another region of the sample shown in Figures 8 and 9. SEM mosaic. Both the aragonite coral (C) and the first-generation aragonite cement (A) are penetrated by numerous endolith (e) borings. Borings have not yet been found in either the overlying dentate (D) or RFC-FOC cement (R). Note the very low angle orientation of the subcrystals in the central crystal of the RFC-FOC fringe.

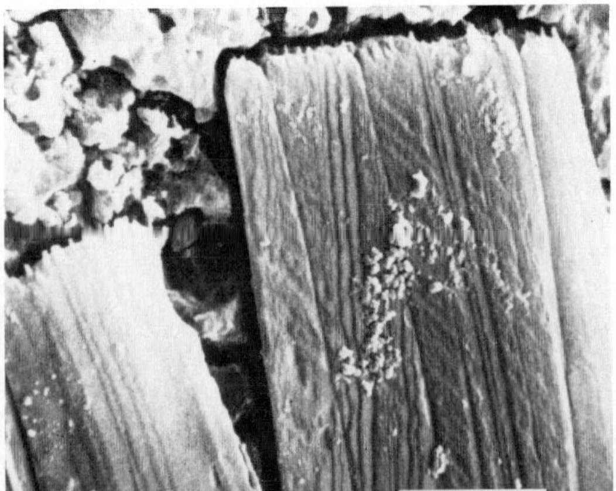

FIG. 11.—Distal ends of two "square-end" aragonite cement crystals (as in Fig. 7D), SEM same sample as Fig. 7. Note the presence of numerous fine needle terminations at the distal surface, suggesting composite crystals.

(Given & Wilkinson note a predominance of calcite in higher latitude or deeper, colder water conditions). During aragonite-inhibiting times, the decreased CO_3^{2-} and generally more uniform conditions globally would mean that aragonite would occur only in rather extreme conditions, if at all.

Because of their large size and distinctive morphology, botryoids should be readily encountered and identified. It is, therefore, significant that they are limited in their distribution, clumped in the interval from Late Mississippian (Davies, 1977a, b) to possibly Early Jurassic (Burri *et al.*, 1973), and in the Early Cambrian (James & Klappa, 1983), middle Precambrian (Grotzinger and Read, 1983) and mid to late Cenozoic (Aissaoui, unpublished).

CONCLUSIONS

1. Modern aragonite cement fringes are characteristically distinct in morphology from ancient fibrous calcite marine cements. The aragonite has more pronounced drusy textures, often with a micritic proximal portion composed of fine, poorly ordered needles. Except where substrate control produces ordered aragonite palisades, even the coarser distal aragonite crystals lack the ordered, often perpendicular arrangement and generally equal lengths typical of isopachous calcite marine cements.

2. Aragonite cement can be preserved unaltered, encased in secondary calcite spar, even in rocks as old as Pennsylvanian. Such preservation occurs even in cases where an underlying aragonite skeletal substrate has been calcitized. Aragonite also occurs as scattered relics in more massive cements (e.g., botryoids) in which most of the original aragonite has been replaced by irregular calcite. Given comparable diagenetic environments, similar relics could well be encountered even in much older calcitized or engulfed aragonites.

3. Textural and compositional properties do exist that allow differentiation of originally calcitic and calcitized, originally aragonitic constituents in ancient limestones. Such textures are, of course, most secure when accompanied by compositional indications of original aragonite, e.g., elevated Sr^{2+} content.

4. Where original aragonite mineralogy of cements inferred to be calcitized aragonite is well substantiated (Criteria 2 & 3, Sandberg, 1983), the diagenctic behavior is like that seen in skeletons and ooids known to have been aragonite. The resulting textures are coarsened, sometimes extremely so, and replacement crystal boundaries irregularly cross-cut the original structure delineated by organic and other non-carbonate inclusions and fluid inclusions.

5. Fine-textured, optically and morphologically radial calcite cements (commonly isopachous) cannot be derived from aragonite precursors. Even if pseudomorphic calcitization were a reasonable interpretation (for which no evidence exists) the morphologies of the two cements are dissimilar, as noted in 1. above.

6. Petrographic data and geometric considerations indicate that RFC and FOC are not derivable by alteration at a front migrating distally from the base of the cement sequence. Evidence from Pleistocene material indicates that RFC and FOC are primary precipitates. This interpretation is supported by marine oxygen isotopic values measured in ancient RFC.

7. Aragonite cements show (along with aragonitic ooids) a clumpled distribution in the Phanerozoic. This distribution appears not to be a function of sampling bias, but rather of oscillatory conditions intermittently facilitating and inhibiting aragonite precipitation.

ACKNOWLEDGMENTS

This paper is part of a continuing study funded in part by the National Science Foundation (DES 7517208 and EAR 7905716) and the Petroleum Research Fund of the American Chemical Society (PRF 1174102). The support of the Institut für Geologie und Paläontologie, Philipps-Universität, Marburg, and particularly Maurits Lindström, during a sabbatical visit is gratefully acknowledged. SEM photos were taken on instruments in the Center for Electron Microscopy, University of Illinois.

I am grateful to R.K. Matthews, and B.N. Popp for their thoughtful reviews of the manuscript. This paper has profited from my discussions with Robin Bathurst, John Hudson, Alan Kendall, Johannes Schroeder, Ian Macintyre, Maurits Lindström, Noel James, Torbjörn Alexandersson, Heinrich Zankl, and William Meyers. Samples were provided by Alan Horowitz, Sal Mazzullo, Jeff Dravis, Joachim Neugebauer, Johannes Schroeder, Cor Wagner, Ian Macintrye, Lynton Land and Harry Cook. Bruce Purser and Phil Heckel guided the field collection of samples in France and Kansas, respectively, as well as some memorable gastronomic experiences in those areas. I am indebted to Brian Popp for assistance with field work in Kansas.

REFERENCES

ALEXANDERSSON, T., 1969, Recent littoral and sublittoral high-Mg calcite lithification in the Mediterranean: Sedimentology, v. 12, p. 47–61.

_____, 1972, Intragranular growth of marine aragonite and Mg-calcite: Evidence of precipitation from supersaturated seawater: Jour. Sed. Petrology, v. 42, p. 441–460.

_____, AND MILLIMAN, J.D., 1981, Intragranular Mg-calcite cement in Halimeda plates from the Brazilian continental shelf: Jour. Sed. Petrology, v. 51, p. 1309–1314.

ALLEN, R.C., GAVISH, E., FRIEDMAN, G.M., AND SANDERS, J.E., 1969, Aragonite-cemented sandstone from the outer continental shelf off Delaware Bay: Submarine lithification mechanism yields product resembling beachrock: Jour. Sed. Petrology, v. 39, p. 136–149.

ASSERETO, R., AND FOLK, R.L., 1976, Brick-like texture and radial rays in Triassic pisolites of Lombardy, Italy: A clue to distinguish ancient aragonitic pisolites: Sed. Geology, v. 16, p. 205–222.

______, AND ______, 1980, Diagenetic fabrics of aragonite, calcite, and dolomite in an ancient peritidal-spelean environment: Triassic Calcare Rosso, Lombardia, Italy: Jour. Sed. Petrology, v. 50, p. 371–394.

AUSTIN, S.A., 1979, Uniformitarianism—A doctrine that needs rethinking: The Compass, v. 56, p. 29–45.

BANNER, F.T., AND WOOD, G.V., 1964, Recrystallization in microfossiliferous limestones: Geol. Jour., v. 4, p. 21–34.

BATHURST, R.G.C., 1964, The replacement of aragonite by calcite in the molluscan shell wall: *in* J. Imbrie and N.D. Newell (eds.), Approaches to Paleoecology: Wiley, New York, p. 319–344.

______, 1974, Marine diagenesis of shallow water calcium carbonate sediments: Ann. Rev. Earth Planetary Sci., v. 2, p. 257–274.

______, 1983, Neomorphic spar versus cement in some Jurassic grainstones: Significance for evaluation of porosity evolution and compaction: Jour. Geol. Soc. London, v. 140, p. 229–237.

BLATT, H., MIDDLETON, G., AND MURRAY, R., 1980, Origin of sedimentary rocks: Prentice Hall, Englewood Cliffs, N.J., 634 p.

BRAND, U., AND VEIZER, J., 1980, Chemical diagenesis of a multicomponent carbonate system—1: Trace elements: Jour. Sed. Petrology, v. 50, p. 1219–1236.

BUCHBINDER, L., AND FRIEDMAN, G.M., 1980. Vadose, phreatic, and marine diagenesis of Pleistocene-Holocene carbonates in a borehole: Mediterranean coast of Israel: Jour. Sed. Petrology, v. 50, p. 395–408.

BUCKLEY, H.E., 1951, Crystal Growth: Wiley, New York.

BURRI, P., DU DRESNAY, R., AND WAGNER, C.W., 1973, Tepee structures and associated diagenetic features in intertidal carbonate sands (Lower Jurassic, Morocco): Sed. Geology, v. 9, p. 221–228.

CHAFETZ, H.S., 1979, Petrology of carbonate nodules from a Cambrian tidal inlet accumulation, central Texas: Jour. Sed. Petrology, v. 49, p. 205–222.

COTTER, E., 1966, Limestone diagenesis and dolomitization in Mississippian carbonate banks in Montana: Jour. Sed. Petrology, v. 36, p. 764–774.

CROSS, T.A., AND KLOSTERMAN, M.J., 1981, Primary submarine cements and neomorphic spar in a stromatolitic-bound phylloid algal bioherm, Laborcita Formation (Wolfcampian), Sacramento Mountains, New Mexico, U.S.A., *in* Claude Monty (ed.), Phanerozoic Stromatolites: Springer Verlag, Berlin, p. 60–73.

CULLIS, C.G., 1904, The mineralogical changes observed in the cores of the Funafuti borings: *in* T. G. Bonney (ed.), The Atoll of Funafuti, Royal Soc., London, p. 393–420.

DAVIES, G.R., 1977a, Former magnesian calcite and aragonite submarine cements in upper Paleozoic reefs of the Canadian Arctic: A summary: Geology, v. 5, p. 11–15.

______, 1977b, Carbonate-anhydrite facies relationships, Otto Fiord Formation (Mississippian-Pennsylvanian), Canadian Arctic archipelago: Amer. Assoc. Petroleum Geologists, Bull., v. 61, p. 1929–1949.

DICKSON, J.A.D., AND COLEMAN, M.L., 1980, Changes in carbon and oxygen isotope composition during limestone diagenesis: Sedimentology, v. 27, p. 107–118.

DRAVIS, J., 1979, Rapid and widespread generation of Recent oolitic hardgrounds on a high energy Bahamian platform, Eleuthera Bank, Bahamas: Jour. Sed. Petrology, v. 49, p. 195–208.

FENNINGER, A., 1968, Das Kalzitgefüge der sparitischen Kalke des Plassen (Tithonium, nördliche Kalkalpen, Oberösterreich): Sedimentology, v. 10, p. 273–291.

______, AND FLAJS, G., 1974, Zur Mikrostruktur rezenter und fossilier Hydrozoen: Biomineral. Forsch. Berichte, v. 7, p. 69–99.

FISCHER, A.G., 1981, Climatic Oscillations in the Biosphere: *in* M. Nitecki (ed.), Biotic Crises in Ecological and Evolutionary Time: Acad. Press, p. 103–131.

FLÜGEL, E., 1982, Microfacies Analysis of Limestones: Springer Verlag, Berlin, 633 p.

FOLK, R.L., 1973, Carbonate petrography in the post-Sorbian age, *in* R.N. Ginsburg (ed.), Evolving Concepts in Sedimentology: Johns Hopkins Univ., Studies in Geol., v. 21, p. 118–158.

______, AND ASSERETO, R., 1976, Comparative fabric of length-slow and length-fast calcite and calcitized aragonite in a Holocene speleothem, Carlsbad Caverns, New Mexico: Jour. Sed. Petrology, v. 46, p. 486–496.

FRIEDMAN, G.M., 1964, Early diagenesis and lithification in carbonate sediments: Jour. Sed. Petrology, v. 34, p. 777–813.

______, 1973, Cementation in reefs: Bull., SNPA, Centre Rech. Pau, v. 7, p. 171–176.

______, AMIEL, A.J., AND SCHNEIDERMANN, N., 1974, Submarine cementation in reefs: example from the Red Sea: Jour. Sed. Petrology, v. 44, p. 816–825.

FÜCHTBAUER, H., 1974, Sediments and sedimentary rocks 1: Stuttgart, Schweizerbart, 464 p.

GEVIRTZ, J.L., AND FRIEDMAN, G.M., 1966, Deep-sea carbonate sediments of the Red Sea and their implications on marine lithification: Jour. Sed. Petrology, v. 36, p. 146–151.

GINSBURG, R.N., AND JAMES, N.P., 1976, Submarine botryoidal aragonite in Holocene reef limestone, Belize: Geology, v. 4, p. 431–436.

______, MARSZALEK, D.S., AND SCHNEIDERMANN, N., 1971, Ultrastructure of carbonate cements in a Holocene algal reef of Bermuda: Jour. Sed. Petrology, v. 41, p. 471–482.

GIVEN, R.K., AND WILKINSON, B.H., in press, Morphology, composition, and mineralogy of abiotic sedimentary carbonates: Control by kinetic reaction parameters and physical environment: Jour. Sed. Petrology.

GOULD, S.J., 1965, Is uniformitarianism necessary?: Amer. Jour. Sci., v. 263, p. 223–228.
GROTZINGER, J., AND READ, J.F., 1983, Evidence for primary aragonite precipitation, lower Proterozoic (1.9 Ga) Rocknest dolomite, Wopmay orogen, Northwest Canada: Geology, v. 11, p. 710–713.
HALLEY, R.B., HARRIS, P.M., AND HINE, A.C., 1983. Bank Margin Environments, *in* P.A. Scholle, D.G. Bebout, and C.H. Moore (eds.), Carbonate Depositional Environments: Amer. Assoc. Petroleum Geologists Mem. 33, p. 463–506.
HATHAWAY, J.C., AND ROBERTSON, E.C., 1961, Microtexture of artificially consolidated aragonitic mud: U. S. Geol. Survey, Prof. Paper 424, p. 301–304.
HECKEL, P.H., 1978, Upper Pennsylvanian cyclothemic limestone facies in eastern Kansas: Guidebook, SEPM field trip 2, Guidebook Series 2, Kansas Geol. Surv., p. 1–69.
HOLLAND, H.D., 1984, Chemical evolution of the atmosphere and oceans, Princeton Univ. Press, in press.
HUDSON, J.D., 1962, Pseudo-pleochroic calcite in recrystallized shell limestones: Geol. Magazine, v. 99, p. 492–500.
______, AND COLEMAN, M.L., 1978, Submarine cementation of the Scheck Limestone conglomerate (Jurassic, Austria): Isotopic evidence: N. Jb. Geol. Paläont. Mh., v. 1978, p. 534–544.
ILLING, L.V., 1954, Bahamian calcareous sands: Amer. Assoc. Petroleum Geologists, Bull., v. 38, p. 1–95.
JAMES, N.P., 1974, Diagenesis of scleractinian corals in the subaerial vadose environment: Jour. Paleontol. v. 48, p. 785–795.
______, 1983, Reef environment, *in* P.A. Scholle, D.G. Bebout, and C.H. Moore (eds.), Carbonate Depositional Environments. Amer. Assoc. Petroleum Geologists, Mem. 33, p. 345–543.
______, AND KLAPPA, C.F., 1983, Petrogenesis in Early Cambrian reef limestones, Labrador, Canada: Jour. Sed. Petrology, v. 53, p. 1051–1096.
______, AND GINSBURG, R.N., 1979, The seaward margin of Belize barrier and atoll reefs: Int. Assoc Sed., Spec. Publ., No. 3, 191 p.
______, ______, MARSZALEK, D., AND CHOQUETTE, P., 1976, Facies and fabric specificity of early subsea cements in shallow Belize (British Honduras) reefs: Jour. Sed. Petrology, v. 46, p. 523–544.
JINDRICH, V., 1983, Structure and diagenesis of Recent algal-foraminifer reefs, Fernando de Noronha, Brazil, Jour. Sed. Petrology, v. 53, p. 449–459.
JORGENSEN, N.O., 1976, Recent high magnesian calcite/aragonite cementation of beach and submarine sediments from Denmark: Jour. Sed. Petrology, v. 46, p. 940–951.
KENDALL, A.C., 1977, Fascicular-optic calcite: A replacement of bundled acicular carbonate cements: Jour. Sed. Petrology, v. 47, p. 1056–1062.
______, AND BROUGHTON, P.L., 1978, Origin of fabrics in speleothems composed of columnar calcite crystals: Jour. Sed. Petrology, v. 48, p. 519–538.
______, AND TUCKER, M.E., 1973, Radiaxial fibrous calcite: a replacement after acicular carbonate: Sedimentology, v. 20, p. 365–389.
KOSTECKA, A., 1972, Calcite paramorphs in the aragonite concretions: Ann. Soc, Geol. Pologne, v. 42, p. 289–296.
______, 1978, Lower Muschelkalk carbonates of the south-western margin of the Holy Cross Mountains (Central Poland): Ann. Soc. Geol. Pologne, v. 48, p. 297–325.
LAND, L.S., 1967, Diagenesis of skeletal carbonates: Jour. Sed. Petrology, v. 37, p. 914–930.
______, AND GOREAU, T., 1970, Submarine lithification of Jamaican reefs: Jour. Sed. Petrology, v. 40, p. 457–462.
______, AND MOORE, C.H., 1980, Lithification, micritization and syndepositional diagenesis of biolithites on the Jamaican island slope: Jour. Sed. Petrology, v. 50, p. 357–370.
LASEMI, Z., AND SANDBERG, P.A., 1983, Recognition of original mineralogy in ancient micrites: Amer. Assoc. Petroleum Geologists, Bull., v. 67, p. 499–500.
LOHMANN, K.C., AND MEYERS, W.J., 1978, Microdolomite inclusions in cloudy prismatic calcites: a proposed criterion for former high magnesium calcites: Jour. Sed. Petrology, v. 47; p. 1078–1088.
LONGMAN, M.W., 1980, Carbonate diagenetic textures from nearshore diagenetic environments: Amer. Assoc. Petroleum Geologists, Bull., v. 64, p. 461–487.
LOUCKS, R.G., AND FOLK, R.L., 1976, Fanlike rays of former aragonite in Permian Capitan Reef Pisolite: Jour. Sed. Petrology, v. 46, p. 483–485.
LUCIA, F.J., 1968, Recent sediments and diagenesis of south Bonaire, Netherlands Antilles: Jour. Sed. Petrology, v. 38, p. 845–858.
MACINTYRE, I.G., 1977, Distribution of submarine cements in a modern Caribbean fringing reef, Galeta Point, Panama: Jour. Sed. Petrology, v. 47, p. 503–516.
______, MOUNTJOY, E.W., AND D'ANGELJEAN, B.F., 1968, An occurrence of submarine cementation of carbonate sediments off the west coast of Barbados, W.I.: Jour. Sed. Petrology, v. 38, p. 660–663.
MACKENZIE, F.T., AND PIGOTT, J.D., 1981, Tectonic controls of Phanerozoic sedimentary rock cycling: Jour. Geol. Soc. London, v. 138, p. 183–196.
MARSHALL, J.D., 1981, Zoned calcites in Jurassic ammonite chambers: trace elements, isotopes, and neomorphic origin: Sedimentology, v. 28, p. 867–887.
______, AND ASHTON, M., 1980, Isotopic and trace element evidence for submarine lithification of hardgrounds in the Jurassic of eastern England: Sedimentology, v. 27, p. 271–289.
MAZZULLO, S.J., 1980, Calcite pseudospar replacive of marine acicular aragonite, and implication for aragonite cement diagenesis: Jour. Sed. Petrology, v. 50, p. 409–422.
______, AND CYS, J., 1979, Marine aragonite sea-floor growths and cements in Permian phylloid algal mounds, Sacramento Mountains, New Mexico: Jour. Sed. Petrology, v. 49, p. 917–936.
MIRSAL, I., AND ZANKL, H., 1979, Petrography and geochemistry of carbonate void-filling cements in fossil reefs: Geol. Rundschau, v. 68, p. 920–951.

MOORE, C.H., 1973, Intertidal carbonate cementation, Grand Cayman, West Indies: Jour. Sed. Petrology, v. 43, p. 591–602.
NEUMANN, A.C., AND LAND, L.S., 1975, Lime mud deposition and calcareous algae in the Bight of Abaco: A budget: Jour. Sed. Petrology, v. 45, p. 763–786.
NEWELL, N.D., 1955, Depositional fabric in Permian reef limestone: Jour. Geology, v. 63, p. 301–309.
OTTE, C. JR., AND PARKS, J.M., JR., 1963, Fabric studies of Virgil and Wolfcamp bioherms, New Mexico: Jour. Geol., v. 71, p. 380–396.
PINGITORE, N.E., 1970, Diagenesis and porosity modification in *Acropora palmata*, Pleistocene of Barbados, West Indies: Jour. Sed. Petrology, v. 40, p. 712–721.
______, 1976, Vadose and phreatic diagenesis: processes, products and their recognition in corals: Jour. Sed. Petrology, v. 46, p. 985–1006.
POPP, B.N., AND WILKINSON, B.H., 1983, Holocene lacustrine ooids from Pyramid Lake, Nevada, *in* T. Peryt (ed.), Coated Grains: Springer-Verlag, Berlin, p. 142–153.
______, ANDERSON, T.F., AND SANDBERG, P.A., in press, Textural, elemental and isotopic variation among constituents in middle Devonian limestones, eastern U.S. and Canada: Jour. Sed. Petrology.
PURDY, E.G., 1968, Carbonate diagenesis: A review: Geol. Romana, v. 7, p. 183–228.
PURSER, B.H., 1969, Syn-sedimentary marine lithification of Middle Jurassic limestones in the Paris Basin: Sedimentology, v. 12, p. 205–230.
RAVN, R.L., AND HECKEL, P.H., 1978, Algal sparite facies in Captain Creek Limestone: Guidebook, SEPM field Trip 2, Guidebook Series 2, Kansas Geol. Surv. p. 70–79.
RICHTER, D.K., AND BESENECKER, H., 1983, Subrecent high-Sr aragonitic ooids from hot springs near Tekke Ilica (Turkey), In T. Peryt (ed.) Coated Grains: Springer-Verlag, p. 154–162.
SANDBERG, P.A., 1975a, Bryozoan diagenesis: bearing on the nature of the original skeleton of rugose corals: Jour. Paleontology, v. 49, p. 587–606.
______, 1975b, New interpretations of Great Salt Lake ooids and of ancient non-skeletal carbonate mineralogy: Sedimentology, v. 22, p. 497–537.
______, 1983, An oscillating trend in Phanerozoic non-skeletal carbonate mineralogy: Nature, v. 305, p. 19–22.
______, in press, Recognition of calcitized skeletal and non-skeletal aragonite: Bull. Amer. Paleontology.
______, AND HUDSON, J.D., 1983, Aragonite relic preservation in Jurassic calcite-replaced bivalves: Sedimentology, v. 30, p. 879–892.
______, AND POPP, B.N., 1981, Pennsylvanian aragonites from southeastern Kansas—Environmental and diagenetic implications: Amer. Assoc. Petroleum Geologists, Bull., v. 65, p. 985.
______, SCHNEIDERMANN, N., AND WUNDER, S.J., 1973, Aragonitic ultrastructural relics in calcite-replaced Pleistocene skeletons: Nature, Phys. Sci., v. 245, p. 133–134.
SCHERER, M., 1975, Cementation and replacement of Pleistocene corals from the Bahamas and Florida: Diagenetic influence of non-marine environments: N. Jahrb. Geol. Paläont., Abh., v. 149, p. 259–285.
______, 1977, Preservation, alteration, and multiple cementation of aragonite skeletons from the Cassian Beds (U. Triassic, southern Alps): Petrographic and geochemical evidence: N. Jahrb. Geol. Pälaont., Abh., v. 154, p. 213–262.
______, AND WENDT, J., 1978, Diagenese oberpermischer Kalkschwämme aus Patch-Reefs des Djebel-Tebaga (S-Tunisien): N. Jahrb. Geol. Pälaont. Abh., v. 157, p. 196–202.
SCHNEIDERMANN, N., 1970, Genesis of some Cretaceous carbonates in Israel: Israel Jour. Earth Sci., v. 19, p. 97-115.
______, SANDBERG, P.A., AND WUNDER, S.J., 1972, Recognition of early cementation of aragonitic skeletal carbonates: Nature, Phys. Sci., v. 240, p.
SCHLAGER, W., AND JAMES, N.P., 1978. Low-magnesian calcite limestones forming at the deep-sea floor, Tongue of the Ocean, Bahamas: Sedimentology, v. 25, p. 675–702.
SCHROEDER, J.H., 1972a, Fabrics and sequences of submarine carbonate cements in Holocene Bermuda cup reefs: Geol. Rundschau, v. 61, p. 708–730.
______, 1972b, Calcified filaments of an endolithic alga in Recent Bermuda reefs: N. Jahrb. Geol. Pälaont., Mh., v. 1972, p. 16–33.
______, 1973, Submarine and vadose cements in Pleistocene Bermuda reef rock: Sed. Geology, v. 10, p. 179–204.
______, 1974, Carbonate cements in Recent reefs of the Bermudas and Bahamas, Keys to the past: Ann. Soc. Geol. Belgique, v. 97, p. 153–158.
______, 1979, Carbonate diagenesis in Quaternary beachrock of Uyombo, Kenya: Sequences of processes and coexistence of heterogenic products: Geol. Rundschau, v. 68, p. 894–919.
SHINN, E.A., 1969, Submarine lithification of Holocene carbonate sediments in the Persian Gulf: Sedimentology v. 12, p. 109–144.
______, 1971, Aspects of diagenesis of algal cup reefs in Bermuda: Gulf Coast Assoc. Geol. Socs., Trans., v. 21, p. 387-394.
______, 1975. Polygonal cement sutures from the Holocene; a clue to recognition of submarine diagenesis [abs.]: Amer. Assoc. Petroleum Geologists, Ann. Mtg. Abstr., v. 2, p. 68.
SIMONE, L., 1981, Ooids: A review: Earth-Science Reviews, v. 16, p. 319–355.
SORBY, H.C., 1879, The structure and origin of limestones. Proc., Geol. Soc. London, v. 35, p. 56–95.
STEINEN, R.P., 1978, On the diagenesis of lime mud: Scanning electron microscopic observations on subsurface material from Barbados, W.I.: Jour. Sed. Petrology, v. 48, p. 1139–1147.
TALBOT, M.R., 1971, Calcite cements in the Corallian Beds (Upper Oxfordian) of southern England: Jour. Sed. Petrology, v. 41, p. 261–273.

TAYLOR, J.M.C., AND ILLING, L.V., 1969, Holocene intertidal carbonate cementation, Qatar, Persian gulf: Sedimentology, v. 12, p. 69–107.

TEBBUTT, G., 1975, Paleoecology and diagenesis of Pleistocene limestone on Ambergris Cay: Amer. Assoc. Petroleum Geologists, Studies in Geol., v. 2, p. 297–331.

TOWE, K., AND HEMLEBEN, C., 1976, Diagenesis of magnesian calcite: Evidence from miliolacean foraminifera: Geology v. 4, p. 337–339.

TUCKER, M.E., AND KENDALL, A.C., 1973, The diagenesis and low-grade metamorphism of Devonian styliolinid-rich pelagic carbonates from West Germany: possible analogues of Recent pteropod oozes: Jour. Sed. Petrology, v. 43, p. 672–687.

VEIZER, J., 1977, Diagenesis of pre-Quaternary carbonates as indicated by tracer studies: Jour. Sed. Petrology, v. 47, p. 565–581.

WAGNER, P.D., AND MATTHEWS, R.K., 1982, Porosity preservation in the Upper Smackover (Jurassic) carbonate grainstone, Walker Creek field, Arkansas: Response of paleophreatic lenses to burial processes: Jour. Sed. Petrology, v. 52, p. 3–18.

WALLS, R.A., MOUNTJOY, E.W., AND FRITZ, P., 1979, Isotopic composition and diagenetic history of carbonate cements in Devonian Golden Spike reef, Alberta, Canada: Geol. Soc. America, Bull., v. 90, p. 963–982.

WARD, W.C., 1975, Petrology and diagenesis of carbonate eolianites of northeastern Yucatan peninsula, Mexico: Amer. Assoc. Petroleum Geologists, Studies in Geol., v. 2, p. 500–571.

WARDLAW, N., OLDERSHAW, A., AND STOUT, M., 1978, Transformation of aragonite to calcite in a marine gasteropod. Canadian Jour. Earth Sci., v. 15, p. 1861–1866.

WARREN, JOHN K., 1982, The hydrological significance of Holocene tepees, stromatolites, and boxwork limestones in coastal salinas in South Australia: Jour. Sed. Petrology, v. 52, p. 1171–1201.

WENDT, J., 1977, Aragonite in Permian reefs: Nature, v. 267, p. 335–337.

WENK, H.R., BARBER, D.J., AND REEDER, R.J., 1983, Microstructures in carbonates: Reviews in Mineralogy, v. 11, p. 301–367.

WILKINSON, B.H., 1979, Biomineralization, paleoecology, and the evolution of calcareous marine organisms: Geology, v. 7, p. 524–527.

______, JANECKE, S.U., AND BRETT, C.E., 1982, Low-magnesian calcite marine cement in Middle Ordovician hardgrounds from Kirkfield, Ontario: Jour. Sed. Petrology, v. 52, p. 47–57.

______, SMITH, A.L., AND JANECKE, S.U., 1983, Low-magnesium sparry calcite marine cements in Phanerozoic hardground grainstones: Amer. Assoc. Petroleum Geologists Bull., v. 67, p. 569.

WOLF, K.H., 1965, Petrogenesis and paleoenvironment of Devonian algal limestones of New South Wales: Sedimentology, v. 4, p. 113–178.

RADIAXIAL FIBROUS CALCITE: A REAPPRAISAL

ALAN C. KENDALL
Sohio Petroleum Co., 50 Freemont Street, San Francisco, California 94111

ABSTRACT

Radiaxial fibrous calcite, a common cavity-fill in ancient reefs and mudmounds has previously been identified as a replacement of acicular carbonate cements. Diagnostic features of the calcite were suggested to be a combination of those inherited from the original cement and those introduced during neomorphism. A more rigorous examination of the proposed replacement theory, and of the evidence employed to support it, raises doubts as to its correctness.

New observations upon cavity-filling radiaxial calcites from Western Australia suggest that most of the features originally considered of neomorphic origin are instead primary. The peculiar, and diagnostic, fabric of distally-convergent fast-vibration directions within each crystal (contrasting with a distally-divergent pattern of subcrystals) is now reinterpreted to have formed by a process of asymmetric growth within calcite crystals undergoing split-growth. Radiaxial crystals are composite and are attempts by these crystals to assume a spherulitic growth form in diagenetic environments which favor the growth of length-slow calcite. The changes from split-crystal spherulitic growth to more 'normal' unit-crystal growth (and visa-versa) that can be observed in the Australian cavity-fills were probably induced by changes in the chemistry of the cavity fluids.

Implications of this study also include (1) that split-crystal growth is a common phenomenon in calcite cements, and (2) calcite, probably both high- and low-magnesian calcite, grew in the form of large, sparry crystals within the marine phreatic environment—a habit and mineralogy not common in present-day comparable settings.

INTRODUCTION

Depositional or early diagenetic cavities in reefs and mudmounds of Paleozoic age are commonly filled by a variety of calcite cements composed of crystals with undulose extinction and abundant inclusions that make them cloudy or turbid (Bathurst, 1959, 1982; Davies, 1977; Kendall and Tucker, 1973; Krebs, 1969; Ross *et al.*, 1975; and Walls *et al.*, 1979). Three types are most commonly recognized: distinction being made upon the nature of the extinction pattern seen under cross-polarized light—itself a reflection of the apparent curvature of calcite crystal lattices.

Radiaxial fibrous calcite (Bathurst, 1959, 1977, 1982; Kendall and Tucker, 1973) is identified by a pattern, within each crystal, of subcrystals that diverge away from the substrate; an opposing pattern of distally-convergent fast-vibration directions (= optic axes, c-axes); and corresponding curvatures of cleavage, twin lamellae and glide-planes (Fig. 1A).

Fascicular-optic calcite (Kendall, 1977) is a less common void filling and is similar to radiaxial calcite, commonly occurring together in the same cavity (Kendall, 1977; Chafetz, 1979; Marshall, 1981). It is distinguished by the presence of a divergent pattern of fast-vibration directions. This pattern coincides with that of the subcrystals (Fig. 1B).

Radial-fibrous calcites (Davies, 1977; Mazzullo, 1980) do not exhibit any marked undulose extinction but resemble other types in being formed of turbid crystals (Fig. 1C). Transitions between radial-fibrous and radiaxial or fascicular-optic calcites occur commonly.

Radiaxial fibrous calcites (and their associates) had long been identified, or were suspected of being, of marine origin (see Krebs, 1969). Explanation of the diagnostic crystal fabrics, however, only came later (Kendall and Tucker, 1973) and this explanation (involving a complex replacement of an earlier acicular cement) has not been challenged, although some misgivings were expressed in Kendall and Broughton (1978). This paper constitutes such a challenge. In large part this revision and recant of much of my previous work is the result of new observations (reported here) on radiaxial cements from cavity fills in forereef limestones from the Devonian of Western Australia.

PREVIOUS EXPLANATION OF RADIAXIAL FABRICS

From a consideration of the occurrences in primary and early diagenetic cavities, its relations with internal sediments, and its gross morphological characters, Kendall and Tucker (1973) interpreted radiaxial fibrous calcite to have been a marine cement. On the other hand, the crystal fabrics (non-planar intercrystalline boundaries, lattice curvature) and distinctive inclusion patterns suggested the calcites were of neomorphic origin and had replaced earlier acicular cements composed of laterally-interfering bundles of radiating crystals. It was inferred that replacement proceeded by a micro-scale solution-reprecipitation process with the migration of a fluid film through the acicular host. It was suggested that replacement occurred most rapidly between bundles so that each radiaxial crystal came to replace parts of several adjacent bundles (Fig. 2). The pattern of fast-vibration directions in each radiaxial crystal was believed to have been inherited from the c-axis orientations of the replaced acicular crystals. Some of the inclusion patterns in radiaxial calcites record characters of the replaced acicular cement but others, particularly those that define calcite crystal terminations, were interpreted to have formed by rearrangement of impurities during replacement and thus reveal successive positions of the replacement front and their form.

Two lines of argument suggest the replacement theory proposed by Kendall and Tucker (1973) is untenable.

Boundaries with Internal Sediment

Boundaries between radiaxial calcite layers and overlying internal sediment may be smoothly-curved to scalloped, suggesting an acicular cement precursor, or may define large calcite-crystal terminations similar to those defined by inclusion zones. Explanation of the terminations by a mechanism involving replacement of overlying internal sediment (itself involving renewed growth of the underlying radiaxial crystals) can commonly be disproven by 1) the occurrence of onlapping relations between laminated internal sediment and the underlying crystal terminations and 2) the absence of quartz silt from the calcite crystals when the immediately surjacent sediment contains this material. The crystal terminations thus imply that replacement of an earlier cement, if indeed this occurred at all, must have proceeded immediately after growth of the precursor acicular cement and before any deposition of internal sediment. In some cavity fills there are many alternations of internal sediment and radiaxial cement layers with crystal terminations (previously noted but unfigured in Kendall and Tucker, 1973, p. 386; figured in Bathurst, 1975, Fig. 311; Tucker, 1973, Fig. 9; and this paper, Fig. 4). Each alternation requires precisely the same sequence of events and also that the acicular cement growth and its later replacement occur in the same general environment. These are possibilities that are now considered unlikely. The simpler, and more probable, interpretation is that the calcite crystal terminations are primary.

Radiaxial Fabrics within Elongate Crystals

If radiaxial calcites replace laterally-interfering bun-

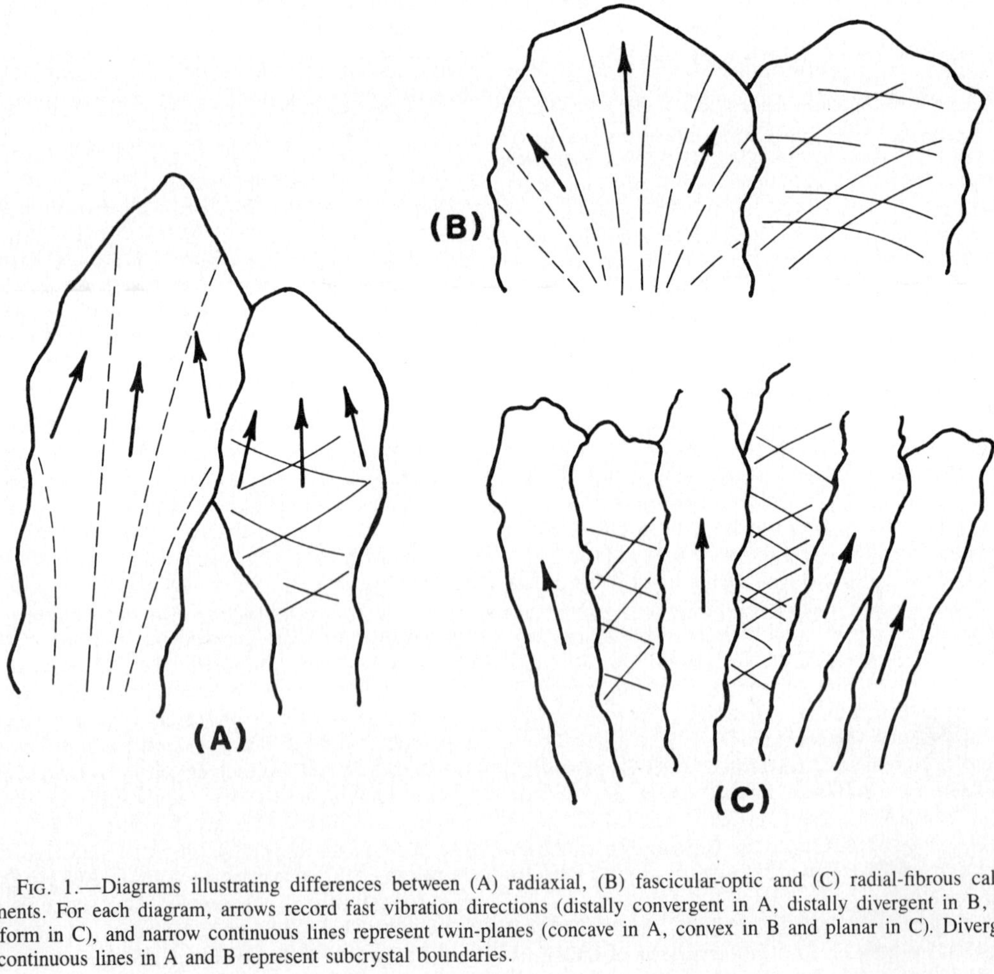

FIG. 1.—Diagrams illustrating differences between (A) radiaxial, (B) fascicular-optic and (C) radial-fibrous calcite cements. For each diagram, arrows record fast vibration directions (distally convergent in A, distally divergent in B, and uniform in C), and narrow continuous lines represent twin-planes (concave in A, convex in B and planar in C). Diverging discontinuous lines in A and B represent subcrystal boundaries.

dles of acicular crystals, it is to be expected that the radiaxial fabrics within the replacement crystals should become less well developed away from the substrate. Laterally-interfering bundles contain progressively less divergent acicular crystals as they grow because the more divergent crystals are soon cut out (Fig. 2). Distal parts of the cement are dominated by nearly-parallel and normally-arranged crystals (the palisade fabric) and replacement calcite developed from this should exhibit little lattice curvature. Radiaxial fibrous calcites, however, exhibit the same degree of lattice curvature throughout their length or this curvature may increase away from the substrate.

A different explanation for the origin of radiaxial fabrics is therefore required. One has been developed from observations made upon cavity-filling cements from Western Australia. It is simpler than that proposed by Kendall and Tucker (1973) and is believed applicable to all radiaxial calcite cements.

RADIAXIAL CALCITE CEMENTS FROM WESTERN AUSTRALIA

The material studied comes from fills of neptunian sills in the Upper Devonian Virgin Hills Formation of McWhae Ridge in the Canning Basin of Western Australia. At this locality, described by Playford (1980),

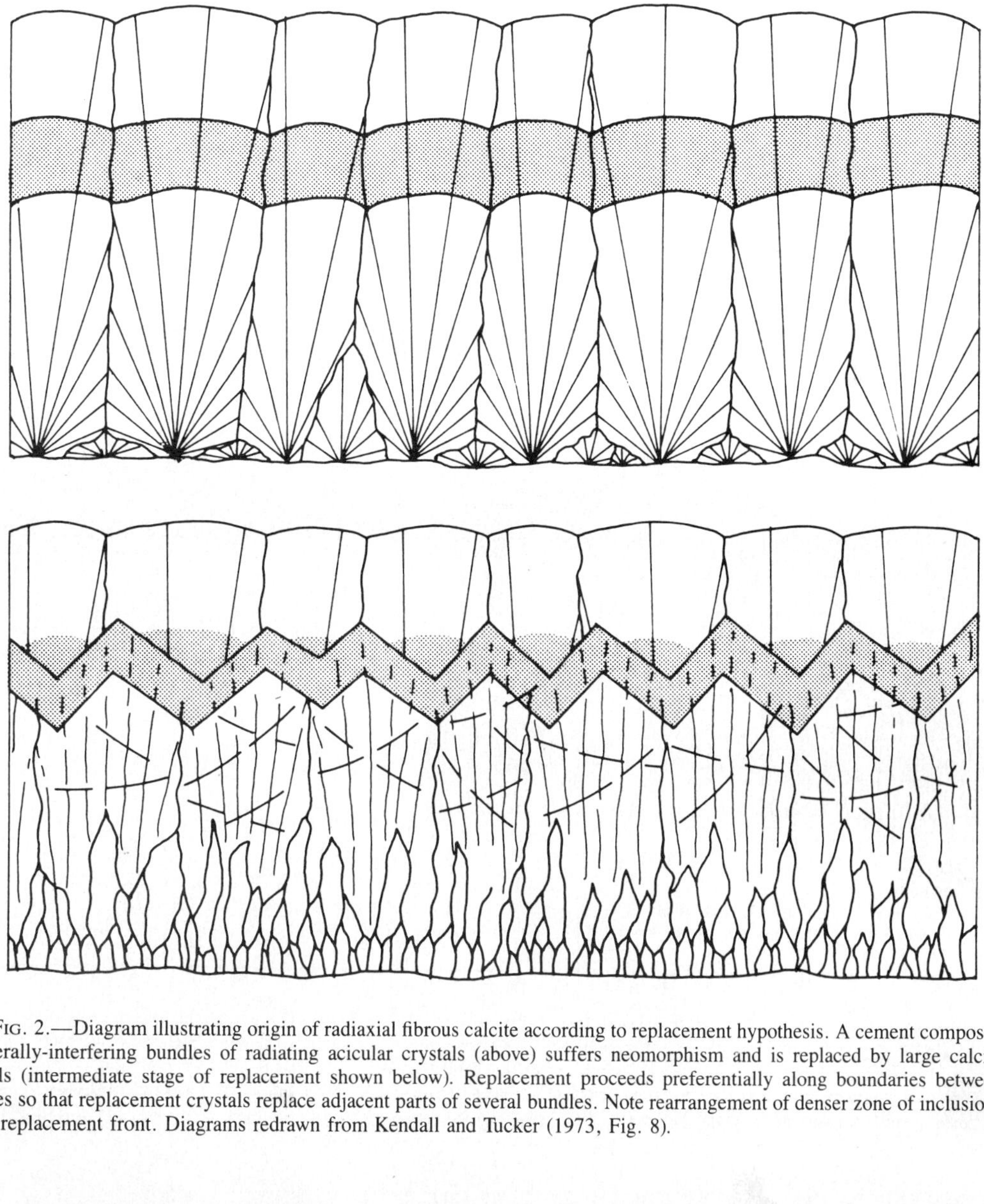

FIG. 2.—Diagram illustrating origin of radiaxial fibrous calcite according to replacement hypothesis. A cement composed of laterally-interfering bundles of radiating acicular crystals (above) suffers neomorphism and is replaced by large calcite crystals (intermediate stage of replacement shown below). Replacement proceeds preferentially along boundaries between bundles so that replacement crystals replace adjacent parts of several bundles. Note rearrangement of denser zone of inclusions along replacement front. Diagrams redrawn from Kendall and Tucker (1973, Fig. 8).

the Virgin Hills Formation is a silty, forereef limestone, red in color, with steep depositional dips that flatten out away from the narrow reefal spine that forms the crest of McWhae Ridge (Fig. 3A). Playford suggests the neptunian sills formed as bedding-plane partings when lithified layers of the forereef moved downslope (pers. comm. 1982). Curvature of the slope caused the lithified slabs to separate, opening up the neptunian sills (Figs. 3B, C). Commonly five or more centimeters wide, sills are traceable along the outcrop for many meters.

The sills are filled by complex alternations of bright red internal sediment and white cement crusts of radial-fibrous, palisade and radiaxial calcite (Fig. 4). Thicker internal sediment layers may contain marine fossils such as echinoderm grains. These indicate the penecontemporaneous origin of the sills and of their fills and suggest the interlayered calcite cements were also of marine-phreatic origin (see also page 74). Crusts line both cavity floors and roofs. Variations in inclusion-density and pattern within the cement crystals allows a cement-stratigraphy to be identified in both floor and roof cements. Different sills, however. have different filling histories.

The most informative cavity fill is shown in Figure 4. Growth of the radiaxial cement crust identified as 'A' in Figures 4 and 5 was terminated by deposition of internal sediment (B). This is largely confined to depressions between crystal terminations of the underlying cement layer. These terminations are markedly elongate and must represent either scalenohedral or acute-rhombohedral forms. They are composed of

FIG. 3.—*A*, McWhae Ridge seen from the north. Note the markedly curved bedding in the forereefal Virgin Hills Formation extending to a narrow reef spine on the left. *B*, Neptunian sills in Virgin Hills Formation. *C*, Cement and internal sediment fill of neptunian sill. A thin-section from this fill is illustrated in Fig. 4. Oiled slab.

FIG. 4.—Photomicrograph of central part of neptunian sill seen in Fig. 3C. A–D are radiaxial calcite cement and internal sediment layers mentioned in the text. 1–6 are junctions between inclusion-rich and overlying inclusion-poor cement zones that can be recognized in both roof and floor layers. Internal sediment (black) fills spaces between high-spired calcite crystal-terminations upon cement layers in the floor. Radiaxial crystals within cement layer C labeled *r* are nucleated directly from the underlying internal sediment. Plane polarized light.

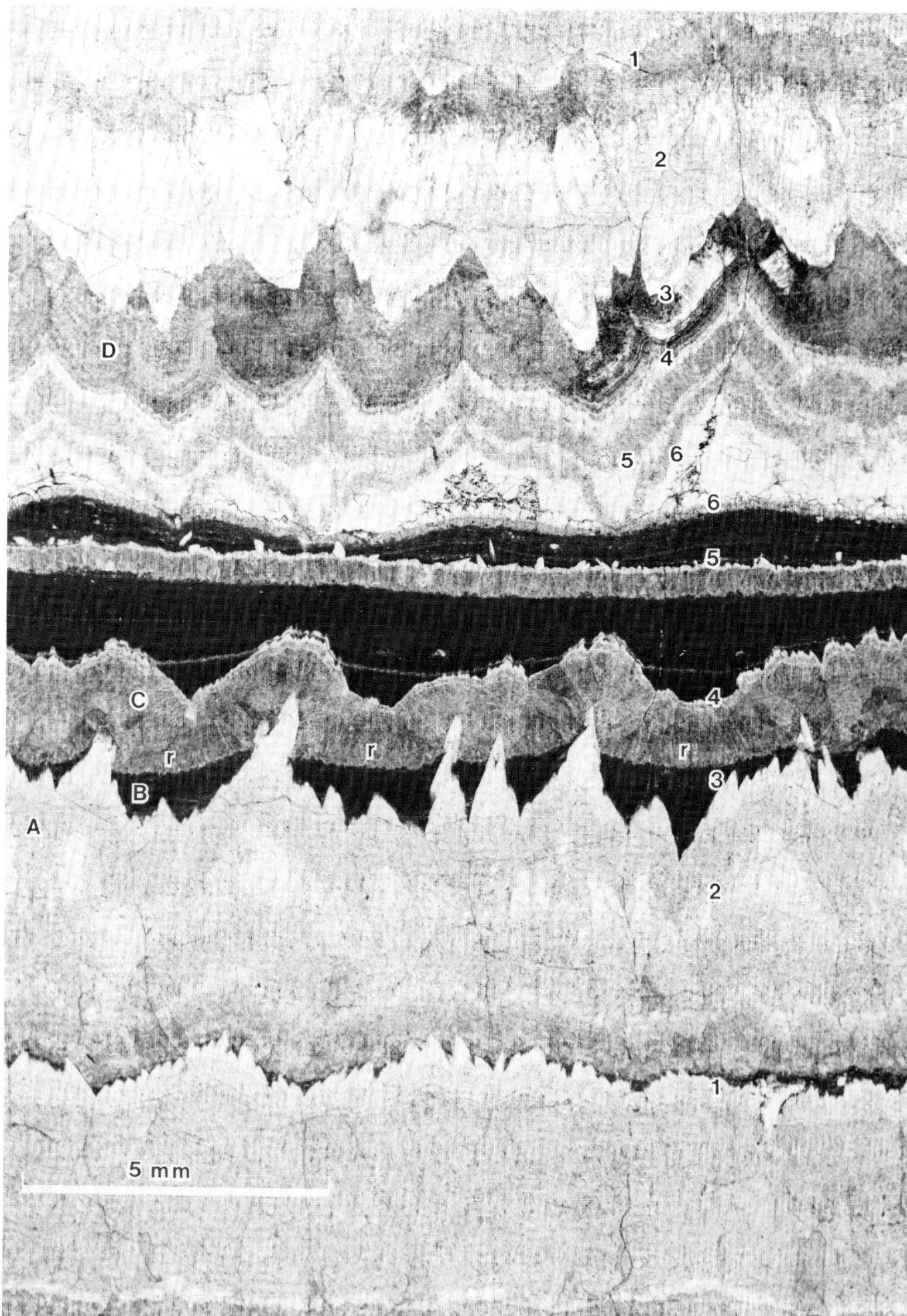
1
2
3
D
4
5
6
6
5
C
4
r
r
r
B
3
A
2
1
5 mm

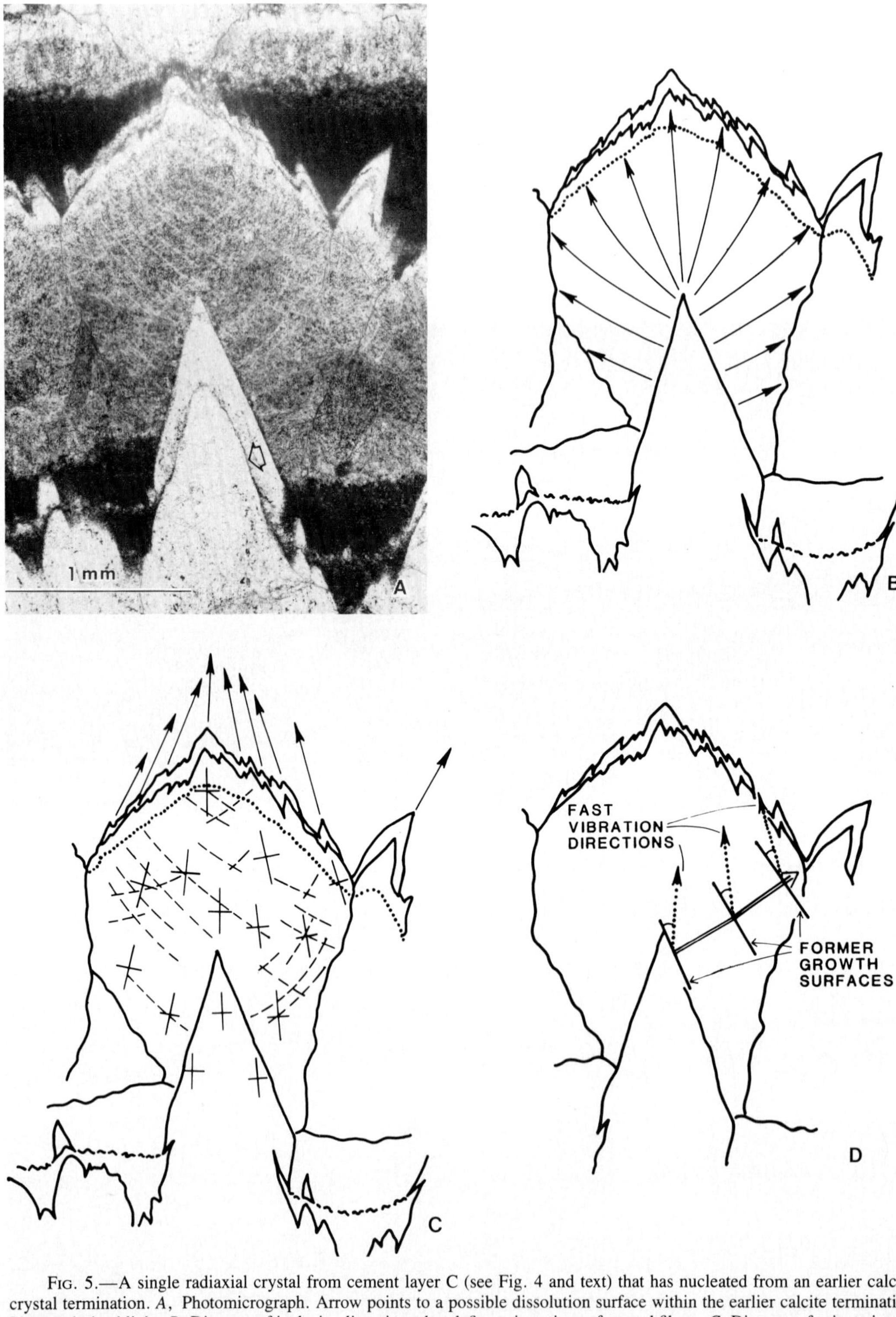

FIG. 5.—A single radiaxial crystal from cement layer C (see Fig. 4 and text) that has nucleated from an earlier calcite-crystal termination. *A*, Photomicrograph. Arrow points to a possible dissolution surface within the earlier calcite termination. Plane polarized light. *B*, Diagram of inclusion lineations that define orientations of curved fibers. *C*, Diagram of orientations of fast-vibration directions as determined from cleavage intersections and the orientations of elongate terminations (arrows). *D*, Diagram showing the changing orientation of fast-vibration directions along the length of a former fiber, and the decreasing divergence between this direction and the former growth surfaces.

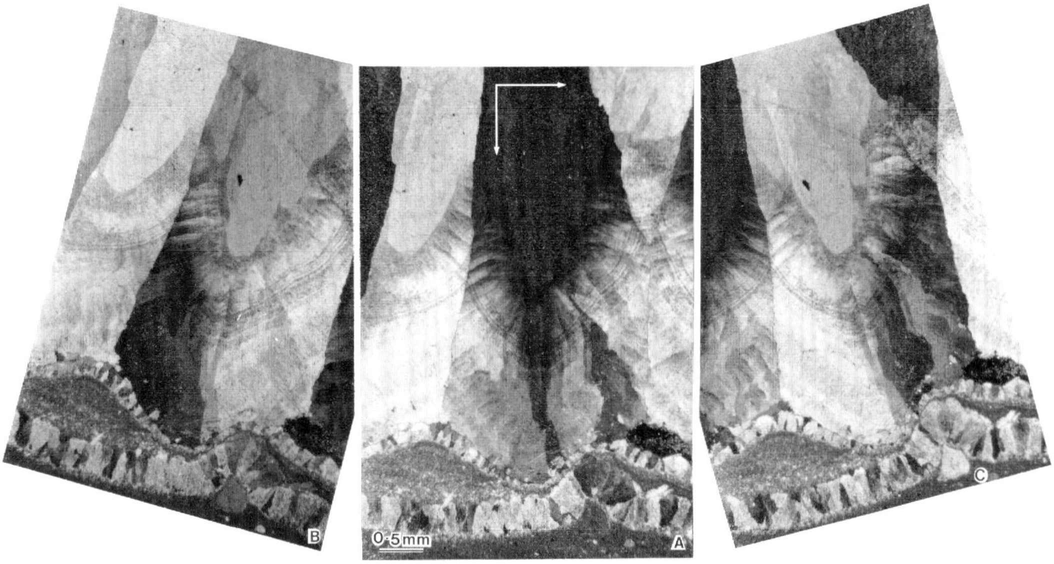

FIG. 6.—Photomicrographs of a radiaxial crystal from a roof cement layer, seen with cross-polarized light and with, at A, the earlier crystal termination placed in the extinction position. This reveals the optical continuity between the crystal substrate and the earliest parts of the overgrowth. It also shows, at B and C, by the changing position of the extinction bar, the gradual appearance of the radiaxial fabric within the inclusion-rich calcite away from the earlier crystal termination. White arrows denote the orientations of polarizer and analyser.

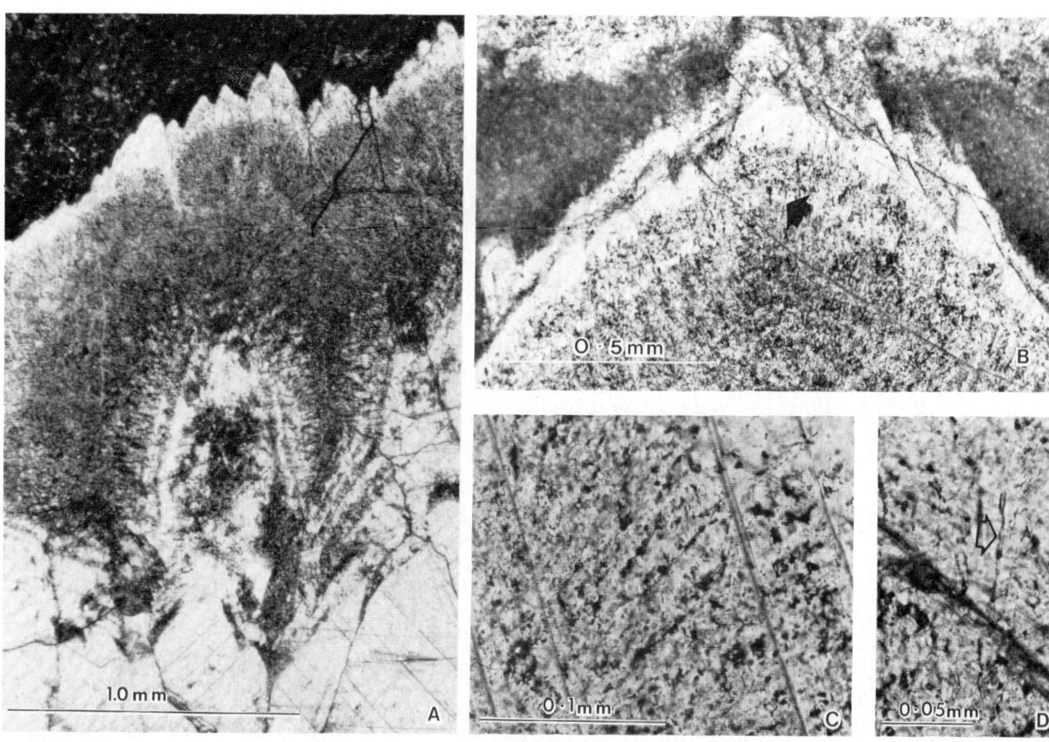

FIG. 7.—Photomicrographs of inclusion patterns. *A*, Radiating inclusion pattern diverging away from the calcite termination. *B*, Radiating linear inclusions from crystal illustrated in Fig. 5. Also note the inwardly-directed and elongate crystal terminations of the latest stage of cement growth (when the calcite was relatively inclusion-free). Arrow points to possible obtuse rhombic terminations which differ markedly from those present in the youngest parts of the crystal. Plane polarized light. *C*, Detail of inclusions showing their irregularity. *D*, Detail of a markedly elongate and fluid-filled inclusion.

almost uniformly-extinguishing and inclusion-free calcite.

Overlying the internal sediment layer, but largely nucleated upon the projecting crystal terminations of the earlier cement, is another cement crust of radiaxial calcite (C); this one containing abundant inclusions. Each exposed termination has given rise to a single inclusion-rich radiaxial crystal. At the boundary between the two cement generations the inclusion-poor and inclusion-rich calcite is in optical continuity (Fig. 6). There can be little doubt that radiaxial crystals in cement crust 'C' have nucleated from the projecting terminations of the earlier calcite cement (A) and therefore that the former is an overgrowth of the latter. Away from the boundary the radiaxial crystals may break up into distally-radiating subcrystals and their lattices become distorted. The crystals exhibit brush extinction (Fig. 6) and curved cleavage and twin-lamellae (Fig. 5A). Fast-vibration directions in each crystal are distally-convergent (Fig. 5C). This pattern is made particularly clear in the most distal parts of the crystals. These parts are composed of inclusion-free calcite and have the form of many elongate crystal terminations. In each crystal these terminations are directed towards each other (Fig. 7B), indicating the c-axes are similarly oriented.

Because the patterns of subcrystals and lattice curvature cross-cut each other, the calcite within any subcrystal exhibits, along its length, a changing fast-vibration direction. This change consistently involves the fast-vibration direction becoming progressively more divergent from the subcrystal elongation.

S.E.M. examination of the inclusion-rich cement zones reveals them to be full of microcavities (presumably originally liquid-filled) and microcrystals of dolomite. These observations, together with the geochemistry of the cements will be the subjects of a future paper.

Inclusions define two important fabrics—growth zones and a radial lineation fabric. Although these are discernable in cement layer 'C', they are better seen in crystals of other crusts, including those of the stratigraphically-equivalent roof crust ('D' in Fig. 4). Crystals of roof crusts also are nucleated on elongate terminations of earlier inclusion-poor cement layers. Inclusions in crystals from cement layer 'C' tend to be more uniformly distributed and thus are more typical of inclusion patterns within radiaxial calcites from other localities.

Variations in inclusion density define growth zones. These reveal that, as growth proceeded, there was a marked change in the shape of the growth front from the initial elongated crystal termination toward an increasingly more botryoidal surface (Fig. 8). This implies faster growth on the flanks of the crystal than on the crystal apex (see Page 68).

Inclusions are individually elongate (Fig. 7B, D) and, within each crystal, are arranged in a radial array (Fig. 5A, B; 7A). This pattern is similar to that in many fascicular-optic cements (Kendall, 1977) but is unusual in radiaxial cements. Each inclusion is elongated normally to the growth layering and is parallel to that of nearby subcrystal boundaries. Because successive growth surfaces exhibit a change in shape toward a botryoidal form, the radiating lines of inclusions and subcrystals display a slight but persistent curvature (Fig. 5B). This deflection is always toward the centerline of the radiaxial crystal.

The inclusions and their patterns are similar in many respects to those described from many calcite crystals in speleothems (Kendall and Broughton, 1978, Braithwaite, 1979) but with one very significant difference. In speleothem calcites the elongation and lineation of inclusions always parallels the c-axis of the host crystal regardless of the orientation of the growing front. In the radiaxial crystals the converse is true: inclusions are oriented normal to former growing surfaces regardless of the c-axis orientation of the host crystal.

Inclusion-defined growth layers show that elongate and large crystal terminations (like those that were present during the last phases of crystal growth in cement crust 'C') were absent. Growth must have occurred on a relatively smooth surface or upon many small terminations.

Repeated alternations of inclusion-rich and inclusion-poor calcite occur within crystals of both floor and

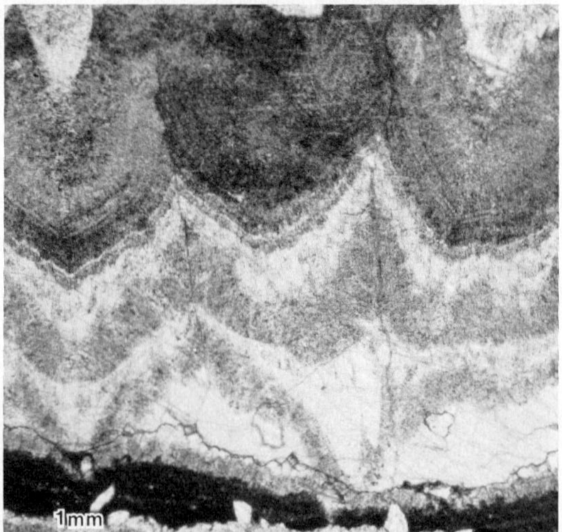

FIG. 8.—Photomicrograph of radiaxial crystals with pronounced growth-zoning revealed by variations in inclusion density. The changing shape of the growth zones suggests a change toward botryoidal (or spherulitic) growth forms. However, the growth zoning still retains the appearance of crystal faces—but faces that are markedly curved. Plane polarized light.

roof cement crusts (Fig. 4). Zones of inclusion-rich calcite never develop large crystal terminations (unless they are thin and simply mimic the form of their substrate). They exhibit a tendency toward smooth botryoidal growth surfaces. In contrast, zones of clear calcite are characterized by large, high-spired crystal terminations. Boundaries between earlier inclusion-free calcite and later inclusion-poor crusts are abrupt. Conversely, boundaries between earlier inclusion-rich calcite zones and inclusion-poor overgrowth are gradational.

INTERPRETATION OF RADIAXIAL FABRICS

Fabrics of the radiaxial cements from Australia will be interpreted first, followed by a discussion of how relevant this is to radiaxial cements elsewhere.

Interpretation of Australian Radiaxial Cements

Together the observations made in the preceeding sections can be used to suggest that radiaxial crystals result from split-crystal growth. Most of the fabrics interpreted to be neomorphic by Kendall and Tucker (1973) are reinterpreted to be primary. This supports a conclusion of Kendall and Broughton (1978) that most fabrics in radiaxial crystals do not definitively prove a neomorphic origin.

Lattice continuity with single crystal substrates.—Each radiaxial crystal was nucleated upon, and initially is in lattice-continuity with, a pre-existing calcite crystal. Implications of this are:

1. Radiaxial crystals cannot have formed by the amalgamation of parts of several neighbouring acicular-crystal bundles.
2. Radiaxial crystals were composed of calcite (probably magnesian calcite because of the presence of dolomite microcrystals, see Lohmann and Meyers, 1977), not aragonite.
3. Growth zones defining calcite crystal terminations are probably primary and do not record the passage of a replacement front through an earlier cement as suggested by Kendall and Tucker (1973) and Marshall (1981).

Growth zones.—The changing form of the growth surface, as revealed by the shape of successive inclusion-defined growth zones (Fig. 8), indicates that the crystal faces of the underlying crystal termination ceased to grow. Overgrowth development must have occurred in a different manner to produce the more botryoidal growth form.

Elongate inclusions.—The radially-arranged elongate inclusions define an acicular-like crystal habit. This may reflect the former existence of true acicular crystals (that later coalesced to form the large radiaxial crystals) or, as in many speleothem calcite crystals (see Kendall and Broughton, 1978), the inclusions may record precipitation on surfaces covered with numerous tiny terminations that grew together just beneath the surface before any acicular fabric could develop. In order to avoid constant repetition of these two alternatives (which differ only in the timing of the lateral coalescence, see Kendall and Broughton, 1978) the term *crystallite* will be used to refer to these separate precipitation sites on the growing surface, and the term *fiber* is used for the material deposited on each of those sites.

The crystallites must initially have grown in lattice-continuity with their underlying single-crystal substrate. The inclusion-rich overgrowth marks the splitting into numerous growing units of the earlier single crystal. The close-spacing of the inclusions, which mark boundaries between the former crystallites and outline the fibers, suggests that crystallites were numerous and of small size. Continued growth on each of the many crystallites, rather than unequal growth on different large crystal faces, eventually led to the formation of botryoidal growth surfaces similar to those found in acicular cements or in speleothems.

Lindström (1979) describes and figures radial-fibrous calcite crystals that contain similar inclusion patterns; similar in that they also do not mimic a lattice curvature in the host crystal. Lindström interpreted the inclusion lineations as marking the growth vectors of the crystal. Each inclusion lineation (or fiber) marks the direction taken by part of the crystal as it grew. The same conclusion is reached here concerning the radiaxial crystals because in them the fibers are consistently oriented normal to the growth layering.

Density of inclusions.—Successive inclusion-rich growth zones have approximately the same spacing of inclusions. From this it may be deduced that crystal-splitting (inferred to have operated when the overgrowth was initiated) must have continued during growth of the radiaxial crystals. If it had not, then as

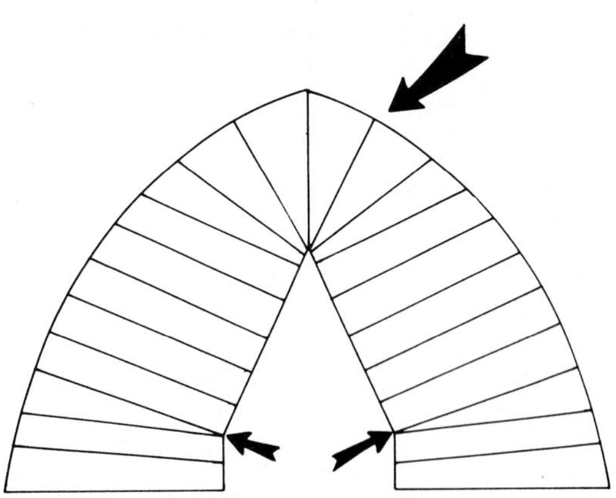

FIG. 9.—Diagram showing that without the presence of crystal-splitting during radiaxial calcite growth the fibers would have to become wider (and intervening inclusions wider spaced). This is most evident at locations where the calcite has overgrown the apex of the underlying crystal termination.

the growing surface expanded (this phenomenon is particularly evident near the acute apex of the former termination, Fig. 9) the fibers would have had to have become wider in order to fill the extra space. Consequently, the inclusions between fibers would have become wider spaced.

The near-uniform spacing of inclusions in inclusion-rich growth zones implies crystallites retained their small size as they grew. Furthermore, small crystallites on an expanding surface must diverge from each other, and other crystallites of near-identical lattice-orientation must have become inserted between them to fill the extra space. These are the hallmarks of products formed by crystal-splitting (see Grigor'ev, 1965).

Relations between fibers and lattice-deformation.—Unlike acicular cements or crystals in many speleothems, the elongation of inclusions in the radiaxial crystals is rarely coincident with the fast-vibration directions of the host calcite. Commonly these two orientations are markedly different and this discrepancy also varies along the fiber length.

The main problem that has hindered interpretation of the distinctive lattice curvature in radiaxial crystals has been caused by viewing the pattern as a whole when, as has just been deduced, the inclusion patterns indicate radiaxial crystals did not grow as single entities. The lattice curvature ought to be examined in relation to the crystallites and to the products of these features—the inclusion-defined fibers.

Although initially parallel with that of the underlying crystal termination the fast-vibration directions become reoriented as they are traced along the fibers (Fig. 5D) so that these directions become progressively more parallel with the former growing surfaces. The fibers become progressively more length-slow as they grew.

At this point it is of interest to note that Lohmann and Meyers (1977) detected a radial-fibrous fabric within radiaxial crystals by employing cathodoluminescence. These authors noted that the fibers were also length-slow.

Another cause of the changing fast-vibration directions appears to be that the orientation of successive growing surfaces changed and the fibers are correspondingly curved (Fig. 5B). If a mechanism could be found that would explain the changing orientations of the fast-vibration directions within individual fibers we would have explained the radiaxial fabric. This mechanism should also affect all crystallites simultaneously and be consistent with the inferred split-crystal growth mechanism.

Curved fibers.—The curved nature of the inclusion-defined fibers implies that the position and orientation of each crystallite on the growing surface progressively changed with time. Since all crystallites initially had the same lattice orientation (all being in continuity with their single crystal substrate) these changes can only be the result of a different growth rate imposed on different parts of the composite crystal surface. Asymmetric growth in crystallite aggregates derived from the repeated splitting of single crystals (spherocrystals) has been analysed by Maleev (1972).

When different parts of a spherocrystal have different growth rates, each crystallite grows more rapidly than its neighbors on one side and more slowly than its neighbors on the other. During growth, each crystallite overtakes the slower growing neighbors and becomes deflected so as to partially occupy the free space (Fig. 10A). But it is also hindered by its faster growing neighbors. As a result all crystallites (and the fibers formed from them) deflect toward the more slowly-growing parts of the overgrowth. If crystal-splitting occurs in addition, daughter crystallites that are inclined toward the more slowly-growing part of the overgrowth are progressively selected because these grow into free space. In contrast, crystallites that are inclined toward faster growing parts of the overgrowth are obstructed by their faster-growing crystallite neighbors (Fig. 10B). In this way, both the elongation of fibers and the crystallographic orientation of the material forming the fibers is deflected toward the slower-growing parts of the overgrowth (Fig. 10C).

Fibers in the radiaxial crystals are consistently curved toward the apex of the composite crystal (Fig. 5B) and this, therefore, must have been the more slowly growing part of the overgrowth. This can be independently deduced from the growth zones which are more closely spaced at the apex of the crystals than along their sides (Figs. 5, 7A).

It should be noted that the proposed mechanism of asymmetric crystallite growth is one that accounts for the changing orientation of the fibers; for the greater rate of change in the orientation of the fast-vibration directions (so that they become more inclined towards the crystal apex and consequently more parallel to the former growing surfaces) and is a mechanism that acts simultaneously on all crystallites. Asymmetric growth also causes the shape of the composite crystal to become progressively modified.

Origin of asymmetric growth.—Two possible explanations can be advanced to explain why radiaxial crystals grew more slowly at the apices of the earlier calcite crystal terminations (= along the c-axis). Either the composite crystals (1) behaved more as spherocrystals and began to assume their most stable form—that of a spherulite, or (2) they behaved more as single crystals—crystals that were changing habit from length-fast to length-slow stable forms.

The first explanation requires little elaboration, but is not favored. Transformation of an elongate calcite termination into part of a spherical aggregate of crystallites clearly involves faster growth rates on the flanks of the former crystal (Fig. 11A). The changing shape of successive growth zones toward a more botryoidal form is consistent with this interpretation. Unfortunately, however, it is more likely that the spher-

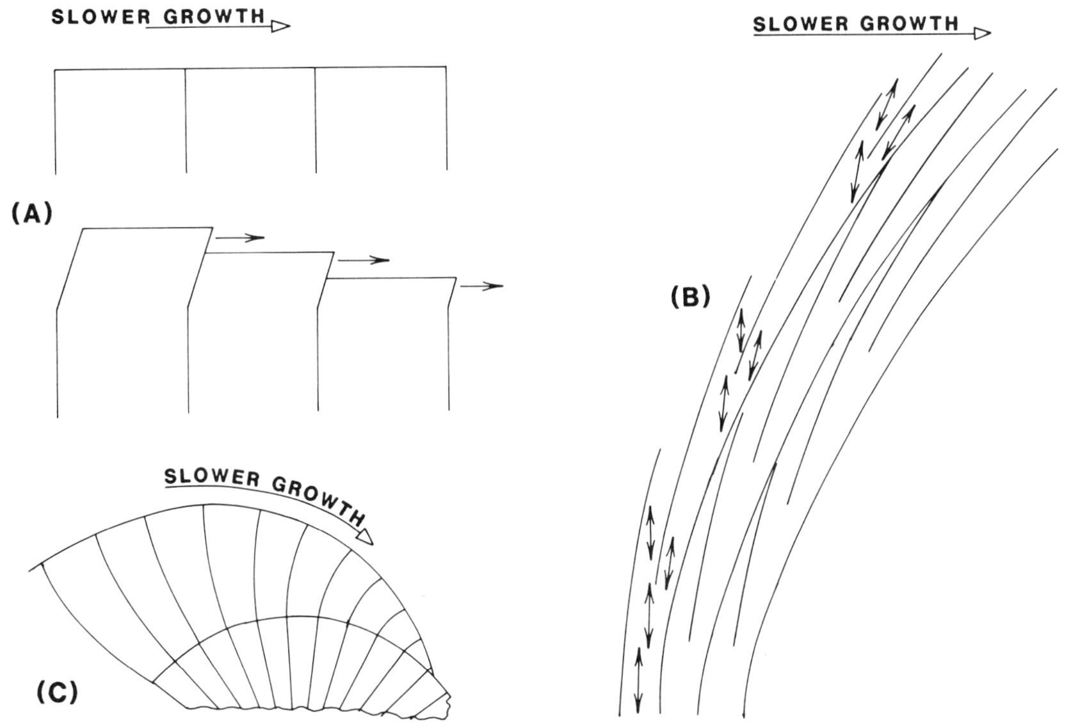

FIG. 10.—Diagrams illustrating asymmetric growth within spherocrystals. *A*, Lateral growth of fibers into the free space above more slowly growing neighbors causes the fibers to become deflected. *B*, Crystal-splitting combined with asymmetric growth causes deflection of fibers and their crystallographic orientations towards the more slowly-growing parts of the overgrowth. For convenience of illustration the crystallographic orientations are shown as being initially parallel with the fiber elongation: this situation is not necessary for the deflection to occur. *C*, Effects of asymmetric growth in spherocrystals are shown in this diagram modified slightly from Maleev (1972). There is both a change in the growth form and the fibers exhibit curvature toward the more slowly-growing parts of the overgrowth.

ical growth form of spherulites is a product of equal crystallite growth rates. The botryoidal-like form of growth-zones presumably therefore mark a tendency toward increasing equality of growth rates over the growing surface than that which had occurred when the crystal termination was forming.

The growth surfaces still retain the form of crystal faces, even though those faces are markedly curved (Fig. 8). It may be deduced from this that all the fibers (which collectively make up a radiaxial crystal) behaved, at least in part, as single crystals. There is reason to believe that these composite crystals grew in environments that promoted growth of length-slow calcite. Since the original termination was markedly length-fast, this change in habit involved a great change in crystal form and a varying growth rate over the former termination (Fig. 11B).

Geometrical considerations indicate that crystallites could not have been bounded by acute rhombohedra or scalenohedra (Fig. 12). These forms, however, were present before and at the end of radiaxial crystal growth, when inclusion-poor calcite was precipitated. Inclusion-rich and inclusion-poor zones must represent different environmental conditions when different crystal habits were stable (see next section).

We may infer that when inclusion-rich calcite was precipitated the environmental conditions did not favor elongate crystallite-terminations. Evidently crystallites were bounded by flatter forms. Calcite crystals with flatter rhombohedral forms have their directions of greatest growth oriented at high angles to the c-crystallographic axis, so that the stable crystal form is length-slow (Dickson, 1978).

From the absence of crystallite terminations that would produce length-fast calcite, it is inferred that length-slow crystal habits were favored by the environmental conditions (however, see later, Page 71). If these conditions caused the whole radiaxial composite crystals to assume a length-slow habit, this would impose faster growth rates on the sides of the former elongate (length-fast) crystal termination. This in turn would lead to asymmetric growth, curvature of the fibers and the development of the radiaxial fabric of convergent fast-vibration directions.

An analogy may be made with echinoderm overgrowths. Most echinoderm grains are single crystals flattened along the c-axis and their overgrowths develop in conditions favoring crystal forms elongated along those axes—forming length-fast calcite (Evamy and Shearman, 1965). The overgrowths, accordingly, grow fastest along the c-axis directions until the stable, length-fast, crystal form is reached. It may be supposed that calcite crystals growing in environments that favor length-slow habits also exhibit preferential growth tendencies upon crystal substrates of the 'wrong' shape. Thus radiaxial overgrowths are interpreted to be composite crystals that are attempting to assume the length-slow stable form upon a length-fast substrate. They therefore display slowest growth rates along the c-axis direction.

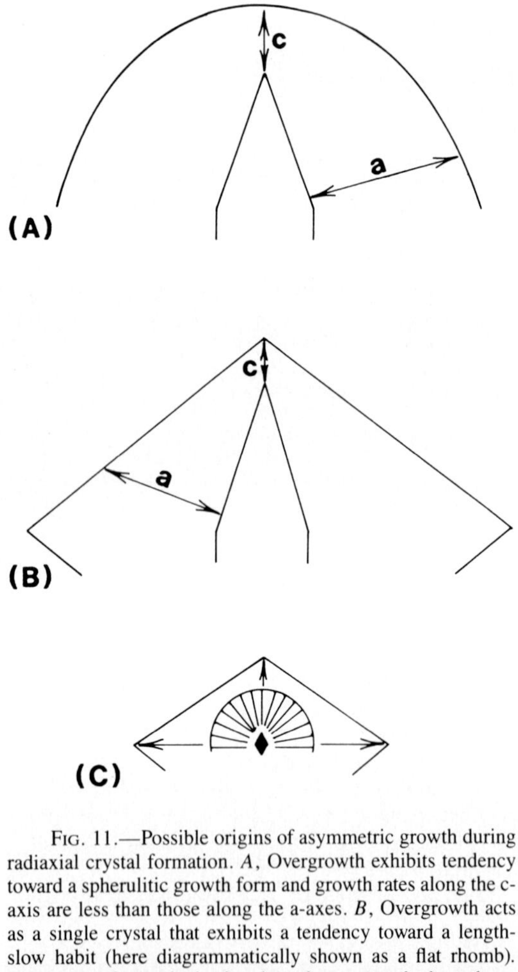

FIG. 11.—Possible origins of asymmetric growth during radiaxial crystal formation. *A*, Overgrowth exhibits tendency toward a spherulitic growth form and growth rates along the c-axis are less than those along the a-axes. *B*, Overgrowth acts as a single crystal that exhibits a tendency toward a length-slow habit (here diagrammatically shown as a flat rhomb). This mode of growth also involves slower growth along the c-axis until the stable form is assumed. *C*, Overgrowth from a small seed crystal (as when radiaxial crystal nucleates from a cavity wall). Crystal-splitting during the initial overgrowth produces a spherulitic growth form which departs significantly from the length-slow calcite crystal habit. Continued crystal growth toward the length-slow crystal habit involves growth asymmetry and the generation of radiaxial crystal fabrics.

Although the explanation for asymmetric growth offered here explains all known features in the radiaxial crystals there is little direct proof that length-slow crystal habits were favored during radiaxial calcite precipitation. Some growth layers do appear to define very flat (rhombohedral ?) terminations (Fig. 7A) but these are far from convincing. The argument rests primarily upon the geometrical inference that crystallites did not possess terminations indicative of length-fast calcite growth.

Crystal terminations and boundaries between cement zones.—Despite the effects of crystal-splitting and reorientation, neighboring crystallites, and their corresponding fibers, still retained near-identical lattice orientations. For this reason they collectively behave as a single, albeit deformed, crystal. The similarity of neighboring crystallite orientations implies that they were always capable of amalgamating to generate larger terminations when the environment changed to that where crystallite growth was no longer favored. The amalgamation process, which generates a more simplified growth surface, is similar to that which occurs in echinoderm overgrowths (Evamy and Shearman, 1965) and in many speleothem calcites (Kendall and Broughton, 1978).

Gradational boundaries from inclusion-rich to inclusion-poor zones are interpreted as recording the gradual elimination of crystallite boundaries as they amalgamate and the growth front simplifies. By way of contrast, the abrupt boundaries between inclusion-poor and overlying inclusion-rich calcite must record an equally-abrupt initiation of crystal-splitting and a complete change in crystal habit.

The two types of calcite are interpreted to reflect calcite precipitation under very different conditions. At times when inclusion-rich calcite was precipitated, growth was accompanied by extensive crystal-splitting and the radiaxial fabrics developed. As we have deduced, these fabrics result from asymmetric growth possibly refecting a change from length-fast to length-slow crystal habits.

In contrast, when inclusion-poor calcite grew, crystal-splitting no longer occurred and the previously separate crystallites amalgamated to form larger and larger crystal terminations. These terminations developed scalenohedral or acute rhombohedral faces—a growth form that could not have been present during episodes of inclusion-rich calcite growth. The large terminations grew in conditions that promoted growth of length-fast calcite. At these times the radiaxial fabrics were no longer being generated but the calcite exhibits brush-extinction because it has overgrown a 'deformed' substrate.

When considering changes in environmental conditions that gave rise to the two calcite types, especially in terms of whether length-fast or length-slow calcite was favored, it is important to recognize that the calcite in question was precipitated in two very different ways—as spherocrystals and as single crystals bounded by large crystal faces. It will now be suggested that the inferred change in favored growth habit (see above) is an apparent one and may be simply a by-product of the presence or absence of split-crystal growth.

Although calcite crystals do exhibit different growth habits it is a misconception to assume that these are thermodynamically equilibrium shapes. Large crystal faces are developed because they are out of equilibrium with the growth medium, due to their proportionately larger kinetic barriers (Dowty, 1976). Relative to faces that disappear, the larger faces (and hence the form of crystals) are those that grow at slower rates (Buckley, 1951). Spherocrystals, on the other hand, are very different from large single crystals because they become covered by numerous crystallite faces that grow at the fastest rate (Maleev, 1972).

Length-fast calcite crystals are bounded by slowly-growing crystal faces (acute rhombohedra or scalenohedra) because the faster-growing flat rhombohedral faces have been selected out. However, if such a crystal (as in the Australian radiaxial cement) is converted to a spherocrystal, by the introduction of split-crystal growth, the faster-growing faces (flat rhombohedral) will dominate and there will be an apparent change toward length-slow crystal growth.

It is also of note that times of inclusion-free calcite precipitation were also those when internal sediment was deposited. Internal sediments always overlie inclusion-poor zones (Fig. 4) and thin layers of this cement type may occur within the internal sediment (Fig. 7A). Episodes of internal sedimentation may also have been associated with dissolution of inclusion-poor calcite as is shown by the rounded nature of some contacts (Figs. 5A, 13). Braithwaite (1979) interpreted similar rounded surfaces in speleothem-like carbonates as resulting from dissolution. Dissolution occurring within episodes when inclusion-poor calcite grew, but not with those when inclusion-rich calcite was precipitated, may indicate the clear calcite type grew more slowly and from less supersaturated waters than did the inclusion-rich cement layers.

The sparcity of dolomite microcrystals in inclusion-poor zones, in contrast to their abundance in inclusion-rich zones, is interpreted to be a reflection of an original inhomogeneous distribution of magnesium within the cement crystals. If this inference is correct, inclusion-poor calcites were precipitated as low-magnesian calcite, whereas inclusion-rich zones were originally precipitated as high-magnesian calcite (with up to 15 mole percent $MgCO_3$; Meyers, pers. comm., 1984). Furthermore, both high- and low-magnesian calcites syntaxially overgrew each other.

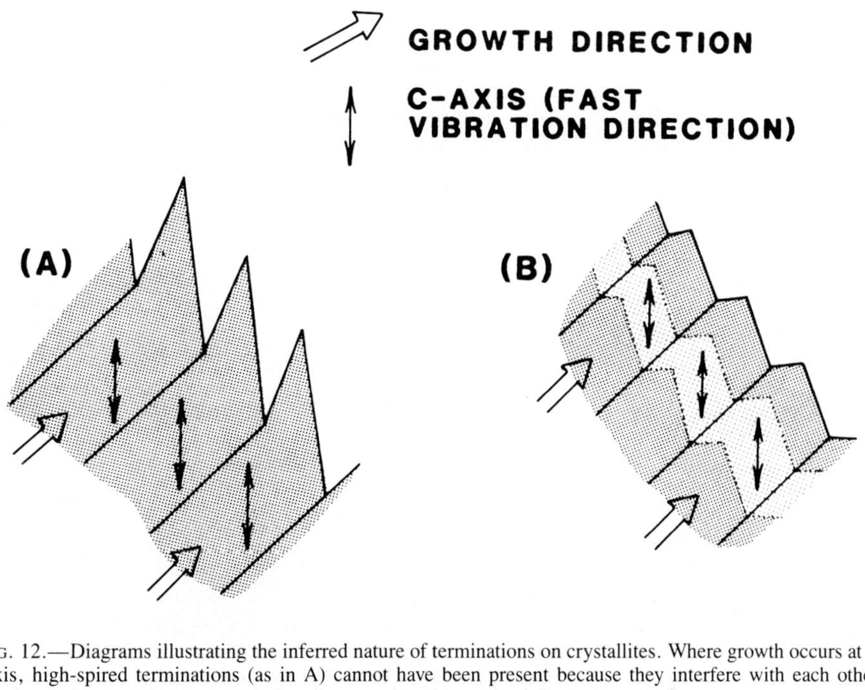

FIG. 12.—Diagrams illustrating the inferred nature of terminations on crystallites. Where growth occurs at a high angle to the c-axis, high-spired terminations (as in A) cannot have been present because they interfere with each other's growth. In contrast, low-spired terminations (as in B) allow continued growth of fibers.

If inclusion-rich calcite layers represent episodes of faster growth, from more supersaturated solutions, this may provide some explanation for the crystal-splitting, which ultimately was responsible for the radiaxial fabrics. Crystal-splitting is commonly caused by crystal poisoning, or when growth occurs from highly supersaturated solutions where solvent molecules interfere with the rapid crystal lattice formation and thus also act as poisons (Buckley, 1951; Grigor'ev, 1965). No other petrographic evidence has been found which would provide clues as to the origin of the crystal splitting. It is hoped that trace-element and isotopic analyses (currently underway) will provide information upon this topic.

Other fabrics.—The irregular intercrystalline boundaries in radiaxial cements were used to infer a neomorphic origin for the crystals (Kendall and Tucker, 1973). However, such boundaries are to be expected when composite crystals compete for growth-space. Identical non-planar boundaries are present between some echinoderm overgrowths (Evamy and Shearman, 1965) and speleothem calcite crystals (Kendall and Broughton, 1978) which are also composed of composite crystals.

Subcrystals are aligned parallel with the inclusion-defined fibers. It may be assumed they are also primary features and originate when some of the crystallites on the growing surface become slightly misoriented during the process of crystal-splitting. The calcite laid down by such a crystallite, and all crystallites that originate from it, has a slightly different lattice orientation from its neighbors and optically may be identified as a subcrystal by virtue of its slightly different extinction position.

Since all parts of radiaxial crystals originate from single crystal substrates and, furthermore, behave as single crystals with strained lattices, the cleavage, glide-planes and twin-lamellae are correspondingly curved. Unusual features are seen when two radiaxial crystals have overgrown neighboring crystal terminations which are themselves parts of a single crystal (Fig. 13). Because the terminations had near-identical lattice orientations, so do their radiaxial overgrowths, and cleavage and twin-lamellae pass from one crystal into the other.

Interpretation of Radiaxial Cements in General

Questions arise as to the relevance of the Australian radiaxial material to an explanation of all radiaxial cements. How typical are the Australian cements and can their interpretation be applied in general?

The Australian radiaxial cements differ from radiaxial calcites elsewhere in two main respects: the nature of their substrate and the pattern of inclusions found within their crystals.

Nature of the substrate.—Most of the Australian radiaxial crystals overgrew large calcite crystal terminations whereas radiaxial crystals in other cements seed directly from cavity walls made of lime mud or exposed skeletal grains. This difference, however, is not complete because some crystals in the Australian cement layers nucleated from the upper surface of internal sediments (Fig. 4). These crystals possess equally well-developed radiaxial fabrics.

The diagnostic fabric of radiaxial crystals—distally-divergent subcrystals cross-cut by a distally-convergent pattern of fast-vibration directions—is such a peculiar one that it is most unlikely that different mechanisms could have formed the same product, especially if they had to form the same products within a single cement layer. It is concluded that all radiaxial calcite crystals originate in the same manner. In most cements, it may be inferred, they nucleate as small crystals on cavity walls and it is these initial seeds that suffer continual crystal-splitting so as to generate composite crystals. Since a small growth asymmetry acting upon a small nucleus will not cause significant departure from the spherulitic shape, the composite crystal is, from the outset, of the 'wrong' shape and will suffer further growth asymmetry and so develop radiaxial fabrics (Fig. 11C).

Inclusion patterns.—Inclusions in the Australian crystals define the 'fibrous' fabric which has not previously been identified as such in other cements. In most radiaxial cements the inclusions are non-linear and do not define fibers. Inclusions may be concentrated along subcrystal boundaries in some cements

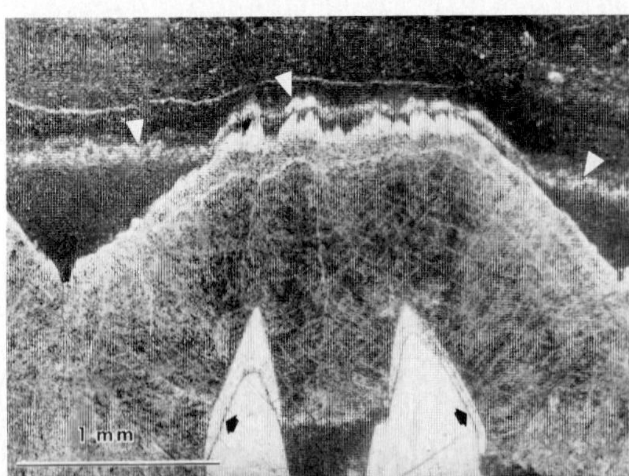

FIG. 13.—Photomicrograph showing the presence of twin lamellae that are continuous across a crystal boundary between two adjacent radiaxial crystals. These crystals are overgrowths of adjoining crystal terminations which are parts of a larger crystal. They and their radiaxial overgrowths thus share a near-identical crystal-lattice orientation. Black arrows point to a possible dissolution surface and white arrows to a cement layer within internal sediment that passes laterally into the radiaxial crystals. Internal sedimentation occurred concurrently with precipitation of inclusion-free calcite and possibly at times of undersaturation when clear calcite was dissolved. Plane polarized light.

and, by analogy with the Australian material, would be parallel with the fiber orientations. One previously published inclusion pattern (Kendall and Tucker, 1973, Fig. 7C) defines a fibrous fabric comparable with that seen in the Australian crystals but was misinterpreted as being of neomorphic origin.

Despite the presence of radiating inclusion patterns in some other radiaxial cements, most lack any vestige of inclusion-defined fibers. If radiaxial fabrics can only be formed in the manner described in this paper, the fibers must originally have been present. In most instances it is believed that the absence of radial inclusion patterns has been caused by a subsequent migration, obliteration or shape-change of the inclusions.

Although turbidity in radiaxial crystals may be caused by dolomite microcrystals (Lohmann and Meyers, 1977), organic material and other solid impurities (Kendall and Tucker, 1973), it is my experience that many inclusions (including most of those in the Australian cement crystals) are fluid-filled microcavities similar to those in stalactites (Kendall and Broughton, 1978). It might have been expected that fiber-defining inclusions in the better-preserved radiaxial crystals would be markedly elongate and spindle-shaped—similar to equivalent inclusions between partially-coalesced fibers in speleothem calcites (Kendall and Broughton, 1978). Instead, most are considerably more irregular (Fig. 7C) and some appear to have formed by a subdivision of more elongate inclusions. Roedder (1977) notes that most inclusions in crystals do not have the shape they had when trapped. Solution on some parts of the inclusion wall and deposition on others is common and causes shape changes or even migration of the inclusions. The fiber fabric that is to be preserved is a very fine one; only the slightest change in inclusion shape and location will cause complete loss of detail and the appearance of a homogeneous inclusion distribution.

From the occurrence of dolomite microcrystals in most radiaxial calcite cements it may be concluded that most were originally composed of magnesian calcite and have suffered some process of unmixing to low magnesian calcite with included dolomite inclusions (Lohmann and Meyers, 1977). This unmixing process is likely to have caused significant damage to the original fluid-filled inclusion pattern and, furthermore, introduces a more homogeneously distributed inclusion pattern (that of the dolomite microcrystals) into the crystals.

The fluid-filled inclusions can also be eliminated by the later growth of calcite within them. This causes a pronounced clearing of the calcite in some of the Australian crystals (Fig. 14) and a false impression that these crystals have suffered neomorphism.

Radiaxial crystals that lack a radial fibrous inclusion pattern could also have arisen, as do some growth layers in speleothems (see Kendall and Broughton, 1978), when the lateral coalescence between crystallites was complete and no intercrystallite space remained to form inclusions. In this type of calcite, or in that, where inclusions have been eliminated subsequently, the inclusion-pattern would be formed of dolomite and other solid materials which are more homogeneously distributed.

It is suggested that all radiaxial cements grew as composite crystals and that many (all ?) possessed a radiating inclusion pattern. This, in most instances, has been obliterated by modifications to the shape and location of fluid-filled inclusions (in part induced by the unmixing of magnesian calcite) and/or the appearance of a more homogeneous inclusion distribution composed of other materials.

If inclusions migrate through the crystals, or are formed within them, they could become trapped (or are formed preferentially) at crystal discontinuities or be lost preferentially at these locations. This could explain the common occurrence of inclusion patterns associated with subcrystal boundaries and twin-lamellae (see Kendall and Tucker, 1973) in many radiaxial cements. Patterns associated with twin-lamellae are common in some of the Australian crystals (Figs. 5A, 13) but only in those with a poorly-developed radial inclusion pattern. This suggests that inclusions have migrated even in some of the Australian cement crystals.

DISCUSSION

Items requiring discussion include 1) explanation of some crystal fabrics previously used to suggest a neomorphic origin for radiaxial calcite after bundled acicular cements, 2) evidence of crystal-splitting in

FIG. 14.—Photomicrograph showing inclusion-rich growth layers passing laterally into inclusion-poor calcite. Much of this change, which appears at first glance to be neomorphic, seems the result of the filling of fluid-filled micro-inclusions by calcite in optical continuity with the host crystal. The reason why only some areas should be affected in this way is not known. Plane polarized light.

other calcites, and 3) the significance of coarsely-crystalline calcite in marine phreatic environments. The first two do not impact directly on the main subject-matter of this paper and are considered in appendices.

Coarsely-Crystalline Marine Calcite

One of the major props of the replacement theory for the origin of radiaxial calcite was the argument that coarsely-crystalline calcite is not found in the marine diagenetic environment (Kendall and Tucker, 1973; Kendall and Broughton, 1978, p. 536).[1] Reinterpretation of radiaxial cements as primary now requires that we accept that this type of calcite could have been precipitated from marine or marine-derived waters.

Inclusion-poor calcite layers with large crystal terminations from marine cements such as those described here and from similar cavity fills from the Devonian of West Germany (Tucker, 1973, Fig. 9) also imply that calcite was capable of growing in the marine environment with the form of large undivided crystals as well as composite crystals. Additionally, in the case of the Australian cements it is suggested the inclusion-poor calcite was originally low-magnesian calcite.

The marine origin of these radiaxial and associated calcites is strongly suggested by their occurrence in locations where the activities of non-marine waters can be precluded. At McWhae Ridge the fills of neptunian sills are interpreted, from the presence of marine fossils in internal sediments and the stratigraphic setting, to have been entirely of marine phreatic origin.

The sills are penecontemporaneous structures within forereef sediments. At McWhae Ridge the Virgin Hills Formation represents a drowning of the original reefal-spine (and its forereef) and contains widening-upward tongues of deep-water stromatolitic limestones. These are interpreted by Playford (1980) to have originally completely capped the older drowned reefal spine (as occurred elsewhere in the Canning Basin in the Elimberrie bioherms; Playford, 1980). The reefal spine, also suffered contemporaneous faulting and is overlain by basinal deposits of the Gogo Formation. The possibility that the cements in the neptunian sills are of freshwater phreatic origin is therefore considered to be remote. Not only were the cements penecontemporaneously emplaced into deepwater sediments (Playford, 1980, estimates water depths greater than 100 m) but the area suffered progressive drowning, not emergence. The presence of any freshwater lens in the McWhae Ridge area during deposition of the Virgin Hills Formation is unlikely and, even if present, would have been ineffective. The reefal spine is less than 200 m wide so that a freshwater lens capable of precipitating calcite cements tens of meters into the forereef sediments and several hundred meters laterally away from the intake area is most improbable.

[1]This was before the discovery of sparry magnesian calcite from the Mediterranean by Frojet (in Purser, 1980, p. 182).

sediments and several hundred meters laterally away from the intake area is most improbable.

Radiaxial calcites from the condensed deepwater limestones upon Devonian sea-mounts in West Germany (Tucker, 1973) are even less likely to have come into contact with meteoric or mixed waters since they were also deposited in deep water initially; were overlain by turbidites; and subsequently were deformed in the Variscan Orogeny before coming into contact with meteoric waters.

Radiaxial calcite and coarsely-crystalline low-magnesian calcite cements have yet to be reported from Recent reefs or mudmounds despite numerous studies of these features. These cement types are common within many carbonate buildups within the Paleozoic and earlier Mesozoic. This discrepancy between Recent and ancient environments is presumably yet another indication that marine water chemistry has changed in the past.

CONCLUSIONS

Radiaxial fibrous calcites do not represent replacements of earlier acicular bundles as was previously thought. Instead they are primary calcite cements, commonly (or always ?) magnesian calcite, that grew as composite crystals in the marine phreatic realm. The composite character is due to split-crystal growth, itself probably caused by crystal poisoning or by growth from highly supersaturated solutions. The distinctive crystal fabrics of a distally-divergent subcrystal pattern cross-cut by a distally-convergent pattern of fast-vibration directions, result from asymmetric growth affecting the split-crystal aggregate. This causes deflection of fiber growth toward slower-growing parts of the composite crystal and the appearance of a lattice-curvature.

Radiaxial crystals are but one product of split-crystal growth in diagenetic calcites. This process is responsible for the distinctive crystal fabrics seen in speleothem calcites (Kendall and Broughton, 1978) and the same process is responsible for the subdivided and fluted character of many artificially prepared and naturally occurring calcite crystals; crystals that may be composed of low- or high-magnesian calcite (see Appendix 2).

The Australian cavity-filling cements provide an example of an interpretation that required the finding of suitable material in order to be developed. I have known for almost a decade that the Kendall and Tucker replacement theory for the origin of radiaxial fabrics was in trouble yet could devise no better explanation. Only when the Australian material was examined (with crystals nucleated upon earlier crystal substrates and the preservation of radial-fibrous inclusion-patterns) was it possible to deduce a possible alternative origin for these most peculiar of cements.

ACKNOWLEDGMENTS

I thank Amoco Canada for supporting my attendance of the A.A.P.G.-sponsored field seminar to Western Australia and Phil Playford for showing so clearly the field relationships at McWhae Ridge. W.J. Meyers (SUNY, Stony Brook) and J.M. Queen (Sohio, San Francisco) kindly reviewed the manuscript and L. Blome redrafted my illustrations.

REFERENCES

ALEXANDERSSON, T., 1972, Mediterranean beachrock cementation: marine precipitation of Mg-calcite, in D.J. Stanley (ed.), The Mediterranean Sea: A Natural Sedimentation Laboratory: Stroudsburg, Pa., Dowden, Hutchinson and Ross, p. 203–223.

BATHURST, R.G.C., 1959, The cavernous structure of some Mississippian "stromatactis" reefs in Lancashire, England: J. Geol., v. 67, p. 506–521.

______, 1975, Carbonate Sediments and Their Diagenesis: Amsterdam, Elsevier, 658 p.

______, 1977, Ordovician Meiklejohn bioherm, Nevada: Geol. Mag., v. 114, p. 308–311.

______, 1982, Genesis of stromatactis cavities between submarine crusts in Palaeozoic carbonate mud buildups: Jour. Geol. Soc. London, v. 139, p. 165–181.

BINKLEY, K.L., WILKINSON, B.H. AND OWEN, R.M., 1980, Vadose beachrock cementation along a southeastern Michigan marl lake: Jour. Sed. Petrology, v. 50, p. 953–962.

BRAITHWAITE, C.J.R., 1979, Crystal textures of Recent fluvial pisolites and laminated crystalline crusts in Dyfed, South Wales: Jour. Sed. Petrology, v. 49, p. 181–194.

BUCKLEY, H.E., 1951, Crystal Growth: New York, Wiley, 571 p.

CHAFETZ, H.S., 1979, Petrology of carbonate nodules from a Cambrian tidal inlet accumulation, Central Texas: Jour. Sed. Petrology, v. 49, p. 215–222.

______, AND BUTLER, J.C., 1980, Petrology of recent caliche pisolites, spherulites, and speleothem deposits from central Texas: Sedimentology, v. 27, p. 497–518.

DAVIES, G.R., 1977, Former magnesium calcite and aragonite submarine cements in Upper Palaeozoic reefs of the Canadian Arctic: A summary: Geology, v. 5, p. 11–15.

DICKSON, J.A.D., 1978, Length-slow and length-fast calcite: a tale of two elongations: Geology, v. 6, p. 560–561.

DOWTY, E., 1976, Crystal structure and crystal growth: 1. The influence of internal structure on morphology: Amer. Mineralogist, v. 61, p. 448–459.

EVAMY, B.D., AND SHEARMAN, D.J., 1965, The development of overgrowths from echinoderm fragments in limestones: Sedimentology, v. 5, p. 211–233.

FOLK, R.L., 1974, The natural history of crystalline calcium carbonate: effect of magnesium content and salinity: Jour. Sed. Petrology, v. 44, p. 40–53.

GARRISON, R.E., AND LUTERNAUER, J.L., 1971, Textures of calcitic cements formed during diagenesis, Fraser Delta, British Columbia, in O.P. Bricker (ed.), Carbonate Cements: The Johns Hopkins Studies in Geology, no. 19, p. 151–154.

______, GRILL, R.D., MACDONALD, R.D., AND MURRAY, J.W., 1969, Early diagenetic cementation of Recent sands, Fraser River delta, British Columbia: Sedimentology, v. 12; p. 27–46.

GRIGOR'EV, D.P., 1965, Ontogeny of Minerals: Jerusalem, Israel Program for Scientific Translations, 250 p.

HANOR, J.S., 1978, Precipitation of beachrock cements, mixing of marine and meteoric waters vs. CO_2 - degassing: Jour. Sed. Petrology, v. 48, p. 489–501.

KENDALL, A.C., 1977, Fascicular-optic calcite: a replacement of bundled acicular carbonate cements: Jour. Sed. Petrology, v. 47, p. 1056–1062.

______, AND BROUGHTON, P.L., 1978, Origin of fabrics in speleothems composed of columnar calcite crystals: Jour. Sed. Petrology, v. 48, p. 519–538.

______, AND SIMPSON, F., 1974, Calcite layers in the uppermost Medicine Hat Sandstone (Upper Cretaceous) of southwestern Saskatchewan: Bull. Canadian Petroleum Geol., v. 22, p. 34–41.

______, AND TUCKER, M.E., 1973, Radiaxial fibrous calcite: a replacement after acicular carbonate: Sedimentology, v. 20, p. 365–389.

KREBS, W., 1969, Early void-filling cementation in Devonian fore-reef limestones (Germany): Sedimentology, v. 12, p. 279–299.

LINDSTRÖM, M., 1979, Diagenesis of Lower Ordovician hardgrounds in Sweden: Geologica et Paleontologica, v. 13, p. 9–30.

LOHMANN, K.C., AND MEYERS, W.J., 1977, Microdolomite inclusions in cloudy prismatic calcites: A proposed criterion for former high-magnesium calcites: Jour. Sed. Petrology, v. 47, p. 1078–1088.

MALEEV, M.N., 1972, Diagnostic features of spherulites formed by splitting of a single crystal nucleus. Growth mechanism of chalcedony: Mineral. Petrog. Mitt., v. 18, p. 1–16.

MARSHALL, J.D., 1981, Zoned calcites in Jurassic ammonite chambers: trace elements, isotopes and neomorphic origin: Sedimentology, v. 28, p. 867–887.

MAZZULLO, S.J., 1980, Calcite pseudospar replacive of marine acicular aragonite, and implications for aragonite cement diagenesis: Jour. Sed. Petrology, v. 50, p. 409–422.

PLAYFORD, P.E., 1980, Devonian "Great Barrier Reef" of Canning Basin, Western Australia: Amer. Assoc. Petroleum Geologists Bull., v. 64, p. 814–840.

PURSER, B.H., 1980, Sédimentation et diagenèse des carbonates néritiques récents, v. 1 Les éléments de la sédimentation et de la diagenèse: Paris, Éditions Technip, 366 p.

ROEDDER, E., 1977, Fluid inclusions as tools in mineral exploration: Econ. Geology, v. 72, p. 503–525.

Ross, R.J., Jr., Jaanusson, V., and Friedman, I., 1975, Lithology and origin of Middle Ordovician calcareous mudmound at Meiklejohn Peak, southern Nevada: U. S. Geol. Survey Prof. Paper 871, 48 p.

Schreiber, B.C., Smith, D., and Schreiber, E., 1981, Spring peas from New York State: nucleation and growth of fresh water hollow ooliths and pisoliths: Jour. Sed. Petrology, v. 51, p. 1341–1346.

Towe, K.M., and Malone, P.G., 1970, Precipitation of metastable carbonate phases from seawater: Nature, v. 226, p. 348–349.

Tucker, M.E., 1973, Sedimentology and diagenesis of Devonian pelagic limestones (Cephalopodenkalk) and associated sediments of the Rhenohercynian Geosyncline, West Germany: N. Jb. Geol. Paläont., Abh., v. 142, p. 320–350.

Walls, R.A., Mountjoy, E.W., and Fritz, P., 1979, Isotopic composition and diagenetic history of carbonate cements in Devonian Golden Spike reefs, Alberta, Canada: Geol. Soc. America Bull., Pt. 1, v. 90, p. 963–982.

APPENDIX 1

Fabrics used to suggest replacement origin for radiaxial calcites

If radiaxial cements and their fabrics are now to be considered primary, it is important to re-examine the evidence that was used to support the interpretation that the calcite replaced earlier acicular cements to produce new fabrics.[1] Some of this has been discussed in the main part of the paper but two fabrics have gone unmentioned.

Replacement of adjacent parts of neighbouring acicular-crystal bundles was suggested by extinction patterns seen in sections cut normal to the radiaxial crystal elongation (Kendall and Tucker, 1973, p. 368). It was expected that pseudo-uniaxial extinction crosses would have been present if optic axes radiated about single axes, as Bathurst (1959) originally supposed. Instead extinction bands were observed which apparently crossed intercrystalline boundaries. Subsequent observations upon other sections cut with this orientation (Kendall, unpublished) have revealed how very difficult they are to interpret. In particular it is difficult to distinguish between crystal and subcrystal boundaries so that extinction bands that appear to cross from one crystal to another may merely be crossing subcrystal boundaries. The conclusion that optic axes radiate from several centers at crystal boundaries (Kendall and Tucker, 1973) cannot be substantiated.

Inclusion-defined zones with scalloped margins were interpreted as revealing positions of former growth stages in the earlier acicular cement. In one example (Kendall and Tucker, 1973, Fig. 5 a) boundaries between adjacent acicular-crystal bundles were interpreted to lie within a radiaxial crystal. This implied that the crystal replaced parts of adjacent crystal-bundles. Re-examination of this photograph (the original thin-section not being available) reveals it to be far from convincing. Study of many radiaxial cements since 1973 has failed to produce a single additional example of this relationship (which should be common if the replacement hypothesis is correct). I must conclude that the original observation and its interpretation were in error.

APPENDIX 2

Crystal splitting in other calcites

Interpretation of radiaxial fabrics is concerned largely with crystal-splitting during calcite growth. If it could be established that crystal-splitting in calcite is relatively common, then the present interpretation gains considerable credibility.

Crystal-splitting does not appear to have been specifically identified as a mechanism affecting the growth of calcite in diagenetic environments. On the other hand, there are many materials that have been figured and described which have features indicating or suggesting they are products of split-crystal growth.

Folk (1974, p. 44–45) noted that some magnesian calcite crystals are fluted and resemble bundles of small parallel fibers aggregating into an overall wheat-shaped to rhombic form. The subdivision of these crystals is a strong indication that they have suffered split-growth. Towe and Malone (1970) artifically precipitated such crystals and they also commonly occur naturally in marine settings (see for example, Alexandersson, 1972, Figs. 13, 14). Folk interpreted the fluting to be the result of crystal-poisoning by magnesium. This, however, cannot be the only cause because similar crystals, but composed of low-Mg calcite, can form from waters with low magnesium contents. Hanor (1978, p. 492, Fig. 4) precipitated low-Mg calcite split-crystals by degassing groundwaters. They were compared with crystals precipitated naturally at water-table locations in beachrocks. The crystals were bounded by numerous offset crystal faces and by multiple parallel c-axis terminations. Identical

[1]It should be emphasized that the reinterpretation of radiaxial fabrics presented here does not imply that radiaxial crystals have not replaced acicular materials. If lateral coalescence of crystallites is delayed (see p. 67) the radiaxial crystals are generated by the fusing together of acicular crystals (= fibers).

compound crystals have more recently been identified from travertine pisolites by Schreiber *et al.* (1981). Binkley *et al.* (1980) have described other composite low-Mg calcite crystals from freshwater beachrocks in Michigan that I would interpret as forming by crystal-splitting. The aggregate crystals are elongate along the c-axis and are composed of many orderly-arranged, partially coalescent, trigonal-prism fibers, between which occur small triangular intercrystallite pore-spaces. Many speleothems are composed of similar crystals which grow from numerous small crystallites (Kendall and Broughton, 1978).

With less confidence I would suggest crystal-splitting was responsible for the crystal fabrics within large low-Mg calcite botryoidal bundles of acicular crystals that were figured by Garrison *et al.* (1969, Figs. 9–10) and by Garrison and Luternauer (1971, Fig. 80). In these bundles adjacent crystals exhibit almost-identical cross-sectional shapes yet also display a gradual and systematic shape-change when traced away from any particular crystal. This behaviour suggests, but unfortunately does not prove, that adjacent crystals have near-identical crystallographic orientations and that all crystals are crystallographically related to each other (in other words, the whole crystal bundle is a single split crystal). It is true that similar fabrics could be generated by combining crystals of different crystallographic orientation (the close packing generating similar cross-section shapes), but this explanation fails to account for the close packing itself, particularly when it is recalled that the crystals were diverging away from each other as the bundle grew in size. Crystal-splitting however, occurring continuously with growth of the bundle, will cause the spaces formed between the divergent crystals to be occupied by daughter crystals of the appropriate crystallographic orientation.

Orientations of crystallites (or fibers) in the split-crystals mentioned so far are crystallographically controlled—most extending along the c-axis. This feature is not that which occurred during growth of radiaxial calcite crystals. In them the crystallite orientations were independent of crystallographic control and, in the case of the Australian crystals, were controlled by the orientation of the substrate surface. Split-crystal growth of this type in calcite is seemingly much less common. Only (1) fibrous crystals described by Lindström (1979) (mentioned on Page 67), (2) fibrous calcite veins figured by Kendall and Simpson (1974, pl. II, Fig. 2), and (3) spherulites examined by Chafetz and Butler (1980, Figs. 6–9), are known to me to be examples of split-crystals that appear to have originated in this fashion. In all three examples, there is a pattern of inclusions that defines a radiating fibrous fabric, but the crystals extinguish uniformly. From this it may be deduced that (1) the crystals are composite, having grown as numerous crystallites, and (2) that the fibers were not elongated along crystallographic directions. Chafetz and Butler, unaware that fibers need not be elongated along crystallographic directions, concluded that their spherulites had suffered replacement by single calcite crystals followed by the development of small terminations on appropriately-oriented crystallites (those elongated along the c-axis).

Through the courtesy of Dr. Chafetz I was able to examine the original thin-sections of these spherulites. I found no evidence supporting a replacement. In fact the manner by which crystal boundaries exactly correspond to spherulite boundaries, even when these have laterally-interfered with each others growth, strongly suggests the calcite crystals are primary. If the crystals were neomorphic it would be expected that crystal boundaries would somewhere cross-cut the original spherulite-boundaries. Furthermore, the fibrous structure is beautifully preserved. I would interpret the absence of terminations upon crystallites that are oriented at large angles to the c-axis as being due to the same geometrical constraints as those that affected crystallites in radiaxial crystals (see Fig. 12). Even though some of the fibers are curved and although changes in crystallographic orientation along the fibers is not present, the spherulites described by Chafetz and Butler (1980) would appear to be the closest analog to radiaxial crystals so far reported.

CARBONATE CEMENTATION—A BRIEF REVIEW

PAUL M. HARRIS, CHRISTOPHER G. ST. C. KENDALL[1] AND IAN LERCHE[1]
Gulf Research & Development Company, P.O. Box 37048, Houston, Texas 77236

ABSTRACT

Porosity prediction in subsurface carbonates remains difficult although our understanding of cementation in carbonate rocks has greatly increased during the past few years. The diagenetic setting is inherited from the depositional setting but continually modified during burial. In the marine setting, cementation occurs in the deep sea, in reefs, on shallow water platforms, and along shorelines. Carbonate and evaporite cementation typifies supratidal flats. During shallow burial, carbonate rocks contact fresh waters and additional cementation occurs along with solution, mineralogical stabilization and possibly dolomitization. Continued burial results in physical and chemical compaction, pressure solution and additional cementation.

Several factors play a role in cementation in carbonates, i.e., the rate and total volume of water movement and timing of cementation relative to hydrocarbon generation; fluid chemistry; and the effect of pH, CO_2 activity, temperature and pressure on mineral solubilities. Petrographic studies coupled with measurements of stable isotopic ratios, trace elements and fluid inclusions are made to assess various stages of cementation and their relation to burial history and fluid flow and composition. Such integrated studies are providing a more detailed understanding of diagenetic events, especially cementation, and ultimately will increase the ability to predict porosity distribution.

INTRODUCTION

With few exceptions, hydrocarbons are reservoired in sandstones or carbonates. In sandstones the porosity is generally interparticle while in carbonate rocks the porosity is associated with a wide variety of depositional and diagenetic fabrics (Choquette and Pray, 1970; Wardlaw, 1976). In efforts to find stratigraphic traps in clastics, despite the local importance of diagenesis, the greatest emphasis is normally placed on interpreting depositional facies and their regional trends. However, with carbonates, cementation and other diagenetic effects are equally important considerations. This paper briefly reviews the environments of cementation in carbonate rocks and many of the techniques commonly used in diagenetic studies.

Depositional porosity in carbonates is usually a function of sediment sorting and shape, which in turn are related to wave and current activity and the biologic community (salinity, turbidity, nutrient supply) at the depositional site. Where currents and waves are particularly active, lime mud is winnowed away leaving porous carbonate sands (Folk, 1962; Dunham, 1962; Enos and Sawatsky, 1981). In contrast, in low energy environments lime muds tend to collect and once the sediment has dewatered, the depositional porosity is much reduced (Shinn *et al.*, 1977; Enos and Sawatsky, 1981). This seemingly simple relation between mud percentage and bottom agitation is complicated since: (1) mud-sized carbonate in reef rock and hardgrounds may actually be micrite cement and not indicative at all of a low energy setting (MacIntyre, 1977); (2) mud-sized carbonate may be replaced during diagenesis by pseudospar and so interpreted as deposited in higher energy settings; and (3) during compaction and pressure solution (Shinn *et al.*, 1977; Wanless, 1979), carbonate grains floating in a mud matrix may be "squeezed" together biasing the texture toward a more grain-rich, mud-free lithology.

Many of the world's larger carbonate reservoirs have porosities which are partly primary in origin. A few examples are the Ghawar field in the Jurassic limestones of Saudi Arabia (Arabian American Oil Company Staff, 1959; Wilson, in press), the Jurassic Smackover limestones in southern Arkansas and northern Louisiana (Moore and Druckman, 1981; Baria *et al.*, 1982), the Devonian Leduc limestones at Red Water in Alberta (Klovan, 1964) and the Mississippian Madison limestones in North Dakota and Saskatchewan (Lindsay and Kendall, in press).

The relationship between porosity and diagenesis is complicated. Apart from mechanical and chemical compaction, three diagenetic processes controlling porosities are solution, cementation and dolomitization (Fig. 1). These processes require a permeable host rock and a mechanism to flush chemically active waters through the rock. Regionally this water movement is controlled by tidal pumping, hydrostatic head, or shale compaction, thermal expansion, smectite to illite conversion and rock fabric of the region.

Partial solution of massive reef limestones in the Miocene of Indonesia and the Cretaceous of the Middle East has led to the development of limestones with excellent reservoir characteristics (Vincellette and Soeparjadi, 1976; Epting, 1980; Frost *et al.*, 1983; Harris and Frost, 1984). However, where dissolution of carbonate sand grains is accompanied by calcite pre-

[1]Present Address: Department of Geology, University of South Carolina, Columbia, South Carolina 29208

cipitation in adjacent primary pores, the end product may be the same amount of porosity but a shift to moldic porosity and variable permeability. Examples of this can be seen in the Smackover Formation in Arkansas (Moore and Druckman, 1981) as well as the Jurassic Haynesville Formation of east Texas where porosities are 15% and higher, but the permeabilities generally are less than 0.5 millidarcies.

The relationship of dolomitization to porosity is complex: dolomitization may reduce, preserve or redistribute porosity. In some carbonate reservoirs, as in the Jurassic Arab limestone of the Ghawar field in Saudi Arabia, dolomite cements partially reduce the primary porosity. Such dolomite cements may be quite common (Kaldi and Gidman, 1982; Sibley, 1982). Contradicting the reputation dolomitization has for porosity enhancement, Halley and Schmoker (1983) have illustrated that at certain burial depths dolomites may have lower porosities than adjacent limestones. In many dolomitized reservoirs such as the Jurassic Smackover formation in Alabama and Florida (Baria *et al.*, 1982) and the Rainbow Lake reef carbonates in Alberta (Langton and Chin, 1968), moldic porosity is also important. Early dolomitization can preserve porosity, particularly in initially porous pelletal lime mudstones, by creating a rigid framework that inhibits compaction of the soft pellets (Roehl, 1967). In these cases the porosity exists either between slightly to moderately compacted soft pellets or between dolomite rhombohedrons. The association of porosity with dolomitized mud-rich carbonates is common in the Mississippian Mission Canyon (Lindsay and Kendall, in press) and Red River formations of the Williston Basin (Roehl, 1967; Longman *et al.*, 1983) along with the Permian San Andres and Grayburg reservoirs of the Permian Basin (Longacre, 1980; Bein and Land, 1982).

Fractures may increase porosity, but more commonly enhance permeability (Stearns and Friedman, 1972; Currie and Nwachukwu, 1974; McNaughton and Garb, 1975; Tillman, 1983). A classic example of this can be seen in the Ekofisk field of the North Sea where overpressured Danian Chalk has been fractured by underlying salt movement (Van den Bark and Thomas,

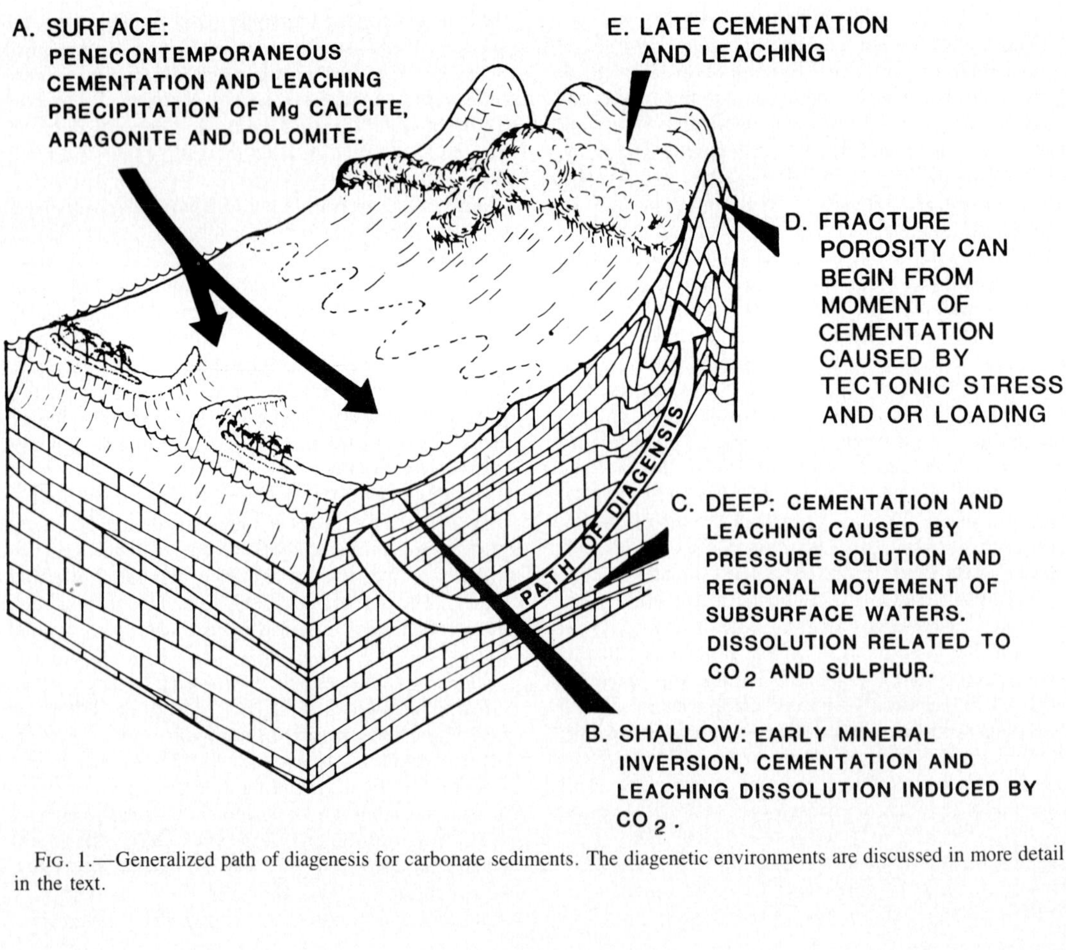

FIG. 1.—Generalized path of diagenesis for carbonate sediments. The diagenetic environments are discussed in more detail in the text.

1981). Here porosities in the reservoir are over 30% but, without fracturing, the permeability would be negligible. From our own observations of interbedded grain carbonates and lime muds, we note that lime muds are more likely to be fractured and the fracturing frequently increases reservoir continuity. Presumably the lime muds are denser and more competent than grain carbonates and so fracture more easily.

Porosity-filling cements in carbonates are characterized by measurements of $^{13}C/^{12}C$, $^{18}O/^{16}O$ and $^{87}Sr/^{86}Sr$ compositions as well as trace elements (see Veizer *et al.*, 1971; Veizer and Compston, 1974; Veizer and Hoefs, 1976; Hudson, 1977; Walls *et al.*, 1979; Campos and Hallam, 1979; Dickson and Coleman, 1980; Moore and Druckman, 1981; Arthur *et al.*, 1982). Microsampling techniques coupled with staining of thin sections (Friedman, 1959; Evamy, 1969) and cathodoluminescence (Meyers, 1974; Pierson, 1981; Frank *et al.*, 1982) are being increasingly used to document the isotopic and elemental signatures of individual cement zones within pores. Quantitative estimates based on mass balance calculations of the minimal amount of fluids that have passed through rocks as recorded by their stable isotopic composition have been presented by Magaritz and Taylor (1976) and Magaritz (1983). The δD composition of aqueous inclusions as well as in clays and chert, $\delta^{15}N$ composition of organic matter and oils, $\delta^{34}S$ composition of associated evaporites and organic matter, together with the uranium-thorium-potassium series of clays and organics can also be used as indicators of the paleo-evolution of the rock.

Although distribution coefficients and fractionation factors may not be precisely known, the trends within co-genetic suites of diagenetic carbonates are used to obtain considerable insight into their diagenetic history (Land, 1980). In addition, examination of covariate trends of $\delta^{13}C$ and $\delta^{18}O$ compositions of components from a single stratigraphic unit is an extremely powerful approach.

The content of fluid inclusions in cements reflects the composition of the fluids during precipitation provided there have been no significant pressure and temperature changes between the time of formation of the inclusion and the time of sampling (Roedder, 1979; Hanor, 1980; Burruss, 1981). When such changes have occurred, corrections can be made to approximate the original composition (Burruss, 1981). The temperature at which the fluid inclusion was formed can be inferred by directly heating the inclusion and seeing at what point the fluid vaporizes; and the salinity of the precipitating fluid can also be estimated by freezing the inclusions (Klosterman, 1981; Moore and Druckman, 1981). Fluorescence studies with respect to different wavelengths of light determine whether the inclusions contain oil, condensate or brine (Burruss, 1981; Burruss *et al.*, 1983). Occasionally, direct composition measurements of the fluid in an inclusion are possible by gas chromatography.

To find efficiently hydrocarbon reservoirs in carbonates, porosity preservation needs to be understood relative to the generation, migration and accumulation of hydrocarbons. An important key to interpreting carbonate diagenesis and cementation is recognition of the respective timing and frequency of diagenetic events and their tie to sea level fluctuations, tectonic events, burial, development of trap and seal and oil migration. The extensive literature on cementation in carbonates is too vast for a complete review, but important studies include Folk (1959), Bricker (1971), Bathurst (1971), Milliman (1974), Zenger *et al.* (1980), Flügel (1982) and summary papers by Folk (1973), Moore (1979, 1980), and Longman (1980, 1981).

ENVIRONMENTS OF CEMENTATION

The cementation and diagenesis of carbonates can take place in many settings (Fig. 2): the marine environment during the deposition of the sediment, in a near-surface position where a freshwater lens penetrates the depositional environment, in a subaerial setting produced by a sea level fall, in a mixing zone between waters of varying chemistries, or in the brines of the deeper subsurface. We cannot apply absolute depths to these diagenetic realms, for the observed processes vary in importance and depth between basins.

The preservation of high porosities and permeabilities at depth within carbonate reservoirs is contrary to current thinking on the importance of cementation in surface environments. This would have the pores of these sediments plugged by cements early in their diagenetic history when unstable mineralogies in sediments are in contact with dynamic near-surface waters. Clearly the Florida and Yucatan aquifers, among many, argue against early porosity loss. Halley and Schmoker (1983) present evidence that in South Florida, despite near surface diagenesis, porosities remain high shallow in the section and decrease slowly with depth. In other settings, perhaps burial was so rapid that carbonates quickly passed through the near-surface phreatic zone and underwent little diagenesis. It is also possible all the carbonate is converted to low magnesian carbonate "upstream" in an aquifer and the supply of $CaCO_3$ originally contributed from metastable mineralogies was exhausted. Another possibility is that evaporites, shales, red beds or dense micrites formed protective impermeable seals updip, over and within the porous carbonates and thus prevented downward penetration of water which might cause cementation and porosity occlusion. A final possibility (suggested by Halley and Schmoker, 1983) is that diagenesis in shallow environments simply does not completely plug porosity, but that tight limestones develop during burial as the result of pressure solution, compaction and additional cementation.

Deep burial processes such as grain to grain interpenetration, stylolitization by physical compaction or pressure solution (Dunnington, 1967; Logan and Semeniuk, 1976; Wanless, 1979; Meyers, 1980; Pratt, 1982) and cementation (Moore and Druckman, 1981) are significant. Studies such as that of the Lake Valley formation (Meyers, 1978; 1980) show that about half of the cement volume is derived from burial processes. It is clear then that to predict porosity in subsurface carbonates one should consider: (1) the depositional environment where depositional porosity is generated, (2) the early diagenetic environments where the porosity can be enhanced, preserved or occluded, and (3) the burial history and associated diagenetic changes.

Marine Setting

Rapid marine cementation has been found to be common and some of the more spectacular examples include: the incorporation of war material and human skeletons into beach rock (Milliman, 1974); the presence of beer bottles in the subaqueous crusts from the Persian Gulf seafloor (Shinn, 1969); and the annual quarrying of beach rock for building purposes in some Pacific atolls, with the rapid formation of further beach rock following the removal of the quarried material (Moresby, 1835). Marine cementation is not restricted by water depth and can be traced from deep water to beaches (Fig. 3).

Deep Sea.—Deep sea cements are formed of magnesian calcite, calcite and aragonite (Gevirtz and Friedman, 1966; Bartlett and Greggs, 1969; Thompson, 1972; Schlager and James, 1978). Deep sea hardgrounds occur in areas where there is good bottom current movement. This is particularly true in crusts and lithoherms on the flanks of carbonate platforms (Mullins *et al.*, 1980) and the margins of submarine channels and canyons (Schlager and Chermak, 1979) where aragonite is precipitated. Deep water cementation may also occur in areas with low sedimentation rate or within enclosed basins with elevated salinities. An example of this latter case are the sediments found in the deeper parts of the Red Sea (Milliman *et al.*, 1969) and the Mediterranean (Fischer and Garrison, 1967) containing hardgrounds cemented by magnesian calcite.

Reefs.—(Fig. 4). In shallower water settings, particularly in reef margins, cementation is more widespread than in deeper water. Reef cements are bladed and micritic magnesian calcite or aragonite; the micritic cement is frequently pelleted (Ginsburg *et al.*, 1971; Friedman *et al.*, 1974; James *et al.*, 1976; MacIntyre, 1977; Mazzullo and Cys, 1979; Shinn *et al.*, 1983).

MINERAL ASSOCIATION TO WATER

MINERAL	CHEMICAL COMPOSITION	SETTING OF PPN
ARAGONITE	$CaCO_3$	MARINE
CALCITE	$CaCO_3$	FRESH WATER AND SUBSURFACE BRINE
MAGNESIAN CALCITE	$CaCO_3$ WITH 5-15 MOLE % $MgCO_3$	MARINE
DOLOMITE	$Ca(Mg)CO_3$	MARINE AND FRESH WATER
SIDERITE	$FeCO_3$	

CRYSTAL SHAPES

CLASS	DIMENSIONS (L TO W)	DIAGRAM
FIBERS	> 6 TO 1	
BLADES	2 TO 1 UP TO 6 TO 1	
EQUANT	2 TO 1 OR 1 TO 1	

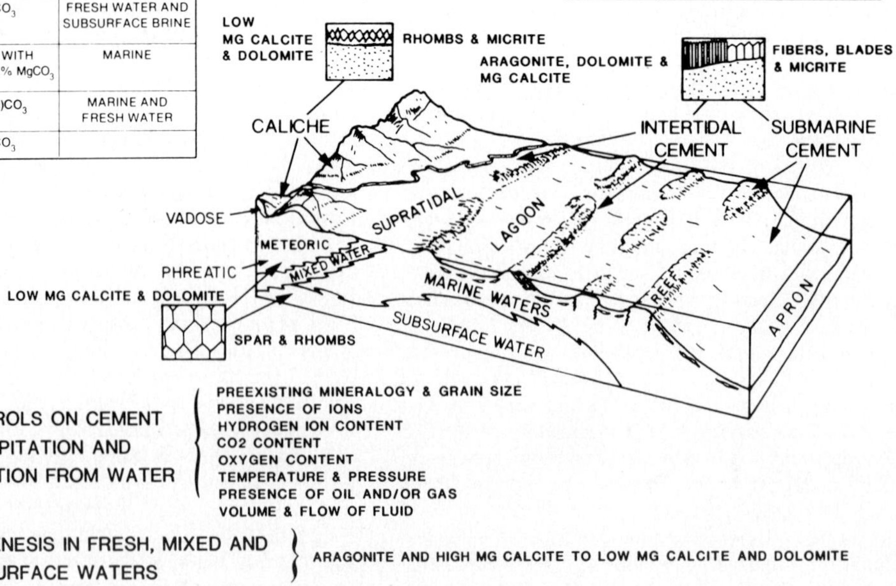

FIG. 2.—Patterns in carbonate cementation for various diagenetic environments. Modified after diagram by R.L. Folk.

The fabric of reef cement is controlled by its access to circulating water; the voids with greater permeability may be filled by isopachous or equant cements whereas the more confined voids commonly show much larger and irregular crystals. In some reef cavities, like those from the reefs of Belize, thick botryoidal crusts of aragonite occur in the larger cavities (Ginsburg and James, 1976; James *et al.*, 1976). The zone of active cementation in reefs appears to be confined in general to the outer few feet of the sediment forming the reef. Evidence of this localization of cementation on a larger scale has been recognized in the modern reefs from Belize (James and Ginsburg, 1979) and Devonian reefs from the Canning Basin (Playford, 1980), which are

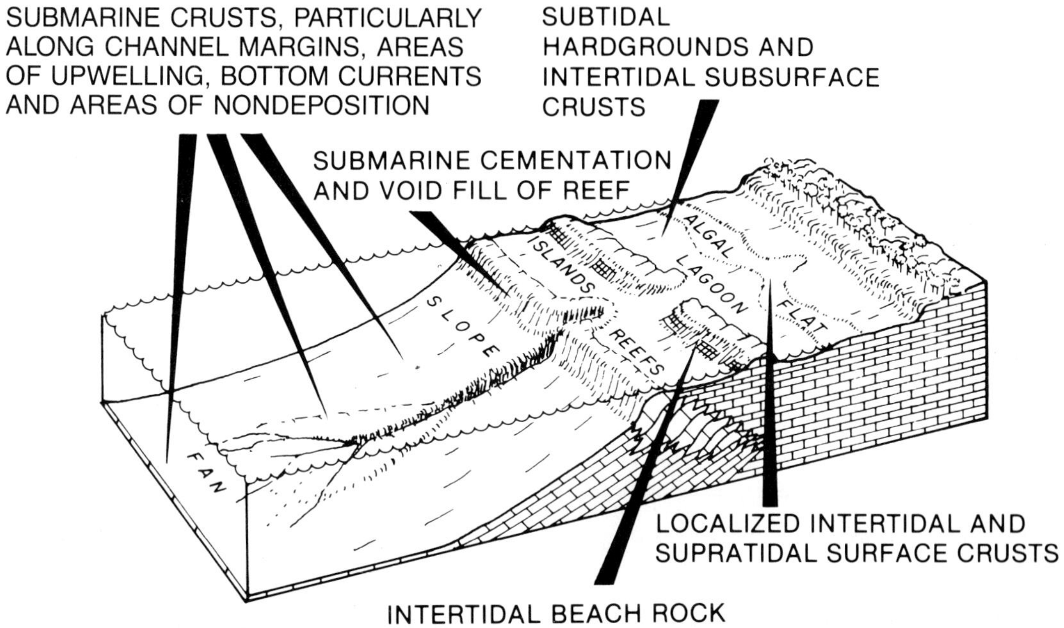

DIAGENETIC EFFECTS OF MARINE WATERS

MICRITIZATION

NEOMORPHIC GRAIN GROWTH

CEMENTATION DURING SEDIMENT DEPOSITION

FORMATION OF { OOLITES, GRAPESTONES, PELLET HARDENING }

SUBMARINE / INTERTIDAL { CRUSTS }

STABLE MARINE CEMENTS

ARAGONITE — FIBERS AND MICRITE
MG CALCITE — FIBERS, BLADES AND MICRITE

FIG. 3.—Environments of marine cementation.

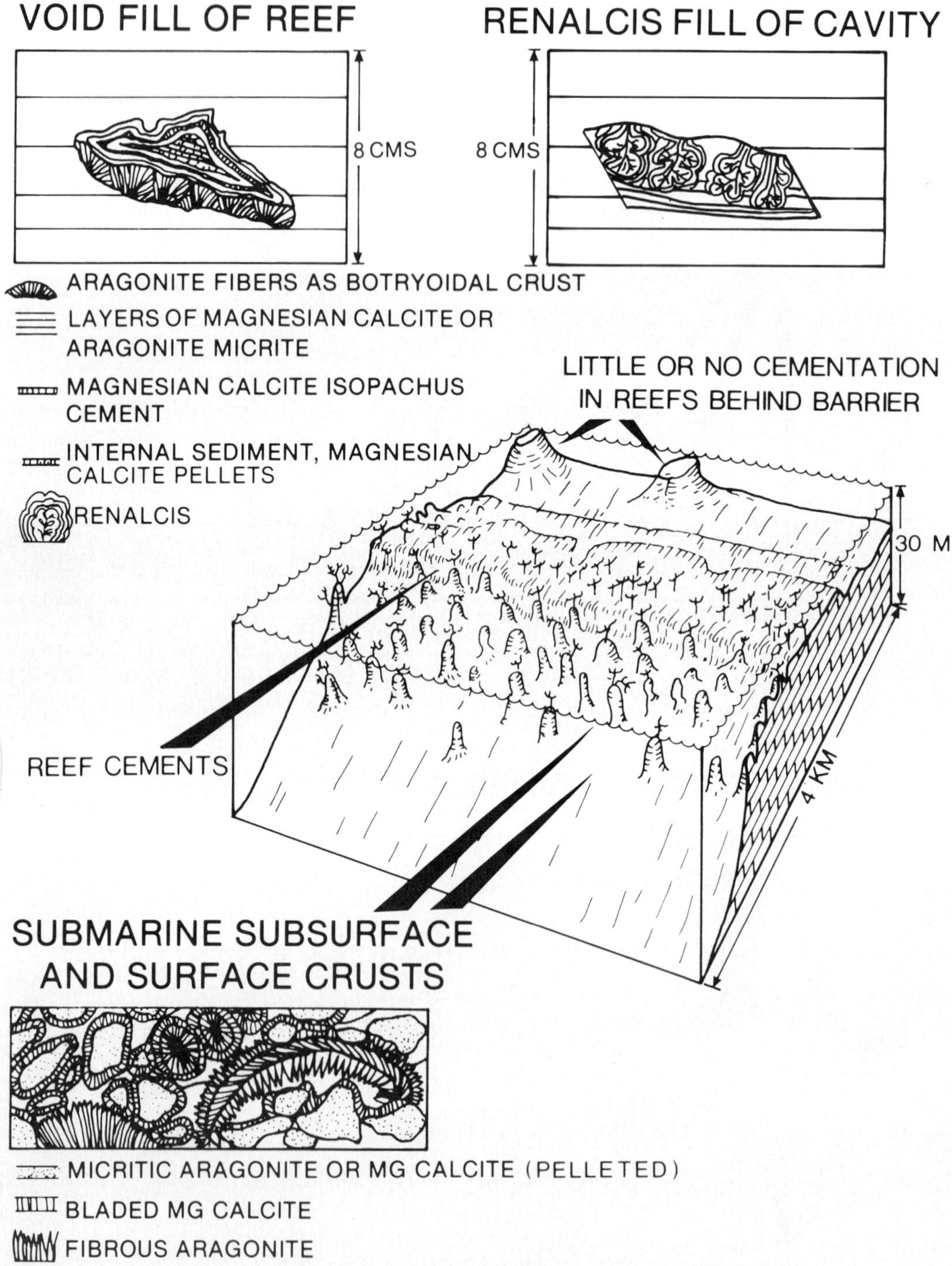

FIG. 4.—Reef cementation.

cemented on their seaward margin whereas the reefs to the lee of the major reef trend show little or no cementation.

Shallow water carbonate platforms.— Cementation penecontemporaneous with deposition of carbonate sediments also occurs across shallow water platforms (Fig. 5). For instance, oolites, which occur in zones of wave and tidal surge, are formed by the precipitation of calcium carbonate around a central nucleus (Illing, 1954; Loreau and Purser, 1973). Similarly grapestones, commonly found on a less agitated seafloor immediately to the lee of oolitic shoals, are formed in part by the precipitation of carbonate cement at points of contact between sandsized gains (Illing, 1954; Kendall and Skipwith, 1969; Winland and Matthews, 1974). Storms also break the cemented crusts that form on oolitic shoals, but do not completely disaggregate the cemented sediment, so partially cemented aggregates

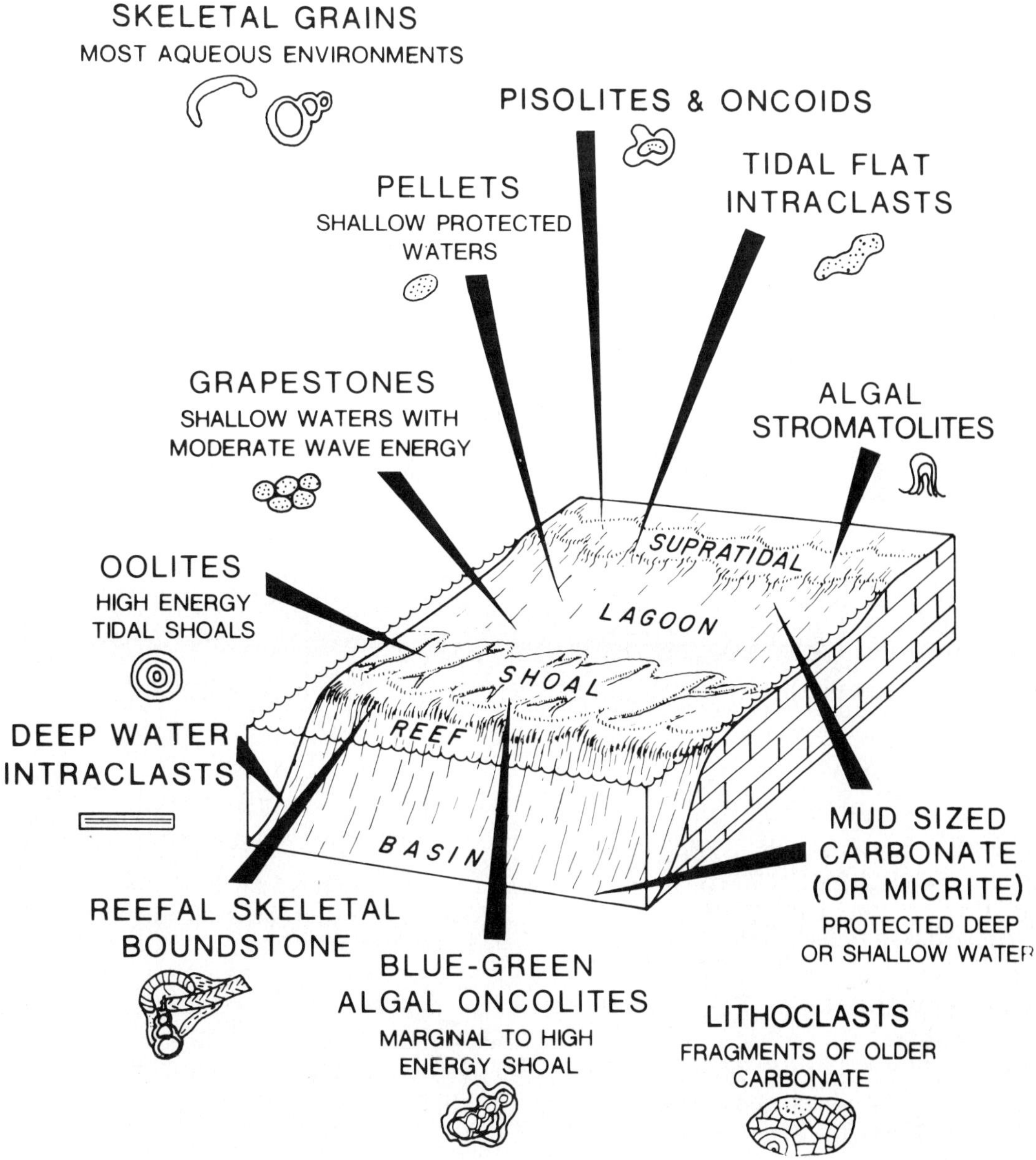

FIG. 5.—Carbonate grain types and their depositional setting.

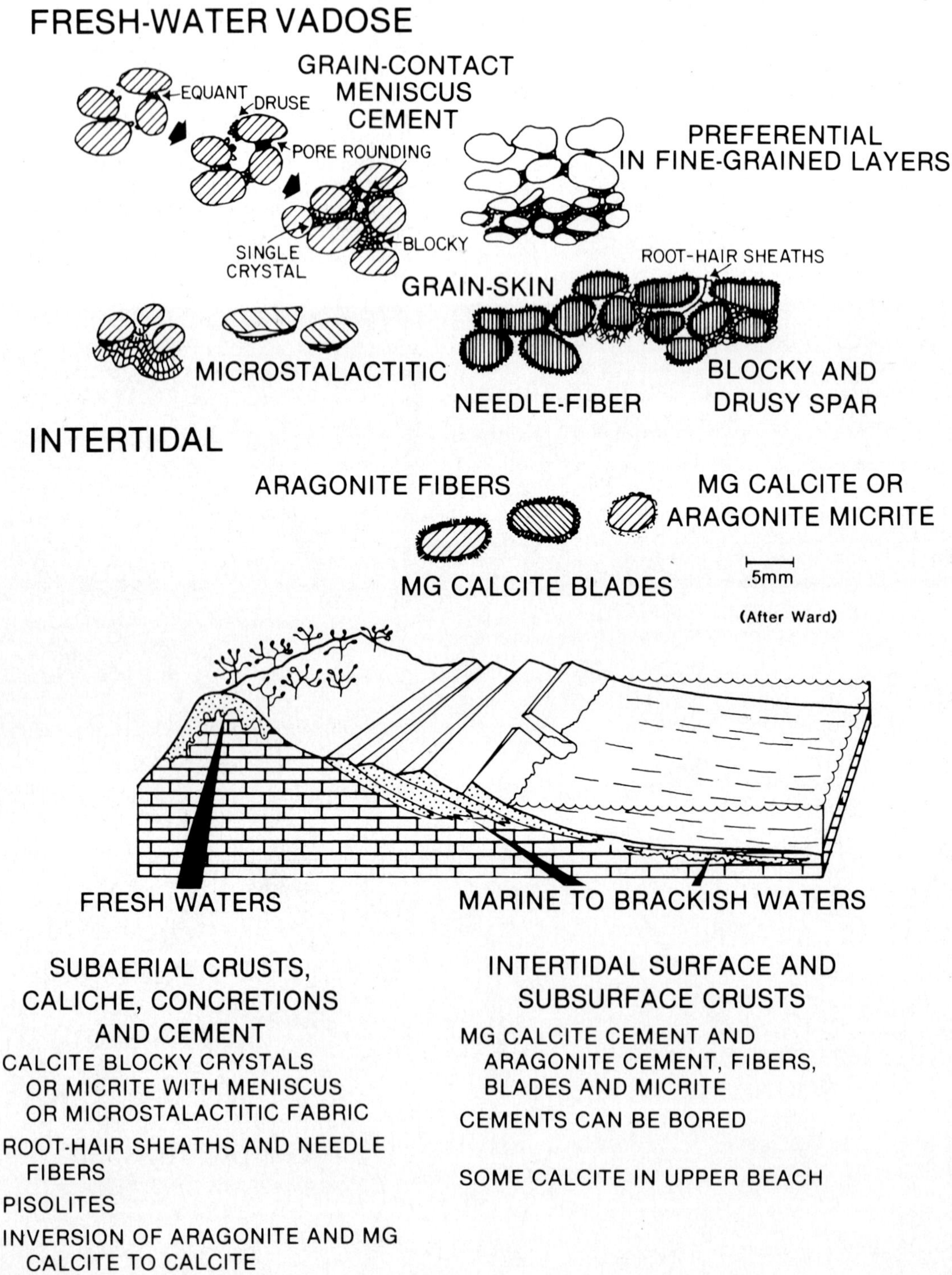

FIG. 6.—Near-shore and freshwater cementation of carbonates.

of the sediments (or intraclasts) remain. Further evidence of cementation can be seen in the way in which soft carbonate pellets excreted by marine organisms are progressively hardened by carbonate precipitation (Illing, 1954; Kendall and Skipwith, 1969). In shallower or more protected water than the hardened pellet zone, the lime mud pellets are not cemented.

Submarine crusts are best developed in grapestone sediments where the seafloor is cemented into thick layers tens of centimeters thick (Taft *et al.*, 1968; Purser, 1969, 1971; Harris, 1978; Dravis, 1979). In some cases, for instance in the Persian Gulf (Shinn, 1969; Assereto and Kendall, 1977), cementation causes the crust to expand and override itself forming compressional ridges or submarine tepees. These submarine crusts or hardgrounds can be recognized in the rock record by organisms encrusted on the hardened surface, borings cut through both sediment and cement and the morphology of the marine cements (Purser, 1969).

The cements involved in the cementation of oolites, grapestones, hardening of pellets and the development of hardgrounds are aragonite and magnesian calcite (Bathurst, 1971). The aragonite may be micritic or fibrous, and the magnesian calcite micritic or bladed. The micritic cements in submarine crusts are more common closer to the water sediment boundary whereas the fibrous and bladed cements are found toward the interior of the crusts and commonly occur as isopachous rims enveloping each grain (Shinn, 1969).

Cementation also occurs during the micritization of carbonate grains, associated with the boring activities of algae and bacteria (Bathurst, 1966; Alexandersson, 1972; Harris *et al.*, 1979). The mineralogy of the grain can sometimes be altered by the processes so that the identification of the altered grain can only be made by its shape. The importance of bacteria and algae in the micritization process may be quite important in that there is probably some organic intermediary involved in the development of marine cements.

Intertidal and Supratidal Marine

Cementation in the intertidal and supratidal zones of carbonate depositional environments is not uncommon (Bathurst, 1971; Milliman, 1974). Cementation in the intertidal zone occurs along beaches where a decrease in CO_2 by degassing due to tidal pumping may form beachrock (Moore, 1973, 1977; Hanor, 1978). The cements are micritic, bladed or fibrous with variable mineralogy (Fig. 6). Magnesian calcite and aragonite cements appear related to marine water tables; calcite cements in contrast seem related to an influx of freshwater from an adjacent landmass into the intertidal sediments.

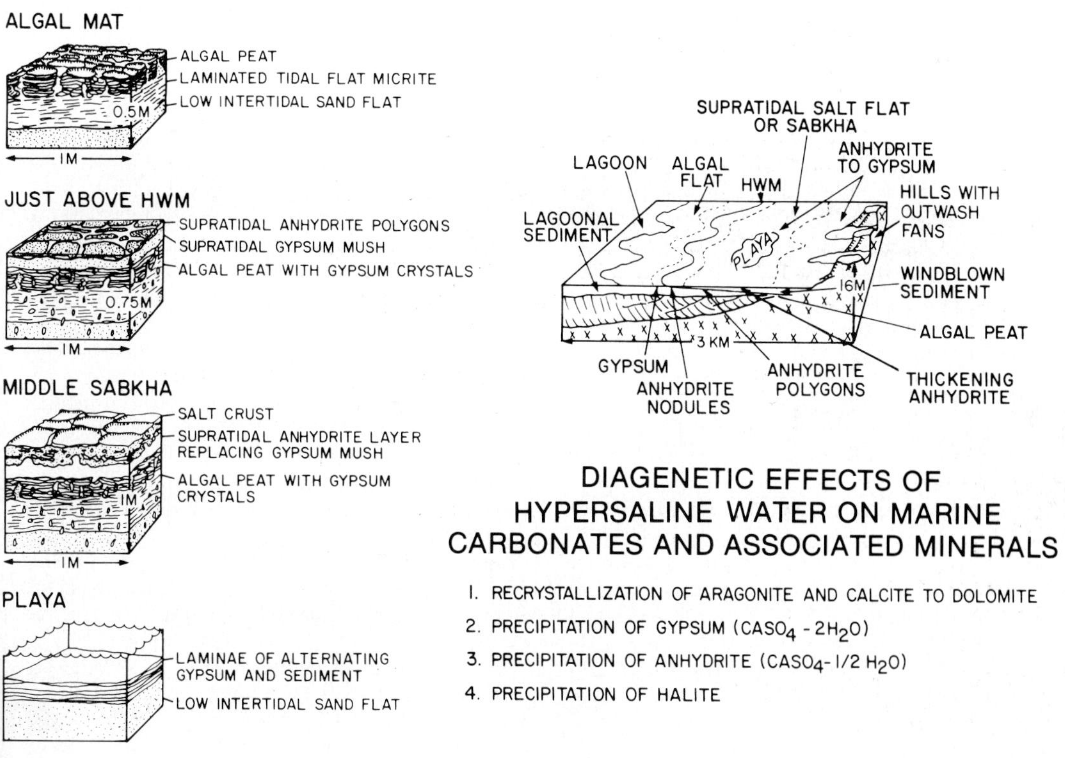

FIG. 7.—Supratidal evaporites and carbonates, their vertical and lateral relationships.

In regions of high evaporation and elevated salinities, the supratidal zone behind quiet water lagoons may be cemented as crusts of variable thickness (Fig. 3). These crusts show evidence of desiccation and expansion (Assereto and Kendall, 1977), and compressional ridges commonly outline polygons with saucer-like depressions. Areas where this type of cemented crust have been described include the southwestern Persian Gulf (Evamy, 1973), Shark Bay and Marion Lake, Australia (Davies, 1970; Warren, 1982) and the Bahamas (Shinn *et al.*, 1965).

Supratidal and intertidal crusts from arid and hypersaline areas commonly develop aragonite travertine coats. Examples of these cements can be seen in the southwest Persian Gulf (Purser and Loreau, 1973), the Permian of the Guadalupe Mountains (Kendall, 1969), Triassic of the Burgamasc Alps (Assereto and Kendall, 1977) and the Jurassic of the Moroccan High Atlas Mountains (Burri *et al.*, 1973). Marine travertines frequently fill the deformed compressional ridges found in these crusts. The deformed crusts are often composed of coated grains or pisolites which formed and/or collected in the splash zone and in the soil horizons of the high intertidal and supratidal environments (Esteban and Pray, 1977).

In addition to the cementation found in the supratidal zone, other diagenetic features are associated with flooding of marine waters. If salinities are elevated or magnesium to calcium ratios are increased, as during the precipitation of anhydrite minerals on supratidal flats of arid regions, the mineralogy of the sediments change (Fig. 7; Curtis *et al.*, 1963; Kendall and Skipwith, 1969; Butler *et al.*, 1982). Aragonite commonly gives way landward across supratidal flats as exemplified by the coastal flats of Abu Dhabi to calcium sulfate in the form of gypsum, anhydrite and finally, by replacement of previously precipitated anhydrite to gypsum. The source of the water may either be storm washover (Butler, 1970) or the evaporative pumping of water derived from the nearby marine environment (Hsü and Schneider, 1973). The precipitation of sulfate is accompanied sometimes by the aragonitic sediment being converted to dolomite (Adams and Rhodes, 1960; Hsü and Schneider, 1973). The dolomitization is initially confined to the finer grained sediments, also eventually replacing the coarser material.

The cements of the supratidal flats show a displacive fabric. The gypsum commonly appears as crystal laths; the anhydrite occurs as contorted ptygmatic layers or nodules. Some cementation and diagenesis may also be related to standing bodies of water in a supratidal setting where marine water is ponded to form supratidal playas. These playas show variable cement stratigraphy and diagenesis depending on their salinites (Logan, 1981; Warren, 1982). Where sulfates have precipitated directly from a standing body of seawater, they form horizontal layers parallel to the sediment water interface.

Shallow Burial

Shallow processes occur at near-surface temperature and pressure in meteoric or marine water. Cementation and dissolution of carbonates initiated in the marine realm continues during contact with freshwater (Fig. 8; Friedman, 1964; Matthews, 1968; Bricker, 1971; Longman, 1980), along with dewatering, grain re-orientation and grain breakage during the earliest burial compaction. The source of the freshwater and the position of the freshwater table is controlled by the relief of the recharge area, climate and the position of the sea with respect to the environment during and immediately after deposition. Freshwater may enter the environment of deposition in the supratidal zone or the intertidal zone through the displacement of marine waters. Freshwater lenses within the sediment extend into the subsurface and may extend out into marine sediments beneath the sea, sometimes spilling into the sea as a freshwater spring. Examples of freshwater lenses and the resultant diagenesis are known from the Bahamas (Halley and Harris, 1979), Barbados (Matthews, 1968, 1971), Bermuda (Land, 1970), Florida (Matthews, 1974) and elsewhere. The thickness of a freshwater lens beneath the land mass with low relief can be substantial. For instance, the Florida peninsula is underlain by a freshwater table that extends 600 meters into the subsurface (Schmoker and Halley, 1982). The greatest cementation is expected at the water table and in the transition zone between freshwater and the marine lens or subsurface brine beneath it (Matthews, 1974; Meyers and Lohman, 1978; Longman, 1980). The mixing zone is important for it is here that the waters have their greatest variation in composition and position. Annual climatic changes cause variations in rainfall and so affect the position of the water table and mixing zone through the year.

The diagenetic overprint is surprisingly uniform throughout the vadose and phreatic zones of Pleistocene limestones from Barbados, the Bahamas and South Florida. Calcite and less commonly dolomite cements form as crusts of equant crystals in the phreatic zone, or may form random mosaics in both the vadose and phreatic zones where their orientation appears independent of the cemented grains. In random mosaics, the size of the crystals range from very fine to medium; the first generation of crystals tends to have more regular crystal facies and is finer than the later crystals. In the vadose zone, cementation takes place within the water meniscae which collect at the points of contact between grains (Fig. 6; Dunham, 1971; Halley and Harris, 1979). Similarily, cements form in the water draped on the lower sides of grains and generate microstalactites (Purser, 1971; Ward, 1975).

The trace element composition of freshwater cements formed at shallow depths is controlled by the temperature, Eh, pH and distribution coefficients of precipitating waters. In well oxygenated groundwaters, ferric ions are common and thus iron is not incorporated into the cements (Evamy, 1969). Manganese is precipitated preferentially to iron in near-surface oxygenated waters (Meyers and Lohmann, 1980), since Mn^{2+} is slightly more soluble in a wider range of Eh conditions than is Fe^{2+} (Garrels and Christ, 1965). Thus Mn^{2+} can occur in more oxidizing waters than Fe^{2+} leading to cements rich in Mn^{2+} but poor in Fe^{2+}. In waters from greater depths, oxygen is less common and in a reducing environment iron and manganese occur as divalent ions in solution which form ferroan and manganese-rich carbonate cements. In strongly reducing conditions both occur in solution. In high sulfide waters, iron is incorporated in pyrite giving preference to $MnCO_3$. $FeCO_3$ forms in low HS^- waters.

The diagenesis associated with freshwater includes solution, the inversion of aragonite to calcite, the exsolution of magnesium from magnesium calcite, cement precipitation and dolomitization. Aragonite and magnesian calcite are unstable in magnesium deficient waters regardless of whether these waters are fresh, brackish or saline. So in the vadose or shallow phreatic zones these minerals tend to change to calcite (Chave, 1962; Land, 1967). In some cases, the original mineral fabrics are completely preserved and in others they are completely destroyed. If the carbon dioxide content of a fluid is increased, the fluid will dissolve more carbonate and hold it in solution. This occurs in soil zones and leads to dissolution and landscape lowering and, where vadose and phreatic waters interact at the water table, to cave formation. The transformation of aragonite to calcite may cause partial plugging of the existing porosity by the formation of calcite cement. In carbonates where the pores are completely occluded by cement, it is generally thought a source of the cement other than the transformation of aragonite to calcite is required, perhaps being derived from the weathering and dissolution of adjacent carbonates (Dunham, 1969; Bathurst, 1971). In the inversion process, strontium is lost from aragonite. Similarly, magnesium may be lost from magnesian calcite to the groundwaters or concentrated to form euhedral crystals of dolomite (Lohmann and Meyers, 1977).

Once the sediment is buried below the freshwater table and comes into contact with shallow subsurface reducing fluids, cementation by calcite occurs at a decreased rate. The crystals tend to be very fine to medium grained and have irregular to equant shapes. Manganese and iron, if present, continue to substitute for calcium (Meyers and Lohmann, 1980), though the quantity of manganese and iron present in the waters is lower than those of the deeper subsurface because of the presence of oxygen. The $\delta^{18}O$ composition of these cements may be enriched in relation to shallower ce-

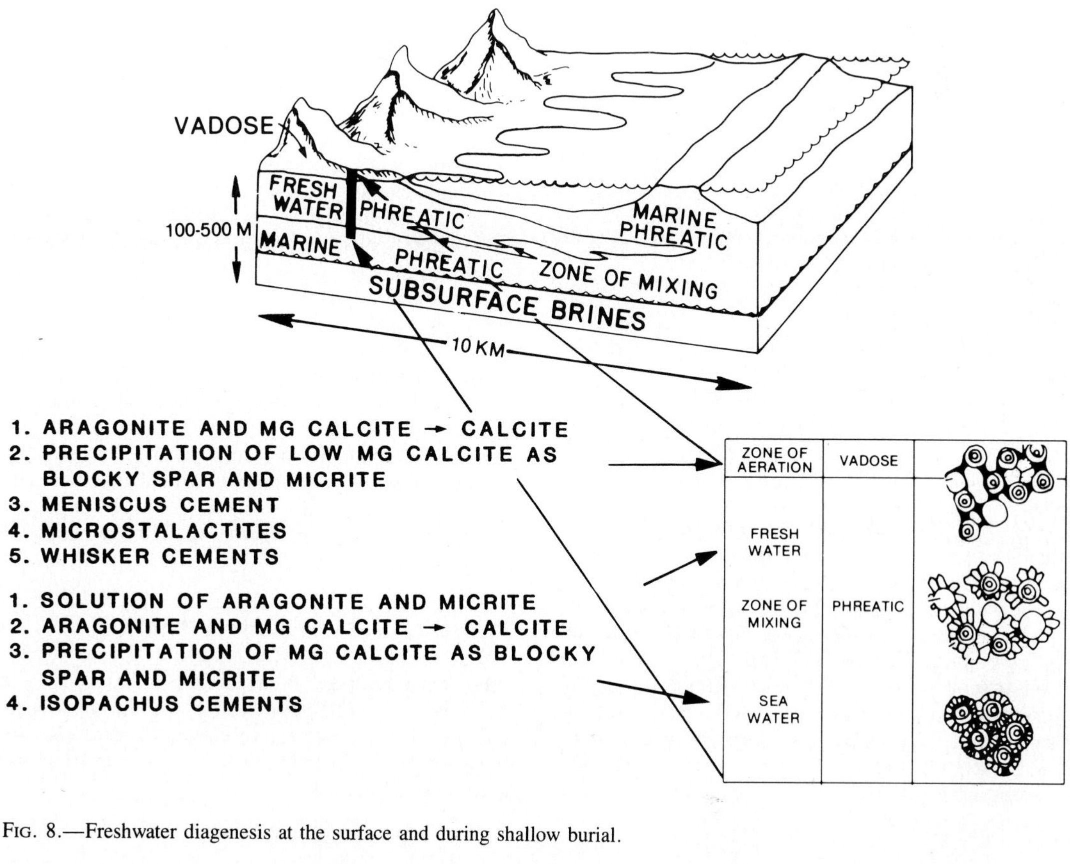

FIG. 8.—Freshwater diagenesis at the surface and during shallow burial.

ments due to contributions from rock oxygen. Conversely the $\delta^{18}O$ composition is depleted once the waters in contact with the sediment rise in temperature.

Deeper Burial

Whereas physical compaction of the kind Shinn *et al.* (1977) described from laboratory experiments obviously begins shortly after a carbonate is deposited and begins to dewater, we are unsure when true chemical compaction or pressure solution begins. Dunnington (1967) places the depth of stylolite formation between 600 to 900 m and Mossop (1972) at around 1000 m, but Schlanger (1964) gives evidence that in some cases burial need not be greater than 90 m, particularly in limestones with a high clay content. Meyers (1980) proves that considerable pressure solution in some carbonate rocks can occur under probably a few 10's of meters overburden, taking the stance that water chemistry may be as important as depth. At one extreme, Logan and Semeniuk (1976) and Wanless (1979, 1982) give evidence that pressure solution is pervasive in some limestone provinces and in some cases, as in the Devonian of the Canning Basin (Logan and Semeniuk, 1976), completely obscures many sedimentary relationships. In contrast, Pratt (1982) feels at least some of the fabrics identified by Wanless (1979) are the result of mechanical compaction. Similarly Playford (1980) persuasively downplays the significance of pressure solution in the Canning Basin. Nevertheless, there are many examples where pressure solution is important in carbonates and is destroying their porosity and permeability (Dunnington, 1967; Coogan, 1970; Mossop, 1972). Probably the best known example is that described by Dunnington (1967) where stylolites following bedding planes form permeability barriers in a producing oil field within Cretaceous limestones in Abu Dhabi.

Pressure solution fabrics take several forms (Park and Schott, 1968; Bathurst, 1971; Mossup, 1972; Wanless, 1979). Wanless (1979) recognized sutured-seam solution (common stylolites and grain-contact sutures), non-sutured seam solution (microstylolites, microstylolite swarms and clay seams) and non-seam solution (pervasive solution thinning of a unit). Solution surfaces are commonly marked by concentrations of dark clay, organic matter or iron oxides.

As Davies (1976), Bathurst (1980), Moore and Druckman *et al.* (1981) have noted, late stage cementation and dissolution of earlier-formed cements by warm subsurface fluids are volumetrically important diagenetic events. These late stage effects are dependent on fabric, texture, grain size and mineral stability, as well as on the pH, salinity and mineral content of the waters. The cement associated with subsurface brines is commonly an equant coarse-grained calcite spar, ferroan and with poikilotopic textures (Meyers, 1974; Grover and Read, 1978, 1983; Moore and Druckman, 1981; Crevello *et al.*, in press). These cements postdate earlier cements and compaction features, often filling cracks in broken grains or containing portions of earlier broken grains or cements. Baroque dolomite (Folk and Assereto, 1974; Radke and Mathis, 1980) is another common burial cement and is characterized by curved crystal faces with sweeping extinction.

Manganese can be an important trace element component of burial cements, but iron rich cements are more common. Subsurface waters are commonly reducing and any iron in the water tends to be incorporated in the calcite or dolomite. The changing manganese and iron compositions of subsurface fluids result in mineral zoning when cements are examined using cathodoluminescent petrography (Sommer, 1972; Meyers, 1974; Glover, 1977; Grover and Read, 1978, 1983; Pierson, 1981). Thermal and pressure effects and chemical solubilities play a dominant role in deeper burial. Oil maturation also proceeds as the temperature increases (Hunt, 1979; Waples, 1980), and hydrocarbons break down into CO_2 and water (Tissot and Espitalie, 1975; Tissot and Welte, 1978) which together may enhance carbonate porosity.

CONCLUSIONS

Carbonate cements form in a wide range of diagenetic environments. Depositional settings play an important role in determining depositional fabrics and the subsequent direction of diagenetic behavior of carbonate rocks. The challenge to the carbonate geologist is to unravel the diagenetic history and assess the importance of each stage of cementation in the alteration of porosity and permeability. Cement fabric and mineralogy reflect the chemistry of their precipitating and presently surrounding waters. The previous sections on environments of cementation point out that similarly appearing cements can form in different diagenetic environments and thus a more accurate interpretation of their origin must rely on geochemical as well as petrographic data. Such measurements of stable isotopic ratios, trace elements and fluid inclusions are being increasingly used in studies of carbonate cements in attempts to determine the composition of fluids and timing of precipitation in various diagenetic environments. Porosity in carbonates may be depositional, diagenetic or tectonically derived; it is lost through a combination of cementation along with chemical and physical compaction with depth.

ACKNOWLEDGMENTS

W.F. Buckingham, R.C. Burruss, W.J. Meyers, D.R. Prezbindowski, B.J. Russell, M. Magaritz and W.D. Wiggins read various drafts of the manuscript and provided numerous valuable suggestions. We thank Gulf Oil Corporation for typing, drafting and permission to publish.

REFERENCES

ADAMS, J.E., AND RHODES, M.L., 1960, Dolomitization by seepage refluxion: Amer. Assoc. Petroleum Geologists Bull., v. 44, p. 1912–1920.

ALEXANDERSSON, T., 1972, Micritization of carbonate particles: processes of precipitation and dissolution in modern shallow-marine sediments: Univ. Uppsala Geol. Inst. Bull., no. 5.3, v. 7, p. 201–236.

ARABIAN AMERICAN OIL COMPANY STAFF, 1959, Ghawar Oil Field, Saudi Arabia: Amer. Assoc. Petroleum Geologists Bull., v. 43, p. 434–454.

ARTHUR, M.A., SCHOLLE, P.A., AND HALLEY, R.B., 1982, Report on the Second Soc. Econ. Paleontologists Mineralogists Research Conference: Stable isotopes in sedimentary geology: Jour. Sed. Petrology, v. 52, p. 1039–1044.

ASSERETO, R.L.A.M., AND KENDALL, C.G.ST.C., 1977, Nature, origin and classification of peritidal tepee structures and related breccias: Sedimentology, v. 24, p. 153–210.

BARIA, L.R., STOUDT, D.L., HARRIS, P.M., AND CREVELLO, P.D., 1982, Upper Jurassic reefs of Smackover Formation, United States Gulf Coast: Amer. Assoc. Petroleum Geologists Bull., v. 66, p. 1449–1482.

BARTLETT, G.A., AND GREGGS, R.G., 1969, Carbonate sediments, oriented lithified samples from the North Atlantic: Science, v. 166, p. 740–741.

BATHURST, R.G.C., 1966, Boring algae, micrite envelopes and lithification of molluscan biosparites: Geol. Jour., v. 5, p. 89–109.

______, 1971, Carbonate Sediments and Their Diagenesis: Developments in Sedimentology, v. 12, Elsevier Pub. Co., New York, 620 p.

______, 1980, Deep crustal diagenesis in limestones: Rev. Inst. Invest. Geol., Diputacion Provincial, Universidad de Barcelona, v. 34, p. 89–100.

BEIN, A., AND LAND, L.S., 1982, San Andres carbonates in the Texas panhandle: Sedimentation and diagenesis associated with magnesium-calcium-chloride brines: The Univ. of Texas-Austin, Bureau Econ. Geology Rep. Inv. 121, 48 p.

BRICKER, O.P. (ed.), 1971, Carbonate Cements: Johns Hopkins Univ., Studies in Geology No. 19, 376 p.

BURRI, P., DUDRESNAY, R., AND WAGNER, C.W., 1973, Tepee structures and associated diagenetic features of intertidal carbonate sands (Lower Jurassic, Morocco): Sed. Geology, v. 9, p. 221–228.

BURRUSS, R.C., 1981, Hydrocarbon fluid inclusions in studies of sedimentary diagenesis, *in* L.S. Hollister, and M.L. Crawford (eds.), MAC Short Course in Fluid Inclusions: Applications to Petrology: Toronto, p. 138–156.

______, CERCONE, K.R., AND HARRIS, P.M., 1983, Fluid inclusion petrography and tectonic-burial history of the Al Ali 2 Well, evidence for the timing of diagenesis and oil migration, northern Oman Foredeep: Geology, v. 11, p. 567–570.

BUTLER, G.P., 1970, Holocene gypsum and anhydrite of the Abu Dhabi Sabkha, Trucial Coast: An alternative explanation of origin: Third Symposium on Salt, Northern Ohio Geol. Soc., v. 1, p. 120–152.

______, HARRIS, P.M., AND KENDALL, C.G.ST.C., 1982, Recent evaporites from the Abu Dhabi Coastal Flats, *in* C.R. Handford, R.G. Loucks, and G.R. Davies (eds.), Depositional and Diagenetic Spectra of Evaporites—A Core Workshop: Soc. Econ. Peleontologists Mineralogists Core Workshop No. 3, p. 33–64.

CAMPOS, H.S., AND HALLAM, A., 1979, Diagenesis of English Jurassic limestones as inferred from oxygen and carbon isotope analyses: Earth and Plan. Sci. Lett., v. 45, p. 23–31.

CHAVE, K.E., 1962, Factors influencing the mineralogy of carbonate sediments: Limnol. and Oceanog., v. 7, p. 218–223.

CHOQUETTE, P.W., AND PRAY, L.C., 1970, Geologic nomenclature and classification of porosity in sedimentary carbonates: Amer. Assoc. Petroleum Geologists Bull., v. 54, p. 207–250.

COOGAN, A.H., 1970, Measurements of compaction in oolitic grainstone: Jour. Sed. Petrology, v. 40, p. 921–929.

CREVELLO, P.D., HARRIS, P.M., STOUDT, D.L., AND BARIA, L.R. (in press), Porosity evolution and burial diagenesis in a reef-debris reservoir, Smackover Formation (Jurassic), Hico Knowles Field, Arkansas, *in* P.O. Roehl and P.W. Choquette (eds.): Springer Verlag, Casebook of Petroleum Reservoirs.

CURRIE, J.B., AND NWACHUKWU, S.O., 1974, Evidence on incipient fracture porosity in reservoir rocks at depth: Bull. Can. Petroleum Geol., v. 22, p. 42–58.

CURTIS, R., EVANS, G.R., KINSMAN, D.J.J., AND SHEARMAN, D.J., 1963, Association of dolomite and anhydrite in the recent sediments of the Persian Gulf: Nature, v. 197, p. 679–680.

DAVIES, G.R., 1970, Algal-laminated sediments, Gladstone Embayment, Shark Bay, Western Australia, *in* B. Logan *et al.* (ed.), Carbonate Sedimentation and Environments, Shark Bay, Western Australia: Amer. Assoc. Petroleum Geologists Mem. 13, p. 169–205.

______, 1976, "Bitumen" in post burial diagenetic calcite: Geol. Survey Canada, Paper 76-1C, p. 107–114.

DICKSON, J.A.D., AND COLEMAN, M.L., 1980, Changes in carbon and oxygen isotope composition during limestone diagenesis: Sedimentology, v. 27, p. 107–118.

DRAVIS, J.J., 1979, Rapid and widespread generation of recent oolitic hardgrounds on a high energy Bahamian platform, Eleuthera Bank, Bahamas: Jour. Sed. Petrology, v. 49, p. 195–208.

DUNHAM, R.J., 1962, Classification of Carbonate Rocks According to Depositional Texture: Amer. Assoc. Petroleum Geologists Mem. 1, Tulsa, p. 108–121.

______, 1969, Early vadose silt in Townsend Mound (Reef), New Mexico, *in* G.M. Friedman (ed.), Depositional Environments in Carbonate Rocks—A Symposium: Soc. Econ. Paleontologists and Mineralogists Spec. Pub. 14, p. 139–181.

______, 1971, Meniscus cement, *in* O.P. Bricker (ed.), Carbonate Cements: Johns Hopkins Univ. Studies in Geology, no. 19, p. 297–300.

DUNNINGTON, H.V., 1967, Aspects of diagenesis and shape change in stylolitic limestone reservoirs: Proc. 7th World Petrol. Cong., Mexico, p. 339–352.

Enos, P., and Sawatsky, L.H., 1981, Pore networks in Holocene carbonate sediments: Jour. Sed. Petrology, v. 51, p. 961–985.

Epting, M., 1980, Sedimentology of Miocene carbonate buildups, central Luconia, offshore Sarawak: Geol. Soc. Malayasia Bull., v. 12, p. 17–30.

Esteban, M., and Pray, L.C., 1977, Origin of the pisolite facies of the upper shelf area, *in* M.E. Hileman and S.J. Mazzullo (eds.), Upper Guadalupian Facies, Permian Reef Complex, Guadalupe Mountains, New Mexico and West Texas, 1977, Field Conference Guidebook, v. 1: Permian Basin Sec., Soc. Econ. Paleontologists and Mineralogists Pub. 77–16, p. 479–486.

Evamy, B.D., 1969, The precipitational environment and correlation of some calcite cements deduced from artificial staining: Jour. Sed. Petrology, v. 39, p. 787–793.

______, 1973, The precipitation of aragonite and its alteration to calcite on the Trucial Coast of the Persian Gulf, *in* B.H. Purser (ed.), The Persian Gulf: Springer-Verlag, New York, p. 329–341.

Fischer, A.G., and Garrison, R.D., 1967, Carbonate lithification on the seafloor: Jour. Geology, v. 75, p. 488–496.

Flügel, E., 1982, Microfacies Analysis of Limestones: Springer-Verlag, Berlin, 633 p.

Folk, R.L., 1959, Practical petrographic classification of limestones: Amer. Assoc. Petroleum Geologists Bull., v. 43, p. 1–38.

______, 1962, Spectral subdivision of limestone types, *in* W.E. Ham (ed.), Classification of Carbonate Rocks: Am. Assoc. Petroleum Geologists Memoir, 1, p. 62–84.

______, 1973, Carbonate petrography in the post-Sorbian age, *in* Evolving Concepts in Sedimentology: Johns Hopkins Univ. Studies in Geology, v. 21, p. 118–158.

______, and Assereto, R.L.A.M., 1974, Giant aragonite rays and baroque white dolomite in tepee-fillings, Triassic of Lombardy, Italy [abs.]: Amer. Assoc. Petroleum Geologists, Abs. with Prog., Ann. Mtg., San Antonio, p. 34–35.

Frank, J.R., Carpenter, A.B., and Oglesby, T.W., 1982, Cathodoluminescence and composition of calcite cement in the Taum Sank Limestone (Upper Cambrian), southeast Missouri: Jour. Sed. Petrology, v. 52, p. 631–638.

Friedman, G.M., 1959, Identification of carbonate minerals by staining methods: Jour. Sed. Petrology, v. 29, p. 87–97.

______, 1964, Early diagenesis and lithification in carbonate sediments: Jour. Sed. Petrology, v. 34, p. 777–813.

______, Amiel, A.J., and Schneidermann, N., 1974, Submarine cementation in reefs—Example from the Red Sea: Jour. Sed. Petrology, v. 44, p. 816–825.

Frost, S.H., Bliefnick, D.M., and Harris, P.M., 1983, Deposition and porosity evolution of a Lower Cretaceous rudist buildup, Shuaiba Formation of eastern Arabian Peninsula, *in* Harris, P.M. (ed.), Carbonate Buildups—A Core Workshop: Soc. Econ. Paleontologists and Mineralogists Core Workshop No. 4, p. 381–410.

Garrels, R.M., and Christ, C.L., 1965, Solutions, Minerals and Equilibria: Freeman, Cooper and Co., San Francisco, p. 450.

Gevirtz, J.L., and Friedman, G.M., 1966, Deep-Sea carbonate sediments of the Red Sea and their implications on marine lithification: Jour. Sed. Petrology, v. 36, p. 143–151.

Ginsburg, R.N., and James, N.P., 1976, Submarine botryoidal aragonite in Holocene reef limestones, Belize: Geology, v. 4, p. 431–436.

______, Marszalek, D.S., and Scheidermann, N., 1971, Ultrastructure of carbonate cements in a Holocene algal reef of Bermuda: Jour. Sed. Petrology, v. 41, p. 472–482.

Glover, E.D., 1977, Cathodoluminescence, Iron and Manganese Content and the Early Diagenesis of Carbonates: Ph.D. Diss., Univ. Wisconsin, Madison, 444 p.

Grover, G.A., Jr., and Read, J.F., 1978, Fenestral and associated vadose diagenetic fabrics of tidal flat carbonates, Middle Ordovician New Market Limestone, southwestern Virginia: Jour. Sed. Petrology, v. 48, p. 453–473.

______, and ______, 1983, Paleoaquifer and deep burial related cements defined by regional cathodoluminescent patterns, Middle Ordovician carbonates, Virginia: Amer. Assoc. Petroleum Geologists Bull., v. 67, p. 1275–1303.

Halley, R.B., and Harris, P.M., 1979, Freshwater cementation of a 1000-year old oolite: Jour. Sed. Petrology, v. 49, p. 969–988.

______, and Schmoker, J.W., 1983, High porosity Cenozoic carbonate rocks of south Florida: Progressive loss of porosity with depth: Amer. Assoc. Petroleum Geologists Bull., v. 67, p. 191–200.

Hanor, J.S., 1978, Precipitation of beachrock cements: Mixing of marine and meteoric waters vs. CO_2-degassing: Jour. Sed. Petrology, v. 48, p. 489–501.

______, 1980, Dissolved methane in sedimentary brines: Potential effect on the PVT properties of fluid inclusions: Econ. Geology, v. 75, p. 603–609.

Harris, P.M., 1978, Holocene marine-cemented sands, Joulters Ooid Shoal, Bahamas: Trans. Gulf Coast Assoc. Geol. Societies, v. 28, p. 175–183.

______, Halley, R.B., and Lukas, K.J., 1979, Endolith microborings and their preservation in Holocene-Pleistocene (Bahama-Florida) ooids: Geology, v. 7, p. 216–220.

______, and Frost, S.H. (1984), Middle Cretaceous carbonate reservoirs, Fahud Field and northwestern Oman: Amer. Assoc. Petroleum Geologists Bull., v. 68, p. 649–658.

Hudson, J.D., 1977, Stable isotopes and limestone lithification: Jour. Geology Soc. London, v. 133, p. 637–660.

Hunt, J.M., 1979, Petroleum Geochemistry and Geology: W.H. Freeman and Co., San Francisco, 616 p.

Hsü, K.J., and Schneider, J., 1973, Progress report on dolomitization-hydrology of Abu Dhabi Sabkhas, Arabian Gulf, *in* B.H. Purser (ed.), The Persian Gulf: Springer-Verlag, New York, p. 409–422.

Illing, L.V., 1954, Bahamian calcareous sands: Amer. Assoc. Petroleum Geologists Bull., v. 38, p. 1–95.

James, N.P., and Ginsburg, R.N., 1979, The seaward marginal Belize barrier and atoll reefs: Int. Assoc. Sedimentologists Spec. Pub. 3, 191 p.

______, ______, MARSZALEK, D.S., AND CHOQUETTE, P.W., 1976, Facies and fabric specificity of early subsea cements in shallow Belize (British Honduras) reefs: Jour. Sed. Petrology, v. 46, p. 523–544.
KALDI, J., AND GIDMAN, J., 1982, Early diagenetic dolomite cements: Examples from the Permian Lower Magnesian Limestone of England and the Pleistocene carbonates of the Bahamas: Jour. Sed. Petrology, v. 52, p. 1073–1085.
KENDALL, C.G.ST.C., 1969, An environmental reinterpretation of the Permian evaporite/carbonate shelf sediments of the Guadalupe Mountains: Geol. Soc. America Bull., v. 80, p. 2503–2526.
______, AND SKIPWITH, P.A.D'E, 1969, Holocene shallow water carbonate and evaporite sediments of Khor al Bazam, Abu Dhabi, south west Persian Gulf: Amer. Assoc. Petroleum Geologists Bull., v. 53, p. 841–869.
KLOSTERMAN, M.J., 1981, Applications of Fluid Inclusion Techniques to Burial Diagenesis in Carbonate Rock Sequences: Louisiana State Univ., Applied Carbonate Research Program, Cont. 7, 101 p.
KLOVAN, J.E., 1964, Facies analysis of the Redwater Reef Complex, Alberta, Canada: Bull. Canadian Petroleum Geol., v. 12, p. 1–100.
LAND, L.S., 1967, Diagenesis of skeletal carbonates: Jour. Sed. Petrology, v. 37, p. 914–930.
______, 1970, Phreatic versus vadose meteoric diagenesis of limestones, evidence from fossil water table: Sedimentology, v. 14, p. 175–185.
______, 1980, The isotopic and trace element geochemistry of dolomite: The state of the art, *in* D.H. Zenger, J.G. Dunham, and R.L. Ethington (eds.), Concepts and Models of Dolomitization: Soc. Econ. Paleontologists and Mineralogists, Spec. Pub. 28, p. 87–110.
LANGTON, J.R., AND CHIN, G.E., 1968, Rainbow Member facies and related reservoir properties, Rainbow Lake, Alberta: Bull. Canadian Petroleum Geol., v. 16, p. 104–143.
LINDSAY, R.F., AND KENDALL, C.G.ST.C. (in press), Depositional facies, diagenesis and reservoir character of the Mission Canyon Formation (Mississippian) of the Williston Basin at Little Knife Field, North Dakota, *in* P.O. Roehl and P.W. Choquette (eds.), Casebook on Carbonate Reservoirs: Springer-Verlag.
LOGAN, B.W., 1981, Modern Carbonates and Evaporite Sediments of Shark Bay and Lake McCloud, Western Australia: Geol. Soc. Australia, Australian Geological Convention, Field Excursion Guidebook, 63 p.
______, AND SEMENIUK, V., 1976, Dynamic Metamorphism, Processes and Products in Devonian Carbonate Rocks, Canning Basin, Western Australia: Geol. Soc. Australia Spec. Pub. 6, 138 p.
LOHMANN, K.C., AND MEYERS, W.J., 1977, Microdolomites in cloudy prismatic calcites, a proposed criteria for former high-magnesium calcites: Jour. Sed. Petrology, v. 47, p. 1078–1088.
LONGACRE, S.A., 1980, Dolomite reservoirs from Permain biomicrites, *in* R.B. Halley and R.G. Loucks (eds.), Carbonate Reservoir Rocks: Soc. Econ. Paleontologists and Mineralogists Core Workshop No. 1, p. 105–117.
LONGMAN, M.W., 1980, Carbonate diagenetic textures from near-surface diagenetic environments: Amer. Assoc. Petroleum Geologists Bull., v. 64, p. 461–487.
______, 1981, Carbonate Diagenesis and Diagenetic Environments: Amer. Assoc. Petroleum Geologists, Fall Education Conference, v. 2, 59 p.
______, FERTAL, T.G., AND GLENNIE, J.S., 1983, Origin and geometry of Red River Dolomite reservoirs, Williston Basin: Amer. Assoc. Petroleum Geologists Bull., v. 67, p. 744–771.
LOREAU, J.P., AND PURSER, B.H., 1973, Distribution and ultrastructure of Holocene ooids in the Persian Gulf, *in* B.H. Purser (ed.), The Persian Gulf, Holocene Carbonate Sedimentation in a Shallow Epic Continental Sea: Springer-Verlag, New York, p. 279–328.
MACINTYRE, I.G., 1977, Distribution of submarine cements in a modern Caribbean fringing reef, Galeta Point, Panama: Jour. Sed. Petrology, v. 47, p. 503–516.
MAGARITZ, M., 1983, Carbon and oxygen isotope composition of Recent and ancient coated grains, *in* T. Peryt (ed.), Coated Grains: Springer-Verlag, New York, p. 27–37.
______, AND TAYLOR, H.P., JR., 1976, Oxygen, hydrogen and carbon isotope studies of the Franciscan Formation, Coast Ranges, California: Geochim, Cosmochim. Acta, v. 40, p. 215–234.
MATTHEWS, R.K., 1968, Carbonate diagenesis-equilibrium of sedimentary mineralogy to the subaerial environment, coral cap of Barbados, West Indies: Jour. Sed. Petrology, v. 38, p. 110–119.
______, 1971, Diagenetic environment of possible importance to the explanation of cementation fabric in subaerially exposed carbonate sediments, *in* O.P. Bricker (ed.), Carbonate Cements: Johns Hopkins Univ. Studies in Geology 19, p. 127–132.
______, 1974, A Process approach to diagenesis of reefs and reef associated limestones, *in* L.F. Laporte (ed.), Reefs in Time and Space: Soc. Econ. Paleontologists and Mineralogists Spec. Pub. 18, p. 234–256.
MAZZULLO, S.J., AND CYS, J.M., 1979, Marine aragonite seafloor growth and cements in Permian phylloid algal mounds, Sacramento Mountains, New Mexico: Jour. Sed. Petrology, v. 49, p. 917–936.
MCNAUGHTON, D.A., AND GARB, F.A., 1975, Finding and evaluating petroleum accumulations in fractured reservoir rock, *in* V.S. Cameron (ed.), Exploration and Economics of the Petroleum Industry, New Ideas, New Methods, New Developments: Matthew Bender & Co., Inc., v. 13, p. 23–50.
MEYERS, W.J., 1974, Carbonate cement stratigraphy of the Lake Valley Formation (Mississipian) Sacramento Mountains, New Mexico: Jour. Sed. Petrology, v. 44, p. 837–861.
______, 1978, Carbonate cements—Their regional distribution and interpretation in Mississippian limestones of southwestern New Mexico: Sedimentology, v. 25, p. 371–400.
______, 1980, Compaction in Mississippian skeletal limestones, southwestern New Mexico: Jour. Sed. Petrology, v. 50, p. 457–474.
______, AND LOHMANN, K.C., 1978, Microdolomite-rich syntaxial cements—Proposed meteoric-marine mixing zone phreatic cements from Mississippian limestones, New Mexico: Jour. Sed. Petrology, v. 48, p. 475–488.

______, AND ______, 1980, Geochemistry of regionally extensive calcite zones in Mississippian skeletal limestones, New Mexico [abs.]: AAPG-SEPM Abst. and Program, Annual Mtg., p. 91.

MILLIMAN, J.D., 1974, Marine Carbonates: Springer-Verlag, New York, 375 p.

______, ROSS, D.A., AND KU, T.L., 1969, Precipitation and lithification of deep-sea carbonates in the Red Sea: Jour. Sed. Petrology, v. 39, p. 724–736.

MOORE, C.H., 1973, Intertidal carbonate cementation, Grand Cayman, West Indies: Jour. Sed. Petrology, v. 43, p. 591–602.

______, 1977, Beach rock origin: Some geochemical, mineralogical, and petrographic considerations: Geoscience and Man, v. 18, p. 155–163.

______, 1979, Porosity in carbonate rock sequences, geology of carbonate porosity: Amer. Assoc. Petroleum Geologists, Continuing Education Course Notes 11, p. A1–A124.

______, 1980, Porosity in Carbonate Rock Sequences: Amer. Assoc. Petroleum Geologists, Fall Education Conference 1980, v. 1, 67 p.

______, AND DRUCKMAN, Y., 1981, Burial diagenesis and porosity evolution, Upper Jurassic Smackover, Arkansas and Louisiana: Amer. Assoc. Petroleum Geologists Bull., v. 65, p. 597–628.

MORESBY, R., 1835, Extracts from Commander Moresby's Report on the Northern Maldives: Jour. Royal Geol. Soc., v. 5, p. 398–404.

MOSSOP, G.D., 1972, Origin of the peripheral rim, Redwater Reef, Alberta: Bull. Canadian Petroleum Geol., v. 20, p. 238–280.

MULLINS, H.T., NEWMANN, A.C., WILBER, R.J., AND BOARDMAN, M.R., 1980, Nodular carbonate sediment on Bahamian slopes: Possible precursors to nodular limestones: Jour. Sed. Petrology, v. 50, p. 117–131.

PARK, W.C., AND SCHOTT, E.H., 1968, Stylolites, their nature and origin: Jour. Sed. Petrology, v. 38, p. 175–191.

PIERSON, B.J., 1981, The control of cathodoluminescence in dolomite by iron and manganese: Sedimentology, v. 28, p. 601–610.

PLAYFORD, P.E., 1980, Devonian "Great Barrier Reef" of Canning Basin, Western Australia: Amer. Assoc. Petroleum Geologists Bull., v. 64, p. 814–840.

PRATT, B.R., 1982, Limestone response to stress-pressure solution and dolomitization—Discussion and examples of compaction in carbonate sediments: Jour. Sed. Petrology, v. 52, p. 323–328.

PURSER, B.H., 1969, Syn-sedimentary marine lithification of Middle Jurassic limestones in the Paris Basin: Sedimentology, v. 12, p. 205–230.

______, 1971, Middle Jurassic syn-sedimentary marine cements from the Paris Basin, France, *in* O.P. Bricker (ed.), Carbonate Cements: Johns Hopkins Univ. Studies in Geol. 19, p. 178–184.

______, AND LOREAU, J.P., 1973, Aragonitic, supratidal encrustations on the Trucial Coast, Persian Gulf, *in* B.H. Purser (ed.), The Persian Gulf: Springer-Verlag, New York, p. 343–376.

RADKE, R.M., AND MATHIS, R.L., 1980, On the formation of saddle dolomite: Jour. Sed. Petrology, v. 50, p. 1149–1168.

ROEDDER, E., 1979, Fluid inclusion evidence on the environment of sedimentary diagenesis, a review, *in* P.A. Scholle and R.R. Schluger (eds.), Aspects of Diagenesis: Soc. Econ. Paleontologists and Mineralogists Spec. Pub. 25, p. 89–108.

ROEHL, P.O., 1967, Stony Mountain (Ordovician) and Interlake (Silurian) facies analogs of recent low-energy marine and subaerial carbonates, Bahamas: Amer. Assoc. Petroleum Geologists Bull., v. 51, p. 1979–2032.

SCHLAGER, W., AND JAMES, N.P., 1978, Low-magnesian calcite limestones forming at the deep-seafloor, Tongue of the Ocean, Bahamas: Sedimentology, v. 25, p. 675–702.

______, AND CHERMAK, A., 1979, Sediment facies of platform-basin transition, Tongue of the Ocean, Bahamas, *in* L. Doyle and O.H. Pilkey (eds.), Geology of Continental Slopes: Soc. Econ. Paleontologists and Mineralogists Spec. Pub. 27, p. 193–208.

SCHLANGER, S.O., 1964, Petrology of limestones of Guam: U.S. Geological Survey, Prof. Paper 403D, p. 1–52.

SCHMOKER, J.W., AND HALLEY, R.B., 1982, Carbonate porosity versus depth: A predictable relation for south Florida: Amer. Assoc. Petroleum Geologists Bull., v. 66, p. 2561–2570.

SHINN, E.A., 1969, Submarine lithification of Holocene carbonate sediments in the Persian Gulf: Sedimentology, v. 12, p. 109–144.

______, GINSBURG, R.N., AND LLOYD, R.M., 1965, Recent supratidal dolomite from Andros Island, Bahamas, *in* L.C. Pray and R.C. Murray (eds.), Dolomitization and Limestone Diagenesis: A Symposium: Soc. Econ. Paleontologists and Mineralogists Spec. Pub. 13, p. 112–123.

______, HALLEY, R.B., HUDSON, J.H., AND LIDZ, B.H., 1977, Limestone compaction—An enigma: Geology, v. 5, p. 21–24.

______, ROBBIN, D.M., LIDZ, B.H., AND HUDSON, J.H., 1983, Influence of deposition and early diagenesis on porosity and chemical compaction in two Paleozoic buildups: Mississippian and Permian age rocks in the Sacramento Mountains, New Mexico, *in* P.M. Harris (ed.), Carbonate Buildups—A Core Workshop: Soc. Econ. Paleontologists and Mineralogists Core Workshop No. 4, p. 182–221.

SIBLEY, D.F., 1982, The origin of common dolomite fabrics—Clues from the Pliocene: Jour. Sed. Petrology, v. 52, p. 1087–1100.

SOMMER, S.E., 1972, Cathodoluminescence of carbonates, I. Characterization of cathodoluminescence from carbonate solid solution: Chem. Geology, v. 9, p. 257–273.

STEARNS, D.W., AND FRIEDMAN, M., 1972, Reservoirs in fractured rock: Amer. Assoc. Petroleum Geologists Mem. 17, p. 82.

TAFT, W.H., ARRINGTON, F., HAIMOVITZ, A., MACDONALD, D., AND WOOLHEATER, C., 1968, Lithification of modern carbonate sediments at Yellow Bank, Bahamas: Bull. Marine Sci. Gulf and Carib., v. 18, p. 762–868.

THOMPSON, G., 1972, A geochemical study of some lithified carbonate sediments from the deep sea: Geochim. Cosmochim. Acta, v. 36, p. 1237–1253.

TILLMAN, J.E., 1983, Exploration for reservoirs with fracture-enhanced permeability: Oil and Gas Jour., February 21, p. 165–179.

TISSOT, B.P., AND ESPITALIÉ, J., 1975, The thermal evolution of organic matter in sediments: Rev. Inst. Fr. Petrology, v. 30, p. 743–777.

______, AND WELTE, D.H., 1978, Petroleum Formation and Occurrence, A New Approach to Oil and Gas Exploration: Springer-Verlag, Berlin, p. 527.

VAN DEN BARK, E., AND THOMAS, O.D., 1981, Ekofisk: First of the giant oil fields in western Europe: Amer. Assoc. Petroleum Geologists Bull., v. 65, p. 2341–2363.

VEIZER, J., AND COMPSTON, W., 1974, Sr^{87}/Sr^{86} Composition of seawater during the Phanerozoic: Geochim. Cosmochim. Acta, v. 38, p. 1461–1484.

______, AND HOEFS, J., 1976, The Nature of O^{18}/O^{16} and C^{13}/C^{12} secular trends in sedimentary carbonate rocks: Geochim. Cosmochim. Acta, v. 40, p. 1387–1395.

______, DEMOVIC, R., AND TURAN, J., 1971, Possible use of strontium in sedimentary carbonate rocks as a paleoenvironmental indicator: Sed. Geol., v. 5, p. 5–22.

VINCELETTE, R.R., AND SOEPARJADI, R.A., 1976, Oil-bearing reefs in Salawati Basin of Irian Jaya, Indonesia: Amer. Assoc. Petroleum Geologists Bull., v. 60, p. 1448–1462.

WALLS, R.A., MOUNTJOY, E.W., AND FRITZ, P., 1979, Isotopic composition and diagenetic history of carbonate cements in Devonian Golden Spike Reef, Alberta, Canada: Geol. Soc. America Bull., v. 90, p. 963–982.

WANLESS, H.R., 1979, Limestone response to stress: Pressure solution and dolomitization: Jour. Sed. Petrology, v. 49, p. 437–462.

______, 1982, Limestone response to stress: Pressure solution and dolomization—Reply: Jour. Sed. Petrology, v. 52, p. 328–332.

WAPLES, D.W., 1980, Time and temperature in petroleum formation, application of Lopatin's Method to petroleum exploration: Amer. Assoc. Petroleum Geologists Bull., v. 64, p. 916–926.

WARD, W.C., 1975, Petrology and diagenesis of carbonate eolianites of northeastern Yucatan Peninsula, Mexico, *in* Belize Shelf-Carbonate Sediments, Clastic Sediment and Ecology: Amer. Assoc. Petroleum Geologists Studies in Geology 2, p. 500–571.

WARDLAW, N.C., 1976, Pore geometry of carbonate rocks as revealed by pore casts and capillary pressure, Amer. Assoc. Petroleum Geologists Bull., v. 60, p. 245–257.

WARREN, J.K., 1982, The hydraulical significance of holocene tepees, stromatolites and boxwork limestones in coastal salinas in South Australia: Jour. Sed. Petrology, v. 52, p. 1171–1202.

WILSON, A.O. (in press), Depositional and diagenetic facies in the Jurassic Arab-C and D Reservoirs, Qatif Field, Saudi Arabia, *in* P.L. Roehl and P.W. Choquette (eds.), Carbonate Petroleum Reservoirs: Springer-Verlag.

WINLAND, H.D., AND MATTHEWS, R.K., 1974, Origin and significance of grapestone, Bahama Islands: Jour. Sed. Petrology, v. 44, p. 921–927.

ZENGER, D.H., DUNHAM, J.B., AND ETHINGTON, R.L., 1980 (eds.), Concepts and Models of Dolomitization: Soc. Econ. Paleontologists and Mineralogists, Spec. Publ. No. 28, 320 p.

PREVENTION OF CARBONATE CEMENTATION IN PETROLEUM RESERVOIRS

CHARLES T. FEAZEL AND RICHARD A. SCHATZINGER
Phillips Petroleum Company, Bartlesville, Oklahoma 74004

ABSTRACT

By the time a carbonate unit has been buried to the depths of most petroleum reservoirs, the significant question is often not "how did the pores originate?", but rather, "why are they still there?" Preservation of porosity, regardless of its origin, is a consequence of one or more of the following mechanisms: (1) minimal burial, (2) reduced burial stress, generally due to overpressured pore fluids, (3) increased framework rigidity, which prevents compaction, (4) exclusion of pore waters by petroleum entry, (5) stable mineralogy, (6) permeability barriers, isolating porous intervals from external diagenetic fluids, and (7) pore resurrection, a consequence of the temporary filling of pores with cement that is subsequently removed. Examples from the stratigraphic record demonstrate that each of these pore-preserving mechanisms may control reservoir quality.

INTRODUCTION

This paper deals not with cements, but with their absence. Exploration for petroleum includes, among its many strategies, prediction of porous reservoirs at depth. Much effort has been expended on studies of the evolution and destruction of pore systems. Although many workers (notably Choquette and Pray, 1970) have catalogued the various types of pores found in carbonate rocks, a question seldom addressed is: "How are the pores preserved in petroleum reservoirs despite burial to depths exceeding five kilometers?" It is generally agreed, despite recent controversy (Moore and Druckman, 1981), that the majority of pore types described by Choquette and Pray (1970) resulted from early, shallow diagenesis. The following discussion concentrates on the processes that preserve such early pores as they pass through successive diagenetic environments.

MECHANISMS OF POROSITY PRESERVATION

We have identified seven mechanisms (Table 1) by which cementation in carbonate rocks may be retarded or prevented: (1) minimal burial, (2) reduced burial stress, (3) increased framework rigidity, (4) exclusion of pore water, (5) stable mineralogy, (6) permeability barriers, and (7) pore resurrection. Petroleum reservoirs from diverse geographic locations and geologic ages are cited in following paragraphs as examples of the control of reservoir quality by one or more of these pore-preserving mechanisms.

Minimal Burial

If carbonate sediments are never buried to significant depth (e.g. below the influence of surface-derived waters) they may retain much of their initial porosity, or have it considerably enhanced by meteoric processes. The shallow burial environment, the site of porosity generation and enhancement, is, however, a dynamic setting in which pore preservation is not favored. In this diagenetic environment porosity is destroyed by recrystallization, infiltration of clays, precipitation of vadose cements, or cements derived from biogenic gas.

The effect of shallow burial on reservoir potential thus depends on a delicate balance between porosity generating and porosity occluding processes. Despite the number of papers that emphasize the importance of porosity destruction during early, shallow diagenesis (Friedman, 1975, Longman, 1980, Bricker, 1971, and others), Halley and Schmoker (1983) and Scholle *et al.* (1983) have shown that significant porosity reduction in some sediments may not begin until carbonate strata are buried to depths of several hundred to nearly a thousand meters. Before these porous rocks can serve as petroleum reservoirs, they must be sealed. The likelihood of competent seals at shallow burial depth is minimal.

Although many examples of porous outcropping limestones that have never been deeply buried might be cited, from Cretaceous to Pleistocene in age (for example: Halley and Harris, 1979, Kerr, 1977, Land, 1973, and Land *et al.*, 1967), our goal is to understand porosity preservation at depth, not at the surface. In the following section we discuss rocks that may be deeply buried, but experience overburden pressures much like those of near-surface units.

Reduced Burial Stress

When pore pressures approach lithostatic pressures, the effective overburden stress resembles that of the shallow burial environment. Pore fluid overpressure reduces stress at grain contacts, and thereby prevents solution-transfer of calcium carbonate (pressure-solution at points of contact and reprecipitation of the dissolved material as pore-filling cement, Durney, 1972). In Maastrichtian and Danian chalk beneath the central North Sea, for example, porosity values of 40 percent are not uncommon at depths of over three kilometers (Van den Bark and Thomas, 1981; Feazel *et al.*, in

TABLE 1.—RELATIONSHIP BETWEEN THE PROCESSES, TIMING, AND RECOGNITION OF SEVEN MECHANISMS OF PORE PRESERVATION

MECHANISM	*PROCESS*	*TIMING*	*RECOGNITION*
Minimal Burial	Meteoric diagenesis Stable tectonics	Early	Stratigraphic setting Cement textures Fluid inclusions Chemical & isotopic signatures
Reduced Burial Stress	Increased pore pressure	Early to late	Pressure tests Log analysis
Increased Framework Rigidity	Lithification Grain support	Pre-compaction	Petrography
Exclusion of Pore Water	Hydrocarbon entry	Early to intermediate	Hydrocarbons in pores Log analysis (Sw)
Stable Mineralogy	Neomorphism Replacement	Early to intermediate	Petrography Trace elements
Permeability Barriers	Depositional encasement Diagenetic barrier	Early to late	Stratigraphic setting Petrography Geochemistry
Pore Resurrection	Occlusion followed by cement removal	Late	Petrography

press). Chalk normally loses porosity rapidly with increasing depth of burial (Fig. 1). The Niobrara chalk of the U.S. midcontinent, for example, has porosities of 40% where exposed, but where it was buried under 1500–2000 meters of Cretaceous and Tertiary deposits shed from the Rocky Mountains, porosity is reduced to 10%. Where the Niobrara is buried to a depth of 2400 meters at Fort Collins, Colorado, porosity is essentially zero (P. Scholle, 1977, and personal communication). The chalk at the Ekofisk Field in the North Sea is unusual because much of its original porosity has been retained. Scholle (1974) believed early dewatering of calcareous ooze reduces the 70–80% initial porosity to 50–60%. The presence in some Ekofisk intervals of porosity values near 50% indicates that little cementation followed mechanical compaction. Scholle (1975, 1977) observed that due to increased pore pressure in the chalk, the mechanism for both mechanical and chemical compaction is much less effective and porosity is preserved. The Central Graben of the North Sea is moderately overpressured, due to the low permeability of the sediments, to the effective seal provided by overlying Tertiary shales, and possibly to retardation of lateral fluid escape by boundary faults. The Ekofisk Field Chalk reservoir is pressurized to approximately 7100 psi at 3000 meters (10,000 feet) (Van den Bark and Thomas, 1981, Fig. 24). Normal pore pressure at that depth is approximately 4300 psi, with a lithostatic load of 9000 psi. Thus almost all the overburden above the Ekofisk structure is supported by the reservoir fluids. Net lithostatic pressure is approximately 1800 psi, which would normally correspond to

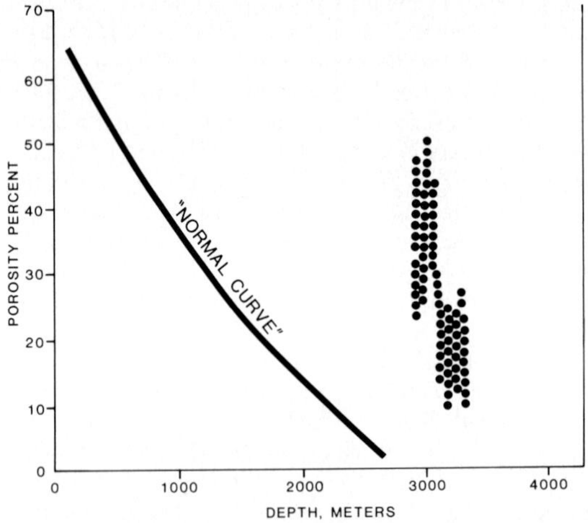

FIG. 1.—Porosity loss with depth for a "normal" pelagic sediment buried with marine or modified marine pore fluids (Scholle *el al.*, 1983). Data at right, from Ekofisk area wells, North Sea, show anomalously high porosity at 3 km burial depth.

a burial depth of only 610 meters (2000 ft.) (Scholle, 1977). The porosity of chalk buried to that depth (Hancock and Scholle, 1975) approximates that of the much deeper Ekofisk section. Without the confining structure and stratigraphy that led to overpressured conditions in the North Sea central graben, the pores would not have been available once the Jurassic source rock reached sufficient thermal maturity to expel petroleum. Oil entrapment as a pore-preserving mechanism will be discussed in a later section.

Increased Framework Rigidity

Any process that increases the strength of a potential reservoir unit, whether through fluid pressure buildup as described above, or by strengthening of the surrounding rock, decreases the probability of mechanical and chemical compaction. Increased rigidity of the rock framework generally results from two processes: (1) achievement of a grain-supported matrix by dewatering and early compaction (Shinn *et al.*, 1977), and (2) early lithification without pervasive pore-filling cementation (Sibley, 1982). The first process marks the transition from mechanical compaction to chemical compaction and may occur early in a rock's history. Figure 2 illustrates the results of the second process: a Holocene beachrock, cemented into a very rigid structure by a fine rim of aragonite crystals fringing the grains, with essentially no loss of primary porosity. Evidence for pre-compaction cement includes spalled fibrous calcite and undeformed delicate skeletal material (Fig. 3). Early dolomitization (pre-compaction: Kaldi and Gidman, 1982, Bebout *et al.*, 1981) of such rim cements (and pore-filling cements as well) leaves a very rigid framework that can withstand burial to depths exceeding five kilometers, even after leaching of the original grains (Fig. 4). The Smackover Formation in the eastern Gulf Coast contains numerous petroleum reservoirs (e.g. Chatom Field, Alabama) that owe their porosity to this mechanism.

Exclusion of Pore Water

Displacement of pore water by oil or gas prevents or significantly retards the precipitation of cement, as most minerals are insoluble in hydrocarbon fluids (Friedman and Sanders, 1978, Epstein and Friedman, 1983). The results of this process may include the preservation of Pennsylvanian aragonitic mollusc shells in

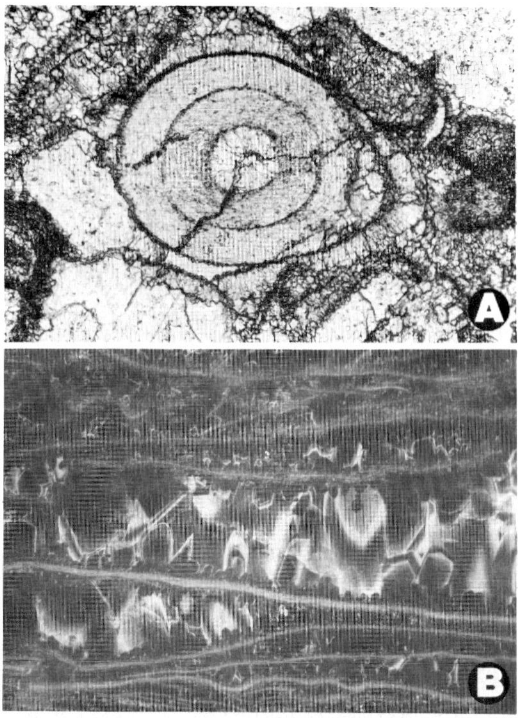

FIG. 3.—Pre-compaction cements. A. Marine calcite cement spalled from ooid by compaction. Pennsylvanian, Horseshoe Atoll, West Texas. Depth approximately 6500 feet. Photo width equals one millimeter. B. Unbroken delicate pelecypod valves preserved from compaction by early luminescent cement. Shublik Formation, Triassic, North Slope, Alaska. Depth approximately 9000 feet. Cathodoluminescence, photo width equals three millimeters.

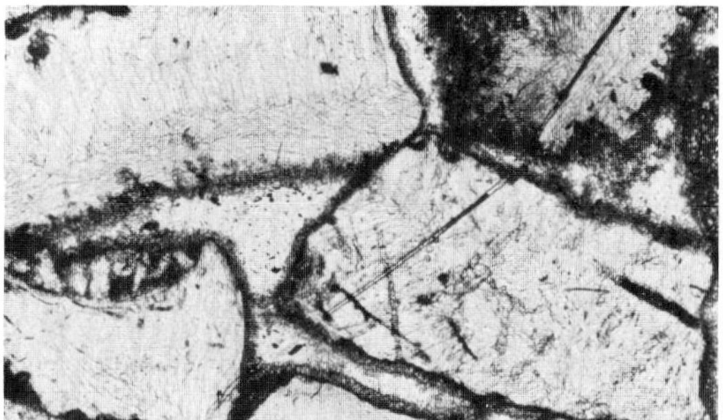

FIG. 2.—Increased framework rigidity caused by early aragonite cement (dark rims on bioclasts). Most primary porosity has been preserved. Only small volumes of cement were required for lithification. Holocene beachrock, U. S. Virgin Islands. Photo width equals one millimeter.

FIG. 4.—Framework rigidity caused by interstitial dolomitization between non-compacted peloids. Moldic pores created by leaching of peloids retain original shape despite deep burial. Smackover Formation, Chatom Field, Alabama, Depth exceeds 16,000 feet. From Feazel (in press).

tar (Squires, 1972). Similar shells of Lower Cretaceous age remain aragonitic where they are in contact with oil, but have recrystallized to calcite where they are in contact with water (Füchtbauer and Goldschmidt, 1964). Organic coatings of many different types on carbonate surfaces have been shown to inhibit diagenesis (Suess, 1970). In most reservoirs, following oil migration there remains a residual water saturation, which may be enough to allow diagenesis to continue after petroleum entry, but at greatly reduced rates.

In some cases (Davies, 1979, oral presentation) the last event before petroleum entry was corrosion of the pore walls. Likely agents for this corrosion were CO_2 and organic acids dissolved in waters moving ahead of the migrating oil. The moving organic fluids may thus have created their own enhanced reservoir porosity (Holland, 1982, Moore and Druckman, 1981). Once the petroleum was emplaced, however, most reservoir diagenesis ceased. An example is found in the Stuart City trend, South Texas (Fig. 5) in which pores contain, in paragenetic sequence: (1) marine cements, (2) meteoric cements, (3) subsurface overgrowths on the meteoric cements, containing hydrocarbon inclusions, (4) oil, forming a near-continuous film around the pore margins, and (5) late diagenetic calcite that apparently precipitated after oil migrated out of the pores (Prezbindowski, 1981).

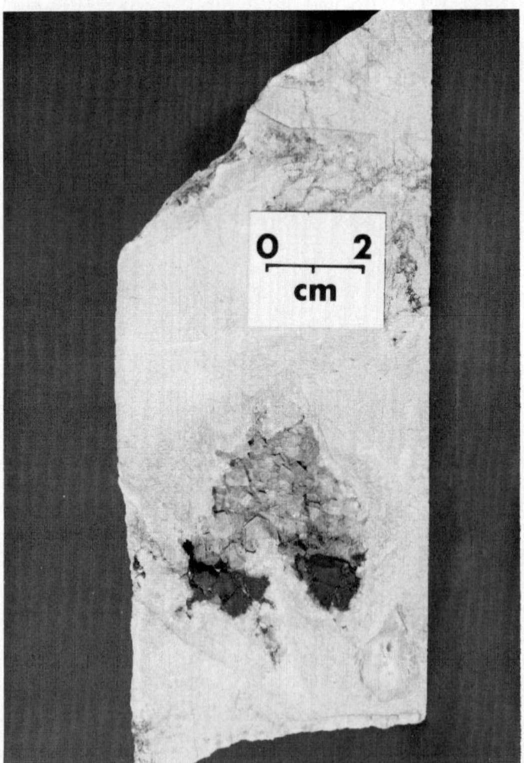

FIG. 5.—Fossil gas-oil contact in large vug. Migrating petroleum displaced pore water, effectively ending diagenesis. The last diagenetic event before petroleum entry was precipitation of coarse equant calcite on vug walls. Stuart City trend, south Texas, depth exceeds 13,000 feet.

Two estimates are possible of the conditions or timing of oil migration (Fig. 6). One involves the use of geochemical or isotopic paleothermometers (Milliken *et al.*, 1981) or fluid inclusion microscopy (Burruss, 1981) to estimate the temperature and salinity of fluids that precipitated cement on pore walls immediately before and during petroleum entry. The other technique involves calculation of the burial history and thermal maturity of petroleum source rocks (Harris *et al.*, 1980; Yukler and Kokesh, in press). Combining these two independent methods gives estimates of the time, depth, and temperature at which carbonate cementation stopped in a particular reservoir (McLimans, 1981).

Below the reservoir zone, however, cementation may continue unabated. Figure 7 illustrates the diagenetic variability found in a North Sea chalk reservoir, with samples from the oil-wet zone, the oil-water contact, and the water-wet zone. Clearly, oil in the pores preserved the chalk from further diagenesis. The abundance of skeletal calcite in the oil zone contrasts with the abundance of authigenic calcite in the water zone. The chalk within the diffuse oil-water contact appears transitional, and contains both visible skeletal elements and authigenic overgrowths. Pore pressure in all three zones is essentially identical. Thus overpressure (Scholle, 1977, cited previously under Reduced Burial Stress) alone is not responsible for porosity preservation in the chalk (Feazel *et al.*, in press). The small net lithostatic load may be sufficient to cause the generation of pore-filling cement by solution-transfer, unless aqueous fluids are prevented from reacting with the pore walls.

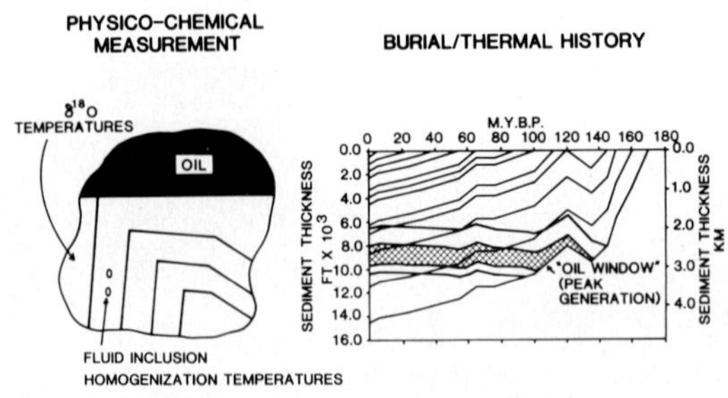

FIG. 6.—Independent techniques for determining the relative timing of hydrocarbon migration and diagenesis include: (a) estimation of paleotemperature and pore fluid composition from geochemical and isotopic signatures and fluid inclusions in cements precipitated before or during hydrocarbon migration; and (b) calculation of source rock maturity based on burial and thermal histories.

Stable Mineralogy

Sediment that starts its burial history with low diagenetic potential (Schlanger and Douglas, 1974) will be little affected by recrystallization. Syndepositional assemblages of stable minerals (low-Mg calcite or dolomite), or early neomorphic assemblages, will suffer less burial alteration than sediments with metastable components. The resistance of the rock framework to chemical alteration preserves the pores from compaction. Mineralogic stabilization is often associated with the increased framework rigidity mentioned earlier. The effects of these two mechanisms are the same; however, one is a chemical process, the other physical.

An example is the porous dolomite from the Smackover Formation shown in Figure 8. It matters little whether the dolomite formed by seepage-reflux, in the mixing zone between seawater and freshwater, or by some other mechanism: dolomitization was apparently early, and was accompanied by leaching of the intercrystalline sediment. Neither chemical alteration nor mechanical compaction altered this early intercrystalline porosity despite burial to a depth of 4.9 kilometers. Such dolomitic reservoirs might be cited throughout the Phanerozoic section (Ordovician, Williston Basin: Longman *et al.*, 1983; Silurian, Michigan Basin: Sears and Lucia, 1980; Devonian, Alberta: Mattes and Mountjoy, 1980; Permian, Permian Basin: Longacre, 1980; Triassic, Italy: Errico *et al.*, 1980; Jurassic, Florida: Ottman *et al.*, 1976; Miocene, East Malaysia: Scherer, 1980). The preservation of porosity in all these dolomite reservoirs is a consequence of simultaneous mineralogic stabilization and increased framework rigidity.

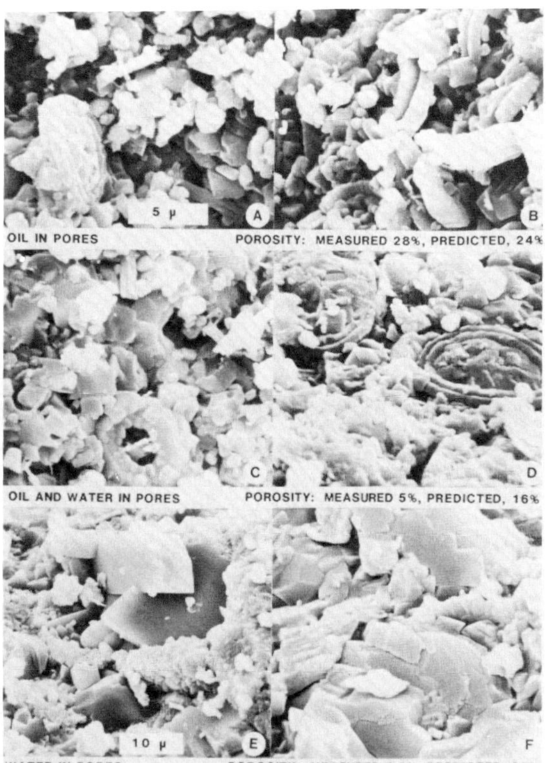

FIG. 7.—Preservation of porosity in chalk by hydrocarbon entry. Chalk from above (top), within (middle), and below base of hydrocarbons. Preservation of calcareous nannofossils suggest that primary porosity was greater than 15% in all three intervals ("predicted" values). Extensive secondary calcite in water-saturated chalk has all but obliterated any trace of skeletal particles. Pore pressures are approximately equal in all three intervals. Tor Field, North Sea, approximate depth 10,000 feet. From Feazel *et al.* (in press).

Permeability Barriers

Barriers to fluid permeability may enclose a lithologic entity, causing it to approximate a closed geochemical system. Truly closed systems, however, probably do not exist in sedimentary basins. The pores in a rock can be effectively sealed against fluid flow (physical barrier), but diffusion during geologic time will permit occlusion of even the most inaccessible pores unless some barrier to diffusion exists (geochemical barrier).

Diffusion may be self-limiting: as porosity decreases, the rate of diffusion (at constant concentration gradient) decreases by an amount equal to the porosity loss squared (Klinkenberg, 1951). Hence cemented margins often isolate porous carbonate reservoirs. (Cemented margins of Devonian reefs: McGillivray, 1977).

Most permeability barriers result from syndepositional processes. Porous carbonates may be surrounded

FIG. 8.—Intercrystalline porosity in dolomite. Replacement of original sediment by geochemically stable dolomite preserved pores from further alteration during burial. Smackover Formation, southeastern Mississippi, depth approximately 20,000 feet.

by impermeable carbonate facies (Permian algal reef with impermeable flanking beds: Mazzullo, 1982). Carbonate reservoirs may be sealed by shales (Cretaceous Stuart City shelf margin buildup: Bebout and Loucks, 1974; Pennsylvanian Horseshoe Atoll: Myers *et al.*, 1956; Miocene Philippine reefs: Longman, 1980; Miocene Indonesian reefs: Friedman, 1982). Evaporites may also encase porous carbonate bodies (Silurian reefs: Petta, 1980; Devonian reefs: Schmidt *et al.*, 1980). These permeability barriers presently seal hydrocarbon accumulations. However, prior to petroleum entry, they insured the preservation of porosity by isolation of the porous interval. The mechanism of porosity preservation is therefore distinct from that of oil entrapment: the former prevents inward migration of cement-precipitating fluids, the latter prevents the escape of hydrocarbons.

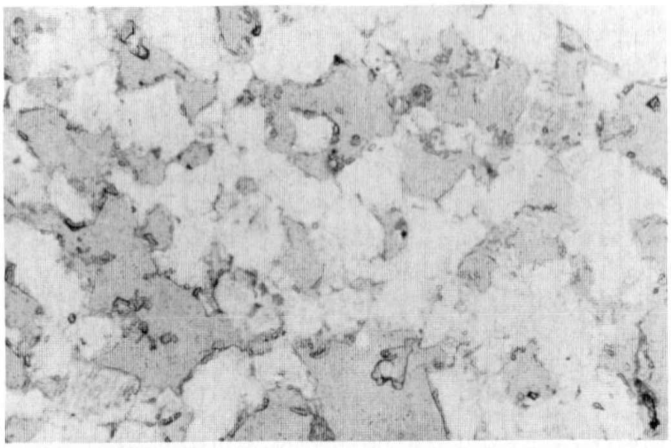

FIG. 9.—Resurrected pores in sandstone. Geometry of quartz grains (light) and epoxy-filled pores (dark) suggests the following sequence: (a) sand deposition, (b) calcite cementation and partial replacement of quartz and (c) resurrection of pore space by leaching of cement. Offshore west Africa, Cretaceous, approximately depth 7000 feet.

If such barriers effectively isolate a porous interval, little cement will precipitate. However, oil and gas may also be excluded if no migration conduit exists between the petroleum source and the potential reservoir. Such permeability barriers need to be breached at the time of petroleum entry, perhaps by fracturing (Longman, 1980). Once within the porous interval, petroleum can displace water by buoyancy. If fractures do not extend into overlying units, the petroleum will remain trapped. Skeletal buildups commonly grew from carbonate platforms. In such cases the preservation of porosity may require only that little cement be derived from the encasing shales or evaporites before oil can enter from below.

Pore Resurrection

A special category of moldic pores might be termed rejuvenated or resurrected. These are voids that formed relatively early in a rock's history, were later occluded by some cementing agent, and subsequently opened a second time by partial or complete removal of cement. Late resurrection increases the likelihood of such

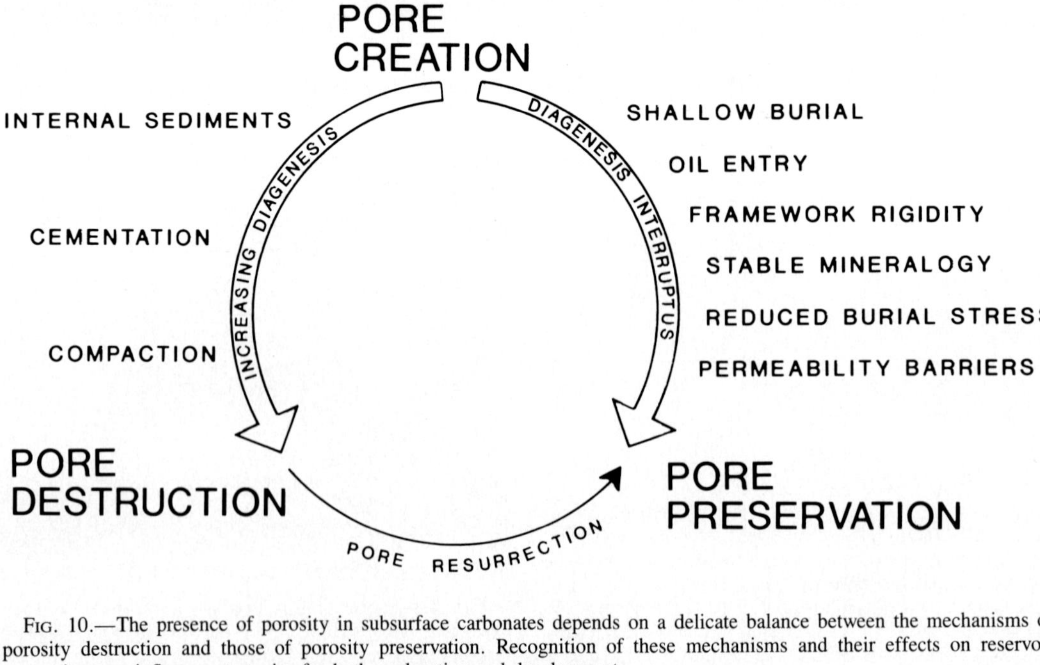

FIG. 10.—The presence of porosity in subsurface carbonates depends on a delicate balance between the mechanisms of porosity destruction and those of porosity preservation. Recognition of these mechanisms and their effects on reservoir properties may influence strategies for both exploration and development.

TABLE 2.—INFERENCES THAT CAN BE MADE ONCE PORE-PRESERVING MECHANISMS HAVE BEEN IDENTIFIED. PRESENCE OF SOME POROSITY ASSUMED.

Mechanism		Implications
MINIMAL BURIAL	Trend	Likely to follow original facies or paleotopography.
	Timing	Early pore preservation. Adjacent source rocks immature.
	Scale	Similar to scale of depositional facies.
	Effect on Reservoir	Bacterial degradation, water washing, or evaporation, probable low reservoir pressure.
REDUCED BURIAL STRESS	Trend	At major depocenters. Early preservation follows depositional patterns. May follow fracture pattern.
	Timing	Overpressure must pre-date compaction and significant cementation. When porosity was preserved compaction was minimal. Overpressure is a transient phenomenon; more common in Cretaceous and younger rocks.
	Scale	Regional.
	Effect on Reservoir	High porosity. High reservoir pressure; pressure maintenance necessary for sustained production. Flow may channel through fractures.
INCREASED FRAMEWORK RIGIDITY	Trend	a) Minor cementation-follows depositional trend. b) Dolomitization/recrystallization—may not follow depositional trend. Permeability variable.
	Timing	Determines trend. Ineffective after compaction or pervasive cementation.
	Scale	Stratigraphically restricted. Geographically restricted or widespread.
	Effect on Reservoir	Fractures and high permeability likely.
EXCLUSION OF PORE WATER	Trend	Structural, stratigraphic, or diagenetic.
	Timing	Pre-compaction; pre-cementation; a) Early: preserves primary pore network. b) Later: porosity enhancement. Secondary porosity. Fractures more important. Mature source rocks prabably deeper.
	Scale	Variable: Controlled by trend of structures or traps.
	Effect on Reservoir	High oil saturation implies good permeability (less post-oil cement). Pervasive cementation possible at base of oil column.
STABLE MINERALOGY	Trend	Depends on stabilization process: a) Initially stable mineralogy-may follow depositional trend. b) Replacement—relates to migration path of diagenetic fluid.
	Timing	a) Pre-burial; b) Early to late diagenetic.
	Scale	a) Initial—field to basin-wide. b) Replacement—cm scale to basin-wide.
	Effect on Reservoir	Depositional permeability may be preserved.
PERMEABILITY BARRIERS	Trend	a) Stratigraphic—follows depositional trend. b) Diagenetic—reflects past geochemical and hydrodynamic gradients.
	Timing	Sequence 1. Isolate porous interval. 2. Either: a. breach barrier so oil can enter, or b. generate oil within isolated interval.
	Scale	Small (diagenetic rims) to intermediate (pinnacle reefs).
	Effect on Reservoir	Fractures likely. Possible formation damage (plugging) as oil is produced and displaced by saturated brine.
PORE RESURRECTION	Trend	Diagenetic—in meteoric zone or at unconformities.
	Timing	a) Early—insignificant b) Late—more likely to be preserved.
	Scale	Dependent on original pore network.
	Effect on Reservoir	Permeability reflects original pore types.

pores' being preserved. This type of porosity is common in clastic rocks from which carbonate cements have been leached (Fig. 9). Its identification in carbonate rocks is less certain, as complete removal of a pore-filling cement is difficult to recognize.

An example was cited by Depowski and Peryt (in press), who described oil and gas reservoirs in the Zechstein dolomite (Upper Permian) of western Poland. They noted that highly productive intervals had been formed by (1) vadose leaching of the centers of ooids, oncoids, and bioclasts, (2) infilling of the pores with sulfate minerals, and (3) later dissolution of the sulfates by hydrothermal solutions.

Other examples (Moore and Druckman, 1981; Fischbuch and Havard, 1977) show that rejuvenated pores are generally solution-enhanced; they extend to some degree beyond their original borders. Such secondary porosity may be due to the migration of corrosive brines along preserved permeability-porosity networks. Therefore, it seems unlikely that resurrection of pores would occur in totally cemented, impermeable host strata.

SIGNIFICANCE TO EXPLORATION AND DEVELOPMENT

A successful exploration or exploitation program requires synthesis of stratigraphy, structure, paleogeography, sedimentation, and diagenesis. As Mazzullo (1982) observed, the best exploration approach asks why certain facies were deposited in particular locations at particular times. It may be equally important to understand the means by which pore systems are preserved following burial. Determination of pore-preserving mechanisms might explain why a pore system has a particular distribution, and represents another step in the hierarchy of models that predict the distribution of potential reservoir facies, whether depositional or diagenetic.

The following questions determine the effect that a given pore-preserving mechanism might have on exploration or development. Table 2 lists possible responses.

1. What trend is the pore system likely to follow?
   a. Is primary porosity preserved or has it been modified? Is preserved porosity facies-controlled?
   b. Is secondary porosity preserved? Is there any relationship to original facies, exposure surfaces, or geochemical gradients?
   c. Is preserved porosity related to structure, paleotopography, or unconformities?
2. What is the timing of pore preservation?
   a. When was cementation interrupted?
   b. When did compaction, leaching, permeability changes, or fracturing occur? What are their effects on the preserved pores?
   c. What is the relative timing of pore generation, hydrocarbon migration and entrapment, and pore preservation?
3. On what scale was the pore-preserving mechanism effective?
   a. Local (e.g. mineral stabilization by dolomitization of a supratidal belt)?
   b. Regional (e.g. mineral stabilization by dolomitization of a massive mid-cratonic platform)?
4. What effect might each pore-preserving mechanism have on reservoir properties?

SUMMARY

We have identified mechanisms by which cementation in carbonate rocks may be retarded or prevented. These mechanisms are responsible for the preservation of pores at reservoir depths, but are distinct from the mechanisms responsible for porosity creation and petroleum entrapment. Pores preserved by these mechanisms are recognized by petrographic, stratigraphic, geochemical, hydrodynamic, and electric log data. We have presented evidence from the rock record that such preserved pores exist, and that the mechanisms responsible for their preservation can be inferred with some certainty. An understanding of the preservation of reservoir porosity at depth has predictive utility to both exploration and development, particularly in terms of the trend, timing, and scale of the mechanisms of pore preservation, and the effect of each on reservoir properties.

REFERENCES

BEBOUT, D.G., AND LOUCKS, R.G., 1974, Stuart City Trend Lower Cretaceous—A carbonate shelf margin model for hydrocarbon exploration: Univ. Texas-Austin, Bur. Econ. Geol. Rept. of Invest. 78, 80 p.

______, BUDD, D.A., AND SCHATZINGER, R.A., 1981, Depositional and diagenetic history of the Sligo and Hosston Formations (Lower Cretaceous) in South Texas: Univ. Texas-Austin Bur. Econ. Geol. Rep. of Invest. 109, 70 p.

BRICKER, O.P. (ed.), 1971, Carbonate Cements: The Johns Hopkins Univ. Studies in Geology, v. 19, 376 p.

BURRUSS, R.C., 1981, Hydrocarbon fluid inclusions in studies of sedimentary diagenesis, *in* L.S. Hollister and M.L. Crawford (eds.), Short Course in Fluid Inclusions: Applications to Petrology: Mineral. Assoc. Canada Handbook 6, p. 138–156.

CHOQUETTE, P.W., AND PRAY, L.C., 1970, Geological nomenclature and classification of porosity in sedimentary carbonates: Amer. Assoc. Petroleum Geologists Bull., v. 54, p. 207–250.

DAVIES, G.R., 1979, Marine to late post-burial diagenesis of carbonate rocks, Arctic Archipelago: Oral presentation at AAPG Research Conf. on Deep Burial Diagenesis in Limestone, Tulsa.

DEPOWSKI, S., AND PERYT, T.M., in press, Carbonate petroleum reservoirs in the Upper Permian Zechstein main dolomite (CA-1), Fore-Sudetic area, western Poland, *in* P.O. Roehl and P.W. Choquette (eds.), Carbonate Petroleum Reservoirs: a Casebook: Springer-Verlag.

Durney, D.W., 1972, Solution-transfer, an important geological deformation mechanism: Nature, v. 235, p. 315–317.
Epstein, S.A., and Friedman, G.M., 1983, Depositional and diagenetic relationships between Gulf of Elat (Aqaba) and Mesozoic of United States East Coast offshore; Amer. Assoc. Petroleum Geologists Bull., v. 67, p. 953–962.
Errico, G., Groppi, G., Savelli, S., and Vaghl, G.C., 1980, Malossa Field: A deep discovery in the Po Valley, Italy, *in* M.T. Halbouty (ed.), Giant Oil and Gas Fields of the Decade: 1968-1978: Amer. Assoc. Petroleum Geologists, Mem. 30, p. 525–538.
Feazel, C.T., Keany, J., and Peterson, R.M., in press, Cretaceous and Tertiary chalk of the central North Sea: A major petroleum reservoir, in P.O. Roehl and P.W. Choquette (eds.), Carbonate Petroleum Reservoirs: A Casebook: Springer-Verlag.
Fischbuch, N.R., and Havard, C.J., 1977, Facies, porosity relationships, and diagenetic processes which modify them, Swan Hills area, Alberta: Canadian Soc. Petrol. Geol. 5th Ann. Core Conf. Proc., Calgary, p. 41–65.
Friedman, G.M., 1975, The making and unmaking of limestones or the downs and ups of porosity: Jour. Sed. Petrology, v. 45, p. 379–398.
______, 1982, Reefs and porosity: Examples from the Indonesian Archipelago: Southeast Asia Petroleum Exploration Soc. Proc.
______, and Sanders, J., 1978, Principles of Sedimentology: New York, John Wiley and Sons, 147 p.
Füchtbauer, H., and Goldschmidt, H., 1964, Aragonitische Lumachelle im bisuminosen Wealden des Emslandes: Beitr. Mineral, Petrog., v. 10, p. 184–197.
Halley, R.B., and Harris, P.M., 1979, Fresh-water cementation of a 1,000-year-old oolite: Jour. Sed. Petrology, v. 49, p. 969–988.
______, and Schmoker, J.W., 1983, High-porosity Cenozoic carbonate rocks of South Florida: Progressive loss of porosity with depth: Amer. Assoc. Petroleum Geologists Bull., v. 67, p. 191–200.
Hancock, J.M., and Scholle, P.A., 1975, The petrology of chalk: Geol. Assoc. London Proc. 86, p. 499–535.
Harris, P.M., Frost, S.H., Kendall, C.G., Shaffer, B.C., and Schneidermann, N., 1980, Cretaceous sea level and stratigraphy, Eastern Arabian Peninsula [abs.]: Amer. Assoc. Petroleum Geologists Book of Abstracts, p. 64.
Holland, M.T., 1982, Reservoir property implications of pore geometry modification accompanying sand diagenesis: Anahuac Formation, Louisiana: 57th Ann. SPE of AIME Fall Conf. (New Orleans), preprint no. SPE-10991, 12 p.
Kaldi, J., and Gidman, J., 1982, Early diagenetic dolomite cements: Examples from the Permian Lower Magnesian Limestone of England and the Pleistocene carbonates of the Bahamas: Jour. Sed. Petrology, v. 52, p. 1073–1085.
Kerr, R.S., 1977, Facies, diagenesis, and porosity development in a Lower Cretaceous bank complex, Edwards Limestone, North-Central Texas, *in* D.G. Bebout and R.G. Loucks (eds.), Cretaceous Carbonates of Texas and Mexico, Applications to Subsurface Exploration: Univ. Texas Bur. Econ. Geol. Rept of Invest. 89, p. 216–233.
Klinkenberg, L.J., 1951, Analogy between diffusion and electrical conductivity in porous rocks: Geol. Soc. America Bull., v. 62, p. 559–564.
Land, L.S.,1973, Contemporaneous dolomitization of middle Pleistocene reefs by meteoric water, North Jamaica, in Coral Reef Project: Bull. Marine Science, v. 23, p. 64–92.
______, Mackenzie, F.T., and Gould, S.J., 1967, Pleistocene history of Bermuda: Geol. Soc. America Bull., v. 78, p. 993–1006.
Longacre, S.A., 1980, Dolomite reservoirs from Permian biomicrites, *in* R.B. Halley and R.G. Loucks (eds.), Carbonate Reservoir Rocks: Notes for Soc. Econ. Paleontologists and Mineralogists Core Workshop No. 1, p. 105–117.
Longman, M.W., 1980, Carbonate petrology of the Nido B-3A core, offshore Palawan, Philippines, in R.B. Halley and R.G. Loucks (eds.), Carbonate Reservoir Rocks: Notes for Soc. Econ. Paleontologists and Mineralogists Core Workshop No. 1, p. 161–183.
______, Fertal, T.G., and Glennie, J.S., 1983, Origin and geometry of Red River dolomite reservoirs, western Williston Basin: Amer. Assoc. Petroleum Geologists Bull., v. 67, p. 744–771.
Mattes, B.W., and Mountjoy, E.W., 1980, Burial dolomitization of the Upper Devonian Miette buildup, Jasper National Park, Alberta, *in* D.H. Zenger, J.B. Dunham, and R.L. Ethington (eds.), Concepts and Models of Dolomitization: Soc. Econ. Paleontologists and Mineralogists Spec. Pub. 28, p. 259–297.
Mazzullo, S.J., 1982, Types and controls of Permo-Pennsylvanian carbonate stratigraphic traps of shallow-marine origin in Permian Basin: exploration models: Oil and Gas Jour. v. 80, p. 124–141.
McGillivray, J.G., 1977, Golden Spike D3A Pool, *in* I.A. McIlreath (ed.), The Geology of Selected Carbonate Oil, Gas, and Lead-Zinc Reservoirs in Western Canada: 5th Canadian Petrol. Geol. Core Conf. Proc., p. 67–88.
McLimans, R.K., 1981, Application of fluid inclusion studies to reservoir diagenesis and petroleum migration: Smackover Formation, U. S. Gulf Coast, and Fateh Field, Dubai [abs.]: Amer. Assoc. Petroleum Geologists Bulletin, v. 65, p. 957.
Milliken, K.L., Land, L.S., and Loucks, R.G., 1981, History of burial diagenesis determined from isotopic geochemistry, Frio Formation, Brazoria County, Texas: Amer. Assoc. Petroleum Geologists Bull., v. 65, p. 1397–1413.
Moore, C.H., and Druckman, Y., 1981, Burial diagenesis and porosity evolution, Upper Jurassic Smackover, Arkansas and Louisiana: Amer. Assoc. Petroleum Geologists Bull., v. 65, 597–628.
Myers, D.A., Stafford, P.T., and Burnside, R.J., 1956, Geology of the late Paleozoic Horseshoe Atoll in West Texas: Univ. Texas-Austin, Bur. Econ. Geol. Publ. 5607, 113 p.
Ottman, R.D., Keyes, P.L., and Ziegler, M.A., 1976, Jay Field, Florida-A Jurassic stratigraphic trap, in S.J. Mazzullo (ed.), Stratigraphic Traps in Carbonate Rocks: Amer. Assoc. Petroleum/Geologists Reprint Series, No. 23, p. 115–125.
Petta, T.J., 1980, Silurian pinnacle reef diagenesis—Northern Michigan: effects of evaporites on pore space distribution, *in* R.B. Halley and R.G. Loucks (eds.), Carbonate Reservoir-Rocks: Notes for Soc. Econ. Paleontologists and Mineralogists Core Workshop No. 1, p. 32–42.

PREZBINDOWSKI, D.R., 1981, Carbonate rock-water diagenesis, Lower Cretaceous Stuart City Trend, South Texas: Ph.D. dissertation, Univ. of Texas, Austin, 236 p.

SCHERER, F.C., 1980, Exploration in East Malaysia over the past decade: in M.T. Halbouty (ed.), Giant Oil and Gas Fields of the Decade: 1968-1978: Amer. Assoc. Petroleum Geologists, Mem. 30, p. 423–440.

SCHLANGER, S.O., AND DOUGLAS, R.G., 1974, The pelagic ooze-chalk-limestone transition and its implication for marine stratigraphy: in K.J. Hsü and H.C. Jenkyns (eds.), Pelagic Sediments; On Land and Under the Sea: Internat. Assoc. Sedimentologists Spec. Pub. 1, p. 177–210.

SCHMIDT, V., MCDONALD, D.A., AND MCILREATH, I.A., 1980, Growth and diagenesis of Middle Devonian Keg River cementation reefs, Rainbow Field, Alberta, in R.B. Halley and R.G. Loucks (eds.), Carbonate Reservoir Rocks: Notes for Soc. Econ. Paleontologists and Mineralogists Core Workshop No. 1, p. 43–63.

SCHOLLE, P.A., 1974, Diagenesis of Upper Cretaceous chalks from England, Northern Ireland, and the North Sea: in K.J. Hsü and H.C. Jenkyns (eds.), Pelagic Sediments; On Land and Under the Sea: Internat. Assoc. Sedimentologists Spec. Pub. No. 1, p. 177–210.

______, 1975, Applications of chalk-diagenetic studies to petroleum—exploration problems (abs.): Amer. Assoc. Petroleum Geologists Bulletin, v. 59, p. 2197–2198.

______, 1977, Chalk diagenesis and its relation to petroleum exploration: oil from chalks, a modern miracle: Amer. Assoc. Petroleum Geologists Bull. v. 61, p. 982–1009.

______, ARTHUR, M.A., AND EKDALE, A.A., 1983, Pelagic environment, *in* P.A. Scholle, D.G. Bebout, and C.H. Moore (eds.), Carbonate Depositional Environments: Amer. Assoc. Petroleum Geologists, Memoir 33, p. 620–691.

SEARS, S.O., AND LUCIA, F.J., 1980, Dolomitization of Northern Michigan Niagara Reefs by brine refluxion and freshwater/seawater mixing, *in* D.H. Zenger, J.B. Dunham and R.L. Ethington (eds.), Concepts and Models of Dolomitization, Soc. Econ. Paleontologists and Mineralogists, Spec. Pub. 28, p. 215–236.

SHINN, E.A., HALLEY, R.B., HUDSON, J.H., AND LIDZ, B.H., 1977, Limestone compaction: An enigma: Geology, v. 5, p. 21–24.

SIBLEY, D.F., 1982, The origin of common dolomite fabrics: Clues from the Pliocene: Jour. Sed. Petrology, v. 52, p. 1087–1100.

SQUIRES, R.L., 1972, Paleoecology, biostratigraphy, and faunal chemistry of a Pennsylvanian age asphaltic limestone, Arbuckle Mountains, Oklahoma: Geol. Soc. America, Abstracts with Programs, v. 4, p. 675.

SUESS, E., 1970, Interaction of organic compounds with calcium carbonate—1. Association phenomena and geochemical implications: Geochim. et Cosmochim. Acta, v. 34, p. 157–168.

VAN DEN BARK, E. AND THOMAS, O.D., 1981, Ekofisk: first of the giant oil fields in western Europe: Amer. Assoc. Petroleum Geologists Bull. v. 65, p. 2341–2363.

YUKLER, M. A., AND KOKESH, F. C., in press, A review of models used in petroleum resource estimation and organic geochemistry: Advances in Petroleum Geochemistry, v. 1, Academic Press.

SECTION TWO: INTRODUCTION

This segment includes examples of early stable and metastable marine cements from a diverse group of environments and geologic ages. Macintyre tackles head-on the highly controversial and often debated question of peloidal textures common to marine magnesium calcite cements. Several factors lead him to conclude that peloidal textures are a result of precipitation rather than of algal, sediment or replacement origin. Friedman discusses the effect on carbonate classification schemes and "energy level" interpretations created by the similar appearance of cryptocrystalline magnesium calcite cements to lithified, mechanically deposited lime muds. Examples of such cements, in an overall framework of reef cementation are given by Lighty. Forereef facies of his early Holocene example from South Florida contain well-cemented layers, whereas reef framework deposits remained poorly cemented until the reef was stranded by continued rise in sea level and the upper portions were transformed into a pavement-like dense crust.

Pleistocene carbonates are usually chosen to study early meteoric diagenetic effects. In an unusual departure, Pierson and Shinn describe a well-preserved sequence of metastable carbonate cements including micritic, bladed and blocky magnesium calcite from Pleistocene reef atoll limestones in the Bahamas. A similarly unusual marine cement of an originally equant spar morphology is interpreted from Jurassic limestones in Wyoming by Wilkinson, Smith and Lohmann. Their predominant marine cement is similar to modern meteoric phreatic low-magnesium calcites, suggesting that generalizations relating cements to diagenetic environments in modern systems may not always be valid when applied to ancient limestones. An extensive history of carbonate cements is well documented by Walls and Burrowes who describe an entire diagenetic sequence in Devonian reefs of Western Canada and discuss the effects of various cement generations on the hydrocarbon productivity of those reefs. The reef margins in their examples are extensively cemented, whereas the reef interiors are characterized by interbedded porous and well-cemented units.

N.S., P.M.H.

SUBMARINE CEMENTS—THE PELOIDAL QUESTION

IAN G. MACINTYRE
Department of Paleobiology, National Museum of Natural History, Smithsonian Institution, Washington, D.C. 20560

ABSTRACT

The origin of peloidal textures common to submarine substrates lithified by magnesium calcite has been a subject of controversy. A review of the proposed origins of this intriguing and diagnostic feature of magnesium calcite submarine cements reveals four suggested sources for these peloids—algal, sediment, replacement and precipitate origins.

A reconsideration of both the petrographic and chemical characteristics of these peloidal textures indicates that they probably originate primarily in chemical processes that may involve repeated nucleation around centers of growth. Several factors support this hypothesis: the peloids have a generally limited size range; their shape is typically spherical; the best-developed peloidal textures occur in restricted microcavities; peloidal textures exhibit well-developed zonation in many cement crusts; peloids and associated magnesium calcite rim cements have a similar chemical composition; and textural and depositional fabric characteristics are similar in magnesium calcite peloids and in precipitated spheroidal aggregates of pyrite. The pelletizing action of cryptic organisms and calcification of algal filaments also play a role—although a somewhat more limited one—in the formation of a peloidal texture in these magnesium calcite cements.

With the recognition that some calcite peloidal textures can be formed chemically, caution is urged in interpreting the origin of "pelleted" limestones.

INTRODUCTION

A dominant—and intriguing—characteristic of magnesium calcite submarine cements is their peloidal texture. The peloids (20–60 μm) consist of a well-developed euhedral rim of sparry, dentate, microcrystalline (4–30 μm) magnesium calcite around a tightly interlocking submicrocrystalline (< 4 μm) nucleus (Fig. 1). The term peloid was originally introduced (McKee and Gutschick, 1969) to describe "ovoid particles of aphanitic limestone . . . believed to have been largely derived from rocks currently being formed nearby" (p. 101). Now, however, it is taken to mean aggregates of micritic limestone of unknown origin (Bates and Jackson, 1980). It is not to be confused with the term pellet, which is generally applied to pellets of fecal origin, as indicated by a characteristic shape or "an internal texture of unoriented silt and clay-size grains" (Milliman, 1974, p. 37).

The purpose of this paper is to review the various proposed origins of peloids and to examine several lines of evidence suggesting that peloidal textures form during the precipitation of magnesium calcite from seawater.

PREVIOUS STUDIES

The controversy over the origin of magnesium calcite peloids associated with submarine lithification dates back to the discovery of submarine cements in

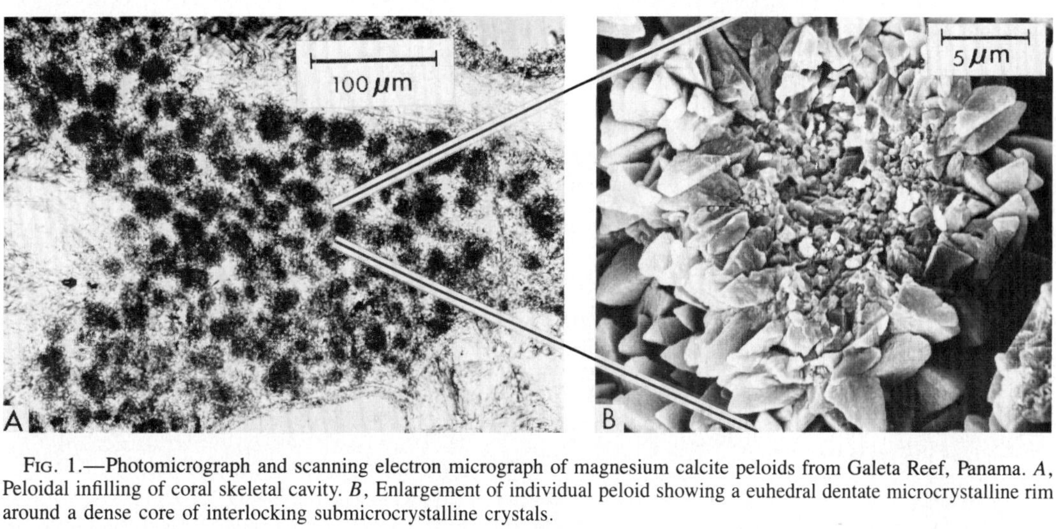

FIG. 1.—Photomicrograph and scanning electron micrograph of magnesium calcite peloids from Galeta Reef, Panama. *A*, Peloidal infilling of coral skeletal cavity. *B*, Enlargement of individual peloid showing a euhedral dentate microcrystalline rim around a dense core of interlocking submicrocrystalline crystals.

the late 1960s. Several explanations of their origins have been presented in the literature:

Pellets Formed by Organisms

One of the earliest reports of contemporary submarine lithification of reef framework (Macintyre *et al.* 1968) noted that magnesium calcite peloids may form either a matrix between sediment grains or a lining in burrows. This occurrence of lithified reef substrate off Barbados, W.I., was considered a local phenomenon associated with the burrowing activity of a pelecypod that lines its burrows with fecal pellets, some of which wash into the surrounding sediment. Subsequent reports (Land and Goreau, 1970; Land, 1971) of similar peloidal textures in the widespread reef cements off Jamaica led Macintyre *et al.* (1971) to revise their earlier hypothesis and to recognize that the submarine cements from Barbados are not confined to areas around burrows. At the same time, they suggested that the well-indurated burrows are indirectly related to the cementation in that they provide pathways for the movement of the waters responsible for the deposition of the cement.

Initially, the submarine cements off Jamaica were said to include "pelletized micrite" (Land, 1971, p. 59), but in later studies the peloidal magnesium calcites from this area were described as pellets "formed by cryptic filter-feeders," including the clionid sponges (Land and Moore, 1980, p. 363), which were thought to pelletize clay-size magnesium calcite that has precipitated and settled within the reef framework.

Replacement of Aragonite Pellets

Another early explanation offered is that peloidal calcites can be formed in tidal flat sediments by the replacement of aragonite fecal pellets or precipitated "clots," as suggested by Taylor and Illing (1969) in their study of intertidal cementation in the Persian Gulf. Shinn's (1969) investigation of subtidal lithification in the same general area supported the idea that geopetal peloidal muds are formed by the replacement of aragonite fecal pellets by magnesium calcite, and he cited textural observations indicating that "pelleted carbonate mud is modified to a more even-textured 5–10 μ 'micro-mosaic'" (p. 133). Shinn further suggested that as recrystallization continues, insoluble organic material is pushed inward, and both the size and definition of the pellets are reduced.

Intermediate Stage or Advancing Front

In a study of intragranular submarine cements in sediment grains from a number of locations in the Caribbean and Mediterranean seas, Alexandersson (1972a) reported 30–60 μm "pellet-like bodies" of magnesium calcite (p. 450). Similar textures in magnesium calcite in beachrock from the Mediterranean were described as "pellet-like globules" (Alexandersson 1972b, p. 215). In both cases, Alexandersson suggested that the peloids appear to be an intermediate stage of development of an advancing front of microcrystalline magnesium calcite cement. That is, after nucleation, the peloids increase in size at the leading edge of a magnesium calcite depositional surface and eventually coalesce to form a "uniform fabric" (Alexandersson 1972a, p. 450).

Calcified Algal Filaments

Still another origin was proposed when Schroeder (1972) noted that cross sections through algal filaments coated with magnesium calcite resemble pelleted magnesium calcite cement. In examining calcified filaments of *Ostreobium* (30–70 μm in diameter) exposed in reef cavities off Bermuda, Schroeder found microcrystalline bladed crystals growing outward from the filament surfaces. The inner surfaces of these filaments were lined with finer submicrocrystalline bladed crystals extending inward, but not always completely filling the filaments. Schroeder noted that in cross sections these calcified filaments appear as double-bladed peloids with euhedral crystals radiating from a well-defined core, in which finer crystals project toward the center of the peloids. He therefore argued that "continuous growth and calcification of these filaments and their periodical fragmentation [could] result in the layered pellet fillings observed in the cavities of these reef rocks" (p. 29). In a subsequent study of reef cements from the same area, Ginsburg and Schroeder (1973) suggested that the magnesium calcite peloids can be attributed both to fecal pellet and calcified algal filament origins. "Pellet-like" structures in magnesium calcite cements from Red Sea reefs have also been associated with calcified algal filaments (Friedman *et al*, 1974).

Detrital Grains Unrelated to Submarine Cement

The hypothesis that peloidal calcites are composed of detrital grains was first proposed by James *et al.* (1976) in their study of submarine cements from the shallow-water reef framework off Belize. James *et al.* noted that "most of the lime silt grains are peloids of unknown origin, discrete ovoid to spherical grains 40–50 μ in diameter composed of Mg-calcite" (p. 535). In emphasizing the detrital origin of the silt-size peloids, they suggested that "wave pumping has flushed suspended, silt-size particles into the interstices" (p. 535). This peloidal magnesium calcite, they argued, is not part of the cement; rather, it promotes the precipitation of "Mg-calcite micrite," which favors infilling "pores less than 15 μ in diameter between lime silt-size grains" (p. 543). Later, James and Ginsburg (1979) examined submarine cements from the deep fore reef off Belize and in their petrographic studies of some of the limestones described "peloids of micrite-sized Mg-calcite, ranging in size from 25 to 70 μm" (p. 127) as silt-size sediments of unknown origin.

Repeated Nucleation

Peloids have also been said to represent an early stage of precipitation in the development of magnesium calcite cements (Macintyre, 1977). That is, "the coarser dentate rims represent a secondary stage of slower and more organized precipitation of coarser, subhedral magnesium calcite" around the previously precipitated cores of a submicrocrystalline "dense mosaic of interlocking anhedral crystals" (p. 508). The somewhat restricted size of peloids was attributed to "repeated nucleation" around centers of growth, and it was suggested that growth ceases when a critical peloidal size range is reached.

Macintyre (1977) further noted that cement crusts and infilled rims of coral skeletons from Galeta Reef, Panama, exhibit a distinct zonation: "the sediment-rich area of the outer edge grades into a peloidal and dense submicrocrystalline to microcrystalline calcite infill which is relatively free of both detrital carbonate and non-carbonate sediment" (p. 508). Both the Galeta Reef study and an examination of extensive submarine lithification of aggregates of serpulid worm tubes in a submarine cave off Belize (Macintyre, 1984) reveal that the best-defined peloid textures occur in restricted or isolated microcavities. Macintyre's (1977, and 1984) contention—which is supported by Marshall and Davies' (1981) study of submarine lithification in the Great Barrier Reef off Australia—is that the peloids and their coarser crystalline rim cements are all part of the lithification process responsible for the precipitation of magnesium calcite.

Coralline Algae

In studying coral-reef diagenesis of Red Sea reefs, Gvirtzman and Friedman (1977) illustrated peloidal magnesium calcite (approximately 50 μm) filling coral skeletal cavities. The similarity between these cavity fillings and peloidal submarine cements from other areas is clear from the textural patterns and the close association of the infilling with "micrite envelopes" and "finely crystalline high-magnesium calcite linings" in Gvirtzman and Friedman's line drawings and a photomicrograph (pp. 363, 365, 375). These authors concluded that the calcite fillings are "coralline algae [that] commonly inhabit interskeletal pores of corals" (p. 375).

DISCUSSION

Magnesium calcite peloids are undoubtedly formed in many ways during submarine lithification. The problem for carbonate sedimentologists is to determine the principal origin of the peloidal textures that are characteristic of magnesium calcite submarine cements. The preceding literature review indicates that these peloids have been attributed to four basic sources. It is useful to consider each of these hypotheses in relation to the extensive development of peloidal textures.

Algal Origins

Both calcified algal filaments and crustose coralline algae have distinctive growth habits and microstructures that are readily recognizable under high magnification. Crustose coralline algae, for example, encrust surfaces, are limited in their ability to fill restricted cavities, and exhibit characteristic growth patterns (Adey and Macintyre, 1973). Crustose coralline algal calcites can be identified on the basis of these features, and generally bear little textural resemblance to peloidal calcites. Cross sections of calcified algal filaments, meanwhile, exhibit a characteristic double-

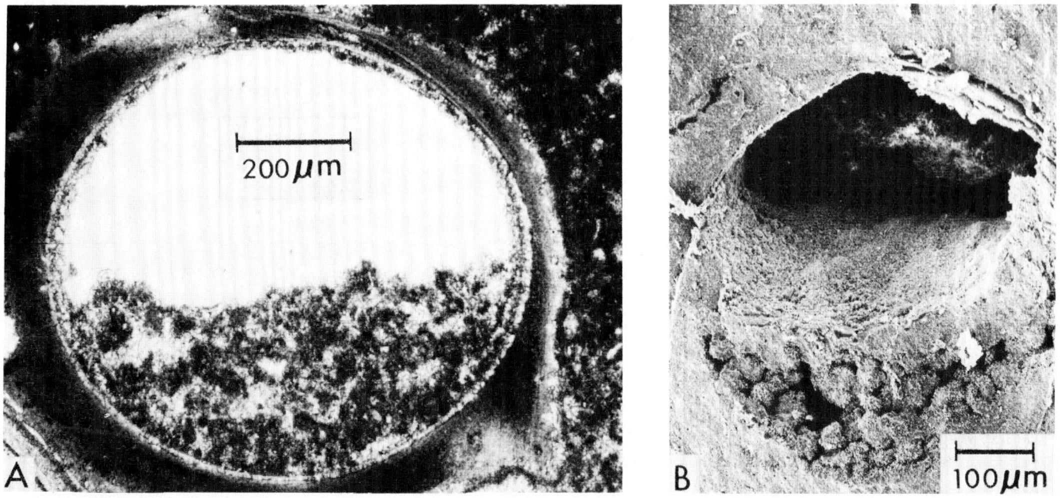

FIG. 2.—Photomicrograph and scanning electron micrograph of peloidal infillings exhibiting a geopetal fabric in serpulid tubes, Columbus Cay Cave, Belize.

bladed crystal arrangement in scanning electron photomicrographs. Although this crystal arrangement may be difficult to discern in thin sections, the calcareous filaments should be recognizable because not all of them will be perpendicular to the section. Another point to consider is that submarine peloidal calcites appear to be generally associated with light-limited, restricted cryptic habitats such as the large submarine cave off Belize (Macintyre, 1984), and thus photosynthesizing algae cannot contribute significantly to the direct formation of peloidal textures.

Sediment Origins

A problem in interpreting peloidal textures is that no standard definitions have been accepted for the terms *precipitate* and *sediment*. James *et al.* (1976), who apparently considered sediment to be any material that settles out of the water column, did not associate peloids that commonly exhibit a geopetal fabric (Fig. 2) with cements. They suggested that peloids are silt-size grains of unknown origin.

Geochemical data indicate, however, that "micritic sediment" including peloids have chemical characteristics similar to crystalline cements. Microprobe analyses have yielded mole percent $MgCO_3$ values of 16.8, 17.9, 16.2 (James *et al.*, 1976), and 16.3 ± 0.4 (Macintyre, 1984) for peloids from Belize, and 17.4 ± 1.1, with a maximum value of 18.4 ± 0.4 (Macintyre, 1977) for peloids from Panama. The sediment-rich "micrites", on the other hand, tend to have lower values of 14.4, 16.8, 15.4 (James *et al.*, 1976) and 14.5 ± 0.7 (Macintyre, 1984) for Belize and 16.6 ± 1.2 (Macintyre, 1977) for Panama. By comparison, mole percent $MgCO_3$ values for the "scalenohedral" sparry rim cements are 18.5 ± 1.0 (Jamaica—Land and Goreau, 1970), 17.0 (Barbados, Jamaica and Cuba—Alexandersson, 1972a) 17.4, 18.6, 18.4 (Belize—James *et al.*, 1976) and 15.2 ± 0.5 (Belize—Macintyre, 1984).

As noted by Land and Moore (1980) this similarity in the chemical composition of micritic and crystalline magnesium calcites indicates that these calcites, within a given depositional site, must have a common source. The fact that peloids settle at the base of cavities does not mean they cannot be considered chemical precipitates (see discussion of pyrite framboids, below). Peloids are merely another form of precipitated magnesium calcite and are not detrital silt-size sediment, as James *et al.* (1976) proposed. If these peloids are sorted detrital grains, then where did the larger and finer peloids settle?

Some "calcite dust" precipitating within a reef framework can, of course, be pelletized by cryptic organisms, as Land and Moore (1980) have suggested. The dominant peloidal textures, however, have two prominent characteristics that are not typical of such biologically formed pellets: their basic shape is spherical rather than ovoid; and detrital debris is absent in their submicrocrystalline cores. The lack of debris within peloids is indicated by the higher mole percent $MgCO_3$ (Macintyre, 1977; Macintyre, 1984) and very low Sr values in microprobe analyses in comparison with sediment-rich calcites. Another significant factor is that the best-developed peloidal magnesium calcites are generally found in very restricted settings (generally, within small skeletal cavities) well away from detrital debris and beyond the range at which the most ambitious cryptic organisms can survive. What organisms might be responsible for the widespread pelletizing of internal reef sediment has not been established. Land and Moore's (1980) suggestion that clinoid sponges are pelletizers is questionable since these gutless organisms have no anatomical characteristics that would enable them to form pellets from clay-size sediment (Per. commn. W.D. Hartman, K. Rützler, 1983).

Replacement Origins

One of the earliest explanations of peloidal textures is that they are formed by the replacement of aragonite clots or pellets (Taylor and Illing, 1969; Shinn, 1969). With the increasing use of electron microprobe analyses and scanning electron microscopy, however, this concept lost hold. For example, Ginsburg *et al.* (1971), noted that the lack of detailed petrographic and geochemical analyses cast doubt on this hypothesis. No further documentation has come to light to support the replacement origin.

Precipitate Origin

Two hypotheses—the "advancing front" and "repeated nucleation" theories—have related peloidal textures to the processes involved in the chemical precipitation of some submarine magnesium calcites. On one hand, Alexandersson (1978) has contended that peloid textures represent "growth structures at the front of advancine Mg-calcite" forming a "featureless cement" (p. 667). This view is not supported by recent observations that the best-developed peloidal textures generally occur in isolated areas (Macintyre, 1977, 1978, 1984) or in a zone marking the inner limit of cement penetration of coral skeletons (Macintyre, 1977). In other words, these are areas in which lithification has been interrupted or inhibited and are not the accreting cement surface, which tends to consist of sediment-rich submicrocrystalline magnesium calcite. This observation was confirmed experimentally when conch shells were attached to the ceiling of a submarine cave and the magnesium calcite collecting on the shell fragments was found to consist of unindurated submicrocrystalline calcite "dust" (Fig. 3) that exhibited no peloidal texture and acted as a matrix supporting sediment grains (Macintyre, 1984). This "dust" appears to be an initial stage of uninhibited magnesium calcite deposition. It should be noted, however, that Pedley (1979) has suggested that a undefined process of "post-depositional aggregation of structureless micrite" (p. 588) could form a peloidal texture in this type of featureless deposit.

Another factor to consider is the degree of uniformity of crystal size following the dense infilling of peloidal calcite. Peloids characteristically exhibit two distinct phases of crystal development--an initial precipitate of submicrocrystalline peloid centers, and a coarser microcrystalline dentate rim (Land and Goreau, 1970; Macintyre, 1977; Marshall and Davis, 1981). These two phases are well documented in dense peloidal magnesium calcites (Fig. 4) in which the size of peloids "does not differ significantly from peloid size of adjacent open peloidal infillings" (Macintyre, 1978, p. 669). In other words, these distinct variations in crystal size prevent the original peloidal texture from being masked, despite dense infilling by later magnesium calcite.

Factors that point to repeated nucleation as the origin of peloids are their spherical shape, general size limitation, the purity of the magnesium calcite, and the isolated sites of deposition. Probably the strongest evidence for a chemical origin, however, is the distinct zonation (Fig. 5A) that occurs in magnesium calcite encrusted rims of coral skeletons (see Macintyre, 1977). The pattern consists of an outer, dense, sediment-rich submicrocrystalline calcite crust that gives way to sediment-free peloidal calcite infillings of coral cavities, which in turn are followed by chalky or submicrosucrosic calcite fillings that commonly exhibit an incipient peloidal texture or clustering of calcite crystals (Fig. 5B). The transition between calcite textures and the upper size limit (70 to 90 μm) for peloids suggest that the peloidal texture is the result of repeated nucleation of submicrocrystalline calcite crystals that stops when crystal growth reaches a specific limit around individual centers of nucleation. The restricted flow of interstitial waters in the isolated sites where peloids are formed indicates that the formation of peloids must be a slow-growth phenomenon.

Similar petrographic characteristics are exhibited by another type of marine precipitate, framboidal pyrite (Fig. 6). In this case, the spheroidal aggregates are believed to be early diagenetic precipitates (Honjo *et al.*, 1965). Although generally smaller in diameter than magnesium calcite peloids—3–25 μm (Honjo *et al.*, 1965) to 40–50 μm (Hudson 1982)—the aggregates of pyrite are within the same general size range. Like magnesium calcite peloids, the framboids are found primarily in isolated skeletal cavities—for example, in ammonite chambers (Hudson 1982) and in radiolarian tests (Honjo *et al.*, 1965), where "80 percent of such framboids lie" (p. 481). In the latter example, the framboids characteristically form geopetal fabrics (Fig. 6A) that suggest "the pyrite was precipitated within the fossil and sank to the bottom of the sphere" (Honjo *et al.*, 1965, p. 480). This same geop-

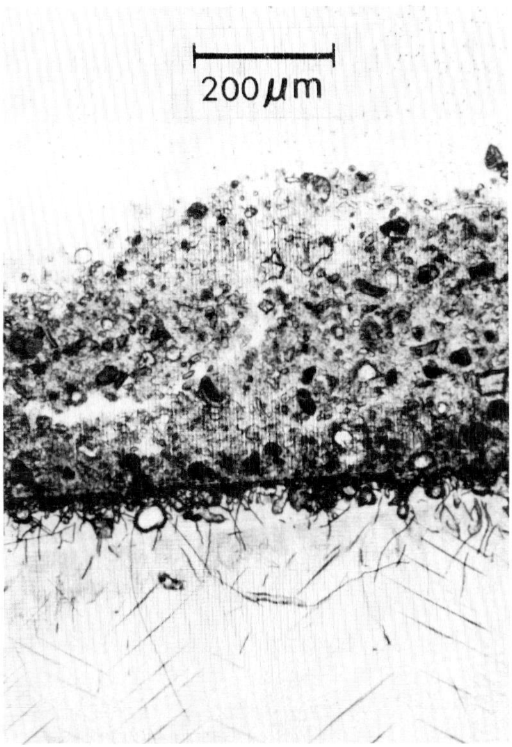

FIG. 3.—Photomicrograph of sediment-rich submicrocrystalline magnesium calcite "dust" on conch shell placed for 3 years on the ceiling of Columbus Cay Cave, Belize. Note fungal microborings on top surface on conch shell.

FIG. 4.—Scanning electron micrograph of dense peloidal filling in a serpulid tube. Note how the differences in crystal size formed in different phases of calcite deposition preserves the peloidal texture.

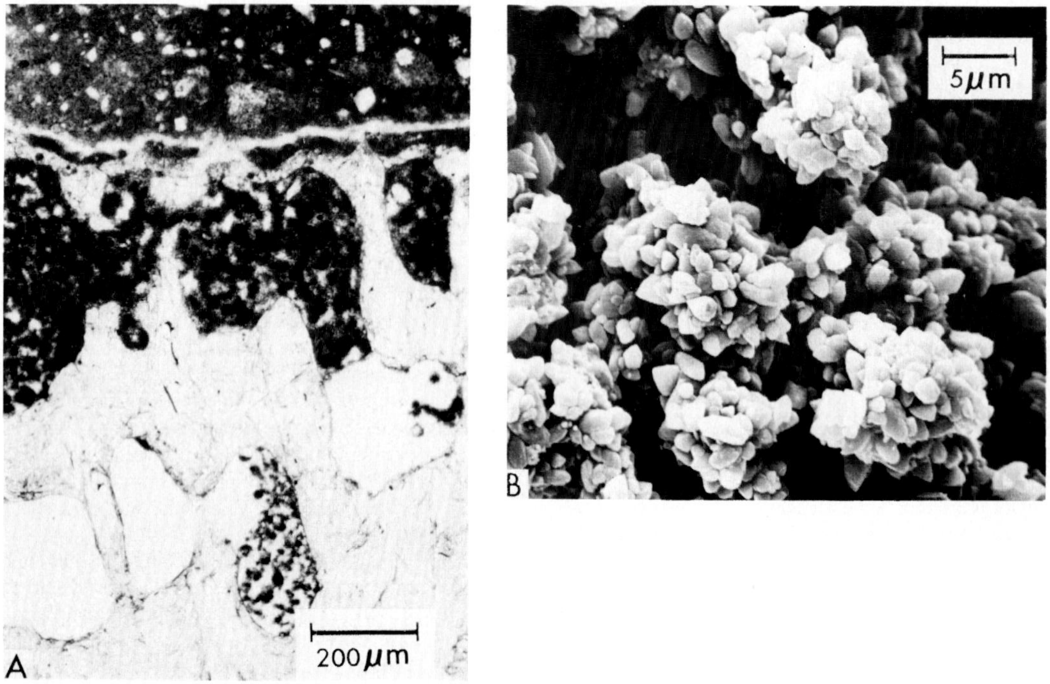

FIG. 5.—Zonation pattern in magnesium calcite cements. *A*, Photomicrograph of outer edge of coral skeleton showing dense sediment-rich crust over well-developed peloidal infilling of skeletal cavities, which is followed by a chalky infilling. This chalky deposit at the inner limit of magnesium calcite penetration commonly exhibits an incipient peloidal texture, Galeta Reef, Panama. *B*, Scanning electron micrograph of chalky or submicrosucrosic magnesium calcite in serpulid tube. Note clusters of crystals forming an incipient peloidal texture, Columbus Cay Cave, Belize.

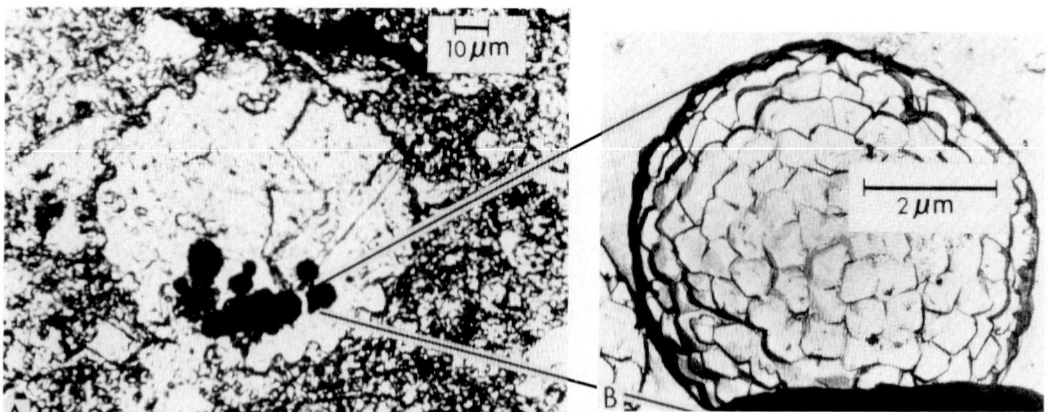

FIG. 6.—Photomicrograph and electron photographs of spheroidal aggregates (framboids) of pyrite in Jurassic limestones from Northern Austrian Alps. (Honjo *et al.*, 1965) *A*, Framboids forming a geopetal fabric within a radiolarian test. *B*, Enlargement of individual framboid showing details of crystal pattern.

FIG. 7.—Burrow with a trail of fecal pellets, Mississippian limestone, Saskatchewan, Canada—rare conclusive evidence in the fossil record of organic origin of pellets.

etal fabric was the basis for James *et al.*'s (1976) suggestion that magnesium calcite peloids are sediments unrelated to the precipitation of magnesium calcite cement.

In the long debate over the origin of framboids, several explanations have been offered for their formation: that they are related to mineralized sulfur bacteria (Schneiderhöhn, 1923; Love, 1958; Fabricius, 1961); that inorganic crystallization of pyrite occurs within an iron sulfide gel (Rust, 1935; Bastin, 1950); that precipitation of pyrite is indirectly controlled by sulfate-reducing bacteria (Vallentyne, 1963); and that framboid structures originate from greigite (Fe_3S_4) spheres formed from an original mackinawite ($FeS_{0.9}$) precipitate (Sweeney and Kaplan, 1973; Rickard, 1975). Whether or not these framboids precipitated inorganically, or in association with sulfur bacteria, their formation within restricted skeletal cavities and their characteristic geopetal fabrics lend credence to the hypothesis that magnesium calcite peloids, which have similar characteristics, are also chemical precipitates.

CONCLUSION

Chemical processes that probably involve repeated nucleation around centers of growth appear to be primarily responsible for the peloidal textures associated with submarine lithified deposits of magnesium calcite. In addition, some peloidal textures in more open cavities can be produced by the pelletizing action of organisms, or, to a lesser extent, by the calcification of algal filaments.

The general spherical shape, size restriction, and isolated sites of deposition (which is well illustrated in the cement crusts on coral colonies) all support a chemical origin for magnesium calcite peloids. The similarity between these peloids and precipitated spheroidal aggregates of pyrite—both in terms of their restricted sites of deposition and geopetal fabrics—provides further support for a chemical origin for these early diagenetic carbonate deposits.

If some peloidal calcites are formed chemically, then the origin of peloidal limestones in the fossil record should be interpreted with more caution. Although some evidence of a fecal pellet origin can be found in the shape or internal texture of peloids, conclusive evidence—such as the pellet-lined burrow from a Mississippian limestone (Fig. 7)—is rarely found in the rock record.

ACKNOWLEDGMENTS

I would like to thank K.M. Towe for his advice and helpful discussions in the preparation of this manuscript. F.W. Beales, R.N. Ginsburg, L.S. Land and E.A. Shinn kindly reviewed this manuscript but in undertaking this task do not necessarily endorse the interpretations presented in it. L.B. Isham assisted with drafting and W.T. Boykins and V.E. Krantz with photography.

REFERENCES

ADEY, W.H. AND MACINTYRE, I.G., 1973, Crustose coralline algae: A re-evaluation in the geological sciences: Geol. Soc. America Bull., v. 84, p. 833–904.

ALEXANDERSSON, E.T., 1972a. Intragranular growth of marine aragonite and Mg-calcite: Evidence of precipitation from supersaturated seawater: Jour. Sed. Petrology, v. 42, p. 441–460.

_____, 1972b, Mediterranean beachrock cementation: Marine precipitation of Mg-calcite: *in* D.J. Stanley (ed.), The Mediterranean Sea: A Natural Sedimentation laboratory: Dowden, Hutchinson and Ross Inc., Stroudsburg, Pennsylvania, p. 203–223.

_____, 1978, Discussion: Distribution of submarine cements in a modern Caribbean fringing reef, Galeta Point, Panama: Jour. Sed. Petrology, v. 48, p. 665–668.

Bastin, E.E., 1950, Interpretation of ore textures: Geol. Soc. America, Memoir 45, 101 p.

Bates, R.L. and Jackson, J.A., 1980, Glossary of Geology: American Geological Institute, Falls Church, Va., 749 p.

Fabricius, F., 1961, Die Strukturen des "Rogenpyrits" (Kössener Schichten, Rät) als Beitrag zum Problem der "Vererzten Bakterien": Geol. Rundschau, v. 51, p. 647–657.

Friedman, G.M., Amiel, A.J., and Schneidermann, N., 1974, Submarine cementation in reefs: Example from the Red Sea: Jour. Sed. Petrology, v. 44, p. 816–825.

Ginsburg, R.N., Marszalek, D.S., and Schneidermann, N., 1971, Ultrastructure of carbonate cements in a Holocene algal reef of Bermuda: Jour. Sed. Petrology, v. 41, p. 472–482.

_____, and Schroeder, J.H., 1973, Growth and submarine fossilization of algal cup reefs, Bermuda: Sedimentology, v. 20, p. 575–614.

Gvirtzman, G., and Friedman, G.M., 1977, Sequence of progressive diagenesis in coral reefs: *in* S.H. Frost, M.P. Weiss, and J.B. Saunders (eds.), Reefs and Related Carbonates—Ecology and Sedimentology: Amer. Assoc. Petroleum Geologists, Studies in Geology No. 4, p. 357–380.

Honjo, S., Fischer, A.G., and Garrison, R., 1965, Geopetal pyrite in fine-grained limestones: Jour. Sed. Petrology, v. 35, p. 480–488.

Hudson, J.D., 1982, Pyrite in ammonite-bearing shales from the Jurassic of England and Germany: Sedimentology, v. 29, p. 639–667.

James, N.P., and Ginsburg, R.N., 1979, The seaward margin of Belize barrier and atoll reefs: International Assoc. Sedimentologists, Spec. Pub. 3, 191 pp.

_____, _____, Marszalek, D.S., and Choquette, P.W., 1976, Facies and fabric specificity of early subsea cements in shallow Belize (British Honduras) reefs: Jour. Sed. Petrology, v. 46, p. 523–544.

Land, L.S., 1971, Submarine lithification of Jamaican Reefs, *in* O.P. Bricker (ed.), Carbonate Cements: The Johns Hopkins Univ. Studies in Geology, No. 19, p. 59–62.

_____, and Goreau, T.F., 1970, Submarine lithification of Jamaican reefs: Jour. Sed. Petrology, v. 40, p. 457–462.

_____, and Moore, C.H., 1980, Lithification, micritization and syndepositional diagenesis of biolithites on the Jamaican Island slope: Jour. Sed. Petrology, v. 50, p. 357–370.

Love, L.G., 1958, Micro-organisms and the presence of syngenetic pyrite: Geol. Soc. London Quart. Jour., v. 113, p. 429–440.

McKee, E.D., and Gutschick, R.C., 1969, History of Redwall Limestone of Northern Arizona: Geol. Soc. America Mem. 114, 726 p.

Macintyre, I.G., 1977, Distribution of submarine cements in a modern Caribbean fringing reef, Galeta Point, Panama: Jour. Sed. Petrology, v. 47, p. 503–516.

_____, 1978, Reply: Distribution of submarine cements in a modern Caribbean fringing reef, Galeta Point, Panama: Jour. Sed. Petrology, v. 48, p. 669–670.

_____, 1984, Extensive submarine lithification in a cave in the Belize Barrier Reef Platform: Jour. Sed. Petrology, v. 54, p. 11–25.

_____, Mountjoy, E.W., and d'Anglejan, B.F., 1968, An occurrence of submarine cementation of carbonate sediments off the west coast of Barbados, W.I.: Jour. Sed. Petrology, v. 38, p. 660–664.

_____, _____, and _____, 1971, Submarine cementation of carbonate sediments off the west coast of Barbados, W.I.: *in* O.P. Bricker, (ed.), Carbonate Cements: The Johns Hopkins Univ. Studies in Geology, no. 19, p. 91–94.

Marshall, F.F., and Davies, P.J., 1981, Submarine lithification on windward reef slopes: Capricorn-Bunker Group, Southern Great Barrier Reef: Jour. Sed. Petrology, v. 51, p. 953–960.

Milliman, J.D., 1974, Marine Carbonates: Springer-Verlag, New York, N.Y., 375 p.

Pedley, H.M., 1979, Miocene bioherms and associated structures in the Upper Coralline limestone of the Maltese Islands: their lithification and palaeoenvironment: Sedimentology, v. 26, p. 577–591.

Rickard, D.T., 1975, Kinetics and mechanism of pyrite formation at low temperatures: American Jour. Sci., v. 274, p. 636–652.

Rust, C.W., 1935, Colloidal primary copper ores at Cornwall Mines, southeastern Missouri: Jour. Geology, v. 43, p. 398–426.

Schneiderhöhn, H., 1923, Chalkographische Untersuchungen des Mansfelder Kupferschiefers: Neues Jahrb. Geol. Paläontol., v. 47, p. 1–38.

Schroeder, J.H., 1972, Calcified filaments of an endolithic alga in Recent Bermuda reefs: N. Jb. Geol. Paläont. Mh., p. 16–33.

Shinn, E.A., 1969, Submarine lithification of Holocene carbonate sediments in the Persian Gulf: Sedimentology, v. 12, p. 109–144.

Sweeney, R.E., and Kaplan, I.R., 1973, Pyrite framboid formation: Laboratory synthesis and marine sediments: Econ. Geol. v. 68, p. 618–634.

Taylor, J.C.M., and Illing, L.V., 1969, Holocene intertidal calcium carbonate cementation, Qatar, Persian Gulf: Sedimentology, v. 12, p. 69–107.

Vallentyne, J.R., 1963, Isolation of pyrite spherules from Recent sediments: Limnol. Oceanog., v. 8, p. 16–30.

THE PROBLEM OF SUBMARINE CEMENT IN CLASSIFYING REEFROCK: AN EXPERIENCE IN FRUSTRATION

GERALD M. FRIEDMAN
Department of Geology Rennsselaer Polytechnic Institute Troy, New York 12181

ABSTRACT

Classifications and "energy" interpretations of limestones are based on the ideas that, in calm waters, micron sized particles of lime mud are (1) available and (2) able to settle on the bottom and remain there, whereas in agitated waters, such tiny particles remain in suspension and are not deposited. Lithified, mechanically deposited lime mud is known as micrite, a contraction of "microcrystalline ooze" which has been inferred to represent a clay-size matrix whose presence signifies a lack of vigorous currents (Folk, 1962, p. 67). Micrite is usually equated to original matrix, hence a mechanically deposited material between particles, as distinct from precipitated cement. As thus interpreted, the loosely termed "micrite cement," used in the literature is a contradiction. Accordingly, physically deposited interparticle material *cannot be* both micrite and cement.

Submarine cryptocrystalline high-magnesium calcite cement in reefs (and beachrock) that is precipitated within millimeters to centimeters of the surfaces of reefrock looks just like micrite. Misidentification of this cement for micrite or non-critical use of the term micrite leads to error in interpretation of depositional facies. Case histories abound where unwary geologists have confused reefrock with a supposed "low-energy" lime-mud facies. Study of reefrock in thin section under an electron beam in which cryptocrystalline high-magnesium calcite cement luminesces and aragonitic particles appear in black presents a dilemma: where the cement : particle ratio is high it is easy to misidentify reefrock, even when one is taking great care.

INTRODUCTION

Since the surge in carbonate research of the 1950's, a wealth of terms has been added to the language of geology. Many of these terms are genetic because a successful geologist, especially a subsurface geologist, must be process-oriented. The search for potential reservoirs in carbonate sequences considers depositional processes and products. Primary porosity and hydrocarbon-reservoir characteristics are commonly related to the energy in the environment of deposition. In general, carbonate sediments deposited in high-energy environments are prone to develop primary porosity and through diagenetic processes these same sediments may develop secondary porosity. With these energy-related concepts in mind, geologists in the 1950's developed classifications based on energy level. As Ham and Pray stated (1962, p. 10) "the concept of energy or water turbulence of the depositional environment provides a significant interpretive parameter for classification of carbonate rocks. 'High-energy' and 'low-energy' carbonate sediments are those which have been deposited respectively in turbulent or quiet waters. The energy concept is well recognized in most modern classifications, although indirectly, by the focus on the relative abundance of grains and matrix." According to Dunham (1962, p. 111) this distinction between sediment deposited in calm water and sediment deposited in agitated water is fundamental. "Inasmuch as calm water is characterized by (lime) mud being able to settle to the bottom and remain there, it seems that the muddy rocks (that is those in which lime mud accumulated between sand-size carbonate particles) deserve to be contrasted with (lime) mud-free rocks" (Dunham, 1962, p. 112–113). The distinction thus parallels the distinction between matrix and particles in sandstone (Pettijohn, 1957, p. 287; Sanders, 1978).

THE TERM MICRITE

Folk (1959) introduced a term known as micrite in reference to matrix which characterizes low-energy carbonate deposits. Micrite is a contraction of "microcrystalline ooze" (Folk, 1962, p. 65–66). He (p. 67) noted that "microcrystalline ooze represents a clay-size matrix whose presence signifies a lack of vigorous currents, just as the presence of clay-size mineral matrix in a sandstone indicates poor washing." Matrix refers to "*mechanically deposited material between particles—as distinct from precipitated cement*" (Bathurst, 1971). Hence micrite may be defined as "lithified mechanically deposited lime mud" (Friedman and Sanders, 1978, p. 565).

Limestones have been classified on the basis of three end members: (1) sand-size and larger carbonate particles, (2) micrite or matrix, and (3) cement (± pore space). Those in which micrite or matrix occupies the space between the sand-size and/or larger particles are interpreted as of low-energy origin, whereas by contrast those in which cement (± pore space) prevails between the particles are considered to be high-energy deposits.

CRYPTOCRYSTALLINE SUBMARINE CEMENT IN REEFROCK OR HOW TO MISTAKE REEFROCK FOR LOW-ENERGY FACIES

A problem in classifying, naming, and interpreting carbonate rocks arises when micrite of matrix origin and cryptocrystalline cement are indistinguishable. I faced this problem for the first time in the early 1960's. At that time, cement was considered always as "sparry", that is according to Folk (1959) a clear crystalline component of the rock having vitreous lustre, that occupies the interstices between the sand-size particles and measuring tens or hundreds of microns in diameter. In my study of lithified carbonate sediments from the Atlantis Seamount, mid-Atlantic Ridge, I was probably the first in the modern literature to face the problem of submarine cement (Friedman, 1964, p. 806–807). Note how confused I was: I mistook this cement for micrite as indicated in the following quotes: "These carbonate cements are well-indurated and consist of lime mud (micrite) in which abundant tests of organisms are embedded" (p. 806), and then further the interpretation: "The presence of abundant shallow-water organisms suggests that these sediments have formed under shallow-water conditions." Yet, I was troubled with this interpretation of micrite occupying interparticle space in a high-energy Atlantic setting and ended up by merely pointing out how these lithified deposits differed from others in that "mineralogically the amount of high-magnesium calcite found in these sediments exceeds that of unconsolidated or lithified carbonate sediments found anywhere else in the world" (p. 806). This paper reported for the first time submarine high-magnesium calcite cement.

The potential for confusion dawned on me soon after. In reference to my experience with the rocks of Atlantis Seamount and newly studied beachrocks of the Red Sea coast, I stated (Friedman, 1968, p. 14) "the high-magnesium calcite fabric is very likely to be confused in ancient marine limestones with an original aragonite micrite fabric."

In following years I studied modern reefrock from many parts of the world, including the Caribbean, Red Sea, Great Barrier Reef of Australia, and Persian Gulf. It soon became apparent that submarine high-magnesium calcite cement is an important constituent in most reefs and that this cement is cryptocrystalline (Friedman, 1973). That is, under the mocroscope, the crystals of this cement are too small to be recognized and separately distinguished and hence this cement may look just like micrite (Friedman and Sanders, 1978, p. 175). Such cryptocrystalline cement in a high-energy reefrock may be easily confused for low-energy lime-mud facies. A cryptocrystalline submarine cement is precipitated within millimeters to centimeters of the surfaces of such high-energy deposits as reefs and beachrocks. Friedman and Sanders (1978, p. 175) have stated that as such a cement can be precipitated within the pore spaces of sediments in even the highest of high-energy environments, non-critical use of what appears to be the particle-micrite ratio as a reliable clue to depositional environment would lead to error. Further words are "caution is necessary."

Many reef cavities of modern reefs are partly filled geopetally with micron-sized carbonate material which I had originally inferred to be lithified lime mud. Yet lime mud was nowhere observed near Red Sea and Australian reef settings, where in studied samples such geopetal fabrics are present. In fact the turbulence on the narrow shelves of the Red Sea precludes lime-mud sedimentation and in the Australian reef environments the surge of the Pacific fetch likewise prevents lime-mud sedimentation. Hence my original interpretation has been abandoned that the reefs in the two above-mentioned geographic areas contain cavities in which lime mud may settle, remain undisturbed, and be cemented. In my re-interpretation (G.M. Friedman *in* Elf Aquitaine *et al.*, 1982, Fig. 2–3C, p. 45) I suggested that the geopetal fill consisting of particles of skeletal fragments accumulated in a cavity of a shell, where it became cemented by cryptocrystalline cement; the upper part of the cavity remained hollow.

PITFALLS IN MISIDENTIFICATION OF SUBMARINE CEMENT: CASE HISTORIES

Two case histories illustrate the confusion in classifying, naming, and interpreting reefrock. Both are based on confusion of mistaking cryptocrystalline cement for micrite (= physically deposited matrix).

Case History 1

In oral presentations and published papers geologists have used the term "micrite cement". To save embarrassment appropriate references will be omitted here. Clearly as micrite has come to be understood, no such usage as "micrite cement" can exist. The material is either micrite (= physically deposited matrix) or it is cement, it cannot be both. For micrite cement, substitute cryptocrystalline cement or else the depositional environment will be incorrectly inferred.

Case History 2

Using dynamite I blasted modern reefrock to determine the internal characteristics and porosity prospects of reefs in their natural settings. In a Ph.D. examination, I asked a candidate to examine the samples from the freshly blasted reefrock and (1) describe the samples and identify their lithology and (2) interpret their depositional environment. The candidate termed the modern reefrock a micrite and identified the depositional environment as "low-energy as in a lagoon," yet the samples had been collected from the reef front, where shallow-water waves surge relentlessly back and forth and storms batter the reef, thus providing the highest energy possible in a marine setting.

These examples illustrate the pitfalls of confusing cryptocrystalline cement for micrite.

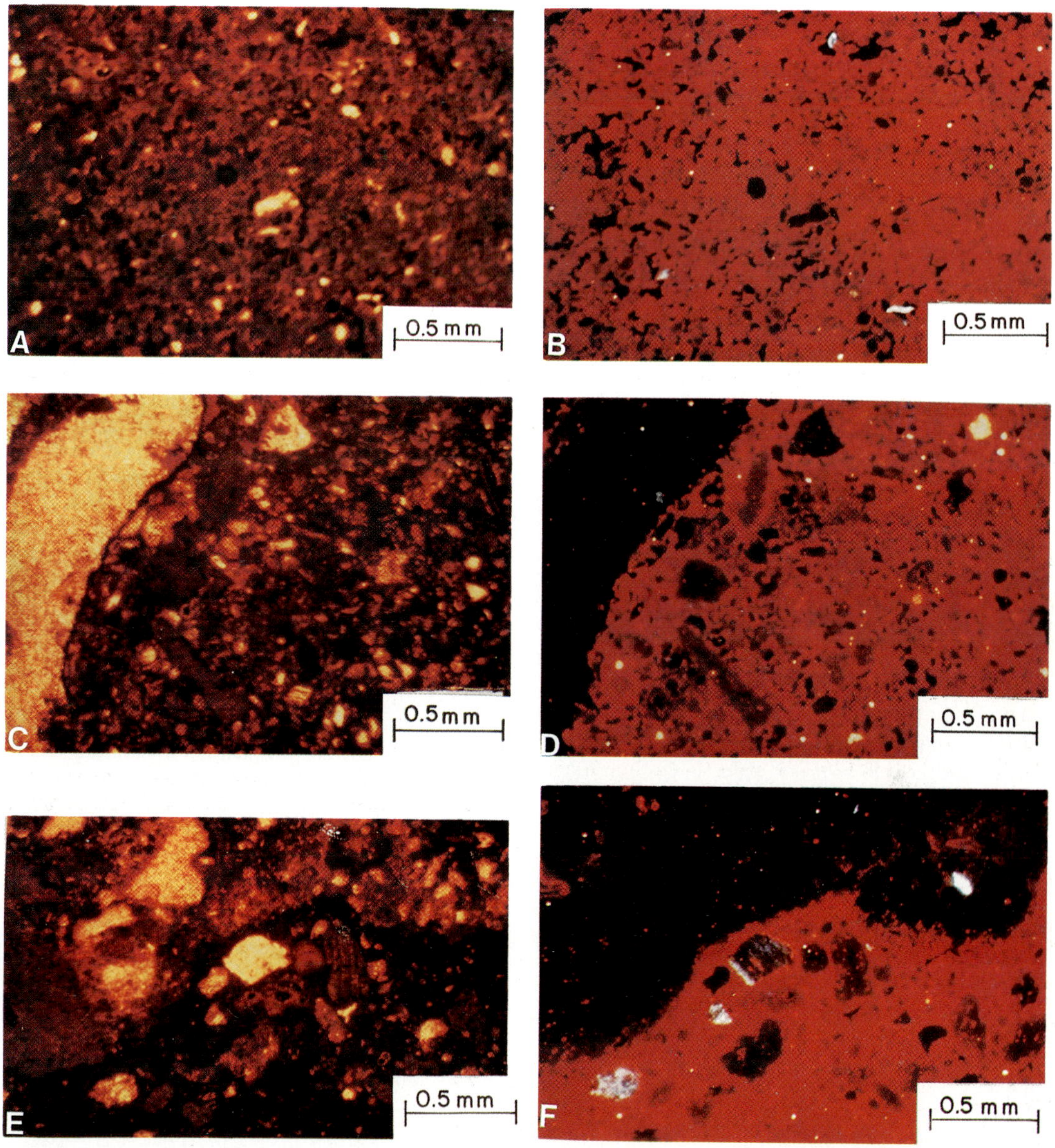

FIG. 1.—*A*, Photomicrograph of thin section of modern reefrock from the Red Sea in which unidentified skeletal fragments are set in a cryptocrystalline high-magnesium calcite cement. Plane-polarized light. *B*, Same view as in Figure 1A, but under an electron beam. Note that the particles, composed of aragonite, do not luminesce, hence may be distinguished from the highly luminescent high-magnesium calcite cement. *C*, Photomicrograph of thin section of modern reefrock from the Red Sea in which various skeletal fragments are cemented by a cryptocrystalline cement. Note large unidentified skeletal particle on left. Plane-polarized light. *D*, Same area of view as in Figure 1C. As in Figure 1B, aragonitic particles do not luminesce (including the large particles on the left), whereas by contrast high-magnesium calcite cement and particles are highly luminescent. *E*, Photomicrograph of thin section of modern reefrock from the Red Sea in which various skeletal and detrital terrigenous particles are set in a cryptocrystalline cement. *F*, Same area of view as in Figure 1E. Note a cement "stratigraphy." The cement in the upper part of the photograph is non-luminescent; probably its high-magnesium calcite has an insufficient concentration of activator ions. The cement in the lower part of the photograph luminesces as in Figure 1B and E. Note the various colors in which the detrital terrigenous particles luminesce.

SUBMARINE CRYPTOCRYSTALLINE HIGH-MAGNESIUM CALCITE CEMENT IN REEFS

Under the microscope, cryptocrystalline submarine cement ranges from translucent to semi-opaque (Friedman, 1975). As revealed by microprobe analysis, concentration of magnesium in the cement, where studied, is slightly enriched (up to 22% mol $MgCO_3$) over that in coralline algae in the same sample (14–90% mol $MgCO_3$) (Friedman *et al.*, 1974). Under the scanning-electron microscope, cryptocrystalline high-magnesium calcite cement consists of micron-size rhombohedral crystals (Friedman *et al.*, 1974). In places, the fabric of cryptocrystalline high-magnesium calcite cement is pellet-like (Friedman *et al.*, 1974, Figs. 12, 15).

To illustrate characteristics of submarine cryptocrystalline cement, not previously published, thin-section views are shown of low-power magnification (1) under a petrographic microscope in plane-polarized light, and (2) of the same view under an electron beam (Fig. 1). Three kinds of submarine cryptocrystalline cement occur in reefs. These are: (1) cements that luminesce brightly, (2) cements that luminesce dimly or faintly, and (3) cements that do not luminesce at all. Figure 1 only illustrates cements that luminesce brightly or those that do not luminesce at all. Under plane-polarized light or crossed nicols these three kinds of cement appear identical. In places, a cement "stratigraphy" can be demonstrated where two kinds of cement appear in sharp lateral stratigraphic contact (Fig. 1E, F). These differences in behavior under an electron beam relate to differences in the concentration of activator (Mn^{2+}) ions. Because the geochemical characteristics of manganese are similar to those of magnesium, manganese replaces magnesium in the high-magnesium calcite lattice in sufficient concentration to produce luminescence. The sharp boundaries between the three kinds of cement indicate variations in concentration of Mn^{2+} ions and different times and/or rates of precipitation.

In contrast to high-magnesium calcite, aragonite does not luminesce and hence appears black under an electron beam. As most particles in some of the studied samples consist of aragonite, a particle-cement ratio is revealed under an electron beam. Figure 1A and B indicate the kind of ratio found in reefs. Honestly, who would take this kind of sample for a reef rock? Figure 1C and 1D show a higher concentration of particles; yet even in this sample, particle-cement ratio may not result in correct reef identification. Views of these rock samples lead to another dilemma: the dilemma of reef identification.

DILEMMA OF REEF IDENTIFICATION

Blasting open a coral reef or drilling through such a reef rarely displays framework-building organisms in growth positions. Biting and boring and rasping or mechanical breakdown convert solid colonies of calcium-carbonate skeletons secreted by the reef organisms into an ever-increasing supply of skeletal particles that accumulate within the reef or in the vicinity of the reef. Among this skeletal debris, coral fragments may be relatively uncommon. Most corals break down into sclerodermites and then further into aragonite needles (Hubbard, 1976). Thus, for instance, contiguous to the Great Barrier Reef of Australia, coral fragments make up only 2 to 15% of the total debris (Bennett, 1971). Boreholes through modern reefs reveal mostly debris; framework builders in a position of growth are sporadic or even absent.

In borehole samples of modern reefs, skeletal debris of corals is underrepresented. The debris present is held together by cryptocrystalline cement which looks like micrite. No wonder reefrock has been described as wackestone, biomicrite, or simply micritic limestone, that has been erroneously referred to a low-energy setting.

CONCLUSIONS

Classifying reefrock may be an experience in frustration: submarine cryptocrystalline high-magnesium calcite cement that is precipitated within millimeters to centimeters of the surfaces of reefrock is identical in appearance with micrite of supposed mechanical origin as matrix. Case histories abound in which unwary geologists have misidentified the reefrock for low-energy facies.

ACKNOWLEDGEMENTS

Thanks are extended to my former graduate students, Constance M. Hill and R.A. Garber who took the photomicrographs. Connie Hill used the photographs in her Master's thesis on cathodoluminescence of carbonate samples. John E. Sanders and R.G.C. Bathurst reviewed drafts of this paper and challenged concepts which resulted in rephrasing of previously made statements. The concepts as presented in this paper, are the sole responsibility of the author. I want to thank the U.S. Department of Energy for funding grant no. DE-ACO-81ER10921 which made this study possible.

REFERENCES

BATHURST, R.G.C., 1971, Carbonate sediments and their diagenesis. Amsterdam, Elsevier Pub. Co., 620 p.

BENNETT, I., 1971, The Great Barrier Reef. Melbourne, Australia, Lansdowne, 183 p.

DUNHAM, R.J., 1962, Classification of carbonate rocks according to depositional texture, *in* W.E. Ham (ed.), Classification of Carbonate Rocks: Amer. Assoc. Petroleum Geologists, Mem. 1, p. 108–121.

ELF AQUITAINE, REECKMANN, A., AND FRIEDMAN, G.M., 1982, Exploration for carbonate petroleum reservoirs. New York, John Wiley & Sons, 213 p.

FOLK, R.L., 1959, Practical petrographic classification of limestones: Amer. Assoc. Petroleum Geologists, Bull., v. 43, p. 1–38.

______, 1962, Spectral subdivision of limestone types, *in* W.E. Ham (ed.), Classification of Carbonate Rocks: Amer. Assoc. Petroleum Geologists, Mem. 1, p. 62–84.

FRIEDMAN, G.M., 1964, Early diagenesis and lithification in carbonate sediments: Jour. Sedimentary Petrology, v. 34, p. 777–813.

______, 1968, The fabric of carbonate cement and matrix and its dependences on the salinity of water, *in* G. Müller, and G.M. Friedman (eds.), Recent developments in Carbonate Sedimentology in Central Europe: New York, Springer-Verlag, p. 11–20.

______, 1973, Cementation in reefs: Bull. Centre Rech. Pau-SNPA, v. 7, p. 171–176.

______, 1975, The making and unmaking of limestones or the downs and ups of porosity: Jour. Sed. Petrology, v. 45, p. 379–398.

______, AMIEL, A.J., AND SCHNEIDERMANN, N., 1974, Submarine cementation in reefs: example from the Red Sea: Jour. Sed. Petrology, v. 44, p. 816–825.

______, AND SANDERS, J.E., 1978, Principles of Sedimentology: New York, John Wiley & Sons, 792 p.

HAM, W.E., AND PRAY, L.C., 1962, Modern concepts and classifications of carbonate rocks, *in* W.E. Ham (ed.), Classification of Carbonate Rocks: Amer. Assoc. Petroleum Geologists, Mem. 1, p. 2–19.

HUBBARD, J.A.E.B., 1976, Sediment budgeting: A timed study of scleractinian necrology [abs.]: 25th International Geol. Congress, Abstracts, v. 3, p. 845.

PETTIJOHN, F.J., 1957, Sedimentary rocks (2nd ed.): New York, Harper & Row, 718 p.

SANDERS, J.E., 1978, Graywacke, *in* R.W. Fairbridge and J. Bourgeois (eds.), The Encyclopedia of Sedimentology: Stroudsburg, Pennsylvania, Dowden, Hutchinson and Ross, p. 389–391.

PRESERVATION OF INTERNAL REEF POROSITY AND DIAGENETIC SEALING OF SUBMERGED EARLY HOLOCENE BARRIER REEF, SOUTHEAST FLORIDA SHELF

ROBIN G. LIGHTY
Department of Geology, Texas A&M University, College Station, Texas 77843

ABSTRACT

Dredging of an offshore pipeline trench across the southeast Florida shelf exposed a well-preserved, porous relict *Acropora palmata* coral reef whose exposed upper framework surface is highly altered into a dense submarine-lithified pavement. Five distinct facies were identified from samples obtained from an excavation ditch perpendicular to the shelf edge 40 km north of Miami: (1) back-reef coral head; (2) back-reef *Acropora cervicornis*; (3) *Acropora palmata* framework; (4) fore-reef coral head; and (5) fore-reef rubble. Reef facies zonation is similar to that of modern Caribbean reefs. Radiocarbon dates of *A. palmata* framework samples range from 7,145 ± 80 to 9,440 ± 85 years BP. Internal structure reveals a growth history of rapid reef-framework accumulation (ave. 6.6 m/1000 yrs) in response to rising sea level, and the transition from a fringing reef to an extensive shelf-edge barrier reef following the submergence of the Florida shelf.

Occasional submarine cements formed during accumulation of reef facies are predominantly microcrystalline magnesium calcite and occur as rim cement, peloidal infill, coatings on grains, and as dense crusts on the surface of corals and reef debris. Syntaxial acicular aragonite cement is present in minor amounts within coral skeletal cavities and predates magnesium-calcite cement where present together. Magnesium-calcite submarine cements generally formed in areas of high agitation (e.g. wave energy) and/or low sedimentation (facies accumulation), preserving most internal skeletal porosity in *A. palmata* and fore-reef coral heads through the formation of dense surficial crusts and thin infill rims around coral skeletons. Submarine cements more easily filled and occluded primary porosity in the delicate branching coral *A. cervicornis* and reef rubble because of smaller sample cross-sectional area. Because reef growth was mostly vertical and somewhat landward, intermittent layers of well-cemented and encrusted rubble formed within fore-reef facies as a result of low sedimentation rates or hiatuses in talus slumping.

Magnesium-calcite cements provide an effective control on primary porosity within the reef deposits, for their limited and crust-like distribution acts to preserve remaining porosity from further cementation and occlusion. Rapid reef growth during a time of rapid sea-level rise (Holocene transgression) prevented reef-framework deposits from being subjected to prolonged periods of submarine cementation, thereby significantly limiting the degree of porosity reduction. The demise of the once flourishing barrier reef was brought about by changing shelf water conditions coincident with the rapid flooding of extensive, relatively flat areas of the back-reef incipient shelf. Following this, the inactive barrier reef became submerged by the continuing rise in sea level, during which time encrusting organisms, and extensive reiterative submarine cementation, bioerosion, and sediment fill transformed the exposed upper reef-framework into a dense, micritic zone of low permeability. This pavement-like reef-framework crust has protected internal reef deposits from additional alteration and cementation, and as an early diagenetic seal may act to further preserve porosity in a potential reservoir facies.

INTRODUCTION

Offshore pipeline construction 40 km north of Miami, Florida (26°15'15" N, 80°03'52" W), sectioned a relict shelf-edge coral reef whose exposed, upper surface is highly altered and extensively infilled by submarine cements. This diagenetic seal has effectively masked the well-preserved underlying *Acropora palmata* (Lamarck) barrier reef. The barrier reef, which extends for at least 95 km along the shelf edge, flourished off southeast Florida during early Holocene time in an area considerably north of present-day reef development (Fig. 1). The upper part of the reef's framework presently forms a low, 10-m relief ridge (Fig. 2) that extends along the shelf break from Palm Beach southwards to at least Miami (Duane and Meisburger, 1969; Macintyre and Milliman, 1970). The crest of this ridge occurs at depths of 15–30 m, and presently supports a deep-water community of alcyonarians, sponges, and scattered coral heads (Macintyre and Milliman, 1970; Goldberg, 1973). Radiocarbon dates indicate that the shallow-water reef "died" about 7,000 years ago and that there has been no significant framework accumulation since that time (Lighty *et al.*, 1978).

As the shelf-edge reef built up with rising sea level, various submarine cements formed within coral skeletons and reef sediments, providing a noticeable facies-specific control on internal reef porosity. Contrary to submarine cement distributions in many modern and ancient reefs, submarine cements in the relict Florida reef are limited in degree throughout most of the reef framework, and so distributed within the reef components as to potentially preserve most primary porosity.

The history of submarine diagenesis in this reef differs from that of most modern reefs because its growth history is different. The early Holocene Florida reef had high rates of reef growth and sediment accumulation during a period of rapid sea level rise, whereas most well-studied modern reefs developed during a pe-

Geological Contribution No. 1, Center for Tropical Marine Research, Box E, College Station, Texas 77841.

riod of slow sea level rise and even stable sea level conditions. Processes controlling submarine cementation in modern reefs most probably acted in much the same manner in early Holocene time, but the final diagenetic product within the relict reef, and its effect on primary porosity, was quite different. A distinct internal reef submarine cement distribution was created because of the reef's particular type of growth history.

This study also examines the special diagenetic history of an inactive, submerged reef structure. Following the barrier reef's demise, the inactive reef framework became further submerged by the continuing rise in sea level of the Holocene transgression, while extensive cementation altered the exposed upper surface of the reef into a dense micritic zone of low permeability. This diagenetic seal has protected internal reef components from further submarine alteration and cementation.

Objectives of this investigation deal with the formation and distribution of submarine cements in reef components that (1) were deposited as reef facies accumulated, and (2) have been exposed at the surface of the relict coral framework since the reef's demise. This study endeavors to explain the circumstances responsible for the rapid demise of a flourishing barrier reef. The character and development of this two-fold diagenetic history are intimately related to the reef's growth history and to the nature of the reef's demise. Other submerged relict reefs, which probably had similar depositional and diagenetic histories, are apparently common today on many continental and insular shelves throughout the Caribbean (Macintyre, 1972), and may be more prevalent in the ancient than has previously been recognized. Submerged (or drowned) reef structures, characterized by transgressive internal reef facies with limited submarine cementation and a framework (or reef core) sealed by a submarine disconformity surface, should have significant early preservation of primary porosity, and through their recognition may provide for a better understanding of basin sea level history.

METHODS

A trench 3 m wide, 450 m long, and having a maximum depth of 10 m was made through the relict shelf-edge reef by use of explosives and dredging, which allowed sample collection from the entire submarine outcrop section along a transect extending from back-reef to fore-reef environments. To avoid bias in sample collection, a grid system was constructed of nylon rope, brass chain, and PVC plastic pipe, with square

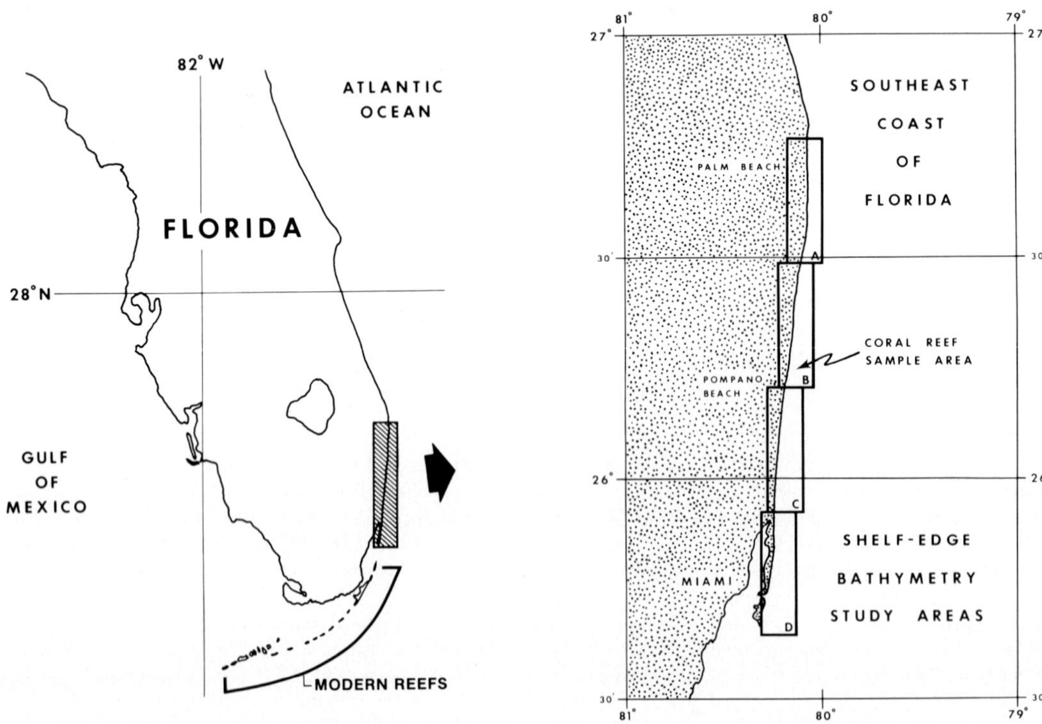

FIG. 1.—Index maps of southeast Florida study area. Details of shelf morphology in study areas A through D are shown in Figure 2. Note that study area of early Holocene barrier reef is located considerably north of present-day Florida reef tract.

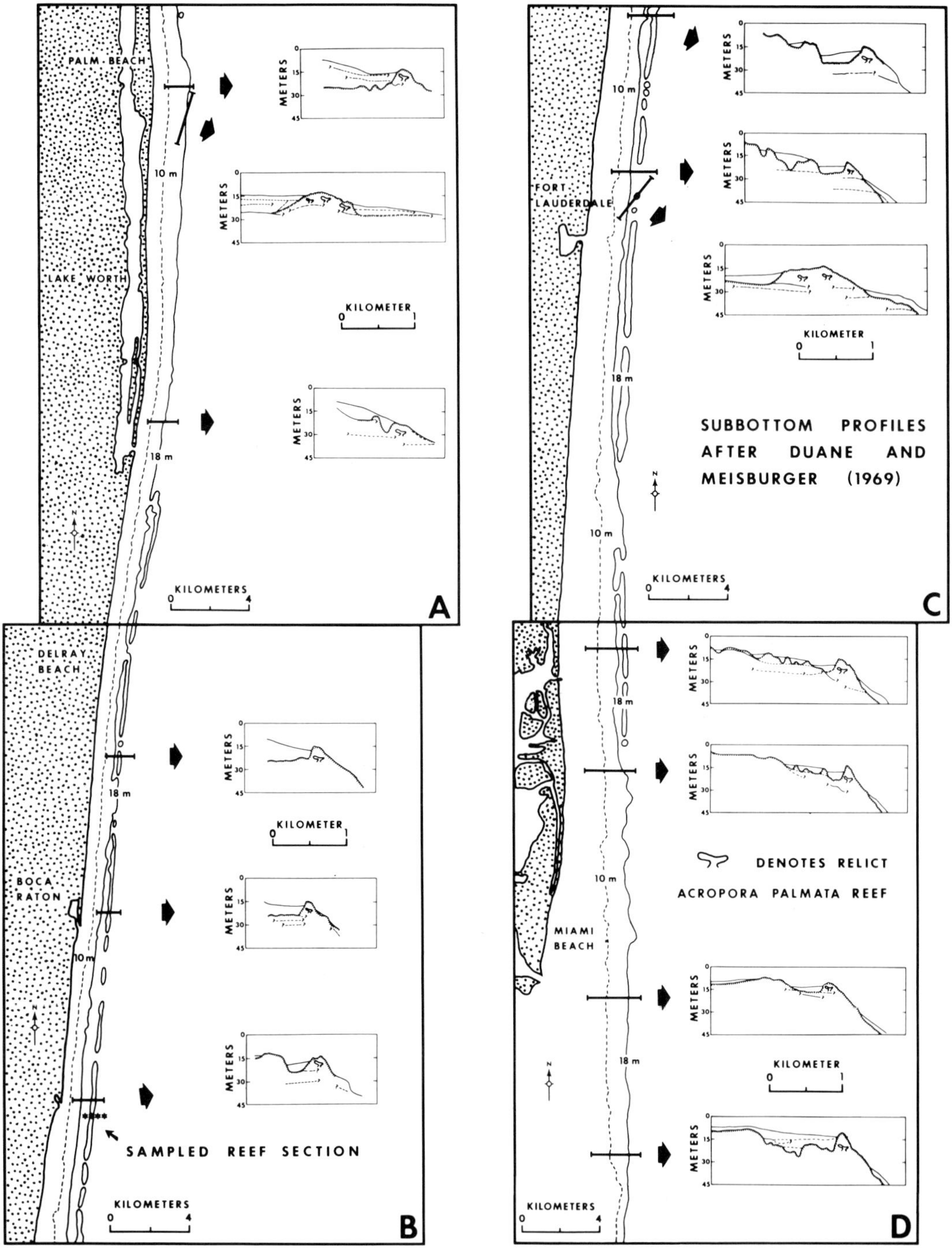

FIG. 2.—Morphology of southeast Florida shelf between Palm Beach and Miami Beach showing shelf-edge relict *Acropora palmata* barrier reef and location of reef section exposed by trench. The nearly continuous 95 km extent of the relict reef is indicated by seismic profile interpretations modified from Duane and Meisburger (1969). Profiles show subsequent partial filling of back-reef areas by sediment.

meter intervals and an overall dimension of 2 m by 10 m. This system was positioned on the trench wall at 25 m spacings along the ditch transect, by unrolling the grid from known coordinates (shelf location, water depth) on the bottom of the ditch up to and over the upper exposed surface of the inactive reef structure. After photographing the location, outcrop samples were collected from the center of each grid square at 110 locations (Fig. 3), with rock (fossil) samples collected at any given location ranging in number from 1 to 75, depending on the size of reef constituents. Although priority was given to collecting coral material, samples of the unlithified predominantly sand-sized sedimentary matrix were also recovered at most back-reef and fore-reef exposures.

Reef samples were plastic impregnated and studied by petrographic thin section analysis and microphotography, with detailed observations made along sample traverses from the center of coral skeletons and reef rubble outwards to surficial cement crusts and encrusting biota. Quantitative X-ray diffraction was run on corals, cement infill and coatings, and matrix sediment to confirm submarine cement mineralogy. SEM samples of poorly-lithified cement infill were glued in slurry mounts and platinum coated twice to prevent their characteristic "charging" phenomena. X-radiographs were taken to reveal skeletal banding in coral skeletons. Radiocarbon ages for unaltered *A. palmata* skeletal material were determined using a half life 5,568 years, and are uncorrected for $^{13}C/^{12}C$ ratios or for secular atmospheric variations.

INTERNAL REEF STRUCTURE

Five distinct facies (Fig. 4) were identified from the trench excavation perpendicular to the trend of the shelf-edge reef feature: (1) back-reef coral head; (2) back-reef *Acropora cervicornis*; (3) *Acropora palmata* framework; (4) fore-reef coral head; and (5) fore-reef rubble. A brief description of these reef facies and their main faunal constituents based on Lighty (1977) is given below to establish the sedimentologic result of this reef's particular growth history and to necessarily identify reef components subjected to early submarine diagenesis.

Back-Reef Coral Head Facies

This facies contains an abundant variety of head corals, that most commonly includes well-preserved heads

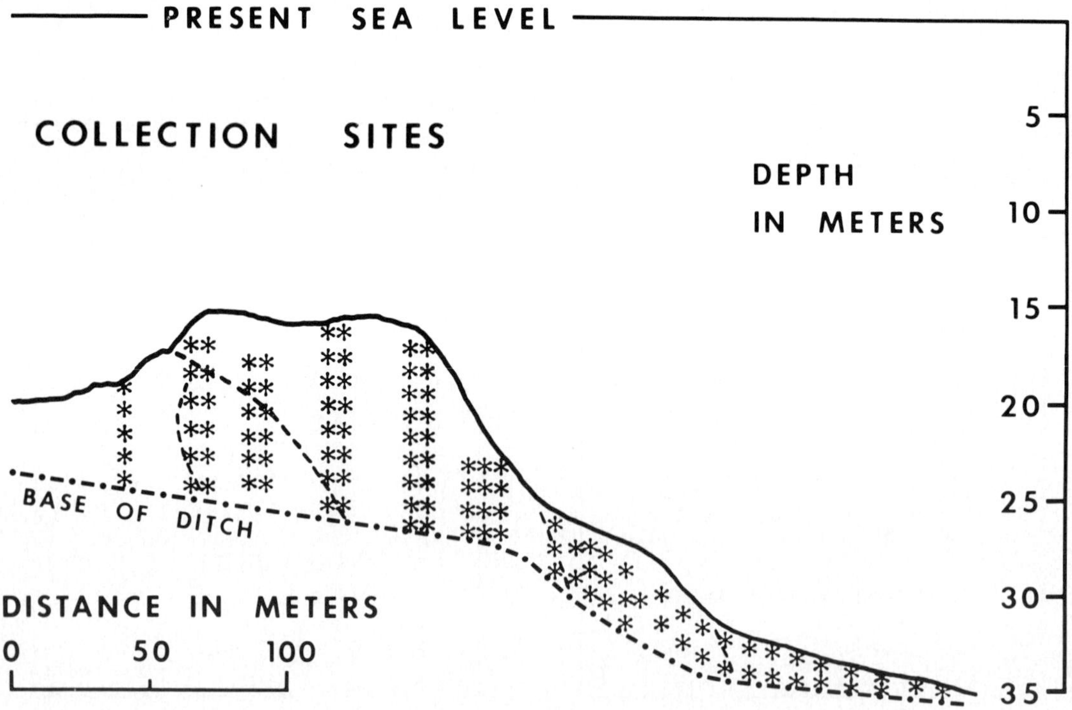

FIG. 3.—Cross section of relict shelf-edge barrier reef along transect of pipeline excavation extending from 1,750 to 2,200 m offshore. Asterisks indicate collection sites on south wall of trench. From 1 to 75 samples were collected at each site, depending on size of the reef constituents.

of *Diploria clivosa* (Ellis and Solander), *Montastrea annularis* (Ellis and Solander), *Stephanocoenia michelinii* (Milne-Edwards and Haime), and *Siderastrea siderea* (Ellis and Solander). These corals are surrounded by an unlithified matrix of sand- to gravel-sized carbonate sediment and reef rubble. Diagnostic coral head growth forms (e.g., closely-spaced corallites in *M. annularis*), the paucity of encrusting organisms, and the lack of extensive macroborings and submarine cement infill, all differentiate these protected shallow-water head corals from their highly-altered, deep-water fore-reef counterparts.

Acropora cervicornis Facies

This back-reef facies is composed almost entirely of poorly lithified, broken *Acropora cervicornis* (Lamarck) branches and heavily encrusted, extensively bored and abraded *Acropora palmata* rubble derived from the reef framework and transported landward by storm action. Included in the rubble are scattered fragments of various *Porites* and *Agaricia* coral species and the hydrocoral *Millepora complanata* (Lamarck), all heavily coated by crustose coralline algae and surficial deposits of magnesium-calcite submarine cement. The sand-sized matrix consists predominantly of skeletal fragments of coral, *Halimeda*, assorted mollusks, and coralline algae.

Acropora palmata Facies

The abundant *Acropora palmata* within this reef facies remains as an in situ rigid framework that is the major structural unit of the relict reef buildup (Fig. 4). Most coral growth forms of *A. palmata* are blades oriented in a landward direction, probably as a result of storms and past hurricanes acting upon this once high-energy reef crest zone on the reef's windward side. Most likely, these same processes resulted in the development of the substantial immediate back-reef rubble deposits. The *Acropora palmata* facies overlaps the *A. cervicornis* facies, with its lateral expansion extending over back-reef rubble and coral deposits as the reef framework attempted to keep pace with rising sea level during the Holocene transgression (Fig. 4). Corals within the *A. palmata* facies are generally well preserved, with only minor amounts of boring by bivalves and sponges, and only a thin covering of encrusting biota and submarine cement. The absence of extensive crustose coralline algal buildups and the lack of sub-

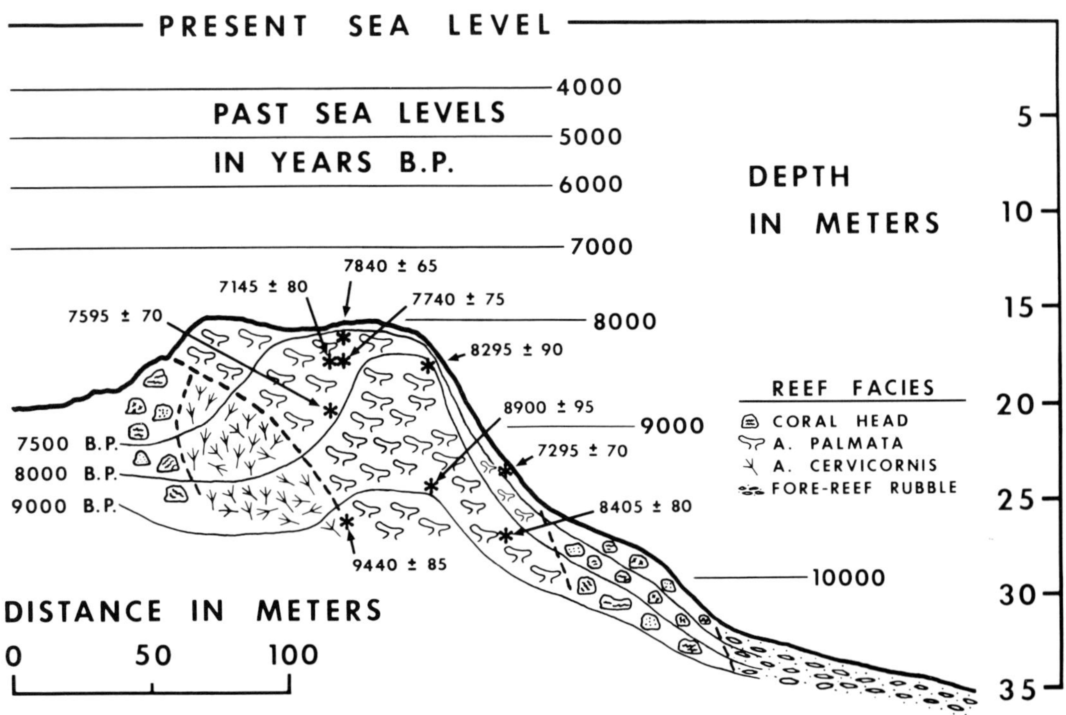

FIG. 4.—Cross section of relict barrier reef showing location of nine radiocarbon-dated coral samples and distribution of five identified reef facies: (1) back-reef coral head, (2) back-reef *Acropora cervicornis*, (3) *Acropora palmata* framework, (4) fore-reef coral head, and (5) fore-reef rubble facies. Time-growth surface lines indicate postulated stages of reef development during early Holocene transgression. The sea level curve used in this paper is Neumann's revision of his 1971 curve, and has been published in Adey (1975), Macintyre and Glynn (1976) and Lighty *et al.*, (1982). Reef growth was rapid and predominantly vertical, with *Acropora palmata* extending into back-reef areas as rate of sea-level rise decreased. There has been negligible reef-framework accumulation over the past 7,000 years.

stantial *Millepora* accumulations indicate that the *A. palmata* community flourished continuously during the growth of the reef, and at no time reached or remained in the intertidal zone long enough to develop a reef-flat facies typical of most modern Caribbean reefs. Nearly every *A. palmata* sample from the upper (most recent) deposits of the framework facies is encrusted by two common cryptic reef organisms, the small solitary corallites of the ahermatypic coral *Astrangia* sp., and the gregarious, small inarticulate brachiopod *Crania* sp. Significantly, these encrusters are absent from all of the coral samples collected in the lower part of this facies, and appear to be latter stage inhabitors of cryptic environments within the open reef framework, following the demise of the reef corals.

Fore-Reef Coral Head Facies

The fore-reef coral head facies is characterized by numerous highly bored and encrusted head corals that most commonly include *Montastrea cavernosa* (Linnaeus), *Montastrea annularis, Diploria* sp., *Stephanocoenia michelinii, Colpophyllia natans* (Muller), *Dichocoenia stokesii* (Milne-Edwards and Haime), and *Siderastrea siderea*. These corals appear to be in growth position and are buried in a matrix of reef-framework coral rubble and as yet unlithified skeletal sediment. There are several isolated pockets of broken *Acropora cervicornis* rubble that represent deposits of small *A. cervicornis* thickets in deeper portions of this fore-reef environment.

Fore-Reef Rubble Facies

This fore-reef facies consists mainly of unlithified deposits of gravel-sized coral detritus that have been so highly bored, encrusted, and infilled by submarine cement that their original character is almost impossible to determine. Several *Porites astreoides* Lesueur and *Dichocoenia stokesii* coral heads were less altered and preserved in growth position, along with the abundant *Agaricia agaricities*. Most reef rubble is heavily coated by crustose coralline algae and the benthic foraminifer *Gypsina plana* (Carter), and mollusks and bryozoans are very common. Skeletal fragments of coral, coralline algae, *Halimeda, Millepora*, and mollusks comprise the sand-sized sediment matrix. Coarser reef rubble is well lithified by extensive submarine cementation and occurs intermittently as discontinuous layers 10–15 cm thick that extend several meters in length down the talus slope.

Accessory Reef Constituents

During sample collection, highest priority was given to the recovery of the diverse and spatially well-defined coral assemblage, as this constituted the major skeletal component of the reef buildup. Sampling and identification of other reef assemblages was somewhat biased, with the majority of these being coral-skeleton encrusters. However, many unattached reef organisms were recovered within the reef rubble and sediment matrix of the back-reef and fore-reef coral head facies, and from the fore-reef rubble facies. Detailed lists and facies distributions determined for accessory reef biota are given in Lighty (1977) and briefly discussed in this section.

Mollusks.—The molluscan assemblage within the relict barrier reef is similar to that found in modern coral reefs of the tropical western Atlantic. A diverse assemblage of bivalves, several gastropods, and a few sessile vermetid gastropods were found in the sedimentary deposits of skeletal sand and rubble. Coral samples collected contained a large variety of boring bivalves that included *Botula fusca* (Gmelin), *Lithophaga antillarum* (Orbigny), *Lithophaga aristata* (Dillwyn), *Lithophaga bisulcata* (Orbigny), and *Lithophaga nigra* (Orbigny). Their macroborings commonly were lined by submarine cement infill within the adjacent coral skeletal cavities.

Bryozoa.—Bryozoans were identified from samples throughout the exposed reef section, and are fairly abundant and extremely well diversified (43 species). All bryozoans were cheilostomes, and their encrusting, cryptic habitats are similar to those reported for many modern coral reefs (Cuffey, 1974). By far the most abundant and ubiquitous species identified was *Smittipora levinseni* (Canu and Bassler). *Celleporaria magnifica* (Osburn) was slightly less abundant, but just as widely dispersed. Some bryozoans only appear to encrust the surface of reef rubble that was deposited in deeper, protected fore-reef areas, whereas others preferred the shallower cryptic environments within the *Acropora palmata* reef-framework. Competition for available reef substrates among bryozoans, as well as with other encrusting biota, resulted in complex distributional patterns and overgrowth sequences within encrustations that vary from sample to sample.

Studies by others on bryozoan distributions within modern coral reefs generally are limited to reef surfaces, and a true picture of bryozoan distribution within most cryptic reef environments has not yet been attained. For example, Schopf (1974), in describing the cheilostome assemblage found within reef environments of a modern *Acropora palmata* reef in the U.S. Virgin Islands, reported a limited cheilostome population for the back-reef *Acropora cervicornis* zone. However, an abundant and well-diversified assemblage of cheilostomes was found within this facies in the relict Florida reef.

The bryozoan assemblage of the relict barrier reef includes *Cigclisula protecta* (Osburn), *Escharoides costifer* (Osburn), *Hippopodina feegeensis* (Busk), and *Figularia contraria* (Lagaaij), all of which have been previously reported only from tropical climates (Osburn, 1940; Kornicker *et al.*, 1959; Lagaaij, 1963). The most abundant of these tropical species, *Hippopodina feegeensis*, presently has its worldwide distribution defined as being entirely within tropical latitudes that do not include the southeast Florida area of this study (Powell, 1969). The presence of these tropical species within the exposed reef facies suggests that warmer water temperatures existed during the growth and development of the now relict shelf-edge barrier reef.

Benthic foraminifera.—The most abundant reef-component encrusters are the benthic foraminifera *Homotrema rubrum* (Lamarck), encrusting all types of substrates, and *Carpenteria monticularis* (Carter), only found on the surface of corals *Acropora palmata* and *A. cervicornis*. *Carpenteria monticularis* previously has been observed living in the modern reefs of Panama and Belize (Macintyre and Glynn, 1976). Other foraminifera also found commonly encrusting coral and reef rubble throughout the entire trench section include *Acervulina inhaerens* (Schultze), *Placopsilina confusa* (Cushman), *Planorbulinoides retinaculatus* (Parker and Jones), and *Sagenina frondescens* (Brady). The deep fore-reef foraminifer *Gypsina plana* occasionally was found encrusting coral debris in the fore-reef rubble facies.

Crustose coralline algae.—Thin encrustations of crustose coralline algae are present on many samples of *Acropora palmata* and *A. cervicornis*. These crustose samples of *Porolithon* and *Lithophyllum* sp. are usually indicative of algal growth in water depths less than 3 m (Steneck and Adey, 1976). There was no extensive binding of coral rubble by crustose coralline algae anywhere within the exposed reef framework. The coral fragments, rubble, and sediment in the fore-reef facies were more commonly encrusted by thin coverings of *Neogoniolithon* and *Lithothamnium*, with rhodoliths being quite abundant. Although encrustations of coralline algae were present as accessory reef components in all reef facies, there was no evidence to indicate the formation of an algal-ridge system, or reef-flat pavement surface, at any time during the growth of the coral reef buildup.

REEF GROWTH HISTORY

Radiocarbon dates for nine samples of reef-framework *Acropora palmata* indicate that the relict reef is early Holocene in age. Ages determined for skeletal aragonite free of borings and magnesium calcite submarine cement range from 7,145 ± 80 years BP near the top of the *Acropora palmata* facies, to 9,440 ± 85 years BP obtained at 10.5 m depth within the same facies (Fig. 4). Growth rates for reef-framework accumulation of the *A. palmata* facies determined from four vertical sequences of dated samples range from 3.6 m to 10.7 m/1000 years, with a mean rate of 6.6 m/1000 years (Lighty *et al.*, 1978). The individual coral growth rates of several modern species have been well documented, but only recently have the reef-framework accumulation rates for several *A. palmata* reefs been determined. The use of submersible coring techniques (Macintyre, 1975) in studying the internal structure of modern reefs has allowed the determination of *A. palmata* framework accumulation rates as high as 10.8 m/1000 years for a fringing reef at Galeta Point, Panama (Macintyre and Glynn, 1976) and 15.2 m/1000 years for an *A. palmata* reef off St. Croix (Adey, 1975). Recent studies on coral-skeletal growth rates have reported axial growth of *A. palmata* measuring from 40 mm/year off Panama (Macintyre and Glynn, 1976) to 8.3 mm/30 days off St. Croix (Gladfelter and Monahan, 1977). These data, along with the facies accumulation rates determined for the relict Florida reef, clearly indicate that *A. palmata* reefs are capable of rapid vertical reef-framework accumulation and should be able to keep pace with a rapidly rising sea level.

Coral radiocarbon dates for the relict Florida reef are compared to Neumann's (1971, revised) sea-level curve for Bermuda (see Fig. 4), and are in some agreement. This curve is believed reliable for this period throughout the tropical western Atlantic (Lighty *et al.*, 1982). *Acropora palmata* dates shown in Figure 4 indicate at least 10 m of rapid, substantial reef-framework accumulation in response to rapidly rising seas of the Holocene transgression approximately 9,440 to 7,740 years BP. Dates also reveal that there has been no apparent framework accumulation during the past 7,000 years. The 95-km long inactive reef structure became further submerged at the shelf edge as the rate of sea-level rise decreased to today's relatively stable position.

The extensive barrier-reef complex initially developed as a fringing reef, most likely situated on a linear shelf-edge topographic high (Pleistocene eolianite ridge?). The fast-growing *Acropora palmata* reef was able to keep pace with the continued rapid rise of sea level, and with the subsequent submergence of back-reef shelf areas, the fringing reef developed into an extensive barrier reef. Depth of penetration of the construction ditch did not allow sampling of either the substrate underlying the reef or the early stages of reef growth. However, it is reasonable to assume that with the Holocene rise of sea level the *Acropora* species quickly established themselves, thereby increasing the reef's potential for rapid reef-framework accumulation.

The distribution of exposed reef facies (Fig. 4) suggests that the ecologic zonation of this reef was similar to that of many reefs in the western Atlantic, including the modern Florida (Ginsburg, 1956; Shinn, 1963), Belize (Rützler and Macintyre, 1982), and Bahamian reef tracts (Newell and Rigby, 1957), and the subaerially exposed Pleistocene reefs of Barbados (Mesolella *et al.*, 1970). It also appears that the observed windward-to-leeward facies zonation, containing deep-water and shallow-water communities, was established early in the reef's development and was maintained during the predominantly vertical growth of the reef complex. The rapid reef-framework accumulation and the constancy of facies relationships through time are functions of the reef's response to a rapidly rising sea, with *Acropora palmata* capable of maintaining rapid growth and development under these ideal conditions. However, there is no evidence to suggest that the *A. palmata* reached or remained in the intertidal zone long enough to undergo a period of stabilization leading to the development of a coralline algal facies (Adey, 1975, Adey and Burke, 1977) or an inactive climax community such as that found in the fringing reef at Galeta, Panama (Macintyre and Glynn, 1976).

The pattern of reef-framework accumulation for the relict reef is similar to reef growth models suggested by Logan (1969) for the reefs of the Yucatan shelf. A comparison of these with the proposed development of the relict Florida reef shows a similarity of predominantly vertical growth (Logan's model 2) as a response to rising sea level. A predominantly vertical growth of *A. palmata* framework has also been interpreted from cores of a shelf-edge barrier reef off St. Croix (Adey *et al.*, 1977). The rapid vertical growth of the shelf-edge Florida reef was followed by a period of substantial *A. palmata* framework construction on the back-reef side of the active framework. This is similar to Logan's model 3, in which the reef-framework extends over a back-reef substrate of off-reef clastic drape. The *A. palmata* framework facies cored in the fringing reef off Panama (Macintyre and Glynn, 1976) was found to extend into back-reef areas in response to available substrates and local sea-level history.

From its initial formation as a fringing reef, vertical growth of *Acropora palmata* framework within the early Holocene Florida reef was in direct response to a rapidly rising sea. Lateral development of the *A. palmata* framework in a shoreward direction was controlled by the rapid accumulation of back-reef deposits, in contrast to fore-reef areas. The high accumulation rates in the back reef are related mainly to the prolific growth of the branching coral *Acropora cervicornis*, augmented to some degree by allochthonous storm deposits. With the continued rise of sea level and the development into a shelf-edge barrier reef setting, the Florida reef was more likely subjected daily to higher energy conditions, and especially during storms and hurricanes. The high unidirectional action of destructive storm events would increase deposition in the back reef, while at the same time prevent substantial fore-reef deposition or seaward extension of the *Acropora palmata* framework.

The destructive forces of hurricanes on the modern reefs of Belize (Stoddart, 1963), Jamaica (Woodley *et al.*, 1981; Tunnicliffe, 1983), and the present-day Florida reef tract (Ball *et al.*, 1967; Perkins and Enos, 1968) have been responsible for significant back-reef deposits of reef-framework debris, as well as restricting the growth of delicate branching corals (e.g. *A. cervicornis*) on the seaward side of reef framework. Conditions would have been favorable for *Acropora palmata* growth on the back-reef side of the reef framework by approximately 8,000 years BP because of the development of a sizable back-reef lagoon, the open-ocean barrier nature of the reef trend, and the close proximity of the swift Florida Current. The establishment of this moderately-high energy environment in the immediate back-reef limited the continued existence of the earlier prolific *Acropora cervicornis* facies, which required a lower energy and more protected environment, and allowed the lateral expansion of the opportunistic *A. palmata* into the immediate back reef. Flourishing *Acropora palmata* commonly inhabits the reef flat and forms interlocking framework in near back-reef areas of the modern Florida reef tract (Shinn, 1963) and the modern Belize barrier reef (personal observation by author, near Ambergris Cay).

Following the period of prolific reef growth prior to 7,000 years BP, the *Acropora palmata* died, and active reef-framework construction ceased. X-radiographs taken of corals collected from upper parts of the framework and back-reef facies reveal pronounced skeletal banding. Similar pronounced banding has been related to periods of high stress on coral growth (Knutson *et al.*, 1972; Buddemeier, 1974; Moore and Krishnaswami, 1974; Macintyre and Smith, 1974; Weber *et al.*, 1975), and possibly may reflect the deleterious effects of turbidity (Dodge and Vaisnys, 1975), and/or an increase in water temperature fluctuations (Hudson *et al.*, 1976). In addition to repeated stress banding in *A. palmata* and *Montastrea annularis* corals, high stress conditions were clearly indicated by the pronounced banding found in a *Porites astreoides* coral head, a species which rarely exhibits prominent banding, even when found living in moderately high stress environments.

In summary, the internal structure of the early Holocene Florida reef reveals a growth history of rapid accumulation of *Acropora palmata* framework in response to a rapidly rising sea level, and the transition from a fringing reef to an extensive barrier-reef complex following the submergence of widespread back-reef shelf areas. The introduction of high stress conditions approximately 7,000 years ago exceeded the tolerance limits of *Acropora palmata* and apparently led to the demise of the shelf-edge barrier reef. Conditions believed responsible for the demise of the reef community are discussed in the following section.

DEMISE OF REEF

The demise of the *Acropora palmata* barrier reef off southeast Florida approximately 7,000 years ago coincides with the extensive flooding of large, relatively flat areas of the Florida continental shelf during the Holocene transgression (Lighty *et al.*, 1978). The early Holocene transgression of the St. Croix insular shelf similarly has been correlated with the demise of a shelf-edge *A. palmata* reef, believed to have been killed by turbidity derived from erosion of shelf soil cover (Adey *et al.*, 1977). In another study, the late

Holocene demise of a 175-km long *A. palmata* barrier reef in the northern Bahamas has been related to initial flooding of back-reef platform areas and the subsequent off-bank transport of turbid and episodically-cooled water masses during storm events (Lighty *et al.*, 1980; Lighty *et al.*, in prep.). Adey (1977) has further speculated that turbidity related to erosion of shelf regoliths has prevented continuous Holocene reef growth on many shelves in the eastern Caribbean. Macintyre (1972) previously noted that many submerged ridges characterize shelf morphology in the eastern Caribbean, and suggested that these features are relict Holocene and Pleistocene reefs.

The absence of a thick soil cover in southern Florida during the Pleistocene (Parker *et al.*, 1955; Perkins, 1977) and the narrowness of the Florida shelf along the relict barrier-reef trend would have limited the source of fine sediments (and turbidity) in close proximity to the barrier reef. However, large quantities of fine sediment were recently transported northward along the shelf edge by the Florida Current, after being placed into suspension by storm effects of Hurricane Donna in the Florida Keys and Florida Bay (Ball *et al.*, 1967). This suggests that during the Holocene transgression a major influx of turbid waters could have originated from wide, restricted areas of Florida Bay and the inner shelf to the south, where soil and lagoonal deposits of fine lime mud were widespread (Enos, 1977). The inception of these episodic conditions would have exceeded the tolerance limits of the early Holocene *Acropora palmata* barrier reef community.

In addition to turbidity, low water temperatures may have contributed to the demise of the shelf-edge reef as a result of atmospheric cooling of the large expanses of shallow water that initially formed during the Holocene transgression of the Florida continental shelf. Episodic cold-water conditions presently occur within large shallow-water areas of south Florida, as documented by minimum water temperatures of 18°C or less recorded for 23 months between 1954 and 1968 in northern Biscayne Bay (near Miami) (Moore, 1969), and for eight winter months between 1968 and 1971 at more than 40 localities in southern Biscayne Bay (Roessler and Tabb, 1974). The residence time of individual cold-water stress conditions may exist over vast shallow-water areas of Florida Bay for several days duration (Roberts *et al.*, 1982).

Furthermore, present-day atmospheric cooling of shallow shelf water north of Miami may be inhibiting *Acropora palmata* reef growth on the many submerged ridges in nearshore areas of the southeast Florida shelf. Although the surfaces of these inner-shelf ridges are shallow enough for the development of *A. palmata* reefs, none were found during extensive underwater surveys of these shelf areas (Raymond, 1972; Goldberg, 1973). Vaughan (1916) similarly concluded that reduced shelf water temperatures currently prevent coral reef growth north of the modern Florida reef tract.

Although pronounced temperature fluctuations and severely cold bottom waters are effectively preventing active reef growth in shallow shelf waters north of Miami, this seasonal atmospheric cooling was probably more effective in reducing water temperatures immediately following the extensive flooding of the Florida continental shelf approximately 7,000 years ago. The large expanses of very shallow, back-reef incipient shelf-water during this period would have been easily cooled by the passage of winter storms and atmospheric cold fronts.

Present-day water temperatures may be more favorable for reef growth along the shelf edge because of seasonal intrusions of the warm Florida Current, but present water depths of 15–30 m over the relict barrier reef are considerably too deep for active *Acropora palmata* reef growth. Only a deep reef fauna of hermatypic coral species was found to inhabit the surface of the relict reef (Goldberg, 1973). Absent were framework builders such as *A. palmata* and *Porites porites* (Pallas) and only minor occurrences of *Acropora cervicornis* and *Montastrea annularis*, the other most common reef-builders in the Caribbean, were observed. The early Holocene radiocarbon ages for samples collected just below the surface of the relict barrier reef (Lighty *et al.*, 1978) indicate that this deeper-water epifaunal community has contributed no appreciable relief to the shelf-edge structure.

In summary, although one cannot conclusively prove the existence approximately 7,000 years ago of low shallow-water temperatures at the shelf edge, or the catastrophic influx of extensive turbidity of carbonate muds, these events are suggested as reasonable causes for the indisputable demise of the early Holocene barrier reef off southeast Florida.

SUBMARINE CEMENT TYPES

Throughout the early Holocene reef's depositional history, submarine cements were precipitated within cavities and skeletal pores, and on the surface of many reef components. Submarine cements have been recognized in many modern reefs (see Friedman *et al.*, 1974 for summary), and several studies on the origin and distribution of submarine cements within the internal structure of modern coral frameworks have been published (James *et al.*, 1976; Macintyre, 1977). Submarine cements are widespread within the framework of the relict Florida reef. These are predominantly microcrystalline (4–30μ) and submicrocrystalline (<4μ) magnesium calcite in the form of rim cement, peloidal fill, coatings on grains, and thin dense crusts on the surface of coral skeletons and reef rubble. Minor amounts of acicular aragonite cement occur throughout the reef as syntaxial overgrowths (typically up to 100μ long) on coral sclerodermites bordering skeletal cavities (Figs. 5, 6, 12A). The amount of aragonite cement within the submerged Florida reef is considerably less than that reported for modern reefs in the Red Sea

(Friedman *et al.*, 1974). However, the dominant role of magnesium calcite as a reef cement has been previously reported for many modern reefs, including the western Atlantic reefs of Barbados (Macintyre *et al.*, 1968), Jamaica (Land and Goreau, 1970), Bermuda (Ginsburg *et al.*, 1971), Belize (James *et al.*, 1976), Panama (Macintyre, 1977), and Brazil (Jindrich, 1983). Magnesium-calcite submarine cements with micritic and peloidal textures also are abundant in fringing reef deposits of Reunion Island, Indian Ocean (Montaggioni, 1977), and ubiquitous throughout fore-reef areas of the southern Great Barrier Reef, Australia (Marshall and Davies, 1981).

Magnesium-calcite cements displaying a variety of growth forms occur throughout samples collected from the reef-framework and fore reef of the relict Florida reef. Magnesium-calcite sometimes forms a sparry rim cement of dentate crystals lining the cavity walls of coral skeletons or reef debris (Figs. 7 and 8), or encrusts and fills chasmolithic algal filaments (Figs. 9 and 10), exhibiting the "palisades form" described in Bermuda reefs by Schroeder (1972). Magnesium-calcite cement more commonly appears as an interlocking mosaic of microcrystalline calcite that has a clotted, or peloidal texture (Fig. 11) and forms a thin, dense crust on coral skeletons and other reef components, or infills cavities within reef components. This peloidal texture, with peloids ranging 20–60μ in size, is a common characteristic of magnesium-calcite submarine cements (Macintyre *et al.*, 1968; Shinn, 1969; Land and

FIG. 5.—Scanning electron micrograph showing skeletal cavity in coral *Acropora palmata* partially infilled by acicular aragonite submarine cement.

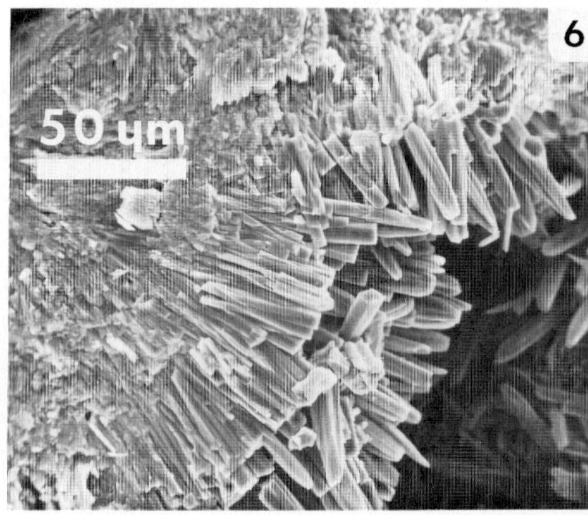

FIG. 7.—Scanning electron micrograph of rim cement of dentate magnesium-calcite crystals lining skeletal cavity in *Acropora palmata*.

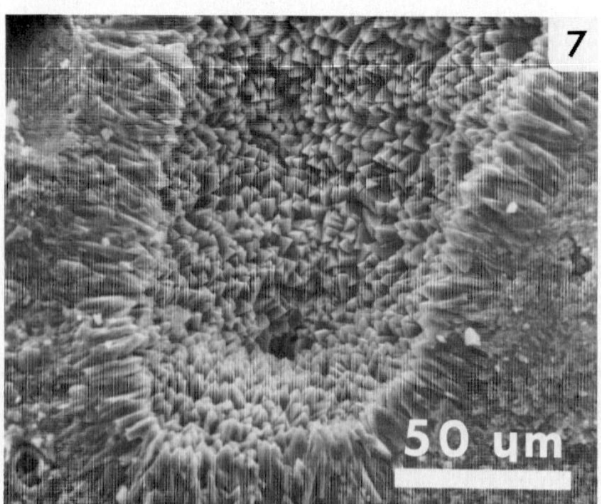

FIG. 6.—Scanning electron micrograph showing syntaxial overgrowths of aragonite cement lining walls of *Acropora palmata* skeletal cavity.

FIG. 8.—Higher magnification SEM micrograph of Figure 7 showing well-developed rhombic crystals of magnesium calcite.

Goreau, 1970; Alexandersson, 1972a), and often exhibits varying degrees of interpeloidal cementation. Cementation ranges from poorly-lithified open mosaics that are more common within internal skeletal cavities (Figs. 11, 12B), to dense well-cemented mosaics that infill near-surface cavities (Figs. 12C–E) and frequently form surficial crusts over reef components (Fig. 12F).

The magnesium-calcite peloids average 40μ in diameter, and consist of a well-developed sparry dentate rim of microcrystalline rhombic calcite (5μ to 10μ in length) around a nondescript anhedral nucleus (Figs. 14–16). This bimodal texture has been described differently by several workers as an intermediate crystal growth stage (Shinn, 1969; Alexandersson, 1972a&b), as crystal growth on a nucleus of silt-size detritus of clastic origin (James *et al.*, 1976) or pelletal origin (Land and Moore, 1980), or as a product of two stages of cement precipitation (Macintyre, 1977). Petrologic relationships and textures determined for over 200 thin sections and nearly 100 hours of SEM observations in this study support the contention that the bimodal internal structure of the peloids represents two stages in the development of these cements—an initial rapid precipitation of anhedral magnesium calcite forms the nuclei for a second stage of slower and more organized precipitation of larger euhedral magnesium calcite, the dentate rims of peloids (Macintyre, 1977, 1978).

The uniform size and textural composition of nearly all peloids, and both the consistently uniform, anhedral microclotted texture of the nuclei and the uniform crystal size and euhedral character of the outer rim, support the hypothesis that the peloids are not an intermediate growth stage, but an end-product of cementation. The absence or limited state of cement crystal growth on silt-size skeletal debris concomitant to the formation of the surrounding well-developed peloids (Fig. 17, 18), suggests that formation of the peloids is not controlled by the availability of a silt-size detrital nuclei. Also, most small primary skeletal pores containing abundant peloidal cement show no evidence of macroboring inhabitants that may have provided pelletal nuclei.

Within the early Holocene reef's components, microenvironmental controls on submarine cementation often result in different diagenetic products inside adjacent cavities (Figs. 12D, 23). It is common to find, within close proximity, aragonite cement in one cavity, magnesium-calcite peloids in another, and no cementation in yet another. Many skeletal cavities contain magnesium-calcite peloids that overlie earlier, well-preserved syntaxial overgrowths of aragonite (Figs. 19–22).

Magnesium-calcite crystals commonly radiate from a single nucleation point on the surface of the peloids, forming small rosettes (Figs. 13, 24) that, along with the generally uniform, spherical shapes of the peloids, may indicate isotrophic concentration (supersaturation) gradients during precipitation. This suggests that the pore waters within the cavity were well mixed as crystals formed on the outer surface of the peloids. Observed petrologic relationships in this study indicate that (1) skeletal cavities devoid of cements were probably always isolated from external marine waters, (2) cavities containing only syntaxial overgrowths of aragonite were initially open to marine circulation, but were quickly isolated by continuous coral-skeleton growth, and (3) cavities that contain magnesium-calcite cements remained open or later became open (by borings or abrasion) to marine circulation for some

FIG. 9.—Scanning electron micrograph of a transverse section showing magnesium-calcite crystals radiating from algal filament substrate ("palisades" growth form described by Schroeder, 1972) within *Acropora palmata* skeletal cavity.

FIG. 10.—Higher magnification SEM micrograph of Figure 9 showing algal filament substrate with magnesium-calcite crystals precipitated on the filament surface (left), and within the filament chamber (right).

amount of time after coral growth ceased. Precipitation of syntaxial aragonite cement appears to be contemporaneous with active coral growth, forming just beneath the surface of active skeletal accretion (Hubbard, 1975; Macintyre, 1977). Magnesium-calcite cements apparently postdate coral growth, generally forming in those cavities near the exposed surface of coral skeletons, or near internal skeleton surfaces exposed by macroboring organisms.

Most individual magnesium-calcite peloids do not appear to have undergone any significant movement after their formation. However, open mosaics of peloids commonly form geopetal structures within coral skeletal and boring cavities (Fig. 25), suggesting that crystal growth occurred as the peloids (including the initial nuclei) remained, or were remixed into suspension within the pore cavity by pulses of marine water, then settled out of suspension as circulation of marine water through the pore ceased. The spherical uniform nature of the peloid's dentate crystal rim, as well as the apparent self-nucleating character of the initial precipitate, also tend to support the hypothesis of peloidal development during periods of prolonged or repeated suspension within the pore cavity. As noted earlier, it appears that magnesium calcite preferred to self nucleate rather than utilize small-sized detrital grains that were present within the same pore space, even those grains having a similar mineralogy such as echinoid and coralline algal fragments. This observation is considered valid because in most small cavities, as peloids developed, they most likely sealed off effective permeability that would have allowed detritus to enter at a later stage, and fully-developed peloids were often too big and numerous to have been easily transported through intercavity channels into pore spaces that contained unaltered skeletal detrital grains. In most cases

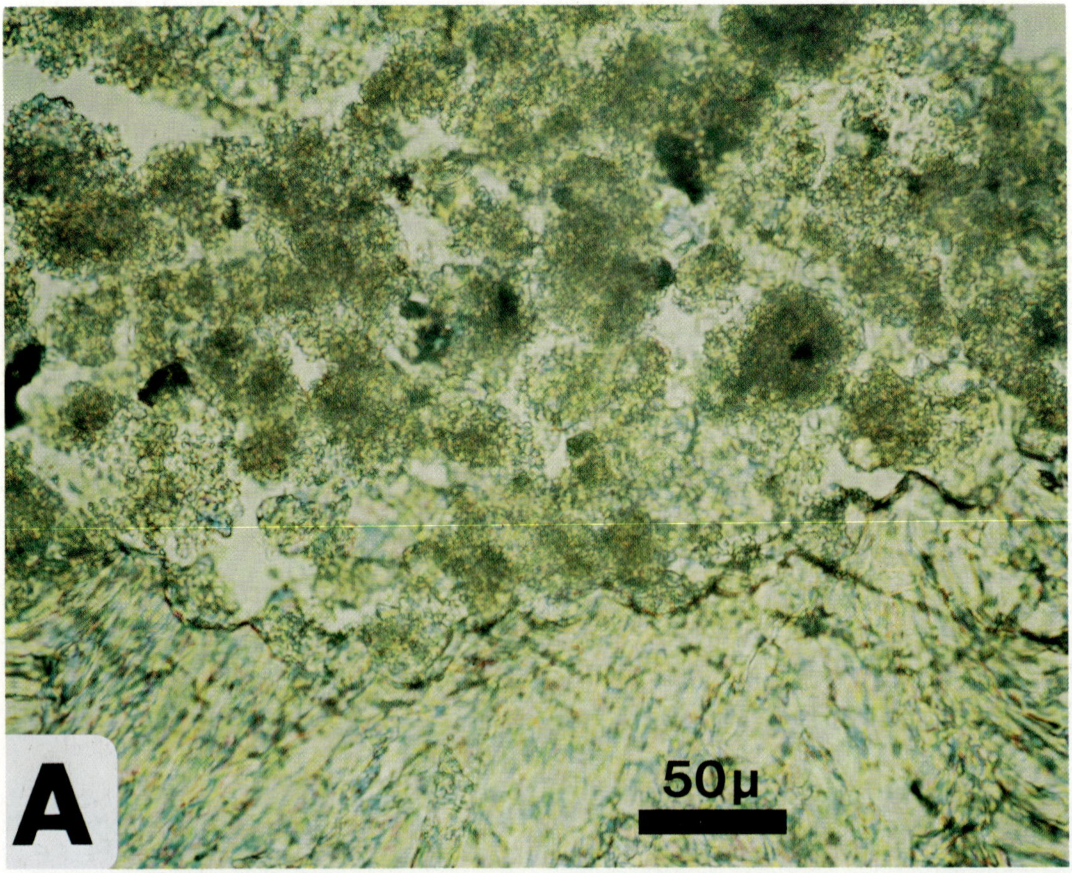

FIG. 11*A*.—Thin section photomicrograph of magnesium-calcite peloidal cement filling sponge boring (scalloped pattern) within *Acropora palmata* coral skeleton. Note the uniform size and texture of peloids, and the open nature of cement mosaic. Magnesium-calcite peloids were most likely transported into, or remixed in suspension, within cavity and deposited as geopetal infill. Brown coloration and dense nucleus, as shown here in plane-polarized light, along with characteristic size and shape, are most common recognition criteria of magnesium-calcite peloidal submarine cement. Interpeloidal areas may be open or completely infilled by additional (nonpeloidal) cementation.

it appears that continued growth of peloids reduced the volume of pore water within the cavities and restricted water circulation within and into the cavity, thereby limiting, or self-defeating, further precipitation of cement.

In areas where the cementation process has continued beyond the formation of peloids, the remaining interpeloidal porosity is infilled by a new and distinctly different stage of magnesium-calcite crystal growth. Alexandersson (1972b) described a similar subsequent precipitation of distinct, randomly-oriented interpeloidal crystals, but he believes this entire cementation process to be a "two-stage peloid-to-paste growth mechanism" in the formation of an "apparently featureless micrite" and "as a valid process for micrite growth" along the front of advancing cementation (Alexandersson, 1978). In the study reported here on this dense microcrystalline cement fabric, using SEM I could only rarely distinguish faint circular outlines of original peloids (Figs. 41–43), but in thin section these same samples almost always retained a clotted or peloidal texture that was still recognizable when examined with plane-polarized light. This extremely dense cement mosaic represents the final and ultimate product formed by peloidal submarine cements, and in most cases can be distinguished from featureless micrite.

DISTRIBUTION OF SUBMARINE CEMENTS

Reef Components and Reef Matrix

The distribution of magnesium-calcite submarine cements within the early Holocene barrier reef is facies related. The most significant amount of cementation, and therefore porosity reduction, generally occurred in those reef environments that had higher rates of agita-

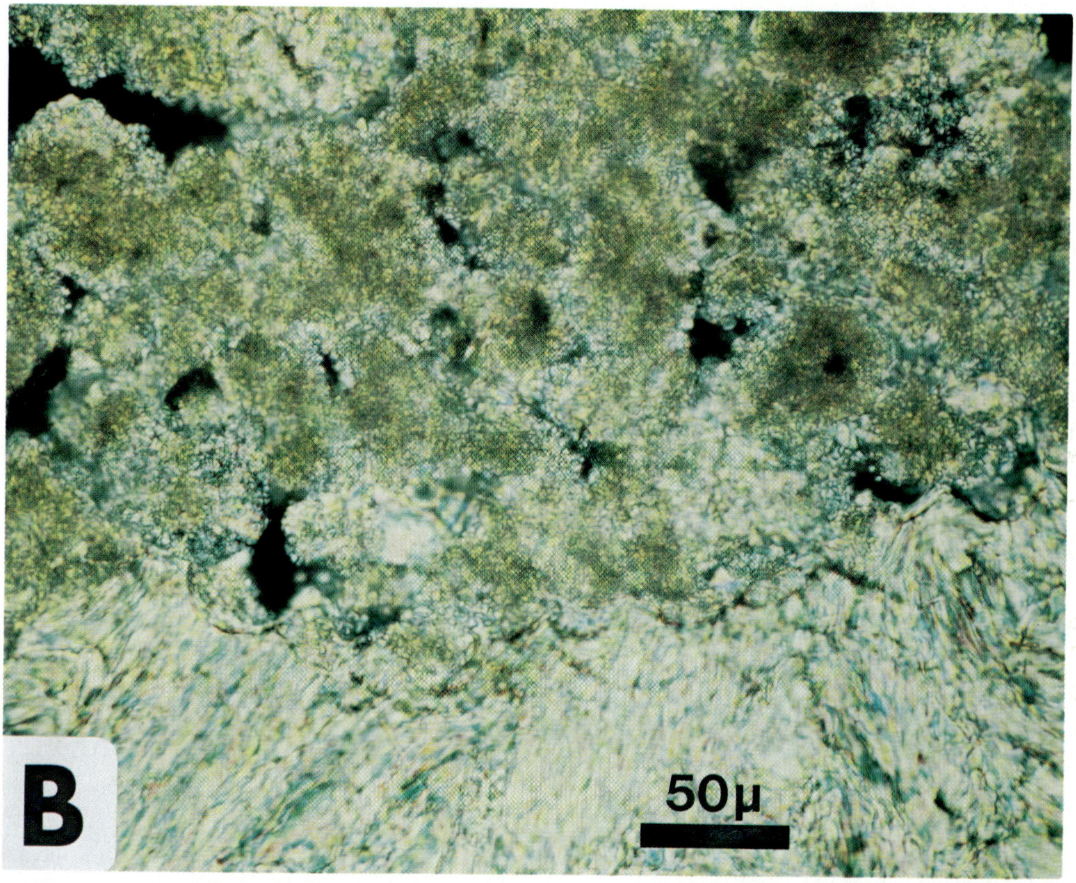

FIG. 11*B*.—Photomicrograph of same view as in Figure 11A, except taken in cross-polarized light. Individual peloids usually are more easily recognized in plane-polarized light, but in cross-polarized light the euhedral crystal rims of magnesium calcite appear as light, transparent fringes surrounding the typically dense, anhedral nuclei. This bimodal internal texture results from two stages of magnesium-calcite precipitation.

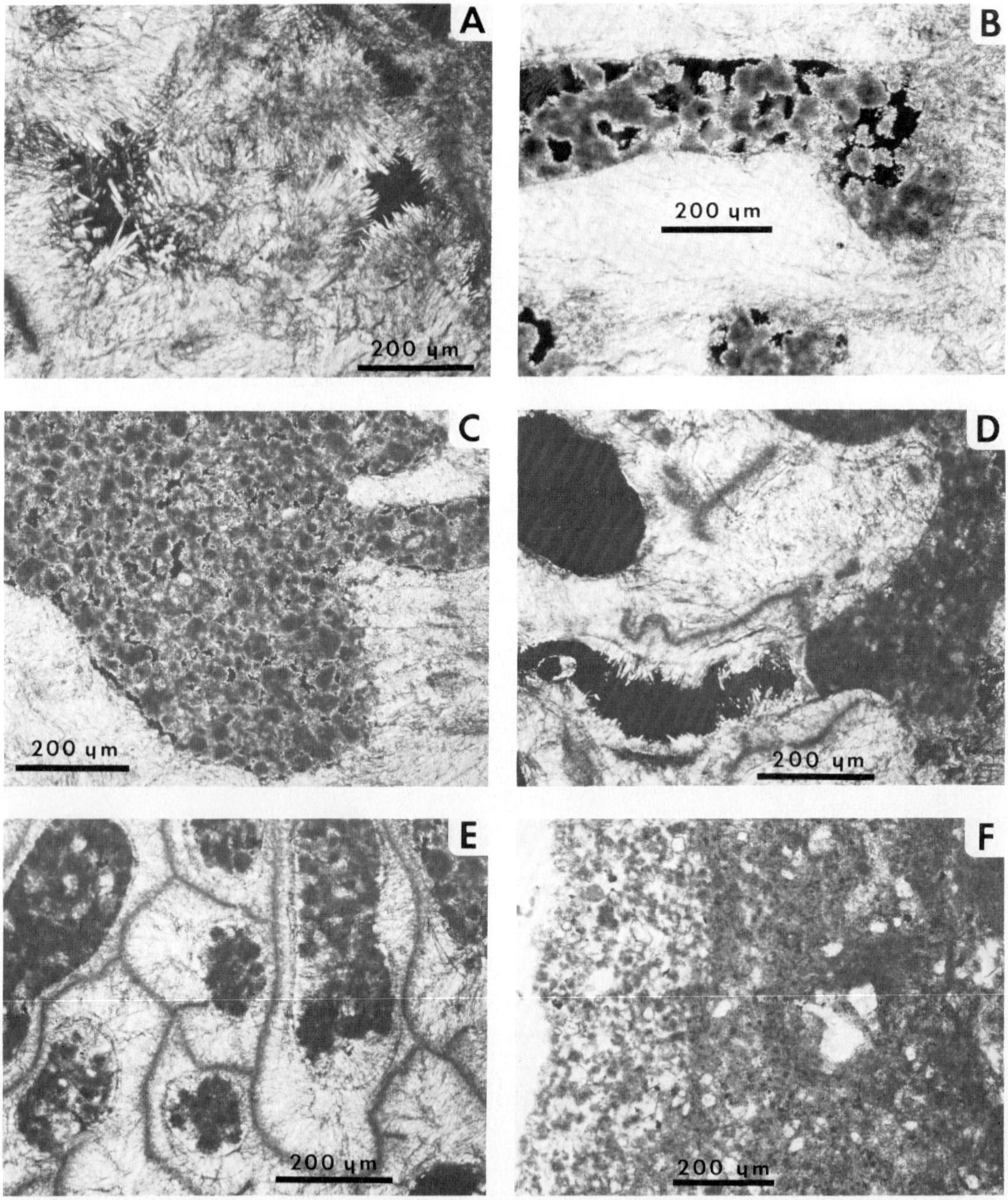

FIG. 12.—Thin section photomicrographs showing distribution of submarine cements within coral skeletal cavities. *A*, Syntaxial overgrowths of aragonite cement filling skeletal cavity in *Acropora palmata*. Cross-polarized light. *B*, Cavity within *Acropora palmata* skeleton partially filled by open mosaic of peloidal magnesium-calcite cement. These cement precipitates often behave as sedimentary clasts and are transported throughout internal pores. Cross-polarized light. *C*, Pore space completely filled by uniform mosaic of peloidal magnesium-calcite cement. Sample located 5 mm beneath surface of *Acropora palmata* skeleton. Cross-polarized light. *D*, Microenvironmental controls on submarine cementation is evident by the different kinds of cements found in adjacent skeletal cavities within *Acropora palmata*. Magnesium-calcite peloidal cement densely fills cavity on right, syntaxial aragonite cement is in lower left, and no cements are present in the upper left cavity. Cross-polarized light. *E*, Complete filling of primary porosity in *Acropora cervicornis* coral by dense cementation of magnesium-calcite peloids. Cross-polarized light. *F*, Crust of magnesium-calcite on surface of *Acropora palmata*, incorporating reef sediment, and ranging (right to left) from dense peloidal mosaic to a less dense, near-surface mosaic, predominantly composed of detrital skeletal fragments. Plane-polarized light.

tion (wave energy, etc.) and/or lower sedimentation rates or framework accumulation rates. Formation of submarine cements is a near-surface phenomenon, with precipitation taking place near the sediment-water interface (Shinn, 1969) and near the reef component-water interface (Land and Goreau, 1970). Magnesium-calcite cements are common within coral skeletons and reef debris in every facies except the back-reef coral head facies, where low agitation and substantial deposition of fine-grained sediment created conditions unfavorable for development of submarine cements. This absence of cementation in protected back-reef areas of fringing and barrier reefs has been documented elsewhere (James *et al.*, 1976; Macintyre, 1977).

Throughout the early Holocene period of prolific reef growth and rapid framework accumulation, magnesium-calcite cements coated and infilled pore cavities in coral skeletons and associated reef rubble in the immediate back-reef *Acropora cervicornis* facies, the *Acropora palmata* framework facies (Figs. 26, 27), the fore-reef coral head facies (Fig. 29), and the fore-reef rubble facies (Figs. 30, 31). Reef sediments deposited around in situ coral within the *Acropora palmata* framework facies and the substantial deposits forming the sedimentary matrix of the back-reef facies were, with rare exception, generally void of cements and unlithified. Intermittant layers of heavily-cemented reef rubble are characteristic of the fore-reef facies

FIG. 13.—Scanning electron micrograph showing magnesium-calcite crystals radiating from a single nucleation point, indicative of well-mixed pore water during precipitation.

FIG. 15.—Scanning electron micrograph of a fractured, average-sized magnesium-calcite peloid found within *Acropora palmata* skeletal cavity. Internal bimodal crystal structure typical of peloids is shown.

FIG. 14.—Scanning electron micrograph of a small magnesium-calcite peloid found within *Acropora palmata* skeletal cavity. Internal bimodal crystal structure revealed by peloid fracturing during sample preparation.

FIG. 16.—Higher magnification SEM micrograph of Figure 15 showing fine-grain uniform texture of peloid center. Rapidly precipitated colloidal magnesium calcite flocculates (self-nucleates) to form substrate for further crystal growth of euhedral magnesium calcite.

(Figs. 28, 30, 31), and probably represent periods of limited or nonexistent fore-reef deposition.

Magnesium-calcite cements form thin dense crusts on the surface of coral skeletons and reef debris, which characteristically incorporate many small encrusting organisms (e.g. bryozoans and the foraminifera *Homotrema rubrum* and *Carpenteria monticularis*), producing an irregular surface that commonly reflects the morphology of the underlying encrusting biota. Sometimes these encrusters are found interlayered with or covering the upper surface of well-developed submarine cement crusts. This indicates that most submarine cement crusts formed before reef components (corals, encrusters, and cement crusts) became buried by the influx or downward filtering of reef sediment. This late-stage sediment fill is still preserved as the generally unlithified reef matrix.

Magnesium-calcite cements were also precipitated within coral skeletal cavities, forming dense infill rims (up to 1 cm thick) within the outer part of the coral skeleton and commonly along the walls of macroborings. The infill rims, consisting of pore-filling magnesium-calcite peloids, appear in sectioned hand samples as a grey-colored zone within the outermost part of the skeleton (Figs. 26, 27, 29), and are light-grey or brown colored in thin section (Figs. 34–36). Where a continuous layer of crustose coralline algae covered the outer surface of a coral skeleton, circulation of marine waters

FIG. 17.—Scanning electron micrograph showing magnesium-calcite peloids filling *Acropora palmata* skeletal cavity. Note absence of extensive crystal growth on detrital skeletal fragment (lower right).

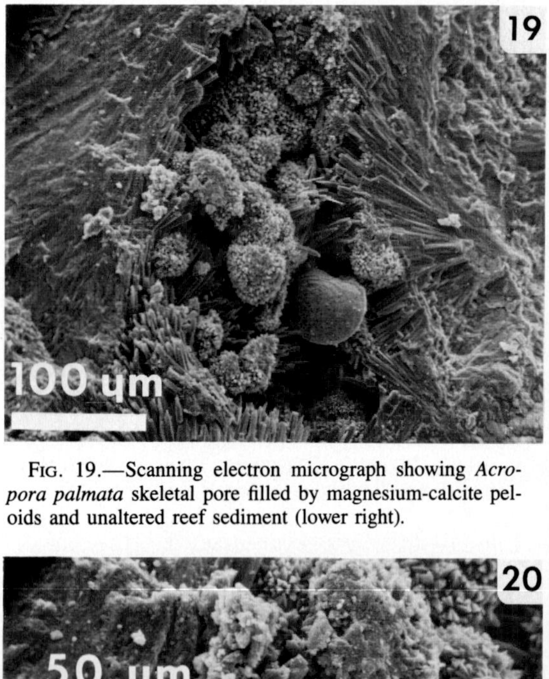

FIG. 19.—Scanning electron micrograph showing *Acropora palmata* skeletal pore filled by magnesium-calcite peloids and unaltered reef sediment (lower right).

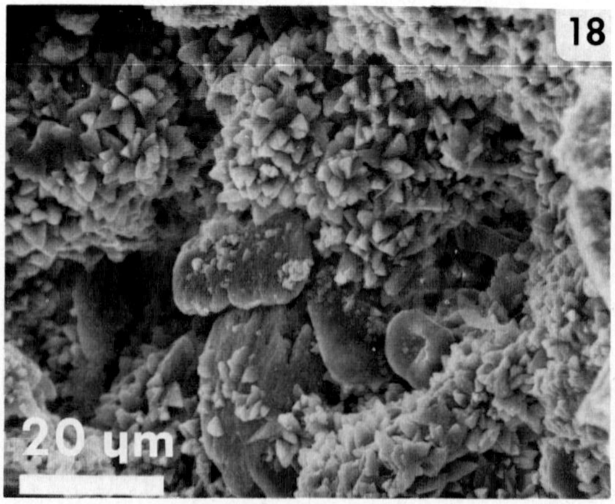

FIG. 18.—Higher magnification SEM micrograph of Figure 17 showing lack of cements on skeletal fragments that filled pore space before or during precipitation of peloidal cement.

FIG. 20.—Higher magnification SEM micrograph of Figure 19 showing peloids overlying earlier syntaxial overgrowths of acicular aragonite.

into the skeleton was effectively cut off, preventing the formation of a magnesium-calcite infill rim (Fig. 34). However, surficial cement crusts often were capable of developing on top of coralline algal encrustations.

The role of permeability barriers, such as encrusting biota, on the distribution of submarine cements within reef components helps to explain the limited penetration of peloidal cement infill in host substrates where there is no apparent restrictive surface deposit and marine waters could have migrated or flushed through unimpeded. Observations made from SEM samples and thin sections, and suggestions by others (e.g. Macintyre, 1977), indicate that the precipitation of peloidal submarine cements within pores of reef components leads to an inevitable reduction in effective permeability. Magnesium-calcite peloids form in the most favorable location, near the surface of the skeleton or debris clast where the influx of marine waters is strongest and more immediate. By their very creation, they become barriers to fluid migration and prevent the inward development of additional cementation and infill. This leads to the commonly observed diagenetic end-product of thin, surface-related cement infill zones or rims. This self-defeating effect of submarine cementation has been noted on a larger scale (thicker zones) for

FIG. 21.—Porosity in *Acropora cervicornis* completely filled by magnesium-calcite cement and earlier precipitation of acicular aragonite; scanning electron micrograph.

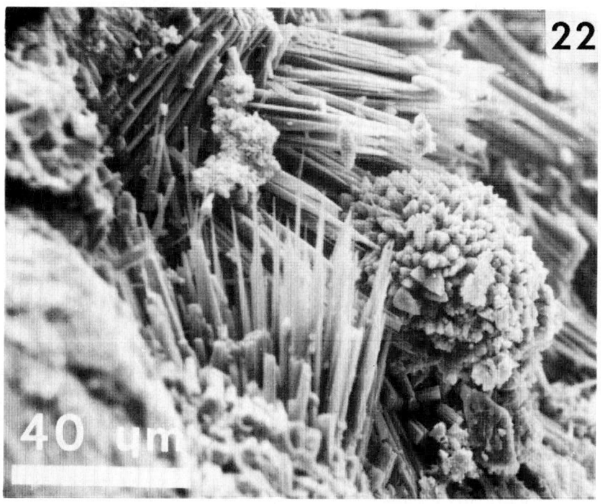

FIG. 22.—Scanning electron micrograph of *Acropora palmata* skeleton showing magnesium-calcite peloid on right and well-preserved acicular aragonite cement on left, indicating precipitation of magnesium calcite was not at expense of aragonite dissolution. Note small, doubly-terminated crystals of magnesium-calcite that developed upon several aragonite crystals (located immediately above peloid).

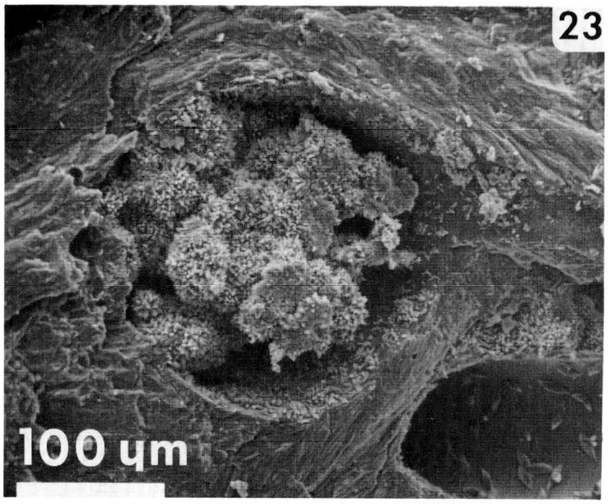

FIG. 23.—Scanning electron micrograph of magnesium-calcite peloidal cement fill within *Acropora palmata* skeletal cavity (in left, but absent in lower right cavity), indicating microenvironment controls on processes of submarine cementation.

FIG. 24.—Higher magnification SEM micrograph of peloids in Figure 23 showing well-developed rhombic crystals of magnesium calcite.

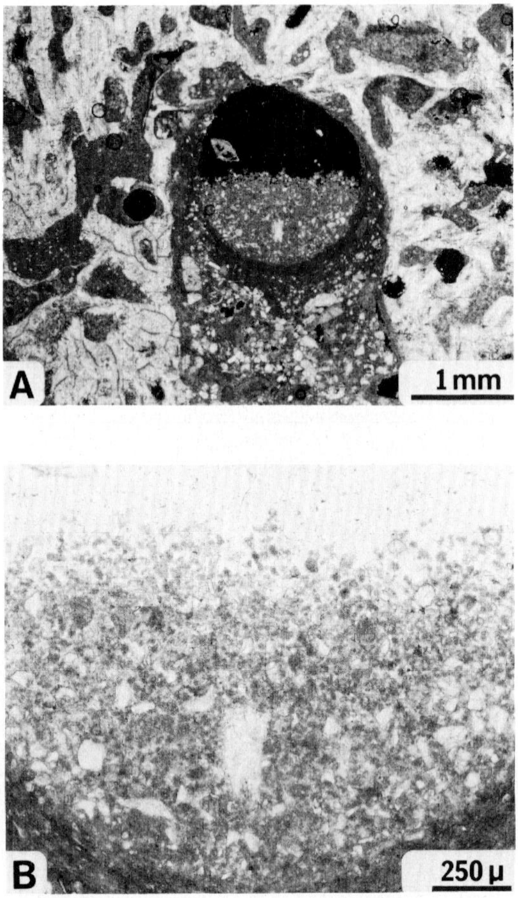

FIG. 25.—*A*, Thin section photomicrograph of magnesium-calcite peloids and internal reef sediment forming geopetal structure within worm tube boring in *Acropora palmata* skeleton. Flat horizontal surface of partial void fill indicates that original sample orientation at time of deposition and cementation was in upwards direction, perpendicular to the upper infill surface. A small miliolid foraminifer is impregnated in left side of geopetal void. Light areas in figure are coral skeleton and dark areas throughout the skeletal material are mainly dense infills of magnesium-calcite peloidal cement. Cross-polarized light. *B*, Close-up photomicrograph of geopetal submarine cement and internal sediment shown in Figure 25A. Individual magnesium-calcite peloids and common carbonate skeletal debris, that in part may be sponge boring chips, are clearly visible. Note that peloids and skeletal grains are well cemented together except near the upper surface of the internal fill. These surface grains and peloids were apparently the last to enter or form within the worm boring, and conditions for further intergranular cementation did not exist or were terminated abruptly. The magnesium-calcite peloids most likely behaved as sedimentary clasts that were transported into the worm boring, or if they were precipitated within the void, at least moved freely about. Plane-polarized light.

submarine hardgrounds formed on nonreefal carbonate packstone-grainstone facies (see Taft *et al.*, 1968; Shinn, 1969). However, within a reef environment (facies) conducive to submarine cementation, and where the size of the reef component is small (minimal cross-sectional area usually less than 4 cm), such as in *Acropora cervicornis* and fore-reef rubble facies, marine waters are able to flow through the entire reef component long enough and strong enough to permit magnesium-calcite formation throughout. The entire reef component can be infilled by cement before the process is terminated by a near-surface localized reduction in permeability. This apparently happened to a wide extent within the early Holocene Florida reef (Figs. 12E, 21, 36).

Macroborings and large cavities within reef components are commonly filled with unaltered skeletal debris and poorly-developed magnesium-calcite cements forming a porous matrix exhibiting a "chalky" texture. These "submicrosucrosic" cements (Macintyre, 1977) consist of small subhedral to euhedral crystals ($<4\mu$), and generally display an extremely fine or incomplete peloidal texture (Figs. 38–40). Submicrosucrosic cements occur where circulation during precipitation is restricted or short-lived. They represent either a last stage of cementation before the pore cavity is completely isolated (Fig. 27), or a final phase of cementation immediately preceding burial of the reef component (Figs. 29, 33).

In the fore-reef coral head and fore-reef rubble facies, and within some well-developed *Acropora palmata* cement crusts, the reef components have undergone extensive or prolonged cementation, and the remaining pore space between well-developed magnesium-calcite peloids is often completely filled by additional magnesium-calcite cement. This forms a highly altered, dense, well-lithified mosaic that still retains a bimodal peloidal texture (Figs. 41–43).

Diagenetic Seal on Reef Framework

In addition to the limited diagenetic alteration of internal reef components during active reef-framework construction, submarine cementation continued to alter nearly the entire, exposed upper surface of the relict reef framework. As the barrier reef became more submerged following its demise, bioerosion, sediment infill, encrusting biota, and extensive submarine cementation transformed the top of the reef into a hard, dark-grey colored rock that is extremely well lithified by a dense micritic matrix of magnesium-calcite cement. This crust-like layer or diagenetic seal is commonly iron stained, has a maximum thickness of one meter, and has undergone so many generations of submarine cementation, boring, and subsequent infill, that the original coral skeletal framework is almost beyond recognition (Figs. 32, 33, 37).

The micritic dark-grey matrix and cement infill, within this laterally extensive zone of pronounced diagenetic alteration, consists of a dense mosaic of magnesium-calcite peloidal cement. Nearly all of the remaining interpeloidal pore space has been filled by a later stage of cementation of usually larger (10μ) magnesium-calcite crystals (Figs. 42 and 43). Macroborings within the diagenetic seal zone commonly are filled by white, friable, submicrosucrosic magnesium-calcite cement (Fig. 33), with minor amounts of unaltered skeletal grains. The characteristic reiterate generations of boring, sediment infill and subsequent lithification, encrusting, and submarine cementation are as complex as those reported for the highly-cemented, slowly-accumulated modern reef rock of Bermuda (Ginsburg and Schroeder, 1973; Focke and Gebelein, 1978) and the Bahamas (Schroeder and Zankl, 1974; Lighty, 1981; Lighty *et al.*, in prep.). Many of the diagenetic aspects of the relict reef's exposed upper surface, when considered with the deep-water reefal community presently living on its surface, are similar to many ancient deep-water carbonate hardgrounds (Purser, 1969; Goldring and Kaźmierczak, 1974; Fürsich, 1979). The main difference is that this hardground surface is covering the top of a submerged relict barrier-reef complex.

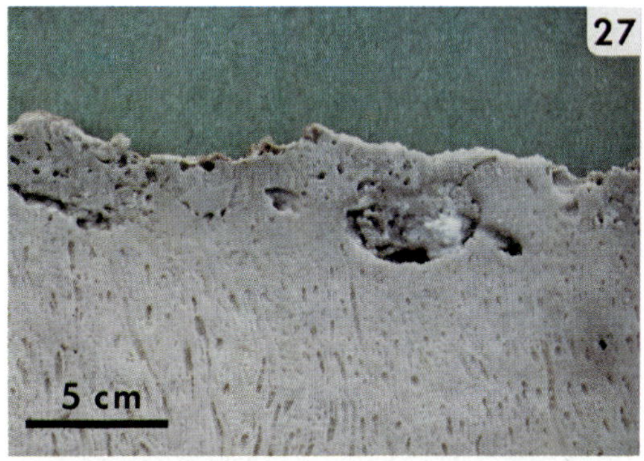

FIG. 26.—Sectioned *Acropora palmata* sample showing grey-colored surficial crust and infill rim of magnesium-calcite submarine cement.

FIG. 27.—Sectioned *Acropora palmata* sample showing 1 cm thick infill rim of magnesium-calcite peloidal cement. Note fresh, unaltered aragonitic coral skeleton beneath the grey-colored zone of submarine cement fill. Red encrusting organism on top surface is common foraminifer *Homotrema rubrum*.

FIG. 28.—Lithified *Acropora cervicornis* rubble and reef sediment, found in fore-reef coral head facies. Red-brown coloration is localized submarine iron staining. Fragments are cemented together by magnesium-calcite cement.

FIG. 29.—Sectioned *Montastrea annularis* coral from fore-reef coral head facies showing extensive penetration and infill of dark grey magnesium-calcite cement along surface of coral skeleton and bordering extensive macroborings, characteristic of head corals from the fore-reef facies. White-colored powder is skeletal sediment and poorly-developed "microsucrosic" magnesium-calcite submarine cement which was the last stage of fill prior to coral's burial by reef sediments.

Diagenetic Model

The total distribution of submarine cements within the relict reef is a function of (1) energy—high agitation being most favorable, (2) time of exposure—low rate of sedimentation or slow framework accumulation being most favorable, and (3) permeability/porosity character—the ability of marine waters to circulate into and through small semi-restricted voids. A qualitative distribution of energy environments and rates of sedimentation/framework accumulation is shown in Figure 44. Ideally, one would expect to find the greatest amount of submarine cementation in a reef environment of high energy and low sedimentation (framework accumulation) and the least amount of submarine cementation in a reef environment of low energy and high sedimentation (accumulation) rates. The observed distribution of magnesium calcite cements within the relict barrier reef is schematically summarized in Figure 45. The cementation pattern depicted emphasizes that the size of the host substrate can be an important over-riding factor in determining the degree of cementation within a reef component, and in some cases even within an entire reef facies. For example, a substantial amount of skeletal porosity is occluded in the immediate back-reef facies—an area subjected to only moderate agitation (wave, tidal, or storm energy) and high rates of facies accumulation—because the small cross-sectional size of *Acropora cervicornis* allowed complete penetration of cement-producing fluids, resulting in extensive development of submarine cements throughout the entire skeleton, destroying a great deal of primary skeletal porosity.

FIG. 30.—Lithified fore-reef rubble of assorted reef components, cemented by crustose coralline algae and magnesium-calcite submarine cement.

FIG. 31.—Lithified fore-reef rubble containing bivalves, serpulids, vermitids, *Halimeda* fragments, and ahermatypic corals deposited with silt- and sand-sized skeletal reef sediment.

FIG. 32.—Submarine iron-stained top surface of dense, micritic reef-framework crust (diagenetic seal, hardground pavement) that has been extensively bored, encrusted, and submarine cemented by magnesium calcite.

FIG. 33.—Side view of lithified reef-framework crust showing dense, micritic dark-grey magnesium-calcite cement and peloidal infill, and pockets of white, friable, submicrosucrosic magnesium-calcite cement formed in cavities with limited access to circulation of marine waters.

In the *Acropora palmata* framework facies, high agitation and an open framework structure during, and immediately following reef growth, counteracted high accumulation rates and allowed the larger-size skeletons to obtain a widespread uniform facies distribution of thin cement crusts and infill rims that may act locally as dense protective seals in preserving primary porosity during later stages of subsurface diagenesis. These rims and crusts have been observed to still retain their dense peloidal texture after alteration (stabilization) to low-magnesium calcite in Pleistocene samples of *A. palmata* and *A. cervicornis* from Barbados (James, 1971, p. 71–74) and in Pleistocene *A. palmata* from St. Croix (Macintyre, 1977, verbal communication). Micritic peloidal textures have been interpreted as products of submarine diagenesis in Miocene bioherms from Malta (Pedley, 1979), a Cretaceous patch reef from Texas (Petta, 1977), and forming surficial crusts and infill rims within a Devonian bioherm from western Canada (Mountjoy and Riding, 1981).

The more or less impermeable cement crusts and infill rims have the potential to preserve a considerable amount of original skeletal porosity or form rigid, diagenetically-stable structural rims around later dissolved skeletons, thereby protecting voids from compaction and infill during burial. One of the key factors in the development of these cement crusts is the associated rapid rate of facies accumulation. Significantly, rela-

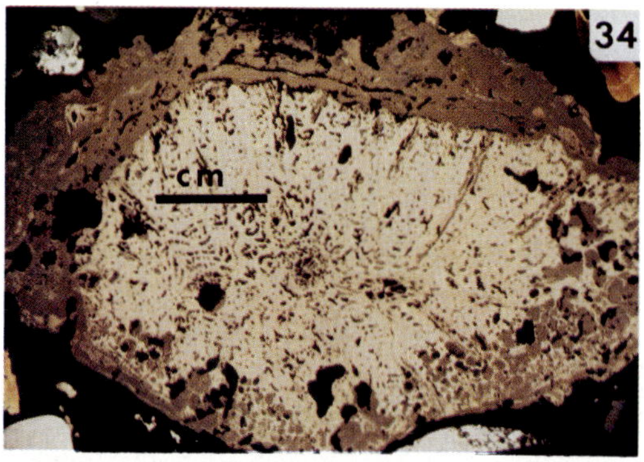

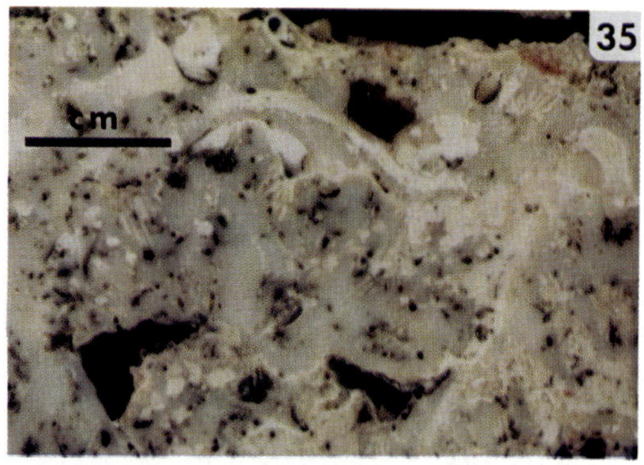

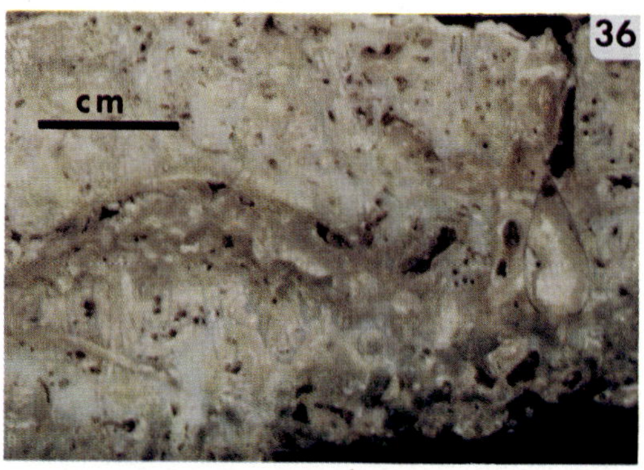

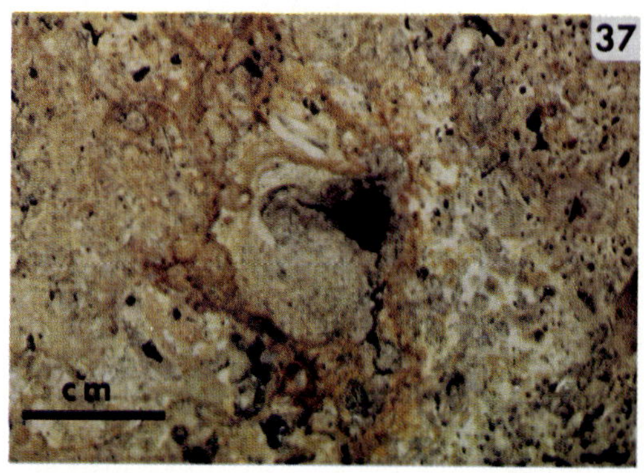

FIG. 34.—Polished, plastic-impregnated slab of sectioned *Acropora palmata* in reflected light showing unaltered, porous aragonitic skeleton in center, and thin infill rim of magnesium-calcite peloidal cement along bottom surface (brown coloration). Note absence of peloidal fill beneath surface that is encrusted by crustose coralline algae (top surface).

FIG. 35.—Polished, plastic-impregnated slab of fore-reef coral *Montastrea annularis* in reflected light showing partial fill of skeletal cavities by tan-colored magnesium-calcite submarine cement.

FIG. 36.—Polished, plastic-impregnated slab in reflected light showing fore-reef coral rubble with extensive fill of light-brown colored magnesium-calcite cement, occluding over 80 percent of the skeletal porosity. Note that submarine cement is also filling shell of boring bivalve.

FIG. 37.—Polished, plastic-impregnated slab in reflected light showing reef-framework crust (diagenetic seal) in which extensive submarine cementation has altered original skeletal composition beyond recognition. Note the iron-stained dissolution (bioerosion?) surface in center of photograph.

tively high accumulation rates are generally interpreted for most volumetrically dominant reef facies in ancient reef complexes.

Within the early Holocene reef, protective, impermeable cement crusts and infill rims are only partially developed in fore-reef coral heads and reef debris because of a high degree of macroboring (Fig. 29) and numerous thick encrustations of coralline algae. This has resulted in the formation of abundant macroporosity, along with the preservation of some primary skeletal porosity and occlusion of much microporosity by well advanced, but spatially limited submarine cementation. Extensive cementation of the upper surface of the submerged relict reef was made possible by extremely low accumulation rates and by sustained interstitial flushing of overlying shelf waters driven by the Florida Current (Gulf Stream), producing moderately-high energy conditions at water depths of 15 m. It is not uncommon to find strong ocean currents and very slow accretion rates resulting in pervasive submarine cementation at even greater depths, such as 50 m in the fore-reef slope of Bermuda (Focke and Gebelein, 1978) and 600 to 700 m in the Straits of Florida (Neumann *et al.*, 1977). This cementation has, in effect, created an extensive low-permeability cement crust over the entire relict reef framework, forming a di-

FIG. 38.—Scanning electron micrograph showing poorly-developed, white submicrosucrosic magnesium-calcite cement and fresh skeletal debris (upper right) that has filled a boring-bivalve shell within a sample of fore-reef rubble.

FIG. 39.—Higher magnification SEM micrograph of Figure 38 showing poorly-developed magnesium-calcite crystals and abundant skeletal fragments.

FIG. 40.—Scanning electron micrograph showing submicrosucrosic magnesium-calcite cement that occurs where pore-water circulation once adequate for precipitation of magnesium-calcite is restricted in early or late stages of peloidal cementation. Sample is from *Acropora palmata* infill rim.

FIG. 41.—Scanning electron micrograph of dense mosaic of magnesium-calcite cement that formed within well-lithified fore-reef rubble. Note large magnesium-calcite crystals filling remaining pore space (center) between peloids.

agenetic seal that has protected the highly porous internal reef facies from further marine alteration and cementation.

The factors that controlled submarine cementation within the exposed upper surface of the reef following the reef's demise are similar to those that control cementation within submarine hardgrounds. During active reef growth, the formation of submarine cements within the reef's internal structure was facies related, but cementation was not only confined to the fore-reef and reef-framework facies, as observed in modern reefs of Belize (James *et al.*, 1976) and Panama (Macintyre, 1977). Submarine cements are also present in the immediate back-reef facies, similar to the cement distribution in St. Croix reefs (Zankl and Multer, 1977). Extensive submarine lithification also has been noted in back-reef lagoonal patch reefs of Enewetak Atoll (Warme *et al.*, 1980) and the Great Barrier Reef (Marshall and Davies, 1981). The fact that back-reef submarine cementation was not recognized in most previous investigations of modern linear reefs probably is because these studies are based on core samples that very rarely recovered the loose rubble and sediment that usually dominates back-reef deposits. In the present study, abundant sample material was collected from the trench wall outcrop of back-reef facies. Analysis of these samples indicates a significant amount of submarine cement infilling within skeletal components and reef debris in the immediate back-reef area, thus effecting significant reef porosity. There was only minor interclastic lithification of coral and sediment deposits, which may account for the poor recovery from this type of facies in reef coring operations.

SUMMARY AND CONCLUSIONS

A 95-km long relict shelf-edge reef is located off southeast Florida in an area considerably north of the present-day Florida reef tract. The internal structure of this submerged barrier reef was exposed during excavation of an offshore pipeline trench 40 km north of Miami. Five reef facies were identified: back-reef coral head, back-reef *Acropora cervicornis*, *Acropora palmata* framework, fore-reef coral head, and fore-reef rubble. Radiocarbon dates indicate that the *A. palmata* reef flourished during the early Holocene (ca. 9,400 to 7,000 yrs. ago), with framework accumulation rates as high as 10.7 m/1000 years. Distribution of reef facies suggests that there was a predominantly vertical reef-framework accumulation (at least 10 m), with some lateral extension of *A. palmata* framework over substantial back-reef deposits as the fast-growing *A. palmata* kept pace with rapidly rising sea level.

An abrupt change in back-reef shelf water conditions brought about the demise of the shelf-edge coral reef during the time of initial Holocene transgression of most of the Florida continental shelf. Turbidity derived from transgressive erosion of soils and lagoonal carbonate muds, and seasonal cold-water stress conditions formed within the vast, shallow incipient shelf waters, apparently were transported off the shelf and caused the rapid demise of the shelf-edge barrier reef. Such that, following a period of rapid vertical buildup the reef was killed off, with its relict structure submerged and unburied at the shelf edge as sea level continued to rise.

Various submarine cements are found within the relict barrier reef. Microcrystalline magnesium calcite

FIG. 42.—Scanning electron micrograph showing dense mosaic of magnesium-calcite cement that comprises a substantial part of the well-lithified reef-framework crust (diagenetic seal). All primary porosity has been occluded. Note bimodal appearance of cement texture.

FIG. 43.—Higher magnification SEM micrograph of Figure 42 showing bimodal texture of cement. Larger magnesium-calcite crystals occur as latter stage cementation, filling remaining void space between peloids. Spherical pattern produced by this last separate precipitation is mainly a function of the shape of remaining pore space between closely-spaced magnesium-calcite peloids.

forms rare dentate rims and abundant peloidal cement that fills near surface voids and forms thin dense crusts on corals and reef rubble. Peloidal cement crusts are often intercalated with skeletal deposits of cryptic encrusting biota, indicating an early (preburial) timing for cementation. Skeletal porosity that is isolated from marine circulation is devoid of submarine cements, while minor amounts of acicular aragonite commonly form overgrowths within skeletal cavities open to marine circulation during coral growth. Magnesium-calcite cements formed in porosity that remained open or later became open to marine circulation (by borings or abrasion shortly after coral growth).

Magnesium-calcite peloids, the predominant submarine cement, exhibit a bimodal texture of dentate rhombic crystals around a nondescript anhedral nucleus of microclotted magnesium calcite. This characteristic bimodal composition is believed to be the result of two stages of precipitation—an initial rapid stage of colloidal precipitation, which then flocculates to form self-nucleating substrates for the next stage of slower and more organized precipitation of euhedral magnesium calcite. Peloidal cementation within reef porosity ranges from poorly-lithified open mosaics, which often form geopetal structures, to dense well-cemented mosaics that have interpeloidal areas further infilled by additional nonpeloidal magnesium-calcite crystal growth.

During reef growth, submarine cements best developed in reef facies with substantial water movement and low sedimentation (framework accumulation) rates. In addition, clast size of the host reef-component was an over-riding factor, with porosity in smaller-sized skeletons and reef rubble often completely filled by submarine cements. For example, submarine cementation in the high-energy *Acropora palmata* facies was limited by the brief time available for extensive submarine diagenesis prior to decreasing energy conditions and burial as reef framework rapidly accumulated above. The low-energy/high-sedimentation conditions in sheltered back-reef environments were unfavorable for precipitation of submarine cements, and none were observed in samples from the back-reef coral head facies. However, the small size of coral skeletons and moderate energy conditions in the immediate back-reef area did allow submarine cements to extensively infill skeletal porosity in the smaller branching coral *Acropora cervicornis*. An unlithified late-stage sedimentary

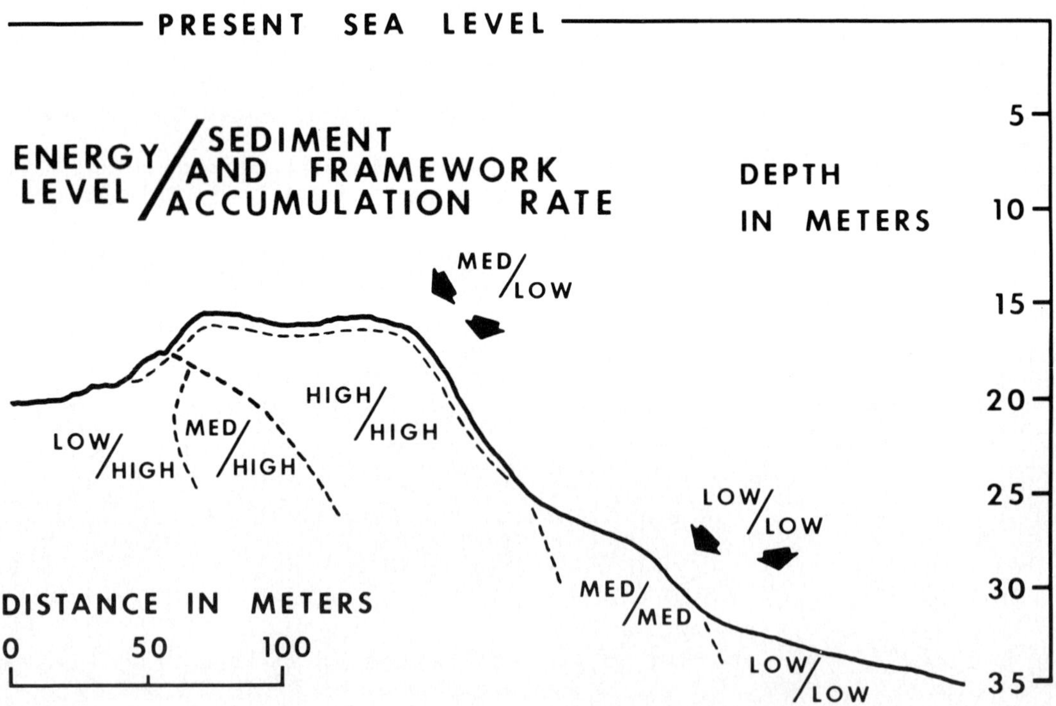

FIG. 44.—Cross section showing variations in energy environments and rates of sedimentation, including reef-framework accumulation, that occurred throughout the early Holocene reef facies and are presently occurring on the exposed upper surface of the relict reef. These parameters directly influence the distribution and occurrence of magnesium-calcite submarine cementation, with most extensive cementation found in areas of high energy/low sedimentation and minimum cementation in areas of low energy/high sedimentation.

matrix encases the *in situ A. palmata* framework and surrounds corals of the fore-reef and back-reef facies. Fore-reef corals are more highly bored and cemented, and intermittent layers of well-cemented reef rubble and sediment are present in the fore-reef talus, both as a result of low rates of sedimentation.

Following demise of the reef, a submarine-lithified crust (hardground) developed over the barrier reef. Extensive cementation by peloidal magnesium-calcite and destructive effects of repeated boring by organisms has altered the original skeletal composition of the uppermost framework almost beyond recognition. If reef growth ever resumed, new reef zonation would most likely mimic relict facies patterns, and this diagenetic hardground pavement over the relict reef structure would most likely represent a disconformity separating two similar facies.

The "cap rock" role of the dense well-lithified pavement over the relict reef, and the facies-related distribution of submarine cements within the relict structure, provide a most effective control on the entire reef's primary porosity. Unlike many modern reefs, the internal structure of this early Holocene reef is characterized by the preservation of a great deal of skeletal porosity and only limited submarine cementation, due to its high facies accumulation rates in response to the rapid rise in sea level. The formation of low-permeability crusts and infill rims on the surface of the dominant coral, *Acropora palmata*, as well as the early-stage formation of a dense, micritic diagenetic seal over the relict reef structure, may ultimately influence late-stage diagenesis of this reef.

The impermeable nature of both the reef-component crusts and the overlying reef-framework crust may act as early seals or impedances to fluid migration, whether it be meteoric or subsurface waters, or hydrocarbons. Because the barrier reef was eventually outpaced by rising sea level, it has never been subaerially exposed and, hence, has never undergone meteoric diagenesis. It is conceivable that this reef may be buried and continually subside with the Florida continental margin to remain within the marine phreatic realm, mainly subject to possible intrusion of basinal subsurface fluids. This reef's depositional history provides a better Holocene analog for many ancient reefs.

Seismic reflection profiling of modern shelves throughout the Caribbean (Goreau and Burke, 1966; Macintyre, 1972; d'Anglejan and Mountjoy, 1973),

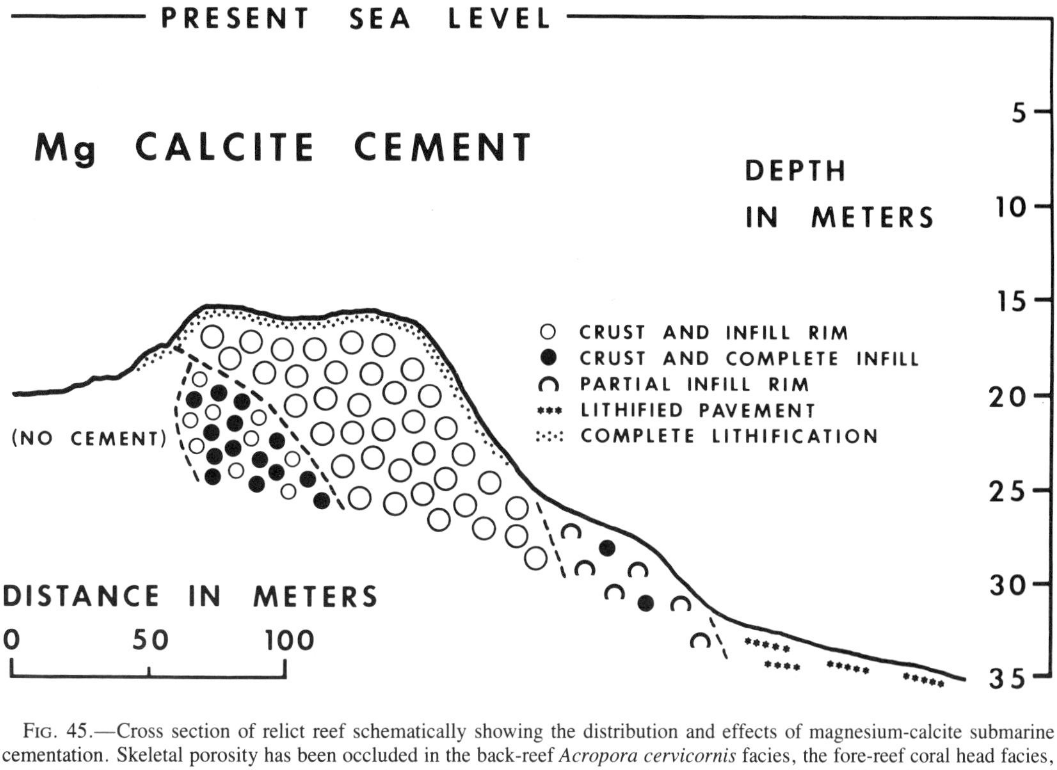

FIG. 45.—Cross section of relict reef schematically showing the distribution and effects of magnesium-calcite submarine cementation. Skeletal porosity has been occluded in the back-reef *Acropora cervicornis* facies, the fore-reef coral head facies, and within the reef-framework crust (diagenetic seal) that has developed on the exposed top surface of the relict reef structure. This diagenetic seal has protected internal reef components and sediment matrix from further submarine alteration. In addition, thin infill rims and surficial cement crusts on corals, especially on the abundant *Acropora palmata*, may act to preserve or influence subsurface diagenesis of substantial primary porosity within coral skeletons.

Bermuda (Stanley and Swift, 1968), and the Bahamas (Hine and Neumann, 1977; Lighty and Neumann, 1979), along with core studies off St. Croix (Adey *et al.*, 1977), have revealed numerous submerged shelf-edge reefs and reef-like structures that most likely experienced a similar history of rapid vertical growth and subsequent diagenetic sealing. Given the number and magnitude of ancient sea level fluctuations (Vail *et al.*, 1977), it is probable that the specific depositional/diagenetic/sea level scenarios necessary for submarine diagenetic sealing of porous carbonate buildups have occurred in the geologic past. The most likely carbonate stratigraphic settings would be developed on rapidly subsiding (cooling) oceanic crust, on tectonically-active tropical margins, and on down-faulted crustal blocks found under slopes and rises of passive continental margins. And indeed, drowned (submerged) reefs (Krebs, 1972; Matthews *et al.*, 1974; Playford, 1980; Read, 1982) and drowned carbonate platforms and shelves (Bosellini, 1973; Krebs, 1974; Bova, 1980; Schlager, 1981) are being recognized in the ancient record.

In conclusion, this study discusses the growth, demise, and submarine diagenesis of a submerged early-Holocene barrier reef on the southeast Florida continental shelf. Major contributions include: documentation of reef facies response to a rapid sea level rise, a new model for termination of reef growth, evidence that submarine cementation in reefs may not always be pervasive, a greater understanding of origin and significance of peloidal cements, and a new model for the development of diagenetic seals in carbonate facies. Hopefully, an understanding of these Holocene processes of formation and postdepositional alteration can aid geologists in recognizing similar depositional and early diagenetic histories recorded in ancient carbonate buildups.

ACKNOWLEDGMENTS

Sample collection supported by a S.E.E.D. grant to Florida Atlantic University Department of Geology and with the cooperation of Powell Brothers, Inc., Ft. Lauderdale, Florida. Special thanks to E.M. Clark, Jr. for assistance in diving operations. Most laboratory work conducted at Duke University Department of Geology, and at Smithsonian Institution Department of Paleobiology while author was a Visiting Research Fellow in Division of Sedimentology. W.R. Brown provided SEM support and coral radiocarbon dates were determined by R. Stuckenrath, Smithsonian Institution Radiation Biology Laboratory. With my deepest appreciation, I thank R.R.H. Lemon, K.A. Lighty, R.G. Loucks, G.W. Lynts, I.G. Macintyre, A.C. Neumann, R.D. Perkins, and R.M. Slatt for discussing and reviewing various evolving versions of this manuscript. Publication funded by the Texas A&M University Department of Geology and the Center for Tropical Marine Research, College Station, Texas.

REFERENCES

Adey, W.H., 1975, The algal ridges and coral reefs of St. Croix: Their structure and Holocene development: Atoll Research Bull. 187, 67 p.

______, 1977, Shallow water Holocene bioherms of the Caribbean Sea and West Indies, *in* Proceedings of Third International Coral Reef Symposium, v. 2: Miami, Florida, Atlantic Reef Committee, p. xxi–xxiv.

______, and Burke, R.B., 1977, Holocene bioherms of Lesser Antilles—geologic control of development: Amer. Assoc. Petroleum Geologists Studies in Geology No. 4, p. 67–81.

______, Macintyre, I.G., Stuckenrath, R., and Dill, R.F., 1977, Relict barrier reef system off St. Croix: its implications with respect to late Cenozoic coral reef development in the western Atlantic, *in* Proceedings of Third International Coral Reef Symposium, v. 2: Miami, Florida, Atlantic Reef Committee, p. 15–21.

Alexandersson, T., 1972a, Intragranular growth of marine aragonite and Mg-Calcite: Evidence of precipitation from supersaturated seawater: Jour. Sed. Petrology, v. 42, p. 441–460.

______, 1972b, Mediterranean beachrock cementation: marine precipitation of Mg-calcite, *in* The Mediterranean Sea, D.J. Stanley (ed.): Dowden, Hutchinson & Ross, Stroudsburg, Pennsylvania, p. 203–223.

______, 1978, Distribution of submarine cements in a modern Caribbean fringing reef, Galeta Point, Panama: Discussion: Jour. Sed. Petrology, v. 48, p. 665–668.

Ball, M.M., Shinn, E.A., and Stockman, K.W., 1967, The geologic effects of Hurricane Donna in south Florida: Jour. Geology, v. 75, p. 583–597.

Bosellini, A., 1973, Modello geodinamico e paleotettonico delle Alpi Meridionali durante il Giurassico-Cretacico: Academia Nazionale Dei Lincei, v. 183, p. 164–190.

Bova, J.A., 1980, Lower Ordovician Chapultepec-Stonehenge sequence: An ancient analogue of a drowned platform: Geol. Soc. America, Abstracts with Programs, v. 12, p. 390.

Buddemeier, R.W., 1974, Environmental controls over annual and lunar monthly cycles in hermatypic coral calcification, *in* Proceedings of Second International Coral Reef Symposium, v. 2: Brisbane, Australia, Great Barrier Reef Committee, p. 259–267.

Cuffey, R.J., 1974, Delineation of bryozoan constructional roles in reefs from comparison of fossil bioherms and living reefs, *in* Proceedings of Second International Coral Reef Symposium, v. 1: Brisbane, Australia, Great Barrier Reef Committee, p. 357–364.

d'Anglejan, G., and Mountjoy, E.W., 1973, Submerged reefs of the eastern Grenadines shelf margin: Geol. Soc. America Bull., v. 84, p. 2445–2453.

DODGE, R.E., AND VAISNYS, J.R., 1975, Corals as environmental indicators: Geol. Soc. America, Abstracts with Programs, v. 7, p. 1055.
DUANE, D.B., AND MEISBURGER, E.P., 1969, Geomorphology and sediments of the nearshore continental shelf, Miami to Palm Beach, Florida: U.S. Army Corps Engineers Coastal Eng. Research Center Tech. Memo 29: Washington, D.C., p. 8–14.
ENOS, P., 1977, Holocene sediment accumulations of the south Florida shelf margin, *in* P. Enos and R.D. Perkins (eds.), Quaternary sedimentation in south Florida: Geol. Soc. America Mem. 147, p. 1–130.
FOCKE, J.W., AND GEBELEIN, C.D., 1978, Marine lithification of reef rock and rhodolites at a fore-reef locality (−50 m) off Bermuda: Geologie en Mijnbouw, v. 57, p. 163–171.
FRIEDMAN, G.M., AMIEL, A.J., AND SCHNEIDERMANN, N., 1974, Submarine cementation in reefs: example from the Red Sea: Jour. Sed. Petrology, v. 44, p. 816–825.
FÜRSICH, F.T., 1979, Genesis, environments, and ecology of Jurassic hardgrounds: Neues Jb. Geol. Paläont. Abh., v. 158, p. 1–63.
GINSBURG, R.N., 1956, Environmental relationships of grain size and constituent particles in some south Florida carbonate sediments: Amer. Assoc. Petroleum Geologists Bull., v. 40, p. 2384–2427.
______, AND SCHROEDER, J.H., 1973, Growth and submarine fossilization of algal cup reefs, Bermuda: Sedimentology, v. 20, p. 575–614.
______, MARSZALEK, D.S., AND SCHNEIDERMANN, N., 1971, Ultrastructures of carbonate cements in a Holocene algal reef of Bermuda: Jour. Sed. Petrology, v. 41, p. 472–482.
GLADFELTER, E.H., AND MONAHAN, R.K., 1977, Primary production and calcium carbonate deposition rates in *Acropora palmata* from different positions in the reef, *in* Proceedings of Third International Coral Reef Symposium, v. 2: Miami, Florida, Atlantic Reef Committee, p. 389–394.
GOLDBERG, W.M., 1973, The ecology of the coral-octocoral communities off the southeast Florida coast: Geomorphology, species composition, and zonation: Bull. Marine Science, v. 23, p. 465–488.
GOLDRING, R., AND KAŹMIERCZAK, J., 1974, Ecological succession in intraformational hardground formation: Paleontology, v. 17, p. 949–962.
GOREAU, T., AND BURKE, K., 1966, Pleistocene and Holocene geology of the island shelf near Kingston, Jamaica: Marine Geology, v. 4, p. 207–225.
HINE, A.C., AND NEUMANN, A.C., 1977, Shallow carbonate-bank-margin growth and structure, Little Bahama Bank, Bahamas: Amer. Assoc. Petroleum Geologists Bull., v. 61, p. 376–406.
HUBBARD, J.A.E.B., 1975, Life and afterlife of reef corals: A timed study of incipient diagenesis: Abstracts IXth international sedimentological congress (Nice), Theme 7, p. 75–80.
HUDSON, J.H., SHINN, E.A., HALLEY, R.B., AND LIDZ, B., 1976, Sclerochronology: A tool for interpreting past environments: Geology, v. 4, p. 361–364.
JAMES, N.P., 1971, Late Pleistocene reef limestones, northern Barbados, W.I., Unpub. Ph.D. dissertation, McGill Univ., Montreal, 282 p.
______, GINSBURG, R.N., MARSZALEK, D.S., AND CHOQUETTE, P.W., 1976, Facies and fabric specificity of early subsea cements in shallow Belize (British Honduras) reefs: Jour. Sed. Petrology, v. 46, p. 523–544.
JINDRICH, V., 1983, Structure and diagenesis of recent algal-foraminifer reefs, Fernando de Noronha, Brazil: Jour. Sed. Petrology, v. 53, p. 449–459.
KORNICKER, L.S., BONET, F., CANN, R., AND HOSKIN, C.M., 1959, Alacran reef, Campeche Bank, Mexico: Pub. Inst. Marine Science, Univ. Texas, v. 6, p. 1–22.
KNUTSON, D.W., BUDDEMEIER, R.W., AND SMITH, S.V., 1972, Coral chronometers: Seasonal growth bands in reef corals: Science, v. 177, p. 270–272.
KREBS, W., 1972, Facies and development of the Meggen Reef (Devonian, West Germany): Geol. Rundschau, v. 61, p. 647–671.
______, 1974, Devonian carbonate complexes of central Europe, *in* L.F. Laporte (ed.), Reefs in Time and Space: Soc. Econ. Paleontologists and Mineralogists Spec. Pub. 18, p. 155–208.
LAGAAIJ, R., 1963, New additions to the bryozoan fauna of the Gulf of Mexico: Pub. Inst. Marine Science Univ. Texas, v. 9, p. 162–236.
LAND, L.S., AND GOREAU, T.F., 1970, Submarine lithification of Jamaican reefs: Jour. Sed. Petrology, v. 40, p. 457–462.
______, AND MOORE, C.H., 1980, Lithification, micritization and syndepositional diagenesis of biolithites on Jamaican Island slope: Jour. Sed. Petrology, v. 50, p. 357–369.
LIGHTY, R.G., 1977, Relict shelf-edge Holocene coral reef: southeast coast of Florida, *in* Proceedings of Third International Coral Reef Symposium, v. 2: Miami, Florida, Atlantic Reef Committee, p. 215–221.
______, 1981, Fleshy-algal domination of a modern Bahamian barrier reef: Example of an alternate climax reef community, *in* Proceedings of the Fourth International Coral Reef Symposium, v. 1: Quezon City, Philippines, Marine Sciences Center, Univ. the Philippines, p. 722.
______, AND NEUMANN, A.C., 1979, Fore-reef shelf and marginal escarpment of northern Little Bahama Bank, Bahamas [abs.]: Geol. Soc. America, Abstracts with Programs, v. 11, p. 187.
______, MACINTYRE, I.G., AND NEUMANN, A.C., 1980, Demise of a Holocene barrier-reef complex, northern Bahamas [abs.]: Geol. Soc. America, Abstracts with Programs, v. 12, p. 471.
______, ______, AND STUCKENRATH, R., 1978, Submerged early Holocene barrier reef south-east Florida shelf: Nature, v. 276, p. 59–60.

______, ______, and ______, 1982, *Acropora palmata* reef framework: A reliable indicator of sea level in the western Atlantic for the past 10,000 years: Coral Reefs, v. 1, p. 125–130.

______, Neumann, A.C., and Macintyre, I.G., in prep., Late Holocene demise of a Bahamian barrier reef: Geologic and ecologic consequences: Submitted to Jour. Sed. Petrology.

Logan, B.W., 1969, Coral reefs and banks, Yucatan shelf, Mexico (Yucatan reef unit), *in* Carbonate sediments and reefs, Yucatan shelf, Mexico: Amer. Assoc. Petroleum Geologists Mem. 11, p. 129–198.

Macintyre, I.G., 1972, Submerged reefs of eastern Caribbean: Amer. Assoc. Petroleum Geologists Bull., v. 56, p. 720–738.

______, 1975, A diver-operated hydraulic drill for coring submerged substrates: Atoll Research Bull. 185, p. 21–25.

______, 1977, Distribution of submarine cements in a modern Caribbean fringing reef, Galeta Point, Panama: Jour. Sed. Petrology, v. 47, p. 503–516.

______, 1978, Distribution of submarine cements in a modern Caribbean fringing reef, Galeta Point, Panama: Reply: Jour. Sed. Petrology, v. 48, p. 669–670.

______, and Glynn, P.W., 1976, Evolution of a modern Caribbean fringing reef, Galeta Point, Panama: Amer. Assoc. Petroleum Geologists Bull., v. 60, p. 1054–1072.

______, and Milliman, J.D., 1970, Physiographic features on the outer shelf and upper slope, Atlantic continental margin, southeastern United States: Geol. Soc. America Bull., v. 81, p. 2577–2598.

______, Mountjoy, E.W., and d'Anglejan, B.F., 1968, An occurrence of submarine cementation of carbonate sediments off the west coast of Barbados, W.I.: Jour. Sed. Petrology, v. 38, p. 660–664.

______, and Smith, S.V., 1974, X-radiographic studies of skeletal development in coral colonies, *in* Proceedings of Second International Coral Reef Symposium, v. 2: Brisbane, Australia, Great Barrier Reef Committee, p. 277–287.

Marshall, J.F., and Davies, P.J., 1981, Submarine lithification on windward reef slopes: Capricorn-Bunker Group, southern Great Barrier Reef: Jour. Sed. Petrology, v. 51, p. 953–960.

Matthews, J.L., Heezen, B.C., Catalano, R., Coogan, A., Tharp, M., Natland, J., and Rawson, M., 1974, Cretaceous drowning of reefs on Mid-Pacific and Japanese guyots: Science, v. 184, p. 462–464.

Mesolella, K.J., Sealy, H.A., and Matthews, R.K., 1970, Facies geometries within Pleistocene reefs of Barbados, West Indies: Amer. Assoc. Petroleum Geologists Bull., v. 54, p. 1899–1917.

Montaggioni, L., 1977, Structure interne d'un récif corallien holocéne (île de La Réunion, Océan Indien), *in* Memoires du B.R.G.M., No. 89, Second International Symposium on Corals and Fossil Coral Reefs: Paris, September 1975, p. 456–466.

Moore, H.B., 1969, Miami sea temperatures and salinities: Inst. Marine Sciences, Univ. Miami Tech. Report ML69039, p. 1–8.

Moore, W.S., and Krishnaswami, S., 1974, Correlation of x-radiography revealed banding in corals with radiometric growth rates, *in* Proceedings of Second International Coral Reef Symposium, v. 2: Brisbane, Australia, Great Barrier Reef Committee, p. 269–276.

Mountjoy, E.W., and Riding, R., 1981, Foreslope stromatoporoid-renalcid bioherm with evidence of early cementation, Devonian Ancient Wall reef complex, Rocky Mountains: Sedimentology, v. 28, p. 299–319.

Neumann, A.C., 1971, Quaternary sea-level data from Bermuda: Quaternaria, v. XIV, p. 41–43.

______, Kofoed, J.W., and Keller, G.H., 1977, Lithoherms in the Straits of Florida: Geology, v. 5, p. 4–10.

Newell, N.D., and Rigby, J.K., 1957, Geological studies on the Great Bahama Bank, *in* Regional aspects of carbonate deposition: Soc. Econ. Paleontologists and Mineralogists Spec. Pub. 5, p. 15–72.

Osburn, R.C., 1940, Bryozoa of Porto Rico with a resume of the West Indian bryozoan fauna. Scientific survey of Porto Rico and Virgin Islands: New York Acad. Sci., v. 16, no. 3, p. 321–486.

Parker, G.G., Hoy, N.D., and Schroeder, M.C., 1955, *in* Water resources of southeastern Florida: U.S. Geological Survey Water-Supply Paper 1255, p. 57–125.

Pedley, H.M., 1979, Miocene bioherms and associated structures in the Upper Coralline limestone of the Maltese Islands: Their lithification and palaeoenvironment: Sedimentology, v. 26, p. 577–591.

Perkins, R.D., 1977, Depositional framework of Pleistocene rocks in south Florida, *in* P. Enos and R.D. Perkins (ed.), Quaternary sedimentation in south Florida: Geol. Soc. America Mem. 147, p. 131–198.

______, and Enos, P., 1968, Hurricane Betsy in the Florida-Bahama area--Geologic effects and comparison with Hurricane Donna: Jour. Geology, v. 76, p. 710–717.

Petta, T.J., 1977, Diagenesis and geochemistry of a Glen Rose patch reef complex, Bandera County, Texas, *in* D.G. Bebout and R.G. Loucks (eds.), Cretaceous carbonates of Texas and Mexico: Univ. Texas-Austin, Bur. Econ. Geol. Report of Invest. No. 89, p. 138–167.

Playford, P.E., 1980, Devonian "Great Barrier Reef" of Canning Basin, western Australia: Amer. Assoc. Petroleum Geologists Bull., v. 64, p. 814–840.

Powell, N.A., 1969, Indo-Pacific bryozoa new to the Mediterranean coast of Israel: Israel Jour. Zoology, v. 18, p. 157–168.

Purser, B.H., 1969, Syn-sedimentary marine lithification of Middle Jurassic limestones in the Paris Basin: Sedimentology, v. 12, p. 205–230.

Raymond, W.F., 1972, A geologic investigation of the offshore sands and reefs of Broward County, Florida: M.S. thesis, Florida State Univ., 95 p.

Read, J.F., 1982, Geometry, facies, and development of Middle Ordovician carbonate buildups, Virginia Appalachians: Amer. Assoc. Petroleum Geologists Bull., v. 66, p. 189–209.

Roberts, H.H., Rouse, L.J., Jr., Walker, N.D., and Hudson, J.H., 1982, Cold-water stress in Florida Bay and northern Bahamas: A product of winter cold-air outbreaks: Jour. Sed. Petrology, v. 52, p. 145–155.

ROESSLER, M.A., AND TABB, D.C., 1974, in Studies of effects of thermal pollution in Biscayne Bay, Florida: U.S. Environmental Protection Agency Report 660/3-74-014, p. 122–129.

RÜTZLER, K., AND MACINTYRE, I.G., 1982, The habitat distribution and community structure of the barrier reef complex at Carrie Bow Cay, Belize, *in* K. Rützler and I.G. Macintyre (eds.), The Atlantic barrier reef eco-system at Carrie Bow Cay Belize: Scientific Reports 1, Smithsonian Contributions to the Marine Sciences No. 12, Smithsonian Institution Press, Washington, D.C., p. 9–46.

SCHLAGER, W., 1981, The paradox of drowned reefs and carbonate platforms: Geol. Soc. America Bull., Part I, v. 92, p. 197–211.

SCHOPF, T.J.M., 1974, Ectoprocts as associates of coral reefs: St. Croix, U.S. Virgin Islands, *in* Proceedings of Second International Coral Reef Symposium, v. 1: Brisbane, Australia, Great Barrier Reef Committee, p. 353–356.

SCHROEDER, J.H., 1972, Calcified filaments of an endolithic alga in recent Bermuda reefs: Neues Jb. Geol. Paläont. Mh., v. 1, p. 16–33.

______, AND ZANKL, H., 1974, Dynamic reef formation: A sedimentological concept based on studies of recent Bermuda reefs, *in* Proceedings of Second International Coral Reef Symposium, v. 2: Brisbane, Australia, Great Barrier Reef Committee, p. 413–428.

SHINN, E.A., 1963, Spur and groove formation on the Florida reef tract: Jour. Sed. Petrology, v. 33, p. 291–303.

______, 1969, Submarine lithification of Holocene carbonate sediments in the Persian Gulf: Sedimentology, v. 12, p. 109–144.

STANLEY, D.J., AND SWIFT, D.J.P., 1968, Bermuda's reef-front platform: bathymetry and significance: Marine Geology, v. 6, p. 479–500.

STENECK, R.S., AND ADEY, W.H., 1976, The role of environment in control of morphology in *Lithophyllum congestum*, a Caribbean algal ridge builder: Botanica Marina, v. XIX, p. 197–215.

STODDART, D.R., 1963, Effects of hurricane Hattie on the British Honduras reefs and cays, October 30–31, 1961: Atoll Research Bull. 95, 142 p.

TAFT, W.H., ARRINGTON, A.H., MACDONALD, C., AND WOOLHEATER, C., 1968, Lithification of modern marine carbonate sediments at Yellow Bank, Bahamas: Bull. Marine Science, v. 18, p. 762–828.

TUNNICLIFFE, V., 1983, Caribbean staghorn coral populations: Pre-hurricane Allen conditions in Discovery Bay, Jamaica: Bull. Marine Science, v. 33, p. 132–151.

VAIL, P.R., MITCHUM, R.M., JR., AND THOMPSON, S., III, 1977, Seismic stratigraphy and global changes of sea level, Part 4: Global cycles of relative changes of sea level: Amer. Assoc. Petroleum Geologists Mem. 26, p. 83–97.

VAUGHAN, T.W., 1916, The results of investigations of the ecology of the Floridian and Bahamian shoal-water corals: Proc. Nat. Acad. Science, v. 2, p. 95–100.

WARME, J.E., SCHNEIDERMANN, N., AND SCHAFERSMAN, S.D., 1980, Contemporaneous reconstruction and cementation of living lagoonal patch reef, Enewetak Atoll, Marshall Islands (abs.): Amer. Assoc. Petroleum Geologists Bull., v. 64, p. 800.

WEBER, J.N., WHITE, E.W., AND WEBER, P.H., 1975, Correlation of density banding in reef coral skeletons with environmental parameters: The basis for interpretation of chronological records preserved in the coralla of corals: Jour. Paleobiology, v. 1, p. 137–149.

WOODLEY, J.D., CHORNESKY, E.A., CLIFFORD, P.A., JACKSON, J.B.C., KAUFMAN, L.S., KNOWLTON, N., LANG, J.C., PEARSON, M.P., PORTER, J.W., ROONEY, M.C., RYLAARSDAM, K.W., TUNNICLIFFE, V.J., WAHLE, G.M., WULFF, J.L., CURTIS, A.S.G., DALLMEYER, M.D., JUPP, B.P., KOEHL, M.A.R., NEIGEL, J., AND SIDES, E.M., 1981, Hurricane Allen: impact on coral reef communities in Discovery Bay, Jamaica: Science, v. 214, p. 749–754.

ZANKL, H., AND MULTER, H.G., 1977, Origin of some internal fabrics in Holocene reef rocks, St. Croix, U.S. Virgin Islands, *in* Proceedings of Third International Coral Reef Symposium, v. 2: Miami, Florida, Atlantic Reef Committee, p. 127–133.

CEMENT DISTRIBUTION AND CARBONATE MINERAL STABILIZATION IN PLEISTOCENE LIMESTONES OF HOGSTY REEF, BAHAMAS

BERNARD J. PIERSON[1] AND EUGENE A. SHINN

University of Miami, Rosenstiel School of Marine and Atmospheric Sciences, Fisher Island Station, Miami Beach, Florida 33139, and U. S. Geological Survey, Fisher Island Station, Miami Beach, Florida 33139

ABSTRACT

Four shallow cores taken at Hogsty Reef atoll, southeastern Bahamas, reach the Early Pleistocene at 30 m below sea level. Unlike in other banks of the Bahamas, Hogsty Reef limestones consist of carbonate grains and cements that have maintained their original mineralogy.

Five types of cement are found in the Pleistocene limestone of Hogsty Reef: (1) micritic Mg-calcite, which is predominant in all four cores; (2) bladed Mg-calcite, which occurs in several thin (<1 m) intervals where it is associated with (3) fibrous aragonite, which is also found in Holocene cemented crust; (4) blocky low-Mg calcite, which occurs in several thin intervals underlying subaerial exposure horizons; and (5) blocky Mg-calcite, an unusual type of cement which consists of clear anhedral crystals 25–100 μm in size and contains less than 10 percent $MgCO_3$. It is found together with bladed Mg-calcite and aragonite.

Preservation of original mineralogy of cements and grains may be attributed to a minor influence of fresh water, resulting from the semi-arid climate of the southeastern Bahamas, and the unlikely formation of a freshwater lens during periods of emergence.

INTRODUCTION

Most shallow marine carbonate sediments and cements are originally composed of the metastable minerals aragonite and magnesian calcite. Inversion of these metastable carbonate minerals to more stable calcite is generally presumed to take place early in the diagenetic history of a limestone. Aragonite-to-calcite transformation through dissolution/precipitation and the loss of magnesium from Mg-calcite occur rapidly in a freshwater environment (Land, 1967; Steinen and Matthews, 1973; James, 1974; Bathurst, 1975, p. 330; Longman, 1980). Halley and Harris (1979) recently documented solution of aragonite and reprecipitation of calcite in a limestone less than 1,000 years old. This mineralogical evolution has been extensively documented in Pleistocene limestones of Florida and the Bahamas (Ginsburg, 1957; Stehli and Hower, 1961; Beach, 1980; Pierson, 1982), of Bermuda (Friedman, 1964; Land, 1967), and of Barbados (Steinen and Matthews, 1973; Pingitore, 1976).

Incomplete or patchy alteration of original Pleistocene age carbonate minerals is not uncommon and has been reported from Barbados (Matthews, 1968; Steinen and Matthews, 1973; Pingitore, 1976, 1978), and Florida (Schneidermann *et al.*, 1972). These authors attributed incomplete mineral stabilization to local variations in climate and fresh water supply. Coral reef limestones with ages exceeding 80,000 years still consist of unstable carbonate minerals in areas of Barbados characterized by low rain fall (1250 mm/year), whereas stable calcite predominates where rainfall is high (1500 to 2000 mm/year; Matthews, 1968).

At Hogsty Reef atoll in the southeastern Bahamas (Fig. 1), all carbonate sediments and cements in the Pleistocene limestone have preserved their original mineralogies, in contrast with the now completely calcite Pleistocene limestone of most other Bahamian banks. The carbonate cements in the Pleistocene limestone of Hogsty Reef atoll are of various types and origins and include marine as well as freshwater cements. Although Hogsty Reef was exposed to subaerial conditions during Pleistocene glacial periods, as shown by soil zone crusts described later, mineral stabilization did not take place.

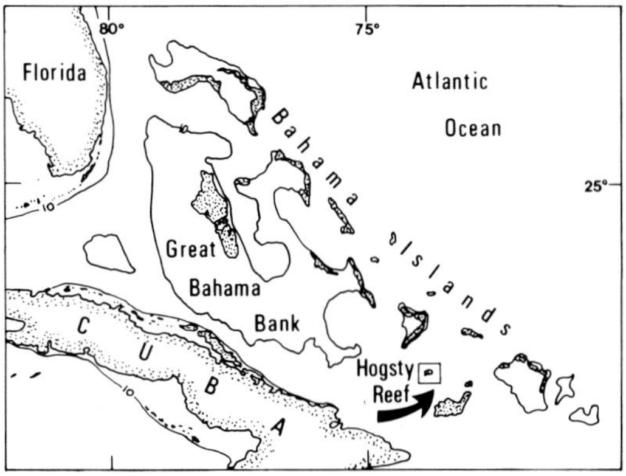

FIG. 1.—Index map showing location of Hogsty Reef atoll in southeastern Bahamas. Contours are in meters.

[1]Present address: Shell Research B.V. (KESPL) 6 Volmerlaan, 2288 GD Rijswijk Netherlands

In this paper we document various types of carbonate cements and speculate on the reasons stabilization has not taken place. We feel that documentation of non-stabilization is important because it contradicts widely accepted beliefs that mineral stabilization always occurs soon after deposition.

METHODS

Rock cores were collected during a cruise to Hogsty Reef in December 1978 on board the R/V CALANUS. Four holes were drilled on the reef (Fig. 2) by the University of Miami in cooperation with the U.S. Geological Survey Fisher Island Station, which provided drilling assistance and its unique underwater drill rig (Fig. 3). The borings were all made at an average depth of 10 m below sea level using a core barrel 1.5 m (5 ft) long. The cores have a diameter of 5 cm for the upper interval (0–12 m) and 3.5 cm for the lower interval (below 12 m). This type of equipment and working conditions limited the depth of penetration to a maximum of 22 m below the surface and significantly reduced recovery for the lower interval.

The percentage of recovery in each of the borings is shown in Figure 4. The intervals of low recovery are believed to consist of friable, poorly cemented limestone, which is difficult to recover with a small diameter core barrel. Holocene uncemented lagoonal sediments from sites HR-1 and HR-4 were not recovered during drilling, but one 2 m long piston core of soft sediment was collected at site HR-1.

The recovered sections were sampled at 30 cm intervals or less. All samples were given an identification number indicating the core number and the sampling depth in feet below the surface; 110 thin sections were analyzed petrographically, after staining with Titan Yellow for determination of magnesian calcite (Choquette and Trusell, 1978) and with Feigl solution for identification of aragonite (Friedman, 1959). X-ray diffraction analyses were conducted at the Rosenstiel School of the University of Miami. Six aragonite samples were dated by C^{14} method at the Radiocarbon Laboratory of the University of Miami, and by uranium^{234}/thorium^{230} method at the U.S. Geological Survey in Corpus Christi, Texas.

The concentrations of major elements (Ca and Mg) as well as of trace elements (Sr, Na, Fe, Mn, Zn) of selected samples were measured with a Cameca electron microscope at the Microbe Analytical Center for Earth Sciences of the University of Louvain-la-Neuve, Belgium.

HOGSTY REEF ATOLL

Located at 21° 40' N and 73° 50' W (Fig. 1), Hogsty Reef is one of the rare atolls in the Atlantic Ocean and the only "true" atoll in the Bahamas. About 9 km long by 5 km wide and shaped like a horseshoe, Hogsty Reef lies between Acklins Island and Great Inagua Island, both of which are more than 75 km away. The bank escarpment plunges to depths exceeding 1800 m, but the atoll lagoon is generally less than 10 m deep. The shallow lagoon is surrounded by a narrow prominent fringing reef (Fig. 2) interrupted by a lagoon pass on the west side. Two small sand cays, called South-

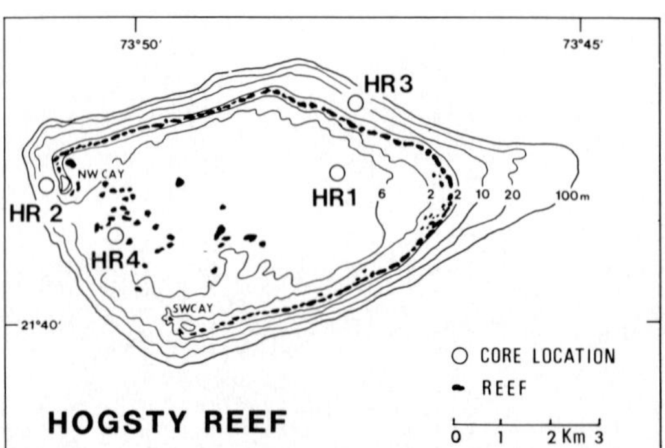

FIG. 2.—Index map showing bathymetry and core location at Hogsty Reef. Contours are in meters.

FIG. 3.—Photograph showing drilling operations at Hogsty Reef in approximately 10 m of water. Underwater drill rig owned and operated by U.S. Geological Survey, Fisher Island Station, Miami.

west Cay and Northwest Cay, mark the extremities of the fringing reef on both sides of the pass. Except for these two cays, Hogsty Reef is totally submerged at present.

Surface sediment composition and distribution were described by Milliman (1967), who defined six different atoll facies. He subdivided the lagoon into an east part, where skeletal grains derived from the windward reef are abundant, and a west part where non-skeletal particles predominate. Recent surface sediments are mostly unlithified, except for scattered buried cemented crusts in the Holocene lagoonal sediments and beachrock on both cays. There is no Holocene coral reef accumulation on the reef front and the gently sloping bottom is practically bare of loose sediment except for sand and gravel in local depressions. Scattered heads and encrustations of corals and red algae grow on the hard, well lithified Pleistocene limestone.

AGE OF THE CORED LIMESTONE AT HOGSTY REEF

Because of the absence of biostratigraphic markers identifiable in thin sections, a search for age diagnostic coccoliths was made in a series of samples of uncemented interstitial mud. Only one sample, HR1-40 (Fig. 5), was found to contain *Gephyrocapsa caribbeanica*, which ranges from the base of the *G. caribbeanica* subzone (approximately 1.6 my) to the present (Hay *et al.*, 1967), indicating that the rock is no older than Early Pleistocene.

Radiometric age determination was possible because original aragonite is preserved. Aragonitic coral skeletons were dated by radiocarbon, as well as by uranium^{234}/thorium^{230} and uranium series methods. All samples dated beyond the range of the C^{14} method (maximum 36,000 yrs.), and the uranium/thorium method gave ages ranging from 110,000 yrs to more than 350,000 yrs (Table 1, Fig. 5). The half-life of U^{234} is approximately 250,000 yrs (Strominger *et al.*, 1958); beyond that age, chances of error increase and 350,000 years is regarded as the lower limit of age determination by the uranium/thorium method (C.W. Holmes, pers. commun.). Ages older than 350,000 years thus remain undetermined.

Corals cropping out at the surface of the indurated reef front (Core HR-3, Fig. 5) are Pleistocene in age and most likely are Sangamon (138,000 ± 13,000 yrs). No younger sediment overlies this Pleistocene reef, although it has been submerged presumably for a long period of time (Pierson, 1980, 1982).

The unconsolidated sediment in the interior lagoon of Hogsty Reef is Holocene and is separated from the Pleistocene limestone by a discontinuity horizon consisting of a thin (4 mm thick) micritic crust.

The recovered Pleistocene rocks of the four borings of Hogsty Reef represent at least two distinct depositional events separated by exposure horizons (Fig. 5). The younger depositional event is Sangamon in age and consists of rock with ages ranging from 100,000 to 150,000 years (Table 1; HR2-12, HR4-24, HR3-0.1). An older depositional event dates beyond 200,000 years and is probably Yarmouthian in age (Table 1; HR1-43, HR2-24). Sample HR1-60, dated as older than 350,000 yrs, could be part of a much older depositional event, but since it is the only available date, it

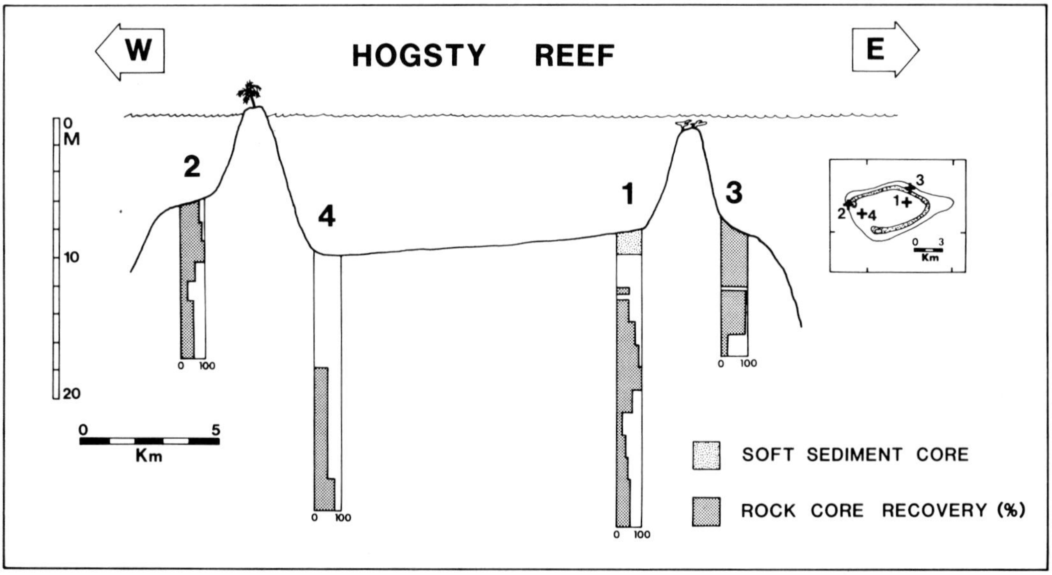

FIG. 4.—Illustration showing intervals of rock and sediment recovery for the four cores taken at Hogsty reef. Recovery factor shown in percent.

will be considered as part of the Yarmouthian deposits.

The cored limestone of Hogsty Reef is thus mostly of Late Pleistocene age and is certainly no older than Early Pleistocene. Several discontinuity zones are found in the cored sections (Fig. 5). These zones are characterized by thin laminated micritic crusts (calcrete), diagnostic of subaerial exposure conditions (Multer and Hoffmeister, 1968; James, 1972; Harrison, 1977; Robbin and Stipp, 1979). At least six periods of emergence of Hogsty Reef bank are recorded in core HR1 (Fig. 5; Pierson, 1982). Three of the discontinuity horizons are clearly visible, whereas three are poorly developed, probably indicating variations in the duration of emergence of the bank or core location. Thin crusts often have patchy distribution and can easily be missed in small diameter cores.

CEMENT TYPES AND THEIR DISTRIBUTION

Six different types of cement have been identified in the Pleistocene limestone of Hogsty Reef, two of which (low-Mg blocky calcite and scalenohedral calcite) may be lumped together. Their respective distribution in the cores is illustrated in Figure 6.

Micritic Mg-Calcite Cement

Microcrystalline Mg-calcite cement is the most common form of magnesian calcite (Alexandersson, 1972; Schroeder, 1973; James *et al.*, 1976; Macintyre, 1977; Longman, 1980) and is by far the most common type of cement found in the Pleistocene limestone of Hogsty Reef (Fig. 6). It consists of a thin (2–5 μm) film composed of individual equant or elongate crystallites 1–2 μm long. This thin film may be overgrown by a 5- to 10-μm-thick rim of elongate microcrystals ranging from 2–8 μm in length (Fig. 7a-b). Based on the intensity of the Titan Yellow staining (Choquette and Trusell, 1978), the magnesium content of the micritic cement is estimated to exceed 10 percent and probably ranges from 12 to 15 mole percent. The micritic cement directly overlies grain surfaces and

TABLE 1.—THORIUM AND URANIUM AGES OF SIX SAMPLES

SAMPLE NO.	Th^{230}/U^{234}	AGE	UPPM	U^{234}/U^{238}
HR1-43	0.86±0.08	213,000±20,000	2.10±0.2	1.14
HR1-60	1.01±0.10	>350,000	2.44	1.08
HR2-12	0.64±0.06	110,800±11,000	3.03±0.2	1.10
HR2-24	0.89±0.08	239,460±24,000	3.13±0.2	1.10
HR3-0.1	0.72±0.07	138,100±13,000	2.72±0.3	1.05
HR4-24	0.72±0.07	138,100±13,000	3.15±0.3	1.10

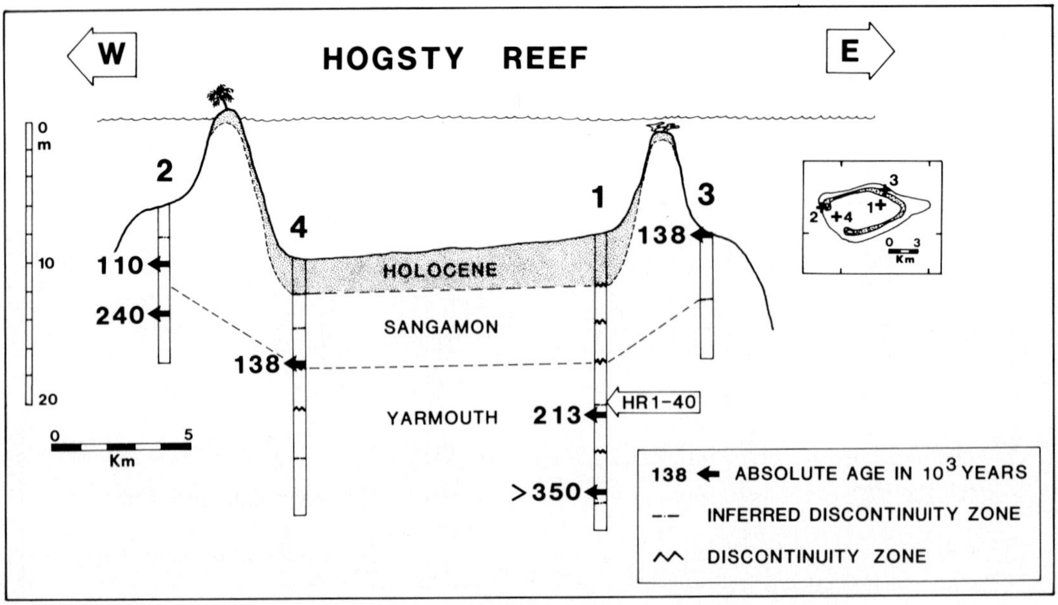

FIG. 5.—Illustration showing the age of the cored limestone of Hogsty reef. Correlations are based on discontinuity zones and absolute ages of selected samples. Age determinations obtained by $uranium^{234}/thorium^{230}$ method.

may overlie a preexisting micritic envelope (Bathurst, 1966; Margolis and Rex, 1971). The magnesian calcite cement constitutes an addition of new material on the grain surface (Macintyre, 1977), and is therefore distinct from micritic envelopes which result from alteration of particle surfaces. The micritic Mg-calcite cement forms a regular isopachous coating around grains and appears as a dark line in thin section, enhancing the outline of particles and their intraskeletal pores (Fig. 7b). Commonly, micritic coatings consist of an irregular zone of superficial micritization overgrown by a regular film of bladed Mg-calcite cement.

Micritic cement forms a coating on the outer surface of the particles and, unlike micritic envelopes, lines the walls of intragranular pores. Lithification by this type of cement alone produces a friable limestone. Mg-calcite cement can grow on any type of particle, regardless of its original mineralogy, although, according to Alexandersson (1972), a substratum of magnesium calcite should induce precipitation of Mg-calcite cement.

Micritic Mg-calcite cement may also occur as small pellets 10–30 μm in diameter (Fig. 7d). Such textures have been described from reef rocks (Shinn, 1969; Land and Goreau, 1970; Shinn, 1971; Friedman, *et al.*, 1974; Macintyre, 1977; James and Ginsburg, 1979), where they seem to be common. Macintyre (this vol.) presents a summary of Mg-calcite pellets.

Micritic magnesian calcite is known to precipitate directly from sea water (Bricker, 1971). Although its occurrence indicates cementation in the marine phreatic environment, Mg-calcite is also known to form in the intertidal zone, as in beachrock (Moore, 1977).

An unusual occurrence of microcrystalline magnesian calcite cement is found in sample HR1-14. A 20-μm-thick coating of micritic cement rims particles and forms meniscus-like growths at points of grain contact (Fig. 7c). The exact origin of this cement remains unanswered, although such cement is known to occur in beachrock (Moore, 1977).

Bladed Magnesian Calcite

Bladed or fibrous high Mg-calcite cement in Hogsty Reef consists of elongated crystals oriented perpendicular to the substrate, growing as an isopachous fringe in inter- and intragranular cavities (Fig. 8a-b). Crystals appear as poorly developed scalenohedra with rhombic terminations 4–8 μm in width and 10–60 μm long.

This type of cement, called bladed spar by James *et al.* (1976), palisade cement by Schroeder (1973), and fibrous Mg-calcite by Shinn (1971), is thought to be

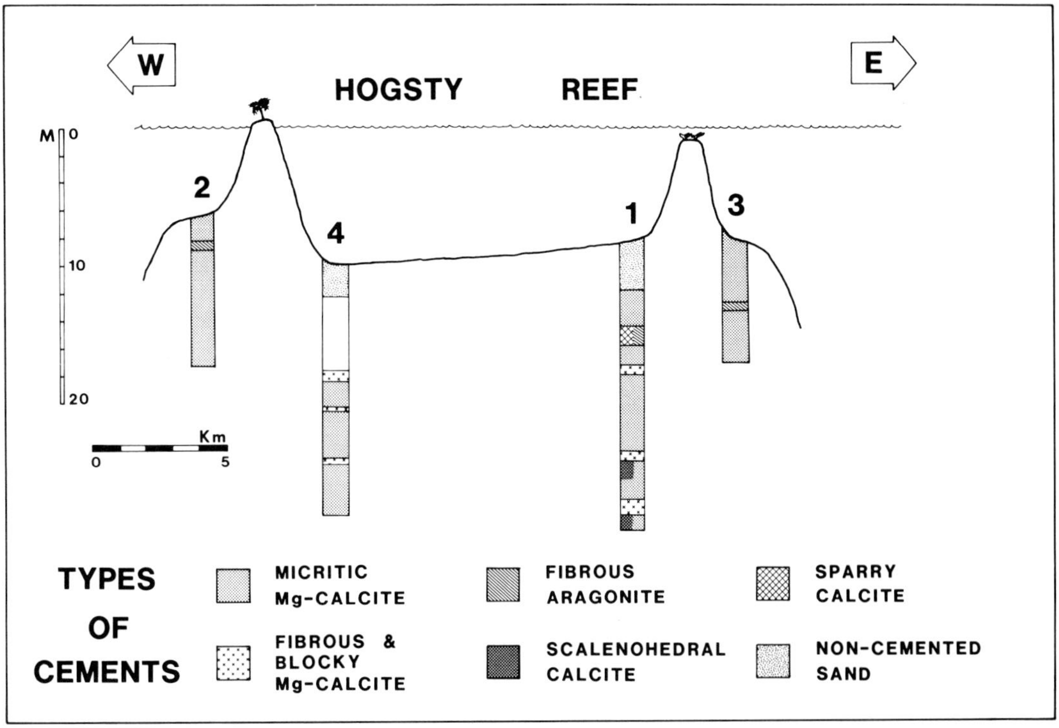

FIG. 6.—Illustration showing cement types and distribution in Hogsty Reef cores.

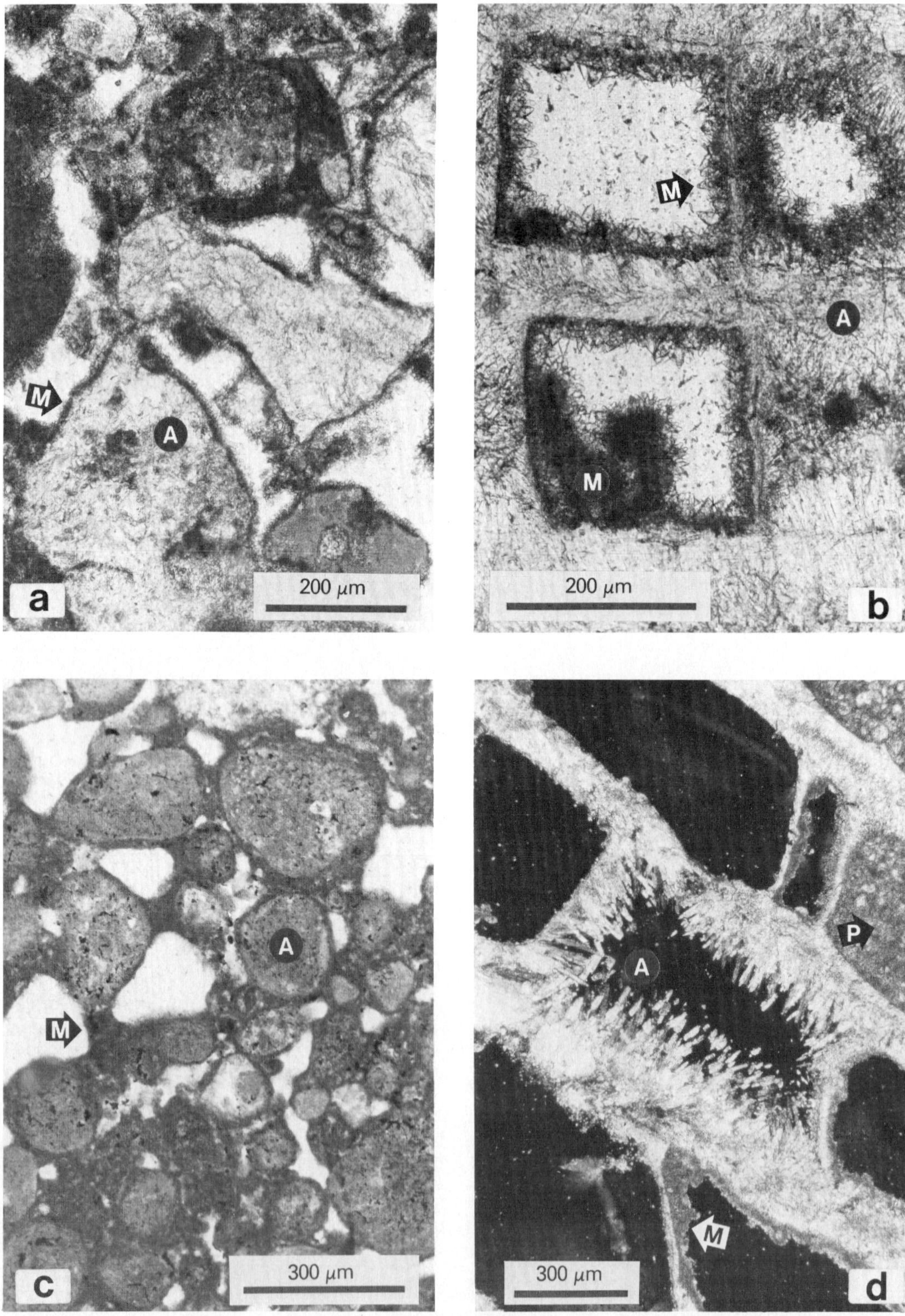
M
A
200 μm
a
M
A
M
200 μm
b
A
M
300 μm
c
P
A
M
300 μm
d

produced by continued growth on the upper faces of rhombs constituting the micritic Mg-calcite cement (Longman, 1980). The magnesium content estimated from the intensity of staining by Titan Yellow (Choquette and Trusell, 1978) ranges from 10 to 15 mole percent. Bladed magnesian calcite is commonly found coexisting with fibrous aragonite (Fig. 6), but the control over the respective distribution of both cement types is not known.

Cementation by fibrous Mg-calcite produces a very hard rock. Bladed Mg-calcite precipitates from marine waters, including intertidal beachrocks (Bricker, 1971).

Fibrous Aragonite

Aragonite cement typically grows as fibrous crystals in a variety of textures (Cullis, 1904; Shinn, 1969; Ginsburg *et al.*, 1971; Alexandersson, 1972; Schroeder, 1973; Friedman *et al.*, 1974; Bathurst, 1975; James *et al.*, 1976; Longman, 1980). Two types of fibrous aragonitic cement commonly occur in the limestone of Hogsty Reef: an isopachous fringe of subequant tightly packed needles 20–80 μm thick, or an irregular coating of needles of various dimensions.

Individual needles average 5–10 μm in width and are on the average 40–60 μm long. Some needles exceed 150 μm in length. Fibrous aragonite crystals rim grains, growing as inter- or intragranular cement (Figs. 7d, 8c). Commonly, the crystals nucleate on grains and grow outward, normal to the substrate. Aragonitic substrates are believed to induce growth of aragonitic cement (Macintyre *et al.*, 1968; Winland, 1971; Alexandersson, 1972), and the common presence of aragonitic void-filling cement in intragranular pores of aragonitic particles, such as coral skeletons (Fig. 7d), substantiates that assertion; however, aragonite is also repeatedly found fringing particles composed of magnesian calcite or growing over a rim of micritic Mg-calcite cement (Fig. 8d). Numerous studies have shown that aragonite precipitates inorganically from seawater at shallow depths and therefore represents a characteristic cement of the marine phreatic and intertidal zone.

Aragonite needles are found in several thin intervals of the subsurface Pleistocene limestone (Fig. 6) and also as the cement of Holocene submarine crusts and beachrock of Hogsty Reef.

Blocky Magnesian Calcite

Clear anhedral crystals of magnesian calcite, ranging in size from 20–50 μm and resembling sparry calcite (Fig. 9a), were occasionally observed in the Pleistocene limestone of Hogsty Reef. The various occurrences of blocky Mg-calcite crystals allow a positive identification of this crystal habit and rule out a misleading observation of transversally cut needles.

Blocky Mg-calcite usually occurs as a last void-filling cement, often post-dating precipitation of fibrous aragonite (Fig. 9b). It grows from the top of the aragonite needles toward the center of the cavity, increasing slightly in size away from the substrate.

Blocky magnesian calcite has rarely been reported in previous studies of carbonate cements (James and Ginsburg, 1979; Schroeder, 1972; Warme and Schneidermann, 1983) and blocky calcite occurring with fibrous aragonite in Figure 28 of Shinn (1969) is probably similar to that reported here. Blocky Mg-calcite cement was detected in the Hogsty Reef limestones by staining thin sections with Titan Yellow. Occurrences of blocky Mg-calcite may have been overlooked in the past because of its resemblance to the more common low-Mg blocky calcite.

Variations in color intensity of the Titan Yellow stain indicate a decrease in the magnesium content of the crystals toward the center of the cavity fill. Results of electron microprobe analyses of the blocky Mg-calcite cement (Table 2) confirmed this systematic variation in the magnesium content (Figs. 10, 11).

A semi-quantitative step-scan analysis across a blocky Mg-calcite zone surrounded by aragonite particles and cements (Figs. 10, 11) showed that the first crystals of magnesian calcite to precipitate are richer in magnesium than the later crystals, implying a gradual depletion of magnesium in the precipitating fluids. Magnesium content ranges from 11.6 mole percent to 3.0 mole percent $MgCO_3$ (Table 2). The decrease in magnesium content appears to be proportional to increase in crystal size.

Blocky Mg-calcite is not common. It was found only in a few samples of the Pleistocene limestone of Hogsty Reef, where it is associated with bladed magnesian calcite and fibrous aragonite. It is believed to have precipitated in the marine phreatic zone.

FIG. 7.—Photographs of thin sections of Hogsty Reef rock (A = aragonite; M = magnesian calcite). *a*, Micritic Mg-calcite cement rimming skeletal particles. A first coating of micritic Mg-calcite (dark line) is followed by formation of incipient bladed Mg-calcite crystals. Thin section stained with Titan Yellow. Sample HR4-58. Normal transmitted light. *b*, Micritic Mg-calcite cement coating intraskeletal pores of coral *Montastrea annularis*. Bladed Mg-calcite followed micritic Mg-calcite (dark line). Thin section stained with Titan Yellow. Sample HR4-26. Normal transmitted light. *c*, Micritic Mg-calcite cement forming meniscus-like connections between grains. Sample HR1-14. Normal transmitted light. *d*, Aragonitic and micritic Mg-calcite cements lining intraskeletal pores of *Montastrea annularis*. No micritic cement preceded fibrous aragonite. The pores lined with micritic cement are commonly filled with micritic pellets (P), also composed of magnesian calcite. Sample HR2-9.6. Cross-polarized light.

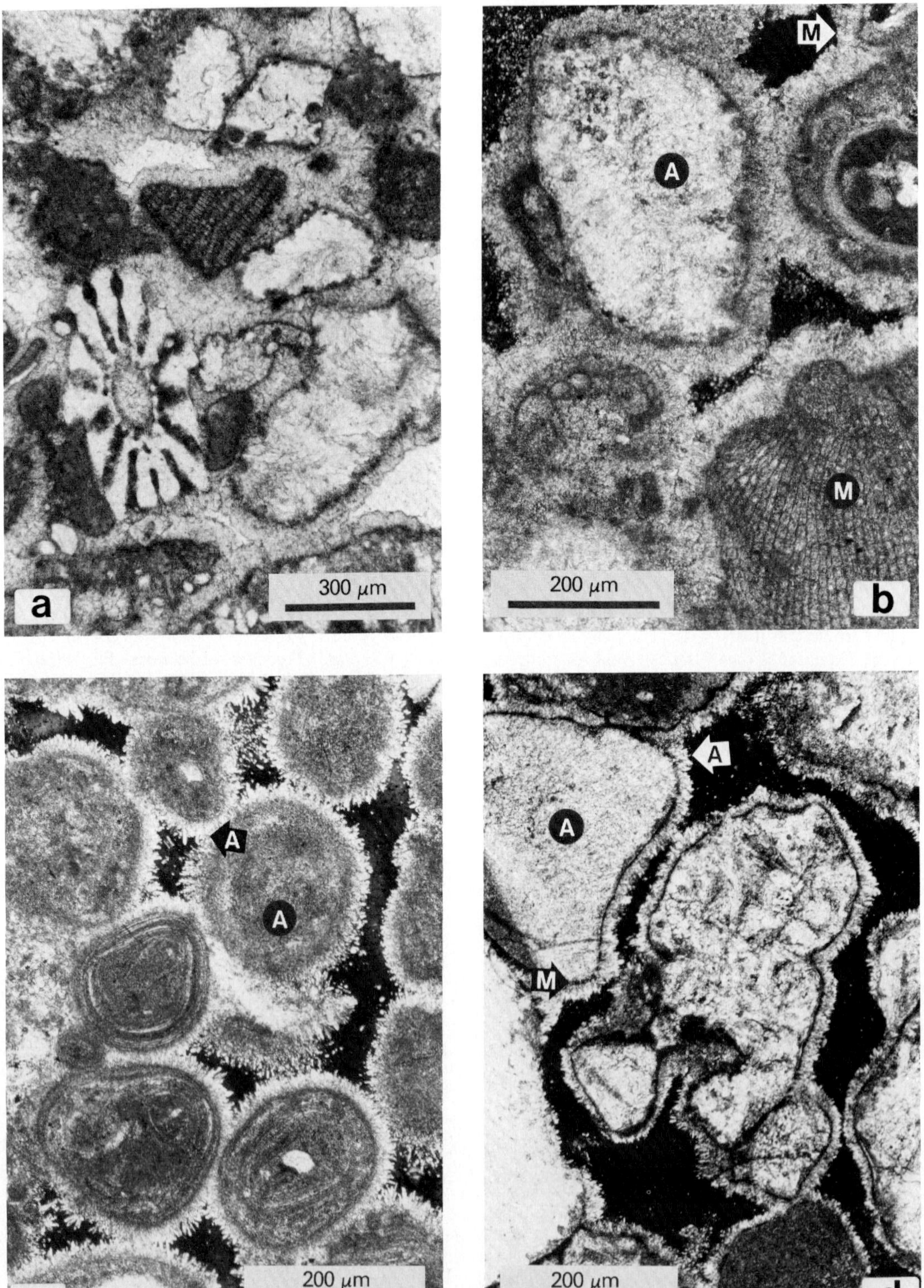
a
300 μm
b
A
M
M
200 μm
c
A
A
200 μm
d
A
A
M
200 μm

Blocky Low-Mg Calcite

Clear rhombs of low-Mg calcite 30–60 μm across locally form isopachous rims around grains (Fig. 9c). Blocky calcite occurs in primary void spaces in the Pleistocene limestone at Hogsty Reef and is probably the last stage of cementation. Preexisting fibrous cements, either aragonite or Mg-calcite, are commonly preserved within the fringes of rhombic calcite (Fig. 9d). The needles appear to be in their original state and have not been affected by neomorphism, as was observed in similar sequences of carbonate cements in Pleistocene limestones of Bermuda (Schroeder, 1973).

Blocky calcite cement distributed equally around grains indicates precipitation within an active freshwater phreatic zone, as shown by Halley and Harris (1979) and Longman (1980). Coexisting with equant rims of blocky calcite, however, are meniscus cements (Dunham, 1971) which are believed to be characteristic of the freshwater vadose zone (Harris, 1979; Halley and Harris, 1979, and many others). Blocky calcite is distributed in patches within several intervals of Hogsty Reef limestone (Fig. 6). A freshwater origin of the calcitic cement is confirmed by the occurrence of a thin calcrete crust directly overlying a 30 cm interval containing this cement. The calcrete crust indicates intense carbonate leaching at the surface of the atoll during periods of emergence. Surface leaching probably provided the calcium carbonate subsequently precipitated as blocky calcitic cement.

The fresh water responsible for the precipitation of the blocky calcite cement did not affect sedimentary particle mineralogy. Grains composed of metastable aragonite or magnesian calcite have not been altered to low-Mg calcite, even though they are rimmed by blocky calcite (Fig. 9c-d). Only occasional traces of aragonite dissolution were observed.

Scalenohedral Calcite

Scalenohedral calcite cement, also called dogtooth cement, consists of elongated crystals of low-Mg calcite shaped as scalenohedra (Schneider, 1977; this paper, Fig. 9b). Crystals 15–30 μm long are common, whereas larger crystals up to 150 μm long occur occasionally. Single crystals grow perpendicular to or slightly inclined on a substrate which could consist of magnesian calcite, low-Mg calcite, or aragonite.

Dogtooth cement may form a regular fringe around grains or grow as isolated crystals on top of a preexisting cement (Fig. 9b). Scalenohedral calcite occurs in two thin intervals (30–50 cm thick; Fig. 6). These intervals are capped by a subaerial exposure horizon marked by a thin calcrete crust. The distribution of dogtooth cement suggests that it formed from supersaturated fresh water during periods of emergence of Hogsty Reef atoll. The fresh water from which scalenohedral calcite cements precipitated, however, left intact that bulk of the sediments composed of metastable carbonate minerals.

DISCUSSION

Preservation of Cement and Grain Mineralogy in Hogsty Reef Limestone

Fresh water left its signature during several periods of Pleistocene emergence at Hogsty Reef atoll as indicated by a number of discontinuity horizons and thin calcrete crusts. Surprisingly, however, carbonate particles and marine cements originally composed of metastable aragonite or Mg-calcite have not undergone mineral stabilization to low-Mg calcite. This observation contrasts strongly with the Pleistocene limestone of most other banks of the Bahamas. Several factors, mainly climatic and hydrologic, may explain this lack of "early" neomorphism (Pierson, 1980).

Rainfall in the southeastern Bahamas: Freshwater supply at Hogsty Reef.—Hogsty Reef lies in a tropical semi-arid to arid climate. Based on data from the Bahamas Meteorological Office, Hogsty Reef probably receives less than 800 mm of rain per year (Fig. 12). Actual precipitation level at Hogsty Reef, however, may be much lower. Bahamian meteorological stations are usually located on larger inhabited islands. On large islands local variations in atmospheric conditions (the "island effect") commonly create a particular rainfall pattern. As oceanic humid air pushed by the dominant easterly winds reaches the east coast of an island, it is carried to higher altitudes by a column of ascending warmer air located above the land mass. Because of the lower temperatures at higher altitudes, condensation occurs and clouds begin to form and grow as they travel westward. Rain is therefore more likely to fall on the west part of an island, whereas the east areas remain more desert-like.

This "island effect" is commonly observed on most Bahamian islands, as for instance on nearby Great Inagua, where the local rainfall pattern determines vari-

FIG. 8.—Photographs of thin sections of Hogsty Reef rock (A = aragonite; M = magnesian calcite). *a*, Bladed Mg-calcite cement forming fringes around grains of various compositions. Micritic envelope and micritic Mg-calcite cement preceded bladed Mg-calcite. Thin section stained with Titan Yellow. Sample HR4-45. Normal transmitted light. *b*, Bladed Mg-calcite forming equant rims (40 μm thick) around skeletal particles. Thin section stained with Titan Yellow. Sample HR4-45. Cross-polarized light. *c*, Fibrous aragonite cement rimming ooids. Isopachous fringes grew on aragonitic substrate. Sample HR4-34. Cross-polarized light. *d*, Fibrous aragonite growing on an earlier micritic cement (M) (dark line). Cement in beachrock. Sample HR-BR2. Cross-polarized light.

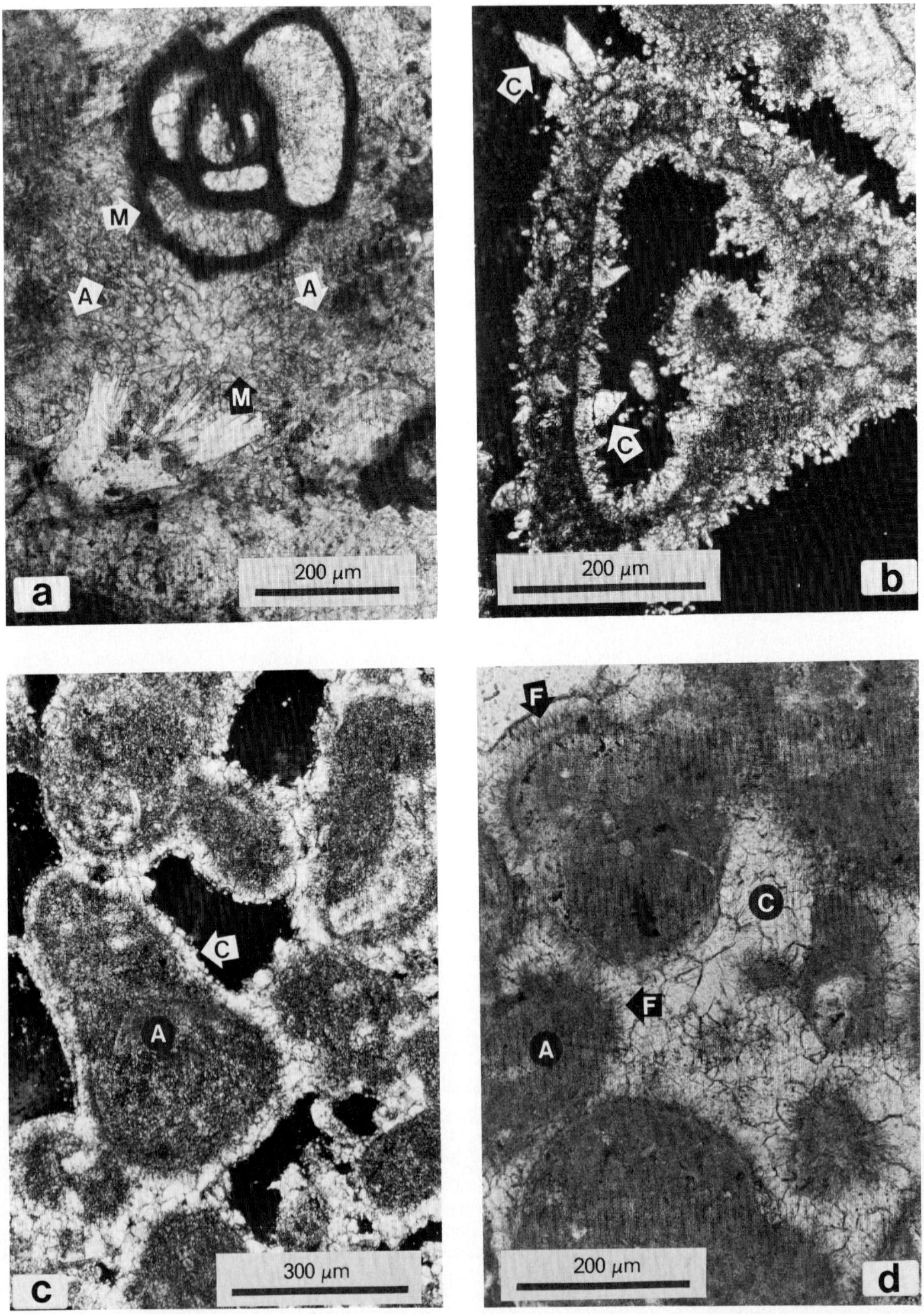
M
A
A
M
200 μm
a
C
C
200 μm
b
C
A
300 μm
c
F
C
F
A
200 μm
d

ous zones of vegetation from an arid type with abundant Cactaceae and low-lying shrubs in the east to a humid type with a dense cover of shrubs and high trees in the west (Pierson, 1982). Since the meteorological station of Great Inagua is located at Matthewtown on the west coast, it records the local high precipitation levels, and as a consequence figures obtained at the station are not representative of the area. Rainfall at Hogsty Reef is therefore considerably lower than the 800 mm/year recorded at Matthewtown, Inagua.

Furthermore, because of the small size of the atoll, atmospheric conditions are hardly modified in the vicinity of Hogsty Reef. A lack of "island effect" which would increase rainfall locally, as observed on larger banks or islands of the Bahamas, accentuates the aridity of the climate at Hogsty Reef.

The present climate at Hogsty Reef is thus relatively arid. The aridity may have been even more severe during glacial periods, coinciding with lower sea-level stands, hence with emergence of Hogsty Reef atoll. Studies of paleotemperatures and paleoclimate in the Caribbean (Schubert and Medina, 1982) showed that the glacial average temperature gradient in the Caribbean during Quaternary glaciations was greater than today, suggesting a drier climate than today. Even if sea level were 100 m lower than today, the size of "Hogsty Island" would not have been very large (see Fig. 2).

Under dry climatic conditions, mineralogical stabilization probably proceeds at a much reduced rate. According to Matthews (1968), sediments subaerially exposed for several hundred thousand years in a dry climate may be virtually unaltered.

A freshwater lens at Hogsty Reef?.—Incomplete mineral stabilization is found in parts of the Pleistocene limestone of Barbados that have always been located in the freshwater vadose zone (Steinen and Matthews, 1973). According to these authors, intervals of the limestone that were occupied at least once by a freshwater phreatic lens show complete mineral stabilization of the matrix to low-Mg calcite and extensive cementation and dissolution.

The state of preservation of carbonate particles and cements in Hogsty Reef limestone therefore suggests that it was never occupied by a freshwater lens. The types and, more relevant, the distribution of the low-Mg calcite cements in the Pleistocene limestone of Hogsty Reef further suggest low-Mg calcite cementation in a freshwater vadose environment.

The lack of a freshwater lens at Hogsty Reef may be partly attributed to location of the atoll in a dry climatic

TABLE 2.—RESULTS OF ELECTRON MICROPROBE ANALYSES OF BLOCKY MG-CALCITE CEMENT

ANALYZED POINT	MOLE % $MgCO_3$	MOLE % $CaCO_3$	PPM Sr	PPM Na	PPM Fe	PPM Mn	PPM Zn	REMARKS
87	3	97	2620	0	230	0	300	Center of cavity-fill
88	4.2	95.7	260	0	440	0	0	Inner zone
89	4	96	0	0	0	0	0	Inner zone
90	6.5	93.5	130	0	100	100	1330	Inner zone
91	1.4	97.5	8860	900	650	170	460	Aragonite cement ± traces Mg-calcite
92	11.6	88	2480	240	170	0	590	Outer zone
93	11.6	88	2220	0	440	0	820	Outer zone
94	8	92	390	370	170	490	0	Outer zone
95	11.65	88	780	90	300	80	0	Mg-calcite (Miliolid foram)
96	4.7	95	0	0	0	300	720	Cement in foram
97	0	98	7560	1530	70	0	0	Aragonite
98	0	98	7430	1620	210	80	150	Aragonite
99	0	98	11200	1960	0	340	0	Aragonite cement

FIG. 9.—Photographs of thin sections of Hogsty Reef rock (A = aragonite; M = magnesian calcite; C = calcite). *a*, Blocky Mg-calcite filling cavity in which fibrous aragonite had already precipitated. A miliolid foraminifer (stained) is composed of Mg-calcite. Thin section stained with Titan Yellow. A decrease in intensity of stain toward center of cavity fill reflects depletion of Mg in the precipitating fluids. Sample HR1-29.5. Normal transmitted light. *b*, Scalenohedral calcite crystals (C) (dogtooth cement) growing on top of preexisting Mg-calcite cement. Sample HR1-66. Cross-polarized light. *c*, Blocky low-Mg calcite (C) forming equant rims around aragonitic particles. Cement shows meniscus-like connections at grain contacts. Sample HR1-21.5. Cross-polarized light. *d*, Blocky calcite forming equant rims around grains. Preexisting fibrous cement (F) is visible in the calcite rhombs. Sample HR1-21.5. Normal transmitted light.

zone but also to the predominantly coarse depositional texture and high porosity and permeability of the rock. The limestone probably did not have the potential to retain a freshwater lens for the period of time necessary for mineral stabilization (Back and Hanshaw, 1970). It should also be noted that if a freshwater lens existed, it would have lasted the longest at and below the 100 m contour (Fig. 2). Freshwater lenses above that level would have been more temporary because they would represent the period when sea level was either rising or falling relatively rapidly. Because our cores only penetrate the upper 25 m, it is impossible to determine if there was a long-lived freshwater lens at the 100 m depth.

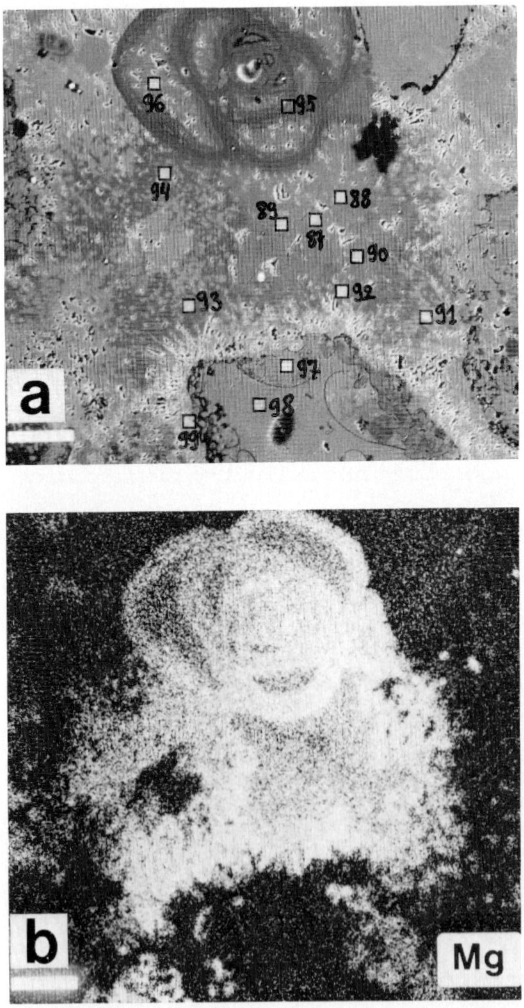

FIG. 10.—Photograph of thin section showing (*a*) the location of points analyzed with the electron microprobe (backscatter electron photo) in a zone of blocky Mg-calcite of sample HR1-29.5, and (*b*) the distribution of magnesium (light areas are rich in magnesium; X-ray element map/Mgk_a). Scale bar is 100 microns. Results of analyses (points 87 to 99) shown in Table 2.

CONCLUSIONS

The Pleistocene limestone of Hogsty Reef contains a variety of carbonate cements, among which micritic Mg-calcite predominates. Bladed Mg-calcite and fibrous aragonite are less abundant but not rare. An uncommon type of cement, blocky Mg-calcite, with a variable magnesium content ranging from 3 to 11.5 mole percent, was found associated with marine cements.

Coexisting with marine cements are freshwater cements of blocky and scalenohedral low-Mg calcite. These cements indicate several stages of freshwater diagenesis during Pleistocene time. In spite of the influence of fresh water, however, metastable minerals that comprise the particles as well as cements have been preserved. This preservation of unstable minerals is unlike the situation on other banks of the Bahamas.

In a few intervals of the Pleistocene limestone of Hogsty Reef, freshwater calcitic cements have grown around particles composed of metastable aragonite and magnesian calcite similar to that reported by Halley and Harris (1979) for Holocene age carbonates. We therefore conclude that the presence of freshwater cements in a limestone does not necessarily imply concurrent neomorphism of the substrate.

The lack of mineral stabilization in Pleistocene limestone is probably unique to Hogsty Reef atoll within the Bahamian archipelago and probably results from arid climate, the lack of "island effect," and the apparent lack of a freshwater lens. It suggests that metastable

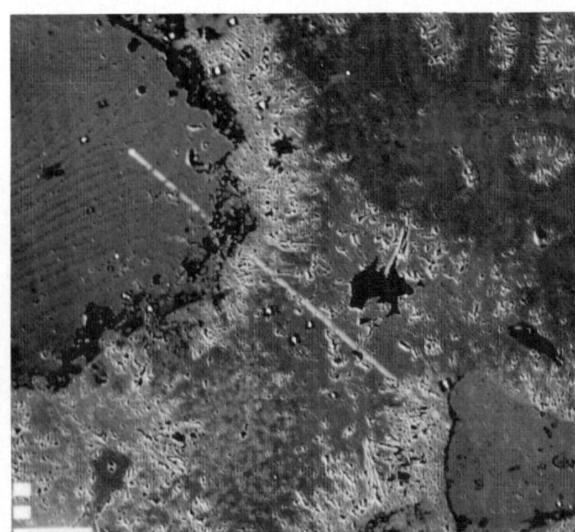

FIG. 11.—Semi-quantitative scan for magnesium across a zone of blocky Mg-calcite. Decrease in magnesium content in center of cavity fill (Fig. 9a) is clearly visible. Path of scan is shown on photo above (back-scatter electron) and is drawn at the scale of the scan curve on opposite page.

carbonate minerals can be preserved and buried for a geologically significant period of time.

ACKNOWLEDGMENTS

We thank R.N. Ginsburg and Wolfgang Schlager of the University of Miami for facilitating the research and for review of the manuscript. We are grateful to J. Harold Hudson of the U.S. Geological Survey and Douglas Lanesky of the University of Miami for their considerable aid in the underwater core boring. We appreciate the willing help of Captain James A. Adamson and First Mate Richard Entwistle of the University of Miami's R/V CALANUS, and Captain Roy Gaensslen of the SEA ANGEL. Barbara Lidz of the U.S. Geological Survey contributed substantially to preparation of the manuscript.

We thank J.J. Stipp of the University of Miami for C^{14} age determinations and Charles W. Holmes of the U. S. Geological Survey in Corpus Christi, Texas (now Minerals Management Service) for the uranium/thorium age determinations. John Wray of Marathon Research in Denver identified the foraminifera. We also thank Jacques Wautier of the University of Louvain-la-Neuve, Belgium, for the electron microprobe analyses, and Willem Poldermans of Shell Research (KSEPL) in

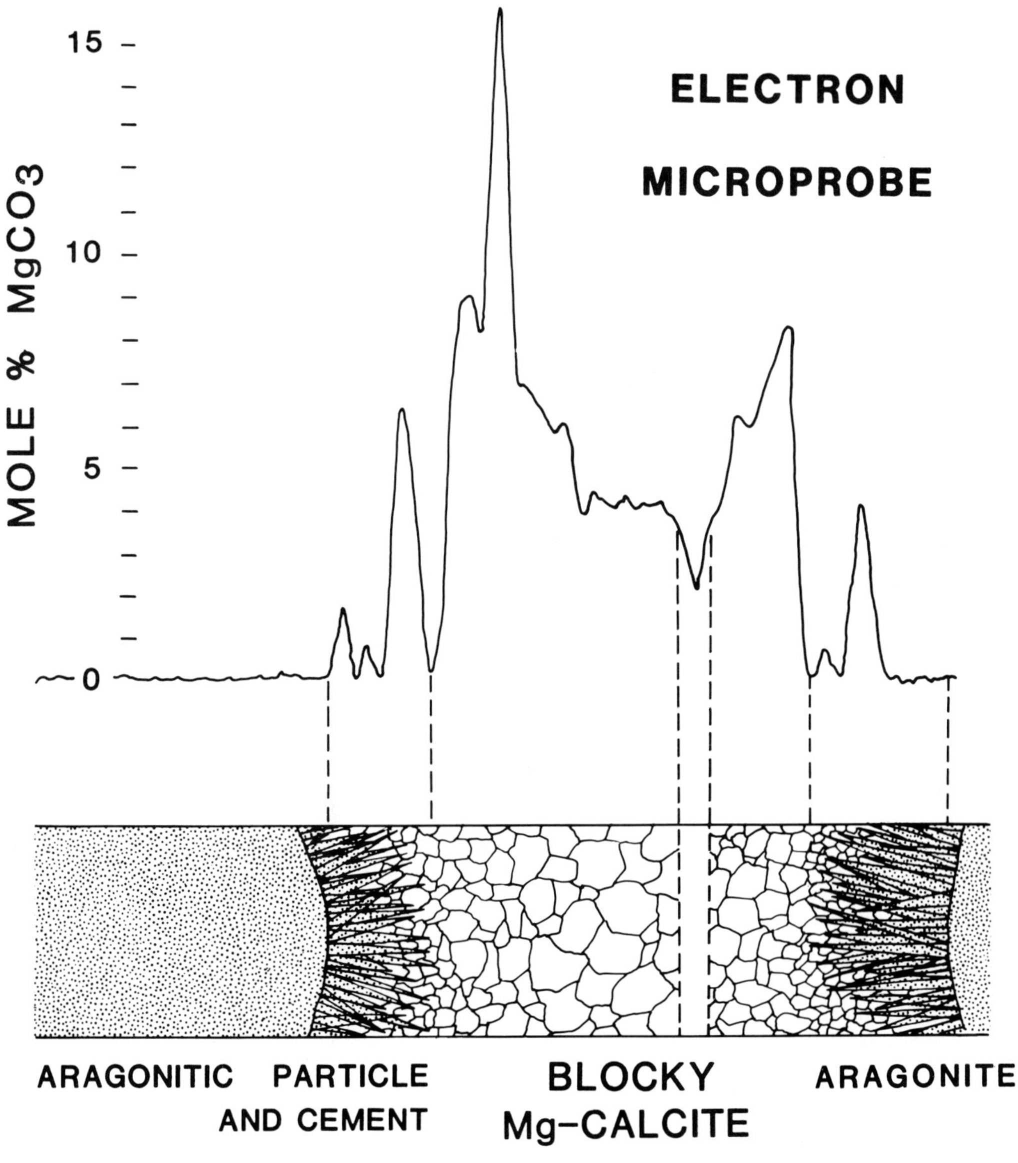

Rijswijk, Netherlands, for microphotography.

We also thank Robert Halley and Linda Pratt of U.S. Geological Survey in Denver for final review of the manuscript and suggestions.

Research was supported by National Science Foundation Grant #EAR 79-27212 and by funds from The Industrial Associates of the Comparative Sedimentology Laboratory.

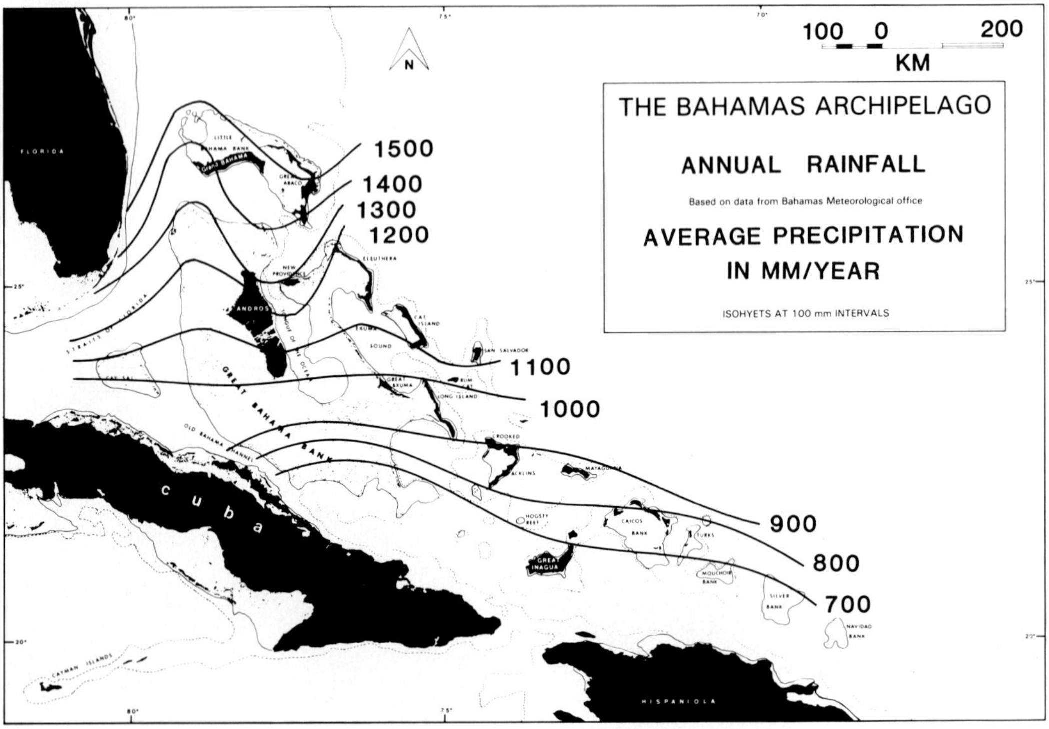

FIG. 12.—Map showing average annual rainfall in the Bahamas archipelago.

REFERENCES

ALEXANDERSSON, T., 1972, Intragranular growth of marine aragonite and Mg-calcite: evidence of precipitation from supersaturated sea water: Jour. Sed. Petrology, v. 42, p. 441–460.

BACK, W., AND HANSHAW, B.B., 1970, Comparison of chemical hydrogeology of the carbonate peninsulas of Florida and Yucatan: Jour. Hydrology, v. 10, p. 330–368.

BATHURST, R.G.C., 1966, Boring algae, micrite envelopes and lithification of molluscan biosparites: Geol. Jour., v. 5, p. 15–32.

______, 1975, Carbonate Sediments and Their Diagenesis: Developments in Sedimentology, v. 12, Amsterdam, Elsevier Publishing Co., 658 p.

BEACH, D.K., 1980, Depositional and Diagenetic History and Stratigraphy of Northwestern Great Bahama Bank: Ph.D. Dissertation, University of Miami, Coral Gables, 235 p.

BRICKER, O.P. (ed.), 1971, Carbonate Cements: The Johns Hopkins Univ. Studies in Geology No. 19, 376 p.

CHOQUETTE, P.W., AND TRUSELL, F.C., 1978, A procedure for making the Titan-Yellow stain for Mg-calcite permanent: Jour. Sed. Petrology, v. 48, p. 639–641.

CULLIS, C.G., 1904, The mineralogical changes observed in the core of the Funafuti borings, *in* T.G. Bonney (ed.), The Atoll of Funafuti: Royal Society of London, p. 392–420.

DUNHAM, R.J., 1971, Meniscus cement, *in* O.P. Bricker (ed.), Carbonate Cements: The Johns Hopkins Univ. Studies in Geology No. 19, p. 297–300.

FRIEDMAN, G.M., 1959, Identification of carbonate minerals by staining methods: Jour. Sed. Petrology, v. 29, p. 87–97.

______, 1964, Early diagenesis and lithification in carbonate sediments: Jour. Sed. Petrology, v. 34, p. 777–813.

______, AMIEL, A.J., AND SCHNEIDERMANN, N., 1974, Submarine cementation from reefs: Example from the Red Sea: Jour. Sed. Petrology, v. 44, p. 816–825.

GINSBURG, R.N., 1957, Early diagenesis and lithification of shallow-water carbonate sediments in South Florida, *in* R.J. Le Blanc and J.G. Breeding (eds.), Regional Aspects of Carbonate Sedimentation: Soc. Econ. Paleontologists and Mineralogists Special Pub. 5, p. 80–100.

_____, MARSZALEK, D.S., AND SCHNEIDERMANN, N., 1971, Ultrastructure of carbonate cements in a Holocene algal reef of Bermuda: Jour. Sed. Petrology, v. 41, p. 472–482.

HALLEY, R.B., AND HARRIS, P.M., 1979, Freshwater cementation of a 1,000-year-old oolite: Jour. Sed. Petrology, v. 49, p. 969–988.

HARRIS, P.M., 1979, Facies Anatomy and Diagenesis of a Bahamian Ooid Shoal: Sedimenta VII: The University of Miami Comparative Sedimentology Laboratory, Miami, 163 p.

HARRISON, R.S., 1977, Caliche profiles: indicators of near-surface subaerial diagenesis, Barbados, West Indies: Bull. Canadian Petroleum Geology, v. 25, p. 123–173.

HAY, W.W., MOHLER, H.P., ROTH, P.H., SCHMIDT, R.R. AND BOUDREAUX, J.E., 1967, Calcareous nannoplankton zonation of the Cenozoic of the Gulf Coast and Caribbean-Antillean area and tranoceanic correlation: Trans. Gulf Coast Assoc. Geol. Soc., v. 17, p. 428–480.

JAMES, N.P., 1972, Holocene and Pleistocene calcareous crust (caliche) profiles: criteria for subaerial exposure: Jour. Sed. Petrology, v. 42, p. 817–836.

_____, 1974, Diagenesis of scleractinian corals in the subaerial vadose environment: Jour. Sed. Petrology, v. 48, p. 785–799.

_____, AND GINSBURG, R.N., 1979, The Seaward Margin of Belize Barrier and Atoll Reefs: International Assoc. Sedimentologists Special Pub. 3, 191 p.

_____, _____, MARSZALEK, D.S., AND CHOQUETTE, P., 1976, Facies and fabric specificity of early sub-sea cements in shallow Belize (British Honduras) reefs: Jour. Sed. Petrology, v. 46, p. 523–544.

LAND, L.S., 1967, Diagenesis of skeletal carbonates: Jour. Sed. Petrology, v. 37, p. 914–930.

_____, AND GOREAU, T.F., 1970, Submarine lithification of Jamaican reefs: Jour. Sed. Petrology, v. 40, p. 457–462.

LONGMAN, M.W., 1980, Carbonate diagenetic textures from near-surface diagenetic environments: Amer. Assoc. Petroleum Geologists Bull., v. 64, p. 461–487.

MACINTYRE, I.G., 1977, Distribution of submarine cements in a modern Caribbean fringing reef, Galeta Point, Panama: Jour. Sed. Petrology, v. 47, p. 503–516.

_____, MOUNTJOY, E.W., AND D'ANGLEJAN, B.F., 1968, An occurrence of submarine cementation of carbonate sediments off the West Coast of Barbados, W.I.: Jour. Sed. Petrology, vol. 38, pp. 660–664.

MARGOLIS, S., AND REX, R.W., 1971, Endolithic algae and micrite envelope formation in Bahamian oolites as revealed by scanning electron microscopy: Geol. Soc. America Bull., v. 82, p. 843–852.

MATTHEWS, R.K., 1968, Carbonate diagenesis: equilibration of sedimentary mineralogy to the subaerial environment: coral cap of Barbados, West Indies: Jour. Sed. Petrology, v. 38, p. 1110–1119.

MILLIMAN, J.P., 1967, Carbonate sedimentation on Hogsty Reef, a Bahamian atoll: Jour. Sed. Petrology, v. 37, p. 658–676.

MOORE, C.H., JR., 1977, Beach rock origin: some geochemical, mineralogical and petrographic considerations: Geosci. and Man, v. 18, p. 155–163.

MULTER, H.G., AND HOFFMEISTER, J.E., 1968, Subaerial laminated crusts of the Florida Keys: Geol. Soc. America Bull., v. 70, p. 183–192.

PIERSON, B.J., 1980, Late Cenozoic facies succession and diagenesis on a Bahamian atoll: International Assoc. Sedimentologists European Mtg., First, Bochum, 1980, Abstracts, p. 203–205.

_____, 1982, Cyclic sedimentation, limestone diagenesis and dolomitization in Upper Cenozoic carbonates of the southeastern Bahamas: Ph.D. Dissertation, University of Miami, Coral Gables, 295 p.

PINGITORE, N.E., JR., 1976, Vadose and phreatic diagenesis: processes, products and their recognition in corals: Jour. Sed. Petrology, v. 46, p. 985–1006.

_____, 1978, The behavior of Zn^{+2} and Mn^{+2} during carbonate diagenesis: theory and applications: Jour. Sed. Petrology, v. 48, p. 799–814.

ROBBIN, D.M., AND STIPP, J.J., 1979, Depositional rate of laminated soilstone crusts, Florida Keys: Jour. Sed. Petrology, v. 49, p. 175–180.

SCHNEIDER, W., 1977, Diagenese devonishcer Karbonatkomplexe Mitteleuropas: Geol. Jahrbuch, D 21, 108 p.

SCHNEIDERMANN, N., SANDBERG, P.A. AND WUNDER, S.J., 1972, Recognition of early cementation of aragonitic skeletal carbonates: Nat. Phys. Sci., v. 240, no. 27, p. 88–89.

SCHROEDER, J.H., 1972, Fabrics and sequences of submarine carbonate cements in Holocene Bermuda cup reefs: Geol. Rdsch., v. 61, p. 708–730.

_____, 1973, Submarine and vadose cements in Pleistocene Bermuda reef rock: Sed. Geology, v. 10, p. 179–204.

SCHUBERT, C., AND MEDINA, E., 1982, Evidence of Quaternary glaciation in the Dominican Republic: some implications for Caribbean paleoclimatology: Paleogeography, Paleoclimatology, Paleoecology, v. 39, p. 281–294.

SHINN, E.A., 1969, Submarine lithification of Holocene carbonate sediments in the Persian Gulf: Sedimentology, v. 12, p. 109–144.

_____, 1971, Aspects of diagenesis of algal cup reefs in Bermuda: Trans. Gulf Coast Assoc. of Geological Societies, v. 21, p. 387–394.

STEHLI, F.G., AND HOWER, J., 1961, Mineralogy and early diagenesis of carbonate sediments: Jour. Sed. Petrology, v. 31, p. 358–371.

STEINEN, R.P., AND MATTHEWS, R.K., 1973, Phreatic vs. vadose diagenesis: stratigraphy and mineralogy of a cored borehole on Barbados, W.I.: Jour. Sed. Petrology, v. 43, p. 1012–1020.

STROMINGER, D., HOLLANDER, J.M., AND SEABORG, G.T., 1958, Table of isotopes: Rev. Mod. Phys., v. 30, p. 585–904.

WARME, J.E., AND SCHNEIDERMANN, N., 1983, Patch-reef cementation: Holocene of Enewetok Atoll and Jurassic of Morocco (abs): Amer. Assoc. Petroleum Geologists Bull., v. 67, p. 566.
WINLAND, H.D., 1971, Nonskeletal deposition of high Mg-calcite in the marine environment and its role in the retention of textures, *in* O.P. Bricker (ed.), Carbonate Cements: The Johns Hopkins Univ. Studies in Geology no. 19, p. 278–285.

SPARRY CALCITE MARINE CEMENT IN UPPER JURASSIC LIMESTONES OF SOUTHEASTERN WYOMING

BRUCE H. WILKINSON, AMANDA L. SMITH, AND KYGER C. LOHMANN
Department of Geological Sciences, The University of Michigan, Ann Arbor, Michigan 48109

ABSTRACT

Multiple horizons within the Redwater Shale Member of the Oxfordian Sundance Formation of southeastern Wyoming exhibit megascopic features recording repeated submarine cementation on the Jurassic seafloor. Some limestone units were bored by endolithic molluscs and encrusted by oysters and serpulid worms prior to the deposition of overlying units. Other sandstone and limestone layers were lithified, fragmented, reworked, and abraided at the sediment surface, giving rise to laterally-continuous layers of sandstone and limestone cobbles which were also densely bored and encrusted by benthic invertebrates. Still other encrusted and bored units contain lithified limestone clasts demonstrating multiple episodes of cementation during the deposition of individual hardground layers.

Marine cement, truncated at lithoclast surfaces or crosscut by endolithic borings, consists of equant calcite crystals, acicular isopachous crusts, and epitaxial rims, but is predominantly equant inclusion-free spar, identical in habit to Holocene meteoric phreatic low-magnesium calcite. Its presence in Jurassic marine grainstones as a synsedimentary marine phase demonstrates that generalizations which relate cement morphology and composition to cementation environments in modern systems may not be valid when applied to ancient limestones. The variable morphology of early cements, even within individual hardground units, also suggests that calcite crystal habit may more frequently record the effect of kinetic factors, such as rates of crystal growth during cement precipitation, than the often-implied effect of Mg/Ca concentrations in ambient pore fluids.

INTRODUCTION

Recent evaluations of carbonate diagenesis have focused on two particular aspects of carbonate cements. First, studies of Holocene and calcitized Pleistocene sequences have documented the relationship between pore water chemistry and the cement phase which is precipitated (e.g. Muller *et al.*, 1972; Folk, 1974). Secondly, such studies have related cement mineralogy and composition to characteristic crystal habits (e.g. Folk, 1974; Longman, 1980). In modern environments the relationships between water chemistry and cement mineralogy and composition are relatively straightforward. Low-magnesium calcite is typically precipitated from supersaturated meteoric water in either phreatic or vadose zones (e.g. Land, 1967; Pingitore, 1970; Bathurst, 1975; Longman, 1980), whereas high-magnesium calcite and aragonite typically develop in shallow marine settings (e.g. Muller *et al.*, 1972). Almost without exception, the mineralogy and composition of Holocene carbonate cement can be directly related to the general chemistry of the water in which it precipitates.

The relationships between cement crystal habit and cement composition and mineralogy are more complex, however, because similar morphologies may be exhibited by different phases. In spite of this similarity, it has been generally concluded, for coarse cement crystals, that low-magnesium calcite is typically equant to slightly elongate with crystal size increasing toward pore interiors, that high-magnesium calcite is acicular, occuring as isopachous crusts, and that aragonite is acicular, occurring as either isopachous crusts or as radiating botryoidal masses (e.g. Folk, 1974; Bathurst, 1975; Longmann, 1980). Employing these generalizations, numerous attempts have been made to infer environments of cementation for ancient carbonate sequences on the basis of the occurrence, morphology, and composition of cement crystals in Holocene sedimentary settings.

In contrast to the strict application of uniformitarian models to ancient carbonates, however, several workers have suggested that some feature of world ocean chemistry has changed during geologic time such that the original composition of shallow marine carbonate components, including synsedimentary cement, has not been constant over geologic time (Folk, 1974; Sandberg, 1975; Mackenzie and Pigott, 1981). Mackenzie and Pigott (1981) further suggested that the relative dominance of aragonite and high-magnesium calcite versus low-magnesium calcite (and possibly dolomite) are closely related to major changes in the position of global sea level during geologic time. Periods of relative continental emergence, such as we are experiencing today, correspond to aragonite and high-magnesium calcite dominance whereas times of relative submergence correspond to low-magnesium calcite (and possibly dolomite) dominance (Mackenzie and Pigott, 1981).

Clearly, determining the validity of applying Holocene depositional and diagenetic models to ancient carbonate sequences is basic to carbonate geology. Do the general habits of aragonite and high-magnesium calcite which predominate as cements in modern shallow marine settings serve as adequate criteria for the discrimination between marine and other cements in the rock record or, in fact, have past ocean chemistries been sufficiently different from Holocene compositions that equant calcite, possibly as low-magnesium calcite, has periodically formed in marine settings as well? In order to evaluate this question, we have examined early cement from Upper Jurassic grainstones now exposed in southeastern Wyoming.

STRATIGRAPHIC SETTING

Marine sedimentary rocks of the Jurassic Sundance Formation are well-exposed throughout the western interior of North America (Imlay, 1947; Pipringos, 1957). In southeastern Wyoming they overlie the Triassic Jelm Formation or the Triassic and/or Jurassic Nugget Sandstone (Pipringos, 1968), and are overlain by the continental Jurassic Morrison Formation (Andersson, 1979). The Sundance Formation has been subdivided into several members, the Lower Oxfordian Redwater Shale Member, which overlies the Pine Butte Member and underlies the Windy Hill Member, is the second from the top (Imlay, 1947; Pipringos, 1957; Andersson, 1979). At least nine, and possibly many more, sandstone or limestone units within this member exhibit features documenting synsedimentary cementation. These have been mentioned in several studies (Pipringos, 1957; Wright, 1971) and have been described in detail by Andersson (1979) insofar as their lateral distribution, depositional environment, general petrology, and origin are concerned. We have relied heavily on this excellent report as the main stratigraphic data base for the Redwater Shale Member and as the source of localities from which samples for this study were collected (Fig. 1).

In southeastern Wyoming, the Redwater Shale Member consists mainly of shale and siltstone, but also

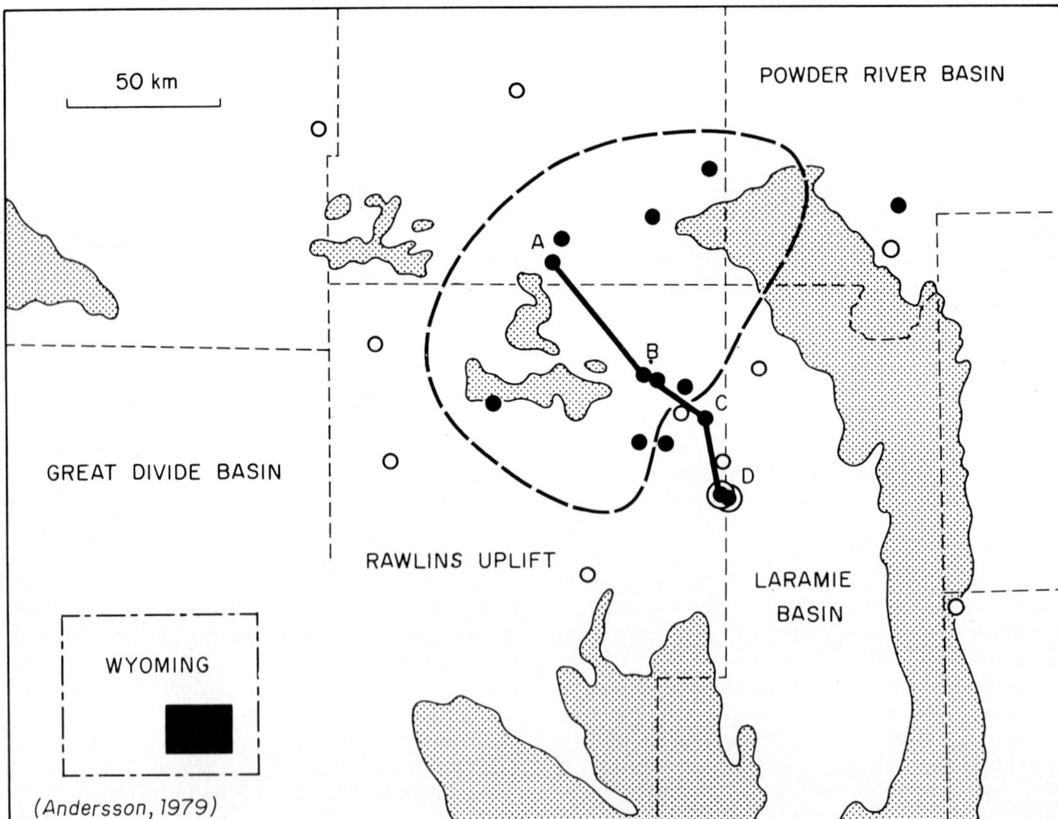

FIG. 1.—Location map of the study area in southeastern Wyoming centered around the Sweetwater Uplift immediately northwest of the Laramie Basin. Features indicating early lithification occur in 14 of the 24 localities examined by Andersson (1979) and are shown as solid dots. One such unit occurs at 9 of the 14 localities (shown within the dashed line), indicating an original areal extent of some 8,000 km². Localities examined during this study are circled dots. A stratigraphic section through the Redwater Shale Member (A–D) is shown as Figure 2. From Andersson (1979).

includes limestone and sandstone, all of which were deposited during an interval of marine transgression in what is commonly refered to as the Sundance Sea (Fig. 2). Units which show evidence for early lithification occur as two types. The first consists of layers of cobble to boulder-sized clasts of limestone or sandstone (Figs. 2, 3C). These are well-rounded, encrusted by oyster valves and/or serpulid worm tubes, and have been bored to varying degrees by endolithic molluscs (Fig. 4). One such sandstone cobble layer now discontinuously crops out over approximately 8,000 km^2 and occurs in sections separated in space by some 100 km (Fig. 1). The second type of unit consists of 0.25 to 1.0 meter-thick oolitic molluscan grainstones that have been encrusted and bored, but were not reworked into clasts on the Sundance seafloor (Figs. 2, 3A). Because this study attempted to evaluate the composition and fabric of early cement, we focused our efforts on boulders and units of coarse grainstone. Initial porosity and the size of early cement crystals are at a maximum within these facies.

GENERAL PETROLOGY

Two features of these grainstones allow for the discrimination between synsedimentary cement and later pore-filling cement. Endolithic pelecypod excavations are the most common. These occur abundantly in the upper surfaces of bedded grainstones and on all surfaces of cobbles (Fig. 4), a distribution which demonstrates the importance of physical rolling of lithified clasts on the Jurassic sea floor (Andersson, 1979). The excavations are several centimeters in length, about one centimeter in diameter, and increase in width from their external surfaces to rounded terminations within

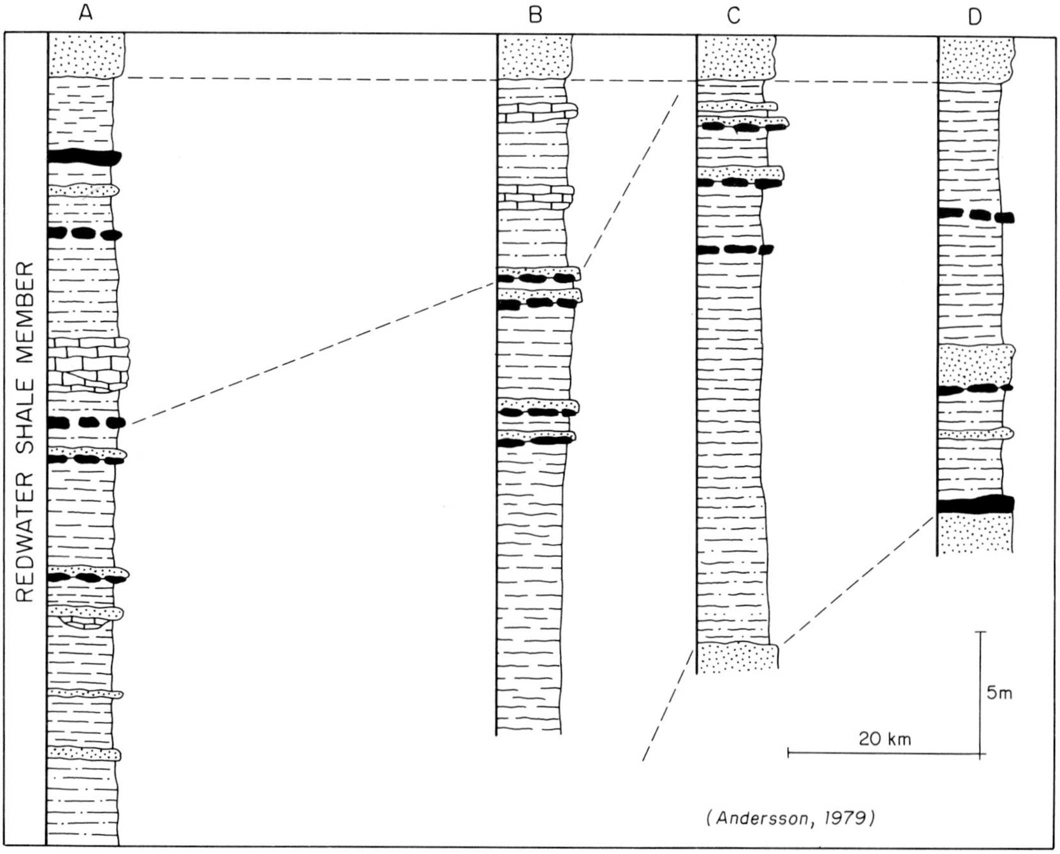

FIG. 2.—Stratigraphic section across the study area showing the general lithology of the Redwater Shale Member of the Sundance Formation and the distribution of units exhibiting features indicative of early cementation. Principal rock types are shale, siltstone, sandstone, and limestone. Layers of encrusted and/or bored sandstone and limestone cobbles are shown as black ovoids; laterally continuous bored-encrusted limestone units are shown as black layers. The three dashed lines, from bottom to top, correlate (1) the top of the Pine Butte Member, (2) the top of the main cobble layer (enclosed in Figure 1), and (3) the base of the Windy Hill Member of the Sundance Formation. Modified from Andersson (1979). See Figure 1 for section location.

grainstone interiors (Fig. 5). They are true borings in that they sharply crosscut both grains and early cement (Fig. 6A).

The second group of features which allow for the identification of early cement are rounded clasts of lithified oolite which occur abundantly in bedded oolitic molluscan grainstone units (Figs. 3B, 6B). Like borings, clast margins truncate both ooids and early cement (Fig. 6C). In addition, units which contain lithified oolite clasts may themselves be bored and encrusted. As such, they record at least two periods of cementation during the deposition of a single grainstone unit: one which lithified the oolite from which clasts were derived, and one which lithified the laterally continuous clast-bearing grainstone prior to boring and encrustation. Cementation, reworking of cemented grainstone, and cementation of lithoclast-bearing grainstone was probably a continuous process during the formation of these units.

Compositionally, Sundance Formation grainstones are low-magnesium calcite containing approximately 1.5 mole% $MgCO_3$. Atomic absorption analyses of representative whole-rock samples (Table 1) indicate that these units are relatively enriched in manganese and iron. Zinc and strontium are present in low concentrations (Table 1), levels typical of many ancient limestones. X-ray diffraction evaluation of grainstone samples further indicates that low-magnesium calcite is the only carbonate phase present. Although Fe occurs in pyrite and other iron-bearing phases, Mg, Mn, Sr, and Zn are probably largely present in carbonate lattice sites.

ALLOCHEMS

Grains which predominate in Sundance limestones are of two types: (1) ooids, which are the dominant grain type in reworked boulders and lithoclasts (Figs. 3C, 6A, B, C) and also make up approximately 50% of the allochems in early-cemented bedded molluscan grainstone units (Figs. 3A, 6B, C), and (2) molluscan debris, which make up the remaining 50% of the grain volume in bedded molluscan grainstone units (e.g. Figs. 3A, 6B, C).

Ooids

Ooids, the dominant allochem within Jurassic grainstones, are 0.3 to 0.5 mm in diameter, consisting of cortical carbonate surrounding nuclei of fine quartz sand and fine biogenic debris (Fig. 6D). The fabric of cortices is exclusively radial with calcite crystals ori-

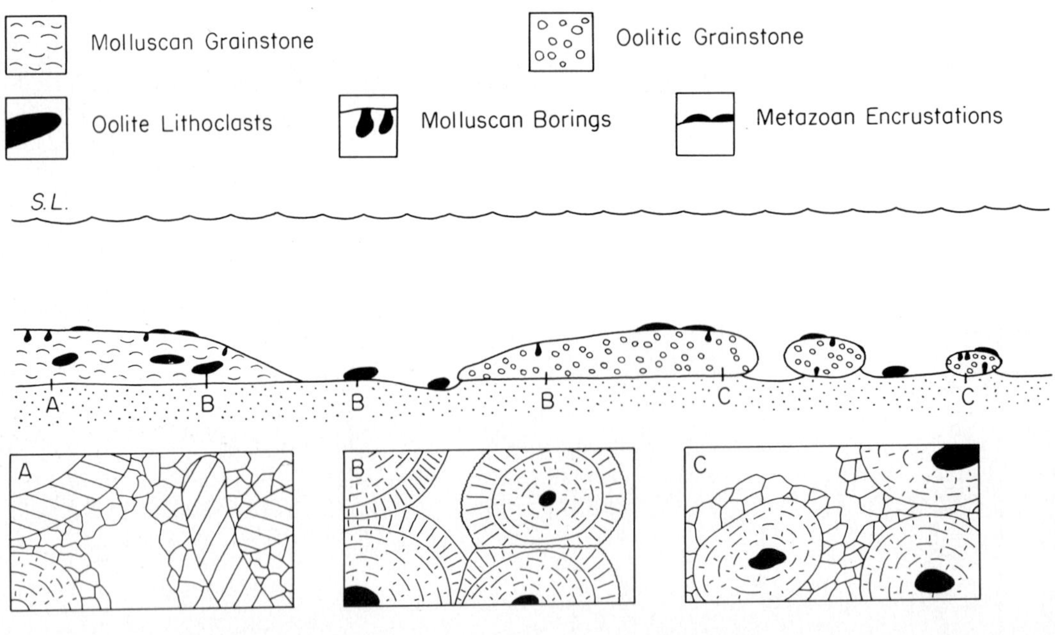

FIG. 3.—Schematic diagram showing local and lateral variation in hardground lithologies and marine phreatic cement morphologies typical of Sundance Formation grainstones. *A*, Bored and encrusted layers of molluscan grainstone which frequently contain lithoclasts of oolitic grainstone. Early cement in these layers is equant crystals of inclusion-free calcite spar. *B*, Lithoclasts of oolitic grainstone included in bored and encrusted molluscan grainstone. Early cement which lithified these oolitic sediments prior to reworking on the Jurassic seafloor consists entirely of isopachous crusts of acicular calcite. *C*, Rounded boulders to cobbles of oolitic grainstone. Individual cobbles were lithified by equant crystals of inclusion-free calcite prior to reworking, boring, and encrustation on the Sundance seafloor. Marine cement in these units is identical to early cement in molluscan grainstones.

ented normal to grain surfaces. Radial crystals are 2–3 μm in width and 5–20 μm in length. Concentric bands typically truncate or crosscut the cortical laths. As first suggested by Sorby (1879), and confirmed by Simone (1974), Sandberg (1975), Wilkinson and Landing (1978), and Heller *et al.* (1980), such radial ooids exhibit primary fabrics and were, in contrast to Holocene marine aragonite ooids, precipitated with a calcite mineralogy. Although the initial minor and trace elemental composition of ancient radial marine ooids is as yet undocumented, they are strikingly similar to high-magnesium calcite ooids from Baffin Bay, Texas (Land *et al.*, 1979).

Radial Sundance ooids are moderately luminescent (dull orange to yellow); variation in luminescence is laminae-specific. Isotopically, they are depleted in both carbon and oxygen compared to Baffin Bay grains and to other Jurassic ooids (Fig. 7, Table 2). Because Holocene radial (as well as other) ooids are non-luminescent and isotopically in near equilibrium with marine water ($\delta^{18}O = -1.3‰$, $\delta^{13}C = +4‰$ PDB, Land *et al.*, 1979), while Sundance Formation ooids are luminescent and isotopically depleted ($\delta^{18}O = -6.8‰$, $\delta^{13}C = +0.2‰$ PDB) with respect to these Holocene and other reported Jurassic radial ooids, it is clear that the Jurassic allochems have been significantly modified since their precipitation on the Jurassic seafloor.

Biogenic Debris

Molluscs were the dominant calcifying benthic invertebrate group in the Sundance Sea (Wright, 1971) and their debris comprise the majority of biogenic allochems in Sundance Formation limestones. Most mollusc fragments have been dissolved; their shape, however, is commonly outlined by thin (0.1 mm) micrite envelopes now filled with calcite spar (Fig. 6B). Among the units examined, molluscan debris are only important in bedded grainstones which were not reworked into limestone clasts following early lithification (Fig. 6C). Echinoderm debris, predominantly as disarticulated ossicles, occur throughout Sundance limestones but are generally rare, making up 2–5% of grainstone allochems.

FIG. 4.—Boulder of oolitic grainstone within a layer of such boulders in the Redwater Shale Member exposed near Como Bluff, Wyoming. Note numerous endolithic mollusc borings, as one-centimeter pits, and several larger oyster valves attached to the boulder surface.

MATRIX ELEMENTS

General

In samples in which allochems and early cement have been crosscut by borings or at the edges of lithoclasts, the volume and characteristic morphology of various matrix phases can readily be determined employing both scanning electron (Fig. 6E) and cathodoluminescence microscopic techniques (Fig. 6F). Cathodoluminescence is particularly useful in discriminating early versus late matrix phases in that respective luminescence characteristics are commonly distinct (Fig. 6F). Early cement crystal size is sufficiently small, however, that cement morphologies are difficult to document by cathodoluminescence alone.

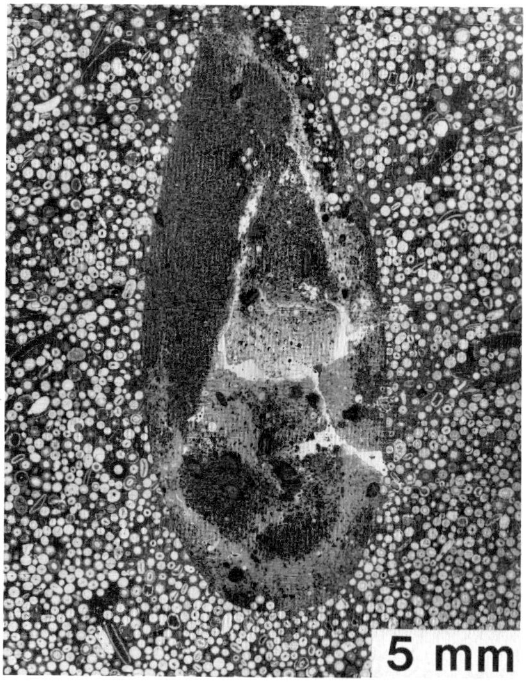

FIG. 5.—Negative print of a thin section through an endolithic mollusc boring in an oolitic grainstone cobble (similar to that in Figure 3C) from a layer of such cobbles exposed near Como Bluff in southeastern Wyoming. The exterior surface of the cobble is immediately above this field of view. The boring was in part filled with uncemented ooids and biogenic debris and subsequently with coarse low-magnesium calcite spar.

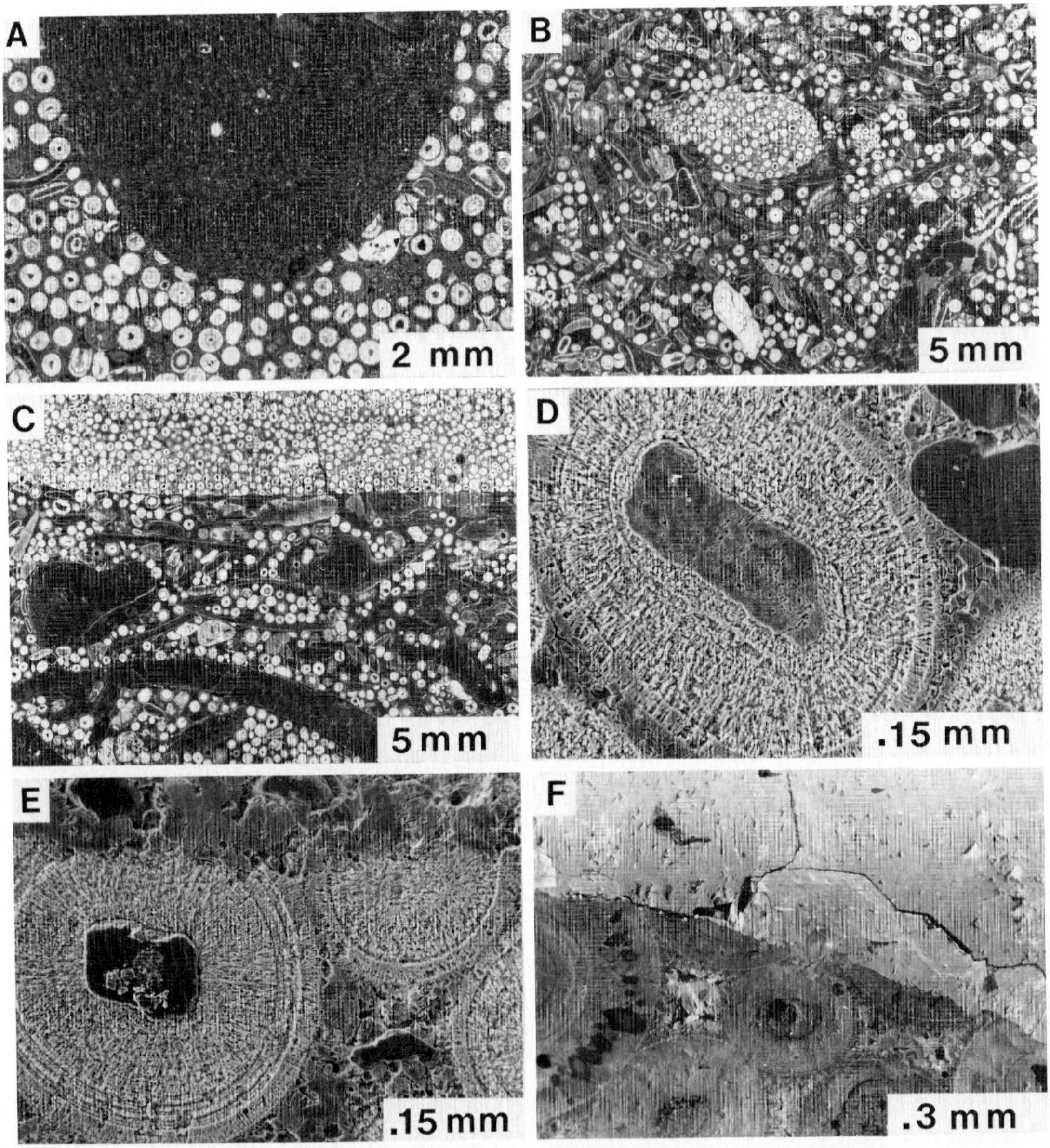

FIG. 6.—Petrographic features of Sundance Formation grainstones. *A*, Negative print of a thin section through the lower portion of an endolithic mollusc boring (similar to that in Figures 3C, 4) showing details of the contact between boring- filling spar (dark) and oolitic grainstone. Note the abruptness of the boring margin truncating grainstone allochems (as well as early cement). *B*, Negative print of a thin section through an oolitic molluscan grainstone (similar to that in Figure 3A) containing a clast of oolitic grainstone (similar to that in Figure 3B) which was lithified prior to reworking into this unit. The unit from which this section was cut was bored and encrusted with oyster valves demonstrating at least two periods of lithification or, more probably, two temporally-separated features recording continuous cementation. Note thin (light) micrite envelopes surrounding now-dissolved molluscan debris. *C*, Negative print of a thin section through the contact between a tabular clast of lithified oolite (top) and (then unlithified) oolitic molluscan grainstone (bottom). The margins of such clasts are formed of abruptly truncated allochems and early cement. *D*, SEM photograph of an ooid from the Sundance Formation. These grains, exhibiting a pronounced radial fabric, are the dominant allochems in early-cemented Jurassic grainstones. *E*, SEM photograph of a boring-grainstone contact showing sharp truncation of ooids and early cement. The dark nucleus within the ooid is a grain of quartz. *F*, Cathodoluminescence image of a boring-grainstone contact. Luminescent boring-filling spar (top) is in sharp contact with more dully luminescent ooids and early cement. A late luminescent spar occludes most intergranular porosity.

SEM images, on the other hand, are more than adequate to discriminate fine details of cement crystal morphology, but only rarely could they precisely locate the contact between early (e.g. pre-boring) and late (e.g. boring-filling) cement (Fig. 8A). Following boring and early cement truncation, late cement grew epitaxially on truncated early cement substrates. As a result, contacts between the two phases are not apparent in SEM images (Fig. 8A). Despite these respective limitations, combined scanning electron and cathodoluminescence microscopy allow both discrimination of early cement and estimation of early cement volumes and characteristic crystal morphologies.

TABLE 1.—ATOMIC ABSORPTION DATA (PPM) FOR THREE REPRESENTATIVE WHOLE-ROCK SAMPLES OF JURASSIC OOLITIC GRAINSTONE SUCH AS THAT IN FIGURES 4 AND 5. EACH VALUE IS AN AVERAGE OF 10 ANALYSES FOR EACH SAMPLE. STANDARD DEVIATIONS ARE IN BRACKETS.

	JSD-1	JSD-4	JSD-5
Mg	3,635(24)	3,595(20)	3,635(24)
Fe	2,210(24)	2,167(24)	2,221(25)
Mn	1,598(9)	1,609(9)	1,620(13)
Sr	80(6)	82(5)	83(6)
Zn	27(8)	53(8)	34(8)

Pore-filling carbonates comprise three distinct phases in Jurassic limestones. These are: (1) early cements which initially lithified grainstone units, (2) carbonate mud which was introduced into pores immediately adjacent to borings during their excavation, and (3) late spar which completely filled remaining intergranular porosity and vacated borings after the deposition of overlying units.

Early Cements

The relative volume of early cements in Jurassic hardgrounds can be estimated in two ways: (1) cathodoluminescence characteristics are commonly dis-

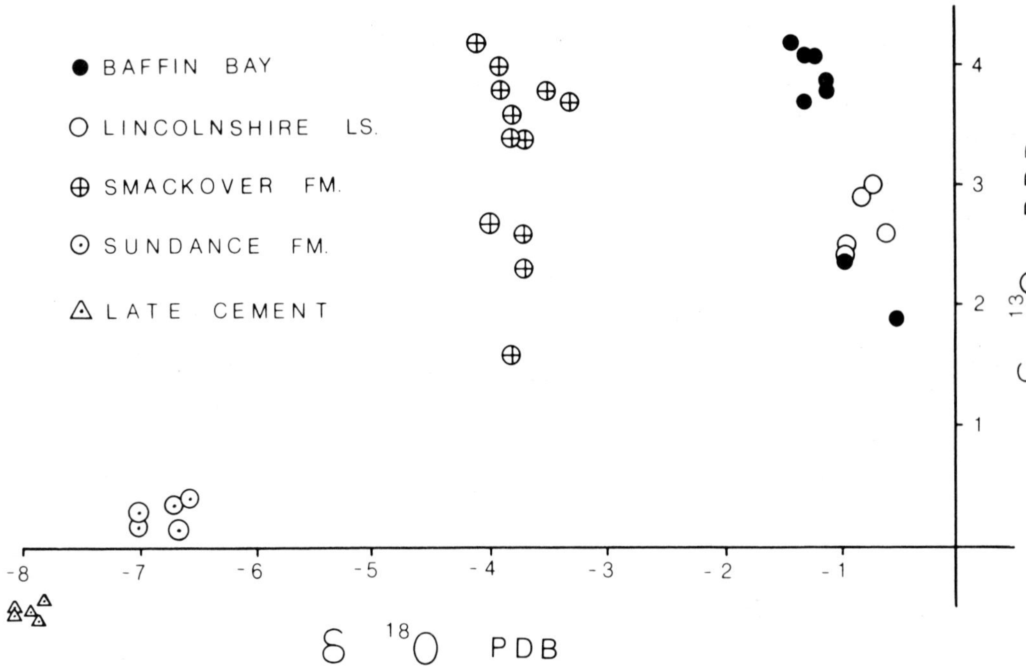

FIG. 7.—Isotopic data for ooids from Baffin Bay, Texas; for Jurassic radial calcite ooids from the Lincolnshire Limestone, the Smackover Formation, and the Sundance Formation; and for Sundance Formation late cement. Data from Baffin Bay is for ooid cortices composed of interlaminated aragonite and radial high-magnesium calcite. Nuclei of these ooids are commonly caliche fragments; the two light carbon data points probably were contaminated with caliche carbonate (from Land *et al.*, 1979). Data from the Upper Jurassic Smackover Formation is for individual ooids (cortex and nucleus) from 9,752 to 11,100 feet in Louisiana and Arkansas (from Moore and Druckman, 1981). Data from the Middle Jurassic Lincolnshire Limestone is for whole rock samples within 5 cm of hardground surfaces. These samples consist entirely of ooids and early cement (from Marshall and Ashton, 1980). Data from the Sundance Formation is for whole ooids (cortices and nuclei) and for late luminescent spar from grainstone cobbles as in Figure 5. Note that radial Sundance ooids are depleted in carbon and oxygen with respect to other radial Jurassic ooids, and that other radial Jurassic ooids are depleted with respect to radial ooids from Baffin Bay. Also note that Sundance Formation late cement is depleted with respect to radial Sundance ooids.

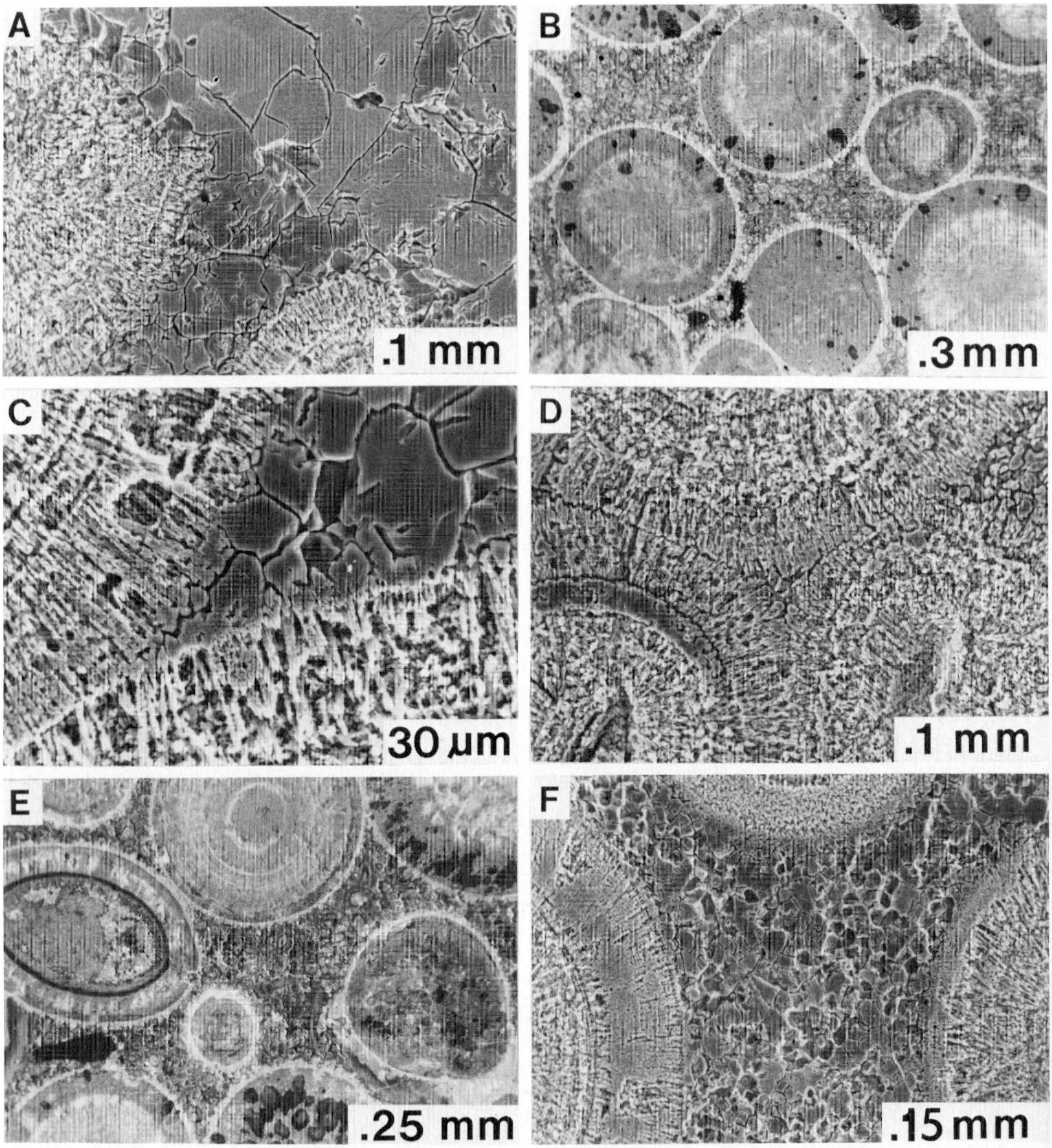

FIG. 8.—Petrographic features of early-cemented grainstones. *A*, SEM image of a boring-grainstone contact which truncates two ooids and early cement. Note that boring-filling cement (upper right) has nucleated on truncated cement crystals. As a result, the boundary between early and late cement is not apparent. *B*, Cathodoluminescence image of the oolitic grainstone in Figure 5. Note that early cement in this sample is a highly luminescent rim around ooids. Other early cement (as in Figure 6F) is only dully luminescent. *C*, SEM image of the dominant early cement morphology in Jurassic grainstones. Here, between two ooids, equant inclusion-free crystals coarsen toward the pore interior in the upper right. *D*, SEM image of early cement in reworked oolite lithoclasts (Figs.6B, C). This early cement is exclusively fibrous, occurring as isopachous crusts. Here, porosity between three ooids was almost completely occluded by this cement. Note that ooid-cement contacts cannot be discerned because cement crystals nucleated on, and grew away from, radial cortical crystals. *E*, Cathodoluminescence image of an oolitic grainstone immediately adjacent to a boring. In this field of view, ooids are surrounded by rims of brightly luminescent early spar. Intergranular pores were filled with carbonate mud, during and/or immediately following boring excavation, which subsequently aggraded to pseudospar now seen as the youngest pore-filling phase. *F*, SEM image of pore-filling pseudospar immediately adjacent to a boring. Note that unlike most cement, this recrystalized carbonate does not increase in crystal size toward pore interiors.

tinctive, and (2) carbonate mud is distinct with respect to its grain size and its luminescence characteristics so that it can be differentiated from early cements, both in thin sections and in SEM samples. Initial intergranular porosity in oolitic grainstones was approximately 35%. Point counting early cements versus later matrix phases (carbonate mud and late cement) demonstrates that 20% of the initial porosity was occluded during early cementation, a volume that now makes up 6% of the grainstones. While not abundant, this volume of cement gave rise to well-lithified substrates prior to encrustation and boring by various invertebrate groups on the Jurassic seafloor.

All early cement is luminescent, ranging from dull orange to bright yellowish orange; the intensity of luminescence varies considerably from cobble to cobble, from unit to unit, and from outcrop to outcrop. Some early cement, for example, occurs as thin isopachous highly luminescent rims (Fig. 8B). In other samples early cement is only dully luminescent (Fig. 6F). None of the early cement exhibits luminescence zoning.

Three morphologies characterize these early cements: equant blocky spar, acicular fibrous crystals, and epitaxial rims on echinoderm debris. The first of these is the dominant morphology in grainstone units where it occurs as nearly equant (10 to 30 μm) subhedral crystals, typically with length-width ratios close to unity, commonly coarsening away from respective substrates (Fig. 8C). They consist exclusively of clear calcite which lack microdolomite and other inclusions.

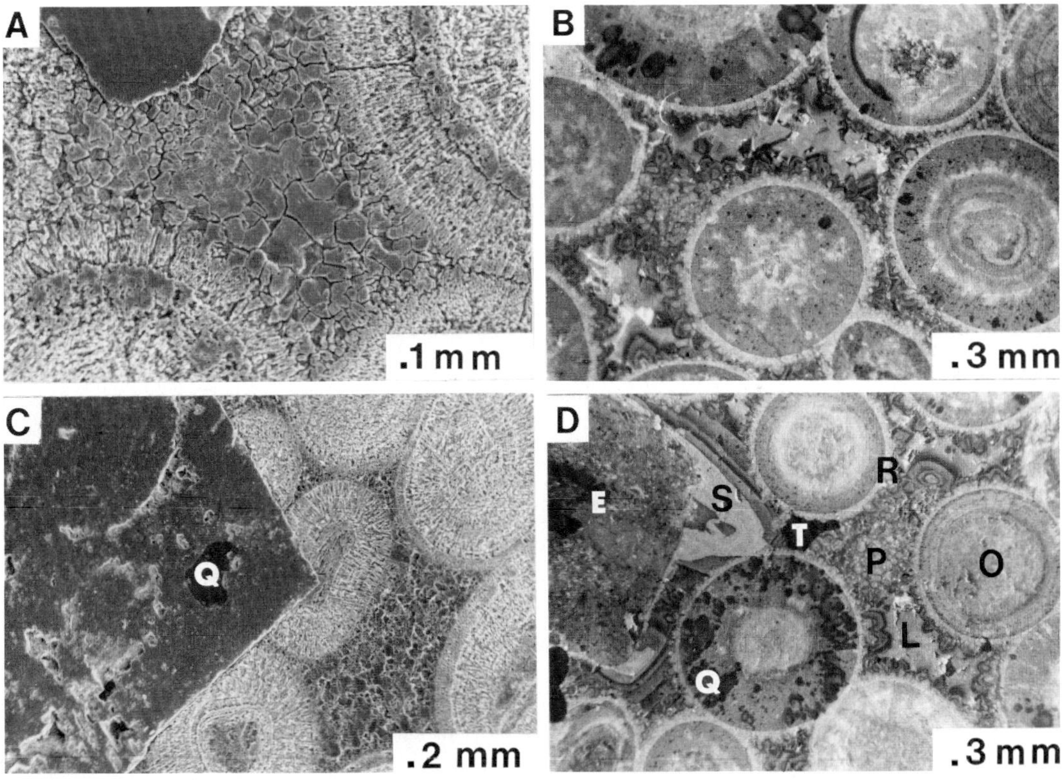

FIG. 9.—Grainstone units from the Sundance Formation. *A*, SEM image of late pore-filling spar in an oolitic grainstone lithoclast. This sample was first lithified by fibrous isopachous cement which encrusted ooids but not terrigenous components (such as the quartz grain shown here). Note that, as in Figure 8D, the contact between fibrous cement and ooid cortices is not discernible. *B*, Cathodoluminescence image of an oolitic grainstone. Note luminescent early cement crusts around ooids, a dully luminescent but strongly zoned generation of late cement, and a final luminescent generation of late cement which filled remaining intergranular porosity. *C*, SEM image of a section through oolitic grainstone and a replacive cubic pyrite crystal. Note ghosts of ooids in the pyrite. An incorporated grain of quartz sand (Q) occurs toward the apex of the pyrite crystal. Such crystals rarely transect grainstone-boring contacts demonstrating that their formation post-dated late cement precipitation. *D*, Cathodoluminescence image of a typical Sundance Formation grainstone. In this field of view note radial ooids with variable luminescence (O), luminescent early cement (R), equant crystals of porefilling pseudospar (P), nonluminescent chalcedonic quartz replacing cortical carbonate (Q), terrigenous quartz sand (T), early to late cement as a zoned epitaxial rim (S) on an echinoderm grain (E), and late luminescent pore-filling spar (L).

The second early cement is fibrous, and occurs only in oolitic grainstone lithoclasts (Figs. 6B, C).

It consists of isopachous crusts of crystals, 1–2 μm wide and 10–20 μm long, oriented normal to substrates (Fig. 8D). Fibrous cement crystals are morphologically identical to cortical carbonate in the ooids which they surround and on which they were nucleated. As a result, the boundary between cortical crystals and fibrous cement crystals cannot be discerned employing either SEM or cathodoluminescence techniques. Planar sutures between coalesced fibrous crusts (which appear as straight lines in section) are common in oolite lithoclasts (Fig. 8D). Early fibrous cement is always dully luminescent. Epitaxial rims on echinoderm debris, like early blocky spar, consist of clear, inclusion-free calcite. They exhibit variable luminescence (Fig. 9D).

Carbonate Mud

Carbonate mud was emplaced into early cement-lined pores immediately adjacent to endolithic mollusc borings, and also partially filled some borings prior to late cement precipitation. Around borings, this allochthonous micritic carbonate occluded all remaining intergranular porosity in a zone approximately 6 mm wide, but is absent from grainstone pores farther removed from mollusc excavations. Where present, both in borings and in intergranular pores, it now exhibits prominent luminescence zones defining crystal faces developed during the subsequent growth of neomorphic euhedral crystals (Fig. 8E). Although the boundaries between the crystals which make up this matrix phase define anhedral 20–30 μm crystals as a pseudospar (Fig. 8F), their prominent zoning demonstrates that the initial carbonate phase was emplaced as finer micritic crystals, that they underwent neomorphism, and that euhedral calcite crystals finally coalesced giving rise to the anhedral crystal forms now characteristic of this phase.

Late Cement

Continued cementation of Jurassic grainstones after deposition of overlying units resulted in the occlusion of almost all remaining porosity. Morphologically, this late cement occurs as equant anhedral crystals on equant and acicular marine cements (Fig. 9A), and as zoned epitaxial rims on early rim cement (Fig. 8D). Early rim cement and late rim cement is morphologically identical and cannot be differentiated employing either scanning electron or cathodoluminescence techniques. Early rim cement, however, is clearly truncated at boring margins whereas zoning in late rim cement commonly extends into boring interiors. As with early cement, the luminescence characteristics of late cement are variable. Strongly zoned dully luminescent crystals commonly give way to brightly luminescent spar toward pore interiors (Fig. 9B). Brightly luminescent late cement is the dominant matrix phase in grainstones, filling 50% of intergranular voids and making up 15% of the rock volume.

Late luminescent cement also fills voids developed during the dissolution of molluscan aragonite (Fig. 10). Thin micrite envelopes underwent various degrees of collapse during compaction following aragonite dissolution. In bedded molluscan grainstone containing tabular clasts of lithified oolite (e.g. Fig. 6C), this compaction was accompanied by fracturing and movement of oolite clast fragments. In several sections it can be demonstrated that fracturing only disrupted lithoclasts and micrite envelopes (Fig. 10), indicating: (1) that the dissolution of molluscan aragonite postdated clast lithification, and (2) that late cement, which infills and surrounds ruptured micrite envelopes, was precipitated after the compactional event.

Along with ooids, late cement was sampled for isotopic analyses (Fig. 7). It is depleted in oxygen and carbon relative to the ooids ($\delta^{18}O = -8.0‰$, $\delta^{13}C = -0.5‰$ PDB). Such values indicate that late cement was precipitated following Jurassic deposition in southeastern Wyoming, either during burial or following Laramide tectonism and exposure to meteoric fluids.

In addition to the allochems and matrix phases discussed above, Sundance grainstones contain minor diagenetic pyrite (Fig. 9C), glauconite, and chalcedonic quartz (Fig. 9D) which replaces cement and/or ooids. Cubic pyrite crystals occur scattered throughout oolitic grainstone, occasionally transecting boring-oolite contacts, replacing both grainstone and late cement that fills borings. Such relationships demonstrate that pyrite formation postdated precipitation of the late cement.

DEPOSITIONAL AND DIAGENETIC HISTORY

On the basis of these relationships, the depositional and diagenetic history of Sundance Formation grainstones was as follows: (1) deposition of oolite sand on the Jurassic seafloor, (2) early cementation of oolitic grainstone by fibrous isopachous cement, (3) local reworking of lithoclasts into (as yet) unlithified molluscan and oolite carbonate sand, (4) continued lithification of these grainstone units by the precipitation of equant spar, (5) local reworking of lithified oolite grainstone into rounded boulders concomitant with boring and encrustation of both oolite boulders and stratified molluscan grainstone units by various invertebrate groups, (6) dissolution of molluscan aragonite with compactional collapse of tabular oolite lithoclasts and micrite envelopes, (7) precipitation of late cement in borings and remaining intergranular pores, neomorphism of intergranular micrite, and (8) minor replacement of ooids and cements by pyrite. The relative timing of the formation of glauconite and chalcedonic quartz are unknown.

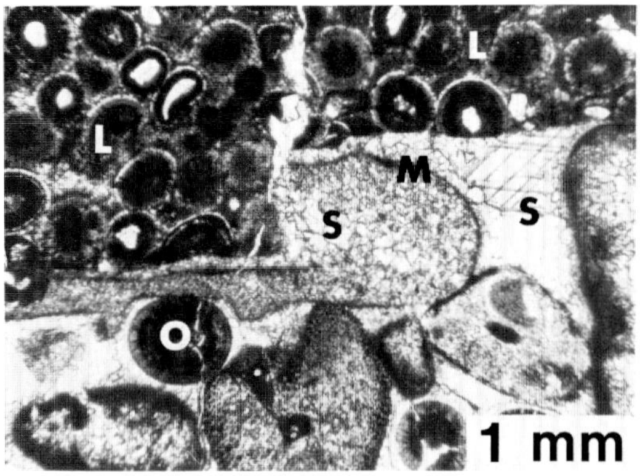

FIG. 10.—Thin section photomicrograph (plane light) through a small fracture which transects an oolite lithoclast (L) and part of a micrite envelope (M) which developed around a now-dissolved aragonite mollusc fragment. Note that while the upper portion of the micrite envelope is ruptured, the lower portion is merely draped over an ooid (O) and a thin layer of early calcite spar. Late spar cement (S) fills the micrite envelope and most intergranular porosity. These relationships require that aragonite dissolution postdated early cement precipitation but predated late cement precipitation.

INTERPRETATION

Cementation Environments

Macroscopic features demonstrate that sandstone and limestone units were repeatedly lithified during deposition of the Sundance Formation. Boring by endolithic molluscs as well as encrustation by oysters and serpulid worms require the formation of well-lithified substrates prior to the deposition of overlying units. Rounding of sandstone and limestone clasts further attests to the repeated development of well-indurated units during deposition. While stratigraphic relationships, such as the thinness and lateral extent of units, and macroscopic features, such as the abundance of marine metazoan borings, strongly suggest that this cementation took place in a marine environment, it might be argued that lithification in fact took place in meteoric waters which infiltrated porous substrates during periodic subaerial exposure, and that cementation events were not related to marine processes. Because acceptance of a marine origin is critical to any further discussion of early cement, the plausibility of meteoric cementation must be addressed in some detail. Several lines of evidence suggest that early cement precipitation took place in marine phreatic settings.

First, where present, early-lithified units occur at approximately 2 m intervals; typical sections contain 3 to 5 such units (Fig. 2). In addition, some relatively thin early-cemented units exhibit great areal extent (Fig. 1), and can be found in coeval sections now separated in space by tens of kilometers with little change in either their thickness or the thickness and lithology of associated sedimentary units. It is difficult to envisage a subaerial setting which could give rise to such closely-spaced pedogenic lithification without the removal of some of the original sequence by erosion. Second, despite the fact that cementation of marine carbonate sediment occurs rapidly when exposed to meteoric waters, the dominant early-cemented lithology in the Redwater Shale Member is, in fact, terrigenous clastic sediment. Although we have focused on carbonate grainstones, most Sundance cobble layers (including the areally extensive unit in Fig. 1) are composed of fine-grained sandstone. Because subaerial, meteoric pedogenic lithification of quartzose sand by calcite spar is a process almost unknown from Holocene settings, it is unlikely that terrigenous Sundance units were repeatedly lithified during subaerial exposure. Lastly, and perhaps most importantly, petrographic relationships demonstrate that early limestone lithification took place prior to the dissolution of aragonite components. Meteoric cementation in Holocene settings, particularly at exposure surfaces, is always accompanied by aragonite dissolution. It is, in fact, the dissolution of aragonite which serves as the source of carbonate for meteoric calcite cementation (e.g. Allan and Matthews, 1982). Because early-lithified grainstone units are thin, but laterally persistent, their upper surfaces must have been nearly coincident with the sediment surface during cementation. The presence of aragonite components in Jurassic units during early cement precipitation, the instability of aragonite at modern exposure surfaces, and the relative stability of Holocene aragonite during marine lithification (e.g. Shinn, 1969; Friedman *et al.*, 1974; James *et al.*, 1976; Macintyre, 1977; Harris, 1978), all indicate that Sundance units were cemented in submarine settings. As such, they are (were) marine hardgrounds in the true sense of the term in that cementation took place at or within a few centimeters of the sediment-water interface.

Cement Mineralogy

Two attributes of carbonate cements frequently are related to their morphologies and the environments in which they precipitate. These are: (1) the mineralogy of the cement phase defined by crystallography and major element composition, and (2) the minor and trace element composition of the cement phase which is determined by the concentration of various ions in precipitating fluids.

What can we infer about the original mineralogy of these Jurassic marine cements? The original mineralogy of a hardground cement, or of any carbonate component for that matter, is either preserved or is replaced by some other mineralogic phase during diagenetic stabilization. In the case of replacement, original mineralogy can only be inferred on the basis of fabrics and/or chemical compositions which may be inherited from the primary phase. In the case of the Jurassic hardground cement, its present calcite mineralogy could be a primary phase or could have formed during the calcitization of primary aragonite. Several lines of evidence demonstrate that the present mineralogy is primary. First, none of the three Jurassic hardground cement morphologies exhibit features typical of calcitized aragonite, which commonly consists of equant to slightly elongate interlocking anhedral crystals of inclusion-rich low-magnesium calcite (e.g. Assereto and Folk, 1976; Davies, 1977; Mazzullo, 1980). Second, the fibrous isopachous early cement in lithoclasts is morphologically identical to fibrous calcite marine cement reported from many modern settings. This fabric is probably characteristic of calcite. Third, with respect to the dominant early cement, equant subhedral spar, crystals typically increase in size away from substrates, a characteristic common to pore-filling calcite cement (e.g. Bathurst, 1975). Similarly, epitaxial rims on echinoderm debris are morphologically identical to primary calcite cement precipitated in numerous marine, meteoric, and deep burial settings; such rims are unknown as either a primary aragonite or a calcitized aragonite phase. Based on the similarity of Jurassic hardground cement morphologies to those reported for primary calcite cements in numerous other settings, there is no reason to doubt that the hardground cements are primary porefilling calcites.

Cement Composition

Given that hardground cements were precipitated as calcite, what then was their original minor and trace elemental composition? The cation which shows greatest variability in Holocene sedimentary calcites is magnesium, ranging from 0–4 mole % (low-magnesium calcite) to 10–18 mole % (high-magnesium calcite). This classic subdivision is purely artificial, of course, arising from the fact that the Mg/Ca ratios of carbonate-precipitating fluids exhibit a pronounced bimodal distribution; magnesium is enriched in marine and hypersaline fluids whereas it is depleted in meteoric and subsurface fluids. Calcites which form in these settings, therefore, are either magnesium-enriched or magnesium-depleted (high-magnesium calcite and low-magnesium calcite respectively). With respect to marine calcites, however, even though Holocene seas exhibit Mg/Ca ratios close to 5, in a context of the Jurassic hardground cement, which may record precipitation from marine water with a composition unlike that of Holocene marine water, there is no reason *a priori* to assume that original magnesium concentrations were either high or low.

If we cannot assume that the Jurassic cements were magnesium-enriched simply because Holocene marine calcite cements are so enriched, how then can original magnesium content be determined? Two possibilities exist. Either the hardground cement was (1) precipitated as the relatively low-magnesium calcite we see today, or (2) was precipitated as a relatively high-magnesium calcite which lost magnesium during transformation to low-magnesium calcite.

Morphological Constraints

If hardground units were largely lithified by crystals of high-magnesium calcite, their dominant morphology as equant spar is seemingly anomalous compared to most reported Holocene high-magnesium calcite which, when coarse, typically occurs as fibrous crusts (e.g. James *et al.*, 1976; Longman, 1980). Ancient examples of former high-magnesium calcites retain this original fabric, commonly occurring as inclusion-rich isopachous coatings with pronounced undulatory extinction (e.g. Davies, 1977; Lohmann and Meyers, 1977). The relationships between crystal habit and magnesium content, however, are almost completely unknown. Even though most Holocene high-magnesium calcite is indeed fibrous whereas most Holocene low-magnesium calcite is typically equant, both compositions may exhibit a range of morphologies. Fibrous low-magnesium calcite is known from numerous Holocene meteoric vadose settings (e.g. Braithwaite, 1979; Kendall and Broughton, 1980; Binkley *et al.*, 1980), and coarse equant high-magnesium calcite spar has now been reported from several Holocene shallow marine settings (Schroeder, 1972; James and Ginsburg, 1979; Pierson and Shinn, 1983; Warme and Schneidermann, 1983). The presence of acicular and equant morphologies as both low-magnesium and high-magnesium calcite in Holocene settings suggest that calcite cement crystal habit may be more closely related to rates of precipitation and other kinetic variables than to ambient Mg/Ca ratios in precipitational fluids. In this context, and bearing in mind that both equant and acicular morphologies occur as early marine cement in the Sundance hardgrounds, there is no apparent reason on the basis of crystal morphology alone why Jurassic hardground cements could not have been precipitated as either a high-magnesium calcite or as a low-magnesium calcite.

If initial compositions were high-magnesium calcite, however, stabilization to low-magnesium calcite must have occurred while preserving the finest details of primary crystal morphology. Although the high-magnesium calcite to low-magnesium calcite transformation front has never been well documented, former high-magnesium calcite components which have stabilized to low-magnesium calcite in other settings, retain virtually all original fabric (e.g. Schlager and James, 1978; Towe and Hemleben, 1976). It is entirely possible, therefore, that Jurassic hardground cements were precipitated as equant and acicular high-magnesium calcites that were subsequently stabilized to low-magnesium calcite during the migration of exceedingly thin fluid fronts, a process which resulted in the retention of original crystal morphology.

Luminescence Constraints

As discussed above, early cements exhibit variable luminescence, ranging from dull (e.g. Fig. 6F) to highly (e.g. Fig. 8B) luminescent. Two interpretations of such luminescence variations are plausible. First, it is possible that these variations are a primary attribute. Holocene marine cements are generally non-luminescent, but may luminesce brightly when precipitated in locally reducing intergranular microenvironments (e.g. Adams and Schofield, 1983). It is possible, therefore, that the early cements record variable degrees of communication between pore fluids and overlying marine water, variable (values of) Eh, and variable degrees of Mn^{2+} incorporation into the growing cement crystals. Although Sundance Formation cementation continuously took place at or near the sediment surface, cement precipitation may have taken place in intergranular micro-environments, from pore waters that were isolated to varying degrees from overlying oxygenated marine water. If manganese content as reflected by luminescence is a primary attribute, however, retention of such a characteristic requires that initial cement compositions were stable in later diagenetic pore fluids, in particular those fluids which gave rise to the subsequent dissolution of molluscan aragonite and the diagenetic stabilization of high-magnesium calcite echinoderm debris to low-magnesium calcite. If luminescence was acquired during initial marine cement precipitation, then initial cement compositions must have been low-magnesium calcites.

Alternatively, variation in early cement luminescence may be a late diagenetic attribute. If cements were precipitated as high-magnesium calcite, stabilization to low-magnesium calcite in manganese-enriched pore fluids would result in the formation of luminescent calcite with spatial variations in luminescence recording spatial variations in Mn^{2+} concentrations and thus redox conditions in pore fluids during stabilization. Compositional data on grainstone ooids and late cement demonstrate the plausibility of such a scenario. The latest blocky cement is the most brightly luminescent and the most depleted in carbon and oxygen (Fig. 7). Ooids are also isotopically depleted, slightly less so than late cement, but significantly so when compared to any reasonable estimates of Jurassic marine values and when compared to high-magnesium calcite ooids from Baffin Bay, Texas (Fig. 7). Because luminescence characteristics and isotopic compositions of Sundance ooids are far removed from what must have been original values, and because they are close to values for late cements, ooids must have diagenetically reequilibrated with manganese-enriched, isotopically-depleted fluids, probably during the precipitation of the manganese-enriched, isotopically-depleted late cement. Early cement is too fine-grained to sample for isotopic analysis, but its luminescence suggests at least the possibility of a similar diagenetic history. High magnesium concentrations in early cement may have given rise to the diagenetic potential for subsequent stabilization which is now recorded by luminescence.

DISCUSSION

Regardless of initial magnesium concentrations, the morphologies of Sundance hardground cements are significant in that the dominant early marine cement morphology, a primary clear equant calcite spar, is seemingly anomalous compared to shallow marine cements reported from Holocene settings. With respect to this spar, even though modern marine cements are almost exclusively fibrous high-magnesium calcite and aragonite, occurrences of equant crystals as both high-magnesium and low-magnesium calcite are also reported. In addition to the reports of equant high-magnesium calcite cited previously, rhombic polyhedral low-magnesium calcite spar is known from Tongue of the Ocean in the Bahamas (Schlager and James, 1978) as well as from shallow seas near England (Al-Hashimi, 1977) and Tasmania (Rao, 1981). These are areas of carbonate precipitation in cold water, conditions which are surely dissimilar to those in which the more nearly equatorial Jurassic grainstones were deposited. They do, however, serve to show that in modern marine settings, cement morphology and habit may be similar to that found in meteoric systems. Ancient hardground units have also been reported from many stratigraphic settings; several of these are oolitic Jurassic grainstones (e.g. Purser, 1969; Marshall and Ashton, 1980). While few studies have attempted to evaluate the habit of early cement crystals, those which have commonly document the presence of equant cement, strikingly similar in habit to the dominant Wyoming hardground cement. These include Ordovician hardground cement from Sweden (Lindström, 1979, 1980), from Oklahoma (Longman, 1976), from the Upper Mississippi Valley (Delgado, 1979), and from Ontario (Wilkinson *et al.*, 1982). Similarly, Purser (1969) reports that the dominant early calcite cement in oolitic Jurassic hardgrounds from the Paris Basin is equant to slightly elongate crystals.

Returning to the question of whether Sundance Formation equant spars as marine cements are indeed anomalous, we can offer no unequivocal answers at this time. Although equant high-magnesium and low-magnesium calcite crystals are now documented from many Holocene marine settings, they are clearly an exceptional morphology in shallow marine environments. Acicular habits are always dominant, even where more equant cement crystals have been co-precipitated. With respect to ancient hardgrounds, however, the few studies to date which have well-characterized marine cement crystal morphologies strongly suggest that equant habits are as common as acicular habits, and may well be the dominant morphology in sequences deposited during relative highstands of global sea level. Coeval ooids in such sequences are characteristically radial, a fabric which documents the dominance of calcite versus aragonite as a primary cortical phase (e.g. Sandberg, 1975). The preponderance of radial ooids, equant marine cements, and a striking lack of primary aragonite (either as marine cement or as cortical coatings) in highstand carbonate sequences all support the suggestion of Mackenzie and Pigott (1981) that the primary mineralogy of physicochemically precipitated carbonate phases has varied cyclically through Phanerozoic time with high-magnesium calcite and aragonite dominant during global sea level lowstands and low-magnesium calcite and (possibly) dolomite dominant during global sea level highstands.

The presence of both equant and acicular cement crystal morphologies in a variety of Holocene meteoric and marine settings, however, in conjunction with reported ocurrences of both high-magnesium and low-magnesium calcites in Holocene marine settings, demonstrates that relationships between fluid chemistry, cement phase, and cement morphology are complex indeed.

Early cements in the Sundance hardgrounds also exhibit variable morphologies, ranging from fibrous to equant. Given that the composition of marine water was probably invariant across the Jurassic shelf, it is not reasonable to invoke differing Mg/Ca ratios (e.g. Folk, 1974) or surface/active cation ratios (e.g. Lahann, 1978) to explain differing crystal forms. Such variation in Jurassic marine cement morphologies, therefore, must relate to the influence of some as yet undocumented kinetic factor such as different rates of crystal growth during hardground lithification.

Clearly, long-term cyclic variations in global atmospheric-hydrospheric chemistries related to variations in the relative positions of global sea level may well have had a profound effect on the mineralogy, composition and habit of sedimentary carbonate phases. Variations in these parameters in Holocene marine cements, however, coupled with uncertainties inherent in inferring similar variations in ancient marine phases such as those in the Sundance hardgrounds, more than demonstrate that simplistically inferred relationships between crystal habit and the chemistry of cementation environments are no more than generalizations. Identification of marine cement employing morphologic criteria alone is to oversimplify the nature of early cementation in various diagenetic settings.

ACKNOWLEDGMENTS

We particularly thank Sandra J. Carlson for assistance during sampling and other aspects of field work. Joyce M. Budai, Karen Rose Cercone, and James Lee Wilson (The University of Michigan) read early versions of the manuscript and made many helpful suggestions. Drafting was done by Derwin Bell, Department of Geological Sciences. Scanning electron microscopy was carried out in the Microbeam Laboratory at The University of Michigan. Research on temporal variation in Phanerozoic marine carbonates at The University of Michigan is supported by the National Science Foundation, NSF Grant OCE-81-17699.

REFERENCES

Adams, A.E., and Schofield, K., 1983, Recent submarine aragonite, magnesian calcite, and hematite cements in a gravel from Islay, Scotland: Jour. Sed. Petrology, v. 53, p. 417–421.

Al-Hashimi, W.S., 1977, Recent carbonate cementation from seawater in some weathered dolostones, Northumberland, England: Jour. Sed. Petrology, v. 47, p. 1375–1391.

Allan, J.R., and Matthews, R.K., 1982, Isotope signatures associated with early meteoric diagenesis: Sedimentology, v. 29, p. 797–817.

Andersson, K.A., 1979, Early lithification of limestones in the Redwater Shale Member of the Sundance Formation (Jurassic) of southeastern Wyoming: Contrib. Geology, v. 18, p. 10–17.

Assereto, R., and Folk, R.L., 1976, Brick-like texture and radial rays in Triassic pisolites of Lombardy, Italy: A clue to distinguish ancient aragonite pisolites: Sed. Geology, v. 16, p. 205–222.

Bathurst, R.G.C., 1975, Carbonate Sediments and Their Diagenesis (2nd enlarged ed.): New York, Elsevier, 658 p.

Binkley, K.L., Wilkinson, B.H., and Owen, R.M., 1980, Vadose beachrock cementation along a southeastern Michigan marl lake: Jour. Sed. Petrology, v. 50, p. 953–962.

Braithwaite, C.J.R., 1979, Crystal textures of recent fluvial pisolites and laminated crystalline crusts in Dyfed, South Wales: Jour. Sed. Petrology, v. 49, p. 181–194.

Davies, G.R., 1977, Former magnesian calcite and aragonite submarine cements in Upper Paleozoic reefs of the Canadian Arctic: A summary: Geology, v. 5, p. 11–15.

Delgado, D.G., 1979, Submarine diagenesis (aragonite dissolution, cementation by calcite, and dolomitization) in Ordovician Galena Group, Upper Mississippi Valley [abs.]: Amer. Assoc. Petroleum Geologists Bull., v. 64, p. 697.

Friedman, G.M., Amiel, A.J., and Schneidermann, N., 1974, Submarine cementation in reefs: Examples from the Red Sea: Jour. Sed. Petrology, v. 44, p. 816–825.

FOLK, R.L., 1974, The natural history of crystalline calcium carbonate: Effect of magnesium content and salinity: Jour. Sed. Petrology, v. 44, p. 40–53.

HARRIS, P.M., 1978, Holocene marine-cemented sands, Joulters Ooid Shoal, Bahamas: Gulf Coast Assoc. Geol. Societies Trans., v. 28, p. 175–184.

HELLER, P.L., KOMAR, P.D., AND PEVEAR, D.R., 1980, Transport processes in ooid genesis: Jour. Sed. Petrology, v. 50, p. 943–952.

IMLAY, R.W., 1947, Marine Jurassic of the Black Hills area, South Dakota and Wyoming: Amer. Assoc. Petroleum Geologists Bull., v. 36, p. 227–273.

JAMES, N.P., AND GINSBURG, R.N., 1979, The Seaward Margin of Belize Barrier and Atoll Reefs: Int. Assoc. Sedimentologists Spec. Pub. 3, p. 191.

______, ______, MARSZALEK, D.S., AND CHOQUETTE, P.W., 1976, Facies and fabric specificity of early subsea cements in shallow Belize (British Honduras) reefs: Jour. Sed. Petrology, v. 46, p. 523–544.

KENDALL, A.C., AND BROUGHTON, P.L., 1978, Origin of fabrics in speleothems composed of columnar calcite crystals: Jour. Sed. Petrology, v. 48, p. 519–538.

LAHANN, R.W., 1978, A chemical model for calcite crystal growth and morphology control: Jour. Sed. Petrology, v. 48, p. 337–344.

LAND, L.S., 1967, Diagenesis of skeletal carbonates: Jour. Sed. Petrology, v. 37, p. 914–930.

______, BEHRENS, E.W., AND FRISHMAN, S.A., 1979, The ooids of Baffin Bay, Texas: Jour. Sed. Petrology, v. 49, p. 1269–1278.

LINDSTRÖM, M., 1979, Diagenesis of Lower Ordovician hardgrounds in Sweden: Geologica et Paläontologica, v. 13, p. 9–30.

______, 1980, Glauconite shrinkage and limestone cementation: Jour. Sed. Petrology, v. 50, p. 133–138.

LOHMANN, K.C., AND MEYERS, W.J., 1977, Microdolomite inclusions in cloudy prismatic calcites: A proposed criterion for former high-magnesium calcites: Jour. Sed. Petrology, v. 47, p. 1078–1088.

LONGMAN, M.W., 1976, Depositional history, paleoecology, and diagenesis of the Bromide Formation (Ordovician), Arbuckle Mountains, Oklahoma: unpub. Ph.D. dissertation, Austin, Univ. Texas, 311 p.

______, 1980, Carbonate diagenetic textures from nearshore diagenetic environments: Amer. Assoc. Petroleum Geologists Bull., v. 64, p. 461–487.

MACINTYRE, I.G., 1977, Distribution of submarine cements in a modern Caribbean fringing reef, Galeta Point, Panama: Jour. Sed. Petrology, v. 47, p. 503–516.

MACKENZIE, F.T., AND PIGOTT, J.D., 1981, Tectonic controls of Phanerozoic rock cycling: Jour. Geology Society, v. 138, p. 183–196.

MARSHALL, J.D., AND ASHTON, M., 1980, Isotopic and trace element evidence for submarine lithification of hardgrounds in the Jurassic of eastern England: Sedimentology, v. 27, p. 271–291.

MAZZULLO, S.J., 1980, Calcite pseudospar replacive of marine acicular aragonite and implications for aragonite cement diagenesis: Jour. Sed. Petrology, v. 50, p. 409–422.

MOORE, C.H., AND DRUCKMAN, Y., 1981, Burial diagenesis and porosity evolution, Upper Jurassic Smackover, Arkansas and Louisiana: Amer. Assoc. Petroleum Geologists Bull., v. 65, p. 597–628.

MÜLLER, G., IRION, G., AND FORSTNER, U., 1972, Formation and diagenesis of inorganic Ca-Mg carbonates in the lacustrine environment: Naturwissenschaften, v. 59, p. 158–164.

PIERSON, B.J., AND SHINN, E.A., 1983, Distribution and preservation of carbonate cements in Pleistocene limestones of Hogsty Reef atoll, southeastern Bahamas [abs.]: Amer. Assoc. Petroleum Geologists Bull., v. 67, p. 534.

PINGITORE, N.E., 1970, Diagenesis and porosity modification in *Acropora palmata*, Pleistocene of Barbados, West Indies: Jour. Sed. Petrology, v. 40, p. 712–721.

PIPRINGOS, G.N., 1957, Stratigraphy of the Sundance, Nugget, and Jelm Formations in the Laramie Basin, Wyoming: Geol. Survey Wyoming Bull., No. 47.

______, 1968, Correlation and nomenclature of some Triassic and Jurassic rocks in south-central Wyoming: United States Geol. Survey Prof. Pap. 594-D.

PURSER, B.H., 1969, Synsedimentary marine lithification of Middle Jurassic limestones in the Paris Basin: Sedimentology, v. 12, p. 205–230.

RAO, C.P., 1981, Criteria for recognition of cold-water carbonate sedimentation: Berriedale Limestone (Lower Permian), Tasmania, Australia: Jour. Sed. Petrology, v. 51, p. 491–506.

SANDBERG, P.A., 1975, New Interpretations of Great Salt Lake ooids and nonskeletal carbonate mineralogy: Sedimentology, v. 22, p. 497–537.

SCHLAGER, W., AND JAMES, N.P., 1978, Low-magnesian calcite limestones forming at the deep-sea floor, Tongue of the Ocean, Bahamas: Sedimentology, v. 25, p. 675–702.

SCHROEDER, J.H., 1972, Fabrics and sequences of submarine carbonate cements in Holocene Bermuda cup reefs: Geol. Rdsch., v. 61, p. 708–730.

SHINN, E.A., 1969, Submarine lithification of Holocene carbonate sediments in the Persian Gulf: Sedimentology, v. 12, p. 109–144.

SIMONE, L., 1974, Genesi e significato ambientale degli ooidi a struttura fibrosoraggiata di alcuni depositi mesozocici dell'area appennino-divarica e delle Bahamas meridionali: Geol. Soc. Italy Bull., v. 93, p. 513–545.

SORBY, H.C., 1879, On the structure and origin of limestones: anniversary address of the president: Quart. Jour. Geol. Soc. London Proc., v. 35, p. 56–95.

Towe, K.M., and Hemleben, C., 1976, Diagenesis of magnesian calcite: evidence from miliolacean foraminifera: Geology, v. 4, p. 337–339.

Warme, J.E., and Schneidermann, N., 1983, Patch-reef cementation: Holocene of Enewetak Atoll and Jurassic of Morocco [abs.]: Am. Assoc. Petroleum Geologists Bull., v. 67, p. 566.

Wilkinson, B.H., and Landing, E., 1978, "Eggshell Diagenesis" and primary radial fabric in calcite ooids: Jour. Sed. Petrology, v. 48, p. 1129–1138.

______, Janecke, S.U., and Brett, C.E., 1982, Low-magnesian calcite marine cement in Middle Ordovician hardgrounds from Kirkfield, Ontario: Jour. Sed. Petrology, v. 52, p. 47–57.

Wright, R.P., 1971, Marine Jurassic of Wyoming and South Dakota: Its paleoenvironments and paleobiogeography: Univ. Michigan Mus. Paleontology, Papers on Paleontology, No. 2, 49 p.

THE ROLE OF CEMENTATION IN THE DIAGENETIC HISTORY OF DEVONIAN REEFS, WESTERN CANADA

RICHARD A. WALLS AND GEOFF BURROWES
KenTex Oil Corporation, 9430 Old Katy Road, Suite 206, Houston, Texas 77055, and Gulf Canada Limited, Gulf Canada Square, 401 9th Ave., S.W., Calgary, Alberta, Canada T2P 3C5

ABSTRACT

Devonian (Givetian and Frasnian) reef reservoirs in Alberta and British Columbia contain 60% of the conventional, recoverable oil and 20% of the recoverable gas in the Western Canada Sedimentary Basin. Although the depositional history of these reefs is well understood, it is the diagenetic "overprint" that is often responsible for their reservoir quality. This paper discusses diagenetic facies common to many Devonian reefs and assesses the role of cementation in controlling reservoir quality.

Frasnian (Woodbend and Beaverhill Lake Group) reefs are characterized by stromatoporoid and coral knoll reef belts deposited near moderately sloping bank edges. Bank interiors are commonly extensive and characterized by cyclic deposition of grainstone shoals and lagoonal and tidal flat sediments. Certain Givetian reefs found in evaporite basins usually occur as areally small, "pinnacle" reefs with steep (>20°) margins and only minor bank interior development.

Givetian reefs studied include the "Presqu'ile" reef complex in northeast British Columbia and Rainbow Member "pinnacle" reefs in northwest Alberta. Frasinian reefs examined in Alberta include: (1) Swan Hills, Judy Creek, Carson Creek, North (Beaverhill Lake Group); (2) Golden Spike, Redwater, Strachan, Ricinus (Woodbend Group); and (3) Nisku reefs at West Pembina (Winterburn Group).

Most reefs have been subjected to diagenesis in essentially three environments: (1) submarine (syndepositional, marine to hypersaline pore waters); (2) subaerial (fresh to marine pore waters); and (3) subsurface (well below phreatic aquifers, saline to brackish pore waters). Fibrous and bladed calcite cements, submarine "cracks", micrite cements, and bored hardgrounds are typical submarine diagenetic fabrics, particularly at bank margins. Submarine cements in reef interiors occur as microcrystalline or finely crystalline, fibrous, or drusy calcite rims on carbonate grains or as isopachous linings of fenestral pores. Subaerial disconformities are common in most reefs, and associated vadose diagenesis produced localized paleosols, microstalactitic cements, and zones with abundant solution porosity. Phreatic cements are predominantly non-ferroan, clear spar calcite with varied Fe^{2+} and Mn^{2+} concentrations. Subsurface cementation produced ferroan and non-ferroan calcites and dolomites which in many cases can be related to stylolite formation. Other subsurface diagenesis includes dolomite and anhydrite replacement, sulfide mineralization, and bitumen formation.

The major diagenetic influences on limestone porosity are cementation and solution. Submarine cements are commonly extensive in reef margins and in Golden Spike and Rainbow reefs accounting for over 70% of the total cement volume. As submarine cements decrease in importance, porosity generally improves.

Alternating cementation and solution associated with fresh water vadose/phreatic environments and localized cementation associated with stylolites is characteristic of most reef interiors. These processes enhance depositionally-controlled layering and produce interbedded porous and well cemented units. This type of "diagenetic stratigraphy" results in stratified reservoirs with numerous permeability barriers which can seriously hamper hydrocarbon recovery.

In the subsurface environment, chemical compaction (pressure solution) and associated calcite and minor dolomite cementation are the most important processes affecting reservoir quality. In Swan Hills and "Presqu'ile" reefs, subsurface cements account for over 55% of the total cement volume. In contrast, subsurface cements represent less than 15% of the total cement volume in Rainbow reefs and certain Leduc reefs (Golden Spike, Redwater). The relative lack of subsurface cements helps to preserve reef porosity that survived early cementation.

INTRODUCTION

The concept of diagenetic environments and diagenetic facies in carbonate rocks received considerable notoriety with the publication of Purdy's excellent review in 1968. Throughout the 1970's numerous articles elaborated on the concept of diagenetic facies. It is now no longer possible to adequately discuss the depositional history of carbonate rocks without also considering the post-depositional, diagenetic overprint.

The importance of diagenesis to the formation of hydrocarbon reservoirs is well established. Unfortunately, exploration and development geologists often underestimate the role of diagenesis in controlling carbonate reservoir quality. In many cases, it can be demonstrated that the basic difference between a productive carbonate reservoir and its "tight" trap rock is the extent of cementation and/or degree of secondary solution.

In Alberta and British Columbia 14.25 billion barrels of oil and 35 trillion cubic feet of natural gas are known to exist in Middle and Upper Devonian carbonate rocks. If petroleum geologists and engineers are to efficiently extract this vast wealth they cannot ignore the basic processes that are responsible for determining reservoir geometry and quality. For example, despite the obvious similarities of depositional facies in different Devonian reefs, reservoir quality is known to vary markedly from one reef to another (Jardine and Wilshart, 1982). Disparities in reservoir quality can be attributed mainly to diagenetic modifications of depositional fabrics. In most carbonate reservoirs it is essential to map diagenetic as well as depositional facies for proper exploitation and reservoir management.

Several key publications over the past 10 to 15 years have addressed in some detail the subject of carbonate diagenesis in the Devonian of Western Canada (Table 1). Although a number of these papers contribute significantly to our understanding of diagenesis in Devonian rocks, they concern themselves largely with local problems. There have been few attempts to examine diagenetic facies within a more regional context, and to compare the diagenetic histories of Devonian reefs of varying ages.

This paper examines the diagenetic history of several Givetian and Frasnian reef complexes in Alberta and British Columbia. The porosity evolution for each of these reef complexes and the similarities and differences in diagenetic history between reefs are described. In particular, we will attempt to clarify the relationship between various stages of cementation and the pathways of porosity destruction through diagenetic time.

Givetian reefs described in this paper include the Slave Point and "Presqu'ile" barrier reef complex in Northeast British Columbia and Rainbow Member "pinnacle" reefs in Northwest Alberta. Frasnian reefs examined in Alberta include: (1) Swan Hills, Judy Creek, Carson Creek, North (Beaverhill Lake Group); (2) Golden Spike, Redwater, Strachan, Ricinus (Woodbend Group); and (3) Nisku reefs at West Pembina (Winterburn Group).

DEVONIAN STRATIGRAPHY AND SEDIMENTATION

Devonian carbonate buildups and reef complexes in Western Canada have been the subject of numerous studies ever since an oil-bearing reef was discovered at Leduc, Alberta in 1947. These carbonate buildups are present in an intracratonic basin that is 1900 km long and 560 km wide (Fig. 1).

The stratigraphic framework used in this paper is illustrated in Figure 2. The major stages of reef building in the basin occurred during the Givetian (Keg River, Rainbow, Slave Point) and Frasnian (Swan Hills, Leduc, and Nisku). General reviews on Devonian stratigraphy and sedimentology in Western Canada can be found in Andrichuk (1958, 1961), McCrossan and Glaister (1964), Belyea (1965), Basset and Stout (1967), Krebs and Mountjoy (1972), and Klovan (1974).

General Depositional History and Facies

Most Devonian buildups are characterized by three general stages of development:

1) Initial marine transgression (with moderate sub-

TABLE 1.—LIST OF KEY PUBLICATIONS DISCUSSING THE DIAGENESIS OF WESTERN CANADA DEVONIAN REEFS.

Author	*Date*	*Subject*
Roedder	1968	Genesis of ore-forming fluids Pine Point, N.W.T.
Dunsmore	1971	Lead-zinc diagenesis in Devonian carbonates
Schmidt	1971	Cementation, Rainbow reefs
Wardlaw and Reinson	1971	Evaporite diagenesis, Middle Devonian
Bebout and Maiklem	1973	Evaporite deposition, Middle Devonian
Wendte	1974	Sedimentation and diagenesis, Cooking Lake Platform
Comby	1975	Sedimentation, diagenesis and reservoir character, Clarke Lake, B.C.
Havard and Oldershaw	1976	Sedimentation and diagenesis, Snipe Lake reef (Swan Hills Fm.)
Fischbuch and Havard	1977	Sedimentation and diagenesis, Swan Hills reefs
Hopkins	1977	Sedimentation and diagenesis of U. Devonian reef slopes, Rocky Mountains
Schmidt and others	1977	Diagenesis, Rainbow reefs
Burrowes	1977	Sedimentation and diagenesis, U. Devonian reef interiors
Mountjoy and Walls	1977	Reef margin cementation, U. Devonian
Walls and others	1979	Diagenesis and isotope history, Golden Spike reef
Wong and Oldershaw	1981	Burial cementation, Kaybob reef (Swan Hills Fm.)
Jardine and Wilshart	1982	Carbonate reservoir characteristics, W. Canada
Mountjoy and Krebs	1983	Comparison of diagenesis in Devonian reefs of Western Canada and Europe.

sidence) and accretion of shallow marine carbonate platforms. These platforms are areally extensive as typified by the Lower Keg River, Slave Point and Cooking Lake Formations.

2) Localized development of reefs and bioherms on these platforms. These reefs and bioherms comprise numerous shallow water facies that develop in response to sediment accretion combined with sea level oscillations (Havard and Oldershaw, 1976).
3) Continued subsidence, reef "drowning", and basin filling with either marine calcareous shales, fine grained limestones or evaporites.

This pattern is repetitive through time with the oldest platform edges and buildups to the north and west (Northeast British Columbia, N.W.T.) and successively younger buildups to the southeast (Fig. 1).

The buildups we will examine vary from broad, layered banks with only minor depositional relief (e.g. Swan Hills) to small "pinnacle" reefs with moderate depositional relief (e.g. Rainbow reefs). The depositional model applicable to most of these buildups is the knoll reef ramp model of Wilson (1975). These knoll reef ramps are characterized by narrow, linear knoll reef belts near the edges of a bank with slopes usually less than 15° (Fig. 3). Knoll reefs are defined as discrete areas (knolls) of organic, frame-built growth at shelf margins (Wilson, 1975, p. 22).

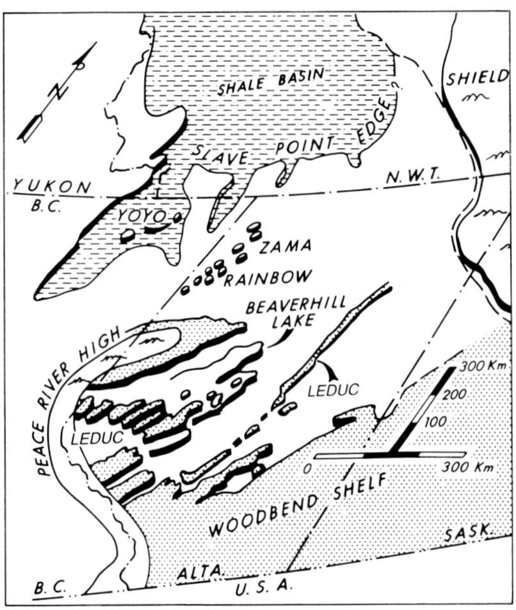

FIG. 1.—Distribution and paleogeography of Middle and Upper Devonian reef carbonates in Western Canada. (modified from Basset and Stout, 1967).

An idealized cross-section (Fig. 4) of a Devonian buildup consists of a narrow belt of bank margin facies enclosing widespread bank interior facies. The general facies mosaic illustrated in Figure 4 varies from buildup to buildup, but the overall pattern is strikingly consistent.

Most of the sediments at the bank margin are composed of skeletal lime grainstones and "reef detritus" (skeletal rudstones). Stromatoporoid-coral framestones are relatively minor. Bank interior deposits vary from grainstones to mudstones and are commonly characterized by cyclic deposition of lagoonal and tidal flat sediments.

The initial stages of reef growth often occur over submarine hardgrounds (disconformities) at the top of the platform (Wendte, 1974). Likewise, subaerial hardgrounds and erosional surfaces are common in bank interior sediments. In Swan Hills reefs for example, Fischbuch (1968) and Havard and Oldershaw (1976) recognize several paleoexposure sufaces some of which appear to be of regional extent.

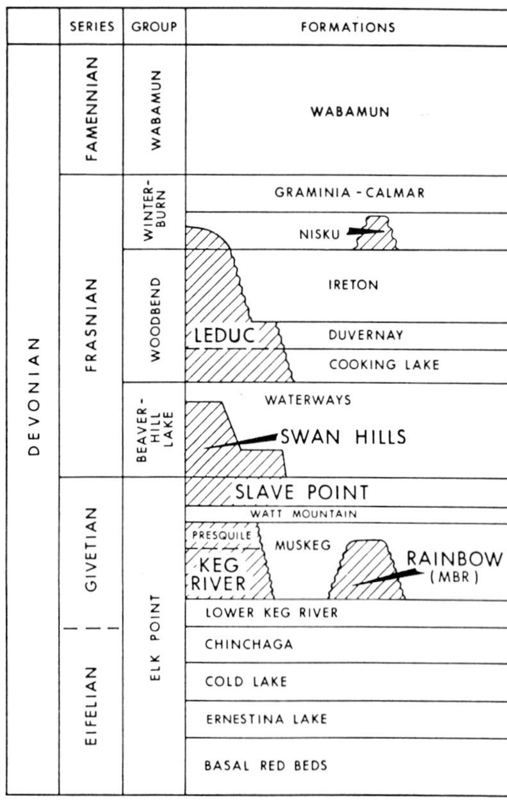

FIG. 2.—Devonian stratigraphy in Western Canada showing stratigraphic position of reef carbonates (modified from Klovan, 1974).

Certain Givetian reef complexes that occur in evaporite basins (e.g. Rainbow Member) do not form knoll reef ramps. Instead they occur as discrete, small, "pinnacle" reefs with steep margins (>25°) and only minor bank interior development.

Givetian reefs.—Givetian reefs discussed in this paper occur in two settings. In Northeast British Columbia, the "Presqu'ile" reef complex (composed, from the base upwards, of the Keg River, Sulphur Point, and Slave Point Formations; see Figure 2) forms a major shelf margin bank or barrier reef complex adjacent to the Hare Indian shale basin (Fig. 5a). Several isolated buildups (e.g. Yoyo) occur on basement highs seaward of the main carbonate bank and isolated within the Hare Indian shales. The "Presqu'ile" is a relatively continuous and partially dolomitized bank that extends from outcrops at Pine Point, N.W.T. to depths of 3000 m in the subsurface of Northeast British Columbia. At Pine Point the "Presqu'ile" is the host rock for one of Canada's largest lead-zinc deposits, while in the subsurface of Northeast British Columbia it contains over 7 trillion cubic feet of natural gas.

In Northwest Alberta, to the east of the "Presqu'ile" reef complex, are evaporite basins that contain numerous, isolated reefs which comprise the Rainbow Member of the Keg River Formation (Fig. 2). These oil bearing reefs were first discovered in 1965 and over 80 productive reefs have been drilled with recoverable reserves greater than 700 million barrels of oil and 850 billion cubic feet of natural gas. In the Rainbow area (Fig. 5b) these reefs occur in the Middle Devonian Black Creek Salt Basin. Restriction of this basin caused intermittent evaporite deposition (Muskeg Formation) and eventual burial of Rainbow reefs in Muskeg anhydrites. Many of the reefs are extensively dolomitized.

Individual Rainbow reefs vary in size from 0.6 km^2 to over 21 km^2 with thicknesses ranging up to 250 m. Several excellent papers provide detailed descriptions of reef geometry, facies, reservoir properties, and production history (McCamis and Griffith, 1967; Langton and Chin, 1968a and b; Barrs *et al.*, 1970; Davis, 1972; Evans, 1972; and Schmidt *et al.*, 1977).

Frasnian reefs.—Frasnian age buildups include

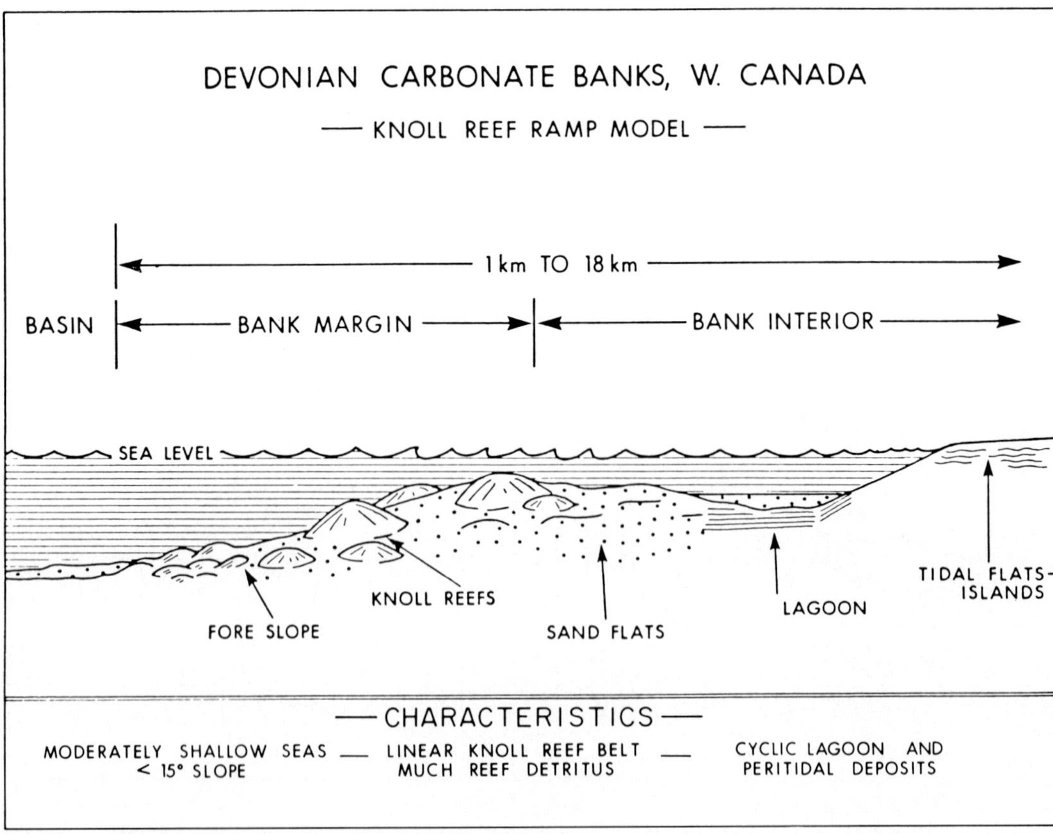

FIG. 3.—Schematic depiction of knoll reef ramp model applicable to most Devonian carbonate reefs and banks.

reefs that develop in the Beaverhill Lake, Woodbend, and Winterburn Groups (Fig. 2). In the Beaverhill Lake Group, the Swan Hills Formation occurs as broad, dominantly limestone, reef complexes on the flank of the Peace River High (Fig. 5c). Swan Hills reefs develop on the Slave Point carbonate platform and attain thicknesses of up to 100 m. Basinal calcareous shales and argillaceous limestones are known as the Waterways Formation.

Most Swan Hills reefs occur in linear patterns (plan view) with individual reefs typically separated by open marine channels or embayments. According to Klovan (1974, p. 794), the broad, arcuate eastern edges of the Swan Hills, Judy Creek, and Carson Creek reef chain (Fig. 5c) represent the windward margins of these reefs.

All Swan Hills reefs are characterized by vertically stratified facies patterns. Reef interiors are composed of widespread subtidal and peritidal deposits exhibiting marked lateral and vertical facies variations. Reefs are commonly coral-dominated at the base changing upwards into stromatoporoid-dominated framestones and bindstones. The uppermost deposits are commonly grainstone and rudstone shoals. Fischbuch (1968) and Leavitt (1968) provide excellent detailed studies of several Swan Hills reefs.

In the Woodbend Group, the Leduc Formation comprises the reef building phase. Subsurface Leduc reefs generally occur as conspicuous, linear reef tracts in the central portion of the Alberta basin (Fig. 5d). For example, the Rimbey-Meadowbrook reef chain trends northeast for a distance of 320 km and contains 15 individual reefs separated by open marine channels. To the east of the Rimbey-Meadowbrook chain and separated by the Ireton shale basin is the broader, northeast trending Wimborne-Bashaw reef tract. With the exception of the Redwater and Golden Spike reefs, the Leduc reefs in Central Alberta are extensively dolomitized.

Redwater is a large, oval-shaped, limestone reef complex that lies to the east of the Rimbey-Meadowbrook trend (Fig. 5d). It is 520 km^2 in area and 245 m thick with 845 million barrels of recoverable reserves trapped in a narrow zone on the eastern, up-dip margin. Golden Spike, located just west of the Rimbey-Meadowbrook trend, is similar to Redwater except that it is only 6 km^2 in area and 170 m thick with recoverable reserves of 208 million barrels.

The Nisku Formation (Winterburn Group) comprises the last reef building stage in the Devonian of Western Canada. Nisku reefs occur as (1) small bioherms or biostromes overlying Leduc reefs; (2) small

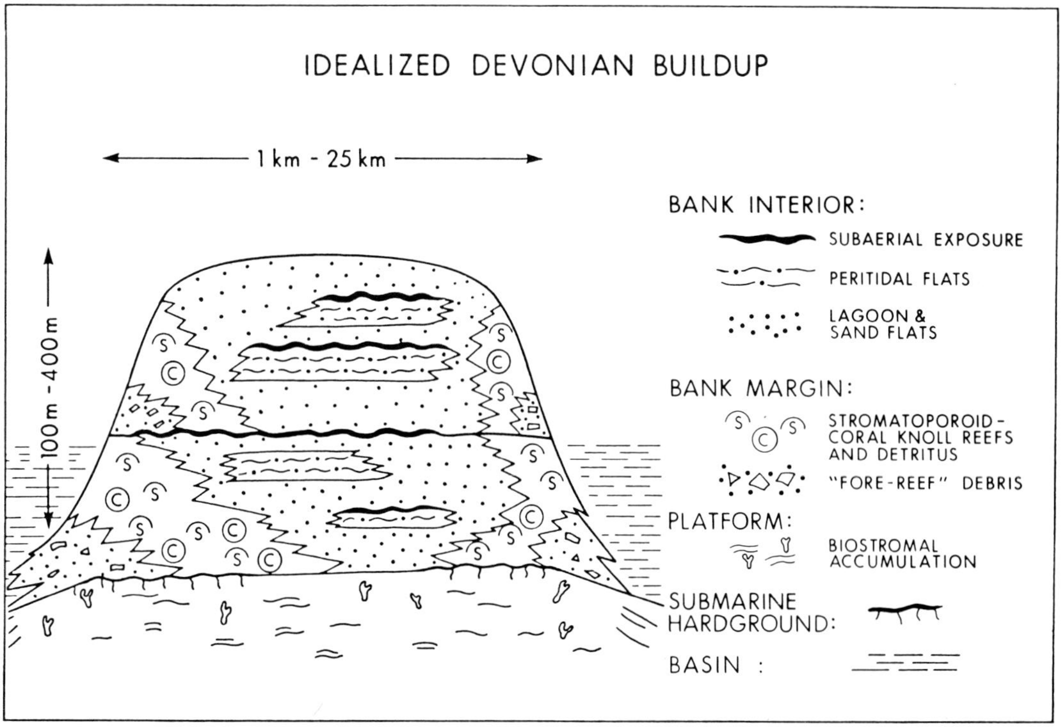

FIG. 4.—Idealized Devonian buildup illustrating major facies.

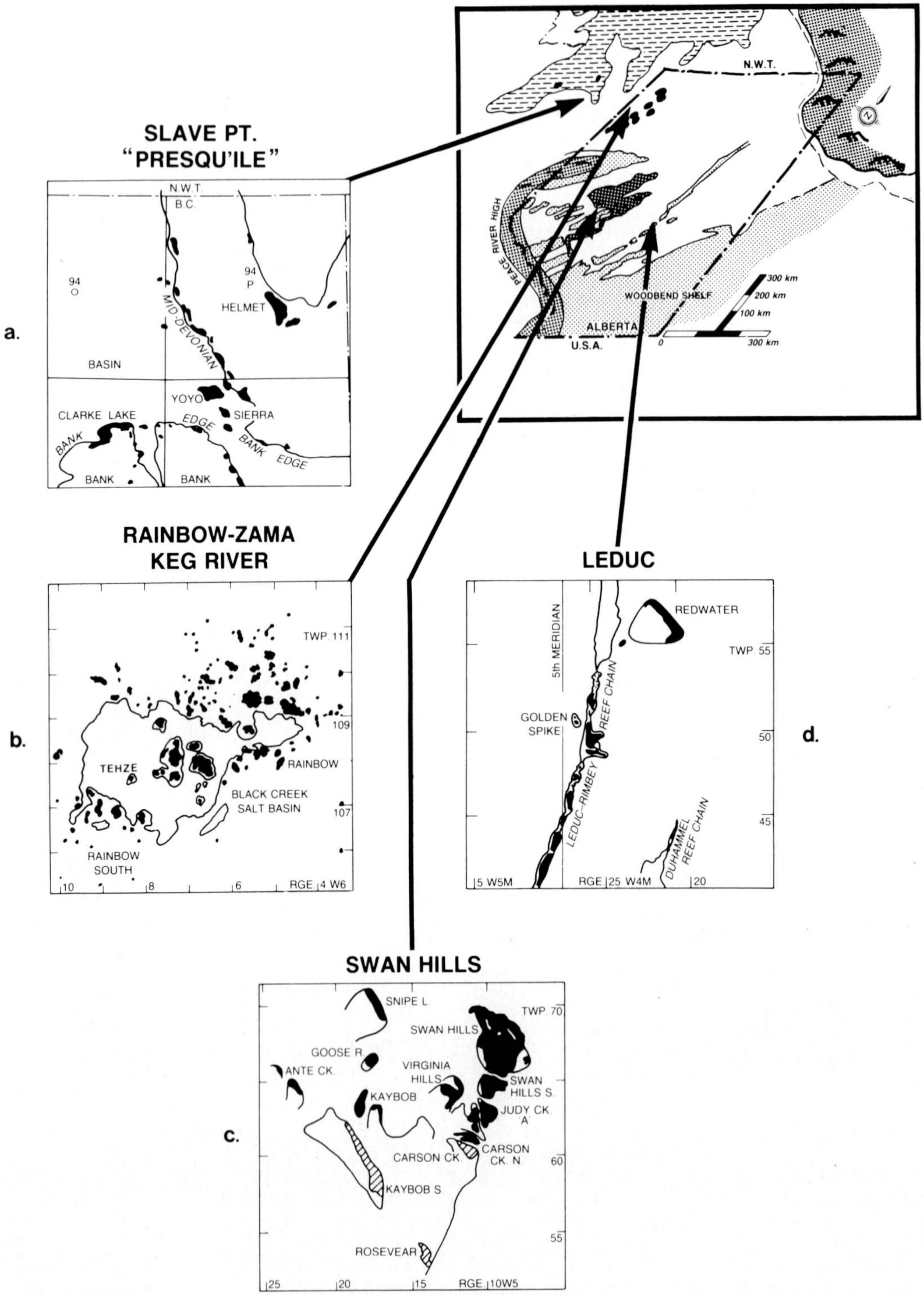

FIG. 5.—Index map showing location and paleogeography of Devonian reefs discussed in this paper.

bioherms isolated in shelf evaporites (Wood River, Alberta); (3) shelf-edge (open marine) bioherms or mounds (Meekwap, Alberta); and (4) shelf-slope stromatoporoid-coral bioherms or mounds (West Pembina, Alberta).

The Nisku (Zeta Lake Member) reefs at West Pembina have received considerable attention since their discovery in 1977. In excess of 40 reefs have since been penetrated with proven reserves estimated at 250 million barrels of oil.

The West Pembina reefs are encased in basinal shales and silty shales of the Cynthia Member (see Fig. 2). These shelf-slope reefs are generally less than 1 km^2 in area and range up to 70 m thick. The main facies are tabulate and rugose coral bafflestones (carbonate mud matrix), tabular and lamellar stromatoporoid wackestones, and skeletal lime grainstones. These reefs generally lack the well developed peritidal facies common in Leduc and Swan Hills reefs.

Excellent descriptions of subsurface Woodbend and Winterburn reefs in Alberta can be found in Downing and Cooke, 1955; Andrichuck, 1958; Bryant, 1960; Klovan, 1964, 1974; Mossop, 1972; Wendte, 1974; McGillivray and Mountjoy, 1975; Reitzel and Callow, 1976; Walls, 1977; Cheshire and Keith, 1977; Stoakes, 1977, 1979; and Chevron Exploration Staff, 1979.

REEF DIAGENESIS

Devonian reefs in Western Canada are characterized by extremely heterogeneous porosity and permeability as demonstrated by Jardine and Wilshart (1982). Much of this heterogeneity can be attributed to depositional facies variations. However, in the following discussions, we will demonstrate that variations in diagenetic processes and products, particularly cementation, are also of considerable importance in the genesis of reservoir heterogeneity.

Processes and Terminology

The three basic diagenetic processes that affect Devonian reef carbonates are: cementation, solution, and replacement by dolomite (Fig. 6).

The importance of cementation in subsurface carbonates is threefold: (1) cements reduce overall effective porosity; (2) they produce permeability barriers

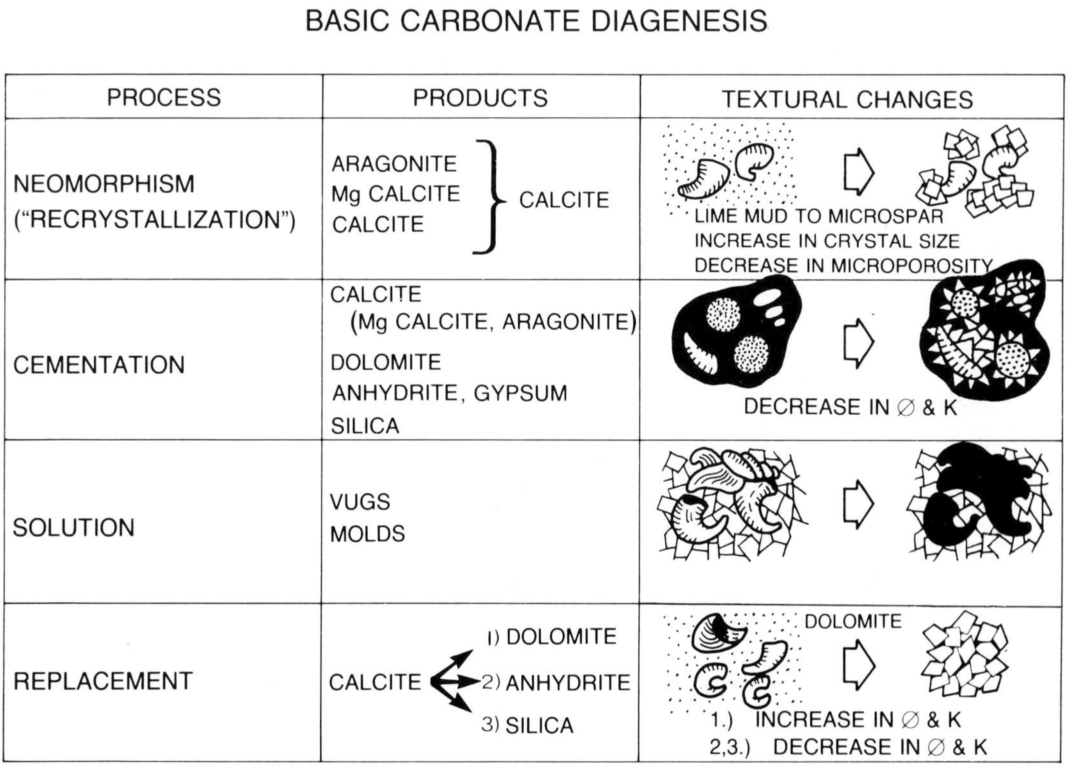

FIG. 6.—Summary of basic processes involved in carbonate diagenesis and their associated mineralogical and textural changes. (ϕ = porosity; K = permeability)

within good reservoirs and otherwise contribute to reservoir heterogeneity, and (3) preferential cementation of a particular facies can result in an up-dip trap for hydrocarbon accumulation. Solution of carbonates occurs primarily in response to fresh water. In ancient carbonates, vug, moldic, and solution-enlarged interparticle porosity are the common expression of fresh water diagenesis associated with subaerial disconformities. Large scale removal of carbonate also occurs during progressive burial by the process of pressure solution.

Dolomite and anhydrite are the most common minerals that replace calcite in Devonian reef rocks. Dolomitization is probably the most important replacement process as it commonly results in an increase in intercrystalline porosity and permeability. However, a comparison of porosities in Devonian reef limestone and dolomite reservoirs of the same age and geological setting reveals little variation in overall porosity. Dolomitization, therefore, has resulted primarily in the reorganization and homogenization of existing limestone porosity with general increase in permeability.

ENVIRONMENTS

Diagenesis of Devonian reefs occurred in three general environments: submarine, subaerial, subsurface (Fig. 7). The submarine environment refers to that area at or slightly below the sediment-sea water interface where diagenetic processes occur contemporaneously with reef growth and sedimentation. Subaerial diagenesis occurs when reef sediments are exposed, either as a result of vertical accretion or low stands of sea level. The subaerial environment includes the vadose (above the water table) and phreatic (below the water table) settings. Fluids involved in subaerial diagenesis may range in salinity from normal marine to fresh water. The subsurface environment is an extensive area occurring below the deeper reaches of the phreatic aquifer in carbonate rocks and is unaffected by near surface processes. The subsurface environment can begin a few tens of meters below the surface and extend to depths of thousands of meters. The transition from the subaerial or submarine environment to the subsurface is usually gradual and difficult to identify.

The following discussion will illustrate the diagenetic fabrics that form in these three general en-

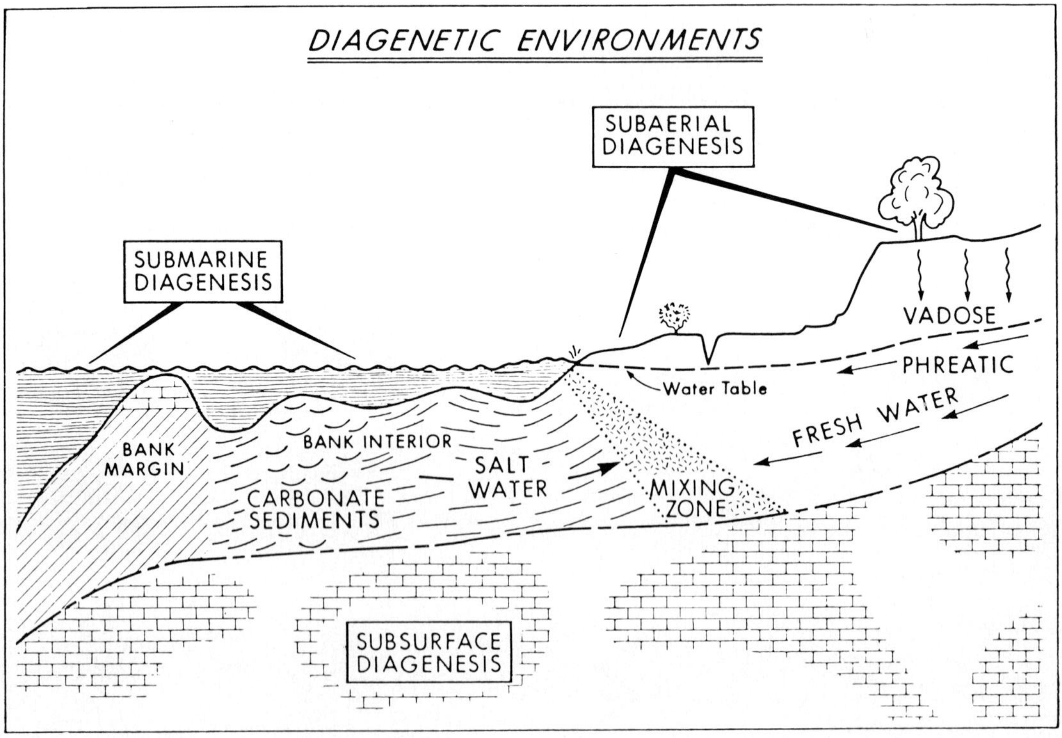

FIG. 7.—Schematic representation of the three major diagenetic environments involved in the diagenesis of Devonian reefs.

vironments for various Devonian reefs. Also, the relative importance of each environment in determining overall reservoir quality will be documented. Emphasis will be placed on the role of cementation in the development and preservation of reservoir rocks.

Submarine Diagenesis

The occurrence and significance of submarine diagenesis and in particular, submarine lithification, has been highlighted in recent years by a number of detailed studies of modern reef complexes (Land and Goreau, 1970; Ginsburg *et al.*, 1971; Schroeder, 1972a, b; Ginsburg and Schroeder, 1973; Friedman *et al.*, 1974; Ginsburg and James, 1976; James *et al.*, 1976; MacIntyre and Glynn, 1976; MacIntyre, 1977; Gerhard, 1977; Mazzullo, 1977; James and Ginsburg, 1979, Mullins, *et al.*, 1980; Marshall and Davies, 1981; and in this volume).

Marine cementation in modern reefs is commonly extensive in near surface portions of reef margins and slopes. Cementation occurs in the form of fibrous, bladed, and finely crystalline (micrite) aragonite and Mg-calcite. These cements greatly enhance the stability of reef margins enabling them to withstand vigorous wave action and maintain steep dips (up to 90° in places). Marine cementation has also been reported from modern subtidal and intertidal lime sands. Cements typically occur as isopachous crusts of aragonite and Mg-calcite.

Petrographic and geochemical evidence collected from numerous ancient reef complexes indicates that submarine lithification processes were also widespread in the past (Krebs, 1967, 1969; Zankl, 1971; Schmidt, 1971; Schmidt *et al.*, 1977; Davies and Krouse, 1975; Davies, 1977; Walls, 1977; Walls *et al.*, 1979; Mountjoy and Walls, 1977; Mazzullo and Cys, 1977, 1979; and others). Many of the diagenetic features identified in ancient reefs are strikingly similar to those attributed to a submarine origin in modern reefs.

In Devonian reefs of Western Canada, evidence of submarine cementation is abundant though not always conspicuous. The three most common manifestations of submarine reef cementation are: (1) hardgrounds; (2) lithified reef-derived debris in allodapic foreslope deposits; and (3) isopachous rim calcite cements in primary cavity systems dominantly at, but not restricted to, reef margins. This paper places emphasis on the latter because it is the most important in affecting reservoir quality.

The sedimentary and diagenetic fabrics of hardgrounds associated with Leduc reefs have been documented by Wendte (1974) and Stoakes (1979). Hardgrounds commonly occur at or near the top of deeper water platforms that underlie the reefs (e.g. Slave Point and Cooking Lake platforms). They are characterized by finely crystalline ($\leqslant 15\mu$) calcite cementation and contemporaneous bioerosion (borings), development of submarine "cracks", and internal (in cracks, cavities, and borings) marine sedimentation (Fig. 8A).

Hopkins' (1977) study of foreslope facies from Devonian exposures in the Rocky Mountains provides excellent descriptions of the types of submarine cementation occurring in this environment.

Submarine cements.—A wide variety of calcite cement fabrics occur in pore systems of Devonian reefs. The predominant fabrics are: (1) isopachous crusts of fibrous, bladed and equant calcite (Figs. 8B–D, 9A, 10); (2) isopachous crusts of banded and nonbanded, prismatic radiaxial calcite (Fig. 9B, C); (3) aggregates of radial fibrous calcite crystals (Fig. 9B) and (4) micrite crusts (Fig. 9B, 10D).

In large, interconnected cavity systems of reef margins, prismatic radiaxal calcite and isopachous crusts of fibrous to bladed calcite are the most common submarine cements (Figs. 8B–D, 9A). In portions of some reefs (i.e. Rainbow and Golden Spike) they account for as much as 50 per cent of the rock volume. The following textural evidence and arguments, when considered collectively, support the interpretation that these cements were precipitated in a marine environment, contemporaneous with deposition:

1) They always predate the earliest stages of compaction.
2) They are largely confined to primary shelter, growth framework, and interparticle pores (Figs. 8B–D).
3) They occur as isopachous linings of voids or as isopachous crusts on skeletal grains (Figs. 8, 9).
4) They are often banded or zoned (Fig. 9A, C) and typically occur in a cloudy (inclusion-rich) fibrous or bladed crystal fabric (Fig. 9B) with long axes of crystals oriented normal to the pore wall or substrate.
5) They predate and/or alternate with geopetal marine internal sediments in several different types of primary cavities (Figs. 8A, B, D; 9B, C).
6) They are locally corroded in response to early subaerial diagenesis (particularly in reef interiors) and overlain by vadose or phreatic cements (Figs. 10, 15B).
7) In many cases, they are texturally similar to submarine cement fabrics described from modern reefs.

Several studies have demonstrated that the crystal habit of a carbonate cement strongly reflects the environment and chemical conditions of precipitation. For example, Folk (1974, p. 41) proposed a model for crystal growth that attributes fibrous or bladed crystals of Mg-calcite and aragonite to precipitation in hypersaline to normal marine waters. Rapid crystallization enhanced by high ionic strength solutions such as sea

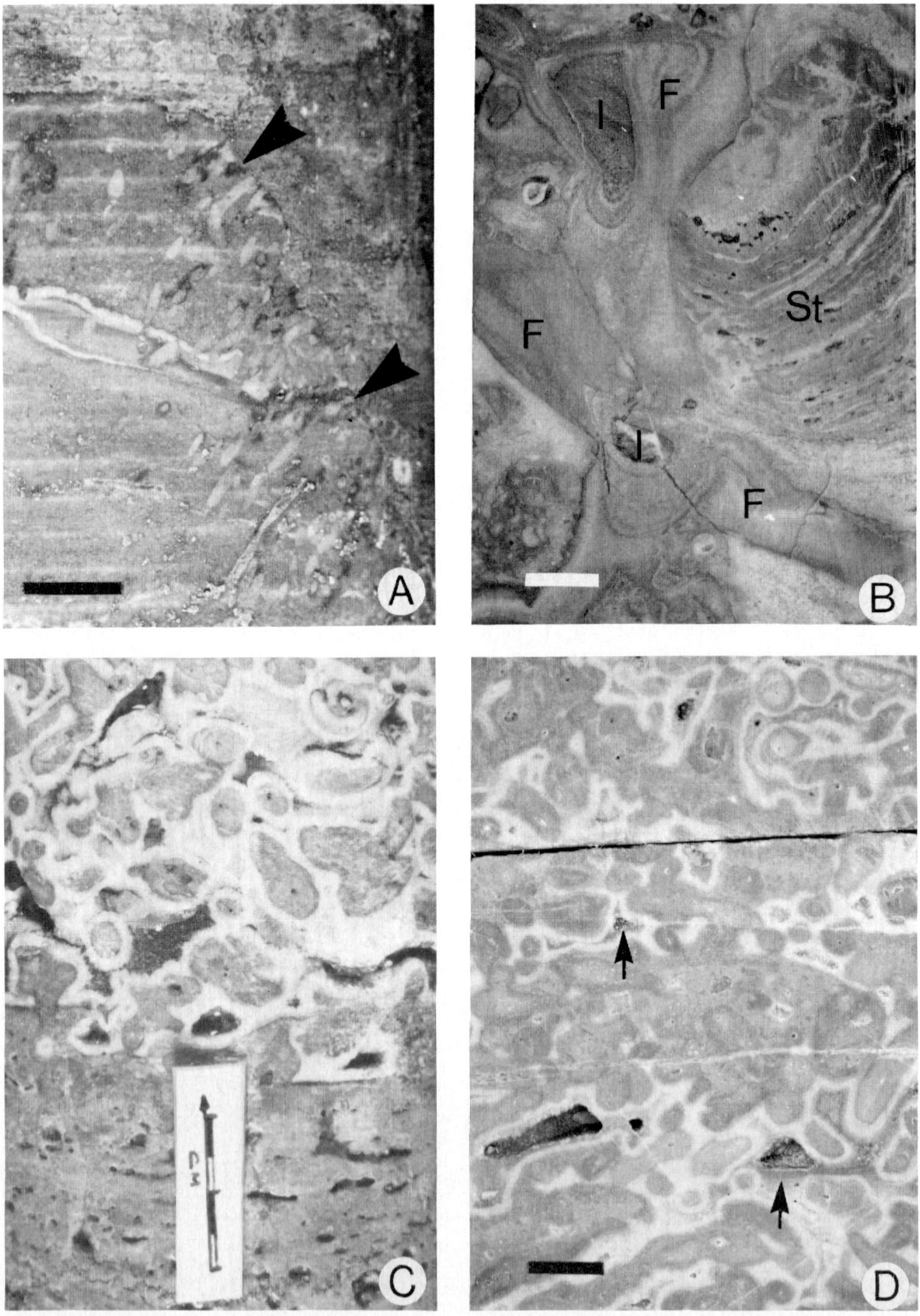

FIG. 8.—Fabrics formed during submarine diagenesis. *A*, Core slab showing extensively bored hardground surface (arrows) near the base of a Swan Hills reef. Laminated geopetal lime mud overlies isopachous submarine cement in associated cavity. Note that cement is truncated by borings. (Carson Creek, North, 6-34-61-12 W5, 8463 ft; scale bar = 1 cm). *B*, Core slab of stromatoporoid (St) rubble from the margin of a Rainbow reef. A banded, fibrous calcite cement (F) fills large interconnected cavity systems and is overlain by a peloidal grainstone internal sediment (I). (Rainbow, 2-27-108-9W6, 6529 ft; scale bar = 1 cm). *C*, Core slab of a *Stachyodes* rudstone from the reef flat facies of a Swan Hills reef. Isopachous, bladed calcite lines and in cases completely fills interparticle porosity. (Carson Creek, North, 6-34-61-12W5, 8518 ft.). *D*, Similar to (B) but with very little porosity surviving the cementation. Geopetal internal sediment (arrows) overlies isopachous cement in some cavities (Judy Creek, 10-16-63-11W5, 8887 ft; scale bar = 1 cm).

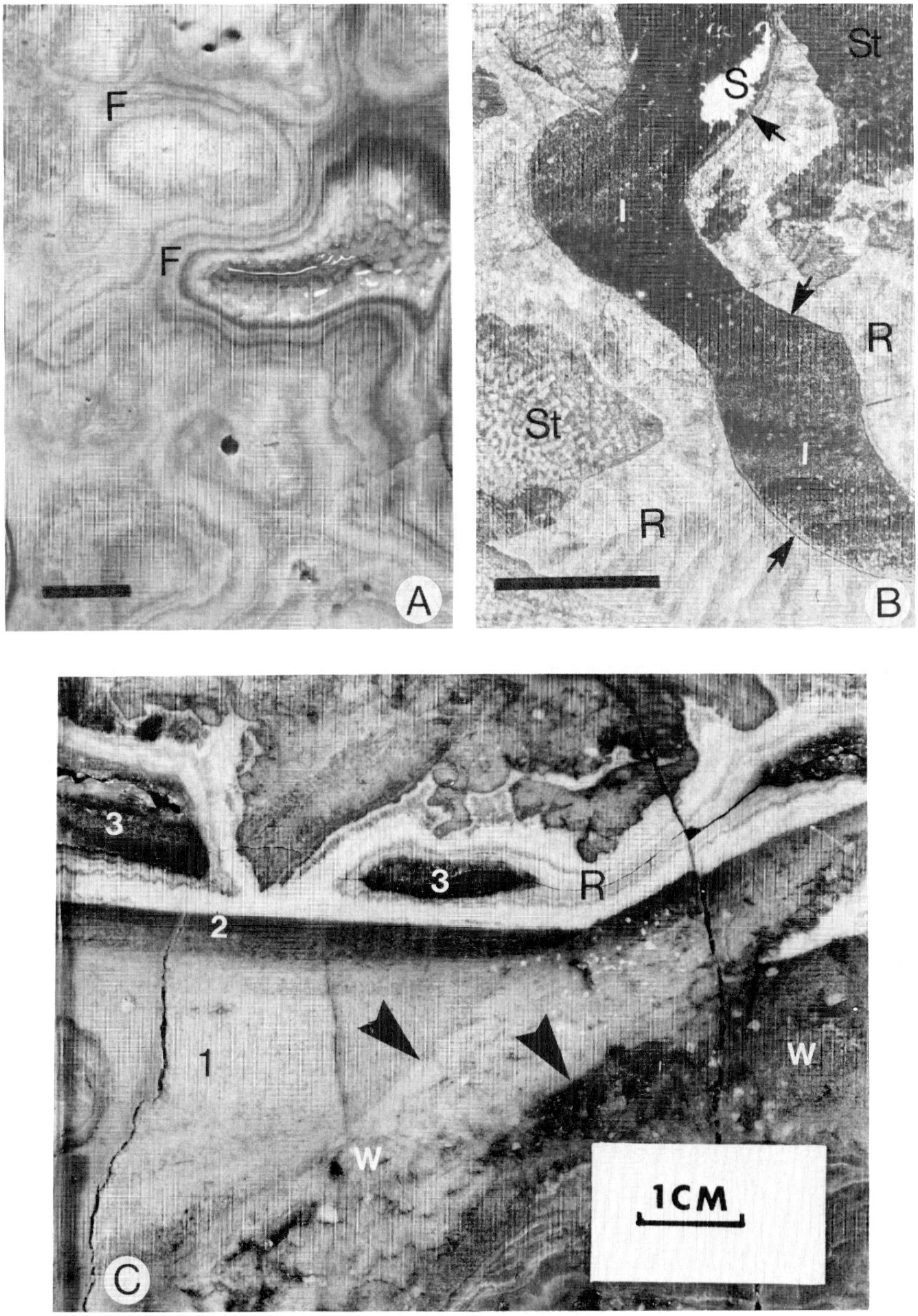

FIG. 9.—Fabrics formed during submarine diagenesis. *A*, Core slab from a Swan Hills reef margin with banded, fibrous calcite cement (F) filling porosity in a stromatoporoid rudstone. (Carson Creek, North, 6-22-62-12W5, 8852 ft; scale bar = 1 cm). *B*, Photomicrograph from a Leduc fore-reef showing a submarine "crack" filled with ostracod, peloidal internal sediment (I). Stromatoporoid fragments (St) are cemented by radiaxial calcite crusts (R). The radiaxial cement is truncated by the "crack" which is likewise lined with a microlaminated micrite cement (arrows). A thin layer of fibrous calcite cement overlies the microlaminated micrite and is postdated by the internal sediment. A non-ferroan sparry calcite (S) fills the remaining pores. (Golden Spike, 12-24-51-27W4, 5829 ft; scale bar = 1 cm). *C*, Core slab from a Leduc fore-reef showing scours (arrows) filled with sediments (1,2). The scours truncate layering in lime wackestones (W). Radiaxial calcite (R) occurs in banded cement crusts on stromatoporoids and geopetal sediments. Lime mud internal sediment (3) overlies the radiaxial calcite cement. (Golden Spike, 12-26-51-27W4, 5941 ft.).

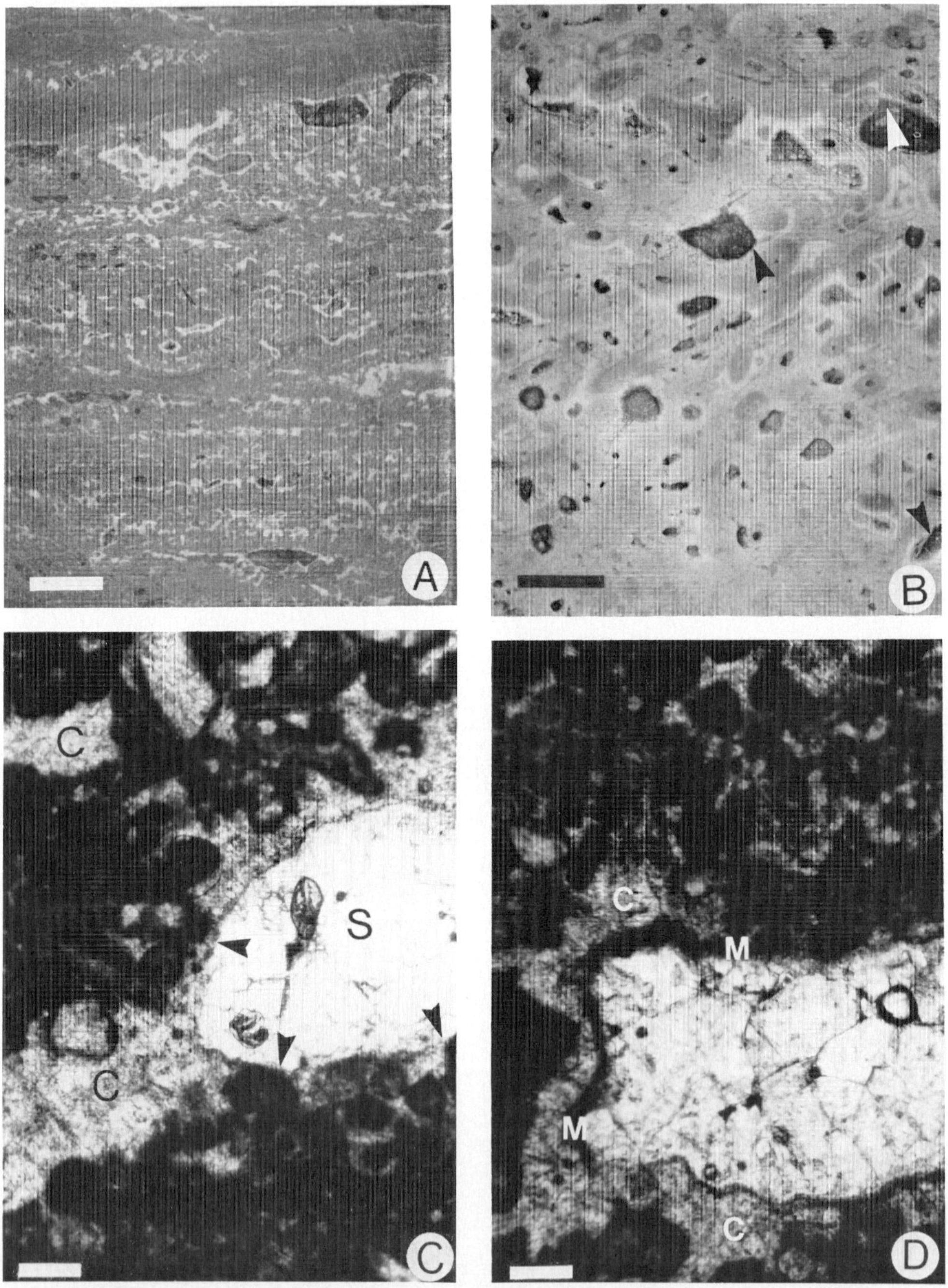

FIG. 10.—Fabrics formed during submarine diagenesis. *A*, Core slab of reef interior, fenestral lime mudstone from a Rainbow reef with well developed isopachous, fibrous, calcite cement lining most pores. (Rainbow, 7-32-109-8W6, 5939 ft; scale bar = 1 cm). *B*, Core slab of *Amphipora* floatstone from a Swan Hills reef interior showing isopachous calcite cements rimming interparticle and shelter cavities. Cements have been partly removed (arrows) by later phase fresh water diagenesis producing solution-enlarged secondary porosity (Snipe Lake, 10-21-70-18W5, 8784 ft; scale bar = 1 cm). *C*, Photomicrograph of skeletal grainstone from a Leduc reef interior showing abundant, inclusion-rich, subequant calcite cement (C). This submarine cement has been partly removed by later dissolution (arrows). Remaining void is filled by clear, spar calcite (S) (Golden Spike, 11-23-51-27W4, 5406 ft; scale bar = 250μ). *D*, Similar to C but with a dark, micrite cement (M) overlying the inclusion-rich, bladed to subequant calcite (C) (Golden Spike, 11-23-51-27W4, 5407 ft; scale bar = 250μ).

water results in precipitation of turbid or cloudy (inclusion-rich) crystals elongated along their c-axes. Cloudy, fibrous and bladed cements are typical of many Devonian reef carbonates. These cements are commonly radiaxial calcites and thus probably represent the neomorphic end-product of original radial fibrous cements (Kendall and Tucker, 1973).

Interesting features observed in fore-reef facies of Golden Spike and Rainbow reefs are "submarine cracks" that locally truncate the initial phases of radiaxial calcite cement (Fig. 9B). The surfaces of these cracks are commonly lined by a microlaminated micrite (Fig. 9B) that either represents a submarine micrite cement similar to that observed by Davies (1977, Fig. 4) in "submarine cracks" in late Paleozoic reefs of the Canadian Arctic, and James and others (1976) in modern reefs, or algal coatings (Krebs, 1969). Microlaminated micrites are subsequently overlain by thin linings of fibrous calcite cements interlayered with marine geopetal sediments (Fig. 9B).

The origin of these "cracks" remains a problem. They may simply be the result of penecontemporaneous, differential settling in partially lithified reef rocks, or caused by "displacive cementation" (Ross *et al.*, 1975; Schmidt *et al.*, 1977) that may occur during the rapid precipitation of early fibrous cements.

Submarine cement fabrics in reef interior facies (lagoons, grainstone shoals, and tidal flats) though less conspicuous than those observed at reef margins, are volumetrically important and play a key role in overall reservoir development. These cements usually occur as finely crystalline bladed to equant calcite rims on carbonate grains or as isopachous linings of fenestral pores (Fig. 10). They apparently predate the onset of any burial compaction and occasionally predate and/or alternate with marine internal sediments. In addition, they are similar to modern (Taylor and Illing, 1969; Shinn, 1969; Sibley and Murray, 1972; Moore, 1973; Dravis, 1979) and ancient (Purser, 1969; Havard and Oldershaw, 1976; Burrowes, 1977; Walls, 1977; and others) intertidal and subtidal cements reported in the geologic literature.

In summary, Figure 11 illustrates the wide spectrum of submarine cement fabrics that occur throughout one Leduc limestone reef complex, Golden Spike.

Subaerial Diagenesis

The episodic subaerial exposure of many Devonian reefs has been documented by a number of authors (Havard and Oldershaw, 1976; Burrowes, 1977; Schmidt *et al.*, 1977; Walls, 1977; Wong and Oldershaw, 1981). Subaerial exposure zones are best preserved in reef interiors and are characterized by the associated occurrences of:

1) Brecciated and fractured lime wackestones and mudstones with microkarst surfaces and green, grey, and brown shale cavity fillings (Fig. 12).
2) Solution-enlarged porosity with truncation of submarine cement fabrics (Figs. 10A–C, 13).
3) Gravitational or microstalactitic rim cements (Fig. 14A, C, D).
4) Clear, spar calcite cements (generally non-ferroan and zoned under cathodoluminescence) (Figs. 14A, B, 15).

Vadose diagenesis.—Fresh water, vadose diagenesis of Devonian reefs produced exposure zones characterized by brecciated limestones and minor shales (usually green colored) that may represent paleosols (Havard and Oldershaw, 1976). However, Wendte and Stoakes (1982) interpret green shales in Judy Creek reef interior limestones as storm transported, basinal muds. Regardless of their origin, these shales are typical of reef interiors and commonly occur at the top of shoaling-upward cycles (Havard and Oldershaw, 1976; Wendte and Stoakes, 1982).

In Leduc and Swan Hills reefs exposure zones typically occur as 30 cm to 2 m thick beds. In Rainbow reefs, exposure zones are much thicker and reach deep into underlying facies. However, exposure in Rainbow reefs occurred under arid, evaporitic conditions (Schmidt *et al.*, 1977) and vadose diagenesis produced fractured and brecciated marine carbonates that are filled with evaporitic muds and anhydrite (Fig. 12D). Minor leaching produced fossil molds and vugs (Fig. 13A).

In Upper Devonian reefs of western and central Alberta (Swan Hills, Leduc, and Nisku Formations) fresh water, vadose diagenesis resulted in the development of secondary vug, moldic, and channel porosity (Fig. 13). Gravitational or microstalactitic, fibrous calcites locally fill both solution pores and primary porosity (Fig. 14A, C, D) in reef interiors.

Microstalactitic cements are characterized by the lobate development of fibrous, cloudy calcite crystals on the undersides of grains or on the uppermost surfaces of pores (Fig. 14C, D). This fabric represents precipitation from waters percolating downward through partially air-filled pores. Microstalactitic cements commonly comprise the first stage of cementation and locally overlie isopachous rims of equant or bladed calcite. The cloudy and radial fibrous crystal habit is similar to many modern supratidal, gravitational cements that precipitate from marine vadose fluids (Taylor and Illing, 1969; Scholle and Kinsman, 1974). A similar origin may explain the common occurrence of these cements in Devonian reef interiors.

Phreatic diagenesis.—Clear, dominantly non-ferroan, spar calcite crystals commonly postdate microstalactitic cements in primary and secondary (solution) voids (Fig. 14A, C and D). These cements do not illustrate gravitational textures but fill voids with a medium to coarsely crystalline, equant mosaic of calcite crystals. The presence of these cements: (1) with brecciated shale and limestone units; (2) overlying carbonate crystal silt; and (3) in solution-enlarged voids, suggests that they formed subsequent to subaerial exposure and vadose diagenesis.

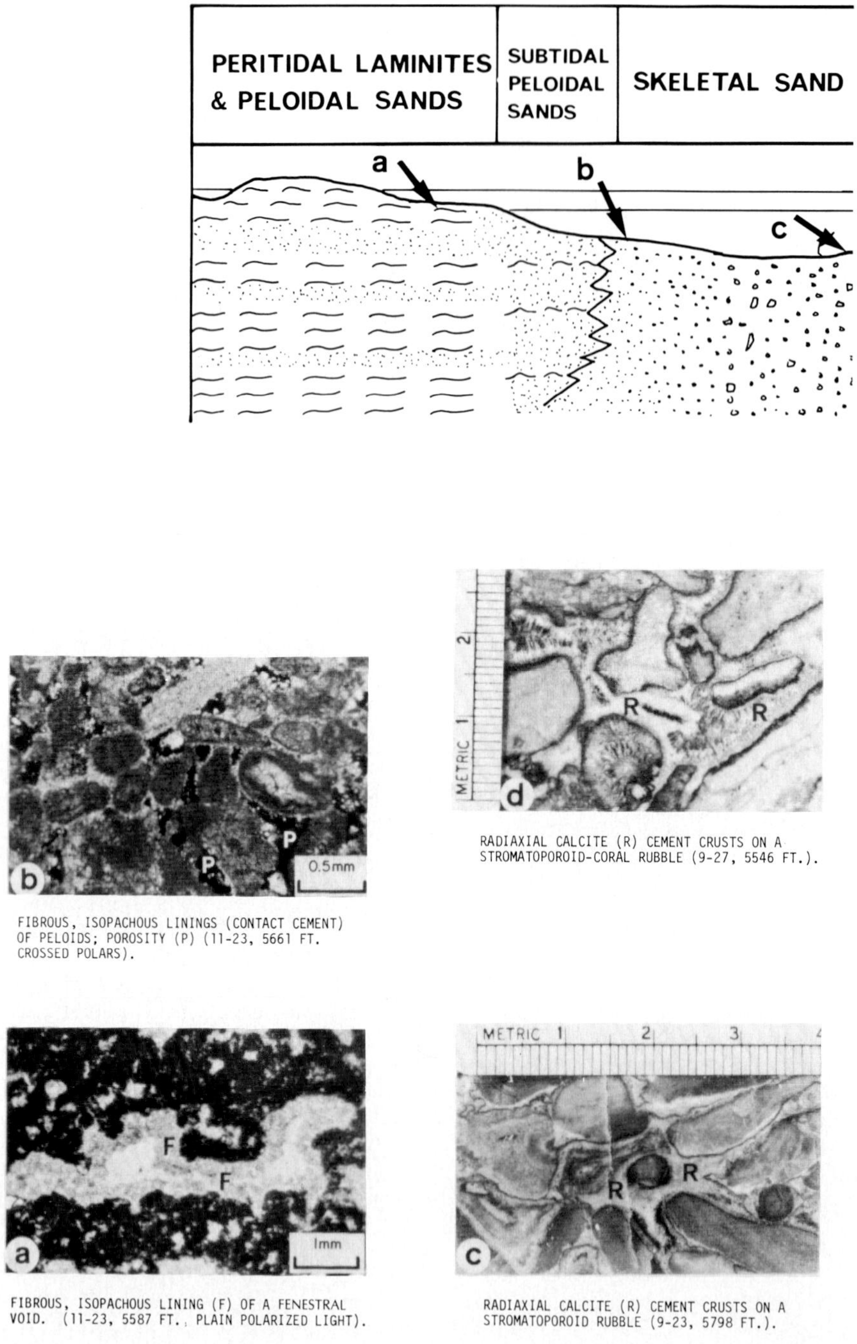

FIG. 11.—Schematic of the Golden Spike (Leduc) reef complex illustrating the common types of submarine cementation occurring in different facies.

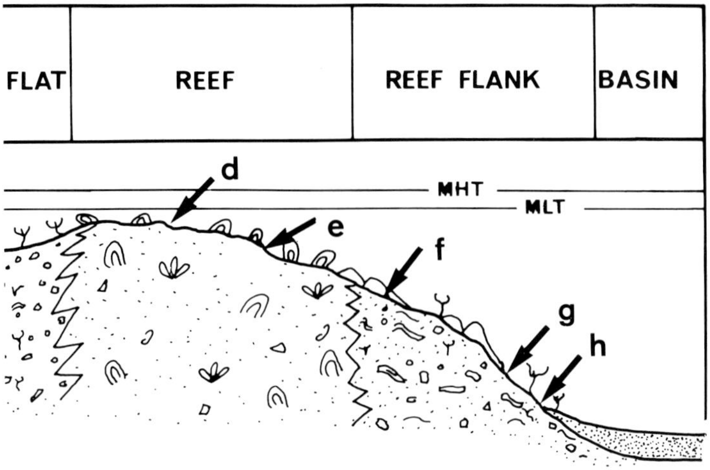

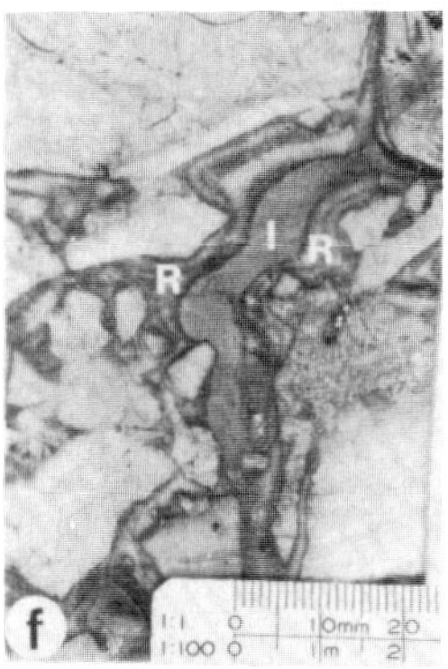

SUBMARINE FRACTURE WHICH TRUNCATES RADIAXIAL CALCITE CEMENT (R) AND IS FILLED BY MARINE INTERNAL SEDIMENT (I) (12-24, 5829 FT.).

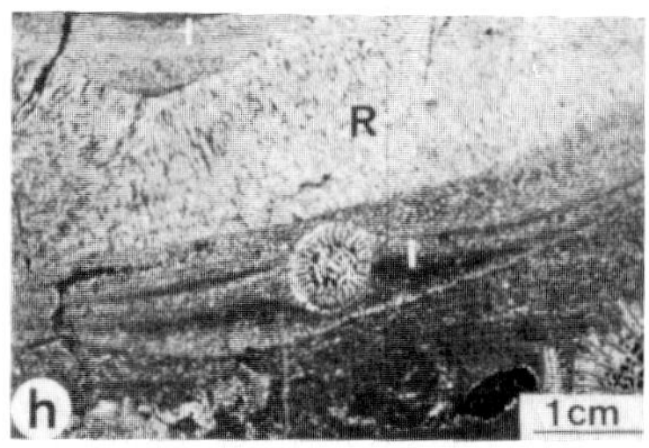

LARGE CAVITY FILLED BY ALTERNATING GENERATIONS OF LAMINATED MARINE INTERNAL SEDIMENT (I) AND RADIAXIAL CALCITE CEMENT (R) (12-26, 5717 FT; CROSSED POLARS).

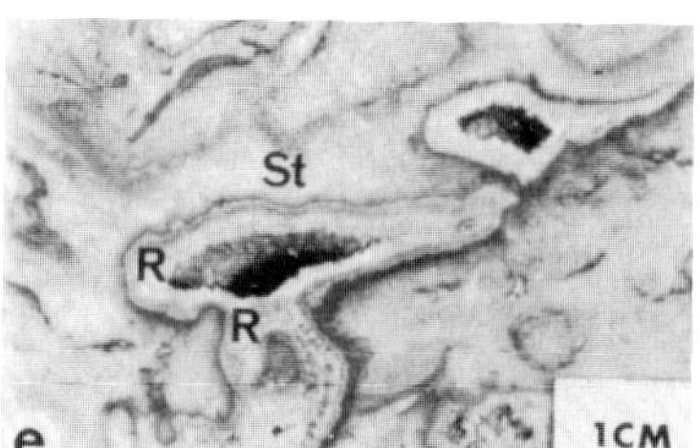

RADIAXIAL CALCITE (R) IN A SHELTER CAVITY UNDER A TABULAR STROMATOPOROID (St) (WELL 3-27, 5600 FT.).

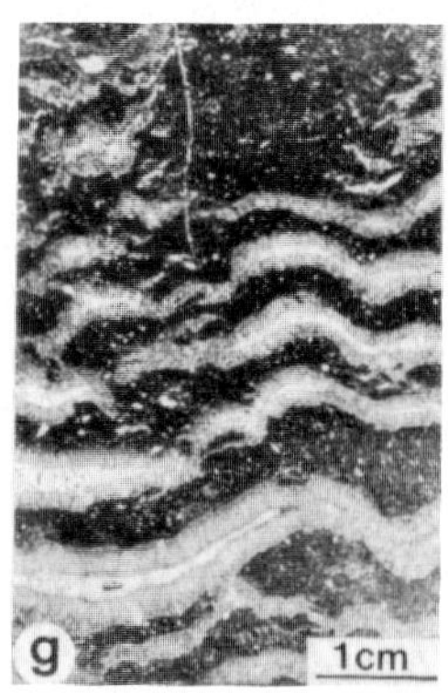

PARALLEL AND INTERCONNECTED CAVITIES FILLED BY RADIAXIAL CALCITE CEMENT (12-26, 5695 FT. PLAIN POLARIZED LIGHT).

Detailed cathodoluminescence petrography of certain Leduc reefs (Burrowes, 1977; Walls, 1977; Walls *et al.*, 1979) has shown that non-ferroan phreatic calcite cements are brightly luminescent and compositionally zones with respect to Mn^{2+} and/or Fe^{2+} (Fig. 15B, D). Strongly zoned cements such as these probably formed under continually fluctuating chemical conditions which are most typical of meteoric waters.

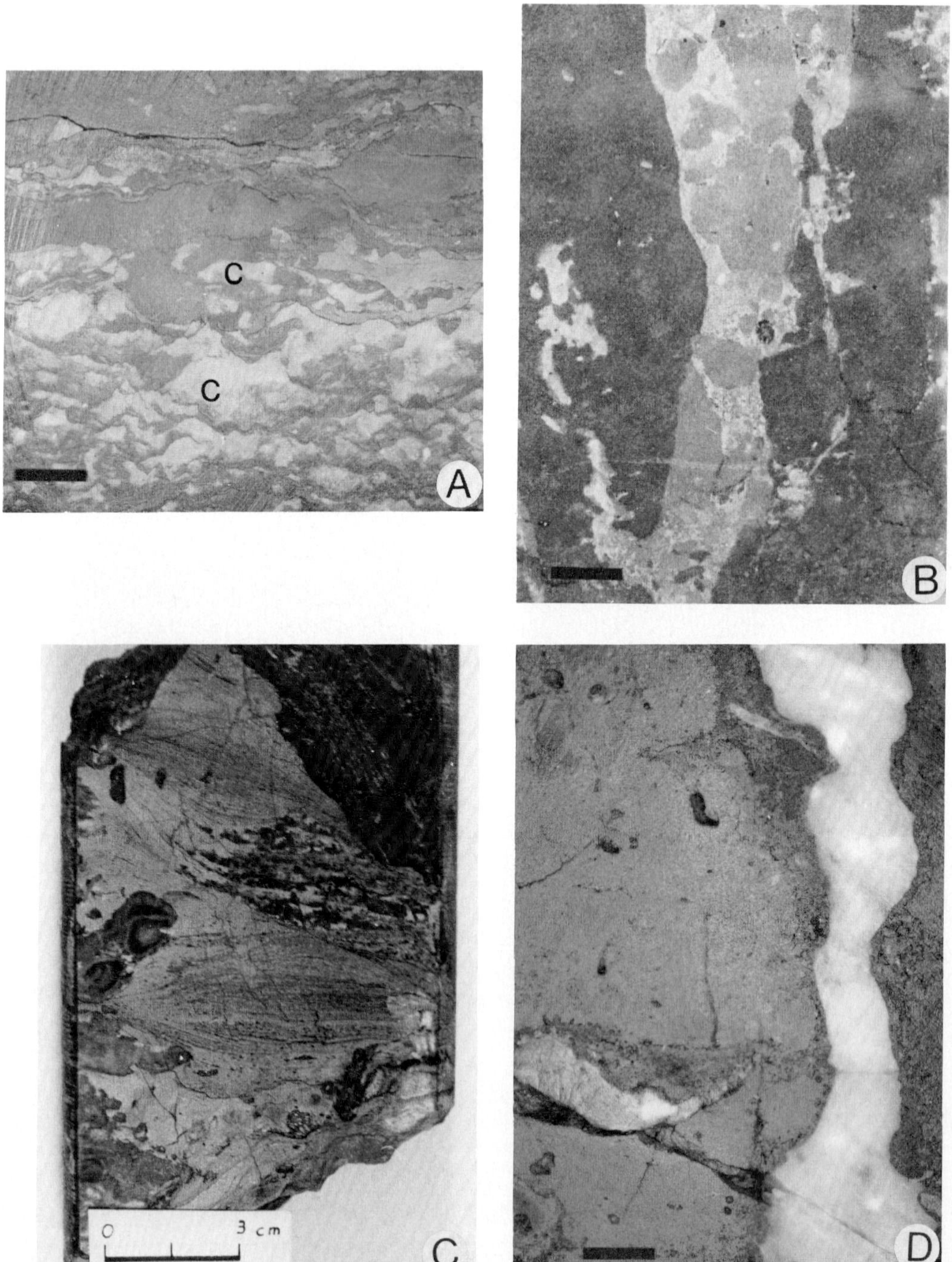

Meyers (1974, p. 857) suggests from existing chemical data that ". . . shallow ground water and surface meteoric waters are more likely media for precipitating Mn^{2+} bearing calcites than deep subsurface waters. . ."

Phreatic calcite cements can be pervasive, filling all primary and secondary pores within a "diagenetic unit" up to several meters thick (Fig. 14A, B) or they can be highly selective as shown in Figure 15A. All degrees of cementation between these two extremes have been observed. It is clear that the presence or absence of these cements can be critical to preservation of reservoir quality (see later discussions).

Subsurface Diagenesis

The subsurface environment is an extensive area that occurs below the deeper reaches of phreatic, meteoric aquifers in Devonian reef rocks and is unaffected by near surface processes. The authors consider the term subsurface synonymous with the deep crustal environment of Bathurst (1980) and mesogenetic environment of Choquette and Pray (1970).

The general burial history of Devonian reefs in the Alberta Basin is fairly well established. In the up-dip margin of the basin (eastern Alberta), Leduc and Nisku reefs have been "breached" by several periods of erosion and are usually overlain by deeply eroded Mississippian rocks rather than Devonian shales (Ireton Fm.). In the Rimbey-Meadowbrook reef trend (central Alberta), Leduc reefs are usually encased in Ireton shales or overlain by Nisku carbonates. Reefs such as Redwater and Golden Spike probably were buried no deeper than 2500 m (O'Conner and Gretener, 1974; Walls *et al.*, 1979). In the down-dip areas of the basin (southwest Alberta), just east of the Foothills, Leduc reefs such as Strachan and Ricinus were once buried to depths greater than 6000 m.

Swan Hills reefs of west central Alberta (Swan Hills, Judy Creek, Carson Creek) are encased in basinal shales and limestones of the Waterways Formation and probably were buried to depths of 3500 m to 4000 m. Rainbow reefs of northwest Alberta are encased in Muskeg evaporites (dolomites and anhydrites) but were never deeply buried. "Presqu'ile" reefs occur in a sublinear "barrier" reef tract that is exposed at its northern end (Pine Point, N.W.T.) but was buried 3000 m or greater at its southern end in northeast British Columbia (Clarke Lake).

A wide spectrum of diagenetic processes occurred in these varied subsurface settings. The most important processes affecting reservoir quality were cementation, chemical compaction (pressure solution), and dolomitization (not discussed in this paper).

Subsurface cementation.—In limestone reefs the most common subsurface cement is medium to coarsely crystalline spar calcite (Fig. 16A, B) which in most cases is either ferroan or contains ferroan zones. Dolomite and anhydrite cements are less common, except in dolomitized reefs.

A subsurface environment is suggested for these calcite cements because they are not restricted to units associated with subaerial exposure features, but are distributed both vertically and laterally throughout the reefs. In addition, these cements often postdate compaction (Fig. 16A) and fill fractures (Fig. 16B) which truncate subaerial as well as submarine cements.

In many Swan Hills reefs there appears a close, causal relationship between the occurrence of stylolites and the occlusion of porosity by coarsely crystalline, calcite cements. Detailed petrographic analysis of the Kaybob reef by Wong (1978) shows that where stylolites are common, cements are pervasive, but in areas where stylolites are absent, cements are less abundant. Likewise, calcite cement crystals decrease both in number and crystal size away from dense, highly stylolitic zones. It is apparent that the fluid source of these cements is the carbonate dissolved during pressure solution (see later discussion). Wong (pers. comm.) estimates that over 75% of the cements in the Swan Hills Kaybob reef may have been derived from pressure solution.

Relationships similar to those described by Wong (1978) were observed by us in other Swan Hills reefs as well as in limestone and dolomite Nisku reefs at West Pembina. For example, Figure 16C illustrates that where stylolites are absent, moldic porosity in tabulate coral mudstones contains only minor dolomite cements. However, with the increasing presence of stylolites, dolomite cements completely fill moldic porosity (Fig. 16D). Thus, identical facies may be porous or tight depending solely on the degree of subsurface cementation.

FIG. 12.—Fabrics formed during subaerial diagenesis. *A*, Core slab of a "paleoexposure" zone in the interior of a Swan Hills reef. Clasts of microkarsted limestone are enclosed in a matrix of green, calcareous shale. (Swan Hills, 10-15-67-10W5, 8039 ft; scale bar = 1 cm). *B*, Core slab of a "paleoexposure" zone in the interior of a Slave Point reef illustrating a solution channel or "crack" in a lime mudstone filled with limestone clasts, green shale and carbonate silt. (Northeast British Columbia, d-47-C/94-P-12, 7132 ft; scale bar = 1 cm). *C*, Core slab of a brecciated lime packstone containing laminated, dolomitic green shale sediment. This sample represents a "paleoexposure" zone at the middle-upper Leduc boundary. (Golden Spike, 9-27-51-27W4, 5589 ft.). *D*, Core slab from the interior of a Rainbow reef showing a vertical channel or "crack" in a dolostone, filled with anhydrite. This fabric represents a "paleoexposure" zone with contemporaneous evaporite deposition. (Rainbow, 7-32-109-8W6, 5983 ft; scale bar = 1 cm).

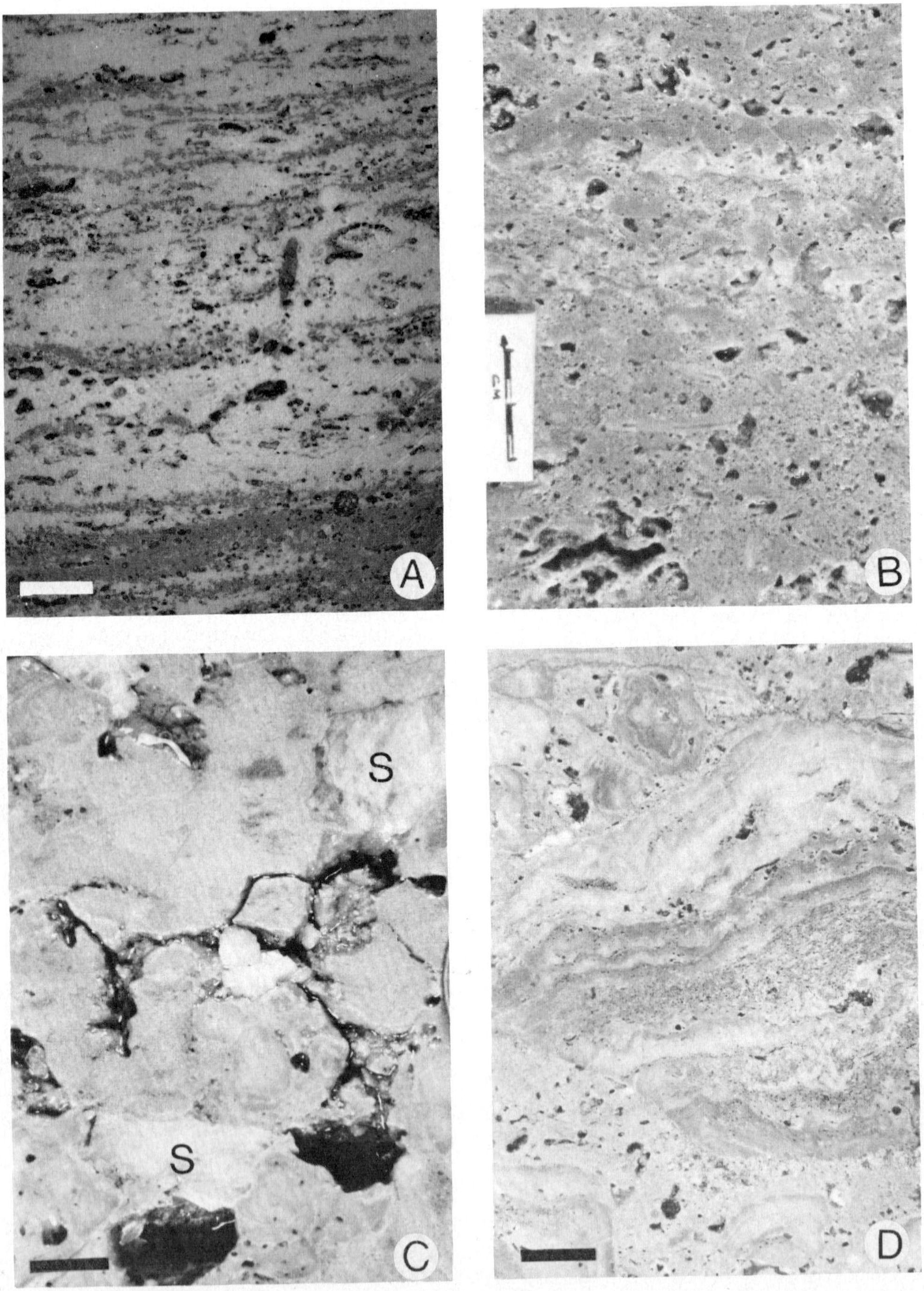

FIG. 13.—Fabrics formed during subaerial diagenesis. *A*, Core slab of a peritidal mudstone from a Rainbow reef interior illustrating primary and solution-enlarged fenestral and moldic porosity. (Rainbow, 7-32-109-8W6, 5931 ft; scale bar = 1 cm). *B*, Core slab of vug and moldic porosity in a Swan Hills reef interior directly underlying a "paleoexposure" zone. (Carson Creek, North, 6-11-62-13W5, 8679 ft.). *C*, Core slab of vug and channel porosity in a *Stachyodes* floatstone from a Swan Hills reef. Vugs are partially filled by late stage, ferroan calcite spar. (Carson Creek North, 16-18-62-12W5, 8827 ft; scale bar = 1 cm). *D*, Core slab showing secondary (solution) porosity in a stromatoporoid rudstone from a Swan Hills reef facies. (Carson Creek North, 1-14-62-13W5, 8941 ft; scale bar = 1 cm).

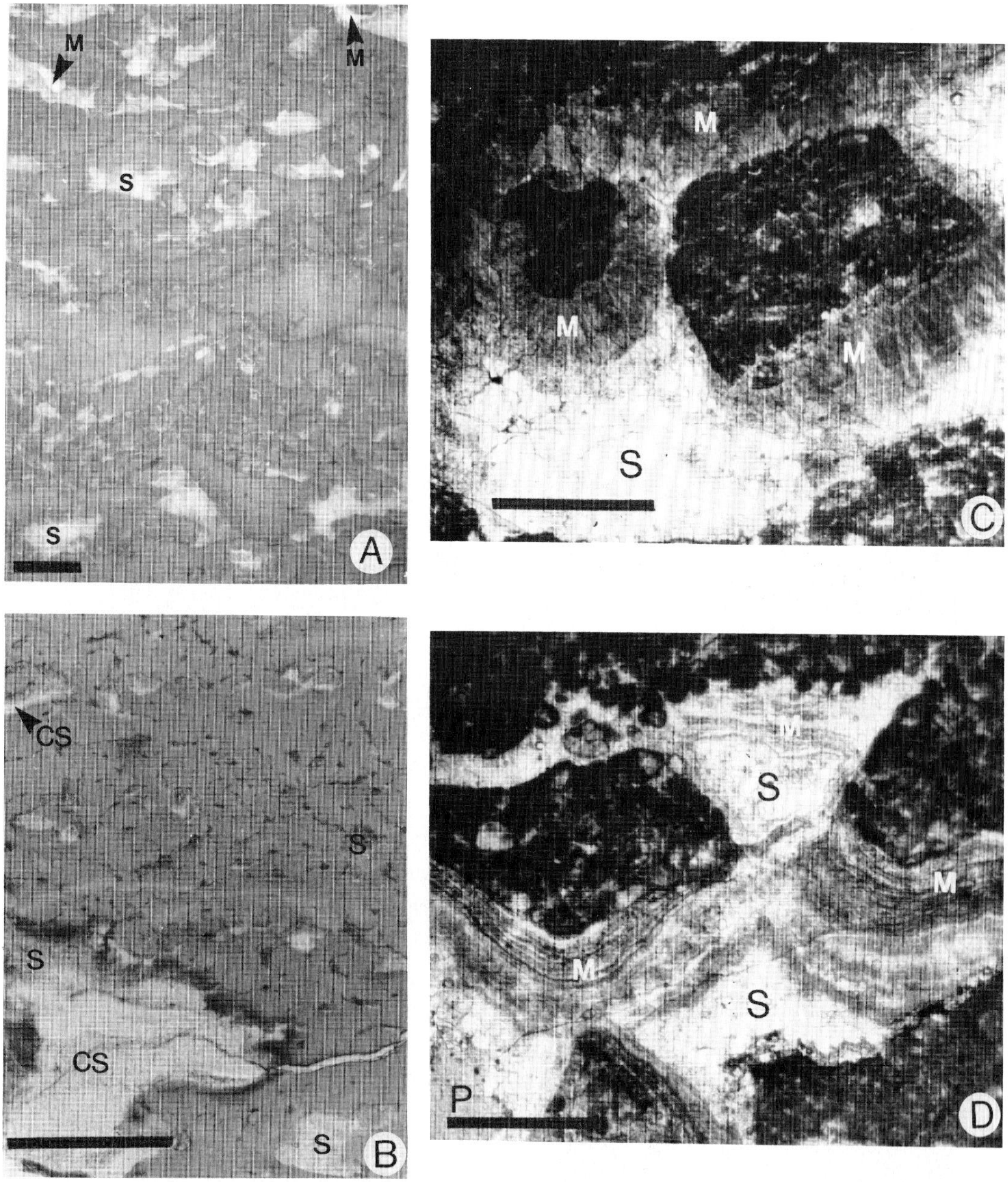

FIG. 14.—Cement fabrics formed during subaerial vadose diagenesis. *A*, Core slab from a Swan Hills reef interior showing primary cavities in *Amphipora* floatstone completely cemented by vadose, microstalactitic cement (M) and later spar calcite (S) (Judy Creek, 4-31-63-10W5, 8545 ft; scale bar = 1 cm). *B*, Core slab of a completely cemented, fenestral lime mudstone from a Leduc reef interior. Fenestral and solution cavities are filled by geopetal crystal silt (light coloured sediment, CS) and clear, non-ferroan calcite spar (S) (Golden Spike, 11-23-51-27W4, 5462 ft; scale bar = 1 cm). *C*, Photomicrograph of radial fibrous, microstalactitic rim cement (M) encrusting undersides of intraclasts. Remaining void space is filled with a non-ferroan spar calcite (S). Occurs in a peritidal laminite facies of a Leduc reef. (Golden Spike, 11-23-51-27W4, 5364 ft; scale bar = 2 mm). *D*, Similar to (C) but fibrous, microstalactitic cement (M) is interlaminated with numerous thin micrite cement bands. The remaining void space is mostly filled with non-ferroan, clear spar calcite (S). Only minor porosity (P) remains. (Golden Spike, 11-23-51-27W4, 5406 ft; scale bar = 2 mm).

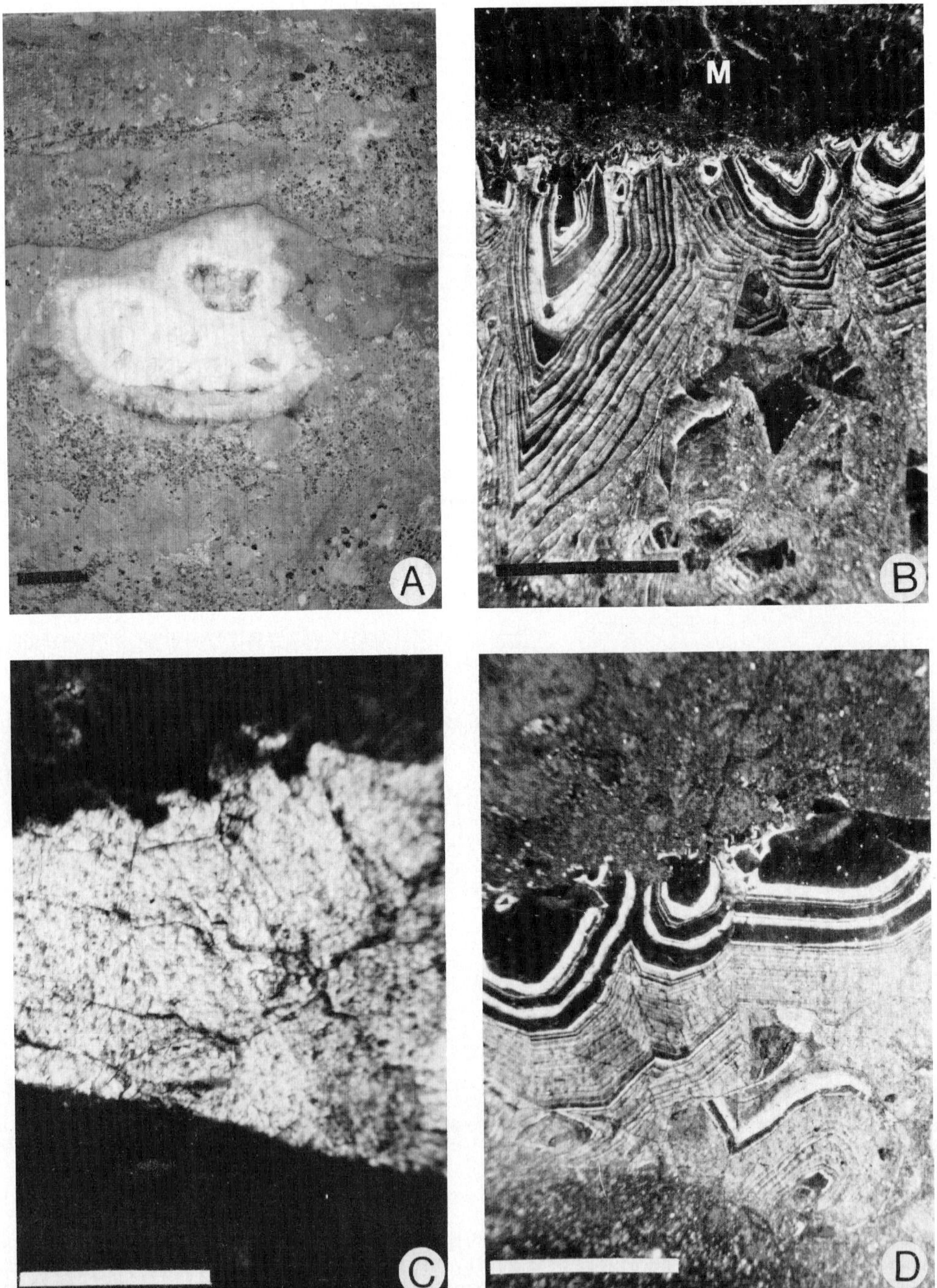

FIG. 15.—Cement fabrics formed during subaerial, phreatic diagenesis. *A*, Core slab of a peloidal grainstone from a Swan Hills reef interior showing a gastropod filled with a clear spar calcite. The calcite cement predates compaction in this facies, and the matrix still retains good interparticle porosity. (Snipe Lake, 10-21-70-18W5, 8721 ft; scale bar = 1 cm). *B*, Cathodoluminescence photograph of cement-filled vug in a Leduc reef interior. Vug is lined by isopachous equant cement (M) of marine origin and infilled by strongly zoned, non-ferroan spar calcite (Golden Spike, 10-27-57-27W4, 5566 ft; scale bar = 2 mm). *C*, Solution cavity from beneath a subaerial disconformity at the Middle-Upper Leduc boundary. Cavity is filled by non-ferroan spar calcite (Golden Spike, 9-27-51-27W4, 5590 ft; scale bar = 2 mm). *D*, Cathodoluminescence photograph of (C) showing strong zoning of phreatic calcite cement.

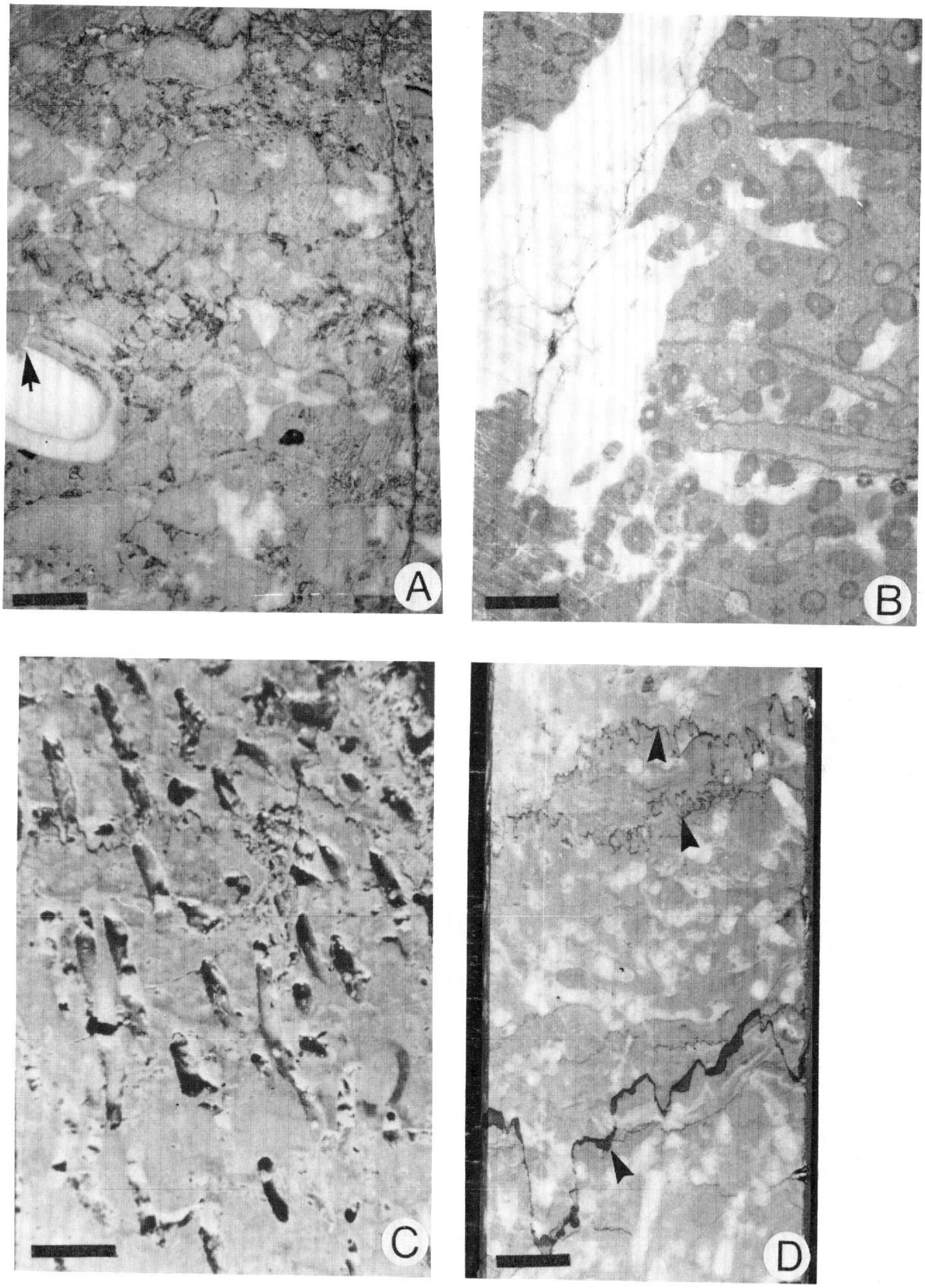

FIG. 16.—Cement fabrics formed during subsurface diagenesis. *A*, Core slab from a skeletal rudstone in a Swan Hills reef flat facies. Most interparticle pores are filled with a coarsely crystalline spar calcite. Cementation apparently postdates compaction (note broken gastropod shell, arrow). (Swan Hills, 10-15-67-10W5, 7959 ft; scale bar = 1 cm). *B*, Core slab of a fracture filled with coarsely crystalline spar calcite. (Judy Creek, 4-31-63-14W5, 8537 ft; scale bar = 1 cm). *C*, Core slab of moldic porosity in a dolomitized, tabulate coral bafflestone from a Nisku reef. A few molds are partially filled with coarsely crystalline dolomite. (West Pembina, 1-32-50-10W5, 8717 ft.; scale bar = 2 cm). *D*, Similar to (C) but all coral molds are filled with a coarsely crystalline dolomite cement. The most complete dolomite cementation in this facies is usually associated with increasing numbers of stylolites (arrows). (West Pembina, 6-25-50-10W5, 8543 ft; scale bar = 1 cm).

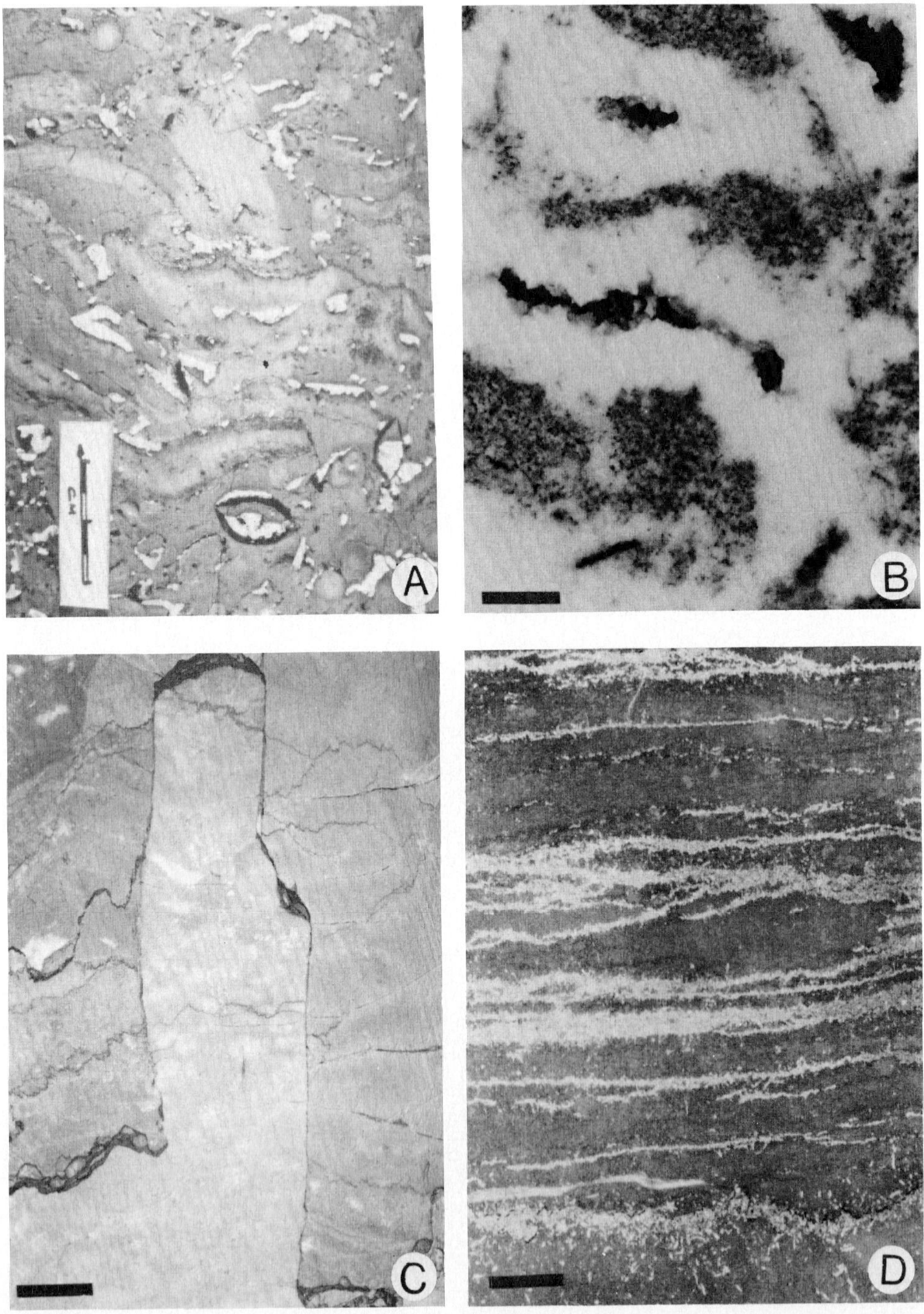

FIG. 17.—Fabrics formed during subsurface diagenesis. *A*, Coarsely crystalline dolomite (white) fills much of the interparticle, intraparticle, and shelter porosity in this tabular stromatoporoid bindstone from a Swan Hills reef. (Carson Creek North, 6-18-61-11W5, 8706 ft.). *B*, Coarsely crystalline dolomite fills a large interconnected cavity system in a dolomitized "Presqu'ile" reef. Finer matrix dolomite is heavily stained by solid hydrocarbons (bitumen). (Northeast British Columbia, d-47-C/94-P-12, 7252 ft; scale bar = 1 cm). *C*, Core slab from a Swan Hills reef interior mudstone showing small, parallel stylolites truncated by a large amplitude stylolite. Notice the contrasting rock types on either side of the large stylolite. (Swan Hills, 10-15-67-10W5, 8044 ft; scale bar = 1 cm). *D*, Core slab of parallel stylolites in a Slave Point lime mudstone. All stylolite seams are lined or filled with finely crystalline dolomite. (Northeast British Columbia, d-47-C/94-P-12, 7220 ft; scale bar = 1 cm).

Dolomite cements generally are rare in Leduc and Swan Hills limestones. However, in some Swan Hills reefs, such as Carson Creek, North coarsely crystalline dolomite occurs as a pore filling in primary porosity of the reef margin facies (Fig. 17A).

As would be expected, dolomite is a common cement in dolomitized Devonian reefs. This paper, however, is concerned primarily with limestone reefs with the exception of the Middle Devonian "Presqu'ile" of northeastern British Columbia. "Presqu'ile" reefs are unique because they have undergone all possible phases of diagenesis from submarine through subsurface, hydrothermal mineralization. Subsurface processes are responsible for the predominant reservoir characteristic of these rocks.

Figure 17B is a typical example of the well known "Presqu'ile" dolomite. Original depositional and early diagenetic fabrics are completely disguised by these subsurface dolomites. Secondary vug, channel, and fracture porosity is partially or completely filled by coarsely crystalline white dolomite cements. These cements are associated with galena, sphalerite, and pyrite minerals and are occasionally overlain by coarsely crystalline ferroan calcite and quartz.

Gray and Kassube (1963), Campbell (1967), Beales and Jackson (1968), Roedder (1968), Comby (1975), Skall (1975, 1977), and others have discussed the origin of "Presqu'ile" dolomites. Most evidence indicates that these coarse dolomites formed during hydrothermal events that occurred discontinuously along a reef margin seated on "deeply rooted" normal faults. Abnormally high bottom-hole temperatures are common today in most wells in northeast British Columbia. Oxygen isotope data from coarsely crystalline dolomites at Pine Point (Skall, 1975) also indicates a hydrothermal origin for these cements.

Anhydrite occurs as a replacement (Fig. 18C) of calcite and dolomite and as a primary cavity filling (Fig. 18D) in a number of reefs. However, with the exception of Rainbow reefs and specific intervals in the Nisku (West Pembina) and deeply buried Leduc reefs (Strachan and Ricinus), anhydrite is generally rare. Limestone reefs such as Swan Hills, Judy Creek, Golden Spike and Redwater contain almost no anhydrite.

Solid hydrocarbons (bitumen), occurring as pore fillings and lining stylolites, are present in varying amounts in almost all reefs. In most cases bitumen is

TABLE 2.—SUMMARY OF THE DOMINANT PROCESSES AND FLUID COMPOSITIONS INVOLVED IN THE SUBMARINE, SUBAERIAL AND SUBSURFACE DIAGENESIS OF DEVONIAN REEFS

CARBONATE DIAGENESIS

SUBMARINE	SUBAERIAL	SUBSURFACE
BIO-EROSION:	NEOMORPHISM:	TEXTURAL NEOMORPHISM
BORINGS	*ARAGONITE* } *CALCITE*	PRESSURE SOLUTION
MICRITIZATION	*Mg-CALCITE* } *CALCITE*	FRACTURING
CEMENTATION:	PEDOGENESIS:	CEMENTATION:
MICRITE	*SOILS*	*CALCITE/DOLOMITE*
FIBROUS	*CALICHE*	*ANHYDRITE*
SYNDEPOSITIONAL FRACTURING	SOLUTION	*quartz, galena, sphalerite*
	CEMENTATION:	REPLACEMENT:
	CALCITE	*CALCITE* { *DOLOMITE*
	microstalactitic, meniscus	*CALCITE* { *ANHYDRITE*
	DOLOMITE	*CALCITE* { SiO_2
	ANHYDRITE	BITUMEN FORMATION
MARINE WATERS	FRESH TO MARINE	SALINE TO BRACKISH

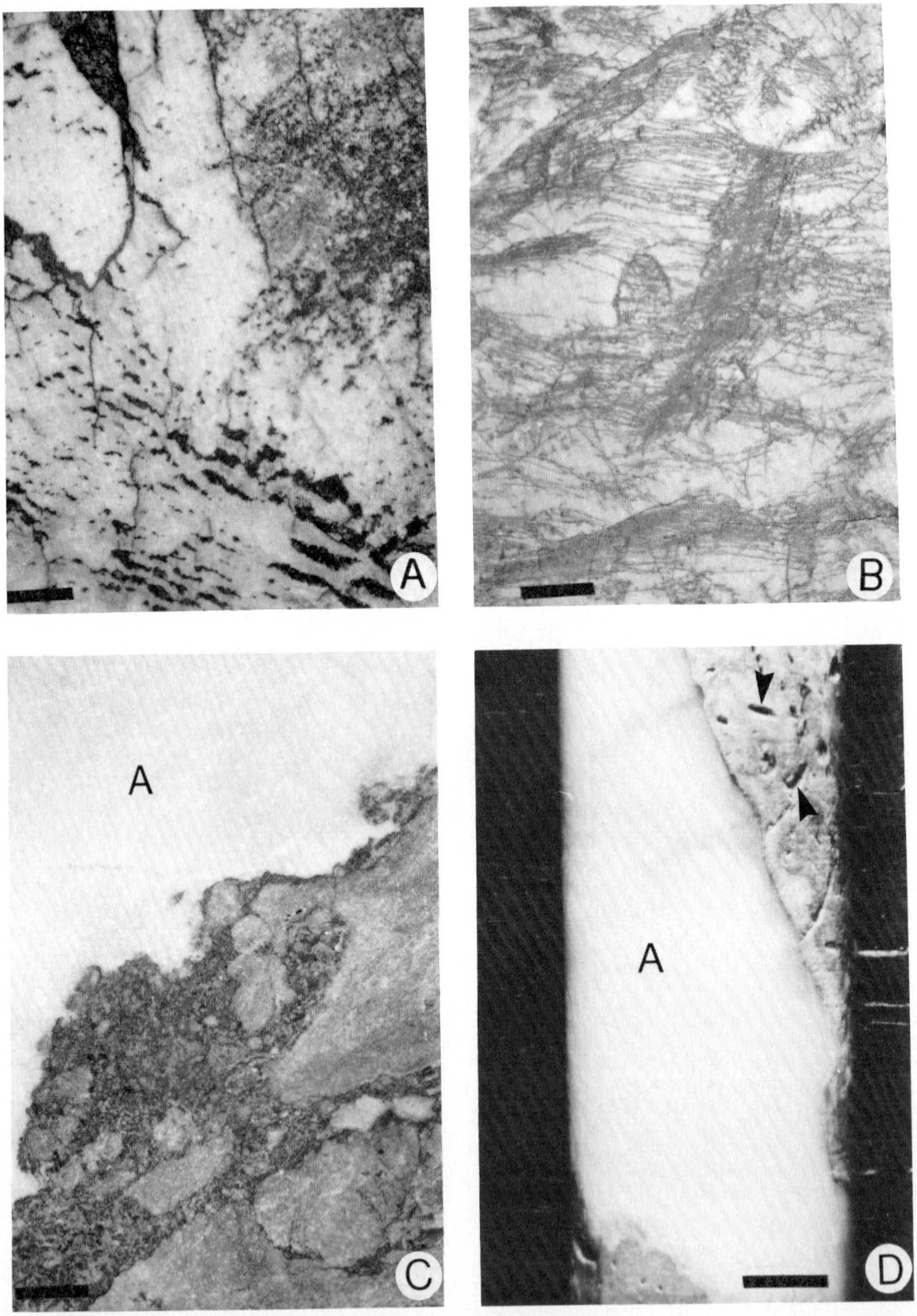

FIG. 18.—Fabrics formed during subsurface diagenesis. *A*, Core slab of completely dolomitized, lime mudstone from a Rainbow reef. This sample contains many small (hairline) fractures as well as vugs which are filled with solid hydrocarbons (bitumen). Rainbow, 2-32-109-9W6, 6256 ft; scale bar = 1 cm). *B*, Core slab from a deeply buried Leduc reef margin limestone showing numerous small (hairline) fractures and stylolites which are completely plugged with solid hydrocarbons (bitumen). (Strachan, 10-31-37-9W5, 13,454 ft; scale bar = 1 cm). *C*, Core slab of anhydrite (A) replacement in deeply buried Leduc reef. (Ricinus, 11-17-35-8W5, 13,908 ft; scale bar = 1 cm). *D*, Nisku reef with massive anhydrite (A) filling a large channel (?) in a dolomitized, tabulate coral floatstone. The dolomitized reef rock retains some of its secondary vug and moldic porosity (arrows). (West Pembina, 14-26-51-9W5, 7851 ft; scale bar = 2 cm).

insignificant and has negligible affects on reservoir quality. However, in specific areas bitumen may completely plug pore throats (for example, reefs in the Zama-Rainbow basin and those in the Southern Alberta Foothills; Fig. 18A, B). In addition to considerably reducing initial permeability and effective porosity, this bitumen, when treated with HCl acid after well bore perforation, forms colloid solutions that flocculate and are known to further plug pore throats and seriously hamper oil recovery.

Chemical compaction and stylolite formation.—The importance of chemical compaction on carbonate sediments and rocks has been demonstrated, most notably, by the recent works of Dunnington (1967), Mossop (1972), Bathurst (1975, 1980), DeBoer (1977), Engelder *et al.* (1981), Meyers and Hill (1983), and Shinn and Robbin (1983). Chemical compaction causes sediments and/or rocks to deform by pressure solution under overburden stress, producing stylolites and releasing $CaCO_3$-rich fluids into surrounding pore waters.

Little is known about the depths of burial required to produce pressure solution. Dunnington (1967) reports that 600 m to 900 m of burial is necessary, but Sellier (1979) observes the effects of pressure solution as shallow as 300 m. In studies of differential compaction within the Woodbend Group along the Rimbey-Meadowbrook trend of central Alberta, O'Conner and Gretener (1974, p. 301) estimate that much of the compaction and "collapse" of the reef cores occurred under only 155 m of overburden. Thus it is possible that some pressure solution, producing small scale stylolites, could have occurred as early as Wabamun time (late Upper Devonian) in such reefs as Golden Spike and Redwater.

In different Devonian reefs, the degree of pressure solution and distribution of stylolites is highly variable. Nevertheless, the following generalizations can be made: (1) stylolites are common in tight rocks (Fig. 17C); (2) stylolites are common in texturally heterogenous rocks (i.e. poorly sorted skeletal wackestone and mudstone) relative to texturally homogeneous rocks (i.e. well sorted grainstones and rudstones); (3) dense concentrations of stylolites act as permeability barriers, particularly in reef interiors. Reductions in bed thickness caused by pressure solution are estimated to be 30 percent in Redwater (Mossop, 1972) and Clarke Lake (Comby, 1975), and somewhat less (20 percent) in Swan Hills reefs such as a Snipe Lake (Wong, 1978). As discussed earlier, it is well established that pressure solution is an important source of $CaCO_3$ for subsurface cements. The marked reduction of porosity by calcite cements associated with stylolites in many Swan Hills reefs is an excellent example of this phenomena.

Stylolites may also provide a source for subsurface dolomitizing fluids (Wanless, 1979). Figure 17D illustrates the commonly observed selectivity of dolomite precipitation within and adjacent to stylolites.

DISCUSSION—DIAGENETIC CONTROLS ON RESERVOIR QUALITY

Like depositional environments, diagenetic environments also repeat themselves through time. The repeated occurrence of particular cement fabrics in reef complexes of different ages and geological settings is well illustrated in Middle and Upper Devonian reefs of Western Canada. However, diagenetic environments are not always easy to map due to our incomplete understanding of the processes and products that occur in these environments, and the common overlap of one environment on another. Nevertheless, it can be demonstrated that processes occurring in each environment have a significant influence on primary, depositional porosity.

The major influences on porosity in limestones are cementation and solution. The relative intensity of these processes will affect reservoir quality and heterogeneity for any given Devonian reef.

In the submarine environment, cementation is far more important than solution. In reefs such as Golden Spike, submarine cementation was so extensive in parts of the reef margin that little porosity remains. In contrast, interior facies of the same reef contain less submarine cement, and primary porosity was affected mainly by solution and/or cementation occurring during subaerial diagenesis.

The subaerial environment was characterized by cementation and solution occurring mainly in vadose and phreatic, meteoric waters. These contrasting processes in reef interiors produced porous units interbedded with well cemented units. This type of "diagenetic stratigraphy", common to Devonian reef interiors, results in stratified reservoirs with numerous permeability barriers which can seriously affect hydrocarbon recovery (Reitzel and Callow, 1976).

Chemical compaction (pressure solution) and associated cementation were the most important processes affecting reservoir quality in the subsurface environment. In reefs where these processes seldom occurred, porosity is well preserved (i.e. Golden Spike, Redwater, Rainbow reefs).

The following discussions will demonstrate the relationship between cementation and reservoir quality for Leduc (Golden Spike), Swan Hills (Swan Hills, Judy Creek, Carson Creek, N.), Rainbow, and "Presqu'ile" reefs. Figures 19 to 22 illustrate the relationship between various stages of cementation and the pathways of porosity destruction through time for reef margins versus reef interiors. Cement volume and porosity percentages are based on visual estimates from numerous drill cores, on core analysis data, and other published data. Although initial porosities certainly varied in different reef facies, for consistency of comparison a value of 50 percent has been assumed in all cases.

Leduc—Golden Spike

Porosity distribution and primary pore system modifications show several general trends throughout

Golden Spike. Porosity in the reef margin, particularly fore-reef facies, is, with few exceptions, consistently low due to occlusion of primary pores by submarine cements. It is estimated that 70 percent of the cement volume in the reef margin formed during submarine diagenesis and that primary porosity was reduced to approximately 20 percent by the end of this diagenetic stage (Fig. 19). Subaerial cements in the reef margin are relatively minor (5 percent) and only about 15 percent of the total cement volume formed in the subsur-

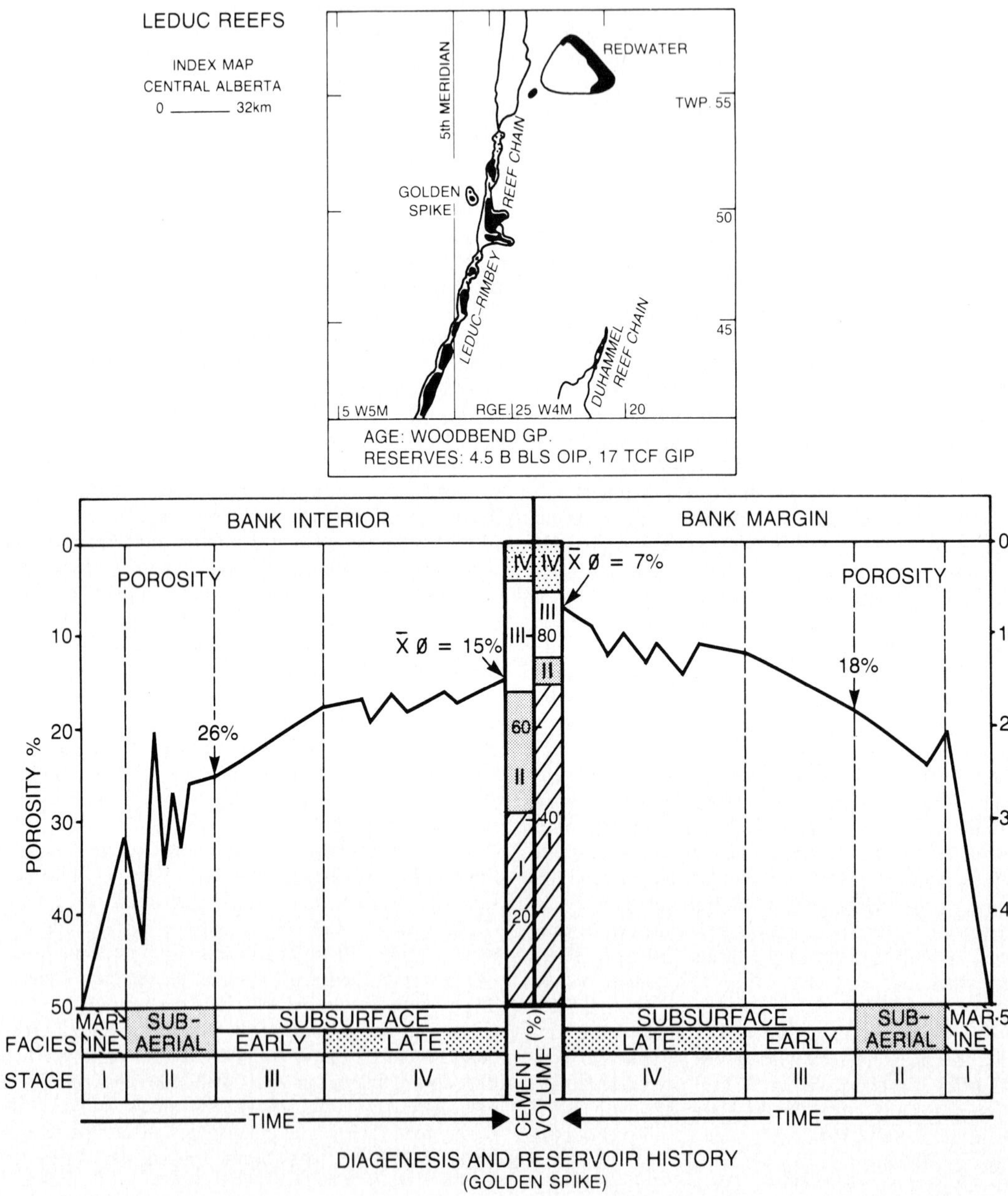

FIG. 19.—Summary of diagenesis and reservoir history for Golden Spike illustrating the pathway of porosity destruction through time, and the estimated cement volume (in percent) formed in each diagenetic facies and stage.

face environment. The average porosity in the Golden Spike reef margin is only 7 percent, due mainly to the abundance of submarine cements (Fig. 19).

Porosity in the reef interior is extremely variable due to the "overlapping" processes of precipitation of submarine, subaerial, and subsurface cements as well as solution, mainly occurring during subaerial exposure (Fig. 19). In contrast to the reef margin, submarine cements account for only 40 percent of the total cement volume. Porosity destruction by subsurface cements in

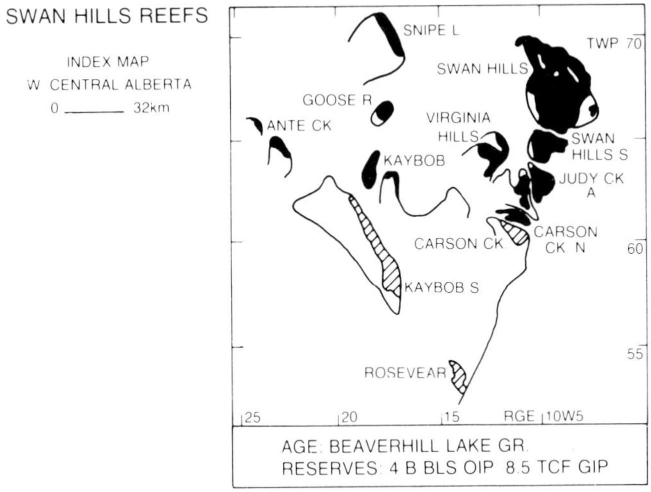

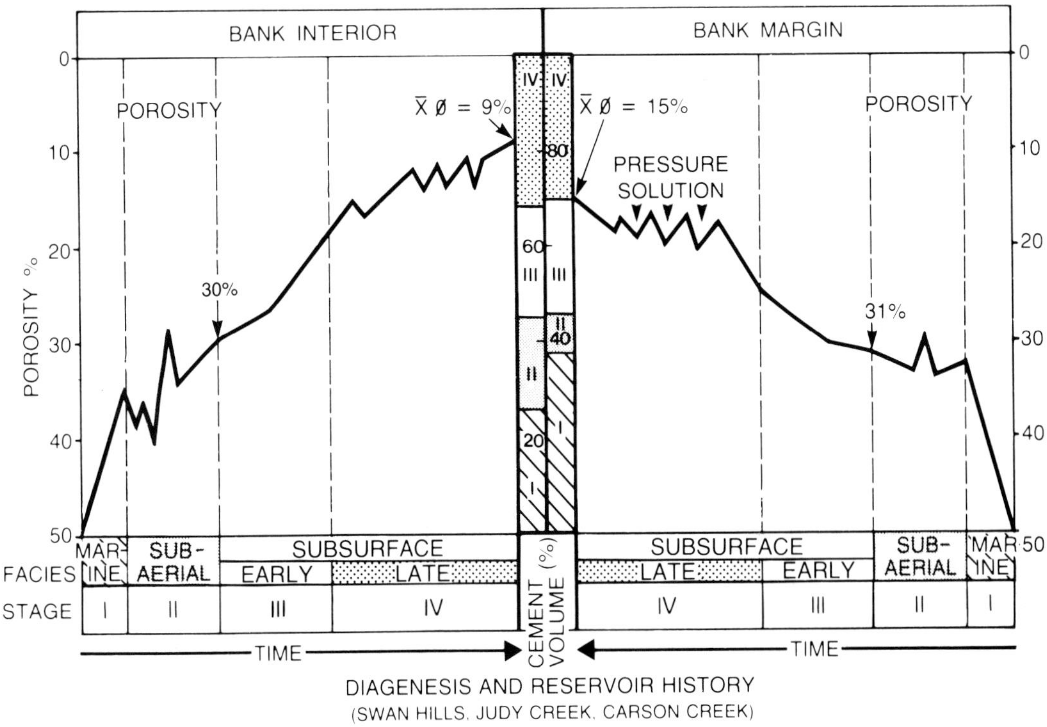

FIG. 20.—Summary of diagenesis and reservoir history for Swan Hills reefs illustrating the pathway of porosity destruction through time, and the estimated cement volume (in percent) formed in each diagenetic facies and stage.

the reef interior is slightly greater than on the margin, but the average final porosity (15 percent) is considerably higher in the interior (Fig. 19). Diagenetic permeability barriers, however, are common in the reef interior and relatively rare in the reef margin.

Swan Hills Reefs

The diagenetic histories of Swan Hills, Judy Creek, and Carson Creek, North are similar. Submarine cements account for about 40 percent of the total cement volume in the reef margin (Fig. 20). It is estimated that reef margin porosity after submarine cementation was approximately 35 percent (Fig. 20).

Porosity destruction in reef interiors followed a similar pattern as observed in Golden Spike. Submarine cements reduced porosity to about 35 percent and alternating stages of subaerial solution and cementation produced porous beds intercalated with well cemented beds. Subaerial cements represent about 20 percent of the total cement volume in the interior. However, reef interior facies still retained approximately 30 percent porosity after subaerial diagenesis. Subaerial cements are much less common (less than 10 percent) in the reef margin (Fig. 20).

Subsurface cementation by ferroan and non-ferroan calcite and some dolomite ultimately reduced porosity in the reef interior to 9 percent. Reef margin facies retained 15 percent porosity after subsurface cementation. Subsurface cements represent 55 percent of the total cement volume in the margin and 60 percent in the interior (Fig. 20).

The deeper burial of Swan Hills reefs relative to Leduc reefs (i.e. Golden Spike and Redwater) may be responsible for the greater abundance of subsurface cements in Swan Hills, Judy Creek and Carson Creek, North. Nevertheless, because of early and pervasive submarine cementation, porosities are much lower in the margin of Golden Spike than in Swan Hills reefs.

Rainbow Reefs

The most striking characteristic of Rainbow reefs is the abundance of submarine cements. It is estimated that 75 percent of the total cement volume in reef margins formed during submarine diagenesis (Fig. 21). These cements created rigid reef rocks penecontemporaneous with reef growth and deposition. In most cases, primary porosity was reduced rapidly to less than 20 percent in reef margin facies (Fig. 21).

In contrast, reef interiors contain about equal amounts of submarine and subaerial cements, but porosity was still reduced to 20 percent by the end of subaerial diagenesis (Fig. 21). As previously discussed, much of the subaerial cementation occurred under evaporitic conditions in Rainbow reefs.

Subsurface cements account for less than 15 percent of the total cement volume in most reefs in the Rainbow Basin (Fig. 21). The relative lack of subsurface cementation helped to preserve porosity that survived early cementation. Thus, reef interiors are generally more porous (13 percent) than margins (8 percent).

Core analysis data reveals that dolomitized Rainbow reefs exhibit slightly less porosity but considerably higher permeabilities than limestone reefs (Fig. 21). In addition, dolomitized reefs show little porosity variation from reef interior to margin. Dolomitization apparently was initiated during early exposure and evaporite diagenesis and continued throughout burial (Fig. 21).

"Presqu'ile" Reefs

Middle Devonian reefs of northeast British Columbia have undergone a complex diagenetic history punctuated by several periods of solution, cementation and mineralogical replacement. It is the subsurface diagenetic history of "Presqu'ile" reefs that sets them apart from other Middle and Upper Devonian reefs discussed in this paper.

Most porosity destruction probably occurred during subaerial cementation and subsurface, hydrothermal diagenesis (Fig. 22). Evidence of submarine cementation is relatively minor in these reefs but this may be due to fabric destruction during later dolomite replacement.

Subaerial dolomite and calcite cements account for about 20 percent of the cement volume in the reef margin and 30 percent in the interior. On the other hand, subsurface cements, many of a hydrothermal origin, account for up to 55 percent of the cement volume in the margin and 40 percent in the interior. The remaining porosity averages 10 percent and 7 percent on the margin and interior respectively (Fig. 22).

Subsurface cements of a hydrothermal origin include, in sequence from oldest to youngest, ferroan dolomite, ferroan calcite, quartz, galena, and sphalerite. Porosity occurs mainly as large vugs, solution channels, and intercrystalline pores.

Summary

Figure 23 summarizes the relative importance of the four cement stages with respect to margin and interior facies of the various reefs. As illustrated, marine cements make up a considerable portion of the cements in all Devonian reeefs of Western Canada. In Golden Spike and most Rainbow reefs, fibrous, bladed, and micritic marine cements are the most commonly occurring cements and significantly reduce primary porosity at reef margins.

Subaerial cements such as fibrous, microstalactitic calcite and clear, non-ferroan spar calcite are much more common in reef interiors than in reef margins (Fig. 23). They usually account for 20 to 40 percent of the total cement volume in reef interiors. The general scarcity of these cements in reef and fore-reef facies indicates that subaerial exposure was uncommon at the seaward margins of many Devonian reefs. The relative lack of subsurface cementation in Golden Spike and

Rainbow reefs resulted in preservation of porosity that survived early, marine and subaerial cementation. Porosity in Swan Hills reefs, however, is affected by subsurface cements which generally account for over 50 percent of the total cement volume (Fig. 23). These ferroan and non-ferroan calcite and minor dolomite cements precipitated from fluids released during chemical compaction (pressure solution).

Subsurface cementation by dolomite and minor amounts of ferroan calcite and quartz, exerts the major

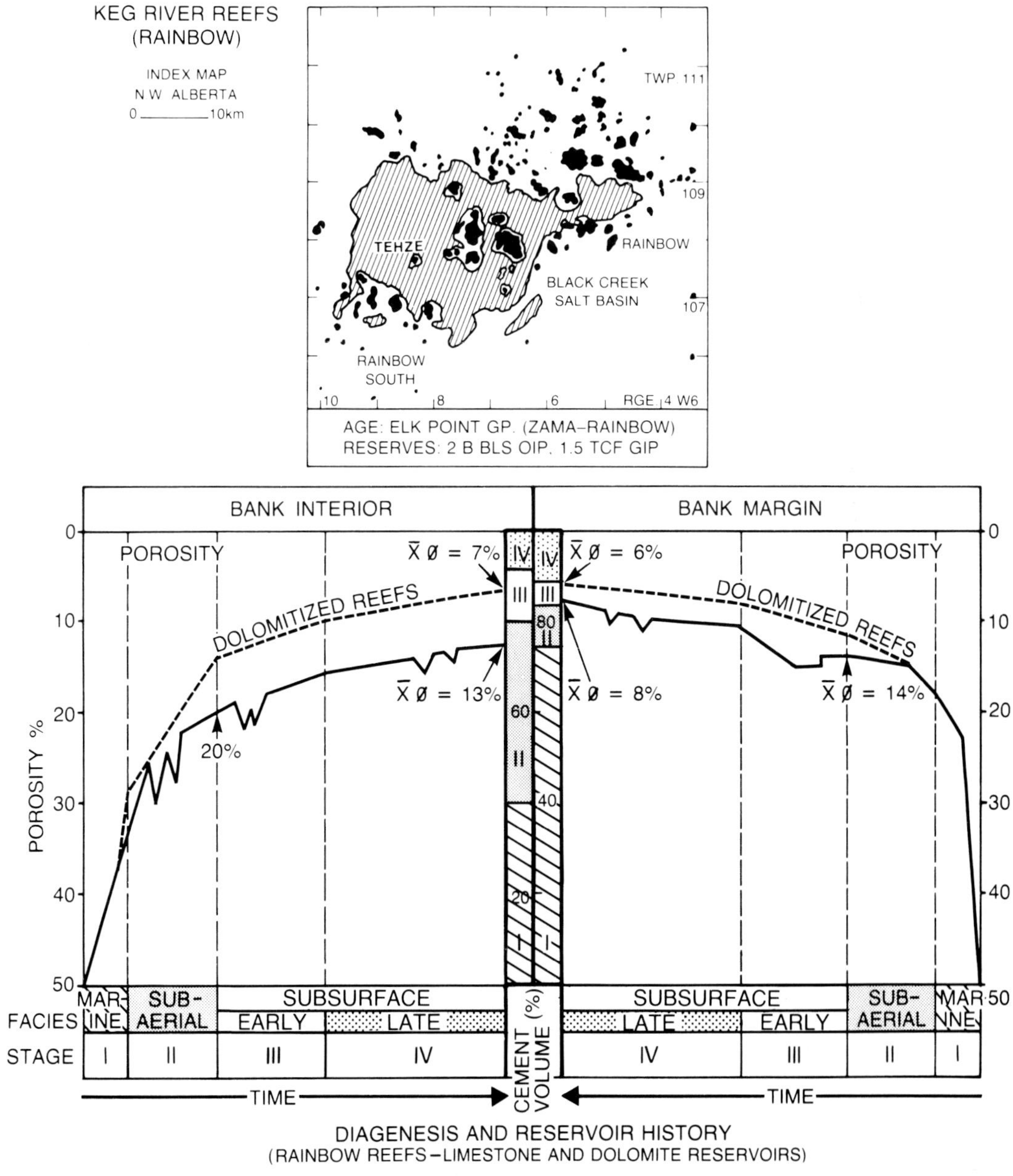

FIG. 21.—Summary of diagenesis and reservoir history for Rainbow reefs illustrating the pathway of porosity destruction through time, and the estimated cement volume (in percent) formed in each diagenetic facies and stage.

influence on reservoir quality in "Presqu'ile" reefs (Fig. 23). These cements account for more than 50 percent of the total cement volume.

As previous discussions have demonstrated, reservoir character and quality varies significantly in Devonian reefs. Table 3 is a summary of porosity, permeability and "reservoir character" for reef margins versus reef interiors. In general, where reef margins have undergone significant submarine cementation (e.g. Rainbow and Golden Spike), average

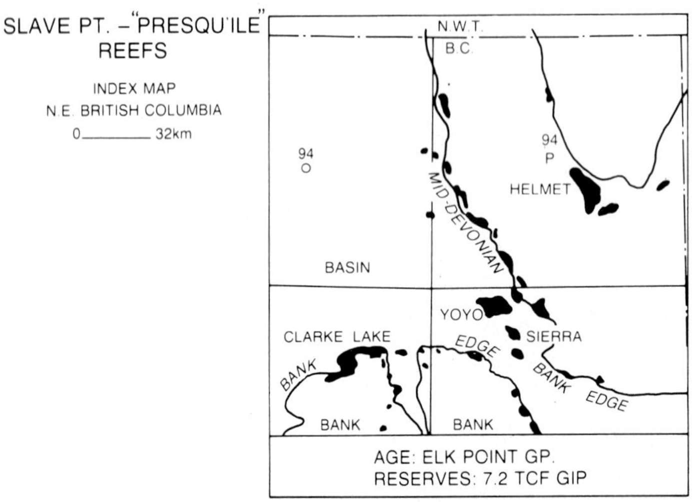

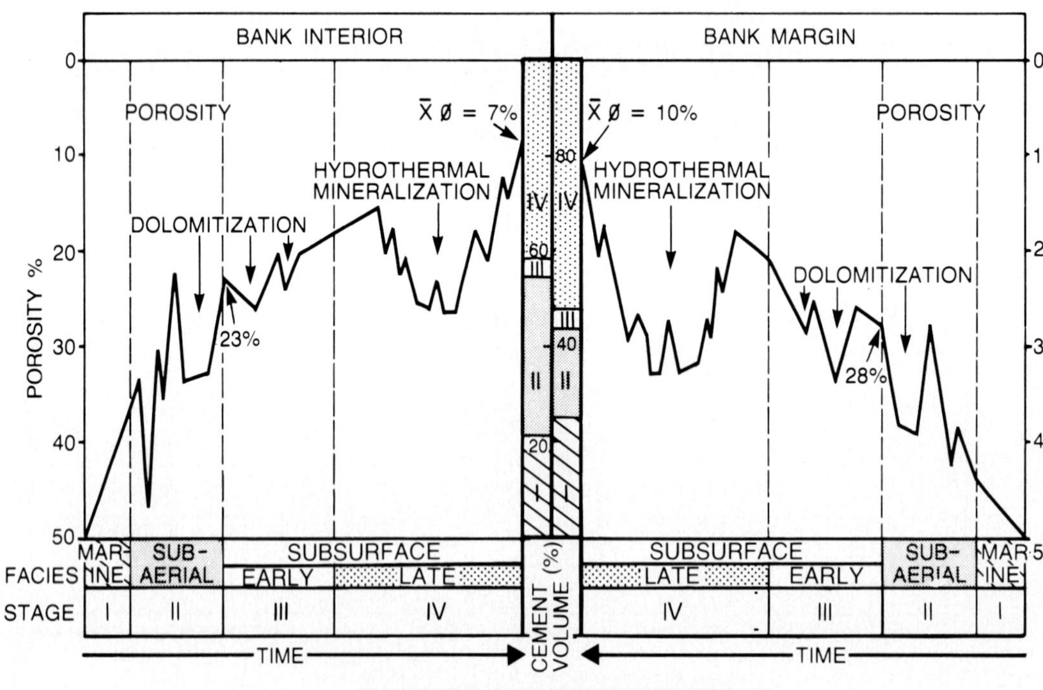

FIG. 22.—Summary of diagenesis and reservoir history for Slave Point—"Presqu'ile" reefs illustrating the pathway of porosity destruction through time, and the estimated cement volume (in percent) formed in each diagenetic facies and stage.

porosities are low. As the amount of submarine cements decrease at reef margins, porosity and permeability improve markedly. For example, most Swan Hills reefs have average porosities of 15 percent and permeabilities in the hundreds of millidarcies in reef margin facies (Table 3).

Reef interiors are characterized by stratified or layered reservoirs. Permeability barriers are commonly the result of cementation that occurred predominantly in subaerial and subsurface environments.

CONCLUSIONS

Reservoir properties of many Middle and Upper Devonian reefs are the result of diagenetic alterations of primary, depositional fabrics. The single most important diagenetic process affecting porosity is carbonate cementation. Cementation in Devonian reefs occurred in three general environments: submarine, subaerial and subsurface.

Submarine cements common in many Devonian reefs include: (1) prismatic, radiaxial calcite (often zoned or banded), (2) isopachous crusts of fibrous to bladed calcite, (3) aggregates of radial fibrous to bladed crystals, and (4) micrite (microcrystalline calcite) rims or crusts.

Types 1 through 3 are more common in reef margin facies, while types 2 and 4 are typical of reef interiors.

Subaerial diagenesis and subaerial exposure zones are best preserved in reef interiors and are characterized by the associated occurrences of: (1) brecciated and fractured lime wackestones/mudstones with microkarst surface and cavities filled with shale and carbonate silt, (2) solution-enlarged porosity, (3) gravitational or microstalactitic calcite rim cements (vadose), and (4) clear, variably ferroan, spar calcites (phreatic). In Rainbow reefs subaerial diagenesis occurred under evaporitic conditions. Therefore, dolomitization and crystalline anhydritization are also common.

TABLE 3.—COMPARISON OF THE RESERVOIR QUALITY AND CHARACTERISTICS OF VARIOUS DEVONIAN REEF MARGINS AND INTERIORS

REEF COMPLEX		∅ MEAN (%)	K MEAN (md)	RESERVOIR CHARACTER
LEDUC (GOLDEN SPIKE)	M	7	25	POOR PERM.; VERT. & HORIZ. PERM. BARRIERS
	I	12	60	"STRATIFIED"; HORIZ. PERM. BARRIERS
SWAN HILLS	M	15	170	GOOD PERM.; FEW PERM. BARRIERS
	I	9	25	"STRATIFIED"; HORIZONTAL PERM. BARRIERS
RAINBOW MBR	M	8–LS 6–DOL.	35 250	VARIABLE PERM. LENSOID RESERVOIRS
	I	13–LS 7–DOL.	90 310	VARIABLE PERM. LIMITED PERM. BARRIERS
SLAVE PT. — "PRESQU'ILE" (DOLOMITE)	M	10	75	GOOD VERTICAL PERM.; LENSOID RESERVOIRS
	I	7	50	LIMITED PERM.

M – REEF MARGIN

I – REEF INTERIOR

In limestone reefs the most common subsurface cement is medium to coarsely crystalline spar calcite (usually ferroan). There is often a close, casual relationship between the occurrence of stylolites (chemical compaction) and the occlusion of porosity by these calcite cements. In the dolomitized "Presqu'ile" reef complex, subsurface (hydrothermal) cements include, from oldest to youngest, ferroan dolomite, ferroan calcite, quartz, galena, and sphalerite.

The major influences on primary porosity are cementation and solution. The relative intensity of either of these processes ultimately determines final reservoir quality. The following conclusions are drawn from analysis of Middle and Upper Devonian reefs.

1) *Leduc Reefs (Golden Spike)*—70 percent of the cement volume in reef margin facies formed in the submarine environment and porosity in these facies is poor (7 percent). Reef interiors are affected equally by submarine and subaerial cementation and porous beds are commonly interbedded with impermeable (well cemented) beds. Average porosity in reef interiors is 12 percent.
2) *Swan Hills Reefs*—Submarine cements account for 25 to 40 percent of the cement volume throughout these reefs. Porosity destruction in reef interiors follows a similar pattern to that observed in Golden Spike. Subsurface cementation ultimately reduced porosity in the reef interior to 9 percent, while reef margin facies retained 15 percent porosity.
3) *Rainbow Reefs*—75 percent of the cement volume in most reef margins formed during submarine diagenesis. These cements created rigid reef rocks penecontemporaneous with reef growth and deposition. Reef interiors contain about equal amounts of submarine and subaerial cements. Subsurface cements are less than 15 percent of the total cement volume. The relative lack of subsurface cementation helped to preserve porosity that survived early cementation. Reef interiors are more porous (13 percent) than margins (8 percent).
4) *"Presqu'ile" Reefs*—It is the subsurface diagenetic history that sets these reefs apart from many other Devonian reefs. Subaerial dolomite

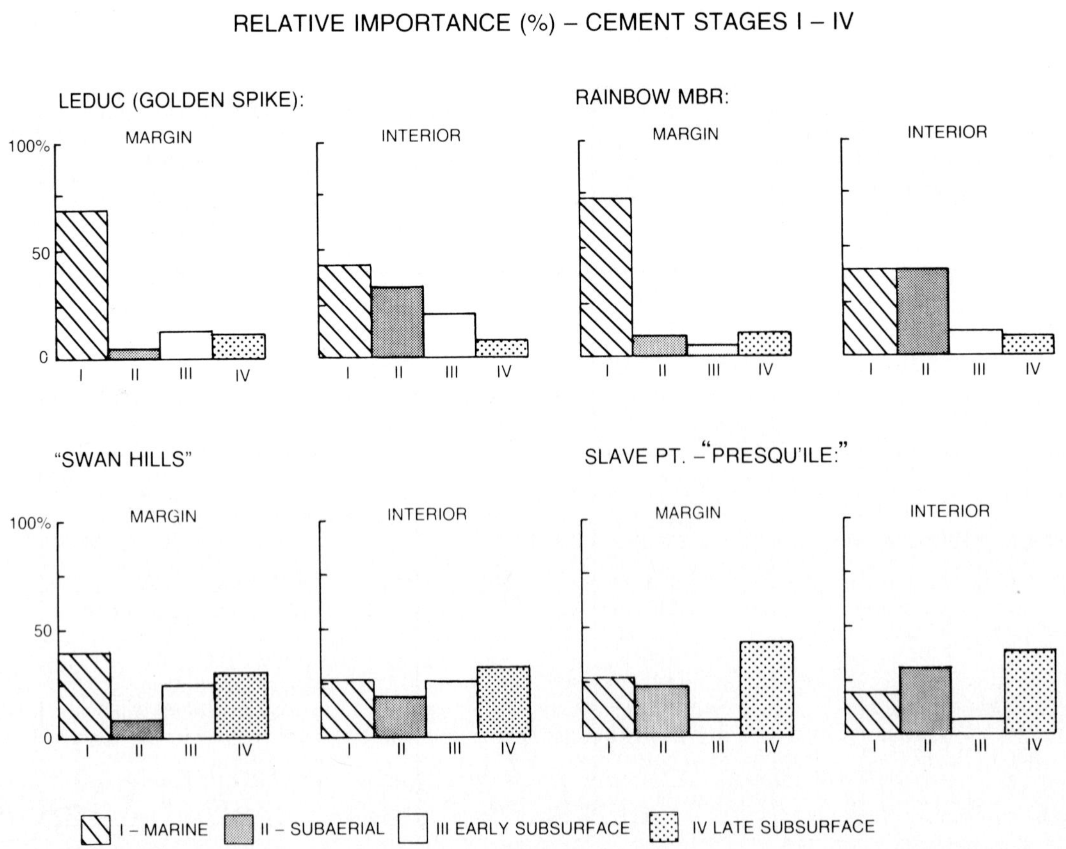

FIG. 23.—Relative abundance of cements formed in each diagenetic stage for reef interiors versus reef margins.

and calcite cement account for 20 percent to 30 percent of the cement volume, while subsurface cements, many of a hydrothermal origin, account for up to 55 percent of the cement volume. These dolomitized reefs have an average porosity of 10 percent and 7 percent in the margin and interior respectively.

ACKNOWLEDGMENTS

This paper is the result of several years accumulated studies conducted while the senior author was at Shell Canada Resources, Ltd., Canadian Hunter Exploration, and Petrosec Exploration, Ltd. We thank all these companies for their support and technical assistance. Gulf Canada, Ltd. provided drafting, clerical, and photographic assistance for the final manuscript. Randy Hughes assisted in much of the drill core photography, and Gilles Morin provided numerous samples, photos, and data on the Carson Creek, North oil field. The continued technical and unending moral support of Hobart W.C. Furbunch is acknowledged. The paper was reviewed by Jack Wendte whose helpful comments and constructive criticisms are greatly appreciated.

REFERENCES

ANDRICHUCK, J.M., 1958, Stratigraphy and facies analysis of Upper Devonian Reefs in Leduc, Stettler and Redwater Area, Alberta: Amer. Assoc. Petroleum Geologists Bull., v. 42, p. 1–93.

______, 1961, Stratigraphic evidence for tectonic and current control of Upper Devonian reef sedimentation, Duhamel area Alberta Canada: Amer. Assoc. Petroleum Geologists Bull., v. 45, p. 612–632.

BARRS, D.L., COPELAND, A.B., AND RITCHIE, W.D., 1970, Geology of Middle Devonian reefs, Rainbow area, Alberta, Canada, *in* M.T. Halbouty (ed.), Geology of Giant Petroleum Fields: Amer. Assoc. Petroleum Geologists Mem. 14, p. 19–49.

BASSETT, H.G., AND STOUT, J.G., 1967, Devonian of Western Canada, *in* International Symposium on the Devonian System, Calgary, v.1: Alberta Soc. Petroleum Geologist, p. 717–752.

BATHURST, R.G.C., 1975, Carbonate sediments and their diagenesis: Elsevier Publ., Amsterdam Developments in Sedimentology 12, 658 p.

______, 1980, Deep crustal diagenesis in limestones: Revista del Instituto de Investigaciones Geologicas Deputacion Provincial, Universidad de Barcelona, v. 34, p. 89–100.

BEALES, F.W., AND JACKSON, S.A., 1968, Pine Point—a stratigraphical approach: Can. Min. and Metal. Bull., v. 61 (675), p. 867–878.

BEBOUT, D.G., AND MAIKLEM, W.R., 1973, Ancient anhydrite facies and environments, Middle Devonian Elk Point Basin, Alberta: Bull. Canadian Petroleum Geology, v. 21, p. 287–343.

BELYEA, H.R., 1965, Upper Devonian of Western Canada, *in* Geological history of Western Canada Basin: Jour. Alberta Soc. Petroleum Geologists, p. 60–88.

BRYANT, J.K., 1960, Stettler field D-2 pool, Oil Fields of Alberta: Alberta Soc. Petroleum Geologists, p. 198.

BURROWES, G., 1977, Sedimentation and diagenesis of back reef deposits, Miette and Golden Spike buildups, Alberta: Unpub. M.Sc. Thesis, McGill University, Montreal.

CAMPBELL, N., 1967, Tectonics, reefs and stratiform lead-zinc deposits of the Pine Point area, Canada: Econ. Geol., Mem. 3, p. 59–70.

CHESHIRE, S.G., AND KEITH, J.W., 1977, Meekwap Field—A Nisku (Upper Devonian) shelf edge reservoir, *in* I. McIreath and R. Harrison (eds.), The Geology of Selected Carbonate Oil, Gas and Lead-Zinc Reservoirs in Western Canada: Can. Soc. Petroleum Geology, p. 107.

CHEVRON STANDARD LIMITED, EXPLORATION STAFF, 1979. The geology, geophysics and significance of the Nisku reef discoveries, West Pembina Area, Alberta, Canada: Bull. Canadian Petroleum Geology, v. 27, p. 326–359.

CHOQUETTE, P.W., AND PRAY, L.C., 1970, Geological nomenclature and classification of porosity in sedimentary carbonates: Amer. Assoc. Petroleum Geologists Bull., v. 54, p. 207–250.

COMBY, O., 1975, Diagenese et compaction mesogenestiques des roches carbonates application au complex recifal de Clarke Lake, Colombie Britannique, Canada: Ph.D. thesis, L'Universite Paris, Paris.

DAVIES, G.R., 1977, Former magnesian calcite and aragonite submarine cements in upper Paleozoic reefs of the Canadian Arctic: A summary: Geology, v. 5, p. 11–15.

______, AND KROUSE, H.R., 1975, Carbon and oxygen isotopic composition of the late Paleozoic calcite cements, Canadian Arctic Archipelago—Preliminary results and interpretation: Geol. Surv. Canada Paper 75–1, Part B, P. 215–220.

DAVIS, T.L., 1972, Velocity-density variation around Keg River Reefs, Alberta: Jour. Canadian Soc. Exploration Geophysicists: v. 8, p. 1–14.

DEBOER, R.B., 1977, On the thermodynamics of pressure solution—interaction between chemical and mechanical forces: Geochem. Cosmochim. Acta., v. 41, p. 249–256.

DOWNING, J.G., AND COOKE, D.Y., 1955, Distribution of reefs of Woodbend Group in Alberta, Canada: Amer. Assoc. Petroleum Geologists Bull., v. 39, p. 189–206.

DRAVIS, J., 1979, Rapid and widespread generation of recent oolitic hardgrounds on a high energy Bahamian platform, Eleuthera Bank, Bahamas: Jour. Sed. Petrology, v. 49, p. 195–207.

DUNNINGTON, H.V., 1967, Aspects of diagenesis and shape change in stylolitic limestone reservoirs: World Petroleum Cong. Proc., 7th, v. 2, p. 339–352.

DUNSMORE, H.E., 1971, Diagenetic model for Middle Devonian Keg River Formation, Rainbow area, northwestern Alberta: unpubl. M.Sc. thesis, Univ. Calgary.

ENGELDER, T., GEISER, P.A., AND ALVAREZ, K., 1981, Role of pressure solution and dissolution in geology: Geology, v. 9, p. 44–45.
EVANS, H., 1972, Zama—a geophysical case history, *in* R.E. King (ed.), Stratigraphic Oil and Gas Fields: Amer. Assoc. Petroleum Geologists Mem. 96, p. 440–452.
FISCHBUCH, N.R., 1968, Stratigraphy, Devonian Swan Hills reef complexes of Central Alberta: Bull. Canadian Petroleum Geology, v. 16, p. 446–587.
______, AND HAVARD, C.J., 1977, Facies-porosity relationships and diagenetic processes which modify them, Swan Hills area, Alberta, *in* I. McIlreath and R. Harrison (eds.), The Geology of Selected Carbonate Oil, Gas, and Lead-Zinc Reservoirs in Western Canada: Canadian Soc. Petroleum Geology, p. 41–65.
FOLK, R.L., 1974, The natural history of crystalline calcium carbonate: Effects of magnesium content and salinity: Jour. Sed. Petrology, v. 44, p. 40–54.
FRIEDMAN, G.M., AMIEL, A.J., AND SCHNEIDERMANN, N., 1974, Submarine cementation in reefs: Example from the Red Sea: Jour. Sed. Petrology, v. 44, p. 816–825.
GERHARD, L.C., 1977, Submarine Cementation of grainstone fabric, St. Croix, U.S. Virgin Islands: Atoll Res. Bull., no. 196, 10 p.
GINSBURG, R.N., AND JAMES N.P., 1976, Submarine botryoidal aragonite in Holocene reef limestones, Belize: Geology, v. 4, p. 431–436.
______, MARSZALEK, D.S., AND SCHNEIDERMANN, N., 1971, Ultrastructure of carbonate cements in a Holocene algal reef of Bermuda: Jour. Sed. Petrology, v. 41, p. 472–482.
______, AND SCHROEDER, J.H., 1973, Growth and submarine fossilization of algal cup reefs, Bermuda: Sedimentology, v. 20, p. 575–614.
GRAY, F.F., AND KASSUBE, J.R., 1963, Geology and stratigraphy of Clarke Lake Gas Field, Northeastern British Columbia: Amer. Assoc. Petroleum Geologists Bull., v. 47, p. 467–483.
HAVARD, C., AND OLDERSHAW, A., 1976, Early diagenesis in back-reef sedimentary cycles, Snipe Lake reef complex, Alberta: Bull. Canadian Petroleum Geology, v. 24, p. 27–69.
HOPKINS, J.C., 1977, Production of foreslope breccias by differential submarine cementation and downslope displacement of carbonate sands, Miette and Ancient Wall buildups, *in* H.E. Cook and P. Enos (eds.), Basin Facies in Carbonate Rocks: Soc. Econ. Paleontologists and Mineralogists, Spec. Pub. 25, p. 155–170.
JAMES, N.P., GINSBURG, R.N., MARSZALEK, D.S., AND CHOQUETTE, P.W., 1976, Facies and fabric specificity of early subsea cements in shallow Belize (British Honduras) reefs: Jour. Sed. Petrology, v. 46, p. 523–544.
______, AND GINSBURG, R.N., 1979, The Seaward Margin of Belize Barrier and Atoll Reefs: Int. Assoc. of Sedimentologists, Spec. Publ. no. 3, Blackwell Sci. Publ., Oxford, 191 p.
JARDINE, D., AND WILSHART, J.W., 1982, Carbonate reservoir description: Soc. Petroleum Engineers, 10010, p. 1–13.
KENDALL, A.C., AND TUCKER, M.E., 1973, Radiaxial fibrous calcite: A replacement after acicular carbonate: Sedimentology, v. 20, p. 365–389.
KLOVAN, J.E., 1964, Facies analysis of the Redwater reef complex, Alberta, Canada: Bull. Canadian Petroleum Geology, v. 12, p. 1–100.
______, 1974, Development of Western Canadian reefs and comparison with Holocene analogues: Amer. Assoc. Petroleum Geologists Bull., v. 58, p. 787–799.
KREBS, W., 1967, Reef development in the Devonian of the eastern Rhenish Slate Mountains, Germany, *in* D.H. Oswald (ed.), International Symposium on the Devonian System: Alberta Soc. Petroleum Geologists, Calgary, v. 2, p. 295–307.
______, 1969, Early void-filling cementation in Devonian fore-reef limestones (Germany): Sedimentology, v. 12, p. 279–299.
______, AND MOUNTJOY, E.W., 1972, Comparison of central European and western Canadian Devonian reef complexes: 24th Int. Geol. Cong., Sect. 6, p. 294–309.
LAND, L.S., AND GOREAU, T.F., 1970, Submarine lithification of Jamaican reefs: Jour. Sed. Petrology, v. 40, p. 457–462.
LANGTON, J.R., AND CHIN, G.E., 1968a, Rainbow Member facies and related reservoir properties, Rainbow Lake, Alberta: Amer. Assoc. Petroleum Geologists Bull., v. 52, p. 1925–1955.
______, AND ______, 1968b, Rainbow Member Facies and related reservoir properties Rainbow Lake, Alberta: Bull. Canadian Petroleum Geology, v. 16, p. 104–143.
LEAVITT, E.M., 1968, Petrology, paleontology, Carson Creek North reef complex, Alberta: Bull. Canadian Petroleum Geology, v. 16, p. 298–413.
MACINTYRE, I.G., 1977, Distribution of submarine cements in a modern Caribbean fringing reef, Galeta Point, Panama: Jour. Sed. Petrology, v. 47, p. 503–516.
______, AND GLYNN, P.W., 1976, Evolution of modern Caribbean fringing reef, Galeta Point, Panama: Amer. Assoc. Petroleum Geologists Bull., v. 60, p. 1054–1072.
MARSHALL, J.F., AND DAVIES, P.J., 1981, Submarine lithification on windward reef slopes: Capricorn Bunker Group, Southern Great Barrier Reef: Jour. Sed. Petrology, v. 51, p. 953–960.
MAZZULLO, S.J., 1977, Synsedimentary diagenesis of reefs: Soc. Econ. Paleontologists and Mineralogists, Permian Basin Sect. Pub. no. 77–16, v. 1, p. 323–354.
______, AND CYS, J.M., 1977, Submarine cements in Permian boundstones and reef-associated rocks, Guadalupe mountains, West Texas and Southeastern New Mexico: Soc. Econ. Paleontologists and Mineralogists, Permian Basin Sect. Pub. no. 77–16, v. 1, p. 151–200.
______, AND ______, 1979, Marine aragonite sea-floor growths and cements in Permian phylloid algal mounds, Sacramento Mountains, New Mexico: Jour. Sed. Petrology, v. 49, p. 917–936.
MCCAMIS, J.G., AND GRIFFITH, L.S., 1967, Middle Devonian facies relationships, Zama area, Alberta: Bull. Canadian

Petroleum Geology, v. 15, p. 434–467.

McCrossan, R.G., and Glaister, R.P. (eds.), 1964, Geological history of Western Canada: Alberta Soc. Petroleum Geologists, Spec. Vol., Calgary, 232 p.

McGillivray, J.G., and Mountjoy, E.W., 1975, Facies and related reservoir characteristics, Golden Spike reef complex, Alberta: Bull. Canadian Petroleum Geology, v. 23, p. 753–809.

Meyers, W.J., 1974, Carbonate cement stratigraphy of the Lake Valley Formation (Mississippian), Sacramento Mountains, New Mexico: Jour. Sed. Petrology, v. 44, p. 837–863.

______, and Hill, B.E., 1983, Quantitative studies of compaction in Mississippian skeletal limestones: Jour. Sed. Petrology, v. 53, p. 595–618.

Moore, C.H., Jr., 1973, Intertidal carbonate cementation, Grand Cayman, West Indies: Jour. Sed. Petrology, v. 43, p. 591–602.

Mossop, G.D., 1972, Origin of the peripheral rim, Redwater reef, Alberta: Bull. Canadian Petroleum Geology, v. 20, p. 238–280.

Mountjoy, E.W., and Krebs, W., 1983, Diagenesis of Devonian reefs and buildups, Western Canada and Europe—a comparison: Z. deutsch. geol. Ges., v. 134, p. 5–60.

______, and Walls, R.A., 1977, Some examples of early submarine cements from Devonian buildups of Alberta: Third Internat. Symp. on Coral Reefs: Western Atlantic reefs, Miami, p. 155–161.

Mullins, H.T., Newman, A.C., Wilbur, J.R., and Boardman, M.R., 1980, Nodular carbonate sediment on Bahamian slopes: Possible precursors to nodular limestones: Jour. Sed. Petrology, v. 50, p. 117–131.

O'Conner, M.J., and Gretener, P.E., 1974, Differential compaction within the Woodbend Group of Central Alberta: Bull. Canadian Petroleum Geology, v. 22, p. 269–304.

Purdy, E.G., 1968, Carbonate diagenesis: An environmental survey: Geol. Roma, VII, p. 183–228.

Purser, B.H., 1969, Syn-sedimentary marine lithification of Middle Jurassic limestones in the Paris Basin: Sedimentology, v. 12, p. 205–230.

Reitzel, G.A., and Callow, G.O., 1976, Pool description and performance analysis leads to understanding Golden Spike's miscible flood: Petroleum Society, C.I.M., Paper n. 7622.

Roedder, E., 1968, Temperature, salinity, and origin of the ore-forming fluids at Pine Point Northwest Territories, Canada, from fluid inclusion studies: Econ. Geol., v. 63, p. 439–450.

Ross, R.J., Jr., Jaanusson, V., and Friedman, I., 1975, Lithology and origin of Middle Ordovician calcareous mudmound at Meiklejohn Peak, Southern Nevada: U.S. Geol. Survey Prof. Paper 871, 48 p.

Schmidt, V., 1971, Early carbonate cementation in Middle Devonian bioherms Rainbow Lake, Alberta, in O.P. Bricker (ed.), Carbonate cements: Johns Hopkins Univ. Geology Studies no. 19, p. 209–215.

______, McDonald, D.A., and McIlreath, L.A., 1977, Growth and diagenesis of Middle Devonian Keg River cementation reefs, Rainbow Field, Alberta, *in* I. McIlreath and R. Harrison (eds.), The Geology of Selected Carbonate Oil, Gas, and Lead-Zinc Reservoirs in Western Canada: Can. Soc. Petroleum Geology, p. 39.

Scholle, P.A., and Kinsman, D.J., 1974, Aragonite and high Mg-calcite caliche from the Persian Gulf—A modern analog for the Permian of Texas and New Mexico: Jour. Sed. Petrology, v. 44, p. 904–916.

Schroeder, J.H., 1972a, Fabrics and sequences of submarine carbonate cements in Holocene Bermuda cup reefs: Geol. Rundschau, v. 61, p. 708–730.

______, 1972b, Calcified filaments of an endolithic alga in Recent Bermuda reefs: N. Jb. Geol. Paläont Mh., v. 1, p. 16–33.

Sellier, E., 1979, Contribution à l'étude pétrologique experimentale de la genèse des stylolithes dans les formations calcaires du basin d'Aquitaine. Thèse non publiée, Univ. Bordeaux III, 300 p.

Shinn, E.A., 1969, Submarine lithification of Holocene carbonate sediments in the Persian Gulf: Sedimentology, v. 12, p. 109–144.

______, and Robbin, D.M., 1983, Mechanical and chemical compaction in fine-grained, shallow water limestones: Jour. Sed. Petrology, v. 53, p. 595–618.

Sibley, D.F., and Murray, R.C., 1972, Marine diagenesis of carbonate sediment, Bonaire, Netherlands Antilles: Jour. Sed. Petrology, v. 42, p. 168–178.

Skall, H., 1975, The paleoenvironment of the Pine Point lead-zinc district: Econ. Geol., v. 70, p. 22–45.

______, 1977, The geology of the Pine Point barrier reef complex *in* I. McIlreath and R. Harrison (eds.), The Geology of Selected Carbonate Oil, Gas and Lead-Zinc Reservoirs in Western Canada: Can. Soc. Petroleum Geology, p. 19–38.

Stoakes, F.G., 1977, The Camrose Member of the Ireton Formation in the Stettler area of Central Alberta *in* I. McIlreath and R. Harrison (eds.), The Geology of Selected Carbonate Oil, Gas, and Lead-Zinc Reservoirs in Western Canada: Can. Soc. Petroleum Geology, p. 89–106.

______, 1979, Nature and control of shale basin fill and its effect on reef growth and termination: Upper Devonian Duvernay and Ireton Formations of Alberta, Canada: Bull. Can. Petroleum Geology, v. 28, p. 345–410.

Taylor, J.C.M., and Illing, L.V., 1969, Holocene intertidal calcium carbonate cementation, Qatar, Persian Gulf: Sedimentology, v. 12, p. 69–107.

Walls, R.A., 1977, Cementation history and porosity development, Golden Spike reef complex (Devonian), Alberta: Ph.D. Thesis, McGill Univ., Montreal, 307 p.

______, Mountjoy, E.W., and Fritz, P., 1979, Isotopic composition and diagenetic history of carbonate cements in Devonian Golden Spike reef. Alberta, Canada: Geol. Soc. America Bull., v. 90, p. 963–982.

Wanless, H.R., 1979, Limestone response to stress: pressure solution and dolomitization: Jour. Sed. Petrology, v. 49, p. 437–462.

Wardlaw, N.C., and Reinson, G.E., 1971, Carbonate and evaporite deposition and diagenesis, Middle Devonian, Winnepegosis and Prairie Evaporite Formations of south-central Saskatchewan: Amer. Assoc. Petroleum Geologists Bull., v. 55, p. 1759–1787.

WENDTE, J., 1974, Sedimentation and diagenesis of the Cooking Lake Platform and Lower Leduc reef facies, Upper Devonian, Redwater, Alberta: Unpubl. Ph.D. Thesis, University of California, Santa Cruz, 221 p.

______, AND STOAKES, F., 1982, Evolution and corresponding porosity distribution of the Judy Creek reef complex, Upper Devonian Central Alberta, *in* W.G. Cutler (ed.), Canada's Giant Hydrocarbon Reservoirs: Amer. Assoc. Petroleum Geologists, Core Conf., Calgary, p. 63–81.

WILSON, J.L., 1975, Carbonate Facies in Geologic History: Springer-Verlag, New York, 471 p.

WONG, P.K., 1978, Sedimentology and diagenesis of the Upper Devonian Kaybob stratigraphic reef: Unpubl. MSc. Thesis, Univ. of Calgary, 207 p.

______, AND OLDERSHAW, A., 1981, Burial cementation in the Devonian, Kaybob reef complex, Alberta, Canada: Jour. Sed. Petrology, v. 51, p. 507–520.

ZANKL, H., 1971, A model of sedimentation and diagenesis in a Triassic reef, *in* O.P. Bricker (ed.), Carbonate cements: Johns Hopkins Univ. Geology Studies, no. 19, p. 189–192.

SECTION THREE: INTRODUCTION

Examples of integrated studies on regional distribution of burial cements are presented in this section. The transition from very early meteoric realms of cementation into deeper burial environments in Mississippian skeletal limestones of New Mexico is described by Meyers and Lohmann who use extensive isotopic and trace element control to characterize four regionally extensive cement zones. Trends in their data reflect progressive stages in the chemical evolution of the water-rock system.

Cementation of the well studied Cretaceous Stuart City trend of south-central Texas is described by Prezbindowski using geochemical and petrographic criteria to discriminate among various cementation events. His studies indicate that marine, near-surface meteoric and shallow to intermediate burial cements significantly reduce porosity. Late-stage cementation is less important, as it occurs only along localized faults and fractures. In a companion paper, Woronick and Land describe late burial cements of various mineralogies in the Lower Cretaceous Pearsall and Lower Glen Rose Formations of south Texas. Using a variety of geochemical data, they identify relative contributions of various subsurface fluids to the cement history.

Combined evidence from fluid inclusions in calcite fracture-fill cements with regional geological considerations and burial histories allows Burruss, Cercone and Harris to develop a sequence of cementation and hydrocarbon migration episodes in Cretaceous limestones from the Middle East. Burial cements in Jurassic Smackover grainstone reservoirs of the central Gulf Coast, their mineralogy and cementation histories are summarized by Moore. Primary porosity in these grainstones was partially filled by two distinct cement types: (1) iron-zoned mosaic calcites related to progressive regional grain-to-grain pressure solution; followed by (2) iron-rich unzoned poikilitic calcites resulting from post-compaction regional subsurface solution. The entire problem of burial diagenesis and the progressive loss of porosity with burial depth is discussed by Scholle and Halley, using data primarily from southern Florida as an example. They suggest important porosity-reducing processes, acting continuously from the surface to great depth, which include physical and chemical compaction and cementation. Calcareous shale sections or marl interbeds act as significant sources of dissolved carbonate which is precipitated as cement in adjacent limestones.

N.S., P.M.H.

ISOTOPE GEOCHEMISTRY OF REGIONALLY EXTENSIVE CALCITE CEMENT ZONES AND MARINE COMPONENTS IN MISSISSIPPIAN LIMESTONES, NEW MEXICO

WILLIAM J. MEYERS AND KYGER C. LOHMANN
Department of Earth and Space Sciences, State University of New York
at Stony Brook, Stony Brook, New York 11794, and
Department of Geological Sciences, The University of Michigan,
Ann Arbor, Michigan 48109

ABSTRACT

Carbonate cements from Mississippian skeletal limestones of southern New Mexico are dominated by echinoderm-syntaxial calcites that comprise four regionally extensive compositional zones. Previous petrography and cement stratigraphy proves that the oldest three zones (zones 1, 2, 3) are pre-Pennsylvanian in age and were precipitated at temperatures around 25°C; they are interpreted as meteoric phreatic cements. The youngest cement, zone 5, is interpreted to be a pre-Permian burial cement precipitated at burial depths less than 1 km and temperatures less than about 60°C.

Originally defined on the basis of luminescent and staining characteristics, each zone has a distinctive assemblage of $\delta^{13}C$, $\delta^{18}O$, Mg^{2+}, Mn^{2+} and Fe^{2+} contents. The low MgO contents (less than 0.25 wt.%) in all zones indicates that seawater was an insignificant component of their precipitational waters over most of the region. Their FeO and MnO contents are compatible with their subsurface interpretation.

Mississippian marine carbonate $\delta^{13}C(\sim +4.0‰$ PDB) and $\delta^{18}O(-1.5‰$ PDB) are estimated from analysis of micrite, synsedimentary marine cement, and crinoid skeletal carbonate. In addition to providing a base line with which to compare non-marine cements, this estimate implies that the composition of Mississippian epeiric seawater was enriched in ^{13}C and comparable in ^{18}O to modern seawater.

The pre-Pennsylvanian cements (zones 1, 2 and 3) are markedly distinct from one another and show a progressive decrease in $\delta^{13}C$ and $\delta^{18}O$ through time ($\delta^{18}O = -1.3, -2.8, -3.7‰$ PDB respectively; $\delta^{13}C = +3.7, +2.4, -0.8‰$ PDB respectively). We propose a model in which zones 1, 2 and 3 precipitated in meteoric phreatic groundwaters largely uncontaminated by seawater. The distinctive trend of pre-Pennsylvanian cements reflects progressive stages in the chemical evolution of the water-rock system. This involved a decrease in the degree of rock-water interaction, which in turn resulted from depletion of metastable mineralogies through recrystallization or through isolation by cementation. Thus, the progressive decrease in ^{13}C reflects equilibration of porewaters with Mississippian skeletal carbonates during early cementation (zones 1 and 2) and increased contribution from soil-gas derived carbon during late cementation (zone 3). Similarly, the progressive decrease in ^{18}O in zones 1 through 3 reflects decreasing contribution of ^{18}O from Mississippian marine carbonates and progressive equilibration with meteoric porewaters. Hence, the isotopic composition of zone 3 cements reflects equilibrium precipitation with meteoric waters. Importantly, decreasing rock-water interaction during pre-Pennsylvanian cementation is supported by decreasing Mg^{2+} contents among zones. High magnesian calcite crinoids are the dominant skeletal component in the Mississippian limestones and likely were the major contributors of carbonate for cementation during diagenesis of all three zones. A progressive decrease in Mg^{2+} content is compatible with a model of diminishing rock-water interaction. Zones 1 and 2 have the heaviest isotopic compositions and highest Mg^{2+} contents. The very low Mg^{2+} content of zone 3 is predicted from very light $\delta^{13}C$ compositions which predicates only minor contribution from crinoidal calcite sources.

Post-Mississippian zone 5 cement has a wide range of $\delta^{18}O$ values (mean $= -7.4‰$ PDB), all more depleted than any pre-Pennsylvanian cement. This light $\delta^{18}O$ probably reflects elevated temperatures in the 40–60°C range. Zone 5 has intermediate $\delta^{13}C$ values which reflect complex predominantly "rock" carbon sources, many of which were probably extraformational as suggested by stratigraphic variations in Mn^{2+} and Fe^{2+}.

INTRODUCTION

A major goal in modern research on carbonate diagenesis is to establish models for cementation on the scale and in tectonic/stratigraphic settings appropriate to widespread epeiric and pericontinental strata. To this end, a particularly fruitful approach has been to analyze trace element and stable carbon and oxygen isotopes of individual limestone components, and further to analyze separately each major diagenetic component and each individual cement zone (e.g., Walls, et al., 1979; Dickson and Coleman, 1980; Moore and Druckman, 1981; Marshall and Ashton, 1980; Meyers and Lohmann, 1980). This paper applies this approach to develop time-sequence reconstructions of the chemistries and flow directions of paleo-groundwater systems that precipitated four major regionally extensive calcite cement zones in Mississippian skeletal limestones of southern New Mexico.

We first summarize background geology and petrography of the Mississippian limestones, briefly discuss the main objectives of our geochemical work, and then present and discuss our results to date.

STRATIGRAPHY OF MISSISSIPPIAN ROCKS

The Mississippian limestones within the study area (Fig. 1) consist of the Caballero Formation (Kinderhookian), the Lake Valley Formation (Osagean), Caloso and Kelly Formations (both Osagean), and the Rancheria Formation (Meramecian-Chesterian) (Laudon and Bowsher, 1941, 1949; Pray, 1961; Armstrong, 1962; Lane, 1974, 1982; Yurewicz, 1977). The stratigraphy is interpreted to represent a lower transgressive interval and an upper progradational interval (Meyers, 1978). The lowermost interval consists mainly of echinoderm-bryozoan limestone of the Andrecito Member of the Lake Valley Formation overlain by deeper-water lime mudstones of the Alamogordo Member. This in turn is overlain by argillaceous echinoderm-bryozoan calcarenites of the Nunn Member, which represent a transition into progressively shallower-water echinoderm-bryozoan packstones and grainstones of the overlying Tierra Blanca Member. The Tierra Blanca represents progradation and depositional shallowing of totally subtidal shallow-water skeletal sand shelf facies.

In the San Andres and Sacramento Mountains a younger deep-water lime mudstone-echinoderm packstone couplet, the Arcente and Dona Ana Members, was deposited on top of the Tierra Blanca. The Rancheria Formation occurs only in the San Andres and Sacramento Mountains. It consists of lime mudstones to fine-grained grainstones (Yurewicz, 1977) and is separated from the Lake Valley Formation by an erosional unconformity (Meyers, 1974).

Mississippian limestones throughout most of the region are overlain by a major pre-Pennsylvanian unconformity that is marked by karstic breccias containing chert and limestone fragments derived from the Lake Valley and Kelly Formations. This unconformity is an extremely important feature in the late diagenetic history of the Mississippian limestones.

The bulk of the Mississippian limestones were deposited in subtidal environments that notably lack cyclic peritidal facies; therefore, their diagenetic history is relatively simple because they were not subjected to multiple diagenetic overprinting by fresh water lenses associated with shoreline migrations.

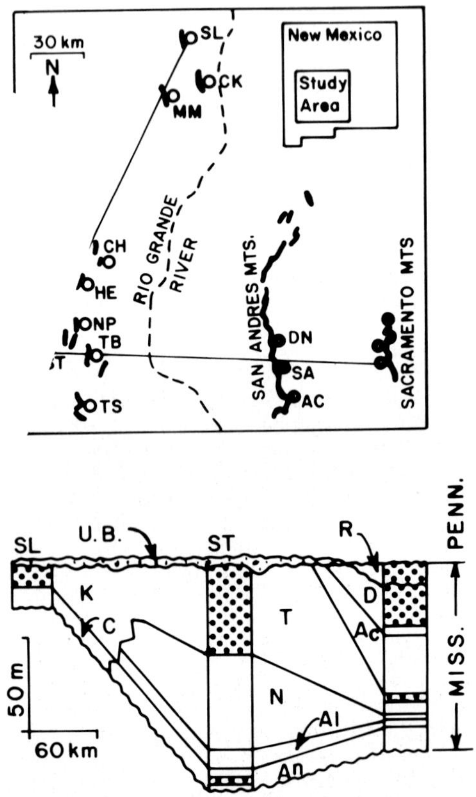

FIG. 1.—Location map and stratigraphic cross section of Mississippian strata in southern New Mexico. Outcrops of Mississippian strata are in black. Open circles on map indicate location of measured sections; closed circles indicate sections of the Sacramento and San Andres Mountains examined in this study. Lines define the line of the cross-section. K = Kelley Fm. (Osagean). C = Caloso Fm. (Osagean). Members of Lake Valley Fm. (Osagean): An = Andrecito, Al = Alamogordo, N = Nunn, T = Tierra Blanca, Ac = Arcente, D = Dona Ana. R = Rancheria Fm. (Meramecian-Chesterian). UB = Pre-Pennsylvanian unconformity breccia containing clasts of Mississippian chert and limestone. Dotted pattern on columnar sections is skeletal packstone and grainstone; unpatterned is skeletal wackestones and mudstones.

CEMENT STRATIGRAPHY

An extensive and detailed cement stratigraphy comprises the essential petrographic framework for detailed geochemical studies (Meyers, 1978). The bulk of the rocks used in this study are echinoderm-bryozoan grainstones and packstones from the Dona Ana, Tierra Blanca, Nunn, and Andrecito Members of the Lake Valley Formation and from the Kelly Formation (Fig. 1). Throughout the several tens of thousands of km^2 of study area, Mississippian limestones have been cemented mainly by syntaxial, inclusion-free calcites on echinoderms which make up about 85% of the total intergranular cements. These inclusion-free, syntaxial cements, which comprise about 26% by volume of the skeletal packstones and grainstones, contain as many as seven compositional zones distinguishable with cathodoluminescence of which four zones are volumetrically important and correlated regionally. These four cement zones, designated zones 1, 2, 3, 5, from oldest to youngest (Fig. 2B), are mappable over all or most of the study area. They are the prime basis of the cement stratigraphy and are the main subject of our geochemical investigations. Of the total inclusion-free intergranular cements, zone 1 comprises about 8%, zone 2 about 40%, zone 3 about 3.5%, and zone 5 about 42% (including 10% granular cement) (Fig. 2A). In addition, about 3% of the intergranular cements comprise zone 2 microdolomite-rich, syntaxial zone 2 cement, interpreted to have formed within a mixing-zone (Meyers and Lohmann, 1978); about 1% comprise microdolomite-rich prismatic calcites interpreted as former magnesian calcite marine cements (Lohmann and Meyers, 1977).

Petrography of the limestones immediately adjacent to pre-Pennsylvanian and pre-Permian unconformities demonstrates that cement zones 1, 2 and 3 are pre-basal Pennsylvanian (pre-Morrowan), and zone 5 is post-Mississippian and probably pre-Permian in age (Meyers, 1978). These age determinations coupled with petrography and cement stratigraphy indicate that zones 1, 2, 3 are shallow meteoric phreatic cements, and zone 5 is a burial phreatic cement. The regional distributions of the pre-Pennsylvanian zones (1, 2 and 3) have suggested a model of cementation in regional north-to-south flowing meteoric phreatic groundwater systems that formed and migrated in response to late Mississippian/early Pennsylvanian eustatic sea-level changes (Meyers, 1978).

Assuming that each cement zone is a "time unit" (Meyers, 1978), this cement stratigraphy provides a unique diagenetic time-stratigraphic framework for examining gradients in the chemistry of individual cement zones over a large region. Furthermore, through the cement stratigraphy and petrography we know the ages of the cement zones and the approximate overburden during their emplacement such that maximum precipitational temperatures can be constrained. Specifically, the pre-Pennsylvanian cements, zones 1, 2 and 3, were precipitated under a maximum overburden of a few hundred meters and zone 5 calcites were probably precipitated under a maximum overburden of about one thousand meters (Meyers, 1978; 1980). With an assumed geothermal gradient of 3.0°C/100 m, this suggests precipitational temperatures of less than about 30°C for zones 1, 2, and 3, and less than about 60°C for zone 5 calcites. These temperature constraints are of obvious importance in estimating trace element and isotopic compositions of precipitational waters from present cement chemistries.

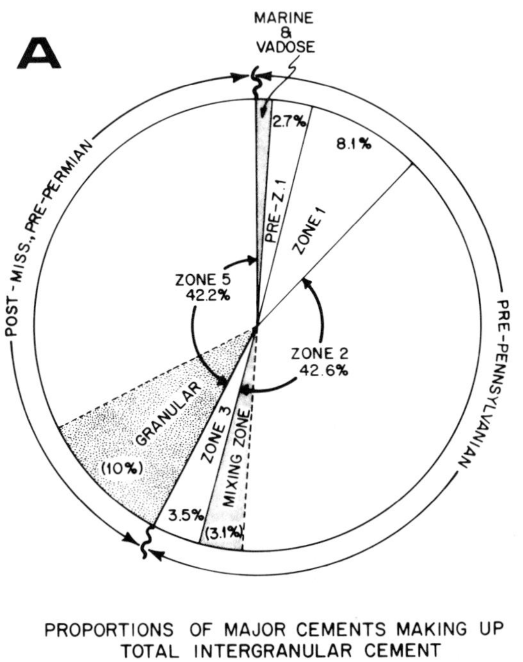

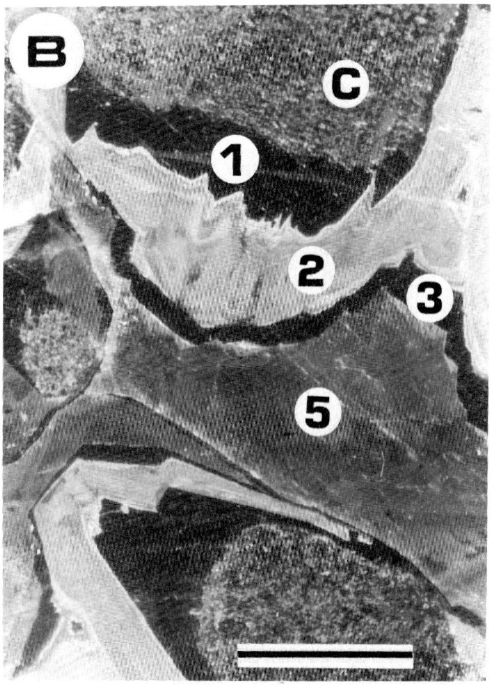

FIG. 2.—*A*, Pie diagram showing mean percentages of major cements, based on point counts of 88 thin sections. All cements are inclusion-free syntaxial cements except those marked "granular", which are inclusion-free granular zone 5 cements; those marked "mixing zone", which are microdolomite-rich syntaxial zone 2 cements; and those labeled "marine", which are microdolomite-rich prismatic cements. *B*, Cathodoluminescent photomicrograph showing the four major regionally extensive cement zones. C = crinoid. 1, 2, 3, 5 = zones 1, 2, 3, 5. Scale bar is 500 microns.

OBJECTIVES OF ISOTOPIC AND TRACE ELEMENT STUDIES

The major goals of our studies on isotope and trace element geochemistry are threefold.

1) The study aims to characterize the isotopic and trace element geochemistry of the four major zones (1, 2, 3 and 5) and to use these data in the context of published equilibrium fractionation factors and homogeneous distribution coefficients to estimate chemistries of precipitational waters. We reason that these cements precipitated from relatively dilute groundwaters, and therefore likely did not experience kinetic fractionation effects. Additionally, we have excellent constraints on precipitational temperatures for zones 1, 2 and 3.
2) This study will address the problem of source(s) of dissolved constitutents in cementing groundwaters through examination of intrasample and intersample variations in trace element and $\delta^{18}O$ and $\delta^{13}C$ composition.
3) This study tests the concept that lateral and vertical gradients in isotope ratios and trace element contents within cement zones reflect similar gradients in the precipitational groundwaters. As such these gradients can be interpreted to infer paleo-flow directions by analogy with chemical gradients in modern groundwaters in carbonate terrains.

This paper focuses mainly on objectives (1) and (2). We do not yet have sufficient data to adequately test for regional or stratigraphic gradients.

TRACE ELEMENT COMPOSITION OF CALCITE CEMENT ZONES

The trace element data is based on electron probe microanalysis of 40 samples from six measured sections. Each sample contains two to four of the major compositional zones; each zone was analyzed by a minimum of five spot analyses, with ten replicates of each spot. Operating conditions for Mg were 10Kv, .015 microamps and about 20 μm spot; and for Fe and Mn, 30 Kv, 0.015 microamps, and 20 micron sized spot. All data were reduced to weight percent oxides using a Bence-Albee procedure modified for fixed stoichiometry and utilizing all carbonate standards. Precision of individual analyses was determined empirically from peak intensity variations among ten replicates of individual spot analyses. Precision at the 95% confidence level for FeO and MnO is approximately 0.005 wt.%, and for MgO is 0.015 wt.%. Volatilization during analysis was evaluated with least squares analysis of peak intensity relative to elapsed time during successive replications.

The regional distribution of our analyses is presently inadequate to test for regional gradients, thus our main purpose in presenting these trace element data is to constrain interpretations of the carbon and oxygen isotopic data. The trace element compositions of the major zones are summarized in Figure 3 and Table 1. The most significant feature is that each zone is chemically distinct when all three trace elements, Mg, Mn, and Fe, are considered. Zone 1 is rich in Mg and poor in Fe and Mn; zone 2 is rich in Mn and intermediate in Fe and Mg; zone 3 is poor in all three trace elements; zone 5 is rich in Fe and Mg, and intermediate in Mn. The fact that each zone is chemically distinct supports the petrographic interpretation that each zone represents a regionally extensive cementation event in a distinct groundwater chemistry. If the zones were lateral yet time-equivalents of one another, there should be more blending and overlap between zonal chemistries.

With these data we can estimate the average aMg^{2+}/aCa^{2+}, aMn^{2+}/aCa^{2+} and aFe^{2+}/aCa^{2+} ratios in their precipitational waters using 25°C for zones 1, 2 and 3 and the homogeneous distribution coefficients from Füchtbauer and Hardie (1976), from Michard (1968), and from Richter and Füchtbauer (1978) for Mg^{+2}, Mn^{+2}, and Fe^{+2} respectively (Table 1). The most significant result of this is the relatively low concentration of Mg in the meteoric porewaters. This we interpret as indicating that the contribution of seawater was absent or insignificant in pre-Pennsylvanian groundwaters. For example, assuming no Mg in the meteoric waters with all the Mg derived from seawater with modern aMg^2/aCa^2 ratio, a mixture of 3% seawater and 97% meteoric water can account for all the Mg in zone 1, the most Mg-rich of the zones. More likely the Mg was derived from crinoids, as the total Mg content of meteoric pre-Pennsylvanian cements can be accounted for by less than 10% loss of Mg from crinoids. Leutloff (1979) measured the present Mg content of crinoids from these rocks by point-counting microdolomite inclusions, and estimated that on average crinoids lost about 25% of their original Mg during diagenesis.

In regard to Mn and Fe, the calculated aMn^{2+}/aCa^{2+} and aFe^{2+}/aCa^{++} ratios are within the ranges of those reported for modern meteoric groundwaters (Hem, 1970; White *et al.*, 1963), and are significantly greater than those for modern oxygenated open marine waters. Similar estimates of element ratios in zone 5 precipitational waters are less certain because of the poorer constraints on temperatures of precipitation, which may have varied from 30°C to 60°C (assuming a pre-Permian age, less than one thousand meters overburden, and 3.0°C/100m geothermal gradient). Assuming a temperature of 45°C, the aMg^2/aCa^2 ratio again is quite low (Table 1) indicating Mg-poor waters.

CARBON AND OXYGEN ISOTOPIC CHEMISTRY

Preliminary examination of carbon-oxygen isotopic chemistry of Mississippian cements concentrated on a suite of fifteen samples from the Sacramento and San Andres Mountains (Figure 1). Paired reflected light and cathodoluminescence photomicrograph mosaics were used as maps to drill out carbonate components from thin-section sized polished slabs employing a micro-

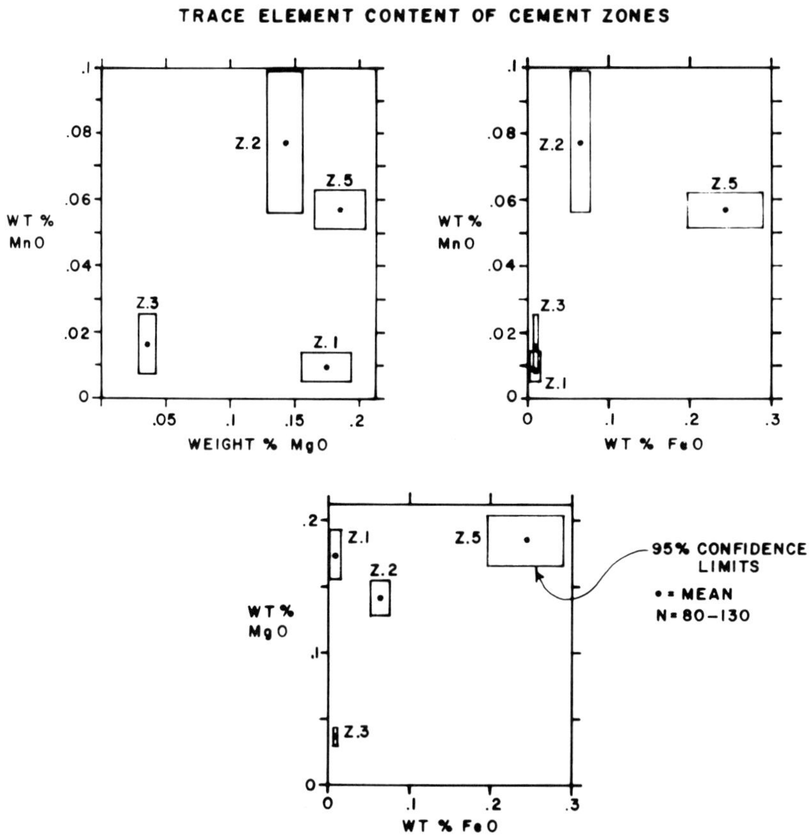

FIG. 3.—Summary of trace element analyses by electron microprobe. Rectangles encompass 95% confidence limits. Z.1, Z.2, Z.3, Z.5 = zone 1, 2, 3, 5.

TABLE 1.—SUMMARY OF TRACE ELEMENT AND ISOTOPE GEOCHEMISTRY OF CEMENT ZONES

	PRE-PENNSYLVANIAN CEMENTS			POST-MISS, PRE-PERMAIN CEMENT
	ZONE 1	ZONE 2	ZONE 3	ZONE 5
Wt. % MgO	0.17 ± .02	0.14 ± .01	0.06 ± .008	0.19 ± .02
Wt. % MnO	0.01 ± .004	0.08 ± .02	0.02 ± .009	0.06 ± .006
Wt. % FeO	0.007 ± .002	0.07 ± .01	0.01 ± .003	0.24 ± .05
δC^{13} PDB	+3.7 ± .2	+2.4 ± .4	−0.8 ± .4	+2.7 ± .5
δO^{18} PDB	−1.3 ± .2	−2.8 ± .9	−3.7 ± .2	−7.4 ± 1.0
ESTIMATES OF WATER CHEMISTRY DURING PRECIPITATION OF CEMENT ZONES				
	ZONE 1*	ZONE 2*	ZONE 3*	ZONE 5**
Mg^{2+}/Ca^{2+}	0.16	0.13	0.06	0.12
Mn^{2+}/Ca^{2+}	3.3×10^{-5}	2.6×10^{-4}	5.7×10^{-5}	5.2×10^{-5}
Fe^{2+}/Ca^{2+}	1.0×10^{-4}	1.0×10^{-3}	1.5×10^{-4}	3.5×10^{-3}
δC^{13} PDB ‰	+0.9	−0.4	−3.6	−0.8
δO^{18} SMOW ‰	+0.6	−0.9	−1.8	−1.8
Mississipian $\delta^{13}C$ HCO_3^- = +2‰ PDB; $\delta^{18}O$ = +0.5‰ SMOW				

*assuming 25°C
**assuming 45°C

scope mounted drill assembly fitted with a drill bit faceted to a twenty micron point. These powders, ranging in size from 0.2 to 0.5 milligrams, were roasted at 380°C for one hour under vacuum to remove volatile organic contamination. Samples were subsequently reacted with anhydrous phosphoric acid at 50°C in a reaction/extraction line coupled directly to the inlet of a VG Micromass 602D ratio mass spectrometer. Isotopic enrichments were converted relative to the PDB carbonate standard and corrected for ^{17}O contribution (Craig, 1957). Analytic precision was monitored through daily analysis of two powdered laboratory standards and through the replication of about 10% of the unknowns; measured precision is 0.08 for $\delta^{18}O$ and 0.05 for $\delta^{13}C$.

Data for crinoids, micrite, synsedimentary marine cements and syntaxial inclusion-free meteoric cements are plotted in Figure 4. Mean compositions and standard deviations (σ) are listed in Table 1. The significant features of these data are: (1) the convergence of trends for crinoids, marine cements and pre-Pennsylvanian cements at a composition of roughly +4.0‰ for $\delta^{13}C$ and −1.5‰ for $\delta^{18}O$; (2) each marine component, crinoids, micrite and marine cement has a distinctive trend reflecting their differing diagenetic reactivities; (3) the isotopic composition of each major non-marine cement is distinct from one another when viewed in carbon-oxygen space, consistent with their unique trace element compositions; (4) the pre-Pennsylvanian cements (zones 1, 2, 3) show a progressive decrease in $\delta^{18}O$ and $\delta^{13}C$ with time; and (5) the post-Mississippian zone 5 has distinctly light $\delta^{18}O$ compared to the pre-Pennsylvanian cements. Calculations of the $\delta^{18}O$ and $\delta^{13}C$ of their precipitating groundwaters are also shown in Table 1, and are based on a constant precipitational temperature of 25°C for all pre-Pennsylvanian cements. Zones 1, 2 and 3, which exhibit a 2.4‰ change in $\delta^{18}O$, could not have formed at depths greater than a few hundred meters. If this change in $\delta^{18}O$ were simply reflecting a change in temperature of precipitation, this would require an increase of about 12°C. However, under the constraints of a shallow burial history, such a temperature increase is geologically unreasonable. Therefore, this variation in $\delta^{18}O$ must reflect a change in the composition of phreatic porewaters. Zone 5 cements formed at much greater burial depths up to 1000 meters, and therefore at estimated temperatures of 30 − 60°C; hence, porewater compositions for zone 5 are calculated for 45°C.

Figure 5 shows variations of carbon and oxygen compositions for crinoids and coexisting cements in the same sample. It is important to note that zone 1 is

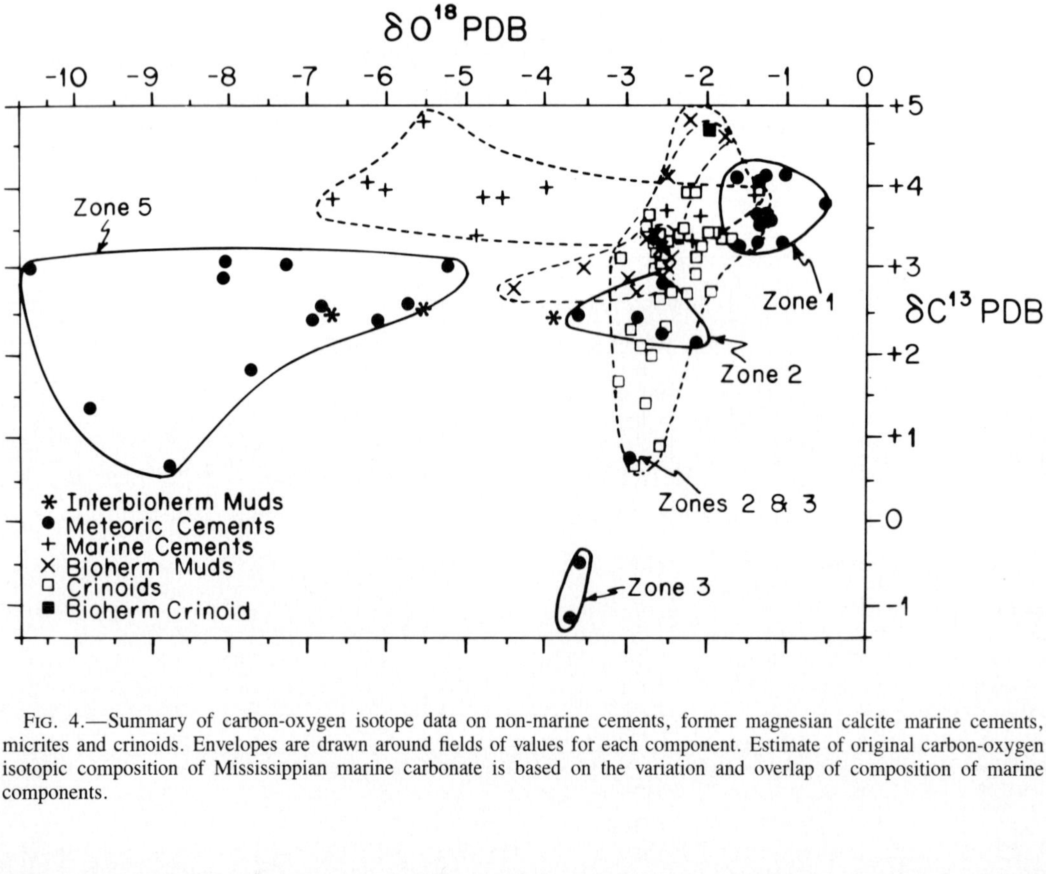

FIG. 4.—Summary of carbon-oxygen isotope data on non-marine cements, former magnesian calcite marine cements, micrites and crinoids. Envelopes are drawn around fields of values for each component. Estimate of original carbon-oxygen isotopic composition of Mississippian marine carbonate is based on the variation and overlap of composition of marine components.

invariably heavier in both carbon and oxygen than coexisting crinoids. Zone 2 is invariably lighter than zone 1, and zone 3 is lighter than zone 2 in both oxygen and carbon, and zone 5 is lighter in oxygen yet heavier in carbon than coeval zone 3.

Estimation of Original Marine Carbonate Composition

In light of numerous studies which have documented significant variation in the isotopic composition of marine carbonates through geologic time (Weber, 1967; Veizer and Hoefs, 1969), an essential first step in all diagenetic studies of ancient carbonate sequences is estimation of the starting composition of the marine carbonates. Such estimates are necessary to provide a baseline relative to which the effects of diagenesis can be evaluated. In this study we have used the overlapping portion of isotopic fields of the marine components, crinoids, micrite and marine cements (Fig. 4) to estimate this composition for the following reasons. Variation in the openness of the diagenetic system during recrystallization of marine cements and skeletal grains result in replacive calcites that range in isotopic composition from primary (closed system) to secondary (open system). Moreover, because skeletal and mineral components have differing diagenetic reactivities, the composition of these secondary calcites reflect a broad range of temperature-porewater conditions. Thus, when numerous marine components are analyzed, an array of isotopic trends will be produced which originate from the primary marine composition and deviate towards numerous secondary calcite compositions. By defining the common field of overlapping composition an initial marine, or near-marine composition can be estimated (Lohmann, in press).

In applying this concept to the data in Figure 4, four samples analyzed from Waulsortian-type bioherms within the Lake Valley possess heavier $\delta^{13}C$ than the majority of marine components. Because of the paucity of data with such heavy compositions, and the dominant convergence of trends at +4.0‰, these samples are interpreted to reflect local perturbations in carbon composition of porewaters, and may be related to anaerobic fermentation-decomposition of organic matter (Irwin *et al.*, 1977). Juxtaposition of all trends indicates a common field of overlap, with the heaviest oxygen and carbon compositions of this field being −1.5‰ $\delta^{18}O$; +4.0‰ $\delta^{13}C$. These values represent our best estimate of Mississippian marine or near-marine carbonate. They are compatible with the heaviest compositions obtained by Brand (1982) for low mag-

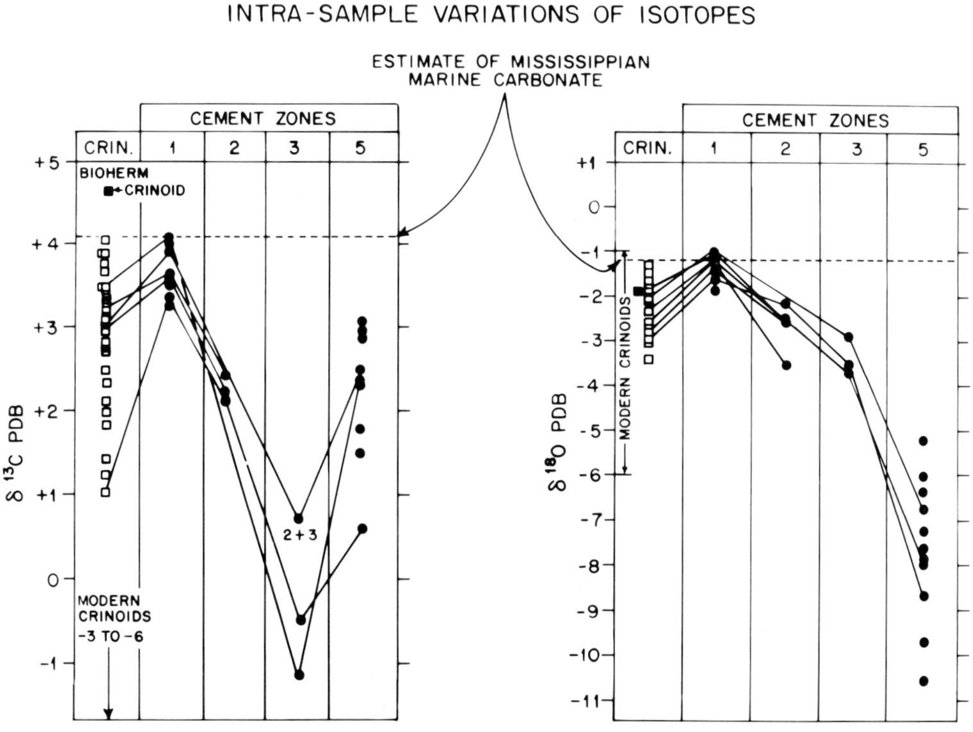

FIG. 5.—Carbon and oxygen isotope date for Mississippian limestone components showing intrasample variations. Lines connect components from the same sample. The dashed line indicates the best estimate for Mississippian marine carbonate composition. Modern crinoid data from Weber and Raup (1968).

nesium calcite (LMC) brachiopods from the Lake Valley Formation (−1.7‰ $\delta^{18}O$; +3.5‰ $\delta^{13}C$), and for crinoids from the Burlington Formation of Iowa (−2.5‰ $\delta^{18}O$; +3.5‰ $\delta^{13}C$) and suggest that Mississippian seawater was more enriched in ^{13}C and similar in ^{18}O to modern marine waters.

Interpretation of Isotopic Trends of Marine Components

Crinoids.—The data in Figure 4 and the previous discussion imply that Lake Valley crinoids originally had ^{13}C and ^{18}O compositions comparable to other Mississippian components, with no marked biological fractionation. The data compare well with ^{18}O of modern crinoids, but the values are considerably enriched in ^{13}C (Fig. 5). Because of this discrepancy with modern data, and because the interpretation of the crinoid isotopic data is important in consideration of sources of cement carbon, several interpretations are addressed below. We rule out the possibility that crinoid skeletal carbonate precipitated in carbon isotopic equilibrium with a seawater that was more depleted in ^{13}C than modern seawater on the basis of our previously discussed estimate of Mississippian marine carbonate compositions.

Regarding the crinoids, one hypothesis is that crinoids precipitated skeletal carbonate in isotopic equilibrium with seawater having a composition shown in Table 1. Their bulk ^{13}C content then decreased as a result of cementation and/or recrystallization in zone 2 or 3 waters (or in a water for which we have no isotopic record). This is consistent with cathodoluminescence which shows that crinoids are plugged with luminescing and non-luminescing calcites. We rule out alteration in zone 5 waters because crinoids are virtually never ferroan. Under this hypothesis, crinoids generally were not altered by zone 1 porewaters because they are typically lighter than coexisting zone 1 cements (Fig. 5). Had they been strongly altered by zone 1 (solution with concommitant precipitation) their values should fall between coexisting zone 1 and our marine values. Crinoids apparently served as a source of carbonate during early meteoric diagenesis, but did not undergo significant intragranular cementation. The interpretation that crinoids were unaffected by major cementation with zone 1 is consistent with the interpretation that the bioherm muds were also unaffected by zone 1 (see subsequent discussion).

A second hypothesis is that crinoids precipitated out of isotopic equilibrium with marine water of the composition shown in Table 1, due to biological fractionation similar to that of modern crinoids. This would imply that they were isotopically altered to more positive $\delta^{13}C$ by cementation and/or recrystallization in zone 1 waters. We rule out major recrystallization by zone 2 and 3 waters in this hypothesis because crinoids are frequently enriched relative to these zones (Fig. 5). However, arguing against this hypothesis, we see no correlation between the $\delta^{13}C$ of crinoids and the presence of zone 1 cements on their exteriors.

In addition, it is unlikely that there was wholesale reconstitution of the chemistry of the crinoids which would be required to change an initial $\delta^{13}C$ composition of −3 to −6‰ (similar to modern crinoids) to the present composition of +2 to +4‰. The argument against such a wholesale change is the fact that crinoids retained most of their Mg and thus were diagenetically altered in an effectively closed system (Leutloff, 1979; Leutloff and Meyers, 1984). In addition luminescing calcite that we interpret as zone 2 cements is a major component plugging the internal skeletal pores of most crinoid particles. Both these features argue against zone 1 waters causing the dominant recrystallization and/or internal cementation of the crinoids.

A further consideration is that the $\delta^{13}C$ data from Mississippian crinoids from England have a range similar to those of the Lake Valley Formation, i.e., +2.1 to 2.6‰ PDB (Dickson and Coleman, 1980). Similar compositions have been reported for Mississippian echinoids (+1.0 to +3.4‰ PDB; Weber and Raup, 1968) and crinoids (+1.8 to 4.0‰ PDB; Brand, 1982) from the Burlington Formation. If all Mississippian crinoids and echinoids originally had isotopic compositions similar to their modern counterparts (about −3 to −6‰ PDB; Weber and Raup, 1968), then they all must have experienced similar diagenetic histories. This seems unlikely considering the wide geographic range of the data.

In summary we suggest that the crinoids precipitated with $\delta^{13}C$ compositions similar to that of our estimated marine value of about +4.0‰, and their bulk isotopic composition was depleted in ^{13}C by cementation and recrystallization in water which produced cement zones 2 and 3. We further point out that the crinoid isotopic data define a distinctive trend which mimics the trend of the pre-Pennsylvanian cements; the significance of this trend will be discussed later.

Micrites.—In contrast to crinoids, micrites define a trend distinct from the pre-Pennsylvanian cements, with biohermal micrites being isotopically heavier than interbiohermal micrites (Cowan, 1980; Meyers *et al.*, 1982). Furthermore, the heaviest biohermal micrites are non-luminescent while the lighter biohermal and interbiohermal micrites are most luminescent.

The interpretation of these data assumes that the micrites had an initial Mississippian marine composition of about +4.0‰ $\delta^{13}C$ and −1.5‰ $\delta^{18}O$ PDB. Based on these data the isotopically heaviest bioherm muds are interpreted to have been "closed" (*sensu* Lohmann, 1978) by marine micritic cementation, thus retaining in large part their marine isotopic signatures. This is supported by their non-luminescent character. The isotopically lighter micrites which trend toward a zone 2 composition are interpreted to have partially recrystallized during precipitation of composition 2 cements which acted as a closure cement. The bioherm micrites with lightest $\delta^{18}O$, and the interbioherm micrites are interpreted as having been influenced by zone 2 waters. This is evidenced by their having $\delta^{13}C$ values comparable to those of zone 2 cements. However, their relatively light $\delta^{18}O$'s indicate that these muds were not closed during zone 2 precipitation, but instead remained open to be influenced by zone 5 waters.

Alternate interpretations are ruled out based on the following reasoning. If zone 1 waters had significantly affected bioherm micrite diagenesis, then the trend of bioherm micrite values would either be directed toward zone 1, or would strongly overlap with zone 1 values. If zone 3 had significantly affected bioherm micrite diagenesis, then at least some of the bioherm $\delta^{13}C$ compositions would be lighter than those of zone 2. If only zone 5 had affected bioherm micrite diagenesis, with no intervening zone 2 effects, then the bioherm micrite values should fall along a straight line trend from the marine composition toward lighter zone 5 compositions.

In summary, most bioherm micrites were cemented to varying degree by micritic marine cements; however, some porosity remained as evidenced by closure of the majority of bioherm micrites by zone 2 cements before the Pennsylvanian. In a small proportion of the bioherm micrites, and in all of the interbiohermal micrites, their intergranular porosity remained high until Pennsylvanian time when zone 5 waters effected their final diagenesis.

Former magnesian calcite cements.—Former magnesian calcite cements from biohermal cavities now consist of prismatic low Mg calcites containing microdolomite inclusions (Lohmann and Meyers, 1977). These marine cements show a wide range in $\delta^{18}O$ composition with only minor variation in $\delta^{13}C$. The patchy luminescence characteristic of these cements is interpreted to result from partial corrosion and replacement of precursor metastable carbonate and intergrowth of secondary LMC (Lohmann and Meyers, 1977; Lohmann, in press). These cements were apparently unaffected by zones 2 and 3 porewaters because the C-O trend lacks any suggestion of convergence with zone 2 or 3 compositions. However, when stained for Fe^{2+} with potassium ferricyanide, local replacement by Fe-rich cements suggests reactivation of the metastable cements during deeper burial in zone 5 porewaters. Reactivation presumably accounts for the range of $\delta^{18}O$ extending toward lighter compositions of zone 5. As in the case of crinoids, marine cements characteristically contain abundant microdolomite inclusions implying at least partial closure of the diagenetic system. Partial closure may be reflected by the analyses with the heaviest $\delta^{18}O$ compositions.

Interpretation of Non-Marine Cements

Pre-Pennsylvanian meteoric phreatic cements define distinct carbon-oxygen fields which exhibit a progressive decrease in carbon, oxygen and Mg with time. As indicated above, variation in oxygen cannot reasonably be attributed to an increase in temperature. Carbon variations similarly, must reflect changes in the source of HCO_3^-, from heavy marine-derived carbonate to light organic-derived bicarbonate. Dissolved Mg in porewaters must be derived from rock sources: the dissolution of metastable marine HMC; dissolution of LMC meteoric calcites; or extraformational sources such as the underlying shales and shaly carbonates of the Caballero Formation and underlying Devonian strata.

Combined variations in trace element and isotopic composition of meteoric cements cannot be accounted for on the basis of temperature changes alone and likewise, must reflect compositional changes of the porewaters. These changes in porewater chemistry may reflect a combination of effects: change in the composition of meteoric waters in response to climatic and vegetational fluctuations; or the evolution of porewater chemistry in response to progressive rock-water interaction. These models are examined more fully in the following section.

It is important to emphasize at this point that compositions of various marine components were evaluated differently from meteoric cements. The composition of the cement phases, zones 1, 2 and 3, reflect equilibrium with porewater compositions and temperatures; thus, their chemistry is a record of the evolution of the porewater system. In contrast, marine- derived cements and skeletal carbonates may have undergone recrystallization within a spectrum of diagenetic environments and as such reflect a mixture of primary and secondary calcites. For example, the HMC crinoidal calcites were probably the principal source of $CaCO_3$ for cementation throughout pre- Pennsylvanian times (Meyers and Hill, 1983), yet underwent a prolonged period of intragranular cementation. As previously discussed, the majority of intracrinoid cements comprise dominantly zone 2 and zone 3 cements with complete infilling prior to precipitation of zone 5. Thus crinoids exhibit a range in composition reflecting the relative proportion of primary marine carbonate and intragranular meteoric cements. This is observed in the compositional trend of crinoidal calcites illustrated in Figure 4 which mimics the trend of pre-Pennsylvanian meteoric cements. Interestingly, the compositions of crinoidal calcites are consistently lighter than coexisting zone 1 cements, emphasizing that while crinoids served as the source of carbonate at sites of active cementation, they underwent negligible internal cementation. Rather, they remained in open contact with porewaters until ultimately infilled by zone 2 and 3 cements.

In contrast to the pre-Pennsylvanian cements, the post Mississippian zone 5 calcites probably record precipitation at elevated burial temperature, and a complex rock source of carbon and trace elements that include intra- and extra-formational sources. Assuming waters in equilibrium with zone 3 cement at 25°C (−2.0‰ SMOW), then apparent temperatures for zone 5 precipitation ranged from 33°C to 56°C which are compatible with pre-Permian burial estimates using a geothermal gradient of 3.0°C per 100 meters. We cannot, however, rule out possible variation in the composition of zone 5 waters. For example, if we assume 45°C, then apparent water compositions would vary from about −0.5 to −5.0‰ SMOW. These lighter values are reasonable for non-marine waters, but the heavier values seem unlikely.

With the above in mind, we discuss two aspects of the interpretation of non-marine cements, with emphasis on the pre-Pennsylvanian cements. First, we consider the constraints on sources of carbon for these cements imposed by the isotopic data, and second we propose a model for the evolution of a porewater-rock diagenetic system.

Sources of carbon.—Crinoids, which were initially composed of metastable magnesian calcite, are the main component of the Lake Valley Formation and as such were probably the major source of $CaCO_3$ for early cements. The interpretation of the isotopic data on crinoids indicates that their initial composition was comparable to Mississippian marine carbonate and that they were internally cemented mainly by zone 2 and 3 waters. Under this interpretation, crinoids yielded $CaCO_3$ for zone 1 by simple dissolution, with only minor intracrinoid cementation. Furthermore, this dissolution yielded most or all of the Mg for zone 1 cements. The marine carbon isotopic composition of zone 1 cements precludes a significant effect of light carbon derived from soil-gas CO_2 or from the oxidation of marine organic matter in these early meteoric porewaters.

The $\delta^{13}C$ of zone 2 is invariably lighter than that of coexisting zone 1 and could have derived partly or wholly from zone 1, from crinoids and other marine components, or from "non-rock" sources (soil-gas CO_2, atmospheric CO_2). Qualitatively zone 2 carbon could have been derived from a mixture of zone 1 or crinoid carbon and organic soil (?) carbon. However, zone 2 is about four times the volume of zone 1, and therefore quantitatively this would have required ⅘ of zone 1 to have been dissolved, the remaining portion representing the undissolved ⅕. We in fact *do* observe chemical etching at the zone 1-zone 2 boundary (Meyers, 1978). However, if there had been large scale dissolution of most of zone 1, then logically it would have been accompanied by widespread dissolution of lime mud in packstones—we do not see petrographic evidence for this, and the above figures seem geologically unreasonable.

Zone 2 carbon could have been derived largely from crinoids only if crinoids had an initial $\delta^{13}C$ heavier than zone 2. In light of the previous discussion, if crinoids had an initial marine $\delta^{13}C$ then zone 2 carbon could have been derived from a combination of crinoid carbon and about 8% organic carbon. This was estimated assuming a $\delta^{13}C$ organic carbon = −25‰ PDB. We see evidence for chemical compaction during pre-Pennsylvanian time (Meyers, 1980; Meyers and Hill, 1983) that could have been one mechanism for generating dissolved carbon from crinoids, the other mechanisms being intraformational dissolution and dissolution at the pre-Pennsylvanian unconformity. Unfortunately, we cannot quantify the first two mechanisms. In regard to dissolution at the unconformity, we estimate that dissolution of about 25 meters of section with 50% porosity and 100% crinoids could supply all the $CaCO_3$ required for zone 2 within a remaining section 70 meters thick.

If crinoids had $\delta^{13}C$'s comparable to those of modern crinoids, it is difficult to see how any significant amounts of carbon for zone 2 could have derived from pure crinoids. Specifically, any mixture of this light crinoid carbon and organic carbon would have resulted in values lighter than −3‰ PDB.

Bryozoans and lime mud, the other two major depositional components, were other potential sources of carbon for zone 2. There is clear evidence of pre-Pennsylvanian chemical compaction of bryozoans at grain margins and within grains (Meyers, 1980). Unfortunately, we cannot quantify this, and we do not know the original isotopic composition of bryozoans.

In regard to lime mud, it is likely that interbioherm micrite had an initial Mississippian marine isotopic composition and a mixture of this and organic carbon could yield carbon of a composition comparable isotopically to that of zone 2. In addition, we see direct petrographic evidence of chemical compaction of lime mud in non-bioherm, micrite supported rocks. In addition, our interpretation of the distinctive trend defined by the isotopes of bioherm plus interbioherm muds (Fig. 4) implies that the interbioherm micrites remained partly uncemented throughout pre-Pennsylvanian times and thus, unlike most bioherm micrite, could have undergone appreciable chemical compaction to yield carbon for zone 2.

In summary, we interpret the carbon of zone 2 as having been derived in large part from crinoids with ^{13}C compositions comparable to Mississippian marine carbonate. This was mixed with light organic carbon possibly derived from a soil zone that was better developed than during zone 1 cementation. Zone 2 carbon also probably had contributions from other marine components (mud, bryozoans) in unknown quantities.

Zone 3 is considerably lighter than coexisting zones 1 and 2, and than coexisting crinoids. Consideration of its source(s) of carbon are similar to those of zone 2 and therefore need not be repeated. By similar reasoning as for zone 2, zone 3 could have contained up to about 19% organic (soil?) carbon.

EVOLUTION OF ISOTOPIC CHEMISTRY OF METEORIC PHREATIC POREWATERS: THEORETICAL AND APPLIED MODELS

Porewaters are the medium in which diagenetic reactions are cycled by dissolution and precipitation reactions; as such it is important to examine factors which control or modify the composition of porewaters and which determine the chemistry of precipitating mineral phases. In a meteoric phreatic system both minerals and fluids are reservoirs of oxygen and carbon and each can be characterized by a specific chemistry and relative sizes.

The important reservoirs for oxygen in a carbonate-meteoric phreatic system include water (H_2O) and rock carbonate ($CO_3^=$), and their relative contribution can be specified by comparing the molar abundance of oxygen contained within equivalent volumes of rock and water. For example, 1 cc of $CaCO_3$ contains 0.0819 moles of oxygen in comparison to 0.056 moles in 1 cc of H_2O, or a relative molar abundance of oxygen rock/water in equal volumes of 1.474. This implies that when evaluating the effects of rock-water interaction, the contribution of oxygen from rock sources cannot be ignored.

Meteoric waters are initially derived by Rayleigh distillation from marine water sources, thus their isotopic composition may vary due to variations in climate (mean annual temperature), distance from marine source (increased fractionation due to multiple episodes of evaporation-precipitation), and to changes in the composition of the marine water source. For example, during an eustatic drop in sea level induced by an increase in polar and glacial ice volume, mean composition of marine waters will become enriched in ^{18}O. This enrichment of marine source waters similarly results in enrichment of derived meteoric waters. In subtropical and tropical coastal environments, meteoric waters typically range to 2‰ lighter than the marine source waters (Dansgaard, 1964).

Carbon is distributed between the rock ($CO_3^=$) and water (HCO_3^-) reservoirs. The relative reservoir sizes are highly discordant, with the rock reservoir being greatly dominant. For example, even in highly CO_2 charged waters (0.1 mg HCO_3^-/kgH_2O) the molar abundance of carbon in $CaCO_3$ is 1.7×10^7 greater than the molar abundance of carbon in an equal volume of water. Thus, in diagenetic systems where significant rock-water interaction has occurred, the isotopic composition of carbon is dominated by the average composition of rock reservoir in equilibrium with porewaters. If metastable carbonates of primary marine composition are actively driving cementation through their dissolution, porewater $\delta^{13}C$ will necessarily converge towards original marine compositions (Allan and Matthews, 1977; 1982). In contrast, dissolved bicarbonate derived from the decomposition of organic matter, although generally not a reservoir of significant size, can produce local perturbations in porewater carbon composition, largely because of the contribution of extremely light carbon (−25‰). Thus, in a vadose recharge zone, where development of an organic-rich soil horizon is extensive, passage of meteoric waters through this soil zone may provide a continuing source of light carbon for phreatic porewaters. This effect is transient however; as light carbon waters are cycled through dissolution-precipitation reactions, the light carbon signatures are rapidly attenuated as equilibrium with rock carbon is attained.

It is necessary to emphasize that dissolution and/or precipitation reactions of mineral phases are an essential requirement to effect exchange between water and rock reservoirs. In meteoric systems, because meteoric waters are notoriously active as diagenetic agents, significant rock-water reaction and ultimate equilibration of their compositions can occur. As $CaCO_3$ is dissolved in a porewater, the isotopic compositions of rock and water are simply mixed in proportion to their isotopic composition, the number of moles of carbon and oxygen they each contribute, and the concentration of the carbon and oxygen molecular species (Lawrence, *et al.*, 1975). For example, when considering oxygen equilibration, oxygen is contained in both the H_2O and dissolved HCO_3^- reservoirs with one mole of oxygen for each mole of H_2O and three moles of oxygen for each mole of HCO_3^- such that during the dissolution of carbonate, oxygen with an isotopic composition of the rock contributes to the total $\delta^{18}O$ composition of the porewater. However, the concentration of dissolved HCO_3^- in the porewater is generally small and minor rock-water reaction will have a negligible effect on the porewater oxygen composition. In contrast, consider the equilibration of carbon. During dissolution of $CaCO_3$, rock-derived HCO_3^- is significant when considering the low concentration of dissolved HCO_3^-. Thus, the relative sizes of the carbon and oxygen reservoirs in rock and water systems define their relative rates of equilibration in response to progressive dissolution-precipitation reactions.

When evaluating how progressive rock-water interaction will be recorded in a carbonate cement record, an important concept which must be considered is the degree to which the diagenetic system is open versus closed. A reaction may occur in open-system equilibrium on a small scale of individual solution-precipitation reactions. However, when considering the cumulative effect on a single volume of porewater proceeding along flow lines in a phreatic system as it is cycled through multiple reactions, the concept of closed system becomes particularly applicable. Although each reaction may involve open-system exchange of isotopic and minor element constituents between mineral and fluid phases, each reaction progressively modifies the composition of the fluid such that a closed-system equilibrium can ultimately be reached. In such systems, the composition of precipitating LMC, when equilibrium is attained, will be controlled by the composition of dissolving carbonate phases which are driving precipitation (Lohmann, 1982).

This becomes apparent when considering variation in isotopic composition of carbon and oxygen in phreatic porewaters. Because of their drastically different relative molar abundances in rock-water reservoirs, carbon and oxygen attain closed-system equilibrium at different rates. For example, even in a soil-gas charged water of abnormally high HCO_3^- concentration and $\delta^{13}C(-25‰)$, reaction of only a small fraction of carbonate is necessary to effectively reach equilibration of water and rock compositions (approximately 20 cc of $CaCO_3$ reacting with 1000 cc water). Oxygen exhibits similar closed-system behavior, yet it requires that a porewater dissolve and precipitate up to 3 equivalent volumes of carbonate rock before rock-water compositions converge.

Utilizing the general mass balance calculations of Lawrence *et al.* (1975) it is possible to model the composition of calcites which would precipitate in equilibrium with a phreatic porewater as its composition evolves in response to rock-water interactions along flowpathways. A distinctive "J-shaped" trend emerges, reflecting the drastically different rates of equilibration of carbon and oxygen (Fig. 6). As soil-gas CO_2 charged porewaters react with the rock system, carbon isotopic composition of porewaters and precipitating cements rapidly converge to the average composition of the dissolving carbonate. The concomitant change in oxygen composition is, however, negligible. These cements exhibit highly variable carbon and relatively invariant oxygen composition (Region A to B, Fig. 6), from which the $\delta^{18}O$ composition of meteoric water can be estimated if the temperature of precipitation can be constrained.

During progressive rock-water interaction as porewaters flow down-dip in a phreatic lens, the oxygen isotopic compositions of rock and water can ultimately equilibrate in what resembles a closed diagenetic system. Because numerous rock-water reactions are necessary to approach a chemically closed system, most meteoric phreatic cements will lie within a region of relatively constant carbon and variable oxygen composition (Fig. 6, region B to C). In carbonate terrains rich in metastable mineralogies, rock-water interaction can effectively approach a closed system equilibrium composition whereby precipitating meteoric cements will be identical in composition to marine carbonates whose dissolution is driving cementation (Fig. 6, convergence of compositions toward C).

From such modelling, it becomes clear that as porewaters progress along flowlines from areas proximal to the vadose recharge toward distal parts of the phreatic lens, calcite cements comprising the full spectrum of compositions can be precipitated as lateral facies within a single phreatic lens. Even though several compositional facies may be simultaneously precipitated, the regional extent of individual cements may be quite restricted. For example, in a metastable carbonate terrain, light carbon cements may occur only as a thin upper horizon in close proximity to vadose recharge, due to the rapid rates of carbon equilibration. Cements of intermediate oxygen composition, however, may extend over large regional expanses due to the slower rate of equilibration between oxygen derived from rock and water sources.

It is also instructive to consider the temporal rather than the spatial evolution of phreatic porewaters. In a mature phreatic system where metastable marine-derived components are depleted or isolated from porewaters through cementation, proximal cement facies

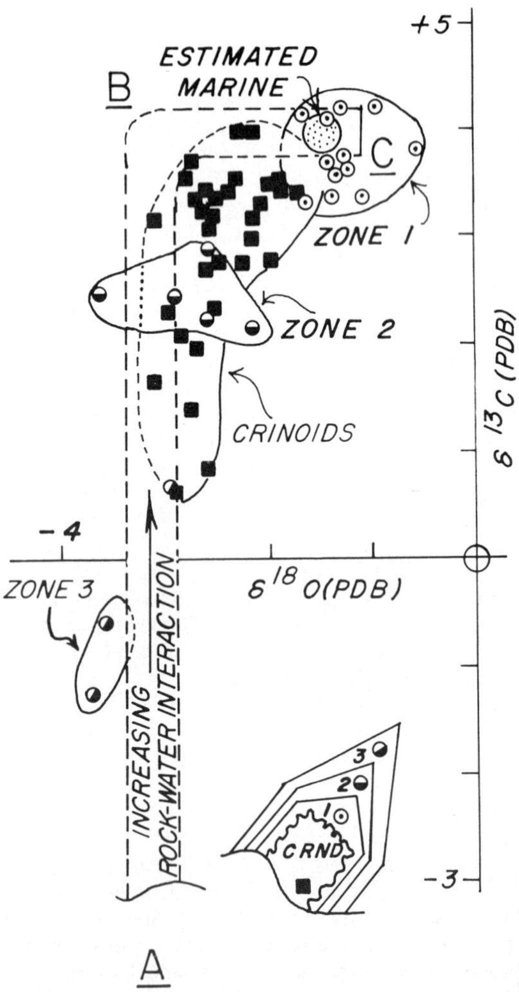

FIG. 6.—Carbon-oxygen isotopic evolution of meteoric phreatic calcite cements. Crinoid and meteoric cement compositional fields are plotted with the predicted "J-shaped" trend assuming progressive rock-water diagenesis during diagenesis. Cements precipitating from light carbon, soil-gas charged meteoric waters proximal to a recharge zone will exhibit compositions near A. As dissolution-precipitation reaction proceeds along porewater flow lines, the composition of meteoric cements will progress along this trend toward B. In the distal part of the phreatic lens, if metastable marine components are undergoing dissolution, meteoric cement compositions will ultimately converge with the composition of these marine carbonates (C). Our estimate of initial marine carbonate composition (+4.0‰ $\delta^{13}C$ and −1.5‰ $\delta^{18}O$ PDB) is indicated by stippled region near C.

can extend down-dip into the distal part of the lens. Open-system light carbon cements, initially restricted to a shallow zone proximal to vadose recharge, would progressively extend deeper and further down dip within the phreatic system as less rock-water interaction occurred. For example, in the groundwaters of South Florida, dissolved HCO_3^- currently have a $\delta^{13}C$ composition of −8‰ or less PDB (Pearson and Hanshaw, 1970)—a carbon composition clearly reflecting significant influence of soil-derived light carbon. If continuous cementation is recorded during the evolution of the porewater composition, the first and oldest cement would represent maximum rock-water interaction (closed-system) and its isotopic composition would approach that of the marine carbonates.

If dissolution of metastable minerals such as HMC was the source of $CaCO_3$ for cementation, we would also expect porewaters to possess their highest enrichment in Mg within the distal part of the phreatic lens and within LMC cements which they precipitated. The youngest cements reflecting least rock-water interaction would approach open-system compositions and have light carbon and oxygen compositions in equilibrium with meteoric waters. Similarly, since carbon equilibration would be achieved if sufficient HMC were dissolved to enrich porewaters in Mg, these porewaters and thus cements should also possess the lowest concentrations of Mg.

The carbon-oxygen compositional trend of pre-Pennsylvanian cements is compatible with such a model. The isotopic composition of zone 1 is virtually indistinguishable from estimated initial marine compositions. Moreover, this zone represents the oldest stage of cementation and reflects the first interaction of meteoric waters with the marine-derived metastable assemblage. The high degree of rock-water interaction is indicated by both the high Mg and heaviest $\delta^{18}O$ compositions. If metastable HMC had an original composition of about 10 mole % $MgCO_3$, a single volume of H_2O would need to dissolve only 1 to 2 volumes of HMC with concommitant precipitation of LMC in order to balance the isotopic and Mg concentrations of zone 1 cement. Similar calculations for zone 3, the youngest stage of meteoric phreatic cementation with extremely light $\delta^{13}C$ and lowest Mg contents, suggests precipitation from a porewater of a mature carbonate terrain in the absence of metastable HMC components.

In contrast, if Mg content of zone 2 cement is estimated based on the rock-water reaction necessary to account for its light carbon isotopic composition, it has anomalously high Mg and intermediate $\delta^{18}O$ compositions which suggest that only minor rock-water interaction has occurred. Enrichment of porewaters in Mg must be derived through the stabilization of HMC marine-derived components, yet $\delta^{13}C$ contents suggest a light soil-gas carbon source. This may be explained, however, by mixing only 8% vadose-derived porewater with a relatively stagnant chemically-equilibrated porewater enriched in Mg. Such mixing would significantly alter the $\delta^{13}C$ of dissolved bicarbonate with only negligible effect on Mg and $\delta^{18}O$ compositions.

Implicit in the geochemical model is the lateral development of the entire spectrum of cement compositions, as cement facies, such that during the precipitation of zone 1 cement, zone 2 and 3 cements would be forming simultaneously in proximal areas of the phreatic lens. This is seemingly incompatible with the petrographic model which suggests that individual zones formed in successive and temporally separate stages of the evolution of the phreatic system. However in a terrain which is rich in metastable carbonates, extensive rock-water reaction in a narrow zone proximal to recharge could rapidly equilibrate porewaters such that the bulk of the phreatic lens would be isotopically homogeneous. In contrast, trace and minor elements should exhibit regional trends as progressive rock-water interaction would continue to modify the porewater minor element concentrations. With depletion or isolation of metastable mineralogies within proximal zones, cement facies exhibiting light carbon and oxygen compositions could effectively expand stepwise through time, to greater depths and further down dip along flow lines within the phreatic system.

On this basis we suggest a working model integrating the cement stratigraphy with the geochemical model. This model, summarized in Figure 7, proposes that each zone precipitated at a different stage in the chemical evolution of the groundwater/rock system which in turn resulted from progressive equilibration of metastable minerals and changing groundwater dynamics.

Under this scenario, zone 1 precipitation occurred during pre-Pennsylvanian falling sea level in a relatively shallow oxygenated(?) meteoric phreatic lens largely uncontaminated by seawater (Fig. 7A). The recharge region for zone 1 waters did not contribute a significant amount of light soil carbon to the regional groundwater system. This was due either to a poorly developed soil zone, or to very rapid equilibration between groundwaters and rock-derived carbon. In the latter case, we suggest that there may be a light-carbon cement facies equivalent in areas proximal to the recharge, a facies that we have not yet sampled or recognized. Zone 1 groundwaters involved the greatest degree of rock-water interaction. Its carbon, oxygen and Mg were derived mainly from simple dissolution of crinoids and perhaps other marine carbonate which had initial $\delta^{18}O$ and $\delta^{13}C$ of about −1.5 and 4.0‰ PDB, respectively. Zone 1 waters cemented crinoids externally but left internal portions of most crinoids and most muds uncemented.

Zone 2 formed in a deep-seated pervasive meteoric phreatic system formed during late stages of sea level drop, stillstand and early stages of sea level rise (Fig. 7B). This system was uncontaminated by seawater except in the south where a mixing-zone "facies" formed

(Meyers and Lohmann, 1978). During zone 2 cementation major sources of Mg and $\delta^{13}C$ included crinoids plus other marine carbonate components (mud, bryozoan) with a significant input of soil-derived light carbon. In addition, the deep-seated nature of the zone 2 groundwater system acted as a widespread gathering system for dissolved constituents, resulting in addition of Mn and Fe to the groundwaters. The dissolved constituents were probably derived from intraformational sources by intergranular pressure-solution and dissolution, especially in recharge regions, and by dissolution at the subaerial unconformity. There may also have been contributions from sub-Mississippian strata. Zone 2 waters represented a major episode of mud lithification and internal cementation of crinoids; by the end of zone 2 precipitation over half of the intergranular cements had been emplaced.

Zone 3 formed in a shallow oxygenated(?) meteoric lens formed during sea level rise at a mature stage of the groundwater evolution. Its groundwaters were also recharged through a relatively well-developed soil zone (Fig. 7C). However, the lack of appreciable remaining metastable carbonates inhibited rock-water reactions and allowed this light carbon signature to be manifested in the precipitating cement. Intraformational sources of heavy carbon and Mg were largely encased or converted to low Mg calcite. This is in contrast to emplacement of zones 1 and 2 where light soil carbon largely equilibrated with rock carbon sources. In addition, the zone 3 $\delta^{18}O$ had negligible input from rock

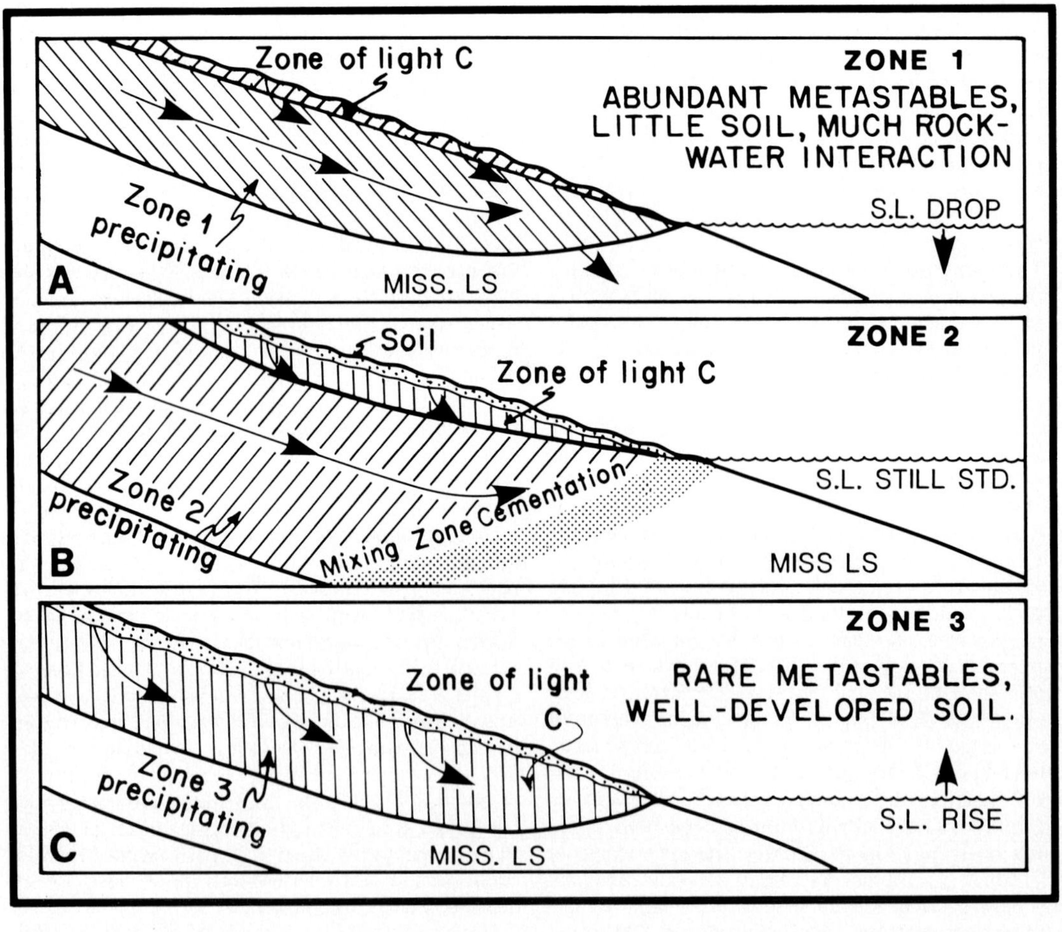

FIG. 7.—Integrated model for trace element and isotope characteristics and regional distribution of pre-Pennsylvanian cement zones. Zones 1, 2, 3 formed at different times, each representing a different stage in the chemical evolution of the groundwater-rock system.

oxygen, and thus its isotopic composition reflects equilibration with relatively pure meteoric waters. During this episode crinoids continued to be affected by internal cementation and/or recrystallization, but muds were largely unaffected.

After zone 3 precipitation the Mississippian limestones were again subaerially exposed and karstified. This was an episode mainly of weathering and disintegrative diagenesis rather than of calcite precipitation. Zone 5 precipitated after Mississippian but probably before Permian times at burial depths up to 1000 m. Zone 5 precipitated at elevated temperatures (but probably less than 60°C). Sources of dissolved constituents were probably largely from extraformational sources as implied by the abundance of Fe and Mn, however, pressure solution could have contributed significant $CaCO_3$ from intraformational sources as well.

CONCLUSIONS

1. Each of the four regionally extensive calcite cement zones (1, 2, 3 and 5) has a trace element and isotope chemistry distinct from the others. Of the pre-Pennsylvanian cements, zone 1 is characterized by relatively high MgO, $\delta^{13}C$ and $\delta^{18}O$ values and low Fe and Mn. Zone 2 is rich in MnO and intermediate in other components, whereas zone 3 is poorest of the three in all trace elements and lightest in $\delta^{13}C$ and $\delta^{18}O$ of all cements.

2. The very low MgO content of all cements (less than 0.25 wt.%) rules out significant input from seawater. The Mg for the pre-Pennsylvanian cements can be accounted for by less than about 10% loss of Mg from echinoderm grains, which is less than half the amount estimated to have been lost.

3. Variation in the isotopic composition of marine-derived components, crinoids, micrite, and marine cement, was examined to estimate the initial composition of Mississippian marine carbonate($-1.5‰$ $\delta^{18}O$; $+4.0‰$ $\delta^{13}C$). These data in comparison with modern marine carbonates suggest that Mississippian epeiric seawater was isotopically similar in oxygen and heavier in carbon than modern open-ocean seawater.

4. On the basis of combined petrographic and geochemical evidence, we interpret crinoids to have had original $\delta^{13}C$ and $\delta^{18}O$ compositions similar to our estimated Mississippian marine values (approximately $+4.0‰$ $\delta^{13}C$ and $-1.5‰$ $\delta^{18}O$ PDB). This is compatible with the $\delta^{18}O$ composition of modern warm water crinoids, but is more enriched in $\delta^{13}C$.

5. The pre-Pennsylvanian zones 1, 2 and 3 show a decrease in ^{18}O and ^{13}C with decreasing age. We interpret this ^{18}O change as due to progressive decrease in ^{18}O of precipitational waters rather than to increasing temperature, as petrographic and stratigraphic data constrain precipitational temperatures to about 25–30°C. The decrease in ^{13}C is interpreted to reflect increasing input of organically-derived light carbon during pre-Pennsylvanian diagenesis. These shifts toward lighter $\delta^{18}O$ and $\delta^{13}C$ during pre-Pennsylvanian cementation probably resulted from decreased interaction of porewaters with marine-derived carbonate and increased development of a soil zone during precipitation of zone 3 cement.

6. Post-Mississippian zone 5 has a wide range of ^{18}O contents, all less than those of the pre-Pennsylvanian cements. These light ^{18}O values reflect, in part, precipitation at elevated burial temperatures, as implied by the estimated burial depths. The ^{13}C contents of zone 5 are more variable than those of the pre-Pennsylvanian cements, but generally show intermediate values. These are interpreted to reflect a complex source of carbon that was dominated by "rock" carbon and probably derived from extraformational strata.

7. Based on petrographic criteria, we propose a model in which cementation of zones 1, 2 and 3 occurred at different times within chemically distinct porewater systems. However, on the basis of modelling of the evolution of porewater chemistry we suggest that the compositional range from zone 1 to zone 3 could be produced as lateral facies by progressive rock-water equilibration along flow lines within a single phreatic system. This apparently contradicts the petrographic model, yet the regional extent of each compositional zone may be quite limited during successive stages of porewater evolution. During zone 1 times, an abundance of metastable carbonate would lead to rapid equilibration of the majority of phreatic lens, with cements of zones 2 and 3 composition being restricted to areas proximal to subaerial recharge. Only during the later stages of diagenesis, when the abundance of metastable carbonates was depleted and rock-water interaction diminished were porewaters reflected in the composition of zone 3. Thus, the compositional trend of pre-Pennsylvanian cements agrees closely with the trend predicted for evolution of cementation during interaction of meteoric waters with a metastable carbonate terrain.

8. The isotopic trends defined by the three main marine components (coarse former HMC marine cements, micrite and crinoids) differ markedly from one another. The crinoid trend mimics the J-shaped trend of pre-Pennsylvanian meteoric cements, reflecting recrystallization/cementation in zones 2 and 3 porewaters. In contrast, the micrites show a distinctive trend reflecting diagenesis in marine waters and zones 2 and 5 porewaters. The coarse marine cements show a trend unrelated to the pre-Pennsylvanian cements reflecting recrystallization through a spectrum of diagenetic temperatures in a system largely closed to extraneous carbon.

ACKNOWLEDGMENTS

This research was supported by NSF grants EAR7412253 and EAR7713133 held by Meyers. The authors gratefully acknowledge the critical reviews of this manuscript by J. Budai, K. Cercone, P. Choquette, C. Kendall and R. Halley. We also wish to acknowledge the cooperation of R.K. Matthews in arrangements for use of the Benedeum Isotope Laboratory facilities at Brown University.

REFERENCES

ALLAN, J.R. AND MATTHEWS, R.K., 1977, Carbon and oxygen isotopes as diagenetic and stratigraphic tools: Data from surface and subsurface of Barbados, West Indies: Geology, v. 5, p. 16–20.

______, AND ______, 1982, Isotope signatures associated with early meteoric diagenesis: Sedimentology, v. 29, p. 797–818.

ARMSTRONG, A.K., 1962, Stratigraphy and paleontology of the Mississippian System in southwestern New Mexico and adjacent Arizona: New Mexico Bur. Mines Mineral Resources Mem. 8, 99 p.

BRAND, U., 1982, The ^{18}O and ^{13}C contents of carboniferous fossil components: Seawater effects: Sedimentology, v. 29, p. 139–147.

COWAN, P., 1980, Diagenesis of lime mud, Mississippian-age bioherms Sacramento Mts., New Mexico: Unpublished MS Thesis, State University of New York, Stony Brook, 159 p.

CRAIG, H., 1957, Isotopic standards for carbon and oxygen and correction factors for mass spectrometric analyses of carbon dioxide: Geochim. Cosmochim. Acta, v. 12, p. 133–149.

DANSGAARD W., 1964, Stable isotopes in Precipitation: Tellus, v. 16, no. 4, p. 436–468.

DICKSON, J.A.P., AND COLEMAN, M.L., 1980, Changes in carbon and oxygen isotope composition during limestone diagenesis: Sedimentology, v. 27, p. 107–118.

FÜCHTBAUER, H., AND HARDIE, L.A., 1976, Experimentally determined distribution coefficients for precipitated magnesian calcites: Application to marine carbonate cements: Geol. Soc. America, Abstracts with Programs, v. 8, p. 877.

HEM, J.D., 1970, Study and interpretation of the chemical characteristics of natural water (second edition): U.S. Geol. Surv. Water-supply Paper 1473, 363 p.

IRWIN, H., CURTIS, C., AND COLEMAN, M., 1977, Isotopic evidence for source of diagenetic carbonates formed during burial of organic-rich sediments: Nature, v. 269, p. 209–213.

LANE, H.R., 1974, The Mississippian of southeastern New Mexico and west Texas—A wedge-on-wedge relationship: Amer. Assoc. Petroleum Geologists Bull., v. 58, p. 269–282.

______, 1982, The distribution of the Waulsortian facies in North America as exemplified in the Sacramento Mountains of New Mexico, *in* K. Bolton, H.R. Lane, and D. LeMone (eds.), Symposium on the Paleoenvironmental Setting and Distribution of the Waulsortian Facies: El Paso, Texas, El Paso Geol. Soc., 202 p.

LAUDON, L., AND BOWSHER, A., 1941, Mississippian formations of Sacramento Mountains, New Mexico: Amer. Assoc. Petroleum Geologists Bull., v. 25, p. 2107–2160.

______, AND ______, 1949, Mississippian formations of southwestern New Mexico: Amer. Assoc. Petroleum Geologists Bull., v. 60, p. 1–88.

LAWRENCE, J., GIESKES, J., AND ANDERSON, T.F., 1975, Oxygen isotope material balance calculations, Leg 35: Initial Reports of the Deep Sea Drilling Project, v. 35, p. 507–512.

LEUTLOFF, A.H., 1979, Regional distribution of microdolomite inclusions in Mississippian echinoderms from southwestern New Mexico: Unpublished MS thesis, State University of New York, Stony Brook, 181 p.

______, AND MEYERS, W.J., 1984, Regional distribution of microdolomite inclusions in Mississippian echinoderms from southwest New Mexico: Jour. Sed. Petrology, v. 54, p. 432–446.

LOHMANN, K.C., 1978, Closed system diagenesis of high magnesium calcite and cements: Geol. Soc. America, Abstracts with Programs, v. 10, p. 446.

______, 1982, "Inverted J" carbon and oxygen isotopic trends: a criterion for shallow meteoric phreatic diagenesis: Geol. Soc. America, Abstracts with Programs, v. 14, p. 548.

______, in press, Covariant carbon and oxygen isotopic trends in calcitized marine cements: Estimation of marine isotopic signatures: Jour. Sed. Petrology.

______, AND MEYERS, W.J., 1977, Microdolomite inclusions in cloudy prismatic calcites: a proposed criterion for former high magnesium calcites: Jour. Sed. Petrology, v. 48, p. 475–488.

MARSHALL, J.D., AND ASHTON, M., 1980, Isotopic and trace element evidence for submarine lithification of hardgrounds in the Jurassic of eastern England: Sedimentology, v. 27, p. 271–289.

MEYERS, W.J., 1974, Carbonate cement stratigraphy of the Lake Valley Formation, Mississippian, Sacramento Mountains, New Mexico: Jour. Sed. Petrology, v. 44, p. 837–861.

______, 1978, Carbonate cements: Their regional distribution and interpretation in Mississippian limestones of southwestern New Mexico: Sedimentology, v. 25, p. 371–400.

______, 1980, Compaction in Mississippian skeletal limestones, southwestern New Mexico: Jour. Sed. Petrology, v. 50, p. 457–474.

______, COWAN, P., AND LOHMANN, K.C., 1982, Diagenesis of Mississippian skeletal limestones and bioherm muds, New Mexico: *in* K. Bolton, H.R. Lane, and D. LeMone (eds.). Symposium on the Paleoenvironmental Setting and Distribution of the Waulsortian Facies, El Paso, Texas, El Paso Geol. Soc., p. 80–95.

______, AND HILL, B.E., 1983, Quantitative studies of compaction in Mississippian skeletal limestones, New Mexico: Jour. Sed. Petrology, v. 53, p. 231–243.

______, AND LOHMANN, K.C., 1978, Microdolomite-rich syntaxial cements: Proposed meteoric-marine mixing zone phreatic cements from Mississippian limestones, New Mexico: Jour. Sed. Petrology, v. 48, p. 475–488.

______, AND ______, 1980, Geochemistry of regionally-extensive calcite cement zones in Mississippian skeletal limestones, New Mexico [abs.]: Amer. Assoc. Petroleum Geologists Bull., v. 64, p. 750.

MICHARD, C.R., 1968, Coprecipitation de L'ion manganeaux avec le carbonate de calcium: Comptes Rendus Acad. Sci. Paris, ser. D., v. 267, p. 1685–1688.

MOORE, C., AND DRUCKMANN, T., 1981, Burial diagenesis and porosity evolution, Upper Jurassic Smackover, Arkansas and Louisiana: Amer. Assoc. Petroleum Geologists Bull., v. 65, p. 597–628.

PEARSON, F., AND HANSHAW, B., 1970, Sources of dissolved carbonate species in groundwater and their effects on carbon-14 dating: Isotope Hydrology. Proceedings Symposium IAEA, p. 271–286.

PRAY, L.C., 1961, Geology of the Sacramento Mts. escarpment, Otero Co., New Mexico: New Mexico State Bur. Mines and Mineral Resources Bull., v. 35, 133 p.

RICHTER, D.K., AND FÜCHTBAUER, H., 1978, Ferroan calcite replacement indicates former magnesian calcite skeletons: Sedimentology, v. 25, p. 843–860.

VEIZER, J., AND HOEFS, J., 1976, The nature of $^{18}O/^{16}O$ and $^{13}C/^{12}C$ secular trends in sedimentary carbonate rocks: Geochim. Cosmochim. Acta., v. 40, p. 1387–1395.

WALLS, R.A., MOUNTJOY, E.W., AND FRITZ, P., 1979, Isotopic composition and diagenetic history of carbonate cements in Devonian Golden Spike reef, Alberta, Canada: Geol. Soc. America Bull., Part 1, v. 90, p. 963–982.

WEBER, J.N., 1967, Possible changes in the isotopic composition of the oceanic and atmospheric carbon reservoir over geologic time: Geochim. Cosmochim. Acta, v. 31, p. 23–43.

______, AND RAUP, D.M., 1968, Comparison of $^{13}C/^{12}C$ and $^{18}O/^{16}O$ in the skeletal calcite of Recent and fossil echinoids: Jour. Paleontology, v. 42, p. 37–50.

WHITE, D.E., HEM, J.D., AND WARING, G.A., 1963, Chemical composition of subsurface waters: U.S. Geol. Survey Prof. Paper 440-F, 67 p.

YUREWICZ, D.A., 1977, The origin of the massive facies of the lower and middle Capitan Limestone (Permian), Guadalupe Mountains, New Mexico, *in* Upper Guadalupian Facies Permian Reef Complex, Guadalupe Mountains, New Mexico and West Texas: Permian Basin Section Soc. Econ. Paleontologists and Mineralogists, Pub. 77–16, p. 45–92.

BURIAL CEMENTATION—IS IT IMPORTANT? A CASE STUDY, STUART CITY TREND, SOUTH CENTRAL TEXAS

DENNIS R. PREZBINDOWSKI
Amoco Production Company, P. O. Box 591, Tulsa, Oklahoma 74102

ABSTRACT

The Stuart City Trend is a shelf-edge buildup of Lower Cretaceous bioclastic and reefal carbonates, now buried at depths from 3300 to 5000 m. Compaction and cementation have generally reduced rock porosities to less than 9%. Sediments were cemented in the marine environment by finely crystalline, bladed, isopachous Mg-calcite and the volumetrically important (14 vol%) coarse to very coarsely crystalline, fibrous to bladed, isopachous, Mg-calcite cement. These cements have been neomorphically altered to low Mg-calcites, forming unusual radial and radiaxial textures. Evidence for a marine origin consists of a relative 1 mol% Mg^{2+} memory, a marine-like isotopic character ($\delta^{18}O \approx -2.5$ and $\delta^{13}C \approx +2.0$) and early relative timing of precipitation. Diagenetic alteration by interaction with meteoric water in lenses formed within topographic highs along the shelf margin changed the initial marine chemical, isotopic, and textural character of the sediments. Secondary porosity formation, mineral stabilization, aggrading neomorphism and equant spar calcite cementation are the important products of meteoric diagenesis. The equant spar calcite cements make up approximately 16% by volume of the limestones studied. They are iron and manganese poor and the majority have a $\delta^{13}C$ composition which falls in the range of modern marine carbonates; i.e., 0.0‰ to +3.5‰. The $\delta^{18}O$ compositions range from −1.3‰ to −6.6‰ relative to the PDB standard.

Oxygen stable isotope and petrographic data suggest that over 7 volume % of the porous sediments was filled by equant spar calcite cements formed in a near-surface meteoric environment. An additional 9 volume % of the pores were filled by equant spar calcite cements formed at shallow to intermediate burial depths (<2,000 meters) in a water-limited system. Thermally induced $\delta^{18}O$ depletion of the equant spar calcites, indicating fluid flow, was of minimal importance. Pyrobitumen pore fillings and inclusions in the outer 0.1 mm-thick rims of the very coarsely crystalline, equant spar calcite cements indicate that only minor amounts of cementation have occurred since the introduction of hydrocarbons. Deep burial diagenesis, i.e., post-hydrocarbon migration, consisted of the precipitation of minor amounts of galena, fluorite and Sr^{2+}-rich equant spar calcites, i.e., <1%. These diagenetic products can be directly related to the present-day formation water.

The chemical, isotopic, and textural characteristics of the Stuart City Trend limestones contain the imprints of their initial marine composition and shallow diagenetic alterations in a hydrodynamic system. Burial diagenesis has not significantly altered these limestones. Fault and fracture control on the movement of formation waters in this system determine the location and intensity of late stage diagenetic events.

INTRODUCTION

The emphasis on studies of Holocene and Pleistocene carbonates during the past two decades is mainly responsible for our understanding of the major surface and near surface (tens of meters in depth) diagenetic processes which alter carbonate sediments to form limestones. However, as carbonate sediments undergo burial, most spend very little time in these well-documented surface and near-surface diagenetic environments. Our ability to recognize and understand diagenetic processes which occur in the deeper subsurface is limited by general inaccessibility. This study is an attempt to understand the general importance of burial diagenesis in altering a specific carbonate package—The Stuart City Trend.

The Lower Cretaceous carbonate shelf system of south-central Texas (Fig. 1) forms a Gulfward-dipping wedge of shallow water carbonates that reaches burial depths in excess of 3000 m downdip. This wedge thickens from 100 m at the outcrop to more than 600 m at the subsurface paleo shelf edge. The Lower Cretaceous Stuart City Trend is an excellent rock unit to document the importance of diagenesis during burial for the following reasons:

1. The stratigraphy and depositional history of these carbonates are generally well known.
2. The structural setting is relatively simple.
3. The carbonate rocks of this shelf system are encased in relatively impermeable Lower Cretaceous and Upper Cretaceous shales and carbonate mudstones.
4. A long and important history of hydrocarbon exploration and production has generated extensive subsurface information and materials.
5. Hydrocarbon production allows the collection of formation-water samples.
6. Burial of this system has continued to present-day and although the rates have varied, these sediments have always moved toward higher pressure and temperature environments.

The major objective of this study was to document the petrographic, chemical, and isotopic character of the individual constructive diagenetic components in the Stuart City Trend and to establish their origin, volumetric importance and timing.

The study is based on approximately 3000 m of core from 17 wells located in 10 counties along the Stuart City Trend of Texas (Fig. 2). The depths at which these cores were taken varied from 3000–5000 m with the majority coming from the upper 100 m of the Stuart City section. The cores are generally restricted to the positive buildups along the Trend.

The core used in this study is biased toward the upper 100 m of section associated with topographic highs along the Trend. There, wells were drilled to find porous hydrocarbon reservoirs. Thus the core material represents the higher energy carbonate facies associated with these topographic highs. The total surface area of the core slabs examined during this study is equivalent to an outcrop only 16 m square, or only nine billionths of one percent of the rock material making up the Stuart City Trend. With these points in mind, a general diagenetic evolution of the porous carbonates associated with the positive highs and the importance of burial diagenesis can be attempted.

METHODS

Standard thin-section staining techniques and luminescent petrography were employed in this study (Dickson, 1965; Sippel and Glover, 1965; Sippel, 1968). No dolomite or iron-rich carbonate components were observed.

Luminescence of individual carbonate components in this study was restricted to the final pore filling, very coarsely crystalline, equant spar calcite cements in only three samples. The vast majority of the blocky spar calcite cement present showed no signs of distinct luminescence. Since iron and manganese were not de-

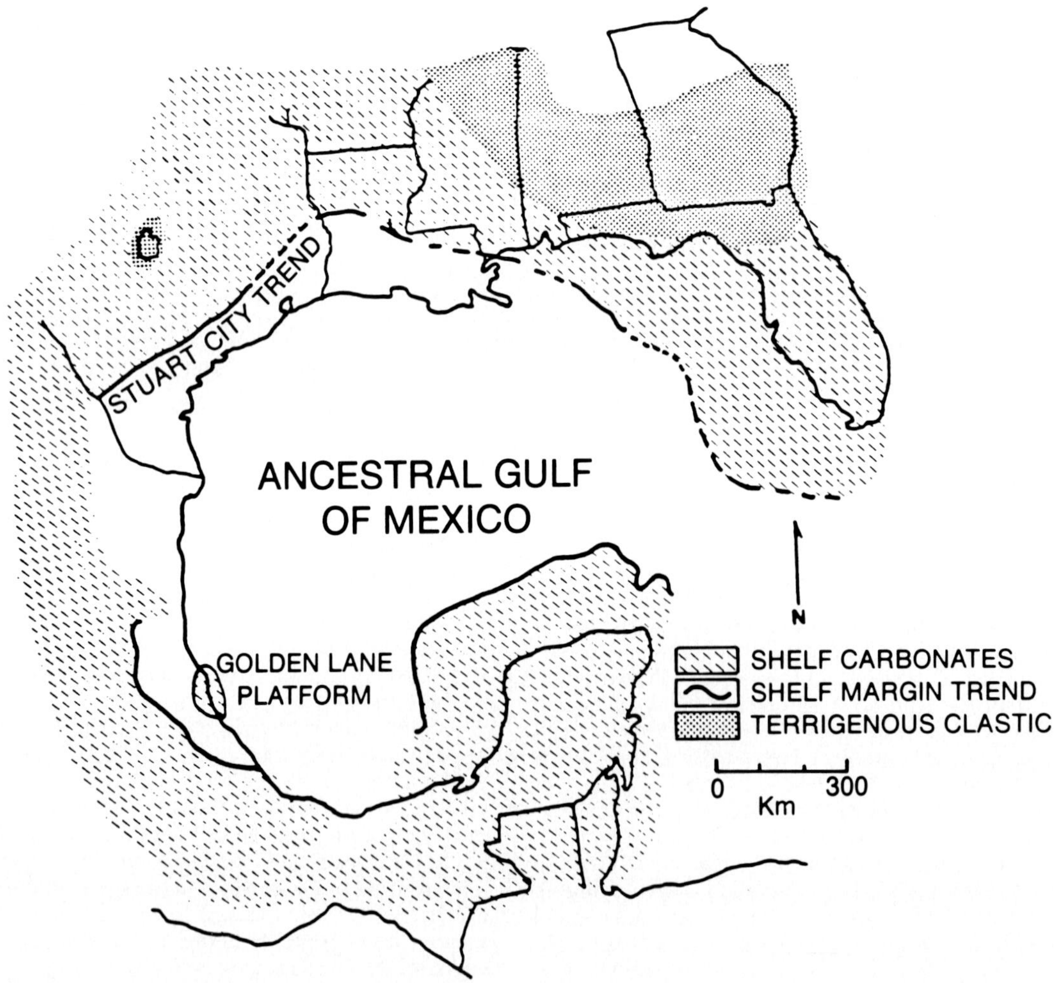

FIG. 1.—Distribution of shelf and shelf-margin facies around the Gulf of Mexico during Lower Cretaceous, Fredericksburg time (from Bebout and Loucks, 1974).

tected in the blocky spar calcites by either staining or microprobe analyses (<370 ppm), nonluminescence of the carbonate cements is attributed to low concentrations or absence of Mn^{2+}.

A three-channel ARL microprobe was used to determine the elemental composition; i.e., Ca^{2+}, Mg^{2+}, Fe^{2+}, Mn^{2+} and Sr^{2+} of individual carbonate components. A 60° Nuclide Mass Spectrometer was used to determine the carbon and oxygen stable isotopic composition of selected samples.

Carbonate grains and individual cements were sampled according to the technique described by Prezbindowski (1980). The carbon and oxygen stable isotopic ratios were determined by the general methods outlined by McCrea (1950). All carbon and oxygen isotopic analyses of carbonates in this study are reported in per mil (‰) units and referenced to the Chicago PDB carbonate standard. Reproducibility was ±0.1‰.

GEOLOGICAL SETTING

Previous Work

The petrology of the Stuart City Trend has been discussed by Bebout and Loucks (1974) and Bebout *et al.* (1977). Bebout and Loucks (1974) provided the first brief survey of pore types, cement types and diagenetic sequences present in the Stuart City Trend. A more detailed study followed (Bebout *et al.*, 1977) emphasizing porosity type and distribution. These works did not characterize the geochemical aspect of the Stuart City Trend carbonates.

General Setting and Stratigraphy

The Lower Cretaceous carbonate rocks of central and south-central Texas form a Gulfward-dipping wedge. This carbonate wedge is divided by the Balcones fault zone into two components; the updip outcropping Glen Rose and Edwards Formations of the Edwards Plateau in central Texas, and the downdip, subsurface Glen Rose and Edwards Formations and their shelf-edge equivalent, the 3 km wide Stuart City Trend in south-central Texas. The Stuart City Trend largely a bioclastic accumulation strikes from the southwest to the northeast across south-central Texas (Fig. 3). It has a gentle northeastward dip along the strike direction. Burial depths increase across Texas from 3000 to more than 5000 m. This system represents a small portion of the Early Cretaceous shelf-edge which nearly encircled the ancestral Gulf of Mexico (Fig. 1; Bebout and Loucks, 1974).

The Stuart City Trend is defined stratigraphically by a lower conformable contact with the Pearsall Shale and an unconformable upper contact with the

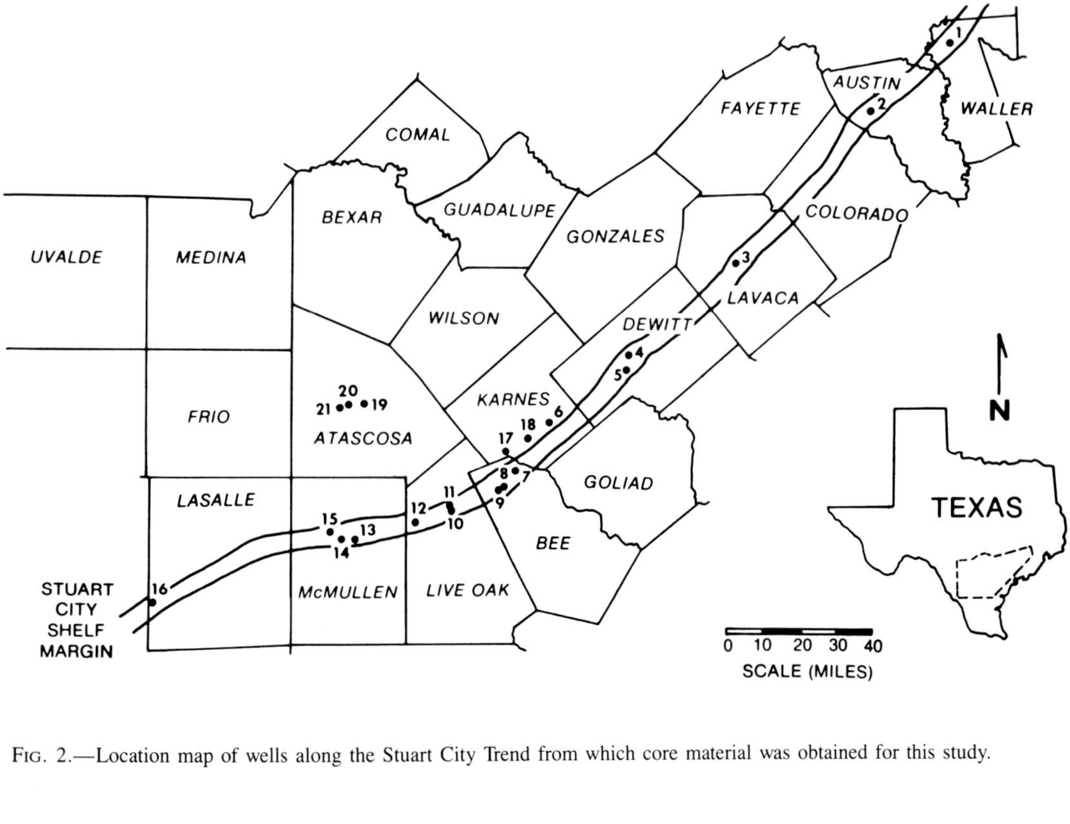

FIG. 2.—Location map of wells along the Stuart City Trend from which core material was obtained for this study.

Georgetown Formation (Fig. 4). This upper contact is difficult to determine in places because of a gradational character and must be picked on the basis of a lithic change from a dark, planktonic foram mudstone to a shallow water lithofacies or by use of electric logs (Winter, 1961).

A precise updip delineation of the Stuart City Formation is difficult to determine because of interfingering of the shallow-shelf carbonate facies of the Glen Rose and Edwards Formations, described by Rose (1968 and 1972), with the shelf-edge facies of the Stuart City Formation. Its limit is best defined by a facies change; i.e., from grainstone and boundstone facies to mud-dominated lagoonal facies (Bebout and Loucks, 1974). Seismic profiles can be used to delineate the backside structural dip of the Stuart City Trend (Wooten and Dunaway, 1977). Basinward, the Stuart City Trend passes into dark, basinal lime mudstone facies of the Atascosa Group defined by Winter (1961).

Bebout and Loucks (1974) and Bebout *et al.* (1977) have described the depositional aspects of the Stuart City Trend. A summary of their findings is discussed below and illustrated in Figure 5. The Stuart City Trend is composed mainly of reef and bank carbonates including rudist pelecypods with lesser amounts of encrusting red algae, stromatoporoids, corals, and other mollusks.

Deposition of the underlying Pearsall Formation and Stuart City Trend sediments appears to have resulted from a cycle of transgression and regression. Progradation began with the deposition of the upper Pearsall but was interrupted during deposition of the lower Stuart City Trend sediments, and resumed as sediments of the middle and upper Stuart City Trend were deposited. This progradational sequence is indicated by the vertical facies succession. The lower part of the sequence begins with a planktonic foraminifera wackestone of open-marine origin and proceeds upwards through the

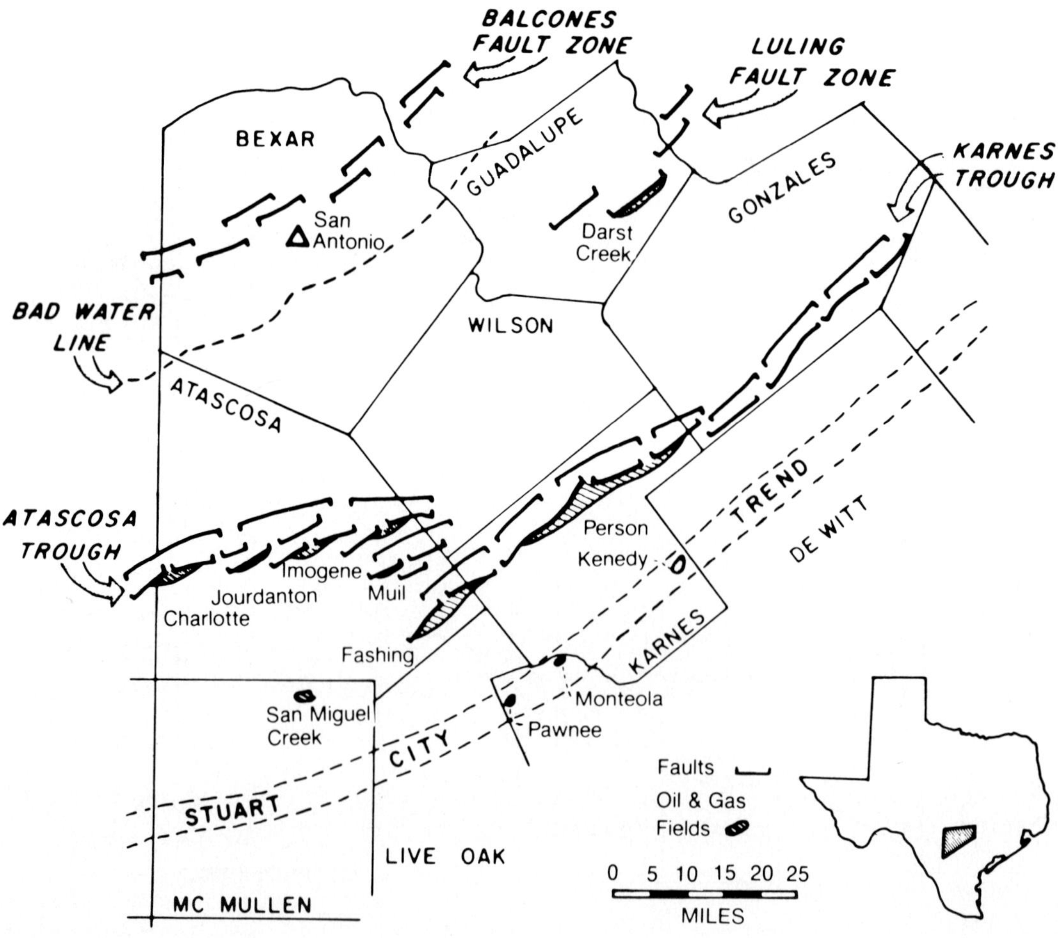

FIG. 3.—Location map of hydrocarbon fields contained in the Lower Cretaceous carbonates, major fault systems and the Stuart City Trend.

following major depositional environments: lower-slope, upper shelf-slope, shelf-margin, and shelf-lagoon. Rapid subsidence of the shelf-edge ended shallow-water carbonate sedimentation on the Stuart City Trend; open-marine, planktonic foraminifera wackestone and mudstone facies reappeared during deposition of the overlying Georgetown Formation.

Structure

Minor faulting of the Stuart City Trend is believed by Holden (1963) to be the result of differential compaction of reef and nonreef sediments. This differential compaction between the sediments of the Stuart City Trend and the adjacent lagoonal and basinal sediments helped accentuate a series of low-relief buildups of about 50 m elevation (Holden, 1963). These closures, although accentuated by compaction, represent topographic highs developed as the result of carbonate deposition. They have a gentle northwest dip on the backreef or lagoonal side and a steep southeast dip on the forereef or deep-water side (Holden, 1963). The majority of the hydrocarbon production from the Stuart City Trend is associated with these structural closures (Wooten and Dunaway, 1977).

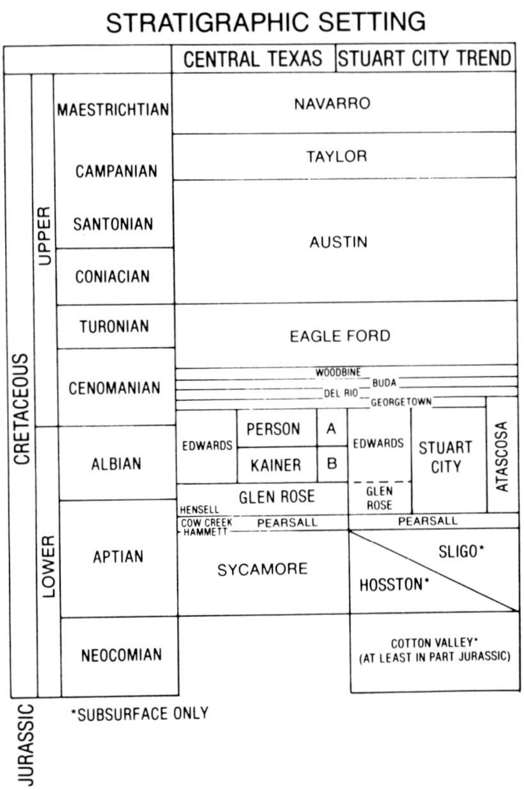

FIG. 4.—Stratigraphic framework of the Cretaceous units in the study area (modified from Bebout and Loucks, 1974).

Large growth faults with displacements of 150 to 300 m have been reported on the basinward side of the Stuart City Trend. They were formed by loading of Tertiary clastic sediments across the Cretaceous carbonate platform and onto the incompetent Cretaceous basinal sediments of the ancestral Gulf of Mexico (Holden, 1963). These growth faults provided avenues for the movement of basinal fluids into the porous sections of the Stuart City Trend (Land and Prezbindowski, 1981).

There are four major fault systems (Fowler, 1956) which affect the diagenesis, hydrology and hydrocarbon accumulations in the Lower Cretaceous Glen Rose and Edwards Formations of the San Marcos platform: the Karnes Trough, Atascosa Trough, Luling, and Balcones fault systems (Fig. 3). They provide important hydrologic pathways for the vertical movement of diagenetic fluids, i.e., basinal brines, meteoric water and hydrocarbons (Prezbindowski, 1981; Land and Prezbindowski, 1981).

DIAGENESIS

Marine

Micrite envelopes.—Micrite envelopes on carbonate allochems are nearly ubiquitous in the carbonates of the Stuart City Trend. Their thicknesses are highly variable, ranging from 5 microns to over 90 microns (Fig. 6A, B and C). A two-step process of micritization leading to the development of micrite envelopes has been documented and discussed by Winland (1971) and Bathurst (1971). These presumable Mg-calcite envelopes are now stabilized to calcite. Microprobe analyses of 11 micrite envelopes found an average Mg^{2+} composition of 1230 ± 95 ppm.

Iron is present in nearly half of the envelopes analyzed. The average iron concentration is 1470 ±700 ppm. These concentrations are similar to those reported by Loucks (1976) for micrite envelopes analyzed from the Lower Cretaceous Pearsall Formation of south Texas.

Bladed isopachous cement.—A finely crystalline, bladed isopachous cement is very common. It is characteristically 30–90 microns in thickness and the individual crystals are bladed with scalenohedral terminations (Fig. 6B). Crystal boundaries are sharp and the transparent appearance of the individual crystals are characteristic.

This cement is thought to have been a high-magnesium calcite of marine origin. The evidence for this interpretation is:

1. It occurs as a first generation cement with a bladed crystal morphology (Fig. 6B, C).
2. It is frequently overlain by a second generation micrite layer suggesting that marine diagenesis continued after cementation.
3. Microprobe analyses demonstrate the existence of a significant magnesium content in the cement

(Fig. 7). A relative increase of 0.8 mol% magnesium exists over the surrounding carbonate components (i.e., equant spar calcite cement and skeletal allochems).

4. The selective occurrence of this cement only on the exterior of allochems that have undergone dissolution (Fig. 6C) support a nonmeteoric-water origin. This type of textural relationship was used as evidence of a marine origin for a similar cement type studied by Kerr (1977) in the Edwards Formation of central Texas.

Micrite cement.—Micrite cements are present as a minor diagenetic component in the porous facies (Fig. 6D). The early diagenetic position of this cement and the even isopachous distribution support a marine origin. This type of cement has been frequently reported from marine and beach rock environments (in Bricker, 1971).

Radial-radiaxial calcite cement.—Radial-radiaxial calcite cement makes up approximately 14%, by volume, of the limestone studied. This important carbonate component is characterized by a "turbid" appearance, isopachous distribution, and large variability in thickness from sample to sample, crystal size, crystal shape and optical orientation. The thickness of this cement ranges from 50 microns to more than 2.5 mm. Crystal size varies from medium crystalline (62 microns to 250 microns) to very coarsely crystalline (1.0 mm–4.0 mm). Although a bladed morphology is characteristic, crystals can be equant in a few cases.

Bebout and Loucks (1974) identified this cement as a "radiaxial calcite cement." Important variations in crystal shape and optical properties make the use of the term "radiaxial calcite cement" undesirable in this study. It is similar to the radiaxial fibrous mosaics originally described by Bathurst (1959) and discussed in detail by Kendall and Tucker (1973). A comparison of these cements is presented in Table 1. It is evident that a great deal of similarity does exist. The important difference, which distinguishes this calcite from a true radiaxial cement, is the large variability in optic axis orientation of the crystals and subcrystals. This type of calcite cement will be referred to as radial-radiaxial cement.

Since its initial description, the origin of radiaxial fibrous texture has been extensively debated. The majority of workers who have studied this texture believe it to be a neomorphic fabric of a synsedimentary marine cement (Kendall and Tucker, 1973; Davies and Krouse, 1975; Davies, 1977; Achauer, 1977). However a wide range of other diagenetic environments have been proposed for its origin. Proposed environments range from early diagenetic, submarine (Schmidt, 1971) to subaerial, meteoric (Friedman and Kolesar, 1971), to late diagenetic (Orme and Brown, 1963).

The most pertinent interpretation is that proposed by Bebout and Loucks (1974) for the origin of "radiaxial

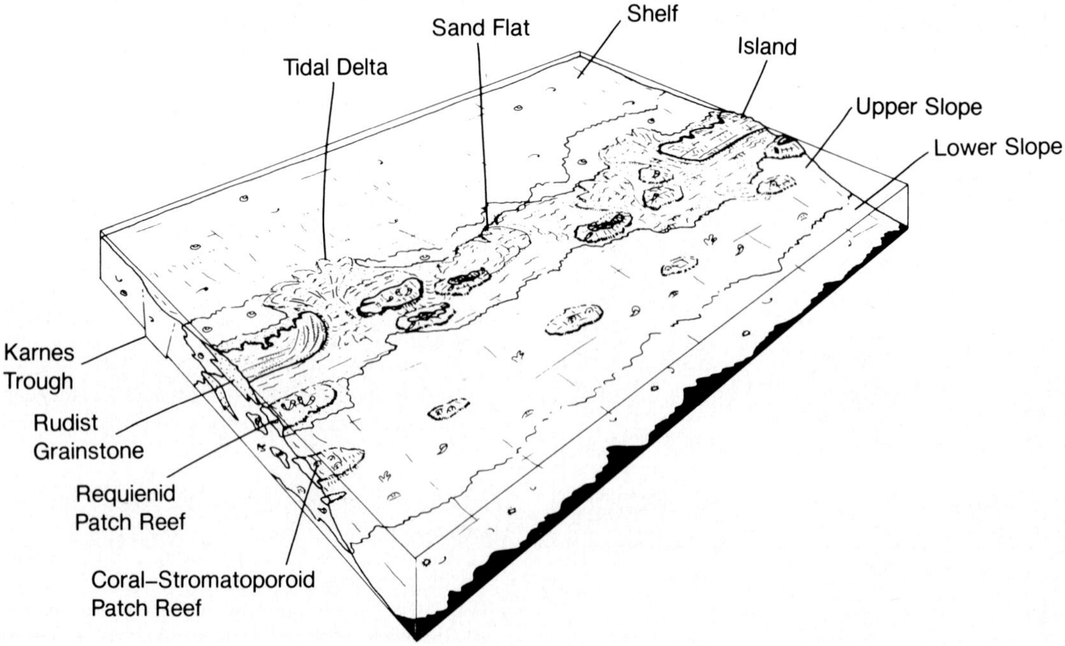

FIG. 5.—Simplified model of Stuart City Trend depositional environments (based on Bebout and Loucks, 1974).

calcite cement" in the Stuart City Trend. They conclude that this cement formed in a shallow subsurface, phreatic meteoric-water environment. This hypothesis is based on petrographic evidence of meteoric-water diagenesis (i.e., meniscus and dripstone cements and secondary porosity formation) predating radial-radiaxial calcite cementation.

This study concludes that the radial-radiaxial calcite cement represents a neomorphic alteration of a synsedimentary, fibrous, marine, high-magnesium calcite

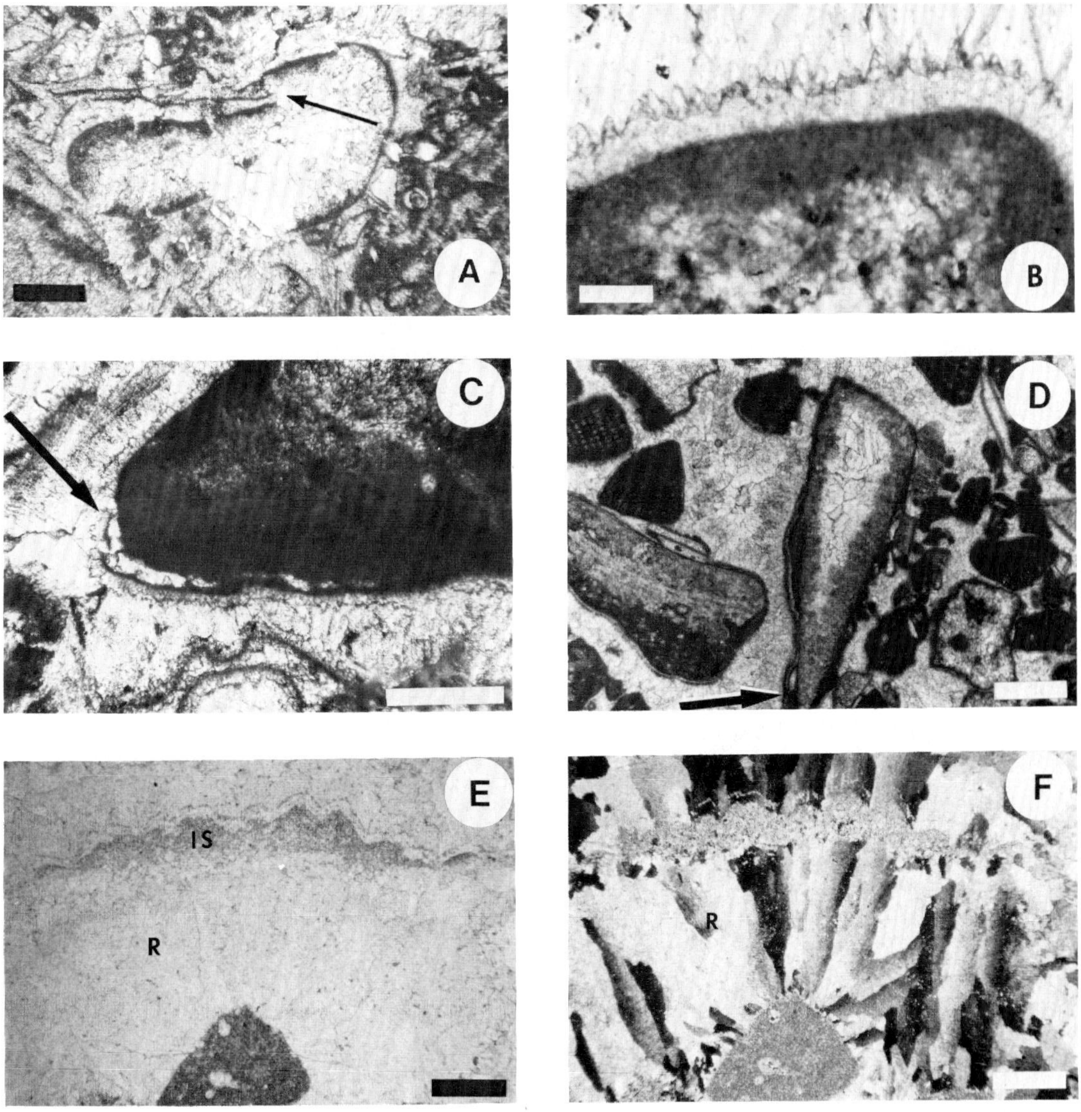

FIG. 6.—Photomicrographs: *A*, Crushed micrite rim of a former mollusk grain. From the Humble No. 1 Dilworth, 11,176 ft. interval. Scale bar = 0.5 mm. *B*, Finely crystalline, bladed, isopachous cement (IC). From the Humble No. 1 Dilworth, 11,497 ft. interval. Scale bar = 50 microns. *C*, Skeletal grain with micrite rim and finely crystalline bladed, isopachous cement separated from the main body of the grain (see arrow). From the Humble No. 1 Dilworth, 11,224 ft. interval. Scale bar = 200 microns. *D*, Mollusk-foram grainstone with a minor amount of micrite cement development (see arrow). Note that the interior of the center carbonate grain is filled by a radial-radiaxial and equant spar calcite cement sequence. From the Humble No. 1 Dilworth, 11,760 ft. interval. Scale bar = 1.5 mm. *E*, Radial-radiaxial cement (R) with interval sediment (IS) unevenly distributed in cement. The rhombic crystal outline within the cement suggest that stabilization of the former high Mg-calcite cement occurred repeatedly after cementation events. From the Shell No. 1 Roehl well, 13,665 ft. interval. Scale bar = 1.5 mm. *F*, Crossed nicols view of area depicted in Fig. 6E.

cement. The evidence that supports this interpretation is:

1. This cement commonly exhibits multiple generations of cementation (Fig. 6E, F). Interlayering of carbonate muds between individual cement layers is common (Fig. 11A). In certain cases this sediment appears current deposited. Such sediment interlayering has been frequently reported in Recent marine cements (Shinn, 1971; Land and Moore, 1980).
2. A significant magnesium content has been found in the radial-radiaxial cements (Fig. 8). Microprobe analyses indicate a significantly higher magnesium content of up to 0.9 mol% is preserved in this cement relative to allochems. Strontium, iron and manganese were not detected. The magnesium content and the lack of detectable strontium suggest a high-magnesium calcite precursor. The magnesium content of this cement is contained in the crystal lattice of the calcite as a replacement of calcium. No evidence was found of micro-dolomite inclusions as reported by Lohman and Meyers (1977) for cloudy prismatic calcites. Relative high magnesium concentrations have been reported in early vadose

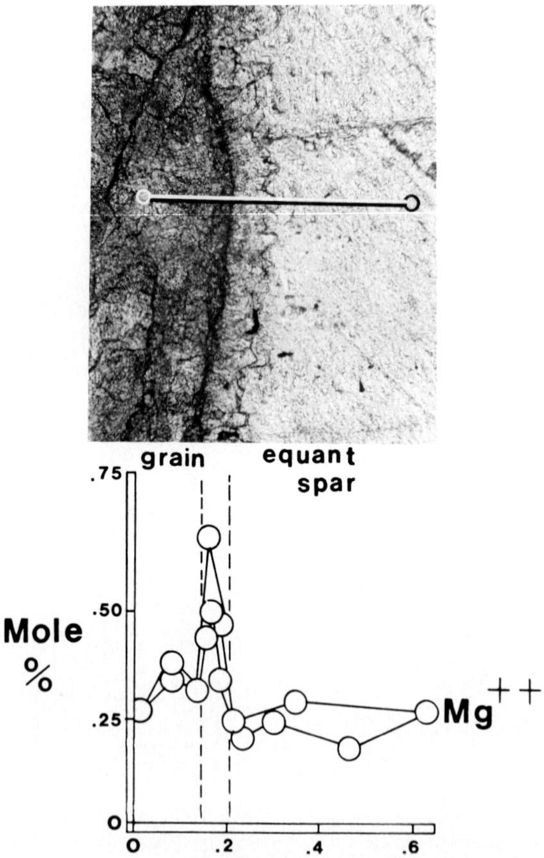

FIG. 7.—Electron microprobe traverses across a fine to medium crystalline, bladed, isopachous calcite cement showing a relative Mg^{2+} memory. Sr^{2+}, Fe^{2+} and Mn^{2+} concentrations were below detectability limits. Circles indicate the approximate two standard deviation relative error for each analysis. From the No. 1 Smith well, 14,098 ft interval.

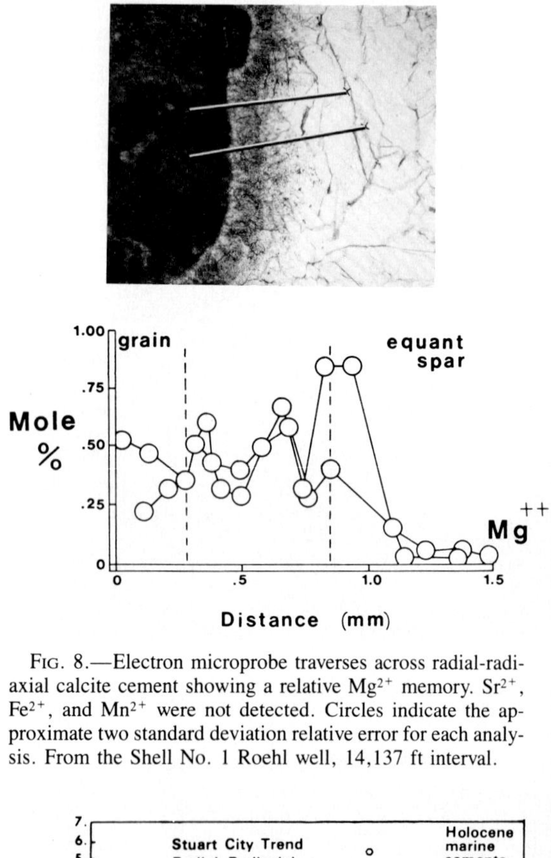

FIG. 8.—Electron microprobe traverses across radial-radiaxial calcite cement showing a relative Mg^{2+} memory. Sr^{2+}, Fe^{2+}, and Mn^{2+} were not detected. Circles indicate the approximate two standard deviation relative error for each analysis. From the Shell No. 1 Roehl well, 14,137 ft interval.

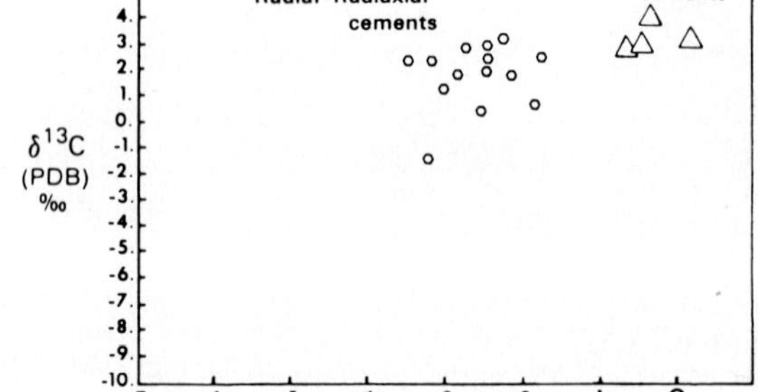

FIG. 9.—$\delta^{18}O$ and $\delta^{13}C$ cross-plot of 15 radial-radiaxial cement samples. Isotopic compositions of the Holocene marine cements taken from Ginsburg *et al.* (1971), Land (1971), Land and Goreau (1970), and Shinn (1969).

cements from the Pleistocene of Barbados, West Indies (Benson and Matthews, 1971). The fibrous to bladed crystal morphology, isopachous geometry, and marine-like stable isotopic character argue against a vadose origin for the radial-radiaxial cements.

3. The stable isotopic composition of this cement, even after neomorphic alteration, is compatible with an origin from a marine water ($\delta^{18}O \approx 0$ to -1) at near surface temperatures (Figs. 9, 10). Lowenstam (1961) has provided evidence that the $\delta^{18}O$ composition of Cretaceous marine water was similar to modern oceans.
4. The occurrence of radiaxial textures in skeletal allochems with remnant wall structures still preserved strongly supports a neomorphic origin. This textural relationship was recognized by Bebout et al. (1977) and termed "neomorphic, radiaxial cement." The radiaxial texture is a product of neomorphic alterations of unstable carbonate components (i.e., cements and allochems). It is not a primary texture.
5. Numerous petrographic features discussed in detail by Kendall and Tucker (1973) support a neomorphic origin (e.g., nonplanar intercrystalline boundaries, lack of competitive growth fabrics, and cross-cutting relationships of crystal boundaries to inclusion patterns).

TABLE 1.—COMPARISON OF THE FEATURES OBSERVED IN THE RADIAXIAL CALCITES DESCRIBED BY KENDALL AND TUCKER (1973) WITH THE RADIAL-RADIAXIAL CALCITE CEMENTS IN THE STUART CITY TREND

Feature	Kendall and Tucker (1973)	Stuart City Trend
Cement Like Morphology	X	X
Neomorphic Alteration Of Allochems		X
Internal Sediment	X	X
Mosaics Composed Of Elongate Crystals With Long Axes Normal To Cavity Walls	X	X
Crystals Composed Of Subcrystals	X	X
Crystal Size Decreases Towards Substrate	X	X
Intercrystalline Boundaries Are Non-Planar	X	X
Crystal Morphology	Fibrous To Bladed	Bladed To Equant
Crystals Are Turbid	X	X
Lattice Curvature Of Crystals	Common	Infrequent
Optic Axes	Convergent Into Pore	Convergent To Divergent
Changes In Inclusion Density Parallel Cavity Walls	X	X
Former Crystal Faces Scalenohedral	X	X

6. The primary evidence used by Bebout and Loucks (1974) and Bebout *et al.* (1977) in proposing a meteoric-water, phreatic origin for the radiaxial cement is the petrographic evidence of meteoric water diagenesis, predating the precipitation of the radiaxial calcite cements (Fig. 6D, 11B and C). Although this diagenetic sequence does occur (i.e., in about 5% of the thin sections studied which contained radial-radiaxial cement), it is not common.

The occurrence of meteoric-water diagenesis prior to the precipitation of some radial-radiaxial calcite cements does not argue against a marine origin. It is likely, given the depositional setting of Stuart City Trend (i.e., shelf-edge system with a broad, shallow marine-shelf system on one side and the ancestral Gulf of Mexico Basin on the other), that sporadic exposure above wave base to allow the influx of meteoric-water and subsequent resubmergence and cementation in a marine environment was possible and did occur.

Exposure and Meteoric Water Diagenesis

Pendulous cements.—Pendulous cements occur in approximately 6% of the thin sections examined. These cements are characteristically 80–400 microns in thickness, have a bladed to fibrous crystal habit, orange color, and characteristic gravity controlled morphology (Fig. 11D). They are always found in carbonates which have undergone extensive neomorphic alteration and early dissolution.

The vadose origin of this cement type is documented (Dunham, 1971 and Müller, 1971). The fibrous to bladed crystal morphology suggests that precipitation occurred from a magnesium-rich fluid (Folk, 1974). This cement is believed to be a beachrock cement precipitated from marine waters.

Meniscus cements.—Only questionable examples of meniscus cements were recognized during petrographic studies of these carbonates (Fig. 11E). Recognition of this cement type was made very difficult by the overprinting of neomorphic and solution processes.

Mineral stabilization.—Original aragonite allochems (e.g., mollusk shell debris) made up a large portion of the carbonate grain population of the Stuart City Trend carbonates. Neomorphism and solution-reprecipitation were both important processes in stabilizing the aragonite component. Nearly 40% of the thin sections examined showed evidence of aragonite dis-

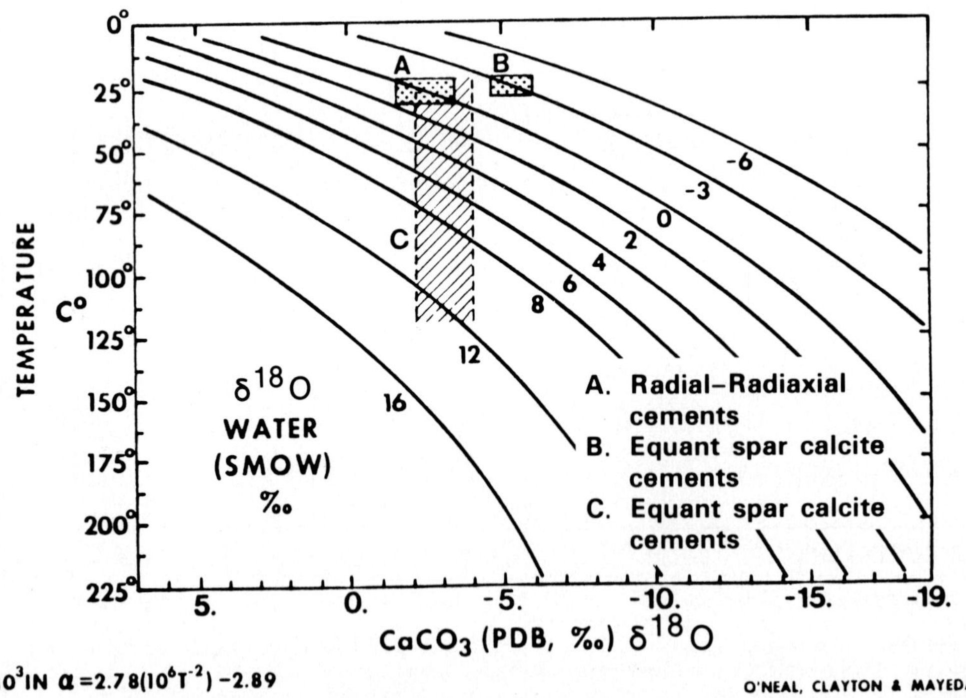

FIG. 10.—Graphic display of the oxygen isotopic equilibrium relationship between temperature, calcite and water. The oxygen isotopic composition of the radial-radiaxial calcite cements (Group A) is in equilibrium with marine-like waters, i.e., ^{18}O relative to SMOW, at near surface temperatures. The ^{18}O-depleted equant spar calcite cements (Group B) are in equilibrium with a ^{18}O depleted water at near surface temperatures. The ^{18}O enriched equant spar calcite cements (Group C) show a wide range of potential environments of formation. Equilibrium relationship from O'Neal, Clayton and Mayeda's (1969) equation as indicated.

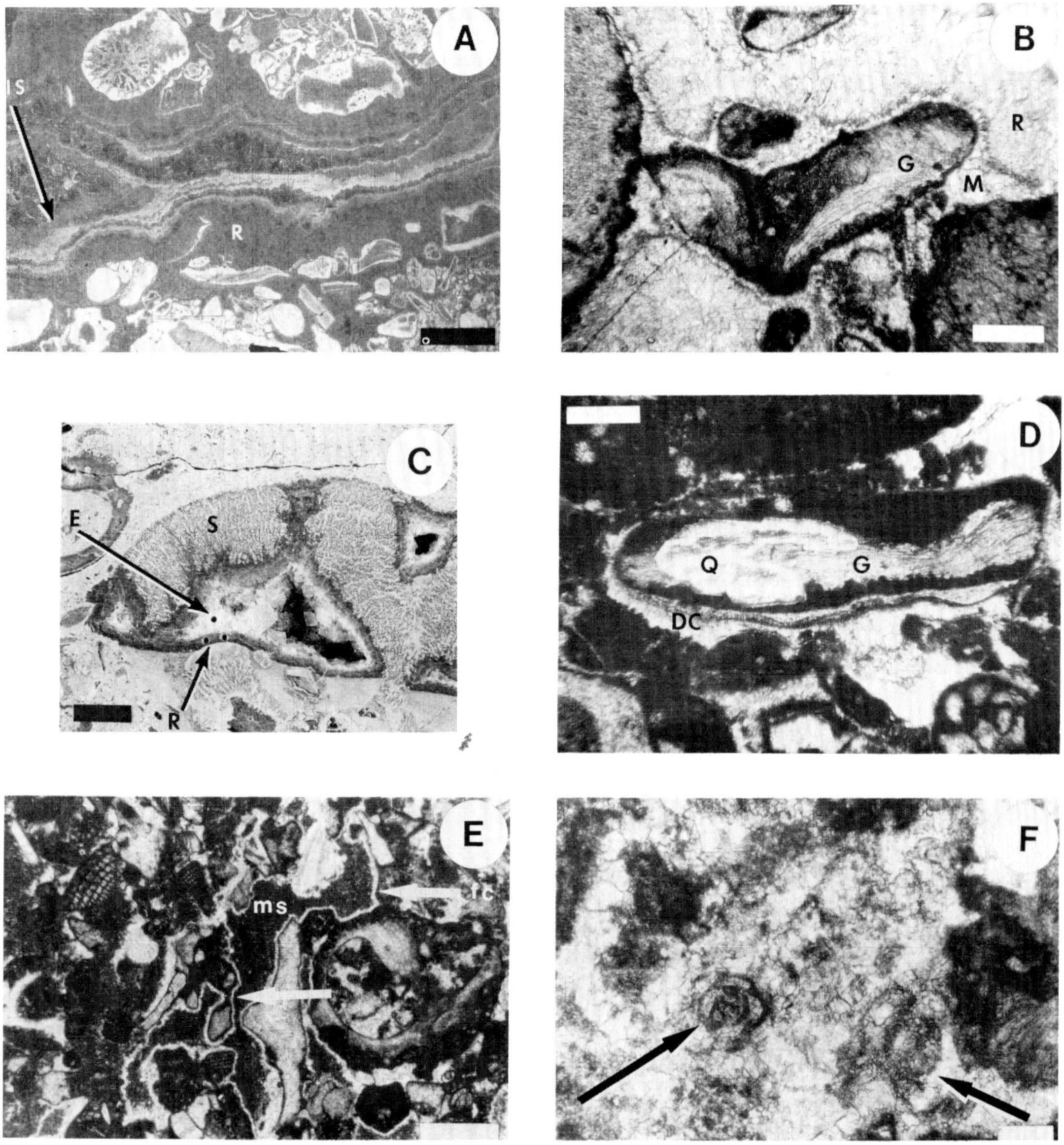

FIG. 11.—Slab photograph and photomicrographs: *A*, Negative print of thin section showing cyclic cementation by radial-radiaxial cement (R) and deposition of internal sediment (IS). From the Shell No. 1 Roehl well, 13,665 ft interval. Scale bar = 3.0 mm. *B*, Meniscus geometry developed in an early equant spar calcite cement (M). This geometry is believed to be the result of partial dissolution of the cement and grains (G). Whether this cement was originally isopachous and has been extensively modified, or meniscus with only a minor amount of modification by dissolution is unknown. Remaining pore space filled with radial-radiaxial cement (R). From the Tenneco No. 1 Alamo Lumber well, 13,595 ft. interval. Scale bar = 500 microns. *C*, Slab photograph of stromatoporoid (S) with solution enlarged intraparticle pore space, partially filled by radial-radiaxial cement (R) $\delta^{18}O$ = −2.2‰, $\delta^{13}C$ = +3.3‰, very coarsely crystalline, equant spar calcite cement ($\delta^{18}O$ enriched), (E) $\delta^{18}O$ = −3.0‰, $\delta^{13}C$ = +3.3‰ (arrows indicate sample holes). From the Atlantic No. 1 Smith well, 14,225 ft. interval. Scale bar = 1.0 cm. *D*, Quartz (Q) replacing mollusk grain (G) and micrite envelope. Dripstone cement (DC) is well developed. From the Humble No. 1 Dilworth well, 11,436 ft. interval. Scale bar = 500 microns. *E*, Foram-mollusk packstone cemented by an early micrite cement (see arrow), finely crystalline, bladed isopachous cement (FC), filled by micritic sediment (MS)? and coarsely crystalline, equant spar calcite cement. From the Tenneco No. 1 Alamo Lumber well, 13,668 ft. interval. Scale bar = 500 microns. *F*, Foram-mollusk packstone, foraminifera (tests) have undergone extensive neomorphic alteration (see arrows). From the Humble No. 1 Dilworth well, 11,620 ft interval. Scale bar = 500 microns.

solution (i.e., moldic and intraparticle pores, broken and floating micrite rims; Fig. 6A). The degree of aragonite dissolution ranged from the isolated development of secondary intraparticle pores to the complete removal of allochem interiors, leaving only the micrite rim.

Neomorphism was also important as evidenced by the preservation of internal shell structures in many former aragonite allochems. It was found to be inversely related to the intensity of aragonite dissolution (Prezbindowski, 1981).

High-magnesium calcites are unstable in Mg^{2+}-depleted waters (Friedman, 1964). Stabilization to a calcite is thought to occur by either incongruent dissolution (Land *et al.*, 1967) or by calcitization (Bathurst, 1971). The preservation of skeletal textures of former high-magnesium calcite allochems (i.e., foraminifer, echinoderm fragments, and coralline algae) indicates the importance of selective Mg^{2+} removal. Although selective Mg^{2+} removal was the most important means of stabilizing high-magnesium calcite to calcite in the Stuart City Trend carbonate sediments, leaching of high-magnesium calcite allochems did occur (Fig. 11F). Dissolution of former high-magnesium calcite allochems was associated with intense meteoric-water diagenesis, indicated by broken and floating micrite rims, abundant microspar formation and depleted $\delta^{18}O$ compositions ($\approx -5.5‰$). These areas of intense diagenesis are normally found closely associated with overlying zones that had been subaerially exposed.

Silica.—Silica is a minor diagenetic component of the Stuart City Trend limestones. It occurs in two forms: (1) monocrystalline, anhedral quartz replacements of mollusk grains and pendulous cements ranging up to 1.5 mm in size (Fig. 11D) and (2) monocrystalline, euhedral quartz replacement of micrite in a pelletal packstone. The anhedral quartz replacements are believed to be an early diagenetic event based on petrographic observations (Prezbindowski, 1981) and $\delta^{18}O$ analysis (Land, 1977). The timing of the euhedral quartz replacements in the pelletal packstone facies is assumed to be similar to the anhedral quartz replacements; i.e., early, although no geochemical and distinctive petrographic data are available to support this interpretation.

Equant Spar Calcite Cements (General Discussion)

Fine to very coarsely crystalline (16 microns to 4 millimeters) equant, spar calcite cements are volumetrically the most important cement types in the Stuart City Trend. These cements make up approximately 15% by volume of the limestones studied. Based on crystal size and stable isotopic composition four general types of equant spar calcite cements are recognized: finely crystalline, coarse to very coarsely crystalline ($\delta^{18}O$ depleted), coarse to very coarsely crystalline ($\delta^{18}O$ enriched) and extremely coarsely crystalline equant spar calcites.

Equant, spar calcite cementation follows one of two general sequences, although there are exceptions. The most common sequence is one in which crystal size increases at a more or less uniform rate toward the center of the filled pore along with a corresponding increase in the number of straight crystal boundaries and triple junctures (Fig. 6D). This general textural evolution was noted by Bathurst (1971) and observed by Loucks (1976) in the Lower Cretaceous carbonates of the Pearsall Formation of South Texas. Loucks (1976) suggested that these alterations in texture can be used to discriminate between equant spar calcites formed in shallow (smaller crystal size, lower frequency of straight crystal boundaries and triple junctions) versus deep subsurface environments. Although this crystallization sequence is common in the Stuart City Trend, it is believed to be a function of the rate of precipitation, which may or may not be related to depth of burial. The $\delta^{18}O$ composition of this cement type is generally enriched relative to the average $\delta^{18}O$ composition of all equant spar calcites analyzed.

The second common sequence consists of a bimodal population of equant spar calcite cement crystals (Fig. 12A). The sequence is typified by an early, fine to medium crystalline equant spar calcite cement with irregular crystal boundaries and a sharp transition to a very coarsely crystalline equant spar calcite cement with well defined linear crystal boundaries. This cementation sequence is generally associated with limestones, which have undergone meteoric-water diagenesis. The $\delta^{18}O$ composition of the very coarsely crystalline equant spar calcite in this sequence tends to be depleted relative to the average $\delta^{18}O$ composition of the equant spar calcites. It may be the result of a dynamic, open meteoric, phreatic system, as opposed to a more restricted phreatic or burial cement sequence with a gradational change in crystal size.

Stable Isotopic Composition

Carbon.—The $\delta^{13}C$ composition of the coarse to extremely coarse crystalline equant spar calcite cements ranged from +4.5 to −7.4‰ relative to PDB (Fig. 14). A significant $\delta^{13}C$ shift away from the normal-marine range of 0.0 to +3.5, as defined by Gross (1964), has occurred in only a few of the samples analyzed.

The $\delta^{13}C$ composition of Holocene marine-carbonate sediments range from 0.0 to +4.0‰ (Allan and Matthews, 1977). The whole rock data from the Stuart City Trend indicates that the marine $\delta^{13}C$ composition of these carbonates during deposition was similar (Fig. 13). The vast majority of the equant spar calcite cements analyzed are within this range (Fig. 14). Depleted $\delta^{13}C$ compositions occur in only six cement samples, four of which are only slightly depleted falling between 0.0 to −1.0‰. The remaining two samples, have significantly depleted $\delta^{13}C$ composition

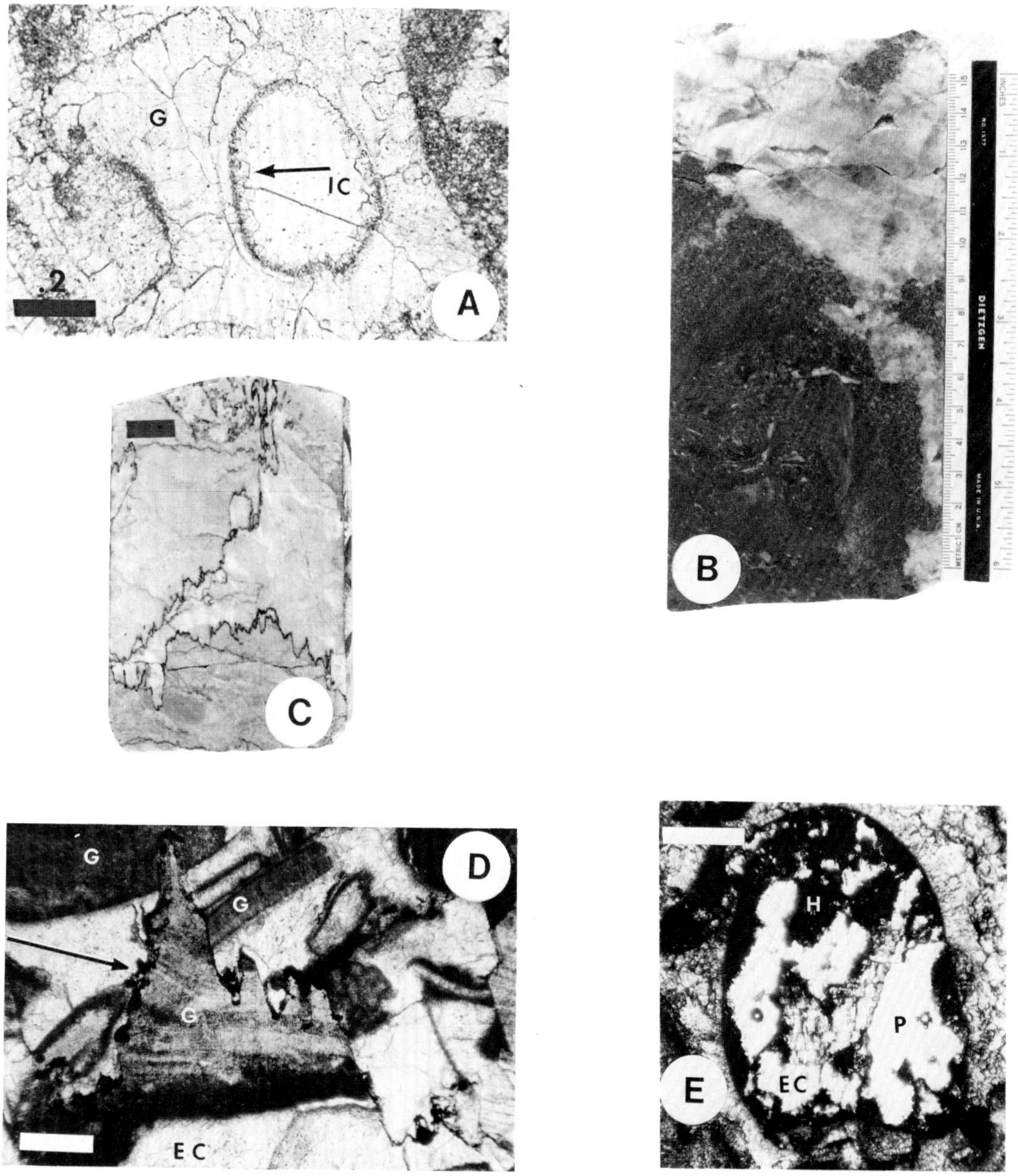

FIG. 12.—Slab photographs and photomicrographs. *A*, Finely crystalline, isopachous, equant spar calcite cement (IC) developed in the aperture of an altered coral grain (G) followed by a coarsely crystalline, equant spar calcite cement ($\delta^{18}O$ depleted). From the Stanolind No. 1 Martin well, 10,353 ft. interval. Scale bar = 200 microns. *B*, Extremely coarsely crystalline equant spar calcite replacement of a mollusk-foram packstone. Isotopic composition of the spar is $\delta^{18}O = -6.5‰$ and $\delta^{13}C = +2.6‰$. The isotopic composition of the surrounding country rock is $\delta^{18}O = -3.9‰$ and $\delta^{13}C = +2.9‰$. From the Amerada No. 1 Horton well, 11,600 ft interval. Scale in centimeters. *C*, Intense stylolitization of a miliolid wackestone. Scale bar = 1.0 cm. *D*, Stylolitic interpenetration of grains (see arrow). From the Atlantic No. 1 Smith well, 14,098 ft interval. Scale bar = 500 microns. *E*, Solution enlarged intraparticle porosity (P). Suggested diagenetic sequence: (1) grain dissolution, infilling with coarse crystalline equant spar calcite (EC) or neomorphic alteration to coarse crystalline equant spar calcite, (2) dissolution, (3) emplacement of hydrocarbons, (4) thermal alteration of hydrocarbons forming bitumen (H). From the Humble No. 1 Dilworth well, 11,908 ft interval. Scale bar = 200 microns.

(i.e., −5.5‰ and −7.4‰, respectively).

Thermal alteration of associated organic material is believed to be the source of the organic carbon incorporated in the equant spar calcite cement. The common occurrence of pyrobitumens within the pore systems from which these samples were taken provides some evidence of such an origin. The relatively enriched $\delta^{18}O$ compositions of the equant spar calcite cement would be the result of precipitation in a restricted carbonate buffered system in the subsurface. There only minor amounts of $\delta^{18}O$ depletion in the precipitated cements would occur.

Oxygen.—Traditionally, the oxygen isotopic composition of a calcite cement has been used to investigate: (1) the nature of the diagenetic fluid responsible for cementation (meteoric, marine, or brine) and/or (2) the approximate temperature of formation (surface, deep subsurface or hydrothermal environments).

There are three important variables in a $^{18}O/^{16}O$ equilibrium carbonate-water system: temperature, $\delta^{18}O$

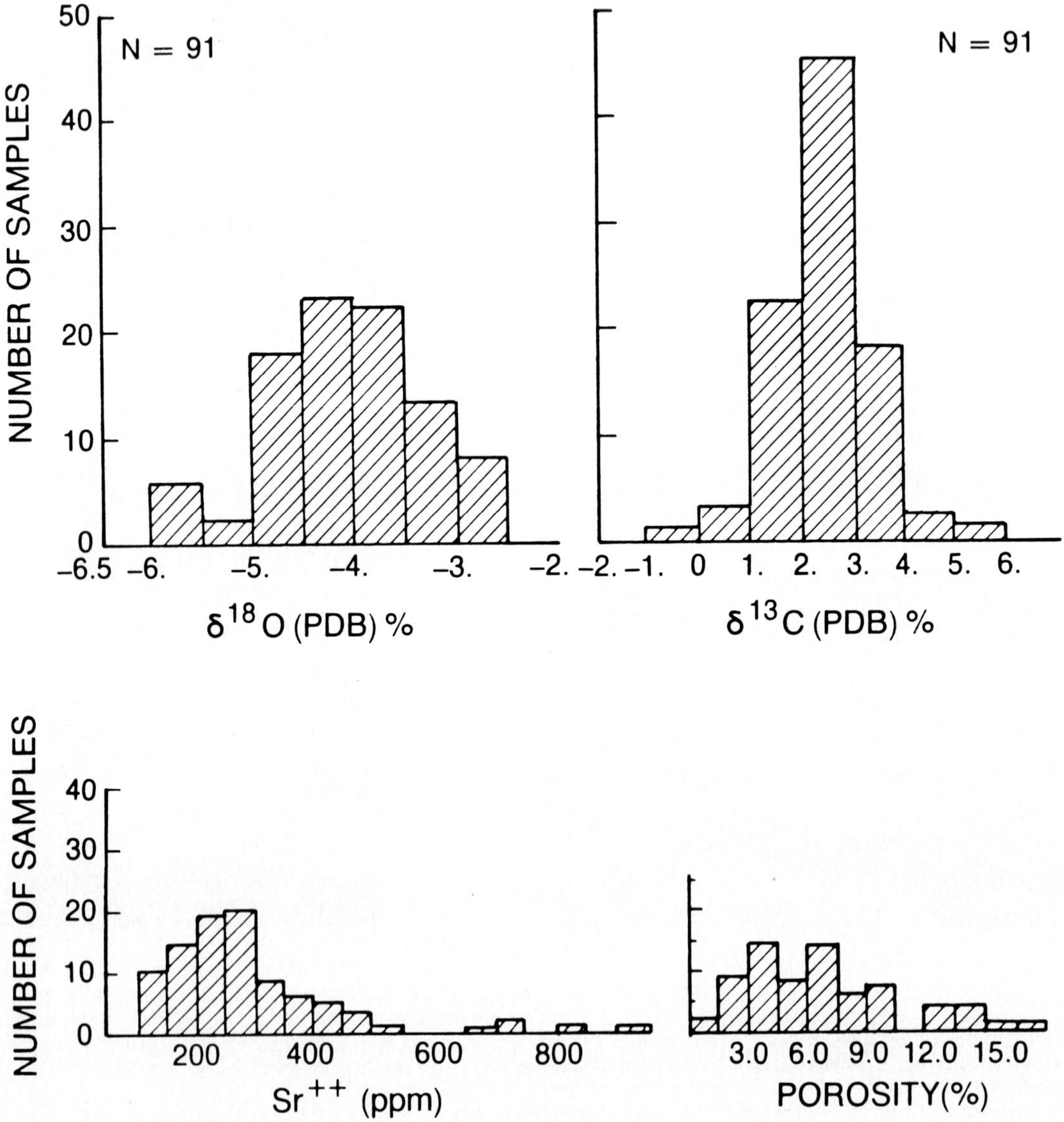

FIG. 13.—Histogram of the $\delta^{18}O$, $\delta^{13}C$, strontium concentration and porosities for 91 whole-rock samples from the Stuart City Trend. Porosities were not determined on all samples analyzed for strontium and stable isotopic composition (from Prezbindowski, 1981).

composition of the water, and $\delta^{18}O$ composition of the carbonate. Knowledge of any two of the three variables allows the determination of the third assuming equilibrium relationships. This relationship allows the testing of specific diagenetic hypothesis.

The importance of the oxygen mole ratio of the fluid to the carbonate country rock in a diagenetic system has been pointed out by Lloyd (1977), Land (1980), and Prezbindowski (1981). The degree of oxygen isotopic shift that occurs in the formation water or precipitated calcite is dependent on the openness and reactivity of the diagenetic system. In an open system, the mole ratio of oxygen in the water to that in the country rock would be very large. Major oxygen isotopic shifts during diagenesis would then be observed in the recrystallized country rock or precipitated cement. This system can be considered a fluid dominated system. If the system is very restricted, the mole ratio of the oxygen in the water to that in the country rock will be small. A major oxygen isotopic shift should be observed in the water during diagenesis but not in the recrystallized country rock or precipitated cement. It is considered a rock dominated system.

The $\delta^{18}O$ composition of the equant spar calcite cements analyzed ranged from −1.7 to −6.6‰, relative to PDB (Fig. 14). These cement types will be discussed individually.

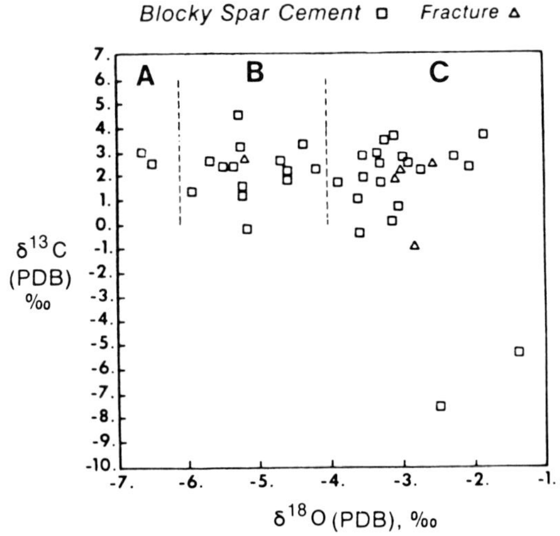

FIG. 14.—Carbon and oxygen isotopic compositional cross-plot for equant spar calcite cements from the Stuart City Trend. Group A, extremely coarse crystalline, equant spar calcite cements. Group B, ^{18}O depleted coarse to very coarse crystalline, equant spar calcite cements. Group C, ^{18}O enriched, medium to very coarsely crystalline, equant spar calcite cements. The square symbol indicates pore filling equant spar calcite cement samples. The triangle symbol indicates equant spar calcite cements taken from fractures.

Finely Crystalline, Isopachous Equant Spar Calcite Cement

This cement is characteristically 30–80 microns in thickness. Individual crystals are transparent with a rhombic morphology (Fig. 12A). It was only found associated with highly altered carbonates; i.e., intense neomorphism and abundant evidence of leaching.

This finely crystalline, equant, isopachous cement is believed to be of fresh-water phreatic origin. The evidence that generally supports this conclusion follows.

1. Microprobe analyses indicate no significant trace elemental content (i.e., Mg^{2+}, Sr^{2+}, Fe^{2+} and Mn^{2+} are below detectability limits). This suggests that precipitation occurred from an oxidized water, depleted with respect to magnesium and strontium (meteoric water?).
2. The rhombic crystal habit of the cement crystals suggests that precipitation occurred in a magnesium depleted water (Folk, 1974).
3. The close association of intense neomorphic alterations, such as micrite rims altered to microspar, suggests interaction with a meteoric type water which was Mg^{2+} depleted.
4. Whole-rock, allochem, and cement $\delta^{18}O$ compositions are nearly identical. They are depleted with a mean value of −5.6‰ (PDB). This average value is compatible with a calcite precipitated from a coastal meteoric-water at near surface temperatures. The lack of a $\delta^{18}O$ distinction between the individual diagenetic components indicates extensive isotopic homogenization has occurred.
5. Morphologically similar cements have been documented from the fresh-water phreatic environment of Joulters Cays in the Bahamas (Halley and Harris, 1979).

However, this cement does not form a "cross-cutting mosaic" with the altered skeletal substrate, a textural characteristic of phreatic-altered corals (Pingitore, 1976). On the other hand, the large, generally equant calcite crystals within the skeletal grain suggest that mineralogic equilibration did occur in a phreatic environment.

Coarse to Very Coarsely Crystalline Equant Spar Calcite Cement (^{18}O Depleted)

This cement type makes up approximately 7% of the porous facies in the Stuart City Trend. It is distinguished by its coarsely crystalline morphology, close association with early dissolution textures and depleted $\delta^{18}O$ composition. It is believed to have formed in a near surface fresh-water phreatic system. The following observations support this interpretation.

1. Microprobe analyses indicated that Mg^{2+}, Sr^{2+}, Fe^{2+} and Mn^{2+} concentrations were below detectable limits.
2. This cement is commonly associated with early dissolution textures, e.g., broken and collapsed

micrite rims, moldic pores, and extensive neomorphism.

3. The $\delta^{18}O$ compositional range of $-5.9‰$ to $-4.2‰$ is compatible with an origin from an open, coastal meteoric water system at near surface temperatures (Fig. 10).
4. It is closely associated with exposure zones as defined by dripstone and meniscus cements and post dates the previously discussed finely crystalline, isopachous, equant spar calcite cement.

BURIAL DIAGENESIS

Coarse to Very Coarsely Crystalline Equant Spar Calcite Cement (^{18}O Enriched)

The cement makes up approximately 9% of the porous facies studied in the Stuart City Trend. It can be distinguished by its coarsely crystalline morphology, association with texturally unaltered limestones and relatively enriched $\delta^{18}O$ compositions. This cement type is believed to have formed in restricted diagenetic systems during shallow and intermediate burial, i.e., <2,000 meters. The following observations support this conclusion.

1. Microprobe analyses of the pre-pyrobitumen equant spar calcite cement failed to detect Fe^{2+}, Mn^{2+} and Sr^{2+}. The apparent lack of detectable amounts of iron and manganese is significant because they are very common late-stage diagenetic components of equant spar calcite cements (Walls *et al.*, 1979; Wood and Armstrong, 1975; Loucks, 1976; Oldershaw and Scoffin, 1967). This absence suggests that diagenetic fluids from the basinal and overlying sediments were not introduced into the porous facies of the Stuart City Trend during shallow and intermediate burial.
2. The $\delta^{18}O$ composition of this cement type ranges from $-3.9‰$ to $-1.8‰$ with an average value of $-2.9‰$. This compositional range is similar to that determined for the radial-radiaxial cements, interpreted to be of marine origin. Precipitation took place in a restricted subsurface diagenetic system where the water was in oxygen isotopic equilibrium with a bulk rock having a marine or near-marine $\delta^{18}O$ composition (rock dominated system).
3. The $\delta^{18}O$ composition of the present day formation waters indicate that the deeper Edwards Formation and Stuart City Trend can be considered a restricted, rock dominated system (Prezbindowski, 1981).
4. The cement is most commonly associated with limestones that have well preserved depositional and early marine diagenetic textures.
5. Pyrobitumen inclusions and strontium enrichment occur in the outer 0.1 mm zone of some pore lining crystals of this type. Documenting a close timing relationship between the final phase of equant spar calcite cementation, hydrocarbon introduction and present day formation waters.

Extremely Coarsely Crystalline Equant Spar Calcite (High Temperature)

This calcite is very coarsely crystalline, having some crystals greater than 4 cm in size. In calcite veins, they cut grains and previously discussed cements placing them very late in the diagenetic sequence. It has a gray to milk-white color and is volumetrically a very minor component (Fig. 12B).

The two samples analyzed have the most ^{18}O depleted isotopic compositions determined from the Stuart City Trend ($\delta^{18}O$ of $-6.6‰$ and $-6.5‰$), respectively. The surrounding unaltered country rock has $\delta^{18}O$ compositions of $-3.5‰$ and $-3.9‰$. The diagenetic fluids responsible for the formation of this calcite have caused a significant $\delta^{18}O$ depletion to occur, relative to the country rock. This suggests an open diagenetic system; i.e., large amounts of oxygen being contributed from the water. It most likely formed by the interaction of the country rock with either of two types of diagenetic waters: (1) an ^{18}O depleted water (meteoric water) at a low temperature (surface or near surface) or (2) an enriched water (positive $\delta^{18}O$ composition) at a high subsurface temperature.

A $\delta^{18}O$ composition of a $-6.5‰$ is too depleted to be explained by a surface or near surface alteration with meteoric water in a coastal setting. Temperatures lower than 25°C would require a meteoric water with a $\delta^{18}O$ composition of $-4.5‰$ to $-6.5‰$ (relative to SMOW) to produce a $-6.5‰$ calcite cement. Present day coastal rains generally have an average $\delta^{18}O$ compositional range of $-2.0‰$ to $-4.0‰$ (Dansgaard, 1964). Since the $\delta^{18}O$ composition of the Cretaceous oceans was similar to today's oceans (Lowenstam, 1961), the $\delta^{18}O$ composition of Cretaceous, coastal meteoric water can be approximated by modern values. As a result, a low temperature meteoric-water origin is believed unlikely.

A high-temperature subsurface water out of isotopic equilibrium with the country rock is capable of producing ^{18}O depleted equant spar calcite cements. Such a diagenetic fluid must be introduced rapidly along a restricted pathway (fractures or faults) to avoid oxygen isotopic and temperature reequilibration with the surrounding country rock. Depending on the initial temperature (25–180°C), a fluid 12.5–28°C warmer than the surrounding country rock is capable of precipitating a calcite that is 1.5‰ more depleted in ^{18}O than the previously precipitated reservoir calcite. The upward movement of heated waters via faults has been documented in the central and south-central Texas area (Woodruff, 1980; Prezbindowski, 1981; Land and Prezbindowski, 1981). This finding lends support to a high-temperature origin for the diagenetic fluids responsible for the formation of this calcite.

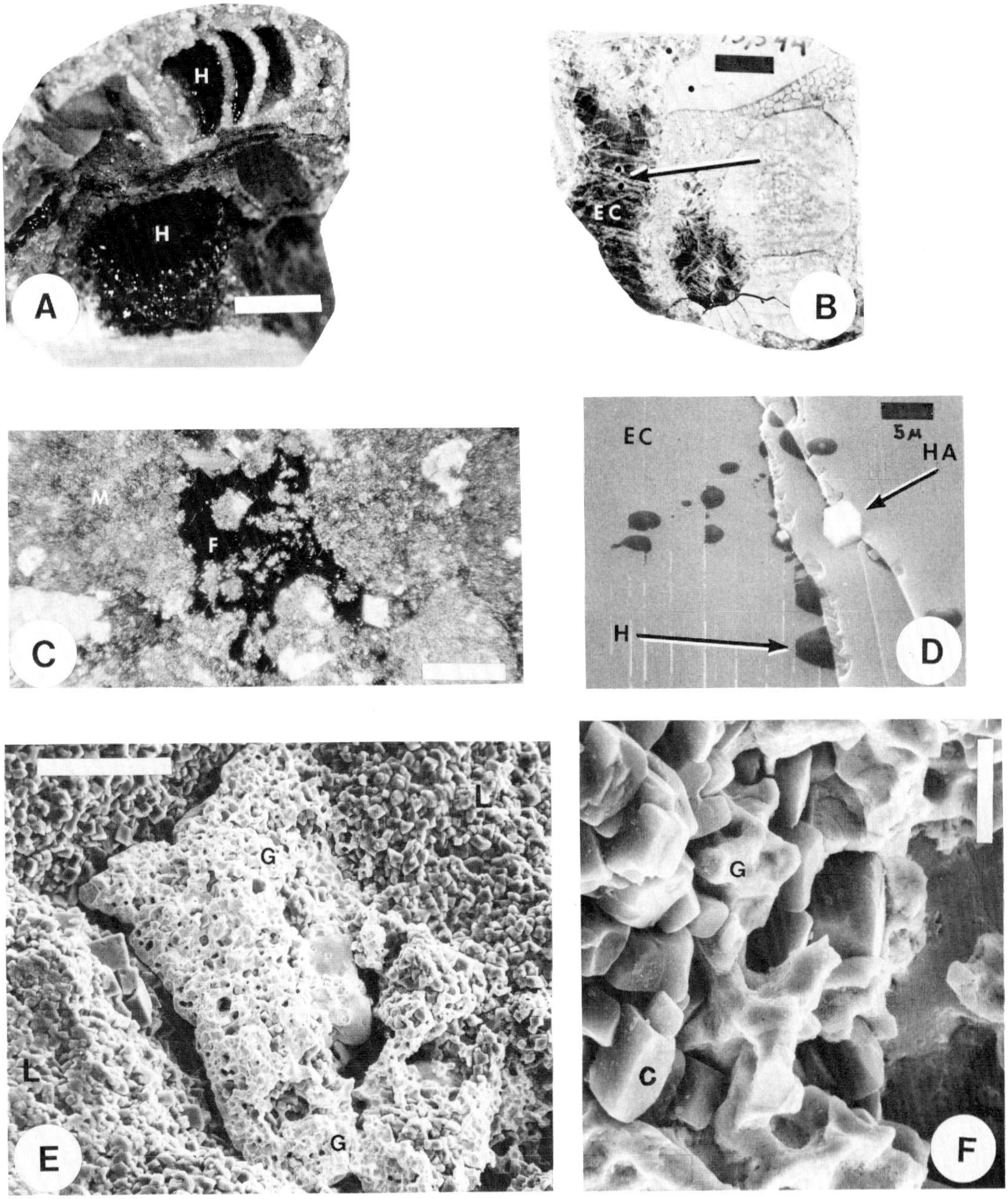

FIG. 15.—Slab, scanning electron micrographs and photomicrographs: *A*, Pyrobitumen (H) with gravity controlled distribution in primary intraparticle pores. Core sample, from the General Crude No. 1 Strawn well, 13,454 ft interval. Scale bar = 1.0 cm. *B*, Cement reduced intraparticle porosity in a caprinid shell. The dark, coarsely crystalline, equant spar calcite cement contains numerous hydrocarbon inclusions. Slab sample from the Tenneco No. 1 Schulz Well, 13,544 ft interval. Scale bar = 1.0 cm. *C*, Flourite (F) replacing calcite (cross-nicols view). From the General Crude No. 1 Wernli well, 13,535 ft interval. Scale bar = 0.5 mm. *D*, S.E.M. photograph of dark, equant spar calcite cement shown in photograph (B). Halite (HA) and hydrocarbons (H) on a freshly broken surface of the cement. Fluids originated from inclusions in the calcite cement crystal. *E*, Scanning electron micrograph of galena (G) replacing calcite (L). Note the sponge-like geometry of the galena. From the General Crude No. 1 Strawn well, 13,541 ft interval. Scale bar = 0.1 mm. *F*, Scanning electron micrograph detailing the galena (g) and calcite (C) textural relationship illustrated in photograph (E).

Pressure Solution

Pressure solution was an important diagenetic process in the Stuart City Trend limestones. Stylolites are present in over 37% of the thin sections studied.

The exact timing and duration of stylolitization in the Stuart City Trend could not be established. Clearly, stylolitization was an active process, which postdated the precipitation of the early, medium to coarse crystalline equant spar calcite cements (Fig. 12C, D). Stylolites that clearly cut across the extremely coarsely crystalline calcite cements of high temperature origin were not observed.

It is likely that pressure solution was an ongoing diagenetic process retarded only with the introduction of hydrocarbons (<2,000 meters of burial). The latter loss of hydrocarbons or cessation of hydrocarbon migration into the porous carbonates may have allowed a reactivation of pressure solution in a water-saturated environment.

Visual observation suggests that pressure solution has mobilized a volumetrically important amount of calcium carbonate. This mobilized carbonate could have been the source of the ^{18}O enriched, coarse to very coarse crystalline equant spar calcite cements believed to have formed in a restricted, subsurface diagenetic system.

Late-Stage Dissolution

A minor late-stage dissolution event occurred prior to and/or during hydrocarbon migration in some sections of the Stuart City Trend (Fig. 12E). The pore-lining coarse to very coarsely crystalline equant spar calcite cements have undergone some late stage dissolution. The surfaces of these crystals are corroded and coated with pyrobitumen. Only a small amount of calcite cementation has occurred after hydrocarbon migration and thermal alteration of the residual hydrocarbons (Fig. 15A-D). This is evidenced by the rare occurrence of pyrobitumen inclusions in the pore-lining equant spar calcite crystals.

The close association of pyrobitumens with late-stage dissolution effects suggests a causative relationship between hydrocarbon migration and dissolution. Whether this dissolution resulted from the

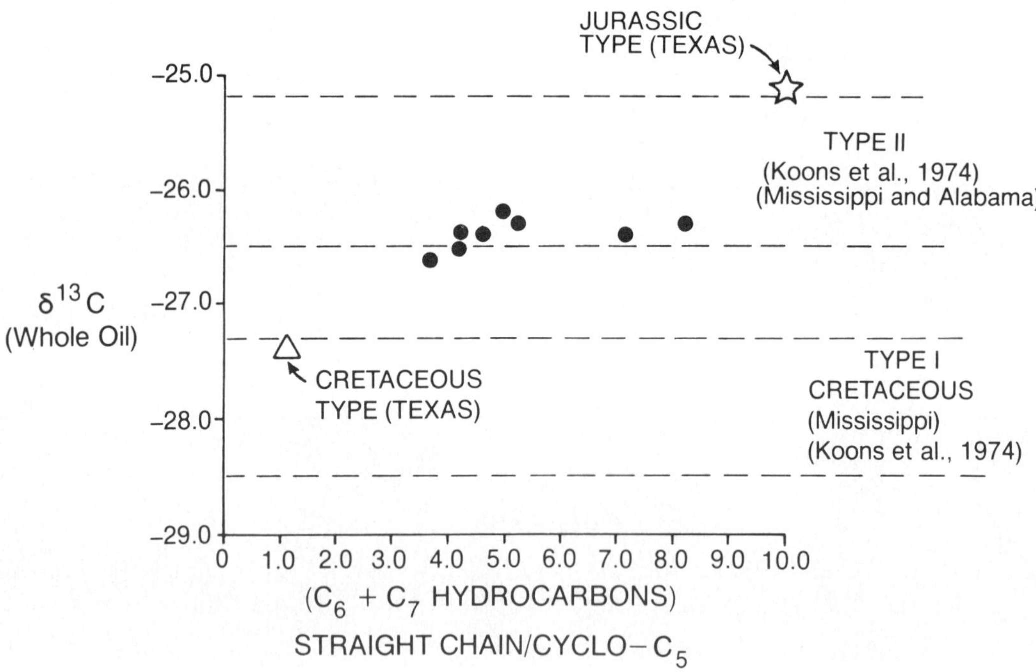

FIG. 16.—Cross plot of the $\delta^{13}C$ whole oil composition against the ratio of the straight chain C_6 and C_7 compounds to the cyclo-C_5 compounds for 8 oil samples from Lower Cretaceous, Edwards reservoirs in the Atascosa and Karnes Trough areas. The typical Jurassic and Cretaceous oil types for the south Texas area are illustrated by the triangle and star symbols (verbal communication, Jack Williams, Amoco Production Company). The $\delta^{13}C$ compositional range determined by Koons *et al.* (1974), for the two major types of oils found in the Lower Tuscaloosa (Cretaceous) reservoirs of Mississippi and Alabama are illustrated by the dashed lines. Koons *et al.* (1974) determined that their type I oils originated from Cretaceous source rocks. They did not state what the source was for their type II oils. Note the agreement between the type Jurassic and Cretaceous oils of south Texas and the Cretaceous and type II oils (source unspecified) from Mississippi and Alabama (Koons *et al.*, 1974). The Edwards oils appear to be a mixture of Cretaceous and Jurassic oils.

introduction of CO_2 and/or organic acids associated with hydrocarbons is unknown at this time. Late secondary porosity formation in many sandstone and carbonate units have been attributed to these processes (Curtis, 1978; Schmidt and McDonald, 1979; Moore and Druckman, 1981).

Hydrocarbons

Migration and source.—The timing of hydrocarbon migration into the reservoir rocks of the Stuart City Trend had not been established. This timing is an important reference point in the diagenetic history of these carbonates. The introduction of hydrocarbons represents the initial stage of deep-burial diagenesis as defined in this paper. In order to determine the approximate time of hydrocarbon migration, the potential source beds must be delineated. It is likely that the organic-poor carbonates of the Stuart City Trend were not source rocks. Explorationists have generally assumed that the bounding Lower Cretaceous basinal sediments of the ancestral Gulf of Mexico provided the source for the hydrocarbons of the Stuart City Trend and the updip Edwards Formation; Rose (personal communication, 1978) suggests a possible Jurassic source for these hydrocarbons. The upward movement of probable Jurassic Formation waters into the Lower Cretaceous carbonates of the Stuart City Trend and Edwards Formation has been documented (Prezbindowski, 1981; Land and Prezbindowski, 1981) and lends credence to a possible Jurassic source for the hydrocarbons.

A study of the pristane/phytane ratios, $\delta^{13}C$ compositions, optical rotations and the relative abundance of the C_6 and C_7 compounds in oil samples from several of the Edwards hydrocarbon fields strongly support a mixed Jurassic and Cretaceous source for the present-day Edwards oils (Fig. 16) (oral communication, Jack Williams).

Hydrocarbons in the Edwards Formation reservoirs and probably the Stuart City Trend gas reservoirs are believed to have been originated from Cretaceous and Jurassic source beds. Using the burial history and present-day geothermal gradient of the Stuart City Trend, a general statement on the maximum depth at which hydrocarbon migration was initiated can be made. Certainly the Jurassic hydrocarbon source beds were buried deeper than the porous reservoir carbonates of the Stuart City Trend and Edwards Formation. The Lower Tertiary sediments of West Africa (Tissot *et al.*, 1974) can provide a roughly comparable burial-temperature model of hydrocarbon generation for these Cretaceous reservoirs. This model suggests that the principal zone of oil formation was initiated at a temperature of 70°C. By analogy, hydrocarbon migration into the reservoir rocks of the Stuart City Trend would have been initiated from deeper Jurassic source beds before reaching a maximum burial depth of 2000 m (>70°C).

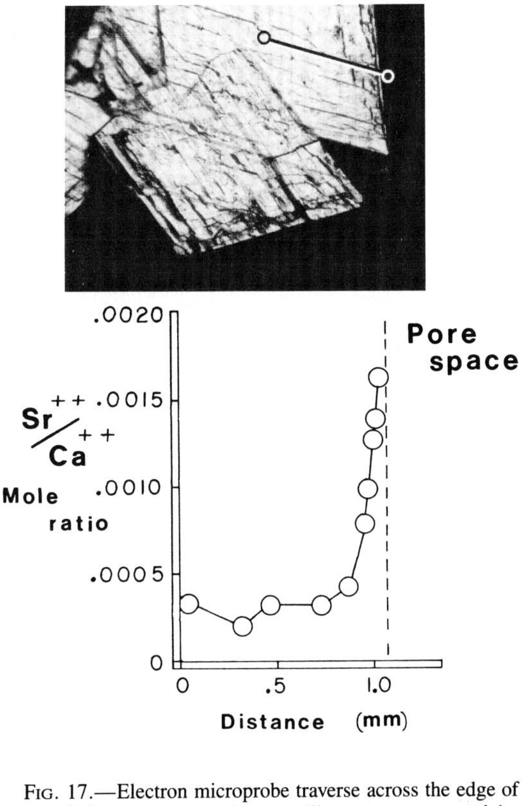

FIG. 17.—Electron microprobe traverse across the edge of a pore lining, very coarsely crystalline, equant spar calcite crystal showing a rapid increase in the strontium concentration toward the pore space. Circles indicate the approximate two standard deviation relative error. From the General Crude No. 1 Strawn well, 13,493 ft interval.

Deep Burial Diagenesis

Pyrobitumen inclusions occur in a small percentage of low iron, very coarsely crystalline $\delta^{18}O$ equant spar calcite cements, and as geopetal pore fillings (Fig. 15). They range in size from less than 50 microns to over 4 cm.

Present-day reservoir temperatures and the late-diagenetic timing of pyrobitumen formation suggest that thermal alteration of liquid hydrocarbons formed the pyrobitumen. The geopetal mode of occurrence evidences its original liquid nature (Fig. 15A). The $\delta^{13}C$ composition of pyrobitumen from the 4,190.5 meter (13,724 ft) interval of the No. 1 Schultz well is −26.0‰ (PDB) very close to the typical Jurassic oil of this area (Fig. 16).

Was equant spar calcite cementation an important process during and after hydrocarbon migration? This is an important question since most section cored in the Stuart City Trend is now water wet. An answer can be given by correlating the strontium concentration in present day formation waters with strontium concentration profiles across pore lining equant spar calcite crystals (Fig. 17). The strontium concentration in formation waters from the Stuart City Trend varies from 4,980 (Sr^{2+}/Ca^{2+} = .08) to 615 (Sr^{2+}/Ca^{2+} = .04) ppm. In the case of the General Crude #1 Strawn Well, high strontium levels in the formation water (4,980 ppm, Sr^{2+}/Ca^{2+} = .08) correlates with the strontium enrichment of the outer 0.1 mm layer of the pore lining equant spar calcite cement (Fig. 17). Not all pore lining calcite crystals show a strontium enrichment toward the pore space. This variability is believed to be a function of the accessibility of basinal fluids rich in Sr^{2+} to the individual pore systems. These fluids were introduced along major faults starting during Eocene time (Prezbindowski, 1981). This time interval correlates with the inferred maximum depth of burial for the introduction of hydrocarbons into the Lower Cretaceous carbonates. The position of the bitumen inclusions and strontium enriched zone indicates that only a minor amount of equant spar calcite precipitation has occurred since Eocene time.

Small amounts of fluorite were noted replacing limestone in one well (Fig. 15C). This replacement is nonselective. Formation waters produced from this well precipitated significant amounts of fluorite scale in the well tubing (Bill Carnegie, written communication, 1977). The nonselective replacement by the fluorite and present-day formation-water chemistry (Land and Prezbindowski, 1981) suggest that fluorite precipitation may be an ongoing process and is associated with the basinal waters introduced into the Stuart City Trend along fault planes.

Galena is a minor late-stage diagenetic component and occurs as small 0.3–4 mm subhedral to euhedral, octahedral appearing crystals along fracture surfaces and replacement of the limestone country rock (Fig. 15E, F). Although high concentrations of lead and zinc have been found in formation waters from several of the wells producing from the Stuart City Trend; only galena has been documented in cores. Like the fluorite, galena precipitation is believed to be directly related to the basinal waters introduced into the Stuart City Trend via fault planes. Woronick and Land (1983) have reported the occurrence of galena and sphalerite in the updip Pearsall Formation and attribute their occurrence to the introduction of similar basinal waters.

SUMMARY

The Lower Cretaceous shelf-edge carbonate complex known as the Stuart City trend has had a relatively simple burial history and is currently near its maximum burial depth (Fig. 18). The current textural and geochemical character of these limestones was largely determined by early diagenetic processes (Fig. 19) consisting of extensive marine cementation in the porous higher energy facies associated with the positive carbonate buildups along the trend. This cement makes up 16% by volume of the porous facies and is believed

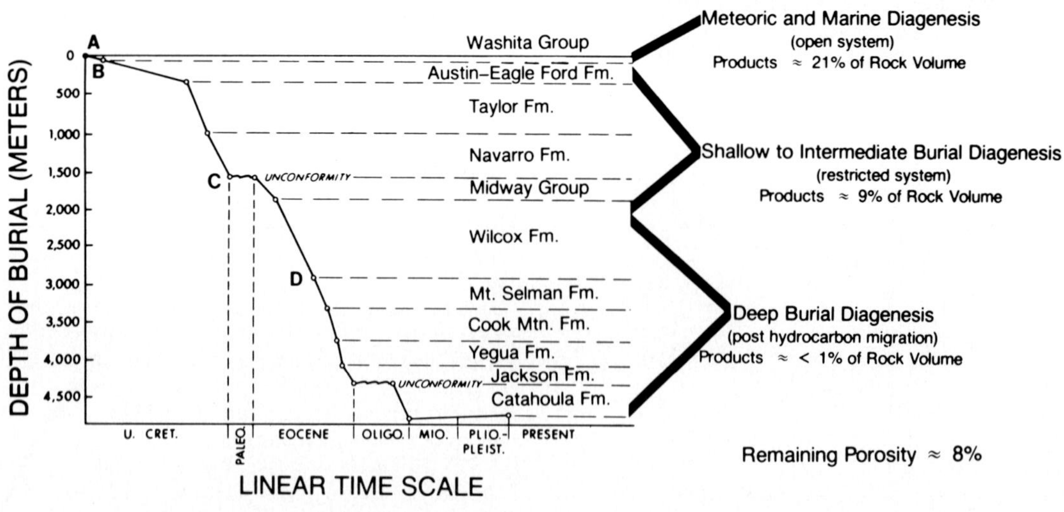

FIG. 18.—Approximate burial curve for the Stuart City Trend in the Karnes County area of Texas. Major growth faulting along the basinward side of the trend was initiated during Eocene time with deposition of the Wilcox Formation. Initial hydrocarbon migration is believed to predate Middle Eocene time, based on analogy with Tissot *et al.* (1974), burial-temperature model for hydrocarbon generation in Lower Tertiary sediments of West Africa. This figure was constructed from information taken from Bebout *et al.* (1976) and Plummer (1966). Intervals A, B, C and D correlate with the cross-sections in Figure 19.

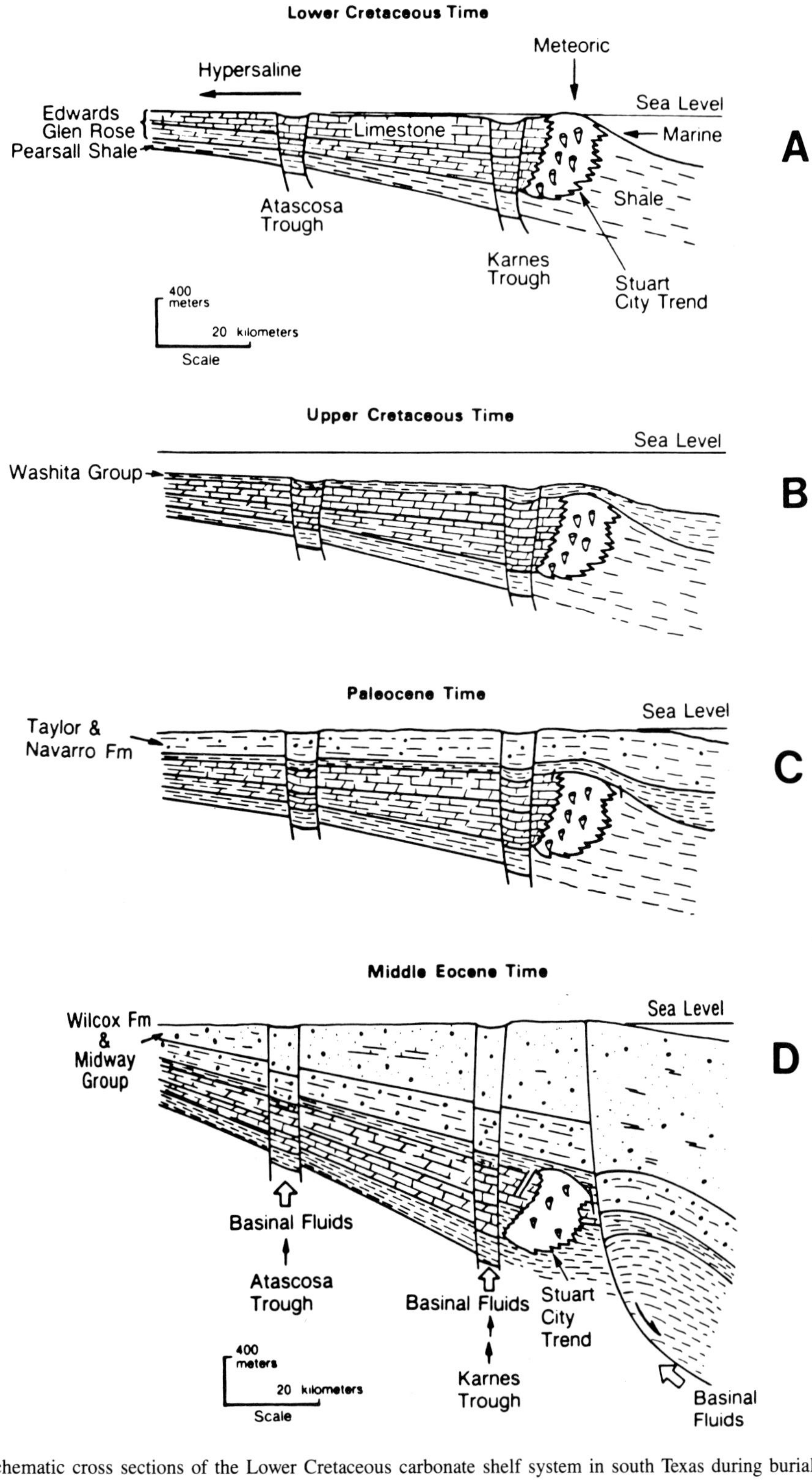

FIG. 19.—Schematic cross sections of the Lower Cretaceous carbonate shelf system in south Texas during burial.

to have originally been a fibrous, high-magnesium calcite. It was neomorphically altered to form the present-day radial-radiaxial cement.

Meteoric water diagenesis was closely associated and sometimes interrelated with marine diagenesis. This relationship was dynamic and controlled by the evolution of small carbonate buildups along the trend, which acted as local meteoric water recharge points during periodic exposure.

The development of this diagenetic system caused extensive textural (secondary porosity formation, neomorphism and equant spar calcite cementation) and geochemical (elemental and stable isotopic) alteration of the marine sediments. The products of these ancient meteoric water systems can be found with varying frequency in the upper portions of the buildups in the Stuart City Trend. Equant spar calcite cements formed in this diagenetic environment reduced overall porosity by approximately 7%.

The relative enriched $\delta^{18}O$ composition and higher strontium concentrations in whole rock samples from the mudstones and wackestones of the interbuildup areas suggest that these sediments were not altered by meteoric diagenesis. Mineral stabilization in these sediments probably occurred during burial with marine waters.

At the end of Stuart City deposition, increased subsidence and the deposition of the deeper water, fine-grained sediments of the Georgetown Formation completed the encasement of the Stuart City Trend carbonate package by relatively fine-grained, low permeability sediments (Fig. 18). Thus a transition was made from an open diagenetic system characterized by high fluxes of marine and meteoric waters to a more restricted diagenetic system isolated from major outside sources of diagenetic fluids. This carbonate package during shallow and intermediate burial underwent internal reorganization (porosity reduction) by pressure solution and the precipitation of iron poor, $\delta^{18}O$ enriched equant spar calcite cements. This process is believed to have further reduced overall porosity by approximately 9%.

The introduction of hydrocarbons was closely associated with a minor, late stage dissolution event. It is likely that this dissolution was caused by products of organic maturation. Whether these products where CO_2, organic acids or some other acid source is unknown at this time. The introduction of hydrocarbons marked the end of volumetrically significant diagenetic alterations. These hydrocarbons were introduced into the Cretaceous reservoirs before burial to depths of 2000 meters from both Jurassic and Cretaceous sources.

Closely associated with the inferred maximum depth of hydrocarbon introduction, growth faulting along the front edge of the Stuart City Trend occurred in Eocene. This faulting was the result of clastic deposition (Wilcox Formation) across the shelf into the ancestral Gulf of Mexico basin. These faults allowed the introduction of hydrocarbons and basinal waters into the porous sections of the Stuart City Trend which were responsible for minor amounts of fluorite, galena and Sr^{2+}-rich equant spar calcite precipitation.

Deep burial diagenesis defined as post-hydrocarbon migration (Fig. 19) was volumetrically unimportant in altering the carbonates of the Stuart City Trend even though much of the rock section is currently water wet.

ACKNOWLEDGMENTS

This paper represents a portion of my dissertation research work carried out at The University of Texas at Austin. The National Science Foundation, NSF EAR-76 17774 supported this research. I am very grateful to the Bureau of Economic Geology of Texas and the oil and gas companies who assisted me in gathering material and information for this study. Don Bebout, Bob Loucks and Rick Schatzinger shared many important geological observations on the study area. I am grateful to Amoco Production Company for their assistance in publishing this work. Robert Handford, Mitch Harris, Merrell Miller and Ed Pittman offered guidance and constructive criticism during the preparation of this paper. I owe them many thanks. I am particularly grateful to Lynton Land for his support, patience and guidance during the formative stages of this work.

REFERENCES

ACHAUER, C.A., 1977, Contrasts in cementation, dissolution, and porosity development between two Lower Cretaceous reefs of Texas, *in* D.G. Bebout, and R.G. Loucks (eds.), Cretaceous Carbonates of Texas and Mexico: Univ. Texas-Austin, Bur. Econ. Geology, Rept. Inv. 89, p. 127–135.

ALLAN, J.R., AND MATTHEWS, R.K., 1977, Carbon and oxygen isotopes as diagenetic and stratigraphic tools: Data from surface and subsurface of Barbados West Indies: Geology, v. 5, p. 16–20.

BATHURST, R.G.C., 1959, The cavernous structure of some Mississippian *Stromatactis* Reefs in Lancashire, England: Jour. Geology, v. 67, p. 506–521.

______, 1971, Carbonate sediments and their diagenesis: Elsevier, Developments in Sedimentology 12, 620 p.

BEBOUT, D.G., AND LOUCKS, R.G., 1974, Stuart City Trend Lower Cretaceous, South Texas (A Carbonate Shelf-Margin Model of Hydrocarbon Exploration): Univ. Texas-Austin, Bur. Econ. Geology, Rept. Inv. 78, p. 80.

______, Luttrell, P.E., and Seo, J.H., 1976, Regional Tertiary cross-sections—Texas Gulf Coast: Univ. of Texas-Austin, Bur. Econ. Geology, Geol. Circ. 76–5, 15 p.

______, Schatzinger, R.A., and Loucks, R.G., 1977, Porosity distribution in the Stuart City Trend, Lower Cretaceous, South Texas, in D.G. Bebout and R.G. Loucks (eds.), Cretaceous Carbonates of Texas and Mexico: Univ. Texas-Austin, Bur. Econ. Geology, Rept. Inv. 89, p. 234–256.

Benson, L.V., and Matthews, R.K., 1971, Electron microprobe studies of magnesium distribution in carbonate cements and recrystallized skeletal grainstones from the Pleistocene of Barbados, West Indies: Jour. Sed. Petrology, v. 41, p. 1018–1025.

Bricker, O.P. (ed.), 1971, Carbonate Cements: Johns Hopkins University Studies in Geology, no. 19, 376 p.

Curtis, C.D., 1978, Possible links between sandstone diagenesis and depth-related geochemical reactions occurring in enclosing mudstones: Jour. Geol. Soc. London, v. 135, p. 107–117.

Dansgaard, W., 1964, Stable isotopes in precipitation: Tellus, v. 16, p. 436–473.

Davies, G.R., 1977, Former magnesian calcite and aragonite submarine cements in Upper Paleozoic Reefs of the Canadian Arctic: A Summary: Geology, v. 5, p. 11–15.

______, and Krouse, H.R., 1975, Carbon and oxygen isotopic composition of Late Paleozoic calcite cements, Canadian Arctic Archipelago—Preliminary results and interpretation: Canada Geol. Survey Paper 75–1, pt. B, p. 215–220.

Dickson, J.A.D., 1965, A modified staining technique for carbonates in thin section: Nature, v. 205, p. 587.

Dunham, R.J., 1971, Meniscus cement, *in* O.P. Bricker (ed.), Carbonate Cements: Johns Hopkins Univ. Studies in Geology, no. 19, p. 297–300.

Folk, R.L., 1974, The natural history of crystalline calcium carbonate: effect of magnesium content and salinity: Jour. Sed. Petrology, v. 44, p. 40–53.

Fowler, P., 1956, Faults and folds of south Texas: Gult Coast Assoc. Geol. Societies, Trans. v. 6, p. 37–42.

Friedman, G.M., and Kolesar, P.T., Jr., 1971, Fresh-water carbonate cements, *in* O.P. Bricker (ed.), Carbonate Cements: Johns Hopkins Univ. Studies in Geology, no. 19, p. 122–126.

Ginsburg, R.N., Schroeder, J.H., and Shinn, E.A., 1971, Recent synsedimentary cementation in subtidal Bermuda reefs, *in* O.P. Bricker (ed.), Carbonate Cements: Johns Hopkins Univ. Studies in Geology, no. 19, p. 54–58.

Gross, M.G., 1964, Variations in the O^{18}/O^{16} and C^{13}/C^{12} ratios of diagenetically altered limestones in the Bermuda Islands: Jour. Geology, v. 72, p. 170–194.

Halley, R.B., and Harris, P.M., 1979, Fresh-water cementation of a 1000-year-old oolite: Jour. Sed. Petrology, v. 49, p. 969–988.

Holden, R.N., 1963, A review of the discovery and development of the Lower Cretaceous Reef in south Texas, in Geology of Peregrina Canyon and Sierra de El Abra: Corpus Christi Geol. Soc., Guidebook, Field Trip, May 23–26, 1963, p. 65–71.

Kendall, A.C., and Tucker, M.E., 1973, Radiaxial fibrous calcite: A replacement after acicular carbonate: Sedimentology, v. 20, p. 365–389.

Kerr, R.S., 1977, Facies, diagenesis, and porosity development in a Lower Cretaceous Bank Complex, Edwards Limestone, north-central Texas and Mexico, *in* D.G. Bebout and R.G. Loucks (eds.), Cretaceous Carbonates of Texas and Mexico: Univ. Texas-Austin, Bur. Econ. Geology, Rept. Inv. 89, p. 216–232.

Koons, C.B., Bond, J.G., and Pierce, F.L., 1974, Effects of depositional environments and post-depositional history on chemical composition of Lower Tuscaloosa oils: Amer. Assoc. Petroleum Geologists Bull., v. 48, p. 1272–1280.

Land, L.S., 1971, Submarine lithification of Jamaican Reefs, *in* O.P. Bricker (ed.), Carbonate Cements: Johns Hopkins Univ. Studies in Geology, no. 19, p. 59–62.

______, 1977, Hydrogen and oxygen isotopic composition of chert from the Edwards Group, Lower Cretaceous, Central Texas, *in* D.G. Bebout and R.G. Loucks (eds.), Cretaceous Carbonates of Texas and Mexico: Univ. Texas-Austin, Bur. Econ. Geology, Rept. Inv. 89, p. 202–205.

______, 1980, The isotopic and trace element geochemistry of dolomite: The state of the art: Soc. Econ. Paleontologists and Mineralogist Special Pub. 28, p. 87–110.

______, and Goreau, T.F., 1970, Submarine lithification of Jamaican Reefs: Jour. Sed. Petrology, v. 40, p. 457–462.

______, Mackenzie, F.T., and Gould, S.J., 1967, Pleistocene history of Bermuda: Geol. Soc. America Bull., v. 78, p. 993–1006.

______, and Moore, C.H., 1980, Lithification, micritization and syndepositional diagenesis of biolithites on the Jamaican island slope: Jour. Sed. Petrology, v. 50, no. 2, p. 357–370.

______, and Prezbindowski, D.R., 1981, The origin and evolution of saline formation water, Lower Cretaceous carbonates, south-central Texas, U.S.A.: Jour. Hydrology, v. 54, p. 51–74.

Lloyd, R.M., 1977, Porosity reduction by chemical compaction stable isotope model: Amer. Assoc. Petroleum Geologists Bull., v. 61, p. 809.

Lohmann, K.C., and Meyers, W.J., 1977, Microdolomite inclusions in cloudy prismatic calcites: A proposed criterion for former high magnesian calcites: Jour. Sed. Petrology, v. 47, p. 1078–1088.

Loucks, R.G., 1976, Pearsall Formation, Lower Cretaceous, South Texas—depositional facies and carbonate diagenesis and their relationship to porosity: unpub. Ph.D. dissertation, Univ. Texas-Austin, 362 p.

Lowenstam, H.A., 1961, Mineralogy, O^{18}/O^{16} ratios, and strontium and magnesium contents of recent and fossil brachiopods and their bearing on the history of the oceans: Jour. Geology, v. 69, p. 241–260.

McCrea, J.M., 1950, On the isotopic chemistry of carbonates and a paleotemperature scale: Jour. Chemical Physics, v. 18, p. 849–857.

Moore, C.H., and Druckman, Y., 1981, Burial diagenesis and porosity evolution, Upper Jurassic Smackover, Arkansas and

Louisiana: Amer. Assoc. Petroleum Geologists Bull., v. 65, p. 597–628.
MÜLLER, G., 1971, "Gravitational" cement: an indicator for the vadose zone of the subaerial diagenetic environment, *in* O.P. Bricker (ed.), Carbonate Cements: Johns Hopkins Univ. Studies in Geology no. 19, p. 302–320.
OLDERSHAW, A.E., AND SCOFFIN, T.P., 1967, The source of ferroan and non-ferroan calcite cements in the Halkin and Wenlock Limestone: Geol. Jour., v. 5, p. 309–320.
O'NEIL, J.R., CLAYTON, R.N., AND MAYEDA, T.K., 1969, Oxygen isotope fractionation in divalent metal carbonates: Jour. Chem. Physics, v. 51, p. 5547–5558.
ORME, G.R., AND BROWN, W.W.M., 1963, Diagenetic fabrics in the Avonian Limestones of Derbyshire and north Wales: Yorkshire Geol. Soc. Proc., v. 34, p. 51–66.
PINGITORE, N.E., 1976, Vadose and phreatic diagenesis: processes, products and their recognition in corals: Jour. Sed. Petrology, v. 46, p. 985–1006.
PLUMMER, F.B., 1966, Cenozoic Systems in Texas, *in* The Geology of Texas, Vol. I Stratigraphy: Univ. Texas Bull. No. 3232, p. 519–818.
PREZBINDOWSKI, D.R., 1980, Microsampling technique for stable isotopic analyses of carbonates: Jour. Sed. Petrology, v. 50, p. 643–644.
______, 1981, Carbonate rock-water diagenesis Lower Cretaceous, Stuart City Trend, South Texas: unpub. Ph.D. dissertation, Univ. Texas-Austin, 237 p.
ROGERS, M.A., MCALARY, J.D., AND BAILEY, N.J.L., 1974, Significance of reservoir bitumens to thermal-maturation studies, Western Canada Basin: Amer. Assoc. Petroleum Geologists Bull., v. 58, p. 1806–1824.
ROSE, P.R., 1968, Edwards Formation, surface and subsurface, central Texas: unpub. Ph.D. dissertation, Univ. Texas-Austin, 301 p.
______, 1972, Edwards Group, surface and subsurface, central Texas: Univ. Texas-Austin, Bur. Econ. Geology Rept. Inv. 74, 198 p.
SCHMIDT, V., 1971, Early carbonate cementation in Middle Devonian bioherms, Rainbow Lake, Alberta, *in* O.P. Bricker (ed.), Carbonate Cements: Johns Hopkins Univ. Studies Geology no. 19, p. 209–215.
______, AND MCDONALD, D.A., 1979, Texture and recognition of secondary porosity in sandstones, *in* P.A. Scholle and P.R. Schluger (eds.), Aspects of Diagenesis—Symposia: Soc. Econ. Paleontologists and Mineralogists Spec. Pub. 26, p. 175–208.
SHINN, E.A., 1969, Submarine lithification of Holocene carbonate sediments in the Persian Gulf: Sedimentology, v. 12, p. 109–144.
______, 1971, Holocene submarine cementation in the Persian Gulf; *in* O.P. Bricker (ed.), Carbonate Cements: Johns Hopkins Univ. Studies in Geology no. 19, p. 59–62.
SIPPLE, R.F., 1968, Sandstone petrology, evidence from luminescence petrography: Jour. Sed. Petrology, v. 38, p. 530–554.
______, AND GLOVER, E.D., 1965, Structure in carbonate rocks made visible by luminescence petrography: Science, v. 150, p. 1283–1287.
TISSOT, B.D., ESPITALIE, J., AND COMBAZ, A., 1974, Influence and nature of diagenesis of organic matter in formation of petroleum: Amer. Assoc. Petroleum Geologists Bull., v. 58, p. 499–506.
WALLS, R.A., MOUNTJOY, E.W., AND FRITZ, P., 1979, Isotopic composition and diagenetic history of carbonate cements in Devonian Golden Spike Reef, Alberta, Canada: Geol. Soc. America Bull. Part 1, v. 90, p. 963–982.
WINLAND, H.D., 1971, Diagenesis of carbonate grains in marine and meteoric-waters: unpub. Ph.D. dissertation, Brown University, Providence, Rhode Island, 320 p.
WINTER, J.A., 1961, Stratigraphy of the Lower Cretaceous (subsurface) of south Texas: Gulf Coast Assoc. Geol. Socs. Trans., v. 11, p. 15–24.
WOOD, G.V., AND ARMSTRONG, A.K., 1975, Diagenesis and stratigraphy of the Lisburne Group Limestone of the Sadlerochit Mountains and adjacent areas, northeastern Alaska: Geol. Soc. America, Prof. Paper 857, p. 1–47.
WOODRUFF, C.M., JR., 1980, Regional tectonic features of Inner Gulf Coast: Oil and Gas Jour., Nov. 10, p. 264–275.
WOOTEN, J.W., AND DUNAWAY, W.E., 1977, Lower Cretaceous carbonates of central south Texas, a shelf-margin study, *in* D.G. Bebout and R.G. Loucks (eds.), Cretaceous Carbonates of Texas and Mexico: Univ. Texas-Austin, Bur. Econ. Geology, Rept. Inv. 89, p. 71–81.
WORONICK, R.E., AND LAND, L.S., 1983, Burial cements in Lower Cretaceous Pearsall Formation and Lower Glen Rose Formation, South Texas [abs.]: Amer. Assoc. Petroleum Geologists Bull., v. 67, p. 572–573.

LATE BURIAL DIAGENESIS, LOWER CRETACEOUS PEARSALL AND LOWER GLEN ROSE FORMATIONS, SOUTH TEXAS

R.E. WORONICK[1] AND L.S. LAND
Dept. Geological Sciences, University of Texas at Austin, Austin, Texas, 78712

ABSTRACT

Lower Cretaceous platform carbonates and shales, buried to depths in excess of 450 meters by the end of Eocene time, have been locally affected by late-stage diagenesis. In the South Texas study area, the rocks are presently buried to depths between 600 and 2800 meters. Burial diagenetic cements include ferroan and non-ferroan calcite, ferroan baroque dolomite, anhydrite, kaolinite, Ba-rich celestite, galena and sphalerite. A subsurface origin is evidenced by the absence of these minerals in outcrop, their occurrence in fractures, and their chemistry.

Carbonate cements are chemically and isotopically zoned; the iron content of baroque dolomite cement varies by as much as 10 mole percent from the center to the edge of single crystals. The lack of stratigraphic or regional trends in the distribution of iron in baroque dolomite suggests that the iron is derived from nearby shales. A strong negative correlation between $\delta^{13}C$ and average iron content of baroque dolomite suggests the simultaneous reduction of iron oxides and oxidation of organic material.

The late subsurface carbonate cements commonly have depleted $\delta^{18}O$ values which must either have formed by precipitation from hot brine or from isotopically depleted (meteoric) water. Both possibilities require the vertical movement of water along faults. Galena and sphalerite occur as minor components in some cores. Brines moving up faults after albitizing feldspars in more deeply buried formations could be the source of lead and zinc for these minerals. $^{87}Sr/^{86}Sr$ ratios for late diagenetic phases are similar to ratios of Edwards Formation brines in this area today, confirming that basinal brines have contributed to late-stage diagenesis. Most diagenetic phases can be qualitatively accounted for by the injection of basinal brines along fractures and faults into a meteoric-dominated aquifer.

INTRODUCTION

Lower Cretaceous carbonates in the subsurface of south-central Texas have been extensively studied (Rose, 1972; Loucks, 1976; Prezbindowski, 1981; Bay, 1980, 1982), yet significant diagenetic problems still remain. Most of the spar calcite cementation in these shelf and shelf-margin wackestones and grainstones occurred prior to hydrocarbon migration, yet the degree to which it was emplaced by essentially syndepositional meteoric processes as opposed to shallow burial processes is argued. Following emplacement of the spar, a sequence of events occurred which resulted in the development of some late secondary porosity and the emplacement of very different kinds of cements, including calcite, baroque dolomite, anhydrite, kaolinite, galena and sphalerite. It is these late diagenetic features on which we intend to focus. Aside from the chemical data, the late origin of these features is indicated by several lines of evidence. Late phases fill both matrix porosity and fractures, and since the fractures cut previous spar cements, the time of emplacement is constrained (Loucks, 1976). In addition, the late diagenetic suite of minerals is found only in the subsurface and the distribution of minerals is not restricted by formational boundaries.

The same suite of late phases, having similar chemistry, is found in both the Pearsall and lower Glen Rose Formations. Both formations are of Early Cretaceous age and have a combined thickness of about 300 meters. The Pearsall Formation was deposited on a shallow marine shelf as shoal-water carbonate complexes with contemporaneous terrigenous sediments (Loucks, 1977), whereas the Glen Rose Formation consists of carbonates which were subsequently deposited on the same evolving shelf as shoal-water carbonate complexes, open-shelf sediments and lagoonal sediments (Bay, 1980, 1982).

Most late diagenetic phases occur primarily as cements in fractures and in interparticle and moldic porosity, but also as replacements of other minerals. The most abundant cements, calcite, baroque dolomite and anhydrite, are best developed in the grainstone facies. Minor late phases include quartz, sphalerite, galena, pyrite, kaolinite and Ba-rich celestite. Nowhere are the late cements volumetrically important, and they typically constitute only a few volume percent of the rock section.

Samples were obtained from cores previously studied by Loucks (1976) and Bay (1980) in south Texas (Fig. 1). Today, these rocks are buried to depths between 600 and 2800 meters. Based on the geothermal gradient derived from bottom hole temperatures (23°C/km), these depths correspond to temperatures between 35 and 91°C. In the outcrop area, which is about 65 kilometers to the north, these formations have never

[1] Present address: Mobil Producing Texas & New Mexico, Inc., Nine Greenway Plaza—Suite 2700, Houston, Texas 77046

been buried more than 450 meters (Keith Young, personal communication). Thus these formations in the subsurface have also probably never been buried more than about 450 meters deeper than their present depth, corresponding to about 15°C higher formation temperature. Higher thermal gradients in this area since Cretaceous time cannot be ruled out, however.

MAJOR CEMENTS: LATE CALCITE

Coarsely crystalline (0.05 mm to 2.0 mm diameter), equant, zoned, ferroan and non-ferroan calcite cement is the most common late phase. In many cases it is difficult to assign a specific coarse spar cement to a specific emplacement time or mechanism. Where it precipitated competitively with baroque dolomite, or fills fractures which cross-cut previously formed cements, the timing is unequivocal. Calcite which replaces anhydrite nodules, some silicified, and which replaces feldspars, especially in the Glen Rose formation, is also of late origin. Equant non-ferroan calcite which formed during earlier diagenesis can be very similar in appearance to late non-ferroan calcite cement, however.

Electron microprobe traverses show ranges in iron content from 0.04 (detection limit) to 1.3 mole percent $FeCO_3$ and in magnesium content from 0.05 (detection limit) to 3.3 mole percent $MgCO_3$. Low-iron calcite is found both preceeding and post-dating baroque dolomite emplacement.

Isotopic analysis of microsamples collected using the technique described by Prezbindowski (1980) also indicates extreme variability in single crystals. $\delta^{18}O$ values range from -3.2 (typical of the early cements) to -14.0‰ PDB (Fig. 2), with up to 4‰ variation within 1 cm-sized crystals. The most negative values are from extremely coarse-grained non-ferroan calcite which formed after baroque dolomite. Calcite more negative in $\delta^{18}O$ than about -10‰ could not have formed at its present temperature from pure meteoric water, assuming that Cretaceous and post-Cretaceous meteoric water in this area could not have been more depleted than about -6‰ (Prezbindowski, 1981).

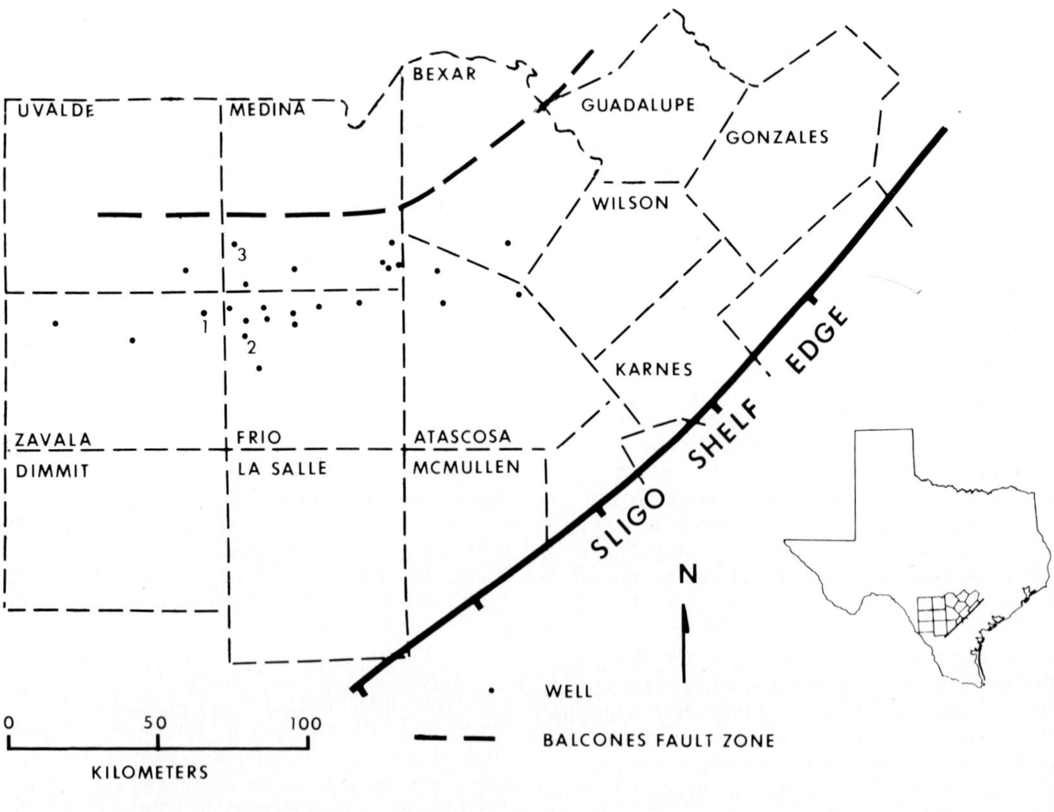

FIG. 1.—Map of the study area showing location of sampled cores, and wells mentioned in the text (1—Rowe #1 Kincaid, 2—Tenneco #1 Mack, 3—Tenneco #1 Ney). During Lower Cretaceous time, shoal-water carbonates and clastics derived from the north were deposited on an extensive shelf along the northern margin of the ancestral Gulf of Mexico.

Thus it is implied that the most negative cements must have formed at temperatures hotter than the rocks experience today. $\delta^{13}C$ values range from about 0 to +3‰ PDB, and do not correlate with any other measured variable.

MAJOR CEMENTS: BAROQUE DOLOMITE

Baroque, or "saddle-shaped" dolomite is typically characterized by undulose extinction, curved or saddle-shaped crystal faces, opaque white color and relatively large crystal size (AGI Glossary of Geology, 1980). In the Lower Cretaceous carbonates, zoned ferroan and non-ferroan baroque dolomite cement, 0.02 mm to 2.5 mm in size, occurs in primary pores, secondary pores and in fractures. Although dolomite *per se* increases in abundance toward the more restricted parts of the carbonate platform which outcrop extensively toward the northwest (Fisher and Rodda, 1969), baroque dolomite is restricted to the subsurface and is a late diagenetic phase everywhere it is found (Loucks, 1977). Baroque dolomite crystals are mostly calcian, containing 51 to 56 mole percent $CaCO_3$. The iron content ranges from 0.04 (detection limit) to 15 mole percent $FeCO_3$. Some crystals are strongly luminescent and zoned with iron; others are relatively homogeneous. The iron content of single zoned crystals can vary by as much as 10 mole percent. We found no correlation between iron content of baroque dolomite and depth, and no regional trends were observed. Rather, the iron content of the dolomite may be related to the lithology of the surrounding rock. Baroque dolomite from parts of the section with high terrigenous content tends to have higher iron content than baroque dolomite from facies containing little terrigenous material. The most iron-rich baroque dolomite came from carbonate units in the Bexar shale, the Pine Island shale, and in the oncolite facies of the Cow Creek Limestone.

Although it is possible that shales supplied magnesium for the dolomite along with iron (Loucks, 1977), we do not find this idea attractive for several reasons. If shales were a major source of magnesium, late dolomite should be associated with terrigenous-rich sections and calcite with terrigenous-poor sections, and this does not appear to be the case. And second, very little magnesium relative to calcium is released as a result of shale diagenesis (Garrels and Mackenzie, 1971, p. 234). Shales are not common in this largely carbonate section and these rocks have not been buried deeply enough for smectite to undergo significant transformation.

$\delta^{18}O$ values for baroque dolomite range from −1.6 to −8.6‰ PDB (Fig. 3). The values do not correlate with depth, $\delta^{13}C$ values, or with iron content. In contrast to calcite, all baroque dolomite could have precipitated from mixtures of basinal and meteoric water at present or inferred maximum burial depths.

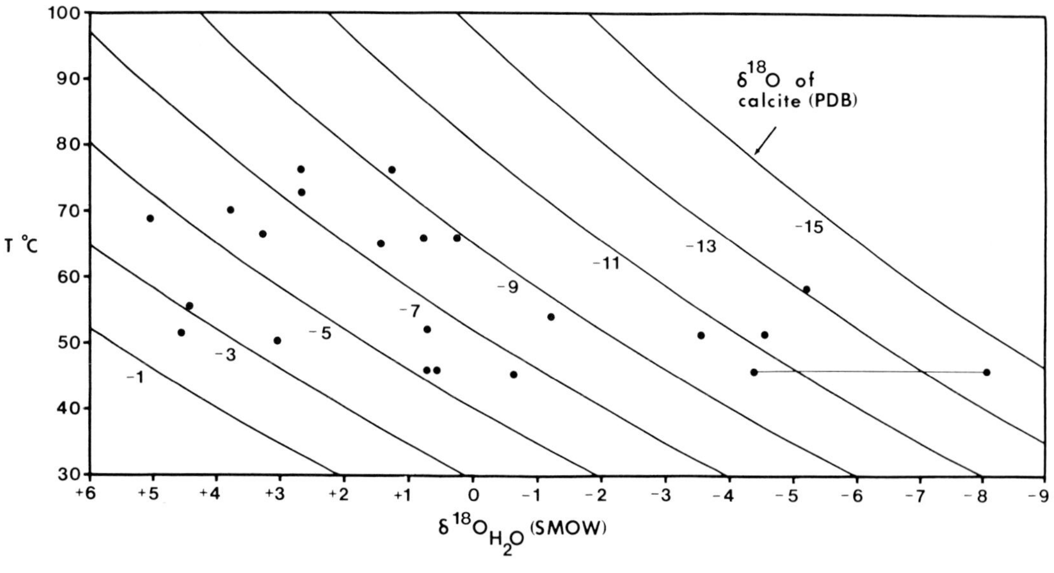

FIG. 2.—$\delta^{18}O$ of late calcite plotted at present burial temperature. Lines for equilibrium precipitation plotted using the relationship $10^3\ln\alpha = 2.78 \times 10^6T^{-2}(°K) - 2.89$. Most calcite could have been emplaced by meteoric water modified by mixing with basinal formation water. Some crystals, including one from which several analyses were obtained (points connected by line), are too depleted to have precipitated from a reasonable meteoric water today. Limited updip movement of heated fluids may thus be recorded by these few samples.

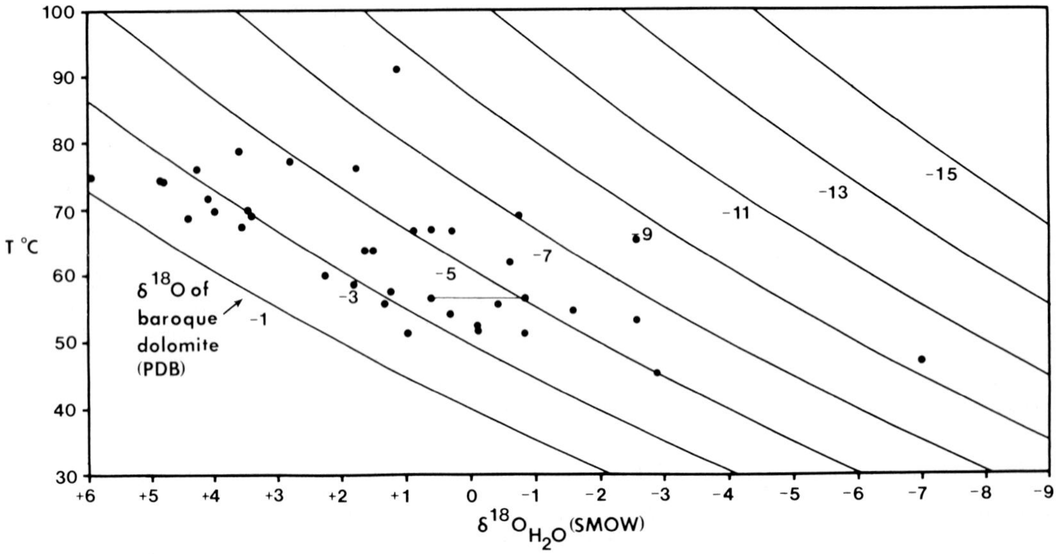

FIG. 3.—$\delta^{18}O$ of baroque dolomite plotted at present burial temperature. Lines for equilibrium precipitation plotted using the relationship $10^3 \ln\alpha = 3.2 \times 10^6 T^{-2}(°K) - 3.3$. All samples could have been emplaced by mixtures of meteoric water with up to about 50% formation water. Line connects analyses from the same crystal.

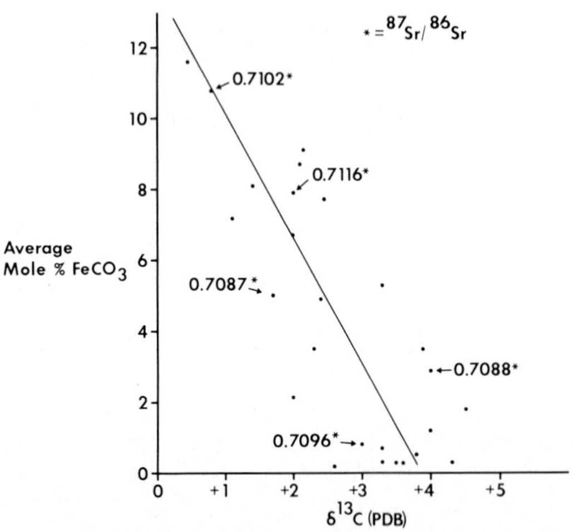

FIG. 4.—Mole percent iron in baroque dolomite plotted versus $\delta^{13}C$. The $^{87}Sr/^{86}Sr$ ratios of five samples (Figure 9) are also plotted. Strong negative correlation between iron and $\delta^{13}C$ suggests that organic-rich solutions transported the iron and contributed ^{13}C-depleted carbon to the pore fluids as a result of bacterial oxidation of their contained hydrocarbons. Vertical injection of ^{87}Sr-enriched saline formation water carrying iron and hydrocarbons into a meteoric aquifer cool enough for bacterial oxidation of the organic matter can account for the observations.

A strong negative correlation exists between iron content and $\delta^{13}C$ of the baroque dolomite (Fig. 4). Iron-rich crystals have more negative $\delta^{13}C$ values, and thus organic-rich fluids seem to have mobilized iron whereas organic-poor fluids apparently did not mobilize or transport iron. As in the case of calcite, baroque dolomite crystals were found to be zoned isotopically, with as much as 1.4‰ variation in $\delta^{13}C$ and 0.7‰ in $\delta^{18}O$ being present.

MAJOR CEMENTS: ANHYDRITE

Bladed to fibrous anhydrite, 0.1 mm to 6.0 mm in length, formed after baroque dolomite. Although anhydrite replaces fossils, matrix material and earlier calcite cement, it occurs primarily as a pore-filling cement. Dissolution of both anhydrite cement and nodules at depths up to 1200 meters is common. Some of this dissolution could have occurred very early, during Cretaceous meteoric activity, but since some late diagenetic phases, especially baroque dolomite, show evidence of corrosion and calcite replacement (Fig. 5), some of the dissolution must be late. These observations, and the replacement of anhydrite by both calcite and quartz indicate that the pore fluids at different times were variously saturated with anhydrite.

Anhydrite was microsampled, dissolved in saturated NaCl solution at room temperature and filtered to remove impurites. $BaSO_4$ was precipitated by addition of 0.5 M $BaCl_2$ in 5% HCl (Lloyd, 1968), then filtered, washed, dried and reduced according to the procedure of Mitzutani (1971). $\delta^{18}O$ values are reported relative to SMOW, and range from +19.5‰ to +22.9‰ (Fig. 6). The $\delta^{18}O$ of two nodular anhydrite samples are within the same range as the anhydrite cements, but are much more enriched than the value of about +15‰ suggested for Cretaceous seawater sulfate by Claypool *et al.* (1980). We interpret these data as proof that the anhydrite nodules have recrystallized in the subsurface and do not record Cretaceous values.

Five $\delta^{34}S$ values for anhydrite and modern Edwards Formation brine are all between +20 and +26‰. Claypool *et al.* (1980) suggest that a $\delta^{34}S$ of less than +16‰ was characteristic of Cretaceous evaporites. Sulfur isotope data thus confirm the assumption that the subsurface dissolved sulfate-water system may have approached oxygen-isotopic equilibrium, given the times and temperatures involved. Simple remobilization of isotopically unmodified Cretaceous sulfate did not occur. Using the fractionation factor for anhydrite-water proposed by Chiba *et al.* (1981), the anhydrite and carbonate cements could all have been emplaced by fluids having similar oxygen isotopic compositions. Fluids ranging in oxygen isotopic composition from +5 to −6‰ could have emplaced virtually all the late calcite, baroque dolomite and anhydrite cements under present burial conditions. Only a few of the calcite values and the single sphalerite fluid inclusion temperature necessitate emplacement under somewhat hotter conditions than are found today.

MINOR CEMENTS

Lutecite and quartzine, two fibrous length-slow varieties of quartz, commonly occur as partial replacements of fibrous molluscs. Previously published analyses (Land, 1977) indicate these phases have $\delta^{18}O$ values similar to those found in outcrop, suggesting emplacement soon after aragonite stabilization (Loucks, 1976). Anhydrite nodules are rarely replaced by lutecite and megaquartz in both formations, and a few occurrences of chert nodules were noted. All these features are also found in outcrop. Quartz overgrowths on detrital quartz grains are locally common in the lower Glen Rose Formation (Bay, 1982), but have not yet been analyzed.

FIG. 5.—Corrosion of baroque dolomite and replacement by calcite (dedolomitization). We interpret this texture as further proof, along with mineral zonation, that fluid compositions varied temporally. Variations could result from periodic basinal fluid loss, or from periodic deep meteoric recharge, or both. Long dimension 1300 microns, plane polarized light.

Kaolinite cement was found in small amounts in about half the cores examined. It formed competitively with and later than anhydrite, and competitively with the late calcite. Kaolinite was also found partially replacing feldspar grains. Detrital feldspar grains are most certainly the source for the aluminum in the kaolinite. Three kaolinite samples have $\delta^{18}O$ values between +16.8 and +17.6‰ (relative to SMOW). Precipitation of kaolinite along with the more depleted calcite and anhydrite is consistent with both the petrographic and isotopic observations, and suggests precipitation under conditions not very different from those resulting in calcite, dolomite and anhydrite emplacement (Fig. 7).

Ba-rich celestite (average $Ba_{.33}Sr_{.66}Ca_{.01}SO_4$, four samples) occurs as a pore-filling cement post-dating baroque dolomite in five of the twenty-five wells studied. Celestite replaces anhydrite nodules and also occurs as a cement. Celestite may have formed as the result of dolomite and/or evaporite recrystallization (Bein and Land, 1983), is known from outcrop, and hence need not be a strictly subsurface phase. Ba-rich celestite, however, is not known from outcrops of these formations.

Gypsum was found to have replaced anhydrite nodules in the Tenneco #1 Ney well at a depth of 994 meters. If this gypsum formed at its present depth, where the temperature (derived from the present geothermal gradient) is about 44°C, it must have formed from fluids of very low ionic strength (e.g. Holser, 1979, Fig. 13). Subsurface fluids in this area are very saline today. Therefore the gypsum must have formed when meteoric water, saturated essentially only with $CaSO_4$ and carbonate phases extended further downdip than it does today.

Minor amounts of sphalerite occur as pore-fillings and as replacement of both calcite and anhydrite in most of the cores examined. Sphalerite makes up as much as a few percent of the Rowe #1 Kincaid core in the 1619 to 1634 meter interval. Sphalerite precedes baroque dolomite in some thin sections and follows it in others. One fluid inclusion homogenization temperature from the Tenneco #1 Mack well at a depth of 2316 meters (77°C) yielded a temperature of 97°C, approximately 20°C higher than present conditions. The freezing point depression was −14°C, indicating precipitation from a very saline fluid.

Galena was found in two of the more basinward

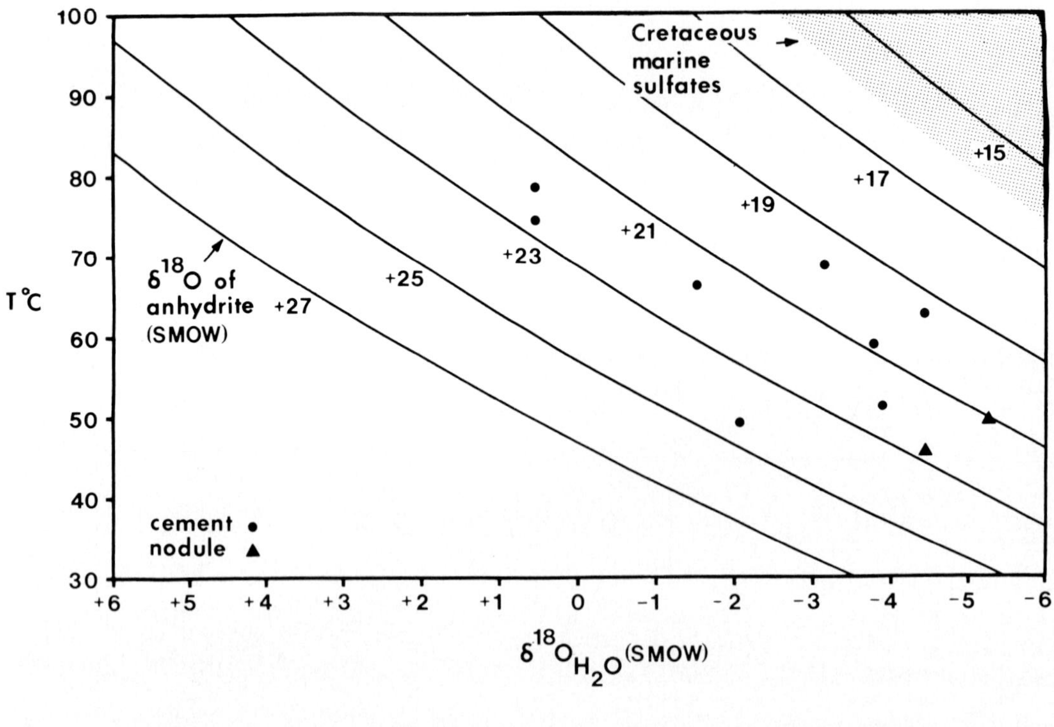

FIG. 6.—$\delta^{18}O$ of anhydrite plotted at present burial depths. Lines for equilibrium precipitation plotted using the relationship $10^3\ln\alpha = 3.21 \times 10^6T^{-2}(°K) - 4.72$. According to Claypool *et al.* (1980), Cretaceous sulfate should have a $\delta^{18}O$ more depleted than about +15‰. Therefore, re-equilibration between the buried sulfate minerals (the ultimate source for the cements) and the pore fluids is indicated. If authigenic anhydrite cements precipitated at equilibrium with the pore fluids, as seems likely, then similar conditions of emplacement are observed as for calcite, dolomite and kaolinite.

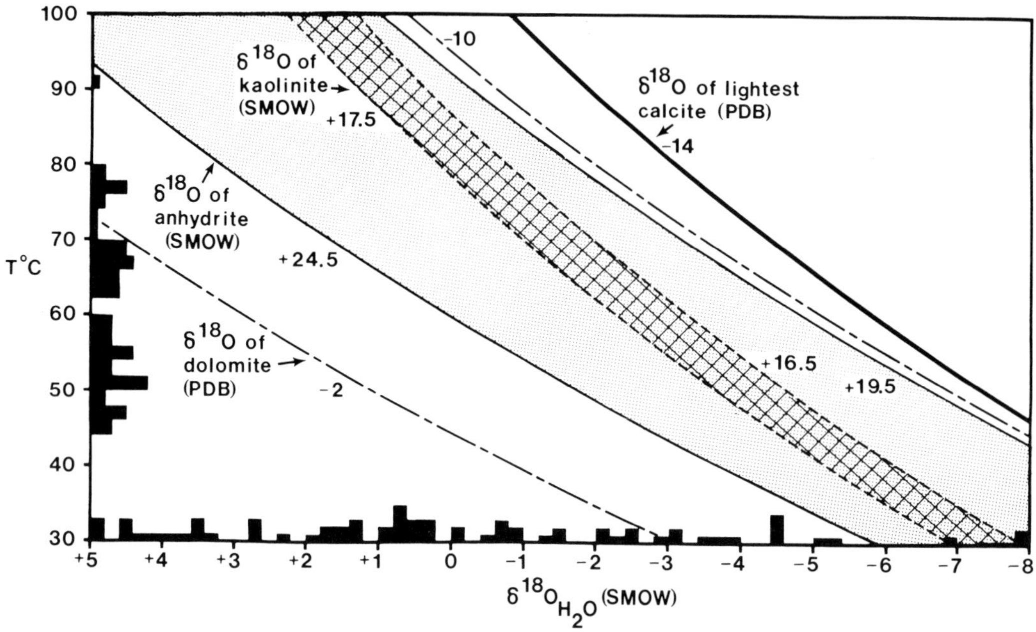

FIG. 7.—Possible conditions of open-system precipitation of pore-filling cements. Only the most depleted calcite is plotted. Temperatures of analyzed samples at present burial conditions are plotted as a histogram on the ordinate. Equilibrium $\delta^{18}O$ values of water, calculated from measured $\delta^{18}O$ values of the phases concerned and the present temperature are plotted as a histogram on the abscissa. The most depleted phases must be interpreted as having precipitated from systems in which meteoric water dominated the oxygen isotopic reservoir. Nearly all phases could have precipitated from mixtures of meteoric water and up to 50% formation water of Edwards-like isotopic composition (+15‰). If precipitation occurred at maximum burial depth (about 15°C hotter than present conditions), then $\delta^{18}O$ values for water would be shifted about +2‰.

wells. It is associated with sphalerite in the Rowe #1 Kincaid well at 1619 meters (Lower Glen Rose) and in the Tenneco #1 Mack well at 2327 meters (Lower Bexar Shale).

FORMATION WATER CHEMISTRY

The Pearsall and Glen Rose Formations are both non-productive of oil and gas in the study area today, so no water samples could be collected for study. Chemical analyses of water from drill-stem tests were made available by Tenneco Oil Company and Border Petroleum Company. These analyses suggest that the formation water in this area today is similar to brines in the nearby Edwards formation studied by Land and Prezbindowski (1981) (Fig. 8).

Brines in the Edwards Formation are very saline Na-Ca-Cl fluids, interpreted to have originated below the Cretaceous section at temperatures greater than 250°C as the result of halite dissolution and albitization (Land and Prezbindowski, 1981). Detrital feldspar grains are almost wholly altered in the Texas Gulf Coast below about 3000 meters (Land and Milliken, 1981; Fisher, 1982; Land, in press, and work in progress). All components incorporated in the detrital feldspars under igneous or metamorphic conditions (except for sodium and potassium, of course), must have been released to the pore fluids as the result of such massive feldspar destruction. A significant fraction of the calcium, strontium, barium, zinc and lead in the brines may have been derived from feldspar destruction, since typical detrital feldspar assemblages contain approximately 200 ppm Sr, 2000 ppm Ba, 20 ppm Zn and 60 ppm Pb (Smith, 1974).

Strontium isotopic data strongly support such a process. Saline formation waters are typically enriched in ^{87}Sr, a decay product of ^{87}Rb. Edwards brines are similarly enriched, having values of about 0.7091 (Fig. 9). The source for the radiogenic strontium is assumed to have been the detrital feldspars which have been destroyed in the deep Gulf of Mexico on a massive scale. Limited analyses of radiogenic strontium in late calcite, baroque dolomite and anhydrite cements clearly indicate that the saline brines have contributed significant amounts of material to the late cements. Limestone wall-rock, in contrast, yields the expected Cretaceous values (Fig. 8).

High Pb and Zn concentrations have been reported from some of the extremely saline, basinward-most Edwards brines (Land and Prezbindowski, 1981). Galena was also found replacing calcite along fractures in

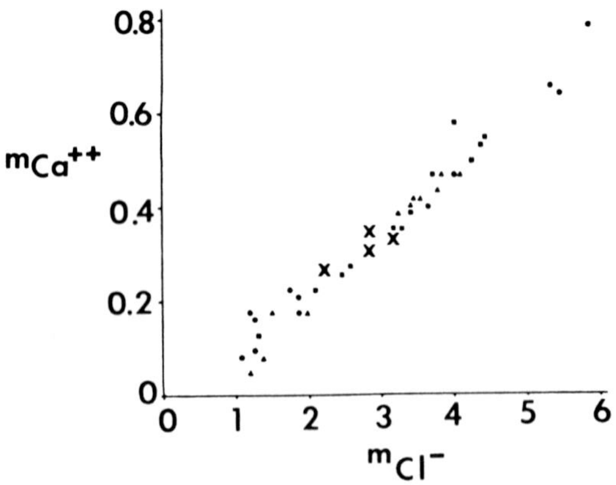

FIG. 8.—Molality of Ca^{++} versus molality of Cl^- in water from four wells in the study area which were analyzed during drilling (× 's). The reliability of these incomplete analyses is questionable, but they do agree with the analyses obtained by Land and Prezbindowski (1981) from the nearby Edwards Formation (dots). Thus we assume that Edwards-type brines have participated in the emplacement of the late diagenetic phases.

one of the most basinward wells in the Stuart City trend studied by Prezbindowski (1981). The oxygen isotopic composition of Edwards brines ranges from +5 to +20‰ today. We will assume, for the sake of a working hypothesis, that similar brines have been present in this area at least since Eocene time when significant burial was achieved. The kinds of fluids from which the late diagenetic phases we have described must have precipitated appear to be limited to only two possibilities: (1) brines similar to those found in this section today, or (2) meteoric water modified by mineral-water reaction as the result of basinward hydrostatic flow. We propose that the late phases we have described formed as a result of the mixing of these two kinds of water.

THE MODEL

The oxygen isotopic data clearly implicate meteoric water in the emplacement of the observed late phases. Extreme temperatures can be ruled out at these relatively shallow burial depths because of rapid heat loss by small volumes of ascending fluids. Precipitation from fluids having $\delta^{18}O$ values between +5 and about −6‰ (SMOW) under present burial conditions is required to explain the range in isotopic composition of

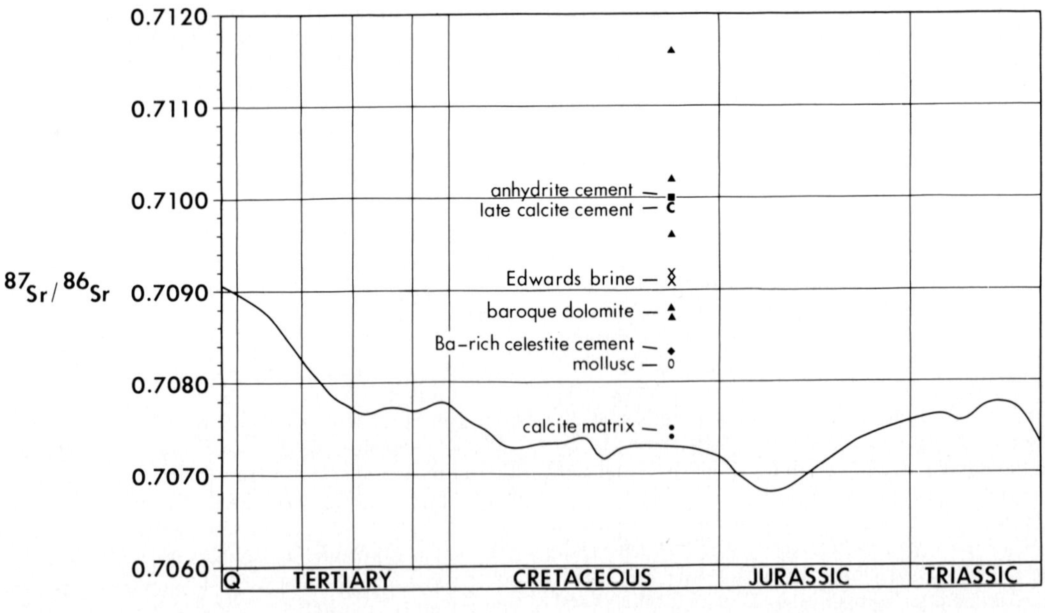

FIG. 9.—Age versus $^{87}Sr/^{86}Sr$ of seawater for post-Paleozoic time (after Burke *et al.*, 1982). Modern Edwards brines are enriched in ^{87}Sr, because, we believe, of massive feldspar destruction in the deep Gulf of Mexico. Late diagenetic phases record the $^{87}Sr/^{86}Sr$ of Edwards-like, or somewhat more radiogenic solutions. The high strontium contents of these brines overwhelmed strontium derived from other sources as the cements formed. Therefore, at least some components from basinal brines have contributed to the late diagenetic phases.

the observed phases (Fig. 7). Thus mixtures of meteoric water with up to about 50% brine (if a brine value of about +15‰ is assumed) is implied. Since Edwards brines contain only a few hundred ppm sulfate (because of the extreme calcium concentrations), sulfate for the precipitation of solid phases cannot have been derived from the brine. Likewise, Edwards brines could not have sourced the magnesium for the precipitation of dolomite since most brines are either undersaturated or just saturated with dolomite today and are at least locally, dedolomitizing (Fig. 5; Land and Prezbindowski, 1981, Fig. 5). Strontium isotopic data and the presence of sulfide phases indicate that other components must have derived from the saline brine, however. We therefore propose that emplacement of the observed phases occurred as a result of injection of Edwards-like brine into a meteoric aquifer which had evolved toward gypsum, calcite and dolomite saturation by down-dip hydrostatic flow (Fig. 10). Injection of a hydrocarbon-bearing fluid of Edwards-like composition caused a series of reactions, including:

1) Bacterial reduction of meteoric-derived sulfate and consequent release of ^{13}C-depleted CO_2 from the oxidized hydrocarbons. Hydrocarbon-rich water transported more iron, resulting in the good correlation between $\delta^{13}C$ and iron content of baroque dolomite. Additionally, iron-rich dolomite tends to be enriched in ^{87}Sr (Fig. 4).

2) Precipitation of galena and sphalerite as the result of sulfide formation,

3) Precipitation of barium, calcium and strontium sulfate phases because of the high cation content of the brine which mixed with a gypsum or anhydrite-saturated solution, and,

4) Precipitation of carbonate phases because of degassing (evasion of CO_2 along with methane) from the deep subsurface fluid. Cations were derived from both the saline brine (Sr), from local shales (Fe) and from the meteoric aquifer (Mg). Gas from Edwards brines today contains several percent CO_2 (P. Lundegard, work in progress). The only mechanism for carbonate precipitation we can suggest is CO_2 loss accompanying methane degassing since simple mixing of the two fluids we have proposed should result in dissolution, not precipitation.

Mineral zonation and both dissolution and precipitation of dolomite and anhydrite suggest periodic fluid movement, possibly as the result of periodic basinal events (Cathles and Smith, 1983), and/or changing recharge in the updip aquifer. It is tempting to relate these events to Miocene faulting which affected the Gulf. But, unfortunately, we have no evidence as to the

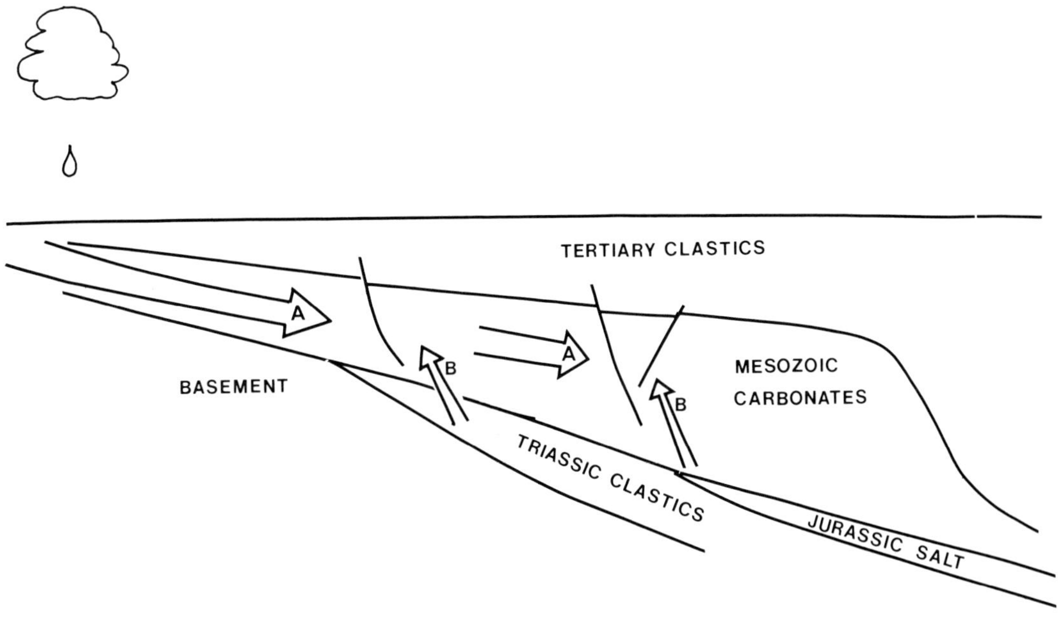

FIG. 10.—Schematic model for precipitation of late diagenetic phases by mixing of meteoric water and saline formation water. We envision the meteoric water aquifer (A), driven by hydrostatic flow, to be essentially saturated with calcite, dolomite and $CaSO_4$, and at a sufficiently low temperature for bacterial activity to be possible. Periodic injection of an Edwards-like fluid (B), containing abundant carbon dioxide, calcium, strontium (radiogenic), and barium, together with small amounts of lead, zinc and hydrocarbons, caused precipitation of the observed phases.

timing of cement emplacement except to be sure that at least some basinal rocks were buried deeply enough to generate hydrocarbons, and that feldspar stabilization reactions must have been operative.

CONCLUSION

The relative paucity of late diagenetic phases in the few cores which are available from this setting (less than 2000 meters of core from six Texas counties, equivalent to the exposure surface of a small outcrop) makes analytical test of our hypothesis very difficult. The reason for the paucity of late phases is, we believe, that communication between the Cretaceous carbonate platform and the adjacent evolving basin was interrupted as terrigenous clastics, beginning with the Eocene Wilcox formation, prograded over the platform and into the basin. Growth faults associated with the platform margin effectively isolated the platform from extensive interaction with basin-derived fluids and the basin dewatered primarily through the offlapping clastics rather than laterally through the Cretaceous platform. Basinal fluids have gradually moved up through the Cretaceous carbonate section with time, however, possibly as the result of slow solution-compaction. Today, highly modified, basinal fluids discharge along the Balcones fault zone as the "bad water" (Pearson and Rettman, 1976), continuing to emplace some of the same phases we find in the deeper subsurface (Longman and Mench, 1978).

ACKNOWLEDGMENTS

This research was supported by the National Science Foundation, EAR-7617774 and EAR-7824081, by the Chevron Oil Field Research Company, Amoco Research Laboratory, and by the Geology Foundation of the University of Texas.

We thank numerous colleagues for stimulating discussion and editorial assistance, including Jim Anderson, Annell Bay, Don Bebout, Dave Budd, Bob Loucks, John Morton, Dennis Prezbindowski and Bob Suchecki.

REFERENCES

BAY, A.R., 1980, Deposition of prograding carbonate sand shoals and their subsequent diagenesis—Lower Glen Rose (Cretaceous), South Texas: Unpub. MA thesis, Univ. Texas-Austin, 189 p.

______, 1982, Evolution and porosity of carbonate shoaling cycles, Lower Cretaceous-Lower Glen Rose, South Texas: Trans. Gulf Coast Assoc. Geol. Soc., 32nd Ann. Meeting, Houston, p. 101–120.

BEIN, A., AND LAND, L.S., 1983, Carbonate sedimentation and diagenesis associated with Mg-Ca-Chloride brines: The Permian San Andres formation in the Texas panhandle: Jour. Sed. Petrology, v. 53, p. 243–260.

BURKE, W.H., DENISON, R.E., HETHERINGTON, E.A., KOEPNICK, R.B., NELSON, H.F., AND OTTO, J.B., 1982, Variation of seawater $^{87}Sr/^{86}Sr$ throughout Phanerozoic time: Geology, v. 10, p. 516–519.

CATHLES, L.M., AND SMITH, A.T., 1983, Thermal constraints on the formation of Mississippi valley-type lead-zinc deposits and their implications for episodic basin dewatering and deposit genesis: Econ. Geology, v. 78, p. 185–203.

CHIBA, H., KUSAKABE, M., HIRANO, S-I., MATSUO, S., AND SOMIYA, S., 1981, Oxygen isotope fractionation factors between anhydrite and water from 100 to 350°C: Earth and Planetary Science Letters, v. 53, p. 55–62.

CLAYPOOL, G.E., HOLSER, W.T., KAPLAN, I.R., SAKI, H., AND ZAK, I., 1980, The age curves of sulfur and oxygen isotopes in marine sulfate and their mutual interpretation: Chemical Geology, v. 28, p. 199–260.

FISHER, R.S., 1982, Diagenetic history of Eocene Wilcox sandstones and associated formation waters, South-central Texas: Unpub. Ph.D. Diss., Univ. Texas-Austin, 185 p.

FISHER, W.L., AND RODDA, P.U., 1969, Edwards formation (Lower Cretaceous), Texas: Dolomitization in a carbonate platform system: Amer. Assoc. Petroleum Geologists Bull., v. 52, p. 55–72.

GARRELS, R.M., AND MACKENZIE, F.T., 1971, Evolution of Sedimentary Rocks. W.W. Norton & Co., 397 p.

HOLSER, W.T., 1979, Mineralogy of evaporites, *in* R.G. Burns (ed.), Marine Minerals: Min. Soc. America, Short Course Notes v. 6, p. 211–294.

LAND, L.S., 1977, Hydrogen and oxygen isotopic composition of chert from the Edwards Group, Lower Cretaceous, Central Texas, *in* D.G. Bebout and R.G. Loucks (eds.), Cretaceous Carbonates of Texas and Mexico, Applications to Subsurface Exploration: Bur. Econ. Geol., Univ. Texas-Austin, Rep. Invest. No. 89, p. 202–205.

______, in press, Frio sandstone diagenesis, Texas Gulf Coast, a regional isotopic study: Amer. Assoc. Petroleum Geologists, Special Publication.

______, AND MILLIKEN, K.L., 1981, Feldspar diagenesis in the Frio formation, Brazoria County, Texas Gulf Coast: Geology, v. 9, p. 314–318.

______, AND PREZBINDOWSKI, D.R., 1981, The origin and evolution of saline formation water, Lower Cretaceous carbonates, South-central Texas: Jour. Hydrology, v. 54, p. 51–74.

LLOYD, R.M., 1968, Oxygen isotope behavior in the sulfate-water system: Jour. Geophys. Res., v. 73, p. 6099–6110.

LOUCKS, R.G., 1976, Pearsall Formation, Lower Cretaceous, South Texas—Depositional facies and carbonate diagenesis and their relationship to porosity: Unpub. Ph.D. Diss., Univ. Texas-Austin, 362 p.

______, 1977, Porosity development and distribution of shoal-water carbonate complexes—subsurface Pearsall Formation (Lower Cretaceous) South Texas, *in* D.G. Bebout and R.G. Loucks (eds.), Cretaceous Carbonates of Texas and Mexico, Applications to Subsurface Exploration: Bur. Econ. Geol., Univ. Texas-Austin, Rep. Invest. No. 89, p. 97–126.

LONGMAN, M.W., AND MENCH, P.A., 1978, Diagenesis of Cretaceous limestones in the Edwards aquifer system of south-central Texas: A scanning electron microscope study: Sedimentary Geol., v. 21, p. 241–276.

MITZUTANI, Y., 1971, An improvement in the carbon-reduction method for the oxygen isotopic analysis of sulphates: Geochemical Jour., v. 5, p. 69–77.

PEARSON, F.J., JR., AND RETTMAN, P.L., 1976, Geochemical and isotopic analyses of waters associated with the Edwards limestone aquifer, Central Texas: U.S. Geol. Survey, Prof. Paper 440-S, 19 p.

PREZBINDOWSKI, D.R., 1980, Microsampling technique for stable isotope analysis of carbonates: Jour. Sed. Petrology, v. 50, p. 643–644.

______, 1981, Burial diagenesis, Edwards Formation, Lower Cretaceous, South-central Texas: Unpub. Ph.D. Diss., Univ. Texas-Austin, 235 p.

ROSE, P.R., 1972, Edwards Group, surface and subsurface, central Texas: Bur. Econ. Geol., Univ. Texas-Austin, Rep. Invest. No. 74, 198 p.

SMITH, J.V., 1974, Feldspar Minerals, V. 2: Springer-Verlag, 690 p.

TIMING OF HYDROCARBON MIGRATION: EVIDENCED FROM FLUID INCLUSIONS IN CALCITE CEMENTS, TECTONICS AND BURIAL HISTORY

ROBERT C. BURRUSS[1], KAREN R. CERCONE[2], AND PAUL M. HARRIS[1]

[1]Gulf Research and Development Company, P.O. Box 37048, Houston, Texas 77236, and [2]Department of Earth and Space Sciences, State University of New York at Stonybrook, Stonybrook, New York 11794

ABSTRACT

The timing of petroleum migration in samples of fractured Cretaceous reservoir limestones from Oman and the United Arab Emirates is determined from observations of hydrocarbon fluid inclusions in calcite cements. Petrography and geohistory analyses of four wells that were variably affected by formation of the Oman Foredeep reveal five stages of diagenesis, fracturing, and fluid migration. (1) Quiet shelf deposition: early cementation associated with regional unconformities; (2) pre-orogenic shelf emergence: fractures cutting Stage 1 cements are healed by very cloudy, cleaved, and twinned calcite containing microfractures with yellow-white flourescent, hydrocarbon fluid inclusions; (3) initial foredeep downwarp: fractures cross cutting Stage 2 fractures are healed with cloudy, cleaved, and sometimes twinned calcite containing dull-blue fluorescent, hydrocarbon fluid inclusions; (4) rapid subsidence and filling of foredeep with sediments including flysch, exotic blocks, and thrust toes: burial and tectonic stylolites crosscut Stage 2 and 3 fractures; and (5) uplift of the Oman Mountains: fractures cross cutting all diagenetic features are filled with clear, untwinned and uncleaved calcite containing only non-fluorescent, aqueous fluid inclusions. By correlating stylolite formation with overburden of approximately 2500 ft. (750 m), the hydrocarbon inclusions in Stage 2 fractures must predate all of Stage 4 and most of Stage 3. In the deepest portions of the foredeep, close to the Oman Mountain front, this correlation limits the presence of oil in fracture porosity to late Turonian–early Campanian time. Farther to the west, in the shallower parts of the foredeep, the timing constraint relaxes, and oil migration occurred as late as early Tertiary.

INTRODUCTION

This paper is an attempt to use observations of hydrocarbon fluid inclusions in calcite cements to document the timing of hydrocarbon migration in carbonate rocks. In particular, we demonstrate that when such observations are placed in the context of the burial and tectonic evolution of a sedimentary basin, they can constrain both the interpreted timing of oil and gas migration and its regional variation across the basin.

The rocks examined in this study were obtained from conventional cores of Lower and Middle Cretaceous reservoir carbonates in Oman and the United Arab Emirates. Samples were taken from four wells (shown in Figure 1) affected to varying degrees by the formation of a deep trough, the Cretaceous Oman Foredeep, and subsequently by the Oman Orogeny during the late Cretaceous and early Tertiary (Saint Marc, 1978; Murris, 1980; Ricateau and Riche, 1980; Soliman and Shamlan, 1982). The interpreted diagenetic history and timing of hydrocarbon migration in the Al Ali No. 2 well, previously described by Burruss et al. (1983) and summarized here, provide the key to interpret observations in the other wells of the present study.

Although all four wells shown on Figure 1 were affected by Late Cretaceous sedimentation and tectonism during the Oman Orogeny (Wilson, 1969; Glennie *et al.*, 1974), the burial and tectonic histories of the rocks varied widely. The Al Ali No. 2 and Bukha No. 1 wells are located in the northern, narrow end of the Oman foredeep, closest to the Oman Mountains and were most strongly affected by tectonism. The Shams No. 1 well is situated on the distal portion of broad central foredeep, farther from the mountain front. The Rashid No. 1 well is entirely out of the Late Cretaceous foredeep basin and was only mildly affected by orogenic events.

The Lower and Middle Cretaceous reservoir rocks of the United Arab Emirates and Oman consist of shallow-water bioturbated packstones and wackestones with associated rudist buildups around intrashelf basins (Saint Marc, 1978; Murris, 1980). The most important of these units (Fig. 2) are the Lower Cretaceous Shuaiba Formation of the Thammama Group (Hassan *et al.*, 1975; Murris, 1980; Frost *et al.*, 1983), and the Middle Cretaceous Mauddud/Mishrif Formations of the Wasia Group (Jordan *et al.*, in press; Murris, 1980; Harris and Frost, 1984). Each of these formations is separated from the overlying rocks by a regional unconformity (Harris *et al.*, 1984). Late Cretaceous sediments filling the Oman Foredeep consist largely of deepwater shales and marls of the Aruma Group. These deposits commonly range from 500 to 3000 ft (150 to 900 m) in thickness, but exceed 10,000 ft (3000 m) in the foredeep basin proper. Adjacent to the mountain front these rocks include flysch and olistostromal blocks shed off advancing thrust sheets during emplacement of the Semail Ophiolite sequence (Wilson, 1969; Glennie *et al.*, 1974; Fig. 1).

Since we are concerned with burial history and sedimentation rates, the biostratigraphic control is quite critical. Figure 3 shows the biostratigraphic zonation used for this part of the eastern Arabian Penninsula by

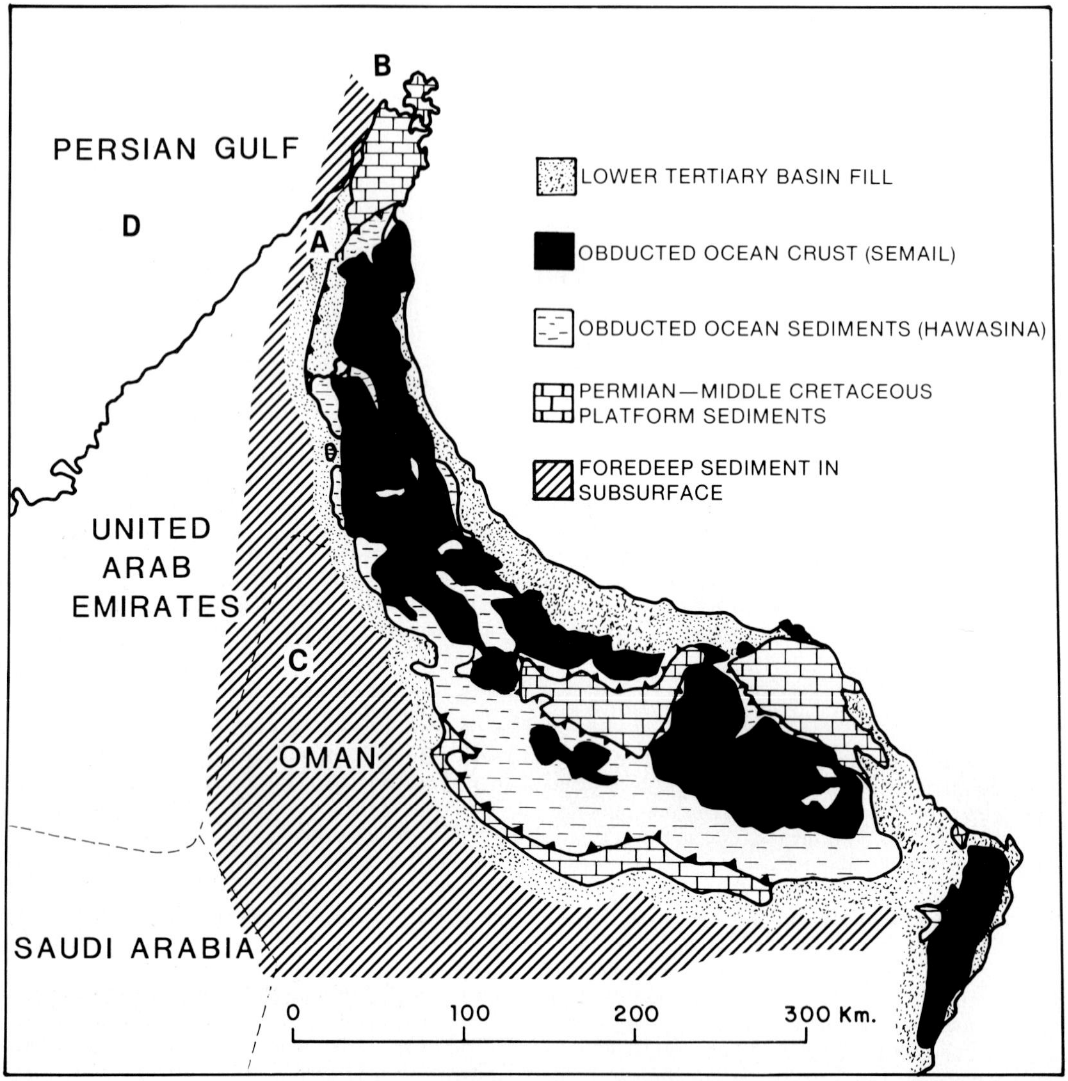

FIG. 1.—Simplified geologic map of northern Oman and Musandam Peninsula showing locations of studied wells. Glennie *et al.* (1974) and Ricateau and Riche (1980) present the details of the geology. The variably wide belt of foredeep sediments recognized in the subsurface is indicated as well as the position of the four wells with respect to the foredeep: (A) Al Ali No. 2 well, (B) Bukha No. 1 well, (C) Shams No. 1 well, and (D) Rashid No. 1 well.

FIG. 2.—Cretaceous stratigraphic units recognized in the subsurface of the Eastern Arabian Peninsula (modified from Harris *et al.*, 1984). The time scale is based on Van Hinte (1976), Kauffman (1977), Childs (1981, COSUNA), and internal data. The Lower Cretaceous Thammama Group, Middle Cretaceous Wasia Group, and Upper Cretaceous Aruma Group are separated by regional unconformities as indicated in black. Shallow shelf reservoir- bearing limestones of the Aptian age Shuaiba Formation and Albian-Cenomanian age Mauddud and Mishrif Formations are examined in this study.

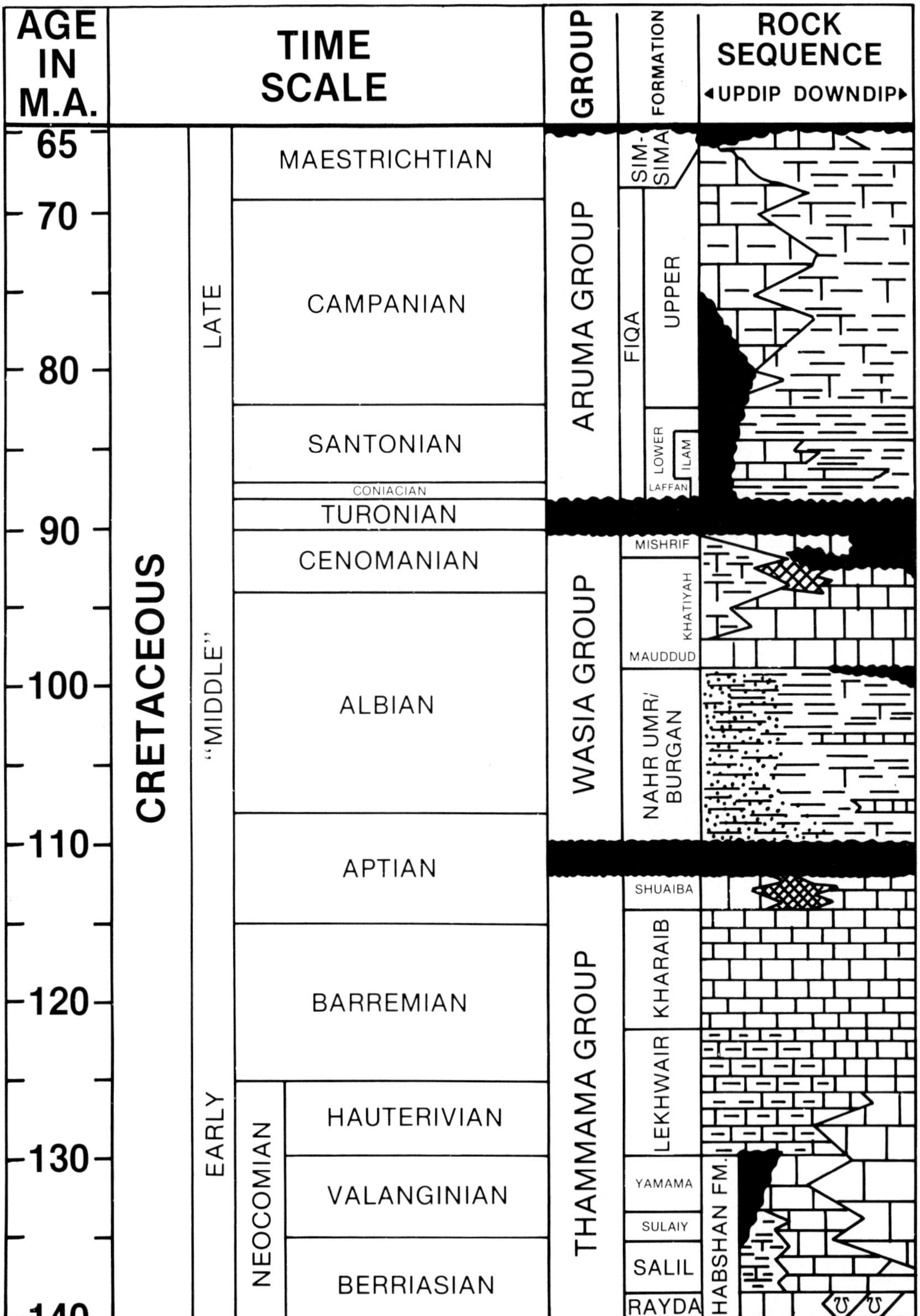
AGE IN M.A.
TIME SCALE
GROUP
FORMATION
ROCK SEQUENCE
◂UPDIP DOWNDIP▸
65
70
80
90
100
110
120
130
140
CRETACEOUS
LATE
"MIDDLE"
EARLY
NEOCOMIAN
MAESTRICHTIAN
CAMPANIAN
SANTONIAN
CONIACIAN
TURONIAN
CENOMANIAN
ALBIAN
APTIAN
BARREMIAN
HAUTERIVIAN
VALANGINIAN
BERRIASIAN
ARUMA GROUP
WASIA GROUP
THAMMAMA GROUP
SIM-SIMA
FIQA
UPPER
LOWER
ILAM
LAFFAN
MISHRIF
KHATIYAH
MAUDDUD
NAHR UMR/ BURGAN
SHUAIBA
KHARAIB
LEKHWAIR
YAMAMA
SULAIY
SALIL
RAYDA
HABSHAN FM.

Harris *et al*. (1984). The shallow shelf reservoir limestones were dated principally by large benthic foraminifera. The detailed zonation of the deeper water deposits of Upper Cretaceous age was based on the occurrence of calcareous nannofossils and planktonic foraminifera and as such, it provided the necessary control on the regional variations in sedimentation rates in the Oman Foredeep.

PETROGRAPHIC OBSERVATIONS

The sequence of diagenetic events affecting Lower and/or Middle Cretaceous carbonates in each well was determined by petrographic analysis of cements and solution features and is summarized in the succeeding sections. Cements are predominantly calcite and fill both pores and fractures. Dolomitization, usually resulting in the formation of zoned rhombic crystals, was minor. Where the carbonates are extensively fractured, generations of fractures were separated by their cross-cutting relations and the appearance of the fracture-filling calcite cement. Attributes of the cements include degree of cloudiness due to abundant mineral and fluid inclusions, mineral cleavage and crystal twinning. Dissolution of these carbonates occurred primarily during stylolitization and interpenetration along grain boundaries. The relative timing of diagenetic mineralization and dissolution features was determined by cross-cutting relationships (Fig. 4).

The fluid inclusions trapped within calcite cements were studied to determine the nature of diagenetic fluids. The presence and characteristics of fluid inclusions are an indication of the type of fluid present in a rock unit at some time in its history. Examples of inclusions observed in this study are shown in Figures 4 and 5. These inclusions contain small volumes (10 to 50 μm largest dimension) of liquid and vapor trapped in minerals either during crystal growth along growth zones or after crystal growth along healed microfractures within the crystal. Fluid inclusions which fluoresce under long wavelength ultraviolet light (366 nm peak of Hg) contain hydrocarbon liquids (Burruss, 1981). The visible color of the fluorescence emission varies with the API gravity of the oil. Condensates have maximum emission at shorter visible wavelengths

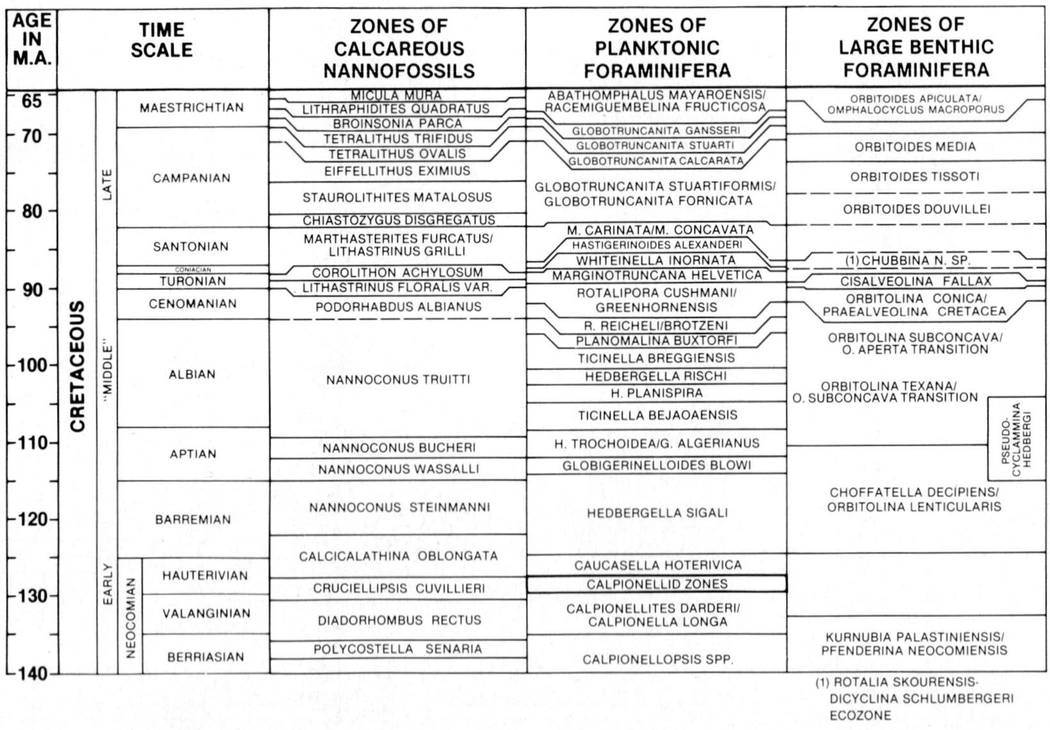

FIG. 3.—Successions of index fossils for the Cretaceous. Each microfossil group has great utility but is limited in some lithologic units due to unfavorable depositional environments. Large benthic foraminifera are the basis for biostratigraphic zones for shallow shelf carbonates of Barremian through Turonian age, whereas calcareous nannofossils and planktonic foraminifers permit zonation of late Cretaceous marls and shales. Identification of calcareous nannofossils by B.L. Shaffer; identification of planktonic foraminifera by G.A. Seiglie, zonation in part based on Bolli (1957) and Van Hinte (1976); identification of large benthic foraminifera by S.H. Frost and G.A. Seiglie.

and fluoresce a deep (or dull) blue, while medium gravity oils have longer wavelength maxima and appear a bright white, yellow, yellow-green, or yellow-orange color (Janezic *et al.*, 1981). Any liquid phase in an inclusion which does not fluoresce can generally be shown to be an aqueous solution (Fig. 5B):

Although we believe that fluid inclusion observations provide reasonable control on the history of diagenetic fluids in this study area, it must be kept in mind that they undoubtably represent a limited record of the formation fluids. Diagenetic mineralization is not a continuous process: cements precipitate only when the fluid chemistry is favorable, fracturing with associated microfracturing occurs during episodically released stresses, and fluid inclusions may only be trapped under certain conditions of crystal growth. If during the presence of a diagenetic fluid, few diagenetic minerals form or very few fluid inclusions are trapped, we will see little evidence of the existence of that fluid.

NORTHERN FOREDEEP

Houston Oil & Minerals Al Ali No. 2 Well, Southeast Umm al Qaiwain

The Lower and Middle Cretaceous carbonates in the Al Ali No. 2 well (Location A, Fig. 1) consist of stylolitized muddy limestones overprinted by several periods of fracturing with all fractures filled by calcite. The diagenetic events affecting these rocks can be broken down into five stages (Fig. 6; Burruss *et al.*, 1983).

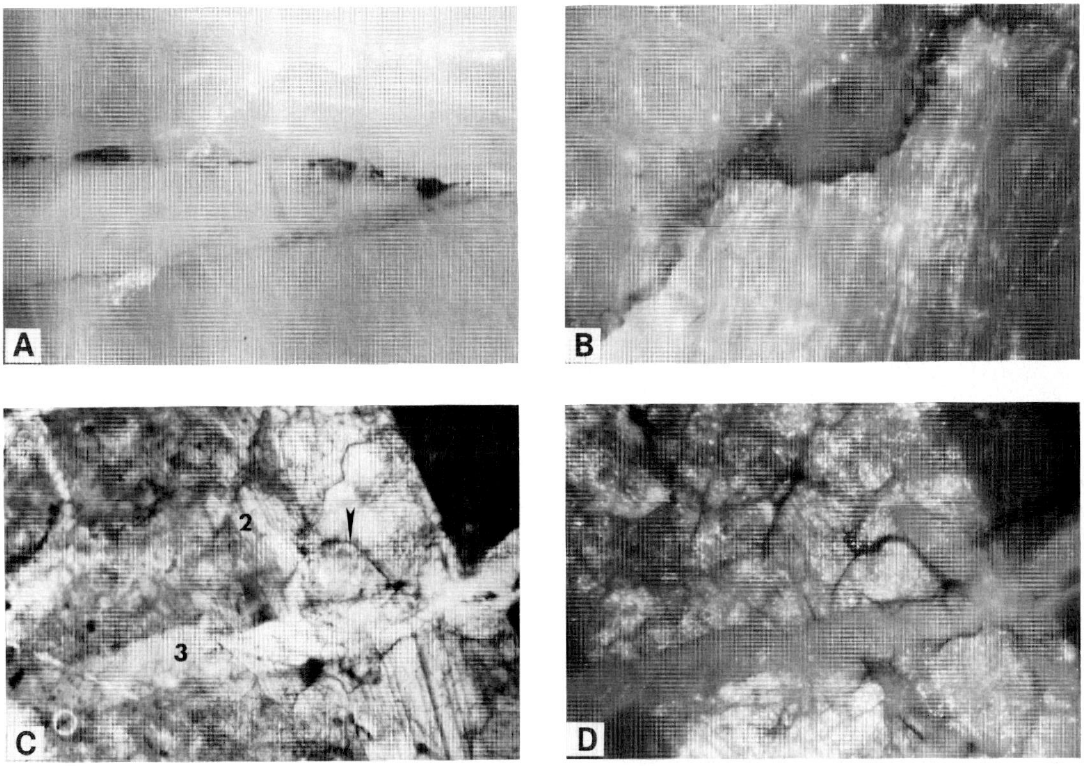

FIG. 4.—Photomicrographs showing cross-cutting relationships between fractures and stylolites. Longest dimension of each photo is 2.4 mm. *A*, Stylolite tooth outlined by insoluble residue penetrates mollusc fragment which contains microfracture with hydrocarbon fluid inclusions. The inclusions appear as white spots fluorescing under vertical, UV illumination (366 nm Hg). Stylolite cross-cuts microfracture and post-dates trapping of fluid inclusions. Top is toward the left of photo. Sample from Shams No. 1 well, 10,032 ft. *B*, Stylolite truncating microfractures in cloudy, Stage 2 calcite cement. Microfractures are decorated by hydrocarbon inclusions shown as white spots fluorescing under vertical, UV illumination (366 nm Hg). Microfractures do not cross stylolite seam indicating dissolution occurred after trapping of inclusions. Sample from Al Ali No. 2 well, 15,192 ft. *C*, Cloudy, Stage 2 (2 on photo) fracture-filling calcite cement cross-cut by less cloudy, Stage 3 calcite (3 on photo) and minor stylolite (arrow on photo) which is also cut by the Stage 3 cemented fracture. Stylolitization and Stage 3 cementation appear to overlap in time, but both events post-date the Stage 2 cements. Transmitted light photomicrograph, sample from Al Ali No. 2 well, 15,829 ft. *D*, Same field of view as C under vertical UV (366 nm Hg) illumination. White spots are fluorescing hydrocarbon fluid inclusions trapped along microfractures in the Stage 2 cement, truncated by stylolites and Stage 3 cements. In other parts of the section the Stage 3 cements contain small numbers of dull-blue fluorescent hydrocarbon inclusions.

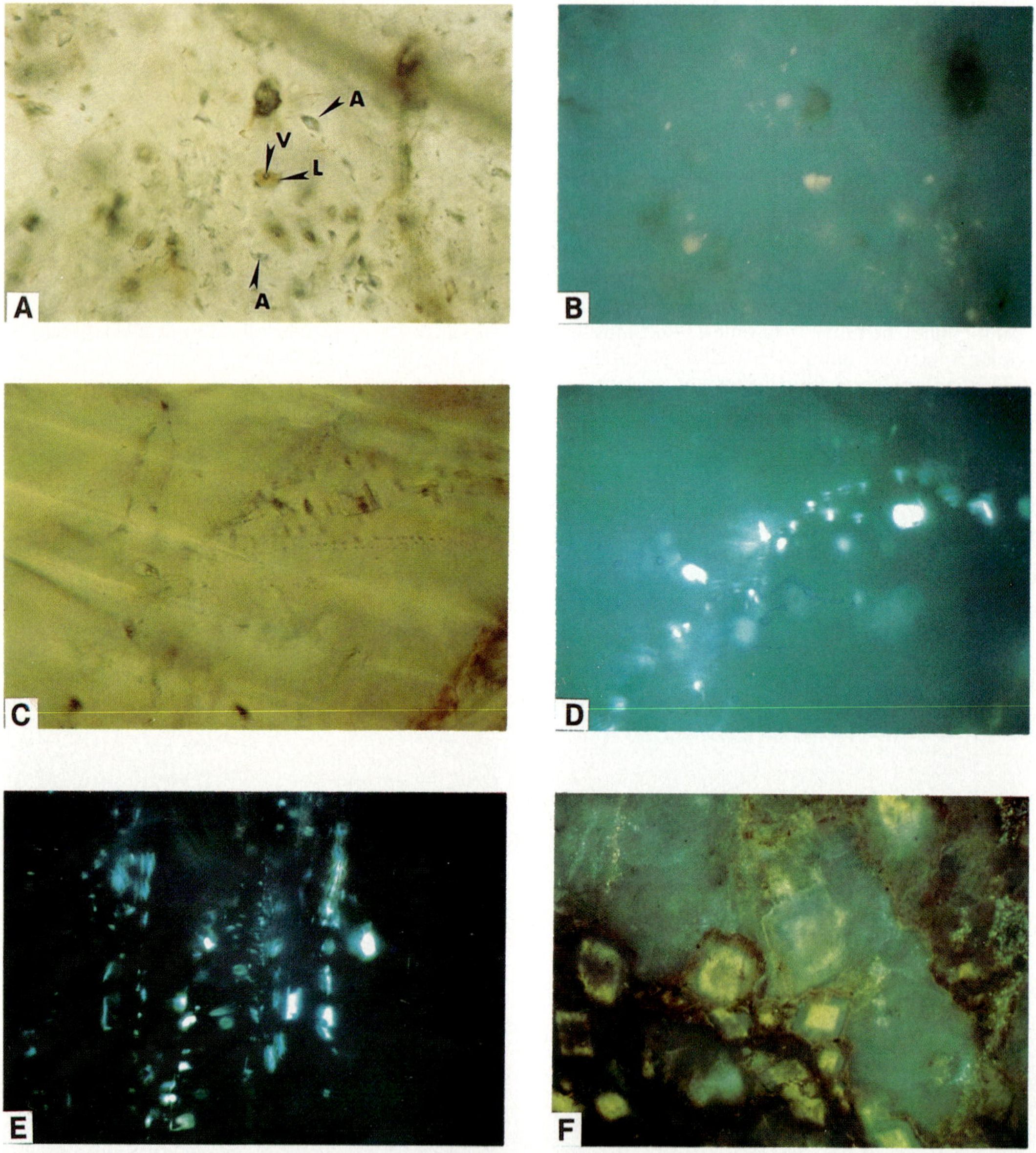

FIG. 5.—Photomicrographs of hydrocarbon and aqueous fluid inclusions in calcite cements. *A, B*, Transmitted light (A) and vertical, UV fluorescence (B), images of hydrocarbon and aqueous fluid inclusions in calcite, Rashid No. 1 well, 9550 ft. Large two-phase, hydrocarbon inclusion shows vapor bubble (V) and brownish, hydrocarbon liquid (L). Examples of small two-phase aqueous inclusions are labeled (A). Note that in fluorescence image, hydrocarbon inclusions fluoresce, but aqueous inclusions do not. In this sample, the hydrocarbon inclusions fluoresce a gold-orange color. Long axis of photomicrographs is 0.6 mm. *C, D*, Transmitted light (C) and vertical, UV fluorescence (D) image of hydrocarbon inclusions trapped along curving, healed microfracture of Stage 2 calcite cement, Al Ali No. 2 well, 15,190 ft. Inclusions fluoresce a yellow-white color. Long axis of photomicrographs is 0.6 mm. (E). Vertical UV fluorescence photomicrograph of hydrocarbon fluid inclusions along healed microfractures in Stage 1, equant calcite from Bukha No. 1 well, 10,555 ft. Although difficult to distinguish in these prints, the color of the fluorescence of these inclusions is blue-white when compared with inclusions from the Al Ali No. 1 well shown in (D). Long axis of photomicrograph is 0.6 mm. (F). Vertical UV fluorescence image of calcite cement and dolomite rhombs caught up in insoluble residue along "microstylolite swarm", Shams No. 1 well, 10,032 ft. Two distinct colors of flourescent, hydrocarbon fluid inclusions occur along healed microfractures in the calcite. The upper right corner contains sub-parallel microfractures with blue-white inclusions and yellow-white inclusions. Dolomite rhombs show yellow fluorescent zoning similar to cathodoluminescence. This is due to trace element activation, not to hydrocarbon fluid inclusions. Long axis of photomicrograph is 2.4 mm.

1. Primary porosity is completely filled with mud or with equant or syntaxial calcite cements. Early cements interpreted to be marine in origin are minor. Early pore-filling is interpreted as predominantly fresh-water derived and formed during regional, subaerial exposure immediately after deposition.
2. An extensive set of fractures cuts Stage 1 cements and is crosscut by all other diagenetic features. These fractures occur as subparallel sets or form braided, spider-web like networks. The fractures are healed with calcite that is commonly cleaved, twinned, and cloudy due to abundant inclusions. Clusters of hydrocarbon fluid inclusions occur along healed microfractures in the calcite crystals (Fig. 5 C and D). All of the inclusions fluoresce bright yellow-white.
3. Stage 2 fractures and trails of yellow-white fluorescent fluid inclusions are crosscut by another fracture set which is filled with calcite that is less cloudy and seldom cleaved or twinned (Fig. 4C and D). This calcite contains isolated fluid inclusions, some of which fluoresce dull blue.
4. Stylolites crosscut Stages 2 and 3 fractures and truncate fluid inclusion decorated microfractures. The stylolites are oriented in all directions, and when intensely developed, obscure the original fabric. Stage 2 fractures are crosscut by stylolites and certainly predate them (Fig. 4B). In contrast, Stage 3 fractures are either clearly crosscut or are only vaguely disturbed by intersecting stylolites, possibly due to an overlap in time between Stage 3 fracturing and Stage 4 stylolitization (Fig. 4C and D).
5. A final fracture set crosscuts all other diagenetic features, and thus post-dates them all. These fractures often occur as networks and can be associated with microfaulting. They are filled with clear, untwinned, and uncleaved calcite containing only nonfluorescent fluid inclusions trapped along growth zones and microfractures.

In the Al Ali No. 2 well, two different liquid hydrocarbon phases (oil-like with bright yellow-white fluorescence, and condensate-like with dull-blue fluorescence) were trapped in the carbonates prior to the period of intense stylolitization. Following pressure so-

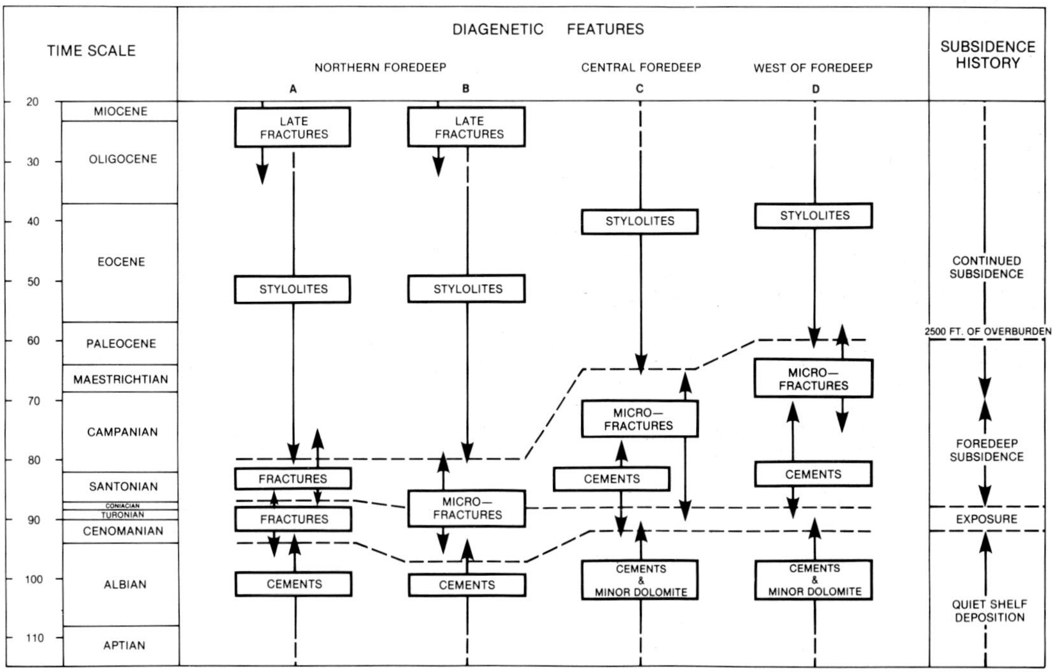

FIG. 6.—Summary diagram of diagenetic features recognized in core samples from four wells in this study. A-D identify wells on Figure 1. The approximate timing of the diagenetic features as well as the important events of the subsidence history are shown. Marine cementation of the seafloor, localized freshwater cementation and solution, and minor dolomitization occurred during quiet shelf deposition. Solution and freshwater cementation typified the regional exposure event and fracturing with subsequent cementation also occurred in the wells located nearer to the foredeep. Foredeep subsidence brought extensive fracturing and subsequent cementation to wells in the northern foredeep and micro-fracturing after cementation to the other wells. Stylolites formed in all wells probably beginning after approximately 2500 ft (750 m) of overburden. They formed earlier in wells from the northern foredeep due to the more rapid burial. A late fracturing and cementation event postdating stylolitization occurs in wells from the northern foredeep.

lution, aqueous fluids were apparently the dominant phase present in the rocks at this location.

Gulf/Elf Aquitaine Bukha No. 1 Well, Offshore Oman

The Lower and Middle Cretaceous carbonates in the Bukha No. 1 well (Location B, Fig. 1) consist of uniform pelleted wackestones and bioclastic packstones, with low rock-matrix porosity but some open fractures. Calcite-filled fractures are rare, but stylolites are common. The diagenetic history of the Cretaceous carbonates in this well (Fig. 6) can be summarized as follows:

1. Cementation, predominantly syntaxial overgrowths on echinoderm fragments, predated other diagenetic events. Less common are neomorphically replaced portions of fossils and equant spar filling pores and fractures. Non-fluorescent aqueous as well as bright yellow-white and bright blue-white fluorescent fluid inclusions were trapped along microfractures in calcite overgrowths, echinoid plates, and equant calcite (Fig. 5E).
2. Stylolites cross-cut calcite cement and neomorphic spar and in some places are so extensively developed that they brecciate the rock. Several sets of stylolites with slightly different orientations and amplitudes can be distinguished in places. Both types of fluorescent hydrocarbon fluid inclusion trails are truncated by stylolites.
3. Late fracturing developed the presently open, partially calcite-filled fractures. These features appear to cross-cut stylolites.

Thus in the Bukha No. 1 well, despite the lack of an extensive sequence of diagenetic events, it again appears that two oil-like phases (bright yellow-white and bright blue-white fluorescent) were present in the rocks prior to a period of extensive stylolitization.

CENTRAL FOREDEEP

Gulf/Quintana/Occidental Shams No. 1 Well, Northern Oman

The Middle Cretaceous carbonates in the Shams No. 1 well (Location C, Fig. 1) are porous grainstones, packstones, and wackestones with both primary and secondary (leached) porosities. The porosity is incompletely filled by calcite cement, and cement-filled fractures and stylolites are minor. The diagenetic history of the Cretaceous carbonates in this well (Fig. 6) can be summarized as follows:

1. Carbonate sediment along with organic-rich laminations were deformed during differential compaction prior to all other diagenetic events.
2. Partial cementation occurred along with localized leaching, and dolomite formed in uncemented organic-rich laminations and pods. Both the dolomite and the cement are cross-cut by all other diagenetic features. Their relative timing cannot be determined because they occur exclusive of each other. Dolomitization of organic-rich zones may have been accompanied by minor solution, creating features which resemble the anastomosing "microstylolite swarms" of Wanless (1979) that form from thin organic laminae.
3. A minor fracture set cross-cuts the calcite cement, and is in turn cross-cut by other diagenetic features. These fractures are filled with calcite in which dull-blue, bright yellow-white, and bright blue-white fluorescent fluid inclusions, as well as non-fluorescent fluid inclusions, were trapped along microfractures (Fig. 4A). Where calcite-filled fractures abut the dolomite-organic laminae, some of this uncemented material fell or was squeezed into the ends of open fractures, and the calcite precipitated around it. Fluorescent fluid inclusions appear to be more abundant in calcite adjacent to the dolomite-organic laminae.
4. Stylolites cut all other diagenetic features in these rocks including the "microstylolite swarms". Stylolites become so intensely developed in places that they appear to brecciate the rock. Stylolites distinctly truncate trails of bright yellow-white and bright blue-white fluorescing fluid inclusions (Fig. 4A and 5F). The relative timing of stylolites and the entrapment of dull-blue fluorescing fluid inclusions could not be determined.

In the Shams No. 1 well, it is evident that two hydrocarbon fluid phases with oil-like characteristics (bright yellow-white and bright blue-white fluorescence) were present in the Cretaceous carbonates prior to a period of stylolitization. The relationship of a more condensate-like hydrocarbon phase (dull blue fluorescent inclusions) to the diagenetic sequence is uncertain.

WEST OF FOREDEEP

Dubai Petroleum Rashid No. 1 Well, Offshore Dubai

Middle Cretaceous packstones and grainstones of the Rashid No. 1 well (Location D, Fig. 1) are more porous and contain more fractures and stylolites than the uniform wackestones of the Lower Cretaceous. Porosity is completely filled in a patchy pattern by calcite cement. The diagenetic history of the carbonate units examined in this well (Fig. 6) can be summarized as follows:

1. Cementation occurred early, as did dolomitization and the formation of "microstylolite swarms" along organic-rich laminae. The relative timing of dolomite and calcite cement can not be determined because both occur exclusively of each other (one in carbonate layers and the other along organic laminae). The cements and

dolomite rhombs are cross-cut by all other diagenetic features.

2. Following lithification, preferential solution formed moldic porosity in the carbonates. This moldic porosity is almost completely filled by equant calcite spar in which gold-orange fluorescing fluid inclusions were trapped along microfractures (Fig. 5A and B). Where mold-filling calcite abuts dolomite/organic layers, some of this uncemented mixture fell or was squeezed into the open mold, and the calcite infill grew around it.
3. Stylolites cross-cut all other diagenetic features (including "microstylolite swarms") and thus post-date them. As in the Bukha No. 1 well, several sets of stylolites can sometimes be distinguished. Trails of fluorescent, fluid inclusions along microfractures in sparry calcite are clearly truncated by these stylolites. Minor fracturing appears to be associated with stylolitization. These fractures (some calcite-filled and some open) are normal to stylolites and terminate further into the rock matrix. The calcite of filled fractures contains no fluid inclusions.

In the Rashid No. 1 well, it is therefore evident that one oil-like (gold-orange fluorescent) hydrocarbon phase existed in Lower and Middle Cretaceous carbonates prior to a period of stylolitization.

REGIONAL RELATIONSHIPS

The similarity and regional extent of the diagenetic sequences observed in the Middle and Lower Cretaceous carbonates of this study suggests those sequences were affected by the tectonic history of the eastern margin of the Arabian subplate. Therefore it should be possible to correlate diagenetic events to tectonic history and thus perhaps date them. Here we briefly review the tectonic history of the area (Wilson 1969; Glennie *et al.*, 1974; Ricateau and Riche, 1980; Searle *et al.*, 1983) and develop the correlation with the diagenetic events described above (Fig. 6).

From pre-Permian to Middle Cretaceous time, the area remained a tectonically quiescent carbonate shelf. Pre-orogenic regional upwarp occurred in late Cenomanian to Turonian time. The orogeny itself took place during Coniacian to Maestrichtian time and consisted of rapid obduction of ocean floor sediments and crust onto the edge of the Arabian craton with associated formation and filling of an orogenic foredeep. Following obduction and foredeep subsidence, thrusting of the carbonate platform edge over the foredeep occurred in the northern Oman mountains during Early Tertiary time and was perhaps reinitiated in the latest Miocene-Pliocene. Post-orogenic activity, doming and differential uplift, took place over the Central Foredeep area in the Mid-Late Tertiary. These tectonic phases, each with its own distinctive kinds of stresses and sedimentation patterns, allows us to isolate the individual effects of each phase and to correlate these to observed diagenetic events.

The fundamental diagenetic/tectonic correlation in each study locality involves stylolitization (Fig. 7). This diagenetic event, unlike the various generations of fracturing, appears to be ubiquitous in the study area. We correlate the formation of nearly-horizontal stylolites (as opposed to vertical or nearly-vertical stylolites formed by tectonic stress) with approximately 2500 ft (750 m) of overburden on the affected unit. This correlation agrees well with the studies of Dunnington (1967) and Mossop (1972), although studies by Schlanger (1964) and Meyers (1980) suggest the burial depth need not be so great. Regardless of the depth correlation, our diagenetic sequence of pre- and post-stylolite events is valid. At a certain time in the diagenetic history of most of these Lower and Middle Cretaceous carbonates, sufficient overburden accumulated to initiate extensive stylolitization which continued for some period thereafter. Since the carbonates were deposited immediately prior to and during the regional Cenomanian-Turonian shelf upwarp and, assuming our correlation between stylolitization and burial of approximately 2500 ft (750 m) is reasonable, are not themselves of sufficient thickness to promote stylolitization, stylolites in areas near the Oman Mountains most likely began to form in the Late Cretaceous (Coniacian-Campanian) when a thick clastic wedge was shed from the orogen (Figs. 1 and 7). In those parts of the study area further away from the orogen, the clastic wedge is thin and stylolites probably did not begin to form until the Early Tertiary (Paleocene-Oligocene) when shallow marine sequences of considerable thickness accumulated in the northeast Rub al Khali basin.

If this primary correlation of stylolites to overburden of foredeep sediment is accepted, then we can correlate tectonic events to various fracture sets with some confidence. Fractures which predate stylolitization probably formed during pre-orogenic shelf upwarp or during early, rapid subsidence of the Oman Foredeep area (Figs. 6 and 7). In the case of the Al Ali No. 2 well where we can distinguish two stages of pre- stylolite fractures, it appears that both tectonic phases (upwarp and rapid subsidence) were recorded (Burruss et al., 1983). Fractures which postdate stylolites are observed only in the northern section of the foredeep (the Al Ali No. 2 and Bukha No. 1 wells) and probably correspond to uplift of the Oman Mountains in the Early Tertiary.

This correlation between tectonic events and diagenetic events puts time constraints on the migration of hydrocarbon fluids. In each well, the fluid inclusions clearly demonstrate that an oil-like hydrocarbon phase was present in the Middle and Lower Cretaceous reservoir carbonates prior to the formation of stylolites. This oil-like phase could have entered the rocks no earlier than Cenomanian (age of the youngest host car-

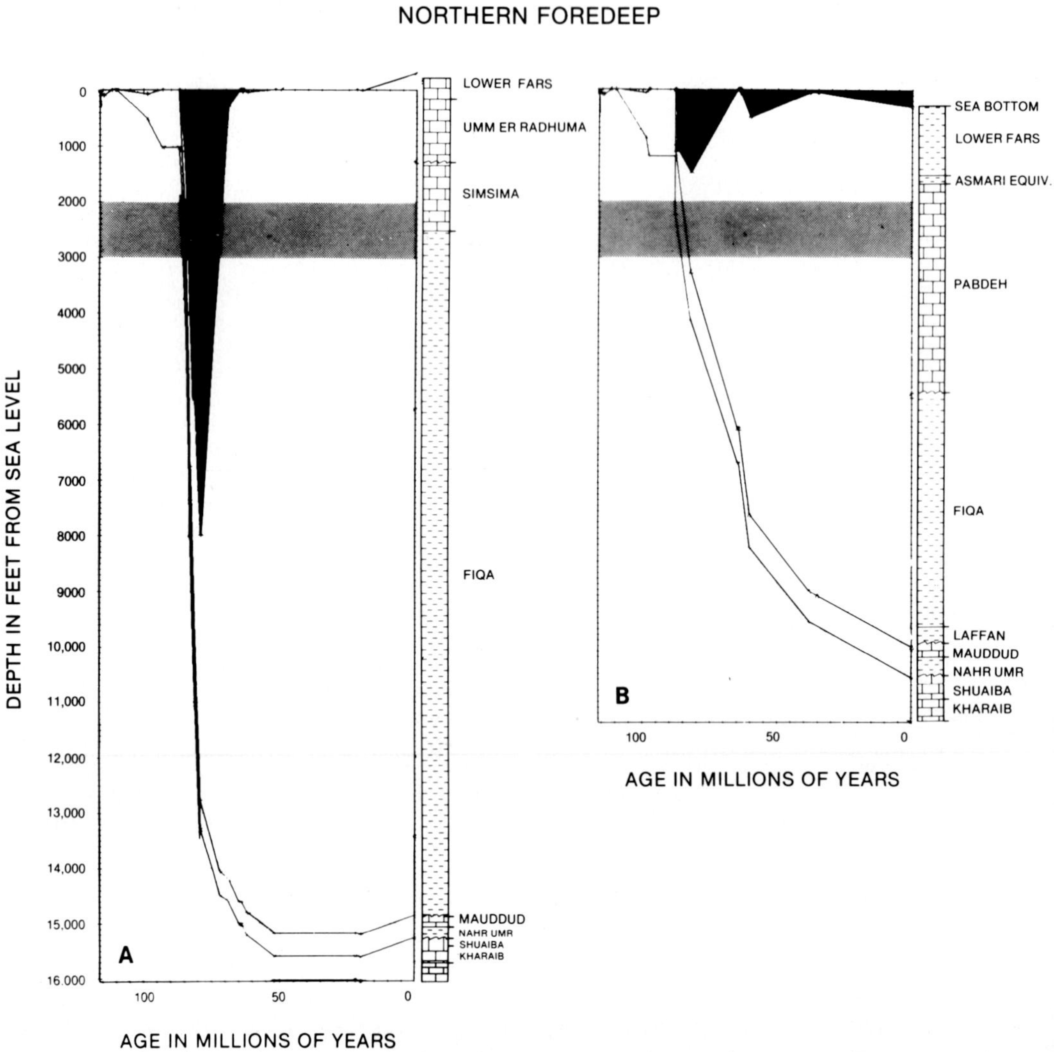

FIG. 7.—Simplified burial history plots of four wells analyzed in this study: A (Al Ali No. 2 well) and B (Bukha No. 1 well) from the narrow northern portion of the Oman Foredeep; C (Shams No. 1 well) from the wider central portion of the foredeep; and D (Rashid No. 1 well) located west of the foredeep. The stratigraphic column penetrated by each well is shown at the right margin of each plot. Age, in increasing amounts from right to left, is plotted on the horizontal axis versus depth in feet penetrated in each well increasing toward the bottom of the vertical axis. The two solid black lines on each plot trace the burial history of important horizons in each well (the top of the Shuaiba Formation by the lower line and the top of the Mishrif/Mauddud Formations by the upper line) from deposition during Cretaceous time to present day burial depth. Differences between the curves represent the differing rates of subsidence and sedimentation due to position with respect to the Oman Foredeep. Paleowater depth is indicated by the black band of variable width along the top of the plot. The large increase in water depth during the late Cretaceous represents development of the Oman Foredeep. The horizontal shaded band across each plot signifies a burial depth sufficient for the onset of stylolitization of the limestones.

bonate units) and no later than latest Cretaceous/earliest Tertiary (probable initiation of the youngest stylolites, although stylolites could have continued to develop throughout much of the Cenozoic). This conclusion is supported by evidence from Abu Dhabi, where Dunnington (1967) states that hydrocarbons charged the Lower Cretaceous reservoirs at Murban (Bab) Field prior to the development of stylolites before the end of Cretaceous time.

If these correlations and time constraints are valid, some interesting diagenetic patterns can be developed. For example, in the eastern edge of the Northern Fore-

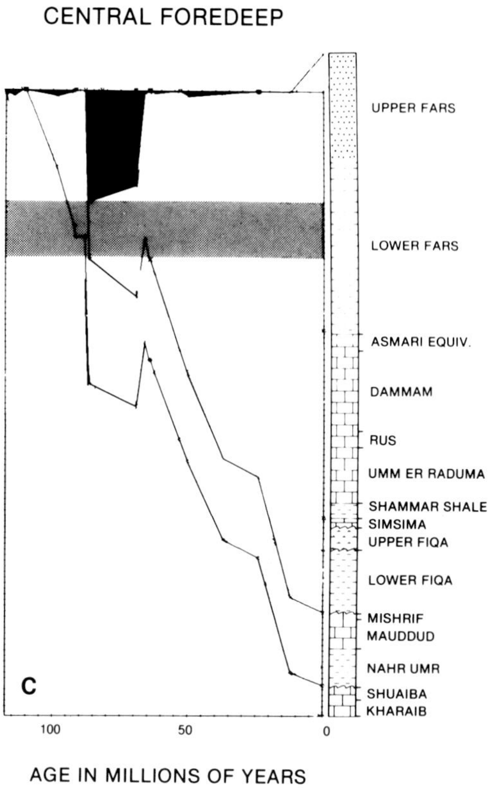

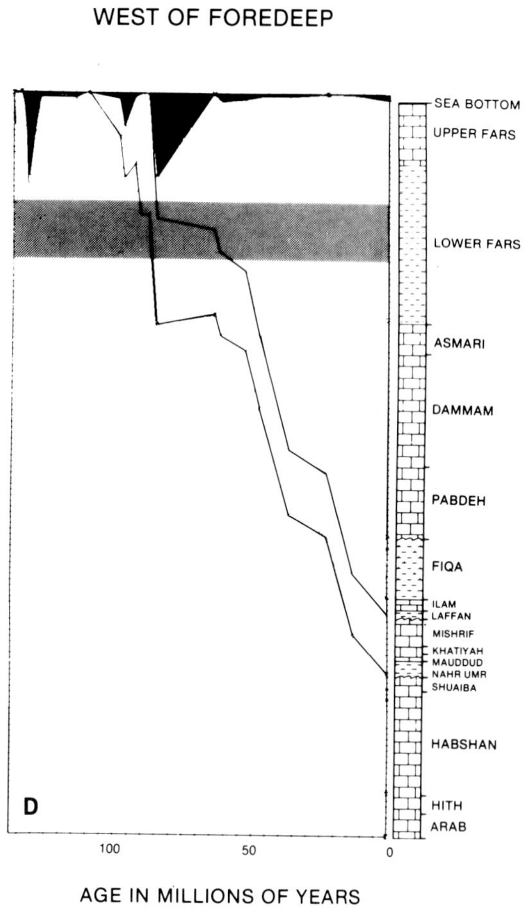

deep (near the Al Ali No. 2 well) the time constraints appear to be much tighter than at the other localities (Burruss *et al.*, 1983). Since thrust emplacement and flysch deposition at this site created an immense amount of overburden in the Campanian, we can reasonably assume this event marks the period of stylolite development and thus the latest possible time in which hydrocarbon liquids could have been present. Also in the area of the Al Ali No. 2 well, the oil-like (yellow-white fluorescent) and the condensate-like (dull-blue fluorescent) hydrocarbon phases, both of which were present prior to stylolitization appear to be chronologically separate. Oil-like hydrocarbon liquids were present during fracturing marking shelf upwarp in latest Cenomanian to Turonian time. Later, in Coniacian to Campanian time, during fracturing probably reflecting the downwarp of the foredeep region, condensate-like liquids were the dominant phase trapped as inclusions. Still later in fractures associated with Early Tertiary thrusting and associated fracturing, only aqueous inclusions are present. Thus in the area of the Al Ali No. 2 well, distinct transitions of diagenetic fluids from an oil-like hydrocarbon phase to a condensate-like hydrocarbon phase to an aqueous phase are recorded.

Slightly further northwest in the Northern Foredeep, at the locality of the Bukha No. 1 well, a limited diagenetic sequence records the presence of two oil-like hydrocarbon phases prior to the Tertiary, but does not permit a distinction of their relative timing if indeed there was one. Even further west, at the location of the Rashid No. 1 well, only one pre-Tertiary oil-like phase was recorded. In this part of the foredeep, oil-like phases could have begun to enter the rocks during a much broader time range (Coniacian-earliest Tertiary) than in the area of the Al Ali No. 2 well. Furthermore, there is no evidence from the Rashid No. 1 well or other nearby wells that hydrocarbon phases were replaced in the reservoir units by an aqueous phase (the wells either produce or tested hydrocarbons). Thus, hydrocarbons could theoretically have continued to enter the reservoirs throughout the Cenozoic, as suggested by McLimans' (1981) conclusion that the Fateh reservoirs were charged during the Middle Tertiary.

In the Central Foredeep, as represented by the Shams No. 1 well, the minimum possible age of oil-like hydrocarbons evident in the rocks seems to be earliest Tertiary. This cannot be shown to apply to more condensate-like liquids, however. Again, as in the northwestern portion of the foredeep, oil-like hydrocarbon phases could have entered the reservoirs during a long period of time. Condensate-like hydrocarbons, on the other hand, could have entered the rock at any time following lithification.

CONCLUSIONS

Several conclusions can be drawn from our diagenetic-tectonic correlations and the deduced patterns of hydrocarbon migration:

1. Calcite cements fill several generations of fractures recognized in Cretaceous limestones. Although the diagenetic sequence varies between wells, two sets of fractures predate stylolitization and one set occurs after it. Cements filling the early stages of fractures are cloudy, cleaved and sometimes twinned calcite. Late stage fractures are healed by clear, untwinned and uncleaved calcite.
2. The examination of fluid inclusions trapped within these cements, especially using fluorescence microscopy, provides a valuable tool for determining the timing of petroleum migration. Calcites filling the early stage fractures contain yellow-white and dull-blue fluorescent, hydrocarbon fluid inclusions, whereas the calcite in late stage fractures contains only nonfluorescent, aqueous fluid inclusions.
3. Oil-like hydrocarbons can be shown to have begun entering reservoir units prior to the Tertiary, therefore the source beds from which these phases were generated must have become mature by Upper Cretaceous time. In the area of the Al Ali No. 2 well, where oil-like hydrocarbons charged the reservoir units prior to the Campanian, the source must have reached thermal maturity by the Cenomanian. Burial history analysis (Fig. 7) suggests that the source of oil-like phases must be pre-Middle Cretaceous. The diagenetic pattern typifying the northern Oman Foredeep as represented by the Al Ali No. 2 well suggests hydrocarbon phases were replaced by an aqueous phase early in the Tertiary. Thus, this local area probably does not have much potential for hydrocarbon accumulations. The extent to which this low potential can be extrapolated to adjacent areas, however, is extremely limited by the complex tectonics of the northeastern Oman Foredeep. On the other hand, in the study localities from the northwestern and central portions of the Oman Foredeep where there has been less tectonic disturbance, it appears that oil-like and condensate-like hydrocarbon phases could have entered reservoirs over a wider span of time continuing to the present.

ACKNOWLEDGMENTS

G.A. Seiglie, B.L. Shaffer, and S.H. Frost provided age determinations. S.H. Frost provided information on the depositional and tectonic setting, and W.D. Wiggins read a draft of the paper. We thank Gulf Oil Corp. for assistance in manuscript preparation, and Gulf Oil Exploration and Production Company and its partners for permission to publish.

REFERENCES

Bolli, H.M., 1957, Planktonic foraminifera from the Oligocene-Miocene Cipero and Lengua formations of Trinidad, B.W.I., *in* A.R. Loeblich and H. Tappan (eds.), Studies in Foraminifera: U.S. National Museum Bulletin, v. 215, p. 97–124.

Burruss, R.C., 1981, Hydrocarbon fluid inclusions in studies of sedimentary diagenesis, *in* L.S. Hollister and M.L. Crawford (eds.), MAC Short Course in Fluid Inclusions, Applications to Petrology, Toronto, Mineralogical Assoc. Canada, p. 138–156.

______, Cercone, K.R., and Harris, P.M., 1983, Fluid inclusion petrography and tectonic-burial history of the Al Ali No. 2 well: Evidence for the timing of diagenesis and oil migration, northern Oman Foredeep: Geology, v. 11, p. 567–570.

Childs, O.E., 1981, Fifth annual progress report of correlation of stratigraphic units of North America (COSUNA).

Dunnington, H.V., 1967, Aspects of diagenesis and shape change in stylolitic limestone reservoirs: World Petroleum Congress 7th, Mexico, Proceedings, v. 2, p. 337–352.

Frost, S.H., Bliefnick, D.M., and Harris, P.M., 1983, Deposition and porosity evolution of a Lower Cretaceous rudist buildup, Shuaiba Formation of eastern Arabian Peninsula, *in* P.M. Harris (ed.), Carbonate Buildups: Soc. Econ. Paleontologists and Mineralogists, Core Workshop No. 4, p. 381–410.

Glennie, K.W., Boeuf, M.G.A., Hughes-Clark, M.W., Moody-Stuart, M., Pilaar, W.F.H., and Reinhardt, B.M., 1974, Geology of the Oman Mountains: Verhandelingen Van Het Koninkliijk Nederlands Geologisch Mijnbouwkundig, Genootschap, 423 p.

Harris, P.M., and Frost, S.H., 1984, Middle Cretaceous carbonate reservoirs, Fahud Field and Northwestern Oman: Amer. Assoc. Petroleum Geologists Bull., v. 68, p. 649–658.

______, ______, Seiglie, G.A., and Schneidermann, N., 1984, Regional unconformities and depositional cycles, Cretaceous of the Arabian Peninsula, *in* J.S. Schlee (ed.), Interregional Unconformities and Hydrocarbon Accumulation: Amer. Assoc. Petroleum Geologists Memoir 36, p. 67–80.

Hassan, T.H., Mudd, G.C., and Twombley, B.N., 1975, The stratigraphy and sedimentation of the Thammama Group (Lower Cretaceous) of Abu Dhabi: Ninth Arab Petroleum Congress, Paper No. 107(B-3), Dubai, p. 1–11.

JANEZIC, G.G., RODGERS, M., BURRUSS, R.C., AND JEFFREY, D.A., 1981, Oil-oil correlations by fluorescence spectrophotometry: (abs.) American Chemical Society Annual Meeting, Atlanta, Georgia, Abs. GEOC 025.

JORDAN, C.F., CONNALLY, R.C., AND VEST, H.A. (in press), Upper Cretaceous carbonates of the Mishrif Formation, Fateh Field, Dubai, U.A.E., *in* P.O. Roehl and P.W. Choquette (eds.), Carbonate Petroleum Reservoirs: Springer-Verlag.

KAUFFMAN, E.G., 1977, Geological and biological overview, Western Interior Cretaceous Basin: Mountain Geologist, v. 14, p. 75–99.

MCLIMANS, R.K., 1981, Applications of fluid inclusion studies to reservoir diagenesis and petroleum migration: Smackover Formation, U.S. Gulf Coast, and Fateh Field, Dubai: Amer. Assoc. Petroleum Geologists Bull., v. 65, p. 957.

MEYERS, W.J., 1980, Compaction in Mississippian skeletal limestones; Southwestern New Mexico: Jour. Sed. Petrology, v. 50, p. 457–474.

MOSSOP, G.D., 1972, Origin of the peripheral rim, Redwater Reef, Alberta: Can. Petroleum Geol. Bull., V. 20, p. 238–280.

MURRIS, R.J., 1980, Middle East, Stratigraphic evolution and oil habitat: Amer. Assoc. Petroleum Geologists Bulletin, v. 64, p. 597–618.

RICATEAU, R. AND RICHE, P.H., 1980, Geology of the Musandam Peninsula (Sultanate of Oman) and its surroundings: Jour. Petroleum Geology, v. 3, p. 139–152.

SAINT MARC, P., 1978, Arabian Peninsula, *in* M. Mollade and A.E.M. Nairn (eds.), The Phanerozoic Geology of the World, II, The Mesozoic: Amsterdam, Elsevier, p. 435–462.

SCHLANGER, S.O., 1964, Petrology of limestones of Guam: U.S. Geol. Survey, Prof. Paper 403D, p. 1–52.

SEARLE, M.P., JAMES, N.P., LOLAN, T.J., AND SMEWING, J.D., 1983, Sedimentological and structural evolution of the Arabian continental margin in the Musandam Mountains and Dibba zone, United Arab Emirates: Geol. Soc. America Bull., v. 94, p. 1381–1400.

SOLIMAN, F.A., AND AL SHAMLAN, A., 1982, Review on the geology of the Cretaceous sediments of the Rub al Khali, Saudi Arabia: Cretaceous Research, v. 3, p. 187–194.

VAN HINTE, J.E., 1976, A Cretaceous time scale: Amer. Assoc. Petroleum Geologists Bull., v. 60, p. 269–287.

WANLESS, H.R., 1979, Limestone response to stress: Pressure solution and dolomitization: Jour. Sed. Petrol., v. 49, p. 437–462.

WILSON, H.H., 1969, Late Cretaceous eugeosynclinal sedimentation, gravity tectonics, and ophiolite emplacement in Oman Mountains, Southeast Arabia: Amer. Assoc. Petroleum Geologists Bull., v. 53, p. 626–671.

UPPER JURASSIC SUBSURFACE CEMENTS: A CASE HISTORY[1]

CLYDE H. MOORE
Applied Carbonate Research Program, Department of Geology, Louisiana State University, Baton Rouge, Louisiana 70803

ABSTRACT

Upper Smackover ooid grainstones, buried to depths of 3400 m across the central Gulf of Mexico, are the hosts for Fe-zoned mosaic calcite and Fe-rich unzoned poikilitic calcite cements. These cements were precipitated into primary preserved pore space in a region where the only early diagenetic overprint was occasional marine cementation (Moore and Druckman, 1981). Petrography indicates that the zoned mosaic calcite cements were formed earlier than the unzoned poikilitic calcites, perhaps overlapping grain-to-grain pressure solution. The unzoned poikilitic calcites were precipitated after compaction and grain-to-grain pressure solution. Both cements contain two-phase fluid inclusions. Assuming the inclusions are primary, the zoned mosaic calcites formed under cooler and slightly less saline conditions than the poikilitic cements. The composition of the fluid phase in mixed zone mosaic and unzoned poikilitic cements at Walker Creek is similar in composition and salinity to present Smackover pore fluids (Klosterman, 1981). The trace element composition of both cement types is generally compatible with present Smackover brines, but because of natural water variation and uncertainties relative to distribution coefficients, these cements may have been precipitated by certain meteoric waters as well. Trace element composition of the mosaic calcite cement is significantly depleted in Mg, Fe and Mn relative to the poikilitic cement, supporting the petrographic and inclusion data, which suggest that the mosaic cement is younger. This relationship indicates that water composition evolved during the formation of the cements. The $\delta^{18}O$ composition of the zoned mosaic calcites becomes progressively lighter from older to younger zones, indicating progressive temperature increase during burial formation, following the interpretation of Dickson and Coleman (1980). The $\delta^{18}O$ composition of the poikilitic calcites (mean 6.4‰) is thought to indicate precipitation from Smackover-type brines under elevated temperatures (Moore and Druckman, 1981).

Elevated levels of $^{87}Sr/^{86}Sr$ in the unzoned poikilitic cements tie these cements directly to present Smackover brines, which have high $^{87}Sr/^{86}Sr$ derived from mixing with siliciclastic-related fluids during early stages of hydrocarbon migration and shale dewatering. The identification of a specific water responsible for the poikilitic cements allows a reevaluation of cement trace element composition, the calculation of paleotemperature of formation from $\delta^{18}O$ composition of the cements, and an evaluation of the two-phase fluid inclusion data by comparison with the $\delta^{18}O$ based paleotemperatures. The low Sr levels in the cements, in the face of elevated Sr/Ca of Smackover brines, call for reevaluation of Sr distribution coefficients under subsurface conditions. The Mg/Ca data indicate that Mg/Ca of the subsurface pore fluids has not evolved appreciably since burial, and supports Katz's (1973) distribution coefficient for Mg in calcite. Fe and Mn cement composition, when compared with the present brine composition indicates that the distribution coefficient for iron in calcite is greater than 1. This data also suggests (see also Michard, 1968), that the distribution coefficient for Mn in calcite probably increases with increasing temperature.

Using the $\delta^{18}O$ composition of the cements and the $\delta^{18}O$ composition of present sea water in Friedman and O'Neil's expression (1977), the minimum temperature of formation for the poikilitic cements is 84°C, while the minimum temperature range of formation for the mosaic cements is from 42°C to 67°C.

When isotopic data are compared with the two-phase fluid inclusion homogenization temperatures, the inclusion data yield consistently elevated temperatures. The poikilitic cements are closer to agreement than the mosaic calcites, suggesting that step-wise physical reequilibration of calcite two-phase fluid inclusions may be a problem.

Using Nunn's (1984) Jurassic geothermal gradient, the poikilitic calcite cements are assumed to have formed after at least 1 km of burial, possibly during Lower Cretaceous Hosston time. This data effectively constrains the time of arrival of hydrocarbons into upper Jurassic reservoirs in the central Gulf to at least Lower Cretaceous Hosston time, since the poikilitic cements clearly formed prior to hydrocarbon arrival (Moore and Druckman, 1981).

The late poikilitic calcite cements are thought to be the result of regional subsurface dissolution of preexisting carbonates by siliciclastic-hydrocarbon related fluids, while the mosaic calcites are thought to be related to progressive grain-to-grain pressure solution.

INTRODUCTION

The relative importance of burial diagenesis in limestones, the recognition of the products of burial diagenetic processes, and the applicability of commonly used petrographic and geochemical concepts and techniques in the deep subsurface environment have become a common concern of carbonate workers (Bathurst, 1980). The present paper will focus these concerns by presenting and evaluating the various lines of evidence used by this writer to conclude that some of the common coarse calcite cements present in Jurassic Smackover limestones of the central Gulf were precipitated after at least 1 km of burial, from brines

[1]Elements of this paper were presented at the Dallas AAPG-SEPM 1983 Carbonate Cements Revisited Symposium with Y. Druckman, J. A. D. Dickson, and A. M. Stueber under the title Jurassic Subsurface Calcite Cementation.

similar in composition to present Smackover pore fluids.

It is hoped that this exercise will serve to illuminate the serious problems facing the petrographer-geochemist in the interpretation of the history of deeply buried carbonate rock sequences.

Geologic Setting and Petrographic Summary of the Subsurface Cements

The host rock for these carbonate cements is late Jurassic Smackover ooid grainstone buried to depths between 3000 and 3400 m in southernmost Arkansas (Fig. 1). These rocks have been the target for intense petroleum exploration across the Gulf Rim for half a century (Collins, 1980). Moore and Druckman (1981) initially studied the petrography, chemistry and distribution of these cements in the context of the regional patterns of diagenesis of the upper Smackover (Fig. 1), and described a strong north-to-south diagenetic gradient in the upper Smackover in Arkansas-Louisiana that affected the sequence shortly after deposition. This gradient was the result of a regional meteoric diagenetic overprint to the north, with a marine-dominated system to the south (Figs. 1, 2). As a result the upper Smackover ooid grainstones of Moore and Druckman's (1981) southern diagenetic zone were buried into the subsurface without significant early diagenetic overprint, except for occasional fibrous marine circumgranular cements. Today, oil production in the southern diagenetic zone is from preserved primary porosity occurring in ooid-dominated sequences of the upper Smackover. The coarse calcite cements under discussion occur across this southern diagenetic zone in oil-wet, as well as water-wet reservoirs, seldom exceeding 10% of the total rock volume.

These cements generally occur as large (up to 2 mm), clear, poikilitic calcite crystals filling one or more intergranular pores. Crystal pore contacts are planar and crystal extinction is uniform (Fig. 3). The calcite generally contains some 3000–4000 ppm Fe in the southern zone, with Fe content of late cement decreasing significantly to the north (Moore and Druckman, 1981). The coarse poikilitic calcites are generally unzoned, or very weakly zoned when stained by the method of Dickson (1966), or observed under cathodo-luminescence.

A second, but far less common fabric is coarse calcite crystal mosaics filling large shelter voids, usually associated with gastropod tests (Fig. 4), or finer mosaics filling preserved primary intergranular porosity. Intercrystalline boundaries are generally straight, but may be irregular. Calcite extinction is uniform (Fig. 4). These crystal mosaics are often complexly zoned, when stained or observed using cathodo-luminescence (Fig. 4). Zoning is related to iron distribution, which varies from 50 to slightly over 1000 ppm.

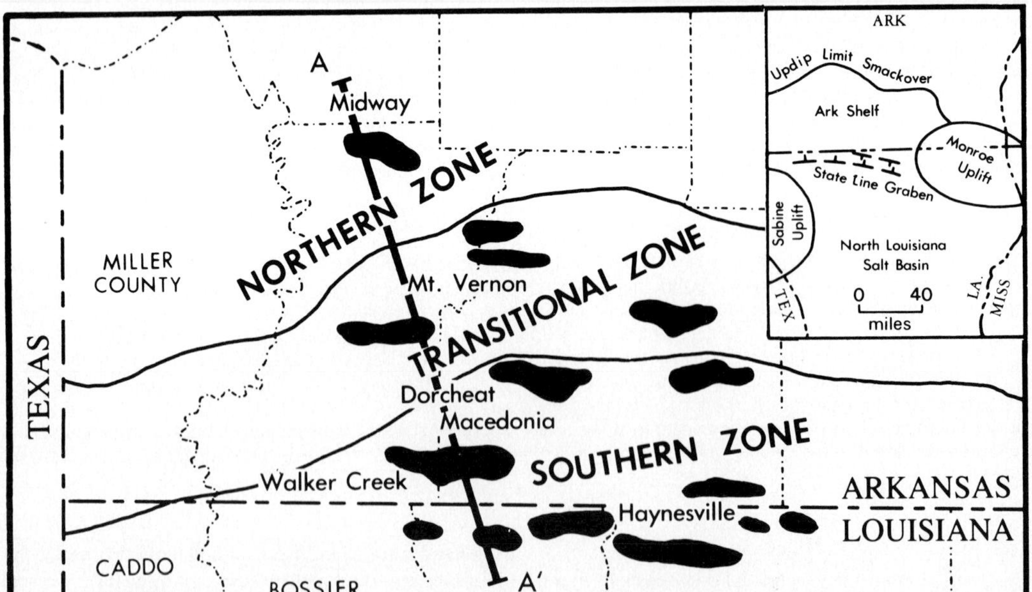

FIG. 1.—Location map of study area showing diagenetic zones of Moore and Druckman (1981) and the location of the diagenetic model of Figure 2.

Moore and Druckman (1981) concluded that these coarse calcite cements were of late subsurface origin and were precipitated from fluids similar to present Smackover brines. Wagner and Matthews (1982), however, in their discussion of the Walker Creek field contended that many of the coarse calcite cements at Walker Creek, and by inference elsewhere in Moore and Druckman's southern zone, were of early meteoric origin.

The following discussion outlines in detail the evidence developed initially by Moore and Druckman (1981), and subsequently by the present author and others supporting the subsurface origin of these cements. The discussion will be structured in approximately the same sequence utilized in the investigations, that have spanned the past five years: petrographic evidence, two-phase fluid inclusions, trace elements, stable isotopes, and finally, strontium isotopes.

PETROGRAPHIC EVIDENCE FOR SUBSURFACE ORIGIN

Perhaps the most compelling argument for the subsurface origin of these cements is the common petrographic evidence that most cementation took place after grain-to-grain pressure solution and after grain fracturing. In many cases early circumgranular crusts of bladed-to-fibrous marine cements were spalled from ooids during compaction. The poikilitic calcite cements incorporated spalled cement as well as the adjacent ooid during precipitation, clearly indicating precipitation after the physical compaction process commenced (Fig. 4). In other cases, ooids and other grains were fractured or the ooid cortex disrupted, and the later cement was precipitated into the fracture or incorporated the disrupted elements of the ooid cortex; both situations indicating that the coarse spar precipitated after enough burial to initiate grain breakage (Figs. 3, 4).

Cementation after significant grain-to-grain pressure solution is perhaps the most common petrographic relationship found in the coarse poikilitic calcite cement. Although significant grain interpenetration with sutured contacts between grains is commonly observed, these sutured contacts stop at the grain-cement boundary, and no evidence of grain-cement interpenetration is noted (Fig. 3). Occasionally, however (as was noted by Wagner and Matthews in 1982) there is evidence that some of the coarse zoned mosaic calcite may have been precipitated before the grain-to-grain pressure solution process was complete. In these cases there is evidence of grain-cement interpenetration (Fig. 3). This relationship is quite rare and has been found to only involve some of the coarse zoned mosaic calcites indicating perhaps that the zoned mosaic calcite may have been formed before the bulk of the coarse poikilitic calcites of the southern zone.

Discussion of the Petrographic Evidence

While the petrographic evidence for the subsurface origin of these cements is clear, the depth in the subsurface at which the cementation took place cannot be determined by petrographic evidence alone. There has been considerable discussion about the processes of pressure solution, factors controlling grain fracture, grain-to-grain pressure solution, as well as porosity loss during burial (see Bathurst, 1980; Fruth *et al.*, 1966; Meyers and Hill, 1983; Coogan, 1970; deBoer, 1977; Manus and Coogan, 1974; Brock and Moore, 1981). There have been few observations, however, relative to the depth at which breakage and/or grain-to-grain pressure solution commences in a mud- or cement-free grainstone. Saller (1984), studying the deep

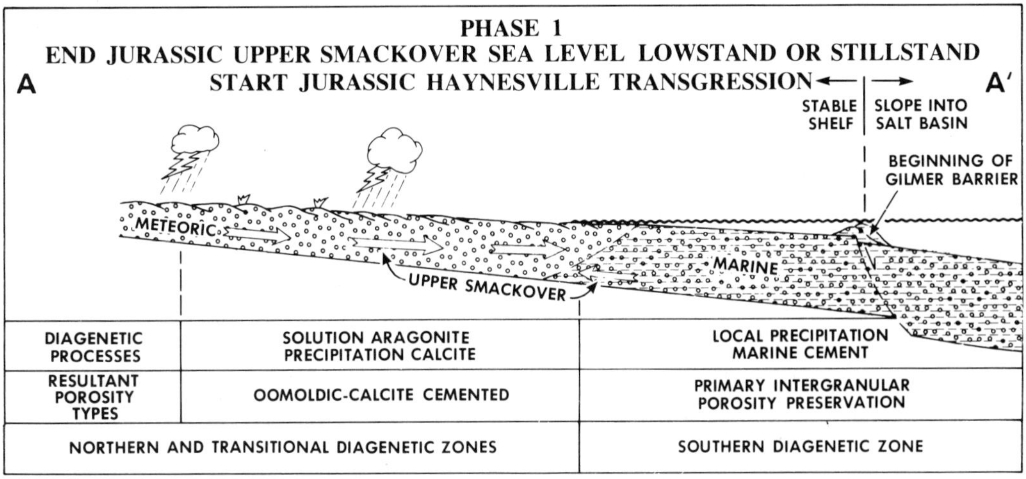

FIG. 2.—Conceptual early diagenetic model of the upper Smackover across southern Arkansas-northern Louisiana prior to significant burial into the subsurface.

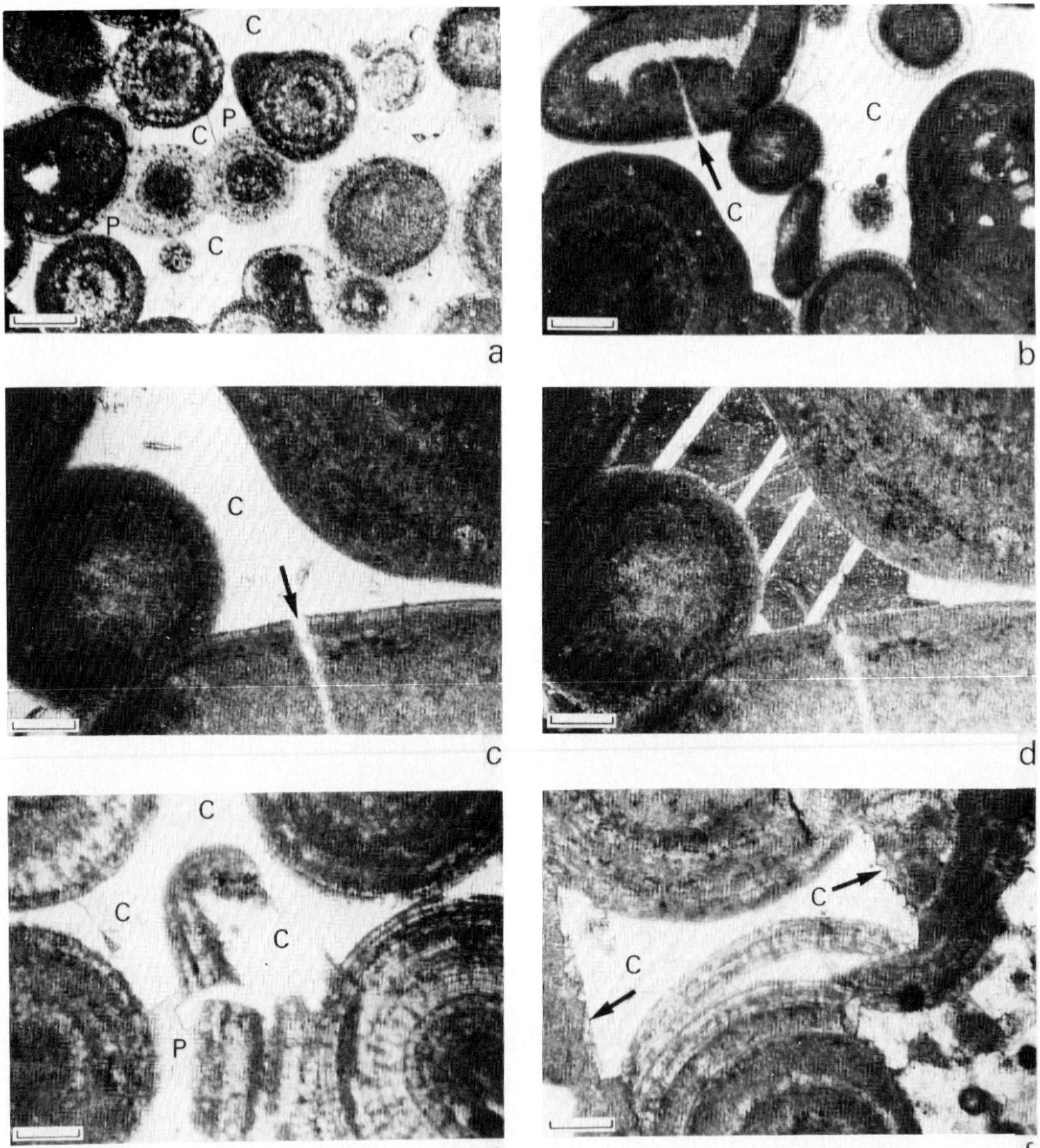

FIG. 3.—*a*, Unzoned poikilitic calcite cement (C): note planar crystal pore contacts (P). Arco Bodcaw #1, 10,873 ft. Columbia Co., Arkansas; plane light. Bar is .25 mm. *b*, Unzoned poikilitic calcite cement (C). Note grain interpenetration without cement involvement. Broken grain healed by cement at arrow. Arco Bodcaw #1, 10,873 ft. Columbia Co., Arkansas; plane light. Bar is .10 mm. *c*, Close view of b showing broken grain (arrow) healed by cement (c). Plane light. Bar is .10 mm. *d*, Same as c with crossed polars. *e*, Broken grain healed by unzoned poikilitic calcite cement (C). Note planar crystal-pore contacts (P). Arco Bodcaw #1, 10,873 ft; Columbia Co., Arkansas; plane light. Bar is .10 mm. *f*, Zoned mosaic calcite cement (C) involved in pressure solution and compaction (arrows). Cement is post ooid rupture because it heals spalled ooid cortex. Arco Bodcaw #1, 10,906 ft., Columbia Co., Arkansas; plane light. Bar is .10 mm.

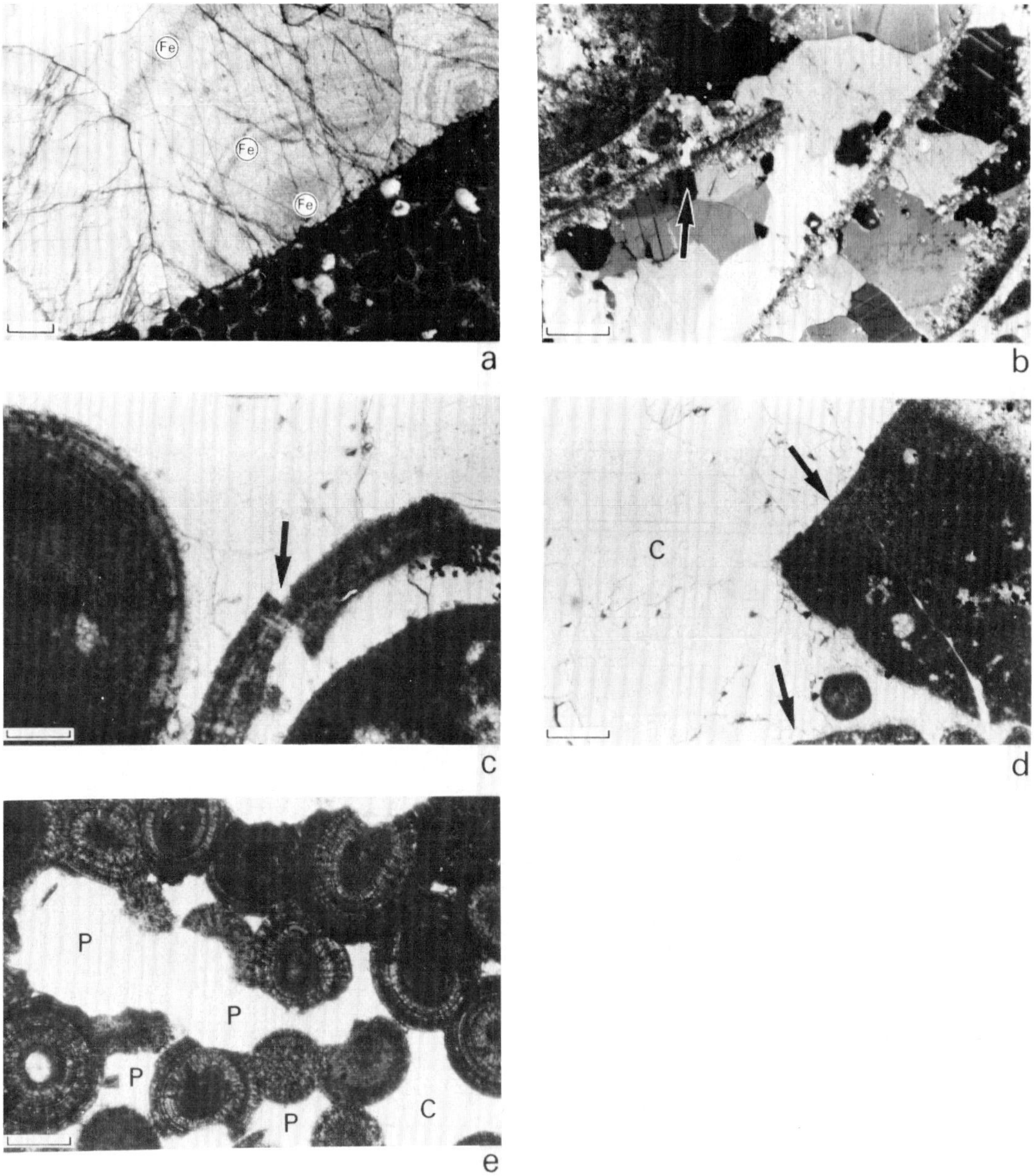

FIG. 4.—*a*, Zoned mosaic calcite cement formed in a large shelter void of a gastropod in the Lear Clement #1 well, Miller Co., at 10,793 ft. Iron rich zones (Fe) displayed after staining using the method of Dickson (1966). Note that crystal boundaries are considerably more irregular than the unzoned poikilitic cements pictured earlier. Bar is .25 mm. Plane light. *b*, Zoned mosaic calcite cement formed in a collapsed gastropod test in the Lear Clement well at 10,700 ft, Miller Co. Note that broken gastropod wall has been healed by the mosaic calcite (arrow). Bar is .25 mm. Crossed polars. *c*, Ruptured ooid cortex (arrow) healed by zoned mosaic calcite cement in the Pennzoil #1 Taylor, Miller Co., 10,793 ft. Plane light. Bar is .1 mm. *d*, Broken clast healed by zoned mosaic calcite cement (C). Note that early marine cement (arrows) in adjacent voids has been cut by the fracture and does not occur on the surface healed by the mosaic calcite (arrow). Lear Clement #1, Miller Co., 10,700 ft. Plane light. Bar is .1 mm. *e*, Late secondary porosity (P). Pore space is filled with colored plastic. Note unzoned poikilitic calcite cement (C) in lower right corner. Arco Bodcaw #1, Columbia Co., 10,894 ft. Plane light. Bar is .25 mm.

borings at Enewetak, the central Pacific, indicated that significant grain fracture in uncemented foraminiferal grainstones was not observed above 600 m. Dunnington (1967) indicated significant pressure solution at 600–900 m for Middle Eastern limestones, while Sellier (1979) reported that pressure solution was important at 300 m burial depth in the Aquitaine Basin. The limestone sequences in both of these study areas contained early cement that no doubt delayed the onset of pressure solution processes. In the Enewetak cores, which are presumably acting as on open system to sea water, significant pressure solution in uncemented foraminiferal sands was not observed above 900 m (Saller, 1984).

Constraints Placed on the Origin of the Cements by Petrographic Data

Both cements were precipitated in the subsurface. The mosaic calcite seems to be older than the poikilitic cement which is clearly post-compaction. No depth constraints, however, can reasonably be applied.

TWO PHASE FLUID INCLUSION EVIDENCE FOR SUBSURFACE ORIGIN

The post-compaction calcite cements under discussion contain abundant two-phase fluid inclusions. Klosterman (1981) compiled a detailed data set of homogenization temperatures and freezing experiments on these cements at the Walker Creek Field, Lafayette and Columbia Counties, Arkansas, located in Moore and Druckman's (1981) southern diagenetic zone (Fig. 1). Klosterman reports a range of homogenization temperatures from 58°C to 158°C, with a mean near 110°C (Fig. 5, this paper). At the time of Klosterman's study, the post-compaction cement in Walker Creek and vicinity was not differentiated into the zoned mosaic and unzoned poikilitic cements described in the introduction of this paper. In 1981, J.A.D. Dickson, during

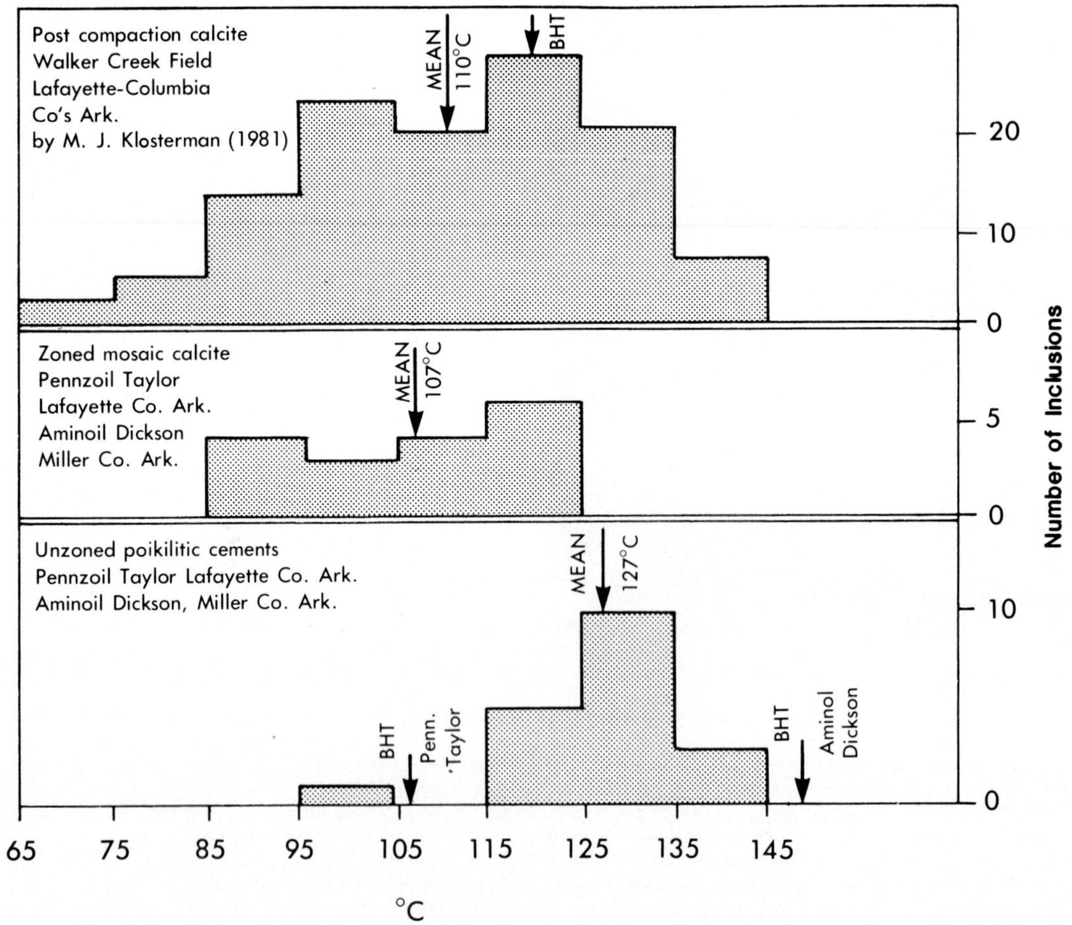

FIG. 5.—Histograms of two-phase fluid inclusion homogenization temperatures taken from zoned mosaic and unzoned poikilitic cements. Bottom hole temperatures (BHT) in study wells are indicated by arrows.

a visit to the Applied Carbonate Research Program, documented these two cement types with the aid of stains. Inclusion data were subsequently generated on the zoned mosaic versus unzoned poikilitic calcite in two wells in the southern diagenetic zone just west of Walker Creek (Fig. 1). This second set of inclusion data was generated by Brian Carter and Terry O'Hearn at L.S.U., using material from Stamatedes (1982).

The zoned mosaic has a wider homogenization temperature range and a lower mean temperature, 108°C, than does the unzoned poikilitic cement which has a mean homogenization temperature of 127°C. A comparison of Klosterman's original data with the present data (Fig. 5) suggests that both temperature populations are represented in her Walker Creek data.

Klosterman (1981) reported that freezing temperature depression, ranging from −15° to −32°C in the inclusions found in Walker Creek post-compaction cements, indicates that the fluids in the inclusions were brines ranging in salinity from 5 times seawater to almost 9 times that of seawater. Initial and intermediate melting temperatures reported by Klosterman strongly suggest the presence of $CaCl_2$ in the trapped fluids.

The later inclusion studies generally confirm Klosterman's gross salinity range (see Table 1), but also indicate that the inclusions of zoned mosaic calcite contain slightly less saline fluid than do those of unzoned poikilitic cements. No conclusions relative to the fluid composition could be reached because the inclusions were too small to observe subtle intermediate melting behavior.

Discussion of Two-phase Fluid Inclusion Evidence

The common occurrence of two-phase fluid inclusions in the coarse post-compaction calcite cements is evidence of their subsurface origin, if the inclusions are primary and are not the result of the physical readjustment of the calcite crystals to increased overburden pressure during burial. If continuous physical reequilibration of the calcite had taken place, all inclusions would reflect bottom-hole temperatures. The results to date show clearly, however, that the unzoned mosaic calcite represents a separate, older population than the unzoned poikilitic cement, paralleling the petrographic evidence developed earlier. This does not rule out the possibility that earlier-formed cements, such as the zoned mosaic calcite of this paper, may have physically reequilibrated in response to loading. Two-phase fluid inclusions of later cements (such as the unzoned poikilitic cement), formed after most physical compaction was complete, may, however, be primary and reflect near true temperatures of formation because they formed in equilibrium with the high temperatures and pressures of the subsurface. It should be noted that Moore and Druckman (1981) reported two-phase fluid inclusions in well-documented early pre-compaction cements in southern Arkansas. These observations would certainly tend to support the concept of a step-wise inclusion reequilibration that could preserve the relative sequence of cementation events in the homogenization temperatures of their two-phase fluid inclusions.

Constraints Placed on the Origin of the Cements by Petrographic and Two-phase Fluid Inclusion Data

The inclusion data generally support the petrographic observations that the zoned calcites are older than the unzoned poikilitic cements. We cannot be certain, however, that the inclusions are primary and reflect true temperatures of formation until independent evidence tying the cement to a particular fluid composition is developed.

TRACE ELEMENT EVIDENCE FOR SUBSURFACE ORIGIN

Table 2 outlines the trace element composition of the post-compaction cements characteristic of Moore and Druckman's (1981) southern diagenetic zone. These data indicate, with the exception of Sr, that the zoned mosaic calcite is chemically different from the unzoned poikilitic cement, containing significantly less Mg, Fe and Mn. Both cements have strontium levels below the detectability limit of microprobe analysis at L.S.U. (100 ppm for Sr). The data for the poikilitic cements parallels the data reported by Moore and Druckman (1981) for their post-compaction cements.

TABLE 1.—FINAL MELTING TEMPERATURES AND CALCULATED SALINITIES OF INCLUSIONS IN POST COMPACTION CEMENTS SOUTH ARKANSAS.

	Zoned Mosaic Calcite	Poikilitic Calcite	M. J. Klosterman, 1981 Post Compaction Calcite Cement Walker Creek
Final Melting Temperature °C Range	− 18 to − 28	−28 to −31	−17°C to −27°C
Mean	−24	−30	−25°C
Wt % NaCl Calculated* From Final Melting Temperature	21	26.	22

*Potter, Clynne and Brown (1978)

Figure 6 represents a detailed microprobe traverse across a single calcite crystal of the zoned mosaic cement, detailing the variation of Mg, Fe and Mn. The crystal was stained and photographed before analysis following the technique of Dickson (1966). The track of the probe traverse, relative to the iron-rich zones revealed by the stain, is also shown in Figure 6. Iron variation faithfully reproduces the stain zonation, with the stain being activated at approximately 300 ppm Fe. There is no significant systematic manganese or magnesium variation. The low levels of manganese are reflected in the very dull luminescence under a luminescope.

Similar traverses were run on the coarse poikilitic calcite cements at Walker Creek (Fig. 7). The pattern of variation of magnesium and iron in the poikilitic cements is strikingly different from the well-developed crisp systematic iron zonation seen in the mosaic cement. Magnesium and iron in the poikilitic calcite seem to have a strong co-varient pattern, while they vary independently in the zoned mosaic cement. The iron variation seen in the microprobe traverses of the poikilitic cements is not organized into a crystal growth zonation that can be detected with luminescence, or with stains.

Discussion of Trace Element Data

The trace element composition of the post-compaction cements should reflect the gross chemistry of the fluids from which they have precipitated (Kinsman, 1969; Land *et al.*, 1975; Bathurst, 1975; Moore and Druckman, 1981; Moore, 1977; and many others). If, as indicated by the fluid inclusion studies, the post-compaction cements were precipitated at elevated tem-

TABLE 2. EQUIVALENT RATIOS OF CERTAIN TRACE ELEMENTS IN PRESENT SMACKOVER BRINES, METEORIC WATER AND SEA WATER WITH THE CALCULATED COMPOSITION OF CALCITE PRECIPITATED FROM THESE WATERS USING INDICATED DISTRIBUTION COEFFICIENTS COMPARED TO MEASURED TRACE ELEMENT COMPOSITIONS OF POST-COMPACTION CEMENTS.

ELEMENT	SMACKOVER BRINE[1]			FRESH METEORIC WATER[1] (N. AMERICAN RIVER SYSTEMS)		
STRONTIUM	Equivalent Ratio Sr/Ca	Calculated[2] Sr (ppm) in calcite ppct. from this water with indicated distribution coefficients at 25°C		Equivalent Ratio Sr/Ca	Calculated[2] Sr (ppm) in calcite ppct. from this water with commonly used distribution coefficients at 25°C	
		.14 (Kinsman 1969)	.054 (Katz et al 1972)		.14 (Kinsman 1969)	.054 (Katz et al 1972)
Range	.0025-.085	307-10,426	118-4022	.0009-.0091	410-1116	43-431
Mean	.025	3000	1350	.0035	420	189
MAGNESIUM	Equivalent Ratio Mg/Ca	Calculated[2] Mg (ppm) in calcite ppct. from this water using Katz's (1973) distribution coefficients at 25°C and 70°C		Equivalent Ratio Mg/Ca	Calculated[2] Mg (ppm) in calcite ppct. from this water using Katz's (1973) distribution coefficient at 25°C and 70°C	
		(.0573) 25°C	(.0973) 70°C		(.0573) 25°C	(.0093) 70°C
Range	0.05-0.52	696-7,244	1183-12,200	0.082-0.89	1142-12119	1939-21,053
Mean	.17	2368	4021	.37	5154	8752
MANGANESE	Equivalent Ratio Mn/Ca	Calculated[2] Mn (ppm) in calcite ppct. from this water using Michard's (1968) distribution coefficient 5.4 at 25°C		Mean Equivalent Ratio[4] Mn/Ca	Calculated[2] Mn (ppm) in calcite ppct. from this water using Michard's (1968) distribution coefficient, 5.4 at 25°C	
Range	2.15 X 10^{-5} - 1.07 X 10^{-3}	64-3174		1.0 X 10^{-3}	2967	
Mean	3.2 X 10^{-4}	949				
IRON	Equivalent Ratio Fe/Ca	Calculated[2] Fe (ppm) in calcite ppct. from this water using Richter and Fuchtbauer (1978) distribution coefficient of 1 at 25°C		Mean Equivalent Ratio[4] Fe/Ca	Calculated[2] Fe (ppm) in calcite ppct. from this water using Richter and Fuchtbauer (1978) distribution coefficient of 1 at 25°C	
Range	2.1 X 10^{-5} to 4.39 X 10^{-3}	92-2450		2.0 X 10^{-3}	1117	
Mean	1.42 X 10^{-3}	793				

peratures from highly saline fluids similar to present Smackover brines, their chemistry should reflect the chemistry of the brines and qualify as positive evidence for their subsurface origin.

Table 2 compares the mean chemical composition of Smackover pore fluids, river water and sea water. Smackover formation waters are $CaCl_2$ brines, with salinities up to 10 times sea water. They have very low Mg/Ca, high Sr/Ca and high Fe and Mn concentrations. There are some serious problems with this comparative approach, however. The chemical characterization of the Smackover brines is, in itself, difficult, because of a wide range of composition as documented by Collins (1974), Carpenter (1978), and others. The same problem of wide variability is encountered when trying to characterize meteoric water systems. Any comparison of the post-compaction cements with the cement chemistry will have to take this variability into account. In Table 2, the low value, high value, and the mean of the trace element equivalent ratios of Sr, Mg, Mn and Fe are presented for Smackover and meteoric waters, where available. These elemental ratios in surface sea water generally act conservatively, so the variability of equivalent ratios is negligible (Gross, 1977).

Experimentally derived distribution coefficients are commonly used to relate the trace element composition of a calcite to the composition of the fluid from which it precipitated. As expressed by Kinsman (1969), in his strontium studies:

$$\frac{\text{trace element in calcite}}{\text{calcium}} = \text{distribution coefficient of the trace element in calcite} \times \frac{\text{trace element}}{\text{calcium}}$$

	MEAN SEA WATER[1]		MEASURED[3] TRACE ELEMENT COMPOSITIONS POST COMPACTION CEMENTS (ppm)	
			Zoned Mosaic	Unzoned Poikilitic
Mean Equivalent Ratio Sr/Ca	Calculated[2] Sr (ppm) in calcite ppct. from this water with commonly used distribution coefficients at 25°C		<100	<100
	.14 (Kinsman 1969)	.054 (Katz et al 1972)		
.0086	1204	464		
Mean Equivalent Ratio Mg/Ca	Calculated[2] Mg (ppm) in calcite ppct. from this water using Katz's (1973) distribution coefficients at 25°C and 70°C		2268 ± 241	3160 ± 498
	(.0573) 25°C	(.0973) 70°C		
5.3	73,833	125,375		
Mean Equivalent Ratio[4] Mn/Ca	Calculated[2] Mn (ppm) in calcite ppct. from this water using Michard's (1968) distribution coefficient, 5.4 at 25°C		143 ± 28	498 ± 85
3.6×10^{-6}	11			
Mean Equivalent Ratio[4] Fe/Ca	Calculated[2] Fe (ppm) in calcite ppct. from this water using Richter and Fuchtbauer (1978) distribution coefficient of 1 at 25°C		356 ± 328	2014 ± 326
1.79×10^{-5}	10			

[1]Equivalent ratios from Moore and Druckman, 1981.
[2]Calculated composition of calcite from expression m Sr/Ca Calcite = Distribution Coefficient x m Sr/Ca of Precipitating Fluid (Kinsman, 1969).
[3]Mean composition in ppm determined by electron microprobe analysis is at the L.S.U. Department of Geology, Microanalysis Laboratory by C.H. Moore. Number of Analyses Mg, Fe, Sr, 100; Mn 55.
[4]Equivalent ratios from Meyers, 1974.

of the precipitating water. It has become increasingly clear in recent years, however, that our understanding of the complexity of the controls on those distribution coefficients in the calcite system is rudimentary at best. One of the main problems involves rate of precipitation, which can have a significant influence on the distribution coefficient (Jacobson and Usdowski, 1976). Since laboratory experiments are conducted under conditions of very rapid precipitation rates, while subsurface cementation occurs at much lower rates, the actual distribution coefficient for a trace element under natural subsurface conditions is presently a matter of conjecture. In addition, there is some controversy concerning the appropriate distribution coefficient for Sr in calcite (Katz *et al.*, 1972), magnesium in calcite (Katz, 1973), and iron in calcite (Land, personal communication).

In Table 2, therefore, we have used the low, high, and mean equivalent ratios for each of the three water systems, combined with various commonly used distribution coefficients, in order to calculate the possible range of trace element compositions of a calcite precipitating from these waters. This range of calculated trace element compositions can then be compared to the measured trace element composition of the Smackover post-compaction cements, to facilitate the evaluation of the cements' compatibility with Smackover brines.

Finally, it should be noted that we are assuming that the brines presently in the pore spaces of the Smackover are not significantly different in chemical composition from the brines, perhaps as old as Cretaceous in age, which precipitated the post-compaction cements.

The very low Sr values seen in both cement types are compatible with meteoric water, using both Kinsman's (1969) and Katz *et al.*'s (1972) coefficients at the low end of the compositional range. The measured cement Sr values are compatible with Smackover brine only if we use Katz *et al.*'s (1972) coefficient at the extreme end of the compositional range.

The magnesium composition of the mosaic cement is compatible with the mean magnesium composition of the Smackover brine at 25°C, while the poikilitic cements may have precipitated from the mean Smackover brine at elevated temperatures (see also Moore and Druckman, 1981). In addition, these cements could have precipitated from meteoric waters either at 25°C or at elevated temperatures, if we use the low end of the meteoric water compositional range. The difference in magnesium composition between the mosaic and poikilitic cements then could be one of

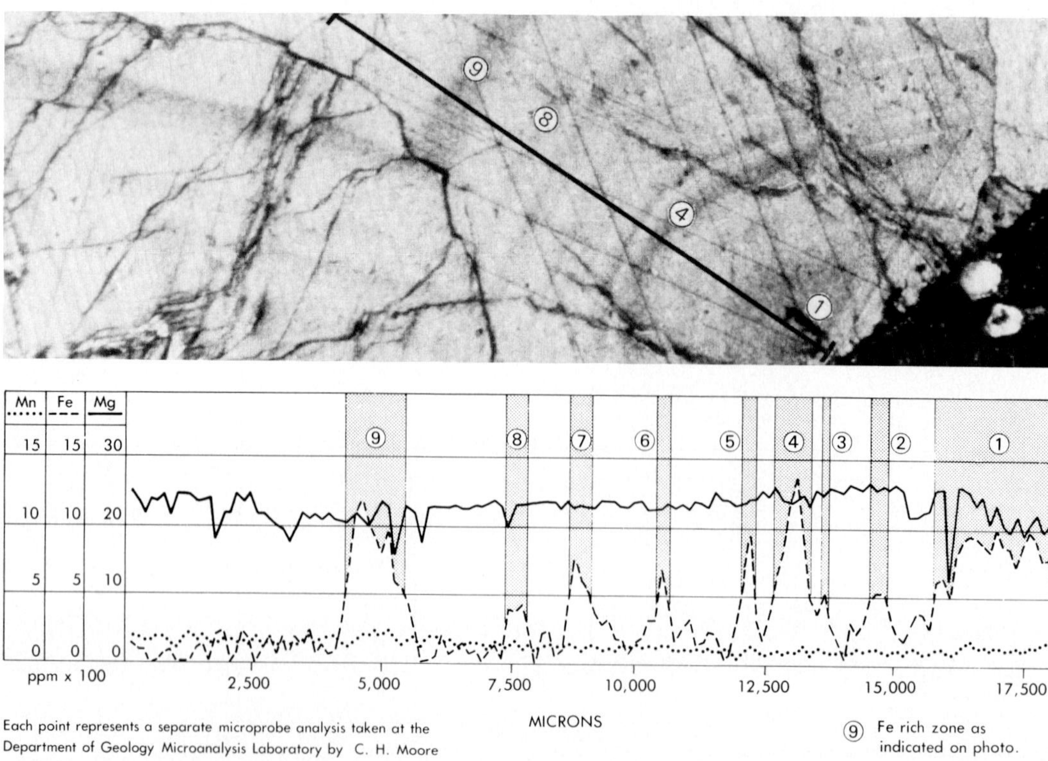

FIG. 6.—Electron microprobe traverse of zoned mosaic calcite cement from Lear Clements #1, Miller Co., Arkansas. Probe traverse, with zones identified on stained calcite shown in photograph at top of figure.

changing water composition or solely the result of temperature.

Both the mosaic and the poikilitic calcites Fe and Mn compositions are compatible with present Smackover brine if we use the low end of the Smackover composition range for manganese and the full compositional range for iron. It would seem that the significant Fe and Mn enrichment of the poikilitic cements relative to the mosaic cements is the result of the change of Fe and Mn composition through time.

Constraints Placed on the Origin of the Cements by Petrography, Two-phase Fluid Inclusions and Trace Element Data

The trace element composition supports the inclusion data, and indicates that the cements precipitated from Smackover brines, only to the extent that the cement composition is generally compatible with Smackover brines under the conditions discussed above. On the other hand, the trace element composition of both cements are probably compatible with the composition of certain meteoric waters as well (Fe and Mn variability data in the meteoric water system are not available, but are assumed to be as great as in the Smackover). The manganese and iron compositional differences between the mosaic and poikilitic cements certainly support the petrographic and inclusion homogenization temperature data, indicating that the mosaic cement is a separate cementation event from the poikilitic cement.

Those data, therefore, are not sufficient to unequivocally tie the post-compaction cements to present Smackover brines and do not qualify at this point as positive evidence for the cements' origin.

STABLE ISOTOPE EVIDENCE FOR SUBSURFACE ORIGIN

The stable isotope composition of the Smackover post-compaction unzoned poikilitic cements was documented initially by Moore and Druckman (1981). The poikilitic cements range in $\delta^{18}O$ composition from -5.5 to $-6.9‰$ (PDB) with a mean $\delta^{18}O$ composition of $-6.3‰$. The $\delta^{13}C$ composition varies from $+2.5$ to $+3.5‰$ with a mean of $+3.1‰$.

The isotope composition of the zoned mosaic calcite cements was investigated by J.A.D. Dickson during a visit to the Applied Carbonate Research Program at L.S.U. in 1982. Dickson carefully used the complex cement stratigraphy of the mosaic calcite, as revealed by staining, to separate progressively younger cement zones for stable isotope analysis much in the manner

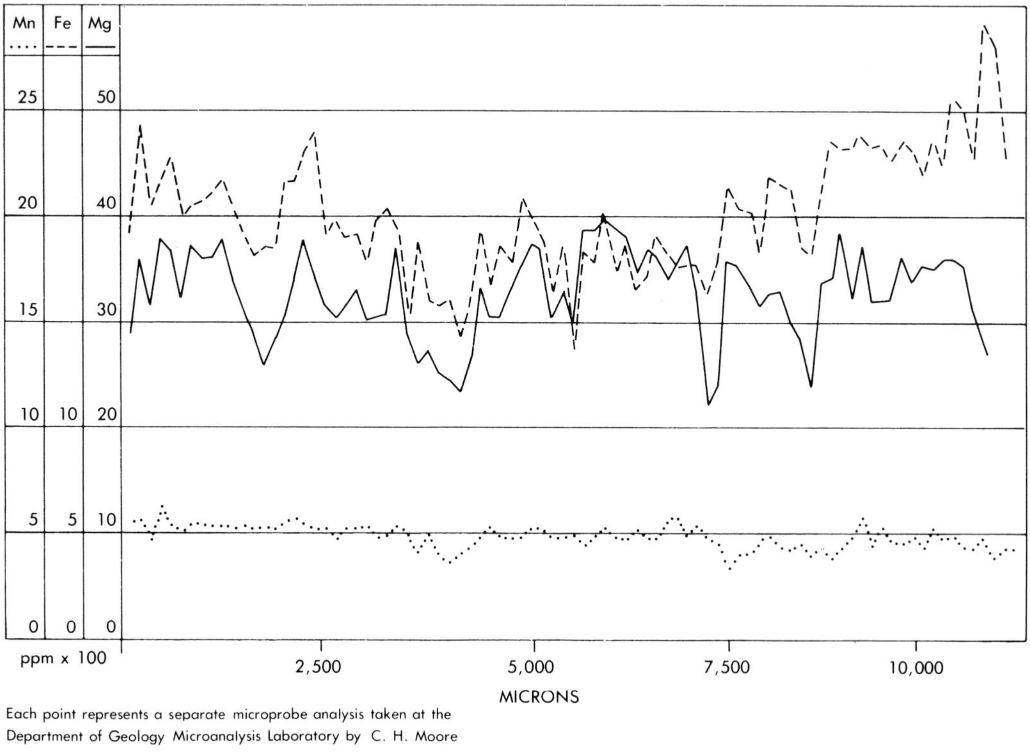

FIG. 7.—Electron microprobe traverse of unzoned poikilitic cement from Walker Creek, Arco Bodcaw #1.

reported by Dickson and Coleman (1980). The results are presented with Dickson's permission in Figure 8. $\delta^{18}O$ becomes progressively lighter through successively younger zones from −4.5 to −8‰, while $\delta^{13}C$ remains relatively constant at approximately +3.0‰ through all zones. The mean $\delta^{18}O$ composition for the mosaic calcites is −5.9‰ while the mean $\delta^{13}C$ is +3‰. The stable isotope composition of the mosaic calcite cements is compared with the isotopic composition of Moore and Druckman's post-compaction cements at Walker Creek in Figure 8.

Discussion of Stable Isotope Data

Moore and Druckman (1981) explained the isotopic composition of the poikilitic cements as the result of calcite precipitation from a water enriched in $^{18}0$ at elevated temperatures. This interpretation was based primarily on the two-phase fluid inclusion data, supported in part by trace element composition of the cement, that indicated precipitation from Smackover type fluids at elevated temperatures. The reported +5.5‰ $\delta^{18}O$ composition of present Smackover brines (Moore and Druckman, 1981) was then used to calculate a precipitation temperature of 90°C. Although this interpretation relies heavily on the assumed primary origin of the two-phase fluid inclusions, the range in $\delta^{18}O$ and $\delta^{13}C$ of these cements lies close to the field of isotopic composition of late sparry calcites from the Devonian Golden Spike of Alberta (Mountjoy and Walls, 1977) ($\delta^{13}C$ +2.9 to +1.4‰ and ^{18}O −8.7 to −14.5‰), and well within the field of the zoned subsurface cements of Dickson and Coleman (1980) ($\delta^{13}C$ +3.8 to +1.2‰ and $\delta^{18}O$ from −2.6 to −12.4‰).

The dramatic ^{18}O-depletion trend related to younger cement zones documented in the Smackover mosaic calcites (Fig. 8) exactly parallels the trends reported by Dickson and Coleman (1980) for Carboniferous cements on the Isle of Man. Dickson and Coleman attributed this trend to increasing temperature and to isotopic evolution of the pore waters during burial, and this would certainly seem to be the most reasonable explanation for the parallel Jurassic trend under discussion here. The small variation in $\delta^{13}C$ in the Jurassic mosaic calcites centered about +3 (Figure 8) supports this interpretation, particularly if we call upon progressive pressure solution as a source for these mosaic cements, which would effectively buffer the carbon isotopic composition.

Since the zoned mosaic calcite cements are demonstrably younger and hence cooler than the poikilitic cements, the close agreement of their mean $\delta^{18}O$ compositions, −5.9 vs −6.3‰ (see also Fig. 8), can be explained by an evolution of pore-water composition during the precipitation of the mosaic and poikilitic cements.

Constraints Placed on the Origin of the Cements by Petrography, Two-phase Fluid Inclusions, Trace Elements and Stable Isotopes

At this point, the web of evidence for the subsurface origin of both generations of cements is convincing indeed, particularly in light of the isotopic trends seen in the mosaic cements. We still cannot, however, estimate with good conscience, the general range of depth over which these cements formed, because of our uncertainty concerning the primary nature of the two-phase fluid inclusions and the equivocal nature of the trace element data. Another line of evidence, independent of temperature, that could tie the poikilitic cements to present Smackover brines would enable us to estimate a temperature of formation and hence depth of burial by using the $\delta^{18}O$ composition of the brines and the $\delta^{18}O$ composition of the poikilitic cements in Friedman and O'Neil's (1977) expression for calcite-water fractionation. Strontium isotopes, as discussed in the next section may well give us this evidence.

STRONTIUM ISOTOPE EVIDENCE FOR SUBSURFACE ORIGIN

The natural variation of the common strontium isotope pair $^{87}Sr/^{86}Sr$ has been studied for some time by geochemists interested in employing strontium isotopes as tracers for the origin and history of water masses (Faure, 1977). ^{87}Sr is the radiogenic product of ^{87}Rb decay, hence its abundance has varied through time. Although the $^{87}Sr/^{86}Sr$ ratio seems to be a constant .7090 in today's ocean water, the sea water ratio has varied systematically through time, as different types of rocks containing distinctly different $^{87}Sr/^{86}Sr$ were exposed to weathering (Faure, 1977). The detailed $^{87}Sr/^{86}Sr$ sea water curve for the Phanerozoic published by Burke *et al.* (1982) gives the geologist reasonably controlled specific reference points for the $^{87}Sr/^{86}Sr$ in the geologic past. Strontium isotopes should be particularly useful in diagenetic studies of carbonate rock sequences because the Sr is a common trace element in carbonates, while rubidium levels are negligible. Once fixed at the time of precipitation in a carbonate, therefore, the $^{87}Sr/^{86}Sr$ will remain the same, and should reflect the $^{87}Sr/^{86}Sr$ of the precipitating fluid because fractionation effects are negligible (Faure, 1977). Alan Stueber, while investigating the utilization of strontium isotopes in oil-field waters as a potential geochemical tracer, determined that the $^{87}Sr/^{86}Sr$ of present Smackover brines (Fig. 9), while exhibiting some variation, were generally more radiogenic than Upper Jurassic sea water (Stueber and Pushkar, 1983; Stueber, Pushkar, and Hetherington, 1984). Since the Smackover brines are commonly thought to be linked directly to Mid-Jurassic Louann salts (Carpenter, 1978), the original brine composition should reflect Mid-Jurassic $^{87}Sr/^{86}Sr$ of .7070 (Burke *et al*, 1982). Stueber, Pushkar and Hetherington attribute the enrichment of the Smackover brines in ^{87}Sr to interaction with clay- and feldspar-rich sequences such as the Norphlet and Bossier Shale during brine evolution and migration (see Fig. 9). Working in conjunction with the author, Stueber analyzed a series of separated rock components from two Arkansas fields in the southern diagenetic zone of Moore and Druckman (1981). In each case, grains including ooids and contiguous post-compaction poikilitic cement and one late saddle dolomite (Moore and Druckman, 1981) were separated. The results of the analyses are shown in Figure 9 (see also Stueber and Pushkar, 1983; Stueber, Pushkar, and Hetherington, 1984). In all cases, grains exhibited $^{87}Sr/^{86}Sr$ that were near Upper Jurassic sea water (.7070), while the unzoned poikilitic cement contained $^{87}Sr/^{86}Sr$ that were significantly higher than the grains, but generally lower (with one exception) than Smackover brines. The one Arkansas saddle dolomite analyzed showed very high $^{87}Sr/^{86}Sr$ at .7099 (Fig. 9).

Discussion of Strontium Isotope Evidence

The high $^{87}Sr/^{86}Sr$ ratio of the unzoned poikilitic cements confirms their clear affinity with present Smackover brines. The fact that most cement $^{87}Sr/^{86}Sr$ values are lower than the Smackover brines may indicate precipitation from the brine before it reached its final $^{87}Sr/^{86}Sr$ composition. If, as indicated earlier, the major sources for the $^{87}Sr/^{86}Sr$ are the Bossier Shale and Norphlet siliciclastics lying beneath and lateral to the Smackover (Moore and Druckman, 1981), than it would be reasonable to assume that the Smackover brines began to receive radiogenic Sr as waters expelled from related siliciclastic units during moderate burial, migrated into the upper Smackover and mixed with brines that may have arrived earlier. Further, these siliciclastic-related fluids may have been the advance guard of the main body of hydrocarbon migration potentially sourced from these siliciclastic units (Moore and Druckman, 1981). Since the unzoned poikilitic calcite cements are demonstrably pre-hydrocarbon, and late saddle dolomites (Moore and Druckman, 1981) have significantly higher $^{87}Sr/^{86}Sr$ than the poikilitic cements (Fig. 9), the author feels that the cements were precipitated from a Smackover brine with an evolving $^{87}Sr/^{86}Sr$ associated with the mixing of these brines with siliciclastic derived fluids (see also

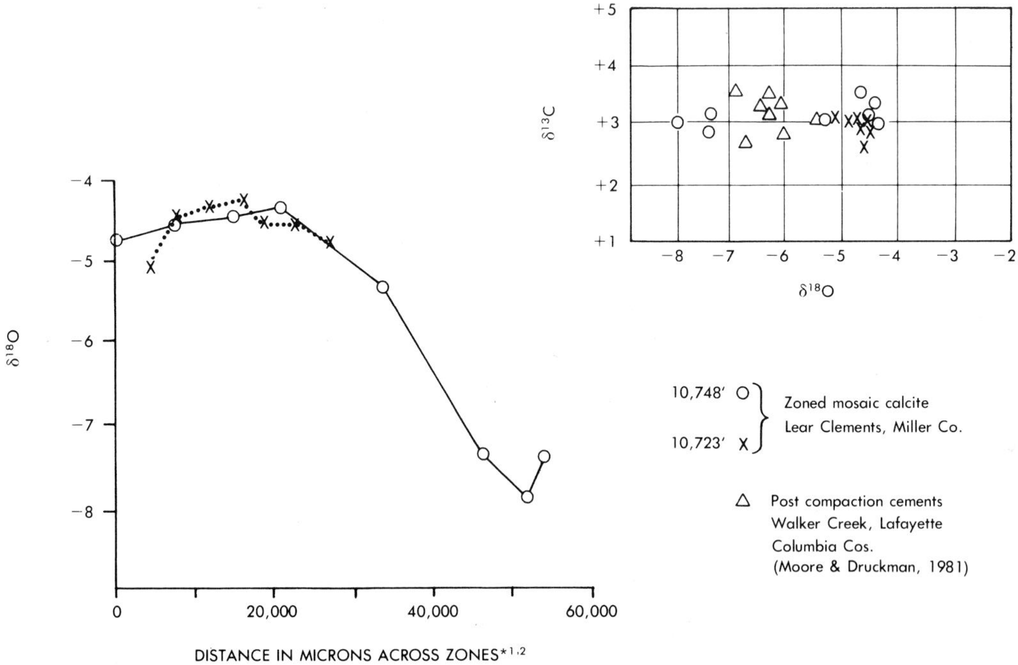

*1 Data generated by J.A.D. Dickson, Cambridge U., presented with permission

*2 *Traverse represents composite based on zone stratigraphy. Good agreement between composite of the two sets of samples from the same well indicates validity of zonal stratigraphy and the technique.*

FIG. 8.—Variation of $\delta^{18}O$ composition across Fe rich zones of mosaic calcite cements with comparison of stable isotope composition of mosaic calcites with bulk isotope ratios of the post-compaction cement of Walker Creek.

Stueber and Pushkar, 1983; Stueber, Pushkar and Hetherington, 1984).

The Jurassic sea water values of the contiguous ooids indicate that ooid stabilization occurred prior to the arrival of the radiogenic components of the Smackover brine. The petrographic and geochemical evidence presented by Moore and Druckman (1981) is compatible with ooid stabilization in contact with Jurassic connate marine pore fluids.

To date, we have not analyzed the zoned mosaic calcite cements for strontium isotopes. If these cements show some enrichment relative to $^{87}Sr/^{86}Sr$ above Upper Jurassic sea water, it will indicate that they have been influenced by Smackover brines. If the analyses show Jurassic sea water values, however, the data will furnish few constraints on the chemistry of the water responsible for the mosaic cement.

Constraints Placed on the Origin of the Poikilitic Cement by Strontium Isotopes

The $^{87}Sr/^{86}Sr$ of the unzoned poikilitic cements tie these cements firmly to subsurface Smackover brines, and hence, in conjunction with the total web of evidence presented, confirm their subsurface origin. We may now proceed to reevaluate our inclusion, trace element and stable isotope data with the certain advantage of knowing the general chemical characteristics of the fluid responsible for at least one of the two post-compaction cements in question.

DISCUSSION OF TRACE ELEMENT COMPOSITION OF POST-COMPACTION CALCITE CEMENTS IN LIGHT OF STRONTIUM ISOTOPE STUDIES

Since the strontium isotope data clearly tie the poikilitic cements to Smackover-like brines, we now have the opportunity to consider how closely the trace element composition of the cement reflects the chemical composition of the mother water, as well as implications of the data relative to the possible evolution of the water after precipitation of the cement. The very low strontium trace element composition of the poikilitic cements, as compared to the general high levels of Sr found in Smackover pore fluids (Table 2) could be considered to reflect a sharp increase in the Sr/Ca of the brines after precipitation of the cement, perhaps in response to the mixing of siliciclastic-influenced waters as indicated by the elevated $^{87}Sr/^{86}Sr$ of the poikilitic cements. If this were the case, the earlier mosaic calcite cements should exhibit a lower strontium trace element composition than the later poikilitic cements. Since both cements have comparably low Sr compositions (Table 2), I am forced to conclude, as did Moore and Druckman (1981), that the commonly used distribution coefficients for Sr in calcite must be re-evaluated for use in the subsurface environment.

The general compatibility of the magnesium data from the cements and the present brine, if a temperature increase is assumed after the precipitation of the mosaic calcite cement (Table 2), may indicate that the Mg/Ca of the Smackover pore fluids reached their present low levels relatively early and have evolved little since. This would indicate that major rock water interactions involving Mg and Ca, such as dolomitization, were complete relatively early in the burial history of the sequence. These data tend to support Katz's (1973) temperature-related range of magnesium distribution coefficients, and indicate the potential importance of magnesium analyses of cement in the study of burial diagenesis.

The increase in iron and manganese in the poikilitic cements versus the earlier formed mosaic cements would seem to indicate a progressive pore-water evolution toward iron- and manganese-rich fluids through time. This increase in iron and manganese could well be related to the arrival and mixing of siliciclastic-derived fluids with brines, as indicated by the elevated $^{87}Sr/^{86}Sr$ of the poikilitic cements. Since the $^{87}Sr/^{86}Sr$ of the poikilitic cement is less than the $^{87}Sr/^{86}Sr$ of the present brine (Fig. 9), it is felt that the brine evolved further relative to siliciclastic-derived waters after precipitation of the cement. This would suggest that the Fe/Ca and Mn/Ca ratios of the fluid at the time of precipitation of the poikilitic cement may have been lower than present levels, and that the distribution coefficient for iron in calcite is greater than 1 as assumed by Richter and Füchtbauer (1978). And finally, as Michard (1968) suggested, the Mn distribution coefficient for Mn in calcite increases with increasing temperature.

TEMPERATURE OF FORMATION OF CEMENTS

Since strontium isotope data ties the poikilitic calcite cements to present Smackover brines, it should follow that we may use the $\delta^{18}O$ composition of these brines and the $\delta^{18}O$ composition of these cements to estimate a cement temperature of formation. We are, however, seriously constrained by our lack of data concerning the variation in the $\delta^{18}O$ composition of the Smackover brines themselves, as well as the assumption that the brines have not appreciably changed isotopic composition in the millions of years since the cement was precipitated. At this time we have one set of samples from the Mount Vernon field in central Arkansas with a mean $\delta^{18}O$ composition of +5.5‰ (Moore and Druckman, 1981). Land (personal communication) reports Smackover brines from the Mississippi salt basin to the east, with $\delta^{18}O$ compositions of +6.2, +5.8 and +5.1‰ respectively. Land and Prezbindowski (1982) report $\delta^{18}O$ compositions of brines from Lower Cretaceous carbonate sequences in south Texas ranging from +3.4 to +19.0 with a mean of +10.7‰. It would seem reasonable to expect, therefore, that the poikilitic cements of the upper Smackover in the central Gulf did precipitate from a fluid enriched in ^{18}O, and that the +5.5‰ $\delta^{18}O$ composition at Mount Vernon could represent a reasonable minimum stable isotope composition since Mount Vernon is located several tens of miles updip from the southern diagenetic zone where the cements in question were collected. Using these data in Friedman and O'Neil's (1977) expression will give us a minimum temperature of cement formation at 84°C.

The temperature of formation for the zoned mosaic cements is difficult indeed to estimate, because we are even less certain of the original isotopic composition of the fluids responsible for these cements, and there is a very real possibility that the fluid was undergoing a rapid chemical evolution because of mixing with subsurface brines. If we assume, however, that the original fluid had a modern sea water $\delta^{18}O$ composition of approximately +1‰, because of the lack of meteoric diagenetic evidence in the southern zone (Moore and Druckman, 1981), then the mosaic cements could represent precipitation at temperatures ranging from 41°C to 67°C based on the progressive change in $\delta^{18}O$ composition of the cement from −4.5 to −8‰ (Figure 8 this paper).

In summary, a reasonable minimum temperature of formation for the poikilitic cements is 84°C, while a reasonable minimum range of temperature of formation for the mosaic calcite cement is 41° to 67°C.

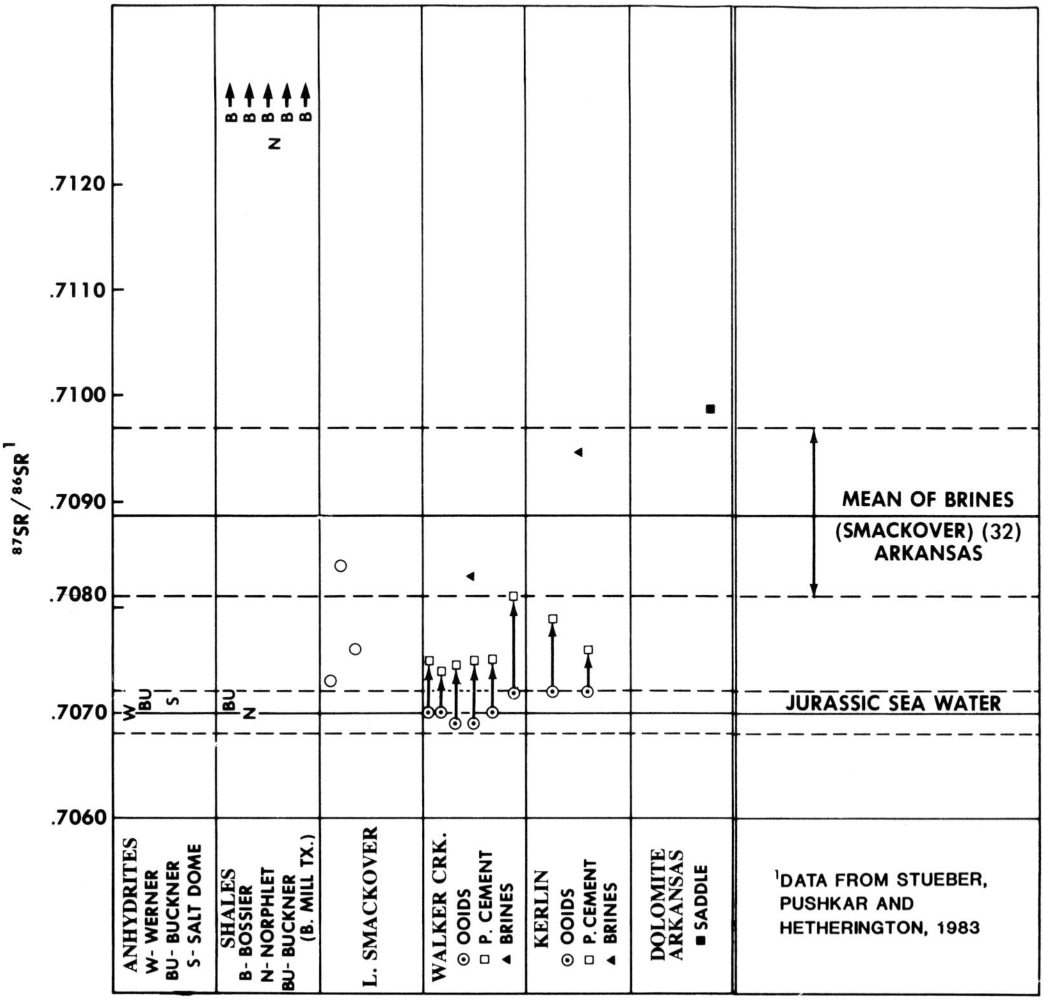

FIG. 9.—$^{87}Sr/^{86}Sr$ composition of brines, evaporites, shales and Jurassic carbonate components, Arkansas. Data from Stueber and Pushkar (1983) and Stueber *et al.* (1984).

COMPARISON OF STABLE ISOTOPE TEMPERATURE ESTIMATES WITH MEASURED INCLUSION TEMPERATURES

Table 3 compares two-phase fluid inclusion homogenization temperatures of both cements under discussion with temperature estimates based on their stable isotope composition. The inclusion temperatures certainly do not compare favorably with the isotope derived temperatures, if the minimum isotope estimates developed in the preceeding section are used. A more favorable comparison for the poikilitic cements is achieved if one applies heavier $\delta^{18}O$ compositions to the original precipitational waters such as have been reported for the Lower Cretaceous of south Texas (Land and Prezbindowski, 1982). It must be realized, however, that the reported homogenization temperatures are not pressure corrected and that some 10°C may well have to be added to these temperatures to correct for ambient pressure at the time of formation (using the pressure correction scheme of Potter (1977) and the calculated Jurassic geothermal gradient of Nunn (1984)). In the case of the zoned mosaic cements, however, the homogenization temperatures are still too high, even if one were to use an original fluid isotopic composition of the Smackover brines at Mount Vernon (+5.5‰).

A full evaluation of the validity of two-phase fluid inclusion temperatures in calcite cements must await a rigorous $\delta^{18}O$ characterization of Smackover brines in south Arkansas. These preliminary data do suggest, however, that while the homogenization temperatures derived from the inclusions of the latest cementation event may come close to true temperatures of formation, step-wise physical re-equilibration may be a serious problem in inclusion studies.

DEPTH OF BURIAL AT TIME OF CEMENTATION AND RELATIONSHIP TO HYDROCARBON MIGRATION

Using an Upper Jurassic geothermal gradient of 75°C/km, based on J. Nunn's (1984) thermal modeling study of the northern Gulf Coast which included burial curves for the Smackover, the unzoned poikilitic cements were precipitated at a minimum depth of 1 km, and hence are approximately Lower Cretaceous Hosston in age. The poikilitic cements in south Arkansas are demonstrably pre-hydrocarbon (Moore and Druckman, 1981), so that hydrocarbons did not arrive in south Arkansas traps, such as Walker Creek, until at least Hosston time.

SOURCE OF $CaCO_3$ FOR THE CEMENTS

There can be little doubt that the unzoned poikilitic cements precipitated from Smackover brines after significant burial, and after grain-to-grain pressure solution had ceased. There is, however, no clear relationship between the relatively rare, high amplitude, stylolites in upper Smackover grainstones and the distribution of the poikilitic cements (Moore and Druckman, 1981). The only reasonable source for these late cements therefore, is the deep subsurface dissolution of existing calcite by corrosive fluids associated with the beginning of hydrocarbon maturation and shale water expulsion. Although Bathurst (1980) indicates that such acidified water would have local and limited influence, significant regionally distributed subsurface porosity generation was documented by Moore and Druckman (1981) in the Smackover of south Arkansas. In those fields studied in detail, primary pore enlargement of more than 10% was not uncommon (see Fig. 4e). The author feels that the carbonate dissolved along the migration pathways was ultimately precipitated nearby as well as upstream, and that the volume of late post-compaction poikilitic cements seen in the southern Arkansas region is consistent with the volume of late subsurface dissolution observed in the same region. The relatively high $^{87}Sr/^{86}Sr$ of the poikilitic cements, indicating a compo-

TABLE 3. COMPARISON OF TWO-PHASE FLUID INCLUSION HOMOGENIZATION TEMPERATURES AND TEMPERATURES CALCULATED FROM MEASURED $\delta^{18}O$ COMPOSITION OF EACH CEMENT USING DIFFERENT POSSIBLE WATER $\delta^{18}O$ COMPOSITIONS.

Cement Types	Range of Homogenization Temperatures (°C)	Mean Homogenization[1] Temperature (°C)	Calculated[2] Temperature Using $\delta^{18}O$ Composition of Cements[3] and Indicated Brine-Water Composition	
			brine $\delta^{18}0$ (+5.5‰)	brine $\delta^{18}0$ (+9‰)
Unzoned Poikilitic Calcite	95°C–145°C	127°C	84°C	117°C
			water $\delta^{18}0$ (+1.0‰)	water $\delta^{18}0$ (+5.5‰)
Zoned Mosaic Calcite	85°C–120°C	107°C	42–67°C	70–100°C

[1]Uncorrected for ambient pressure.
[2]Using Friedman and O'Neil's (1977) expression.
[3]Using mean $\delta^{18}0$ composition of poikilitic calcite of −6.3% as reported by Moore and Druckman in 1981 and range of $\delta^{18}0$ of the mosaic cements as illustrated in Figure 8, this paper.

nent of siliciclastic related waters in the originating brines, clearly supports this view.

The observation that the zoned mosaic calcite cements were formed while pressure solution was still in progress would indicate that the $CaCO_3$, for these cements, may have come from grain-to-grain pressure solution during the earlier stages of subsurface compaction. The cement's relatively small volume, its occurrence in large shelter voids formed by thick-walled gastropods, the progressive change in trace element chemistry, perhaps in response to water mixing reflected by intricate zonation, structured isotopic variation, and finally no evidence of early meteoric water influence, are all consistent with such an origin. Strontium isotope analysis of mosaic calcite cements may well resolve the question in the future.

CONCLUDING STATEMENT

This case history of subsurface diagenesis clearly outlines the serious problems facing the carbonate petrologist-geochemist, when dealing with rock sequences that have encountered the increased pressures, temperatures and unusual fluids during relatively deep burial. While I have, for obvious reasons, emphasized the importance of burial diagenesis in this paper, the reader should not interpret this emphasis as a gauge of my position concerning the importance of deep burial diagenesis relative to early meteoric-dominated diagenetic processes. I fully appreciate the unusual setting of the upper Smackover of southern Arkansas that allowed the rapid burial of a thick carbonate sand into the subsurface without an appreciable early diagenetic overprint to obscure and negate the effects of burial diagenetic processes. I feel, however, that this setting has given us an unprecedented opportunity to extend our view of diagenesis into the subsurface realm. Finally, this paper is intended to be a strong statement in favor of a reasoned integrated approach to carbonate diagenesis in which all available data and techniques, including petrology-geochemistry as well as regional geology and stratigraphy, are utilized to constrain the ultimate diagenetic model.

ACKNOWLEDGMENTS

Because of the complexity, and relatively long time span encompassed by this study, to date, the list of individuals and groups that have contributed significantly to this effort is necessarily long, and if I have inadvertently left someone out, I apologize. First the author would like to thank the industrial associates of the Applied Carbonate Research Program whose continued financial support has made the study possible. I wish to thank the following individuals specifically: my friend, Charlie Druckman, Geological Survey of Israel with whom this project was started, and who continues in so many ways to influence its directions; Lynton Land, the University of Texas, for his encouragement and help in all aspects of carbonate geochemistry, his stimulating discussions, and specifically, for his thoughtful criticism of the manuscript; Mary Jo Klosterman for her perseverance in tackling and seeing through a difficult project, and her subsequent help and discussion of T.P.F.I.; Tony Dickson for many hours of stimulating discussion, friendly encouragement, and for providing the isotope data so painfully extracted from the zoned cements; Alan Stueber, Southern Illinois University, for bringing the application of Sr isotopes to my attention, for his patience and concern in providing the best possible analyses, and who, with Paul Pushkar of Wright State University, provided laboratory facilities for Sr extraction; Mobil Field Research Laboratory, Ernie Hetherington and Dick Koepnick of Mobil, for furnishing many of the strontium isotope analyses; Phil Choquette and Marathon Oil, for furnishing stable isotope analyses on some of the very small samples carved out by Tony Dickson; Frank Brock and Pennzoil, for providing water samples for Sr isotope analysis at Walker Creek; Brian Carter and Terry O'Hearn of L.S.U., for a sterling job on the two-phase fluid inclusions; Barbara Delaville, "Dupe" Duplechin, Mary Eggart, James Kennedy, Danny Peace, Ellen Tye and Art Saller for help in manuscript preparation; the Department of Geology, L.S.U., for logistical support; and finally, my wife Marlene, for her continuous support and patient good humor during its preparation.

REFERENCES

Bathurst, R.G.C., 1975, Carbonate Sediments and Their Diagenesis, 2nd enlarged edition: Amsterdam, New York, Elsevier, 658 pp.

______, 1980, Deep crustal diagenesis in limestones: Rev. Inst. Invest. Geol. Diputacion Provincial Univ. Barcelona, v. 34, p. 89–100.

Brock, F.C., Jr., and Moore, C.H., 1981, Walker Creek revisited: a re-interpretation of the diagenesis of the Smackover formation of Walker Creek field, Arkansas: Trans. Gulf Coast Assoc. Geol. Soc., v. 31, p. 49–58.

Burke, W.H., Denison, R.E., Hetherington, E.A., Koepnick, R.B., Nelson, H.F., and Otto, J.B., 1982, Variation of sea water $^{87}Sr/^{86}Sr$ throughout Phanerozoic time: Geology, v. 10, p. 516–519.

Carpenter, A.B., 1978, Origin and chemical evolution of brines in sedimentary basins: Oklahoma Geol. Surv. Circ. 79, p. 60–77.

Collins, G.A., 1974, Geochemistry of liquids, gases and rock from the Smackover formation: U. S. Bur. Mines Rept. Inv. 7897, 84 p.

Collins, S.E., 1980, Jurassic Cotton Valley and Smackover reservoir trends, east Texas, north Louisiana, and south Arkansas: Amer. Assoc. Petroleum Geologists Bull., v. 64, p. 1004–1013.

COOGAN, A.H., 1970, Measurements of compaction in oolite grainstone: Jour. Sed. Petrology, v. 40, p. 921–929.
DEBOER, R.B., 1977, Pressure solution: theory and experiments: Tectonophysics, v. 39, p. 287–301.
DICKSON, J.A.D., 1966, Carbonate identification and genesis as revealed by staining: Jour. Sed. Petrology, v. 36, p. 491–505.
______, AND COLEMAN, M.L., 1980, Changes in carbon and oxygen isotope composition during limestone diagenesis: Sedimentology, v. 27, p. 1–12.
DUNNINGTON, H.V., 1967, Aspects of diagenesis and shape change in stylolitic limestone reservoirs: World Petrol. Cong. Proc., Mexico, v. 2, p. 339–352.
FAURE, G., 1977, Principles of Isotope Geology: New York, John Wiley and Sons, 464 p.
FRIEDMAN, I., AND O'NEIL, J.R., 1977, Compilation of stable isotope fractionation factors of geochemical interest: U. S. Geol. Survey Prof. Paper 440K, 12 p.
FRUTH, L., JR., ORME, J., AND DONATH, F., 1966, Experimental compaction effects in carbonate sediments: Jour. Sed. Petrology, v. 36, p. 747–754.
GROSS, M.G., 1977, Oceanography, A View of the Earth: New Jersey, Prentice Hall, Inc., 497 p.
JACOBSON, R.L., AND USDOWSKI, H.E., 1976, Partitioning of strontium between calcite, dolomite and liquids: Contr. Mineralogy Petrology, v. 59, p. 171–185.
KATZ, A., 1973, The interaction of magnesium with calcite during crystal growth at 25–90°C and one atmosphere: Geochim. Cosmochim. Acta, v. 37, p. 1563–1586.
______, SASS, E., STARINSKY, A., AND HOLLAND, H.D., 1972, Strontium behavior in the aragonite-calcite transformation: an experimental study at 40–98°C: Geochem. et Cosmochim. Acta, v. 36, p. 481–496.
KINSMAN, D.J.J., 1969, Interpretation of Sr^{+2} concentrations in carbonate minerals and rocks: Jour. Sed. Petrology, v. 39, p. 486–508.
KLOSTERMAN, M.J., 1981, Application of fluid inclusion techniques to burial diagenesis in carbonate rock sequences: unpublished M.S. thesis, Louisiana State University, 102 p.
LAND, L.S., AND PREZBINDOWSKI, D.R., 1982, The origin and evolution of saline formation water, Lower Cretaceous Carbonates, south-central Texas, U.S.A.: Jour. Hydrology, v. 54, p. 51–74.
______, SALEM, M.R.I., AND MORROW, D.W., 1975, Paleohydrology of ancient dolomites: geochemical evidence: Amer. Assoc. Petroleum Geologists Bull., v. 59, p. 1602–1625.
MANUS, R.W., AND COOGAN, A.H., 1974, Bulk volume reduction and pressure solution derived cements: Jour. Sed. Petrology, v. 44, p. 466–471.
MEYERS, W.J., 1974, Carbonate cement stratigraphy of the Lake Valley formation (Mississippian) Sacramento Mountains, New Mexico: Jour. Sed. Petrology, v. 44, pp. 837–861.
______, AND HILL, B.E., 1983, Quantitative studies of compaction in Mississippian skeletal limestones, New Mexico: Jour. Sed. Petrology, v. 53, p. 231–242.
MICHARD, C.R., 1968, Coprecipitation de L'ion Maganeux avec le carbonate de calcium: Comptes Rendus Acad. Sci. Paris, ser D., v. 267, p. 1685–1688.
MOORE, C.H., 1977, Beach rock origin: some geochemical, mineralogical and petrographic considerations: Geoscience and Man, v. 18, p. 155–163.
______, AND DRUCKMAN, Y., 1981, Burial diagenesis and porosity evolution, upper Jurassic Smackover, Arkansas and Louisiana: Amer. Assoc. Petroleum Geologists Bull., v. 65, p. 597–628.
MOUNTJOY, E.W., AND WALLS, R.A., 1977, Some examples of early submarine cements from Devonian buildups of Alberta: Proceedings of the Third International Coral Reef Symposium, University of Miami, p. 155–161.
NUNN, J.A., 1984, Subsidence and temperature histories for Jurassic sediments in the Northern Gulf Coast: A Thermal-Mechanical Model, *in* Jurassic of the Gulf Rim: Special publication of the Gulf Coast Association of Geological Societies.
POTTER, R.W., 1977, Pressure corrections for fluid inclusion homogenization temperatures based on the volumetric properties of the system NaCl-$H_2$0: Jour. Research U. S. Geol. Surv., v. 5, p. 603–607.
______, CLYNNE, M.A., AND BROWN, D.L., 1978, Freezing point depression of aqueous sodium chloride solutions: Econ. Geol., v. 73, p. 284–285.
RICHTER, D.K., AND FÜCHTBAUER, H., 1978, Ferroan calcite replacement indicates former magnesium calcite skeletons: Sedimentology, v. 25, p. 843–860.
SALLER, A.H., 1984, Petrologic and geochemical constraints on the origin of dolomite: subsurface Enewetak atoll: An example of dolomitization by normal sea water: Geology (in press).
SELLIER, E., 1979, Contribution à l'étude pétrologique experimentale de la genèse des stylolithes des les formations calcaires du bassin d'Aquitaine, Thèse non publiée, Univ. Bordeaux II, 300 p.
STAMATEDES, M.R., 1982, Dolomitization of the upper Smackover in Miller County, Arkansas and adjacent areas: unpublished M.S. thesis, Louisiana State University, 164 p.
STUEBER, A.M., AND PUSHKAR, P., 1983, Application of strontium isotopes to origin of Smackover brines and diagenetic phases, southern Arkansas [abs.]: Amer. Assoc. Petroleum Geologists Book of Abstracts, 1983 Annual Meeting, Dallas, p. 166.
______, ______, AND HETHERINGTON, E.A., 1984, A strontium isotopic study of Smackover brines and associated solids, southern Arkansas: Geochim. Cosmochim. Acta. (in press).
WAGNER, P.D., AND MATTHEWS, R.K., 1982, Porosity preservation in the upper Smackover (Jurassic) carbonate grainstone, Walker Creek Field, Arkansas: Response of paleophreatic lenses to burial processes: Jour. Sed. Petrology, v. 52, p. 3–19.

BURIAL DIAGENESIS: OUT OF SIGHT, OUT OF MIND!

PETER A. SCHOLLE[1] AND ROBERT B. HALLEY
U.S. Geological Survey, P.O. Box 25046, Denver, Colorado 80225

ABSTRACT

Examination of porosity data from Holocene and Pleistocene carbonate strata indicates that there generally is little or no porosity loss in the zone of near-surface water circulation [that is, in the vadose, meteoric-phreatic, or mixing zone(s)]. Thus, the transition from very porous carbonate sediments to well-cemented, low-porosity carbonate rocks is a dominantly subsurface process. Indeed, both shallow- and deep-marine carbonate strata show a continuous loss of porosity with depth, indicating that porosity-reducing processes act continuously from the surface to depths in excess of 4 km.

Experimental, observational, and geochemical data show that porosity loss through burial diagenesis results from both physical and chemical compaction and from cementation. In near-surface sections, dewatering, grain reorientation, grain breakage, and other mechanical processes lead to sediment/rock porosities as low as 30 percent. Continued porosity loss requires mechanical compaction, chemical dissolution at grain contacts and along solution seams or stylolites, and/or re-precipitation of dissolved calcite as intergranular cement. Calcareous shales or marl seams (donor beds) can act as significant sources of dissolved carbonate which is precipitated as cement in adjacent limestones (recipient beds). Quantitantive studies of stylolites and solution seams commonly underestimate the total magnitude of pressure solution because they ignore contributions from associated thick calcareous shale sections and from thinner, regularly and irregularly distributed marl interbeds. Through these mechanisms, carbonate rock porosity may be reduced to values near zero in "semi-closed" systems without significant introduction of allochthonous cementing material.

In many young, subsiding basins, patterns of porosity loss with depth are crudely predictable. These patterns provide standards against which individual case studies of diagenesis may be compared and provide predictive tools for estimating porosity prior to drilling. In other areas, the standards allow identification of anomalously high porosity and focus attention on specific mechanisms which would act to preserve primary (or early diagenetic) porosity or to create secondary porosity at depth. Comparisons of oil field porosities with standard curves will allow further refinement of our understanding of diagenetic processes.

Predictive models are still in their infancy, however. There is a critical need to independently assess how rates of porosity loss with depth are affected by time, temperature, depositional setting, early diagenetic history, maturation history of organic matter, and other factors. In addition, overpressuring, early oil migration, dolomitization, and hydrothermal alteration are known to affect porosity-depth relationships. Refining our understanding of these factors may help geologists use their general knowledge of basin history to make valid predictions of carbonate reservoir quality in frontier areas.

INTRODUCTION

"Much (and perhaps most) cementation and formation of secondary porosity (except fractures) in carbonates occurs at relatively shallow depths in one of four major diagenetic environments: the vadose zone, meteoric phreatic zone, mixing zone, and marine phreatic zone" (Longman, 1980, p. 461). This statement from a recent summary of carbonate diagenesis represents a common approach to the understanding of limestone porosity. The reason for the widespread acceptance of this approach is that, for the past three decades, most studies of diagenesis have concentrated on early diagenetic processes (cementation and leaching) in Holocene and Pleistocene sediments. Because early diagenetic processes occur in near-surface settings (commonly on warm, tropical isles), they are accessible for study and in many cases are not obscured by later, complicating events. Many Holocene or Pleistocene limestones have undergone significant diagenetic alteration as a consequence of events acting in the four near-surface environments listed above, and it is natural (but not necessarily correct) to extrapolate this limited experience to the geologic record. A survey of the literature on carbonate diagenesis for the past decade will attest to the frequency with which that extrapolation is made.

There exists, however, a large body of data from rocks of all ages which can be used to argue forcefully against the concept that cementation and leaching of carbonate rocks are dominated by near-surface processes. We would not deny that early diagenetic events may leave an imprint which affects later diagenesis. However, we would argue here that, particularly for rapidly subsiding sedimentary basins, "most cementation" and much leaching take place in moderately deep to very deep subsurface settings (mesogenetic environments).

Observations by Pray (1960) and Zankl (1969) that compaction (largely in the form of breakage of fossils) is rare in limestones convinced many that cementation of carbonate rocks had to precede burial. Indeed,

[1]Present address: Gulf Research and Development Company, P.O. Box 37048, Houston, Texas 77236

Bathurst (1975, p. 439) wrote, "We must therefore infer: (1) that cementation begins early, before compaction, and (2) that about half the volume of $CaCO_3$ in many limestones has been carried into the sediment from outside."

More recently, experimental work by Shinn *et al.* (1977) and Shinn and Robbin (1983) has shown that significant compaction and sediment dewatering can take place with little or no grain breakage. Furthermore, several workers (e.g., Bathurst, 1975) have noted radical differences between the average porosities of near-surface limestones, which may have experienced some or all of the diagenetic environments listed by Longman, and the porosities of ancient limestones. These differences, which average about one order of magnitude, have led geologists in the past decade to search for evidence of burial cementation.

This search has been largely successful. In this paper we summarize data from regional studies, core and outcrop examination, and petrographic and geochemical analyses which document major deep-subsurface compaction fabrics and cements in carbonate rocks.

SOME GENERALITIES

Diagenesis in its broadest sense is any change in a sediment which occurs after its deposition. Burial diagenesis, then, is any change in a sediment which occurs below the zone of near-surface ground water circulation and above the zone of low-grade metamorphism. The depths of these transitions vary, of course, between basins and even within basins. Burial diagenetic processes largely are a consequence of the accumulation of overburden detritus in a subsiding sedimentary basin and the related changes in temperature, pressure, basin hydrology, and pore-fluid chemistry. This concept corresponds with the term "anadiagenesis" as used by Fairbridge (1967) and includes those changes which occur in the "mesogenetic zone" or "zone of deeper burial" as used by Choquette and Pray (1970). Burial diagenesis encompasses a range of processes including mechanical compaction, chemical compaction (pressure solution), cementation, replacement and other neomorphic processes, dissolution, and fracturing. We believe that burial diagenesis plays an important role in subsurface cementation and that understanding its principles will help us identify the sites of major porosity loss in carbonate strata.

The Mass Balance Problem

Dunham (1969) and Bathurst (1975) pointed out that cementation models based on a remote supply of calcium carbonate have difficulty in deriving the volume of cement required to reduce a carbonate sediment with 40 to 50 percent porosity to an ancient limestone with only a few percent pore space. Because calcium carbonate is only slightly soluble in water under typical subsurface conditions, several tens of thousands to several hundreds of thousands of pore volumes of water are required to fill pore space with cement. Pray (oral commun. 1966) estimated that between 10,000 and 50,000 pore volumes of solution were required to fill pore space completely with calcium carbonate. Dunham (1969, p. 161) calculated that between 10,000 and 100,000 pore volumes were needed, based on efficiencies of precipitation of 100 percent and 10 percent, respectively. Even these latter calculations are based on unrealistically high efficiencies of precipitation and do not take into account the constantly decreasing permeability of strata and the increasing difficulty of supplying water and cement to the pore space as cementation proceeds. If the scale of the problem is expanded from that of a single pore to an entire formation with an initial average porosity of about 50 percent, then it would take enormous volumes of water (50,000 times or more the volume of the formation) to cement this unit to near zero porosity. These problems led Friedman (1975, p. 395) to conclude that ". . . the problem of mass balance remains open. The amount of cement needed to convert carbonate sediment into limestone seems larger than can be presently accounted for. . .".

The above calculations were based on water compositions similar to those of near-surface waters, such as sea water or fresh water in carbonate aquifers. A calculation using typical subsurface brines yields similar results. The amount of dissolved calcium carbonate in subsurface waters is generally limited by the concentrations of carbonate and bicarbonate, which generally are on the order of a few hundred parts per million (Collins, 1975). Because carbonate makes up 60 percent of the weight of calcium carbonate, the total dissolved calcium carbonate in subsurface waters typically is less than 1 g per liter. This concentration is similar to concentrations in most near-surface waters. Assuming the ability to precipitate 10 percent of the dissolved calcium carbonate from solution, it would take more than 27,000 cm^3 of brine to precipitate 1 cm^3 of calcite (2.71 g). Thus, even in the subsurface, very large volumes of brine are required to reduce porosity in carbonate rocks if the calcium carbonate must be transported into the system.

The large volume of water required for cementation by importation of dissolved calcium carbonate places limits on the hydrologic settings in which this process can take place. Such cementation could only be effective in areas of active water flow. Such settings include the splash zone on beaches (Ginsburg, 1953), the intertidal zone (Moore, 1973), turbulent areas near reefs and carbonate sand shoals, and major aquifers, especially along their recharge zones and marginal mixing zones where ground water enters the sea (Back and Hanshaw, 1971; Runnells, 1969).

Flow models indicate that waters of compaction in subsiding basins provide insufficient volumes of fluid for major cementation (Bonham, 1980). In models of subsiding basins, formations may be envisioned as simply settling through water previously deposited with the underlying strata. Although the channeling of large volumes of compaction water through volumetrically small areas of permeable sediment may account for local well cemented zones, there is simply not enough compaction water available to cement formations on a regional scale. Therefore, the cementation of units during burial in subsiding basins presents a significant mass balance problem. This is true particularly for basins like South Florida which have a geologic history dominated by subsidence and which have not yet entered a phase of uplift. In basins which have undergone (or are undergoing) uplift, active aquifer systems can be formed which aid in cementation by moving large volumes of head-driven waters through the sedimentary section.

The mass balance problem for subsiding basins is largely circumvented by a model that uses a local supply of cementing material derived during chemical compaction. In this model, some or most of the cement is derived by pressure dissolution from nearby sources. This material is transferred to cementing sites by a combination of diffusion and fluid flow; in fact, diffusion alone may be sufficient where the distances between sites of dissolution and sites of reprecipitation

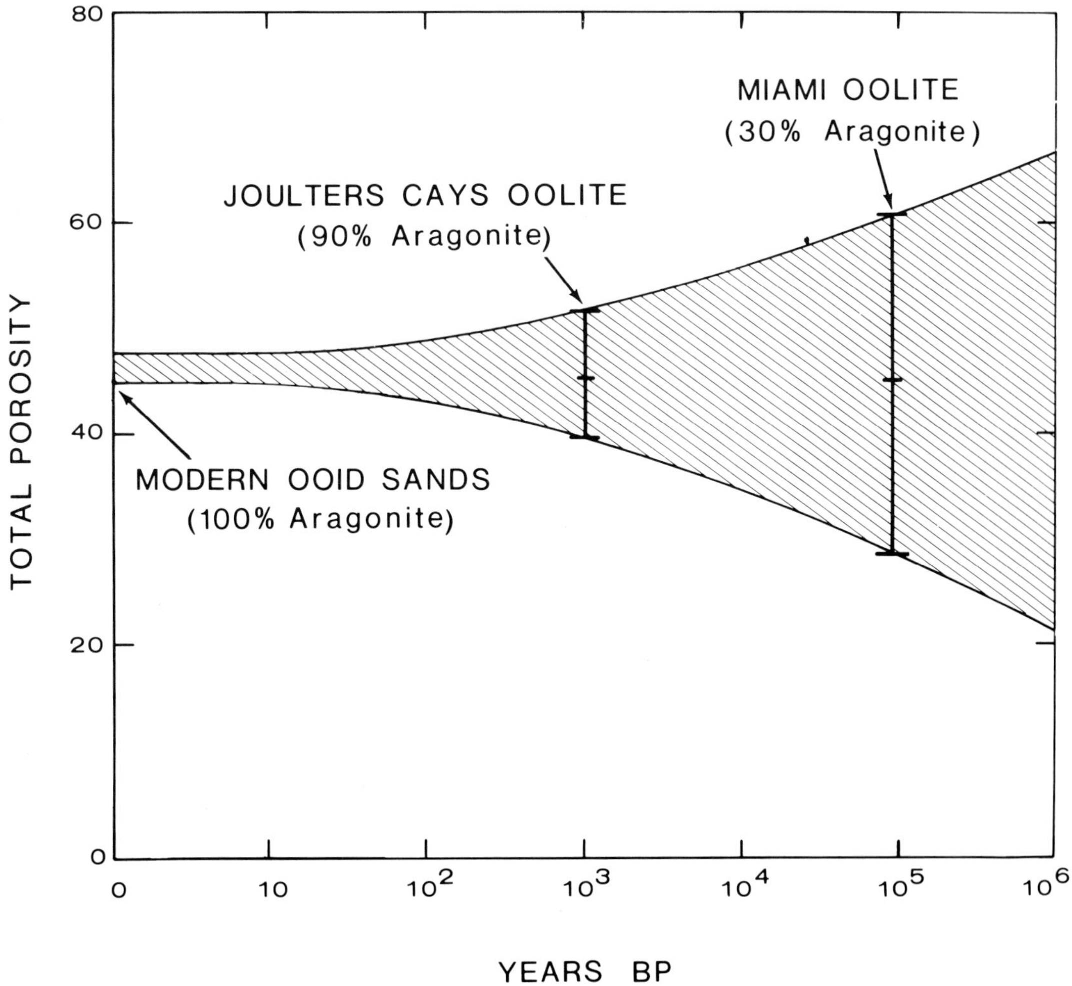

FIG. 1.—Diagram illustrating lack of porosity change during mineralogical stabilization of oolites in Florida and the Bahamas. A modern unconsolidated ooid sand is illustrated at the left of the diagram with 45 percent porosity and 100 percent aragonite. On Joulters Cays, about 10 percent of the aragonite has been converted to calcite cement, causing a wider range of porosity measurements, but the average porosity is still about 45 percent. The oolite facies of the Miami Limestone retains only about 30 percent aragonite, the other 70 percent having been converted to calcite. It has also developed vuggy porosity, causing individual porosity measurements to span a wide range, but the average porosity is still about 45 percent (from Halley and Evans, 1983).

are small. Diffusion is particularly important during late stages of cementation, when porosity is low and permeability is miniscule.

EVIDENCE FOR BURIAL DIAGENESIS

Porosity-Depth Curves

Perhaps the most direct evidence that near-surface diagenesis is not the major cause of porosity loss in carbonate strata comes from empirical examination of porosity values of carbonate rocks of all ages. Porosity data from modern carbonate strata are not abundant. Values for mudstones generally range from 60 to 70 percent (Ginsburg, 1956), and values for grainstones typically range from 40 to 50 percent (Enos and Sawatsky, 1979, 1981; Halley and Harris, 1979). These numbers compare well with those obtained in experimental studies by Vinopal and Coogan (1978).

Recent studies (Halley and Harris, 1979; Beach, 1982; and Halley and Evans, 1983) have shown that Holocene and Pleistocene carbonate sediments of the Florida-Bahamas area that have undergone progressive near-surface vadose and phreatic alteration retain high porosity (Fig. 1). Grainstone porosities in these diagenetically altered sediments still average about 45 percent with maximum porosity values of 59 percent. Data from other areas is sparse, but Harrison (1975) described mineralogically stabilized Pleistocene limestones from Barbados which averaged about 30 percent porosity and similar values were noted for exposed Pleistocene limestones of Bonaire (F. J. Lucia, oral commun., 1983).

Although the carbonate material in these sediments has been moved from one site to another through dissolution and reprecipitation, and although their permeability may have been greatly altered, their average porosity has remained high and, in some cases, essentially constant. The range of porosity values increases with the intensity of diagenesis (as measured by the conversion of aragonite to calcite), but the mean porosity values do not change. This indicates that there has been little or no net addition of carbonate material to the section despite as much as 120,000 years of near-surface alteration.

Sediments which are cemented in marine settings (at the seafloor or in marine phreatic environments) may undergo greater porosity loss due to the net addition of $CaCO_3$ from seawater. Individual thin units can show virtual obliteration of pore space by this process. However, these submarine cement fabrics are commonly recognizable in thin section, and they appear to affect only a small fraction of both modern and ancient strata.

Studies of thick carbonate sections have confirmed the idea that near-surface carbonate strata retain most of their pore space until they are more deeply buried. In chalk sections from both deep marine (Schlanger and Douglas, 1974) and shelf settings (Scholle, 1977), average porosities in excess of 40 percent were typically found in strata which had never been buried deeper than 500 m (Fig. 2). Only in rare, submarine-cemented hardground sections were lower porosities encountered in shallowly buried chalk sections. Burial, with increased lithostatic stress and increased temperature, is the only event which has lead to widespread porosity loss in these carbonate strata.

Some geologists have argued that chalks are atypical sediments because of their initial, stable, low-Mg calcite composition and their consequent low diagenetic potential. There is certainly some validity to this argument. However, it can also be reasoned that if typical shallow-water carbonate sediments are diagenetically stabilized in near-surface settings and still retain their original porosities, then they should also respond as stable strata with low diagenetic potential during burial. The relative scarcity of pre-Neogene aragonite and high-Mg calcite in platform strata supports the concept of near-complete diagenetic stabilization of shelf limestones before deep burial. Only in shaly sections, impermeable carbonate rocks, concretions, oil-saturated strata, or other unusual settings do unstable carbonate minerals persist in older strata.

Recently published data (Schmoker and Halley, 1982; Halley and Schmoker, 1983) have confirmed that chalks and shallow marine limestones behave similarly during burial. These data (Fig. 3) are based on borehole gravity measurements and interpretation of electric logs (as opposed to perm-plug measurements for the chalks) and thus represent the averaging of porosities of much larger volumes of rock, a significant advantage in vuggy carbonate sections. The data were collected from the more or less continuously subsiding Florida Platform and include Quaternary to Cretaceous strata. These limestones and dolomites have gone through the entire spectrum of near-surface diagenetic processes described by Longman (1980), because they were deposited during times of major eustatic sealevel fluctuations and are currently in an area of deep ground-water circulation. Nevertheless, near-surface (0 to 100 m depth) porosity values average about 40 percent, and values down to about 500 m generally exceed 30 to 35 percent. These values are not really so very different from those of chalks when one considers the differences in average grain size and the expected greater dewatering of chalks in near-surface settings. These data also align reasonably well with curves of porosity versus depth compiled by numerous authors for muddy terrigenous sediments (see review in Brown, 1969, p. 489).

Both chalks and shallow-marine carbonates show consistent and generally smooth patterns of porosity loss with burial (Figs. 2, 3). Rather than showing unusual stability, chalks apparently lose porosity at a faster rate than shallow-water carbonates. This presumably is a consequence of the finer grain size, lack of early cement framework, and virtual absence of dolomite in chalks. The fact that curves of porosity versus depth for individual chalk units are reasonably smooth (that is, there are no discrete steps) indicates that processes which destroy pore space act continuously from the surface to depths in excess of four kilometers. These processes are not tied in any way to near-surface circulation of water but rather relate to the inexorable increases of temperature and pressure with depth. Likewise, the data from platform carbonates of Florida indicate a generally continuous decrease in porosity from surface to depth with only minor discontinuities. These discontinuities may be related to inhomogeneities in primary facies, mineralogy, and early diagenesis, to variations in pore water chemistry as a function of depth, or to strong aquifer flow at intermediate depths.

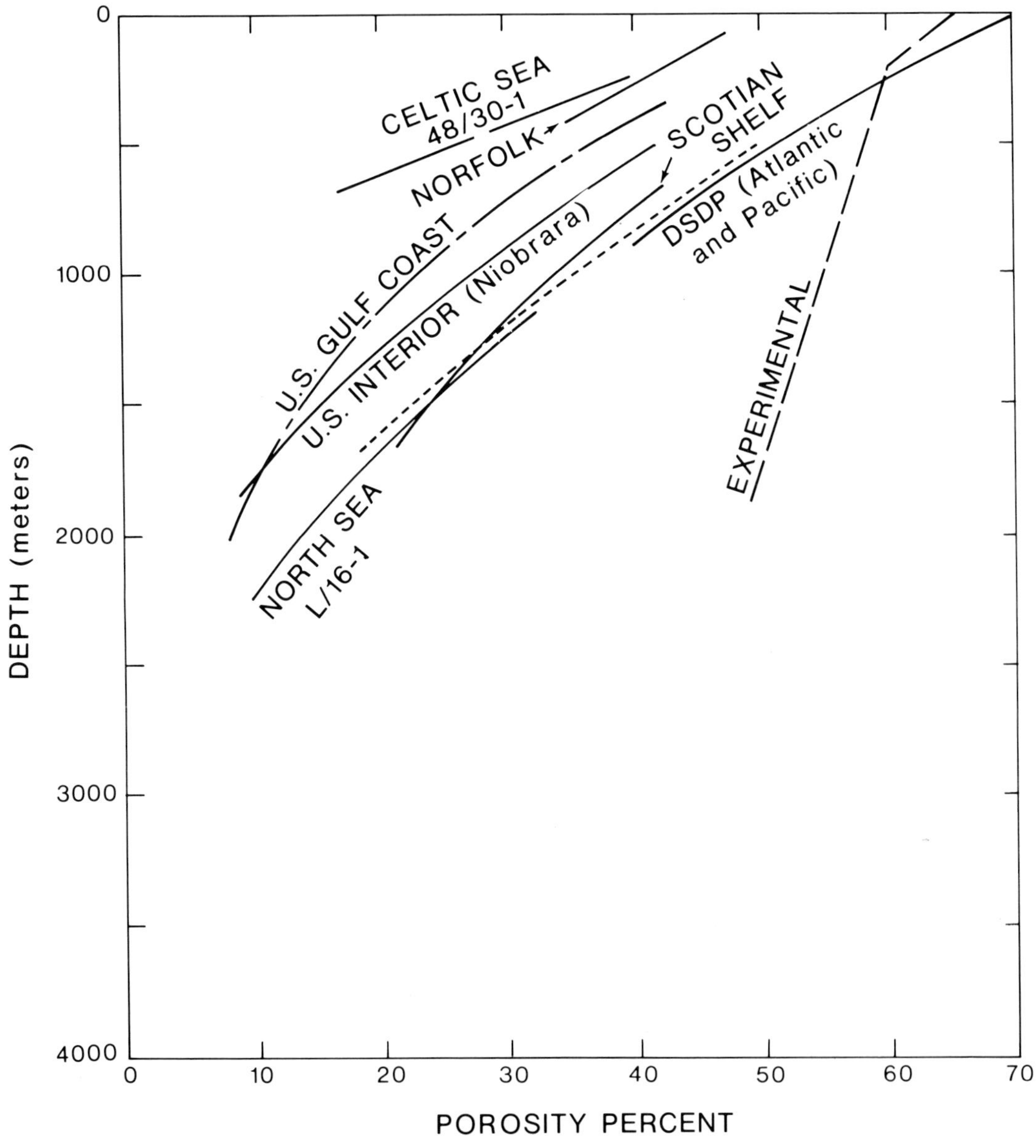

FIG. 2.—Plot of porosity changes as a function of maximum burial depth for onshore chalk sections from North America (Niobrara Limestone) and Europe (Norfolk Chalk, Scotian Shelf, North Sea, and Celtic Sea wells); Deep-Sea Drilling Project (DSDP) holes; and experimental studies (modified from Scholle, 1977). DSDP results from Schlanger and Douglas (1974); experimental work for Hamilton (1959).

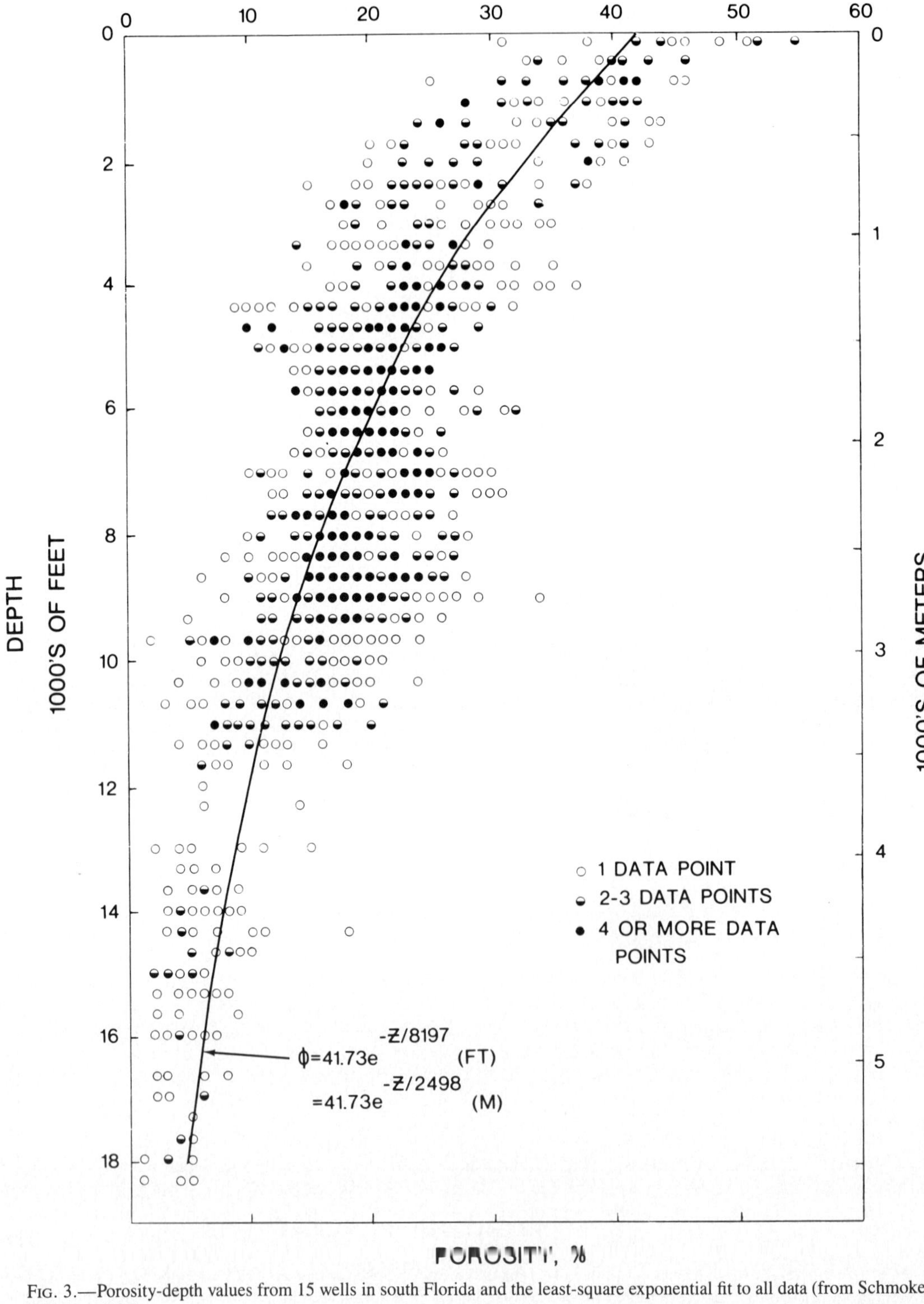

FIG. 3.—Porosity-depth values from 15 wells in south Florida and the least-square exponential fit to all data (from Schmoker and Halley, 1982).

Both chalk and platform carbonate data show rates of porosity loss which decrease with increasing depth but which eventually reach porosity of less than 5 percent (pore space), values typical of ancient carbonate rocks. In chalks, such values generally are reached at depths in excess of three kilometers in sections with marine pore fluids and at one to two kilometer depth in sections with Mg-poor ("fresh") pore fluids (Scholle, 1977; Lockridge and Scholle, 1978). In southern Florida, average porosity values of 5 percent or less were encountered in platform limestones at or just below four kilometers, whereas the average values for platform dolomites at that depth were still above 10 percent (Schmoker and Halley, 1982, Fig. 9).

By presenting this analysis of porosity data, we do not mean to imply that a fundamental or predictive model of porosity-depth relations is yet in hand. The data shown in Figures 2 and 3 were collected from a limited number of relatively young basins, most of which have had a continuous subsidence history with little or no tectonic uplift. Clearly, data exist for low-porosity strata from older basins which were never deeply buried, which were tectonically deformed, or which have had major artesian pore-fluid systems. Furthermore, porosity loss with depth is not irreversible. Several processes (discussed later in this paper) can lead to generation of secondary porosity and can act in deep, basinal settings.

Nevertheless, the porosity-depth data from Figures 2 and 3 indicate that answers to the problem of carbonate cementation in subsiding basins must be sought mainly by gaining an understanding of subsurface processes. Undoubtedly, as our knowledge in this area grows, we will find increasing complexities. Effects of variations in primary depositional facies, carbonate mineralogy, early diagenetic patterns, basin hydrodynamics, geothermal gradients, pressure gradients, organic carbon contents and types, geologic age (time) and many other factors will probably be shown to affect the rates and timing of subsurface porosity loss. Indeed, it may yet be found that variations in early diagenesis can account for significant differences in the response of strata to burial diagenetic effects, thereby confirming a correlation between near-surface diagenetic facies and hydrocarbon reservoirs. Yet even such conclusions will not obviate the fact that near-surface diagenetic phenomena are of significance only in the way they modify deeper-seated processes. Porosity-reducing diagenesis is primarily a subsurface process.

Compaction

Progressive changes of porosity with depth are accompanied by a series of changes in the physical appearance of the strata. These changes include the development of features attributable to both mechanical and chemical compaction. In most instances, mechanical compaction effects—that is, dewatering, grain reorientation, and even grain breakage—dominate the early burial history of carbonate strata. Chemical compaction (solution-reprecipitation) effects such as stylolitization and grain interpenetration generally are most significant later in the diagenetic history. However, in many cases, both processes can be shown to have been active throughout the burial of carbonate strata.

Mechanical compaction.—Dewatering of sediments and mechanical reorientation of grains to produce material with lower porosity is a particularly important process in fine-grained carbonate sediments. Such sediments commonly have initial porosities in excess of 60 percent, with some especially watery sediments ranging from 70 to 80 percent (Richards and Hamilton, 1967; Brown, 1969; Schlanger and Douglas, 1974; Enos and Sawatsky, 1981). Experimental data (Terzaghi, 1940; Hathaway and Robertson, 1961; Fruth *et al.*, 1966; Robertson, 1967; Ebhardt, 1968; Morelock and Bryant, 1971; Shinn *et al.*, 1977; Bhattacharyya and Friedman, 1979; Shinn and Robbin, 1983) and field data (Ginsburg, 1957; Richards and Hamilton, 1967; Schlanger and Douglas, 1974; Enos and Sawatsky, 1981) have shown that extensive loss of porosity can occur with the equivalent of only a few tens to hundreds of meters of burial. Indeed, Ginsburg (1957) and Brown (1969) described decreases in porosities of Florida Bay and Bahamian muds from 80 percent to 65 percent within the "first few inches" of burial; porosities then remain relatively constant to the surface of the Pleistocene. Much of this near-surface porosity loss appears to result from simple dewatering and grain reorientation and repacking as indicated by the lack of chemical changes observed in compaction experiments and the rapid rate at which these significant porosity changes are achieved (Shinn and Robbin, 1983). A common conclusion of these experimental and field studies on carbonate muds (or mud-supported sediments) is that, at porosity values of 30 to 40 percent, mechanical compaction either becomes ineffective or requires very great differential stresses and extensive grain breakage. This holds true at least on the relatively short-time scale of the existing laboratory studies. Presumably at porosity values of about 40 percent, a reasonably strong grain-supported fabric is established, even in micritic sediments. In coarser grained sediments which were grain supported from the time of deposition, physical compaction effects generally are less severe, especially where grains have significant mechanical strength.

Mechanical compaction effects, of course, are not necessarily terminated once a grain-supported fabric is achieved. Where grains are fragile, where little early-diagenetic cement has formed to weld grain contacts, where rates of overburden loading are rapid, or where significant tectonic stresses act on sediment which is only slightly lithified, extensive plastic or brittle grain deformation may take place. The rate of sediment loading appears to be especially important in this context. For example, chalks that show the most extensive flattening of foraminiferal tests or calcispheres generally are those which come from areas of extremely rapid overburden accumulation or significant tectonic deformation. The strongly compacted chalks of Northern Ireland, described by Wolfe (1968), come from an area in which as much as two kilometers of basalt was extruded over the chalk section in only a few million years. Other European chalks which were buried at slower rates, although commonly to greater maximum depths, show far less evidence of mechanical compaction (Scholle, 1977). Similar grain deformation effects and major porosity losses through mechanical compaction were reported by Mimran (1975; 1977) from deformed chalks of southern England. In those studies a direct correlation was established between intensity of deformation and degree of compaction.

In the Western Interior of North America, as well, chalk deposition was widespread during the Late Cretaceous. However, it is mainly in areas of very rapid burial that mechanical compaction effects are strikingly visible. Figure 4 shows a sample of the Niobrara Formation from near Ft. Collins, Colorado, an area where nearly three kilometers of Pierre Shale was laid down atop the Niobrara within a few million years after Niobrara deposition. This area lies along the major boundary faults of the Front Range of the Colorado Rockies, a zone where major normal fault development took place relatively soon after Pierre deposition. Compactional flattening of the microfossil constituents is pronounced, in our experience, mainly in areas which combine rapid sediment loading with later faulting. Such profound crushing of grains presumably is a consequence of application of tectonic or overburden stresses at rates which effectively exceed the ability of the sediment to respond through chemical dissolution and reprecipitation. This may be particularly prevalent where rapid sediment loading resulted in overpressuring, which may have delayed both mechanical and chemical compaction. Where overpressures were released rapidly (through faulting or other mechanisms) the effective rates of differential stress may have been even greater than could have been achieved simply by high initial rates of deposition or deformation.

The absence of grain breakage in most fine-grained carbonate units has led a number of workers to dismiss or downplay the general importance of mechanical compaction (Weller, 1959; Pray, 1960; Zankl, 1969; Bathurst, 1970). Yet experimental work reviewed by Brown (1969) and that of Shinn *et al.* (1977), and Shinn and Robbin (1983) has shown that modern carbonate muds and mud-supported carbonate sands can be compacted rapidly from original porosities of about 70 percent to values less than 40 percent (in some cases to 30 percent, as cited in Bathurst, 1980) without development of strong deformation fabrics. Compactional flattening of burrows and development of wispy, organic-rich seams were noted, but only minor breakage of fossils was observed in these experiments (Shinn *et al.*, 1977; Shinn and Robbin, 1983). Other experiments by Fruth *et al.* (1966) and Bhattacharyya and Friedman (1979) have shown that the higher the content of muddy matrix, the less noticeable are compaction effects. As long as allochems are not in contact with one another they show little or no effect of mechanical deformation, presumably because the micritic matrix responds to stress by minor flowage, providing an essentially nondirectional stress field around the allochems.

Mechanical compaction accounts for between one third and one half the porosity reduction needed to convert mud-supported sediments to ancient limestones. Dewatering, grain reorientation, and minor grain breakage combine to produce a stable framework of interlocking and self-supporting particles capable of resisting additional overburden pressure. Thus, by these mechanisms a significant part of Bathurst's (1975) volume problem in carbonate cementation is eliminated.

Mechanical compaction in coarser, grain-supported rocks is probably less extensive. Because grainstone and packstone sediments have an initially grain-supported fabric, there is little chance for porosity reduction through dewatering. Grain reorientation can sometimes achieve higher packing densities and thus reduce pore space, but in most cases grain deformation (plastic or brittle) is necessary for significant porosity reduction. Most field studies of grain-supported rocks do not (and cannot) distinguish between mechanical and chemical compaction, because these processes tend to work concurrently in grain-supported materials. Nevertheless, these studies commonly note plastic deformation of allochems resulting in planar rather than point contacts between grains and brittle deforma-

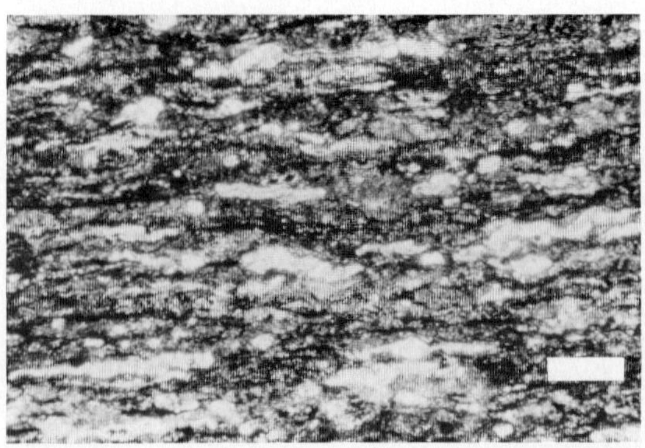

FIG. 4.—Thin-section photomicrograph of the Niobrara Limestone from the Ideal Cement Quarry near Ft. Collins, Colo. Note pervasive compactional flattening of foraminiferal tests. Scale bar is 0.11 mm.

tion of grains (shattering at grain contacts). These features are most readily noted in oolitic strata because the primary grain shapes are uniform and predictable. Observations by Kahle (1966), Coogan (1970), Coogan and Manus (1975), Conley (1977), and many others have demonstrated that plastic deformation and brittle deformation of oolitic grains are both common and produce significant loss of pore space.

Specific studies of mechanical compaction in non-oolitic, grain-supported rocks are scarcer, because mechanical and chemical compaction effects are so closely interrelated. Meyers (1980) and Meyers and Hill (1983) have presented excellent quantitative data on the extent and timing of compaction in Mississippian strata of the southwestern U.S. Interestingly, studies of this same area convinced Pray (1960) that compaction did not occur in Pennsylvanian calcilutites. Meyers and Hill, on the other hand, found that compaction was "a major process of porosity destruction in more than 90 percent of all coarse grainstones and packstones" (1983, p. 231). Compaction, including both mechanical effects and intra- and inter-granular dissolution but not stylolitization, accounted for as much as a 38 percent reduction of pore space, leaving as little as 24 percent pore space to be filled by intergranular cements. Furthermore, stratigraphic evidence indicates "that compaction began under a few tens to a few hundreds of feet of overburden and was completed under less than 6,500 feet (2,000 m) of overburden" (Meyers, 1980, p. 457). Plastic and brittle deformation of bryozoans is quantitatively most important, with breakage of brachiopods and other elongate shells being of lesser significance. Bioherms and strata which had been extensively cemented in marine phreatic environments or in meteoric waters escaped mechanical compaction (Meyers, 1978; 1980). Strata which had been exposed to "microetching" below subaerial exposure surfaces showed the most intense crushing. It should be noted, however, that mechanical compaction was considered to be less important than chemical compaction in bringing about pore space reduction in these Mississippian units (Meyers, 1980).

Evidence for physical compaction in shallow-water carbonates has been used as an exploration tool by Beach and Schumacher (1982). In this work, uncompacted island sediments forming part of Stanley field in the Williston basin were recognized to be surrounded by compacted marine carbonates. Beach and Schumacher (1982) concluded that compaction was important in the development of Stanley field and influenced the reservoir in three ways. First, only marine sediments were mechanically compacted, thus isopach maps of the upper Mission Canyon and overlying Charles Formations reflect distribution of depositional facies in the upper Mission Canyon. Second, compaction implies lack of early cementation, thus suggesting why primary porosity is preserved in the reservoir facies. Third, compaction reduced interparticle pore volume in marine sections, thus reducing porosity in marine reservoir facies and improving the effectiveness of the updip trap. To our knowledge, this is the first direct application of mechanical compaction in shallow-water carbonates to an exploration play.

In the fore-reef facies of the Permian (Guadalupian) Capitan Formation we have observed major mechanical compaction effects. At the top of the slope, all sediment was extensively bound by organic coatings and massive marine phreatic cements; in this zone there is virtually no evidence of mechanical compaction. In the middle slope environment, marine cements are largely lacking, and coarse grainstones display extensive shattering of grains at or near grain contacts (Fig. 5). These fractured grains are cemented by coarse, equant crystals of clear, "late" calcite. In lower slope settings, finer grained sediments were deposited and both mechanical and chemical compaction features are common. Thus, the nature and extent of compaction features depends, in part, on the amount and fabric of the early diagenetic cement framework, the grain size and packing of the sediment, and the amount and rate of overburden addition.

Chemical compaction.—Chemical compaction, solution transfer, or pressure solution is the second major factor in subsurface porosity reduction. Chemical solution phenomena in limestones have been studied for many years (see excellent historical discussion in Bathurst, 1975, p. 468–473), but the specific term "chemical compaction" was introduced by Lloyd (1977) to describe porosity loss in limestones which takes place without the addition of new or allochthonous carbonate material. In the extreme, chemical compaction can yield a completely cemented rock with no introduction of externally-derived cement. All that is required is the expulsion of existing pore fluids

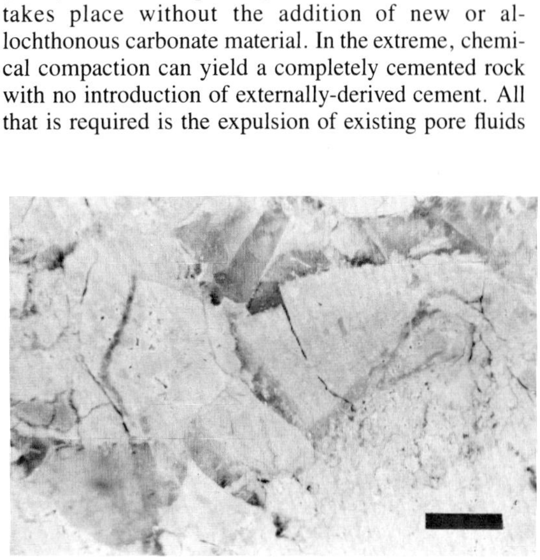

FIG. 5.—Polished slab of Upper Permian (Guadalupian) Capitan fore-reef limestone from McKittrick Canyon, Tex. Sample comes from about 600 ft below the crest of the reef, below the zone of intense penecontemporaneous submarine cementation, and thus shows extensive grain fracturing and compaction due to rapid burial. Scale bar is 1.5 cm.

in order to accomodate the volume of cement. The theoretical basis for such chemical compaction has been reviewed by Durney (1972), Bathurst (1975), de-Boer (1977), and Robin (1978).

Quantitative estimates of the importance of pressure solution really started with the detailed studies of stylolites by Stockdale (1922; 1926) and were greatly advanced by the studies of Dunnington (1967) and Mossop (1972). The importance of non-stylolitic pressure solution phenomena has been recognized in recent years, however, and has led to major improvements in our ability to evaluate the magnitude of chemical compaction.

Pressure solution features have been classified by numerous workers in recent years (Park and Schot, 1968a, 1968b; Trurnit, 1968; Mossop, 1972; Logan and Semeniuk, 1976; Garrison and Kennedy, 1977; Wanless, 1979; Buxton and Sibley, 1981). Three types of chemical compaction features are common, however, to all carbonate rocks: intra- or inter-granular solution features, solution seams, and stylolites. Intra- and inter-granular solution features are found at contacts between grains and at sites within grains where lithostatic stresses are concentrated. These features may be only weakly developed or may eventually lead to the "fitted fabrics" of Buxton and Sibley (1981), in which essentially all the grains in a rock have irregular to linear sutured contacts. Such fabrics apparently form during early stages of chemical dissolution in porous grainstones or in packstone and wackestone fabrics where there is partial matrix support. Solution seams (also termed horsetail groups, residual seams, wispy stylolites, clay seams, microstylolites, or non-sutured seams) represent discrete surfaces of dissolution which extend continuously across many grains and generally form in fine-grained limestones and clay-bearing carbonate rocks. Solution seams are commonly associated with lumpy or nodular fabrics (stylonodular fabric of Logan and Semeniuk, 1976; flaser structure of Garrison and Kennedy, 1977; and fitted nodular fabric of Wanless, 1979). Stylolites are similar to solution seams but have higher amplitudes and generally form in coarser-grained carbonate rocks and/or at lithologic discontinuities in carbonate sections (e.g., Elliott, 1982). Stylolites apparently can form throughout burial, but large-amplitude throughgoing stylolites are commonly formed during late stages of diagenesis but before complete loss of rock porosity (e.g., Dunnington, 1954; Gillett, 1983; Bathurst, 1980). All types of solution surfaces are marked by concentrations of dark, insoluble material (clay, organic matter, iron oxides, etc.).

Qualitative studies showing the wide distribution of chemical compaction features are too numerous to attempt any systematic listing. As even casual visitors to marble-walled lavatories can attest, stylolites are everywhere. Quantitative studies of precisely how much porosity loss is associated with chemical compaction are less common, however. Such studies are done by measuring and totalling the amplitudes of stylolites and other solution features, or by measuring the amount of insoluble residue accumulated along stylolite surfaces. Several factors make such quantitative studies difficult. First, measurement of stylolite amplitudes (or solution seam thicknesses) provides only a minimum estimate of the amount of material removed (Stockdale, 1926; Mossop, 1972). Some estimates use the amount of insoluble material at a solution contact and an average insoluble content for the whole rock to derive the thickness of removed carbonate, but these estimates must assume an initial homogeneity for the rock which is doubtful. Weyl (1959) and Oldershaw and Scoffin (1967) confirmed by observation that stylolites form preferentially in clay-rich layers. Thus, estimates based on measurements of insoluble residues at solution contacts must be considered to be unreliable or maximum values. These values are especially unreliable in cases where the "insoluble" material which accumulates along solution surfaces (the stylocumulate of Logan and Semeniuk, 1976) is not truly insoluble but merely less soluble than the associated limestone. "The irregular distribution of stylocumulate along pressure-solution surfaces and the textural evidence of corrosion of quartz by carbonate idens, suggests that relatively insoluble stylocumulate also is dissolved under appropriate chemical conditions" (Logan and Semeniuk, 1976, p. 137). Finally, the interrelationship of mechanical and chemical compaction makes distinguishing between them difficult in many cases. The distinction is an important one, however, because mechanical compaction simply moves grains closer together, increasing bulk density and reducing porosity. Chemical compaction, on the other hand, not only reduces porosity to some degree by changing packing density (see Manus and Coogan, 1974) but also may reduce porosity by liberating dissolved $CaCO_3$ for local cementation at sites of lesser stress, generally in inter- or intra-granular pore spaces (solution-reprecipitation or solution transfer).

Dunnington (1967) has provided some very interesting and careful observations on chemical compaction of carbonate rocks, primarily from the Middle East. He reported that solution features formed as a consequence of both overburden and tectonic stresses, and that stylolites parallel to bedding (overburden related) resulted in thinning of carbonate sections by as much as 40 percent, with 20 to 35 percent reductions being common. "Volume reductions of up to 35 to 42 percent having resulted from stylolitization" were reported by Park and Schot (1968a, p. 72). Coogan (1970) compiled extensive data from North American oolitic grainstones and concluded that as much as 90 percent (more commonly 50 to 70 percent) of initial porosity had been destroyed by compaction (both mechanical and chemical) in the Jurassic Smackover Formation at approximately three kilometers depth in the

Gulf Coast. Slightly lower values were obtained for deeply buried Mississippian limestones of the Appalachian Basin, whereas the relatively shallowly buried Ste. Genevieve Formation of Indiana showed only about 10 percent compaction. Mossop (1972) demonstrated as much as 24 percent volume reduction through pressure solution in lagoonal strata of the Upper Devonian Redwater atoll of Alberta. Atoll margin carbonates (which were coarser grained and may have had early marine phreatic cements) were compacted by an average of 13 percent. Finally, as noted earlier, Meyers and Hill (1983) have measured as much as 38 percent porosity reduction (from about 42 percent to 27 percent pore space) in Mississippian shelf carbonates of southern New Mexico.

The data clearly show that compaction is effective in reducing pore space in carbonate strata. Bulk volume reduction coupled with pressure solution-derived carbonate cementation can account for the entire change in porosity during the transition from sediment to indurated rock in many sections. The more pertinent and less answerable question, however, is "How universal is this process?" To answer this question, it is necessary to evaluate available evidence on factors which control the rate and extent of chemical compaction.

Controls on compaction.—The amount of overburden needed to produce significant pressure solution is poorly understood. Dunnington (1967) quoted a figure of 2,000 to 3,000 feet (600 to 900 m) as the depth at which pressure solution became "effective". Stratigraphic reconstructions by Buxton and Sibley (1981) allow the conclusion that extensive chemical compaction has taken place in Devonian limestones of Michigan at burial depths of no more than 1,500 m. Depths of "tens to hundreds of meters" were determined by Meyers and Hill (1983, p. 231) to have been adequate for the initiation of chemical compaction in areas with non-marine pore fluids. Similarly shallow depths of formation have been noted in a variety of sources quoted by Bathurst (1975, p. 473). Beall and Fischer (1969) observed that the transition from predominantly mechanical to chemical compaction occurred at about 250 m in North Atlantic DSDP cores. Finally, Neugebauer (1973, 1974) presented experimental data which showed that under the least favorable conditions solution transfer was ineffective at depths less than 300 m, was of only minor significance at depths less than 1000 m, but became extremely important at depths between 2,000 and 4,000 m.

Water chemistry, as noted earlier in this paper, affects both timing and rates of solution transfer processes. Theoretical and experimental work (Neugebauer, 1973; 1974) and field studies (Scholle, 1977; Meyers and Hill, 1983) all indicate that meteoric (Mg-poor) waters accelerate the effects of pressure solution when compared with marine or marine-derived pore fluids. In this context, it is worth noting that basins typically go through at least two phases. By definition, the early history is dominated by subsidence. During this phase, pore waters are dominantly marine-derived and sediments typically follow the porosity-depth curves shown in Figures 2 and 3. Many basins of the world, however, particularly pre-middle-Mesozoic ones, have undergone deformation and/or uplift of the basin margins. In this situation, a completely different hydrologic system typically will evolve. This system involves head-driven artesian fresh water being injected from the uplifted margins towards the basin center. Clearly, this has a major effect on the burial diagenesis of carbonate units for the burial depth vs. porosity curves of Figures 2 and 3. These curves are unrealistic for most, if not all, basins which have such a multi-stage history. These burial depth-porosity curves remain important, however, because they are applicable to the early development of all basins.

There are other effects of pore-fluid composition on burial diagenesis. The exclusion of all waters from the pore space or the prevention of contact between carbonate grains and pore fluids (as in oil-wet reservoirs or in strata in which hydrophobic organic compounds coat grains) greatly reduces stylolite formation, as noted by Dunnington (1967) in platform carbonates and by Scholle (1977) for basinal chalks. Early introduction of oil or gas into the pores can prevent solution transfer entirely, because chemical compaction requires a water phase for the removal of ions. Thus, stylolites may be common in the water zones but absent in the gas and oil zones of a reservoir (Dunnington, 1967; Feazel and Schatzinger, 1983). Carbonate reservoirs which were filled with hydrocarbons before deep burial may preserve porosity while the surrounding carbonate units are thinned and cemented by chemical compaction.

Pore fluid pressures can play an obvious role in affecting the rates of both mechanical and chemical compaction. Mechanical compaction clearly depends on differential stresses to bring about bulk volume reduction. Chemical compaction is just as dependent on differential stresses, however. Because stressed minerals are generally more soluble than unstressed ones, dissolution takes place at sites of maximum stress such as grain contacts (Riecke Principle). As hydrostatic pressures begin to approach lithostatic pressures, differential stresses are reduced, regardless of burial depth. Thus, in overpressured settings, chemical compaction can be greatly retarded. This has been demonstrated in the North Sea (Harper and Shaw, 1974; Scholle, 1977), where chalk porosities as high as 42 percent have been preserved at burial depths in excess of 3,000 m in areas of strong overpressuring. The differential stresses and compaction pressures on these chalks are like those at only about 1,000 m of burial, despite their much greater actual depth. The rest of the stresses are borne by the non-directional hydrostatic pressure.

The related effects of time and temperature variations on compaction are much more poorly understood. Studies on chalks have not shown a major effect of variations in regional geothermal gradients on rates of porosity loss with depth. In the North Sea, chalks are subject to geothermal gradients 1.5 to 2 times as high as those for equivalent age chalks in the U.S. Western Interior. Yet the porosity-depth curves for the two areas (where they are normally pressured) are virtually identical (Lockridge and Scholle, 1978).

On the other hand, continuing studies by J.W. Schmoker (oral comm., 1983) indicate that there is a consistent correlation between porosity-depth curves and time/temperature index in several basins he has examined. In these studies, older units and ones subjected to higher temperatures had lower porosities than otherwise similar strata. In experimental studies as well, elevated temperatures resulted in significantly increased rates of compaction (Ebhardt, 1968). However, data isolating the specific effects of either time or temperature are sparse. It is virtually impossible to find basins of early Mesozoic or older age which have remained under constant burial and temperature conditions and in which one can eliminate the possible effects of meteoric pore fluids. As with many chemical reactions, the effects of time and temperature can substitute for one another in a complex fashion. That is, application of low temperatures for a long time may result in the same degree of compaction as high temperatures applied for a short time, all other conditions being equal.

Primary depositional facies and early diagenetic alteration can also play a significant role in controlling rates of chemical compaction. Grain size and sorting, grain types, mineralogy, and bedding (or lithologic contrast) features inherited from depositional environments will affect the relative importance of mechanical versus chemical compaction as well as the rates of porosity loss with burial.

Enos and Sawatsky (1981) showed that porosity and permeability are negatively correlated in modern carbonate sediments. The highest porosities and the lowest permeabilities are found in carbonate muds, whereas the reverse is true for carbonate sands. In most ancient rock sequences this porosity-permeability relationship is reversed. That is, the most porous ancient limestones are usually the most permeable ones. Part of the explanation for this reversal is the tendency for fine-grained carbonate rocks to become cemented during burial diagenesis more quickly than their coarser-grained counterparts. Although the reasons for this selective cementation are not fully understood, they presumably involve effects related to grain size, such as surface area and reactivity. In addition, clay minerals and organic matter in finer-grained carbonate strata accelerate pressure solution and porosity loss. The result is that, as a crude approximation, those limestones which contained the largest pores and had the highest permeability originally tend to be the ones which retain porosity and permeability in the subsurface. Thus, reefs, carbonate sand bodies, talus deposits, and other originally coarse-grained facies tend to retain porosity preferentially in the subsurface. Finer-grained facies such as lagoonal or basinal muds, on the other hand, preferentially lose porosity during burial diagenesis. Furthermore, the nature or style of burial diagenesis can be influenced by the primary sediment composition and texture. Buxton and Sibley (1981) and Elliott (1982), for example, demonstrated that the style of chemical compaction (intergranular, solution seam, or stylolite dissolution) varies to a considerable degree with the starting lithology.

Effects of variations in early diagenesis on the style and magnitude of burial diagenesis are more difficult to evaluate. Small amounts of early cementation, as shown in Figure 6, are incapable of preventing intergranular dissolution (see also the discussion in Bathurst, 1975, p. 465–466). Nevertheless, variations in the nature and intensity of early dissolution and cementation must be expected to exert some influence on the rate and timing of burial diagenesis (e.g., Meyers and Hill, 1983; Heckel, 1983). Indeed, much of the variance in the porosity at any given depth (Fig. 3) is probably attributable to such early diagenetic variations and their effects on burial diagenesis.

Little is known about the detailed response of various early cement fabrics to burial stresses. However, any primary sedimentologic or early diagenetic property of a limestone which retards late mechanical or chemical compaction (or burial cementation) will help to preserve porosity in the subsurface. Purser (1978) noted the correlation of carbonate reservoir rocks and evidence for early diagenesis. This led him to conclude that early lithification tends to retard or prevent subsequent chemical compaction and thereby reduces the volume of calcium carbonate released for cementation.

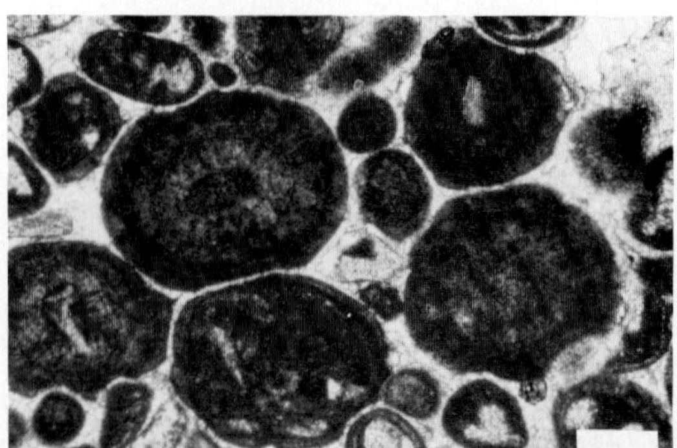

FIG. 6.—Thin-section photomicrograph of the Upper Mississippian Ste. Genevieve Limestone from Tennessee illustrating chemical compaction postdating cementation. Thin rims of submarine cement surround grains, yet ooids show extensive grain interpenetration due to pressure solution. Scale bar is 0.38 mm long.

A similar conclusion had been reached by Becher and Moore (1976) for a Jurassic oolitic reservoir in Arkansas.

Although the particular mechanisms which retard later porosity loss are not known, four consequences of early fresh-water alteration are well documented: (1) mineralogical stabilization; (2) lithification; (3) decreased permeability; and (4) preserved porosity. These factors presumably retard subsurface diagenesis in the following ways: (1) mineralogical stabilization decreases the average grain solubility and thus the potential for chemical compaction of these sediments; (2) lithification increases the resistance to mechanical compaction; and (3) low permeability decreases water flow through the rock, thus limiting the amount of allochthonous cement introduced.

The influence of early diagenesis on late diagenesis appears to be a fruitful topic for future experimental and field research. The literature is filled with papers claiming that porosity is preserved in (and hydrocarbon production obtained from) rocks cemented, leached, or diagenetically stabilized in one environment or another. However, if the data in Figure 3 are to be believed, what makes deposits from one early diagenetic environment untimately "favorable" for oil production as compared with those from another environment is how these early diagenetic fabrics affect the response of the strata during burial. In near-surface settings nearly all strata have adequate porosity and permeability to act as hydrocarbon reservoirs. By the time these units have been buried to depths at which abundant hydrocarbons are generated, however, only a small fraction still have characteristics of reservoir rocks. We do not yet know how near-surface variations in porosity (and perhaps even more importantly, permeability) are translated into the deep subsurface. It may be a "you can pay me now or you can pay me later" situation. That is, strata with significant early cementation and low near-surface porosities may be more resistant to mechanical and chemical compaction than strata which retained high porosities during near-surface alteration. Thus, porosities of these two types may converge or even cross at some depth.

An example of this is provided by the data on dolomites from Schmoker and Halley's (1982) study of Florida Platform carbonates (shown here in Fig. 7). Dolomites have lower porosities than age- and depth-equivalent limestones in near-surface sections. Porosities of the two rock types are similar at about 1,700 m, and at greater depths limestones become progressively less porous than dolomites. Thus, the prevalence of porous hydrocarbon reservoirs in dolomites apparently results less from the dolomitization process or the diagenetic environment of dolomite formation and more from the greater resistance of dolomites to burial diagenetic loss of pore space when compared with limestones. With further study, we will likely find that the limestone data set (Fig. 3) can be broken down into subsets by depositional and early diagenetic environments and that these subsets will show separate and distinct trends.

As noted earlier, the presence of clay minerals can significantly enhance formation of pressure solution features. Many carbonate strata, especially those deposited in outer shelf to abyssal environments, originally have a strong primary bedding rhythmicity consisting of couplets of limestone and marl (e.g., Fischer, 1980; Scholle *et al.*, 1983). These rhythmic beds range in thickness from a few millimeters to as much as a meter and represent between 20,000 and 100,000 years of deposition (Gilbert, 1900; Fischer, 1980). Superficially similar cycles are found in many platform carbonate sections (see papers in Merriam, 1964, and Einsele and Seilacher, 1982, for example).

These calcareous shale or marl interbeds in otherwise pure carbonate sequences are significant because they commonly are sites of intense pressure dissolution. In young pelagic sediments or in shallowly buried sections of older, rhythmically-bedded carbonate strata, carbonate contents of these shaly intervals commonly exceed 50 or 60 percent. Yet in more deeply buried sections, pressure solution may have removed most or all of this carbonate. The marly intervals of chalk sections generally show a progressive increase in the number of wispy solution seams and a progressive decrease in thickness and carbonate content with increasing depth of burial. Complicating geologic factors have never allowed a quantitative evaluation of pressure dissolution, but qualitative observations (e.g., Scholle, 1977) on sections throughout the world indicate that the process is ubiquitous and volumetrically important, at least in pelagic limestones. The stratigraphic effects of this process have been modeled by Hallam (1964) and Arthur (1979), and the results are shown graphically in Figure 8. Basically, there is a progressive "unmixing" of lithologies with a net export of carbonate material dissolved along solution seams from the marly layers (donor beds of Bathurst, 1975). The material derived from these zones is presumably added to adjacent carbonate layers as pore-filling cement (discussed further in the next section). Eventually the material becomes segregated into interbedded pure shales and carbonates. This process acts not only in thin shale interbeds in carbonate sections but also in thick sections of calcareous shales. Perry and Hower (1970), for example, demonstrated significant loss of $CaCO_3$ (shown as CaO) from Tertiary shales in the U.S. Gulf Coast. These shales have CaO contents in excess of 10 percent at depths of about 7,000 ft (2,100 m); by 12,000 ft (3,600 m) the average CaO value has decreased to about 2 percent, suggesting massive export of calcium carbonate.

There are several major consequences to the processes noted above. First, pressure solution can cause significant changes in lithology during burial. Interpretations of paleo-compensation-depths (e.g., Hesse

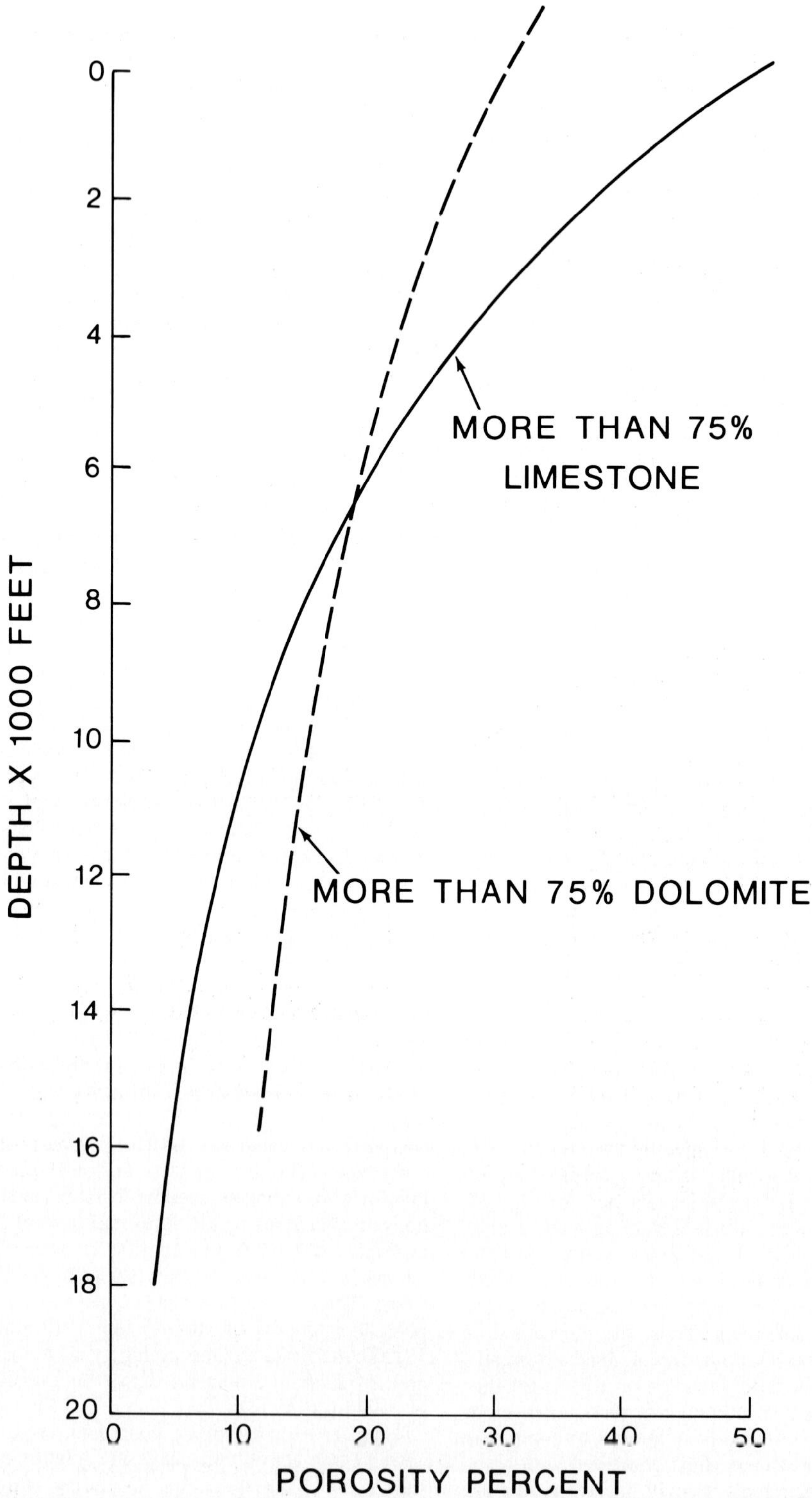

FIG. 7.—Average porosity versus depth for dolomite and limestone in south Florida (from Schmoker and Halley, 1982).

and Butt, 1976; Scholle, 1971) commonly are based on the presence or absence of carbonate material in hemipelagic interlayers between carbonate or clastic turbidites. These relationships clearly can be altered by complete diagenetic removal of carbonate minerals from the hemipelagic interlayers, making environmental interpretations difficult in deeply buried or tectonically deformed sections. Second, the amount of pressure solution in a carbonate-shale section may be seriously underestimated if the shales have been completely purged of their carbonate components. Because of coalescence, individual solution seams may no longer be recognizable, and yet significant amounts of carbonate may have been liberated to act as porosity-reducing cement in associated carbonate beds. Thus, although carbonate rocks in the section may show little or no sign of physical or chemical compaction, associated thin shale interbeds may have been sites of major dissolution, stratigraphic thinning and carbonate export.

It should be clear from the above discussion that we are not now (and indeed we may never be) in a position to prove the quantitative significance of pressure solution and solution transfer as mechanisms for subsurface reduction of porosity. The inability of mechanical compaction alone to account for widely observed and previously discussed porosity loss with increased burial depths, the abundance of pressure solution features, and the apparent impossibility of moving hundreds of thousands of pore volumes of water through most basins in order to introduce externally-derived cement indicate that pressure solution is the predominant porosity-reducing mechanism in burial diagenesis. But the obvious inability to obtain quantitative data in most sections, the fact that much of the dissolution can be taken up in shaly layers of unknown primary carbonate content, and the fundamental difficulty in assessing the world from limited and small-scale data all point toward a permanent inability to provide absolute proof of the efficacy of pressure solution and solution transfer.

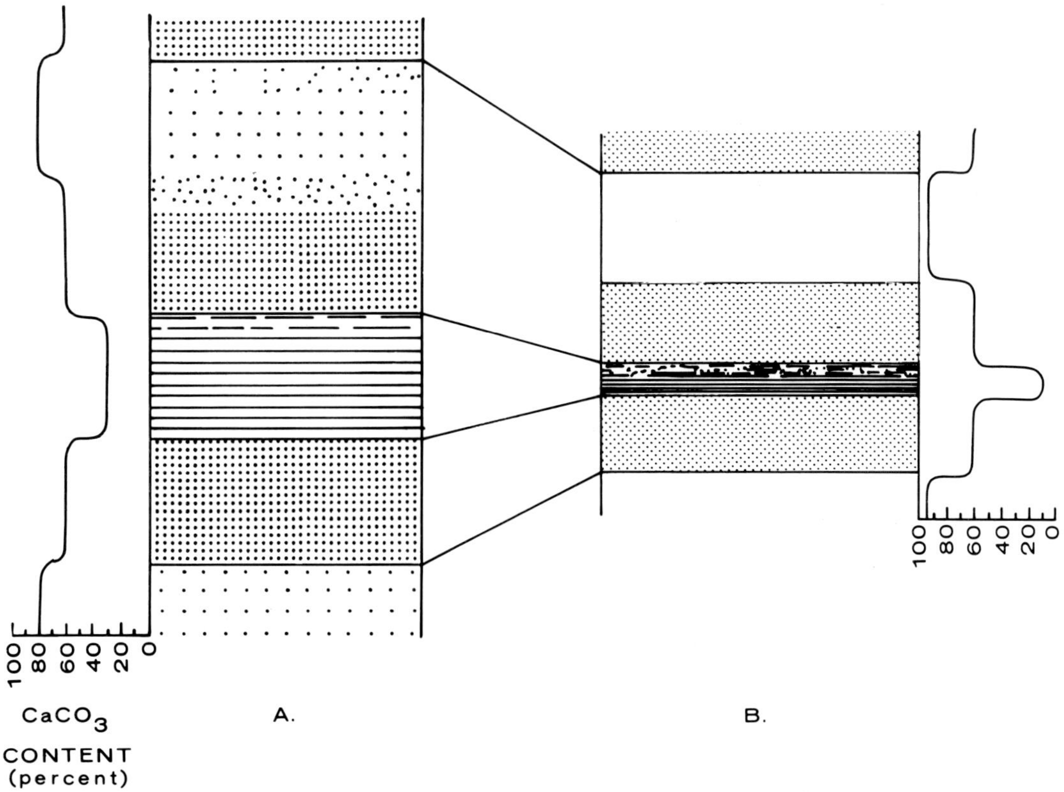

FIG. 8.—Diagrammatic section of typical rhythmically bedded carbonate-shale strata before burial (A) and after burial (B). Note thinning of section due to porosity loss, and note progressive loss of calcium carbonate from shalier units (lined) and increase of carbonate (added calcite cement) in originally more calcareous layers (stippled). Modified from Hallam (1964) and Arthur (1979).

Large-Scale Isotopic Data

As noted earlier, isotopic analyses of cements can provide specific evidence for the temperature and timing of burial cementation. Examination of the entire assemblage of isotopic analyses from carbonate rocks of diverse ages and locations can provide an additional perspective on diagenesis, however. Keith and Weber (1964) and Hudson (1975) noted the fact that most Phanerozoic marine limestones "have the same range of ^{13}C as modern marine shells and carbonate sediments, around 0 per mil on the PDB scale." (Hudson, 1975, p. 20).

Considerable data have now been gathered to show that near-surface meteoric diagenesis typically results in a significant enrichment of ^{12}C in limestones. In some cases, carbon isotopic shifts of as much as 6 per mil have been noted in association with soil zones and vadose zones (Gross, 1964; Hudson, 1975, Allan and Matthews, 1982), although this effect decreases rapidly with depth. Thus, Hudson (1975, p. 22) concluded that the vast majority of ancient limestones had not been significantly affected by near-surface vadose cementation but rather were "cemented mainly during the compactional phase of burial diagenesis. . . The source of this cement is in other limestones or in shales of the sedimentary basin; it is released by pressure solution along stylolites and microstylolites and by solution of aragonite shells."

Although this approach outlined above is useful in demonstrating the overall importance of burial cementation, it only shows that most cementation is accomplished without the introduction of large amounts of CO_2 from external sources. It does not distinguish between possible internal sources of carbonate cement. However, in recent years far more numerous carbon and oxygen isotopic analyses of bulk limestones have been completed on carefully measured stratigraphic sections (Scholle and Arthur, 1980). These data show that there have been major variations in the carbon isotopic composition of sea water through time. These variations are reflected in correlatable changes in the carbon isotopic signature of pelagic limestones. From a diagenetic viewpoint, what is significant, however, is that these isotopic variations occur over intervals of centimeters to meters and are essentially equally well represented in porous and non-porous sections. Figure 9, for example, shows a comparison of four chalk sections of equivalent (Late Cretaceous) age. Although the average porosity values for these sections range from less than 1 percent (Peregrina Canyon, Mexico) to more than 35 percent (Norfolk, England), the isotopic signatures are virtually identical in shape and intensity. Because these two- to three-per mil variations in isotopic composition take place over distances of meters to a few tens of meters and because the variations remain unaffected by diagenetic porosity loss, there is a strong implication that the sources of cement are not only internal but are probably nearby. That is, cements in the carbonate beds most likely are derived from adjacent or very nearby marl-shale layers. If significant volumes of cement had been introduced from outside the local system, this would have resulted in a damping of the primary fluctuations in the well-cemented sections.

Although these carbon isotopic variations are not identical for every section measured, they are sufficiently uniform that one can generalize that in most cases studied to date burial cements are of very local derivation. The fact that there are exceptions (such as the Gubbio section discussed in Scholle and Arthur, 1980) indicates that this isotopically-based approach may be used more fully in the future to refine our understanding of differences in styles of cementation in different basins.

Cementation

Significant loss of porosity occurs during grain-contact dissolution alone, or in most cases, during redeposition of some or most of the removed material as cement. Cement redeposition is particularly important where the porosity is below about 20 percent (Manus and Coogan, 1974; Mitra and Beard, 1980). A number of authors have noted pressure-solution features are often near zones of low porosity (Dunnington, 1967; Harms and Choquette, 1965; Nelson, 1981). A particularly well-documented example from the Devonian of Canada described by Wong and Oldershaw (1981) demonstrates not only the spatial link between pressure dissolution and cementation, but also documents the timing between stylolite growth and cement precipitation using zoned cements to illustrate temporal relationships. However, there are very few examples in which the logical links between pressure solution, cementation, and porosity loss can be demonstrated. Pressure solution features are widespread in many rocks, as is evidence of late cementation, but connecting the two processes in most instances requires a certain leap of faith.

Part of the reason for these difficulties is the rather poor understanding of the controls governing the dissolution and reprecipitation of carbonate minerals in the subsurface. The chemistry of these reactions is exceedingly complex. The reactions typically occur at elevated temperatures in water of widely varying composition and typically proceed very slowly over geologically significant periods of time. Likewise, the effects of grain size, fabric, mineralogy, pore fluid composition, and many other variables are known in only a general manner. For example, Neugebauer (1973) has suggested that magnesium in solution, because it interferes with the precipitation of calcite, retards the effects of pressure solution in chalks. In contrast, experimental studies by Baker et al. (1980), who studied the pressure solution and hydrothermal recrystallization of fine-grained carbonates, showed that clay, diatomite, and basaltic glass inhibited stabilizing reactions. Baker *et al.* (1980) found that variations in calcium, magnesium, alkalinity, and pH of the reacting solutions have only a small effect on the rates of alteration. Interestingly, Baker *et al.* (1980) also found that the presence of clay inhibited pressure dissolution, a result which appears opposite that noted by many earlier workers.

Land (1980) attempted to model changes in ^{18}O, ^{13}C, and strontium in deep-sea carbonates during burial diagenesis. He found particular difficulty modeling the strontium content of deep-sea oozes. In particular, he noted a lack of knowledge of the factors that control values of distribution coefficients. This lack of knowledge was partly remedied by Lorens (1981), who described a rather strong effect of precipitation rate on the distribution coefficient of strontium in calcite. However, Mucci and Morse (1983), monitoring experiments that lasted from two to twenty-four hours, found that the concentrations of magnesium and strontium in calcite were independent of precipitation rate. The question of factors affecting distribution coefficients remains unresolved. Nevertheless, Baker *et al.* (1982) successfully modeled the distribution of strontium in pore water and sediments; in their model, lithification is essentially isochemical, and the cement is derived from within the formation. Advective flow was unnecessary to explain its lithification. To explain the magnesium distribution of lithified calcite, however, required a distribution coefficient many times smaller than previously published values.

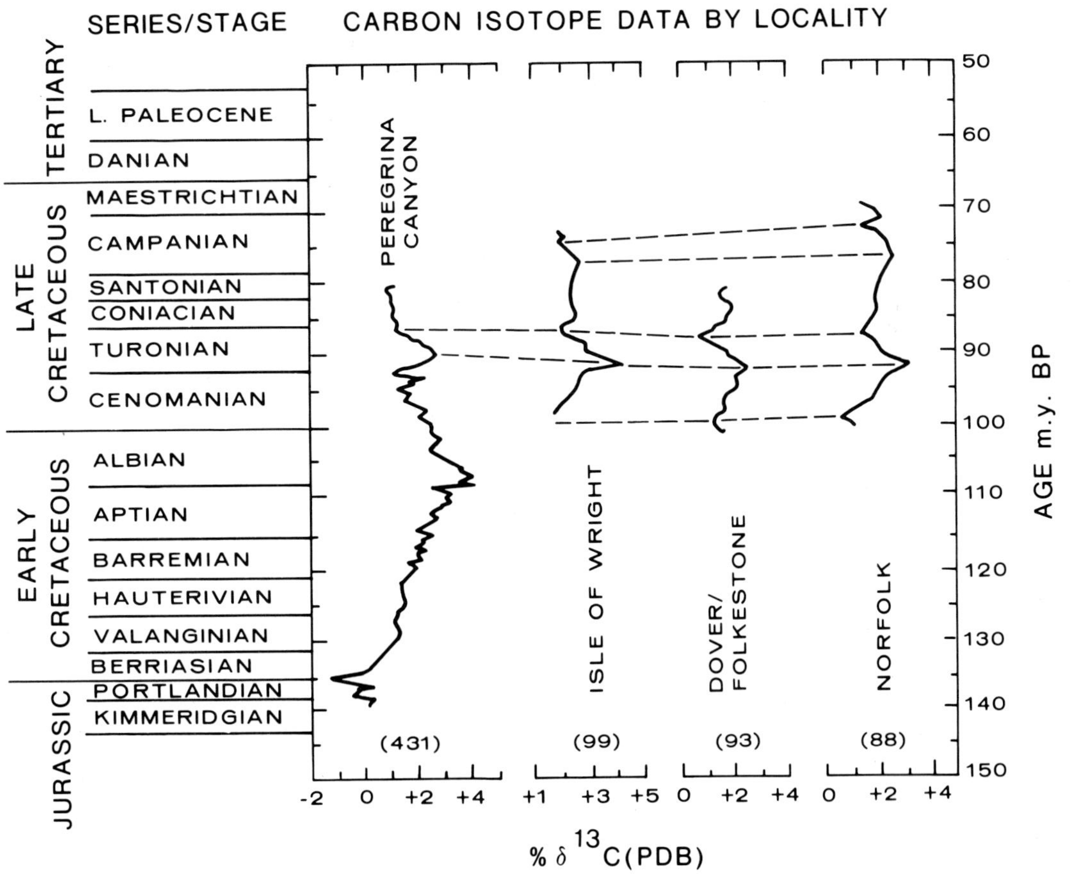

FIG. 9.—Carbon isotopic ratios of whole-rock samples of pelagic limestone from 4 localities (Peregrina Canyon, Mexico; and the Isle of Wight, Dover-Folkestone, and Norfolk, England). Porosities for these units (all cemented during burial) are less than 1 percent for Peregrina Canyon; 10–20 percent for the Isle of Wight; 25–40 percent for Dover-Folkestone; and 35–45 percent for Norfolk. Despite these great variations in porosity, the details of the isotopic cycles are virtually identical in shape and magnitude for all 4 sections, indicating that little or no cement was introduced from outside; rather, cements were apparently derived on a very local scale. Modified from Scholle and Arthur (1980).

Despite our relatively rudimentary understanding of subsurface cementation and the difficulties of cementation models, our ability to recognize burial cements has increased significantly during recent years. The increased use of special petrographic and geochemical methods is adding to a growing list of well documented examples of burial cementation.

CRITERIA FOR RECOGNITION OF BURIAL CEMENTS

The recognition of burial cements hinges on the identification of petrographic relationships which suggest precipitation after other recognizable events or processes in the burial history of a given rock. Burial cements post-date early cements and compaction features, and may post-date stylolites and fractures. Although these relationships do not necessarily establish a depth of burial for cements, or perhaps any burial in special cases, they commonly indicate precipitation some time after the onset of diagenesis and after some burial.

If pore spaces are sufficiently large, burial cements are commonly coarse grained and poikilotopic, because they grow slowly at relatively low levels of supersaturation. Occasionally, grains incorporated by growing burial cements may be seen to be more closely packed near the periphery of the cement crystal, indicating progressively tighter packing of the surrounding sediment during burial and cementation. Recent studies of burial cements have used a number of increasingly advanced techniques for the identification of burial cements and estimation of temperatures and depths at which they form. These include combinations of staining and cathodoluminescent petrography, analyses of fluid inclusions, and stable isotope analyses.

Ferroan Cements

Shortly after burial, often within a few meters or tens of meters below the sediment-water interface, dissolved oxygen is depleted from pore waters in marine sediments by bacterial respiration and organic matter oxidation (Claypool and Kaplan, 1974). Continued oxidation and sulfate reduction may cause pore water to become reducing, which promotes the co-precipitation of ferrous-iron carbonate within calcite and/or dolomite. Friedman (1959) describes staining techniques for the identification of ferroan calcites and dolomites.

Although burial cements are commonly observed to be ferroan, iron content alone is not sufficient for their identification. There is no theoretical reason why ferroan cements could not form early in near-surface anoxic waters (Evamy, 1969). However, such occurrences seem to be rare; most ferroan cements described from carbonate rocks have been attributed to late or burial cementation. Ferroan calcite cements have been described by Oldershaw (1971), Meyers (1974), Loucks (1977), Grover and Read (1978), and Wong and Oldershaw (1981), among others. Calcite cements formed during burial may be iron-free because there may be no source for iron. This can be the case in very thick carbonate sequences or carbonates partially isolated by evaporites. Examples of iron-poor burial calcite cements are illustrated by Moore and Druckman (1981).

Saddle dolomite (Radke and Mathis, 1980), also called baroque dolomite (Folk and Assereto, 1974), is another burial carbonate cement which is often ferroan. It is characterized by curved crystal faces and sweeping extinction and frequently contains iron-rich and iron-poor zones. Examples of late-stage, burial dolomite cements are illustrated by Choquette (1971), Radke and Mathis (1980), Moore and Druckman (1981), and Wong and Oldershaw (1981). Radke and Mathis suggest that saddle dolomite is indicative of temperatures between 60 and 150 degrees Centigrade.

Cathodoluminescence

Fluctuating oxidation states (redox potentials) and changing manganese and iron ratios in subsurface fluids cause some burial cements to exhibit marked zonation when examined using cathodoluminescent petrography (Meyers, 1978; Pierson, 1981; Frank and others, 1982; Fairchild, 1983). These zones can sometimes be mapped and correlated with regional stratigraphic discontinuities, as in the classic study of cement stratigraphy within the Lake Valley Formation of New Mexico by Meyers (1974). In this manner, the age of the cement zones may be bracketed and approximate depths of burial during formation may be estimated. Similarly, Grover and Read (1978, 1983) identified burial cements in Middle Ordovician carbonate ramp limestones from Virginia.

Luminescent cements in the Lake Valley Formation are shown by Meyers (1978) to have formed at relatively shallow depths, generally less than 500 meters. Grover and Read (1983) consider luminescent cements to form at intermediate depths, deep enough not to be influenced by oxidizing meteoric waters, but shallow enough to avoid strongly anoxic waters that allow incorporation of ferrous iron in calcite. In their opinion, luminescent cements represent burial cements separating early meteoric cements from late, deep burial cements. Grover and Read (1983) describe dull luminescence in iron-bearing cements interpreted to have formed at burial depths of 2,000–3,000 meters. Even deeper cementation (4,000 to 7,500 m) is linked to heavy metal mineralization during the late Paleozoic. These two studies (Meyers, 1974, 1978; Grover and Read, 1983) illustrate the growing role of cathodoluminescent petrography in studies of burial cementation.

Cathodoluminescent cements are often studied in conjunction with isotope and fluid inclusion data to best understand their origin. As an example, Figure 10 shows cements which are homogeneous in plane light but exhibit distinct luminescent zones. These zones

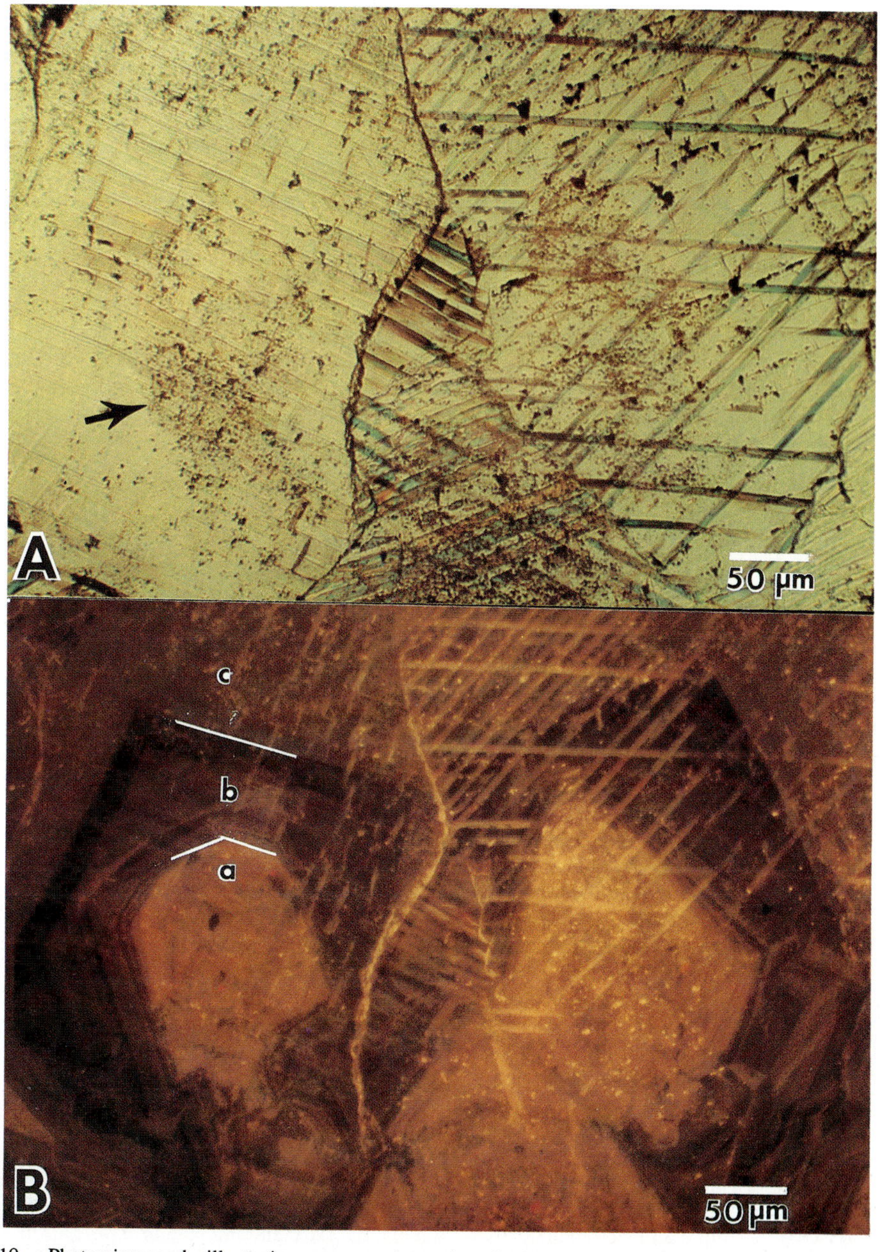

FIG. 10.—Photomicrographs illustrating pronounced zonation of calcite cement visible during cathodoluminescent petrography. Inclusion-rich zone visible in plane light (arrow in A) luminesces brightly under the cathode ray (zone a in B). Clear calcite overgrowths in A are divisible into 2 zones in B, a first zone (b) of alternately light and dark bands, and a second zone (c) of uniformly dull luminescence. ^{18}O analyses for the three cements are approximately −2, −4, and −6 percent (relative to PDB standard) from oldest to youngest, suggesting precipitation from successively lighter water or at successively higher temperatures. The latter interpretation is supported by fluid inclusions from the youngest cement, which homogenize near 175°C. Sample is from Upper Permian sponge-algal reef, near Ziyun, Guizhou Province, Peoples Republic of China.

have characteristic oxygen isotope ratios as well as luminescent characteristics, and most likely record calcite cementation at progressively higher temperatures. The brightly luminescing cores (zone a) of these cement crystals are cloudy with inclusions, have ^{18}O of about −2, and probably represent an early marine cement. Zone b contains alternately light and dark bands and has ^{18}O of about −4 per mil (relative to PDB standard). Zone c has rather uniform dull luminescence, ^{18}O of about −6, and contains fluid inclusions that homogenize at about 175°C. Combining these observations leads to the recognition of at least three distinct episodes of cementation, each at successively deeper burial depths and higher temperatures.

Fluid Inclusions

Primary fluid inclusions in carbonate cements have been analyzed in an attempt to determine the temperature of cement formation. By reference to a reasonable geothermal gradient, the temperature leads to a depth of burial at the time of cement precipitation. Roedder (1979), in a review of fluid inclusion studies in sedimentary rocks, cited only two studies from calcite or dolomite, both unpublished. Several fluid inclusion studies in carbonate rocks have been completed during the last ten years, most as unpublished Master's theses. Nelson (1974) performed over fifty homogenizations in a study of fluid inclusions of the Fernvale Limestone (Ordovician) of northern Arkansas. Homogenization temperatures ranged from 85 to 173 degrees Centigrade. Nelson concluded that two-phase fluid inclusions were trapped in the limestone during recrystallization of an earlier cement. Although homogenization determinations often took several hours to perform, technological advances in fluid inclusion stages have shortened the time required for a determination and have improved their accuracy.

Klosterman (1981) measured 222 homogenization temperatures from coarse calcite cements in the Smackover Formation (Jurassic) of Arkansas. She found a wide range of fluid inclusion homogenization temperatures (58°–158°C) and interpreted the range to represent the gradual precipitation of cement and entrapment of inclusions during changing subsurface temperatures. Depressed freezing temperatures and low initial melting temperatures of the inclusions indicated the cement precipitated from high-salinity brines. Klosterman (1981) cited several difficulties with the fluid inclusion analyses, including determination of inclusion origin, pressure corrections, small inclusion size, metastable phases, and lack of phase data for organic-inorganic systems. To these problems, we might add difficulties associated with stretching or leaking of inclusions, inadvertent operator selection of an unrepresentative sample population of inclusions, and alteration of inclusions during sample preparation, particularly when dealing with low-temperature inclusions (less than 100 degrees Centigrade). The possibility exists that inclusions may increase in volume by stretching when they are heated above their entrapment temperature (Poland, 1982) or during freezing analyses (Lawler and Crawford, 1983). Stretching results in artificially increased homogenization temperatures and has yet to be thoroughly investigated for calcite or dolomite. Stretching in flourite has been studied by Poland (1982), and her study indicates that small inclusions, such as those typically found in sedimentary calcites, may resist stretching at elevated temperatures and pressures in sedimentary basins. She also suggests that the ease at which stretching occurs may be related to the Mohs hardness scale, with soft minerals like calcite and anhydrite near the most stretchable end of the scale. Despite these shortcomings, fluid inclusion analyses are a developing key data source for the recognition of burial cements, as exemplified by the work of Moore and Druckman (1981).

Stable Isotopes

Stable isotope studies have played a key role in the recognition of burial cements. The fractionation of oxygen isotopes between water and precipitating calcite at a given temperature is well known (Epstein *et al.*, 1953). Stable isotopes of oxygen have been particularly useful in interpreting the temperature during the precipitation of calcite, given reasonable constraints on the oxygen isotopic composition of water. Conversely, with ^{18}O of calcite and temperatures of precipitation derived from fluid inclusions, the ^{18}O of precipitating waters may be estimated. Examples of the use of oxygen isotopic data have been previously cited and the reader is referred to Hudson (1975), Scholle (1977), Dickson and Coleman (1980), and Moore and Druckman (1981).

Stable carbon isotopes are not fractionated as extensively by temperature as oxygen isotopes and therefore are not as useful in determining temperature during cement precipitation. However, as suggested earlier, they may provide a key to the distance of migration of calcium carbonate during subsurface cementation.

Variations in the stable strontium isotopic composition of seawater have been documented by Veizer and Compston (1974) and Burke *et al.* (1982). These data provide an original strontium isotopic ratio for carbonate sediments of a given age, which may then be compared with strontium isotopic composition after diagenesis. From this comparison, inferences may be drawn about the strontium composition of waters involved with cementation and about their previous history. For example, Stueber and Pushkar (1983) have shown that some cements in the Smackover Formation yield a ratio of $^{87}Sr/^{86}Sr$ which is too large for cements precipitated from unaltered, buried Jurassic seawater. Rather, the precipitating waters have picked up large quantities of radiogenetic Sr (^{87}Sr), perhaps during feldspar dissolution within the underlying or laterally equivalent formations. Feldspar dissolution, a well documented late diagenetic process forming secondary porosity in sandstones (Schmidt and McDonald, 1979), preceded or coincided with the precipitation of burial cements in portions of the Smackover Formation. Stable strontium isotope analyses show promise for determining the timing of some of the subsurface processes that occur during burial diagenesis of carbonate rocks.

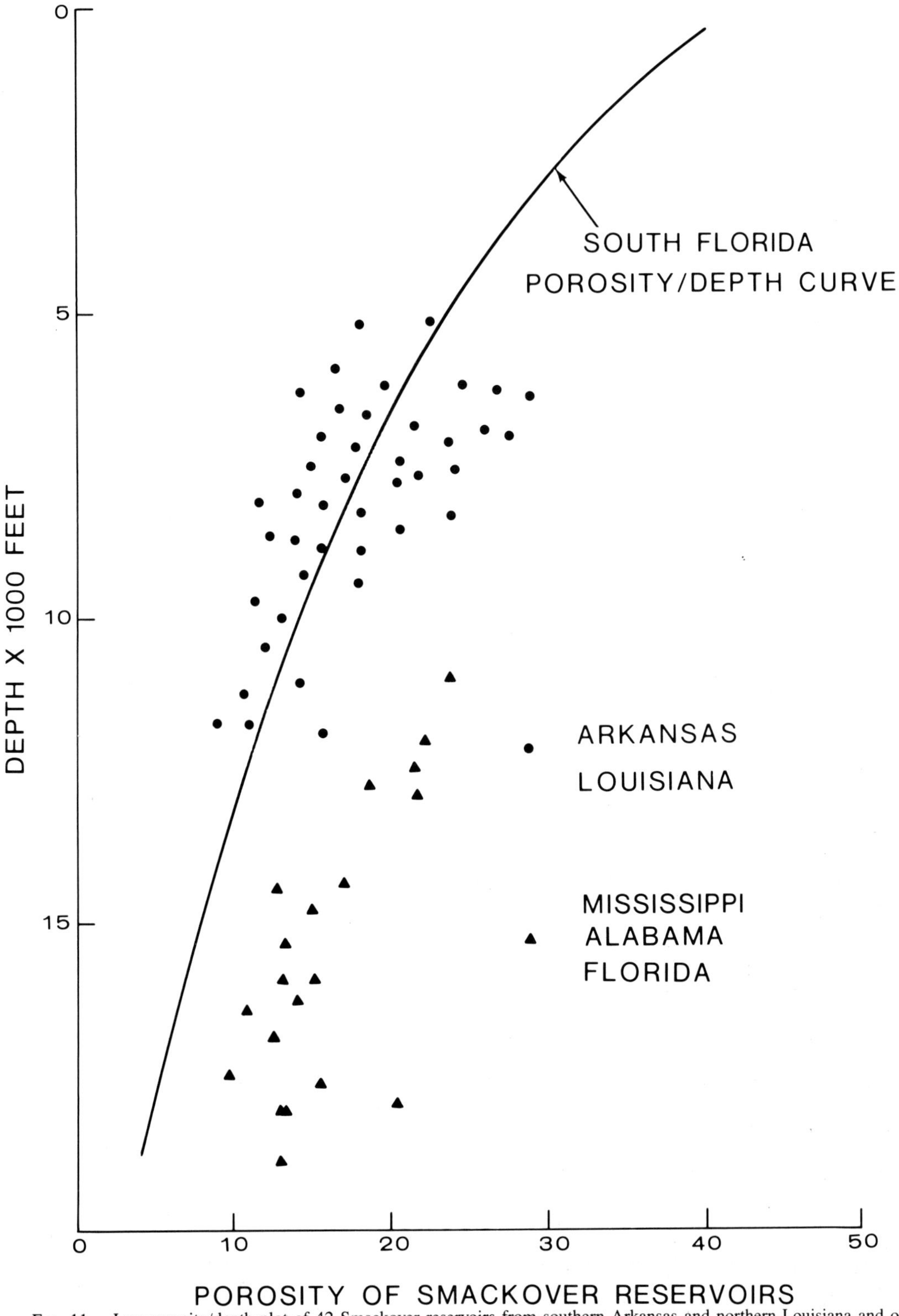

FIG. 11.—Log-porosity/depth plot of 42 Smackover reservoirs from southern Arkansas and northern Louisiana and of 19 deep Smackover reservoirs in Mississippi, Alabama, and Florida. Smooth curve is porosity/depth relation for south Florida from Figure 3. Note that porosity in these deep reservoirs is much higher than would be predicted by the Arkansas-Louisiana data or by the South Florida trend.

ESTABLISHING NEW PERSPECTIVES

The depth-porosity trends discussed in this work have value as predictive tools for estimating average porosity for large volumes of carbonate rock within some sedimentary basins. Undoubtedly, these predictive models will become more useful as controls on subsurface diagenesis are better defined. Beyond the goal of immediate porosity prediction, these models may offer insights, first, to the evolution of porosity within a basin and, second, to unusual and perhaps poorly understood but potentially important geologic processes controlling porosity distribution in carbonate rocks.

In the first case, the depth-porosity trends reveal isolated instants in the history of a subsiding basin. However, to the extent that similar porosity-reducing processes have been operative in the basin, the trend becomes a predictive line tracing the change in porosity, with respect to time, of any unit within the basin. A formation with little porosity may be traced back in time (upward) and an estimate made of the time when it was porous. This information can be combined with knowledge of source rock maturation and hydrocarbon migration (for example, Waples, 1980) to determine if there might be reservoirs in the unit which were charged with hydrocarbons at some time in the past. It is interesting to note, for instance, that the porosity-depth trend in Smackover reservoirs of northern Louisiana and southern Arkansas is very similar to that of non-reservoir rocks in south Florida (Fig. 11; also see Halley, 1981). This might be interpreted to suggest that hydrocarbons moved into Smackover reservoirs when the formation had a porosity-depth profile like that found in south Florida today and effectively blocked any further porosity loss. In this way, the depth-porosity trend of the reservoirs was preserved while the rest of the formation may have undergone continued alteration.

In the second case, porosity-depth trends provide tools for recognizing unusual porosity occurrences. As an example, compare the porosity-depth relationship for the Smackover Formation in Arkansas and Louisiana with similar data from the deeply buried Smackover of Florida, Mississippi, and Alabama (Fig. 11). Clearly, these deep Smackover reservoirs are much more porous than might be expected from the data for shallower reservoirs (Arkansas and Louisiana) or for non-reservoir rocks (Fig. 3). As yet, there is no clear consensus about what factors are responsible for the unusually porous reservoirs in the deep Smackover. Several studies have considered the origin of Smackover reservoirs in some detail. Chimene (1976), Moore and Druckman (1981), and Wagner and Matthews (1982) described the setting, facies, and diagenesis of several fields in Arkansas and Louisiana. Some of the deeper Smackover fields of Florida, Mississippi, and Alabama have been described by Ottman *et al.* (1976), Oglesby (1976), and Mancini and Benson (1981). The deep fields include many which are largely dolomite, some of which are limestone, and a few of which are largely sandstone with carbonate matrix. Some of the deep reservoirs are overpressured, many are not. What unusual happenstance has endowed these deep reservoirs with such porosity at these depths? Has something in their geologic history led to the preservation of early porosity at depth? Or has porosity been secondarily created at depth? These and other questions await thorough documentation and testing. Perhaps the reasons for porosity at depth will turn out to be as numerous as the reservoirs themselves, each existing for its own set of unique circumstances. However, it is our belief that this is not the case. Figure 11 presents a surprisingly consistent relationship and suggests that only a few basic causes underlie the existence of these deep reservoirs. The existence of deep reservoirs in carbonate rocks is promising, and such reservoirs might reasonably be postulated for other areas once the underlying causes are understood. It is our belief that the processes of burial diagenesis hold the keys to this knowledge and that the prediction of porosity at depth in carbonate rocks is controlled as much by burial diagenesis as by facies and early diagenesis.

ACKNOWLEDGMENTS

The authors wish to thank D.L. Gautier, E.A. Shinn, and W.D. Wiggins for reviewing earlier versions of this manuscript and for offering suggestions that substantially improved the final product. This paper was presented at the Abu Dhabi National Reservoir Research Foundation Seminar on Stylolites, November 7th and 8th, 1983, and a version is published in the proceedings for that symposium.

REFERENCES

ALLAN, J.R., AND MATTHEWS, R.K., 1977, Carbon and oxygen isotopes as diagenetic and stratigraphic tools: Surface and subsurface data, Barbados, West Indies: Geology, v. 5, p. 16–20.

ARTHUR, M.A., 1979, Sedimentologic and geochemical studies of Cretaceous and Paleogene pelagic sedimentary rocks: Pt. 1, the Gubbio Section: Ph.D. Thesis, Princeton University, Princeton, New Jersey.

BACK W., AND HANSHAW, B.B., 1971, Rates of physical and chemical processes in a carbonate aquifer: Advances in Chemistry, v. 106, p. 77–93.

Baker, P.A., Gieskes, J.M., and Elderfield, H., 1982, Diagenesis of carbonates in deep-sea sediments—evidence from Sr/Ca ratios and interstitial dissolved Sr data: Jour. Sed. Petrology, v. 52, p. 71–82.

______, Kastner, M., Byerlee, J.D., and Lockner, D.A., 1980, Pressure solution and hydrothermal recrystallization of carbonate sediments—An experimental study: Marine Geology, v. 38, p. 185–203.

Bathurst, R.G.C., 1970, Problems of lithification in carbonate muds: Proc. Geol. Assoc., v. 81, p. 429–440.

______, 1975, Carbonate sediments and their diagenesis: Amsterdam, Elsevier Scientific Publishing Co., 658 p.

______, 1980, Deep crustal diagenesis in limestones: Rev. Inst. Invest. Geol. Deputacion Provincial Univ. Barcelona, v. 34, p. 89–100.

Beach, D.K., 1982, Depositional and diagenetic history of Pliocene-Pleistocene carbonates of northwestern Great Bahama Bank; evolution of a carbonate platform: Ph.D. dissertation, University of Miami, Coral Gables, Florida, v. 1, 447 p.

______, and Schumacher, Ann, 1982, Stanley field, North Dakota: A new model for a new exploration play, *in* J.E. Christopher, and John Kaldi (eds.), 4th International Williston Basin Symposium, p. 235–243.

Beall, A.O., and Fischer, A.G., 1969, Sedimentology, *in* Initial Reports of the Deep-Sea Drilling Project, v. 1: U.S. Govt. Printing Office, Washington, p. 521–593.

Becher, J.W., and Moore, C.H., 1976, The Walker Creek field: A Smackover diagenetic trap: Gulf Coast Assoc. Geol. Socs. Trans., v. 26, p. 34–56.

Bhattacharyya, A., and Friedman, G.M., 1979, Experimental compaction of ooids and lime mud and its implication for lithification during burial: Jour. Sed. Petrology, v. 49, p. 1279–1286.

Bonham, L.C., 1980, Migration of hydrocarbons in compacting basins: Amer. Assoc. Petroleum Geologists Bull., v. 64, p. 549–567.

Brown, P.R., 1969, Compaction of fine-grained terrigenous and carbonate sediments—A review: Bull. Canadian Petroleum Geology, v. 17, p. 486–495.

Burke, W.H., Denison, R.E., Hetherington, R.B., Koepnick, R.B., Nelson, H.F., and Otto, J.B., 1982, Variation of seawater $^{87}Sr/^{86}Sr$ throughout Phanerozoic time: Geology, v. 10, p. 516–519.

Buxton, T.M., and Sibley, D.F., 1981, Pressure solution features in a shallow buried limestone: Jour. Sed. Petrology, v. 51, p. 19–26.

Chimene, C.A., 1976, Upper Smackover reservoirs, Walker Creek field area, Lafayette and Columbia Counties, Arkansas, *in* J. Braunstein (ed.), North American Oil and Gas Fields: Amer. Assoc. Petroleum Geologists Memoir 244, p. 177–204.

Choquette, P.W., 1971, Late ferroan dolomite cement, Mississippian carbonates, Illinois Basin, U.S.A., *in* O.P. Bricker (ed.), Carbonate Cements: Baltimore, Johns Hopkins University Studies in Geology, no. 19, p. 339–346.

______, and Pray, L.C., 1970, Geologic nomenclature and classification of porosity in sedimentary carbonates: Amer. Assoc. Petroleum Geologists Bull., v. 54, p. 207–250.

Claypool, G.E., and Kaplan, J.R., 1974, The origin and distribution of methane in marine sediments in natural gases in marine sediments: New York, Plenum Press, p. 99–139.

Collins, A.G., 1975, Geochemistry of Oilfield Waters: New York, Elsevier Scientific Publishing Co., 496 p.

Conley, C.D., 1977, Origin of distorted ooliths and pisoliths: Jour. Sed. Petrology, v. 47, p. 554–564.

Coogan, A.H., 1970, Measurements of compaction in oolitic grainstone: Jour. Sed. Petrology, v. 40, p. 921–929.

______, and Manus, R.W., 1975, Compaction and diagenesis of carbonate sands, *in* G.V. Chilingar and K.H. Wolf (eds.), Compaction of Coarse-grained Sediments, I. Developments in Sedimentology—18A: New York, Elsevier Scientific Publishing Co., p. 79–166.

de Boer, R.B., 1977, Pressure solution: Theory and experiments: Tectonophysics, v. 39, p. 287–301.

Dickson, J.A.D., and Coleman, M.L., 1980, Changes in carbon and oxygen isotope composition during limestone diagenesis: Sedimentology, v. 27, p. 107–118.

Dunham, R.J., 1969, Early vadose silt in Townsend mound (reef), New Mexico, *in* G.M. Friedman (ed.), Depositional Environments in Carbonate Rocks: A Symposium: Soc. Econ. Paleontologists and Mineralogists Spec. Pub. 14, p. 139–181.

Dunnington, H.V., 1954, Stylolite development post-dates rock induration: Jour. Sed. Petrology, v. 24, p. 27–49.

______, 1967, Aspects of diagenesis and shape change in stylolitic limestone reservoirs: 7th World Petroleum Congress, Proc., Mexico, v. 2, p. 339–352.

Durney, D.W., 1972, Solution transfer, an important geological deformation mechanism: Nature, v. 235, p. 315–317.

Ebhardt, G., 1968, Experimental compaction of carbonate sediments, *in* G. Muller and G.M. Friedman (eds.), Recent Developments in Carbonate Sedimentology in Central Europe: New York, Springer-Verlag Press, p. 58–65.

Einsele, G., and Seilacher, A. (eds.), 1982, Cyclic and Event Stratification: New York, Springer-Verlag Press, 536 p.

Elliott, T.L., 1982, Carbonate facies, depositional cycles, and the development of secondary porosity during burial diagenesis: Mission Canyon Formation, Haas Field, North Dakota, *in* J.E. Christopher and J. Kaldi (eds.), 4th International Williston Basin Symposium: Saskatchewan Geol. Soc. Spec. Pub. 6, p. 131–151.

Enos, Paul, and Sawatsky, L.H., 1979, Pore space in Holocene carbonate sediments [abs.]: Amer. Assoc. Petroleum Geologists Bull., v. 63, p. 445.

______, and ______, 1981, Pore networks in Holocene carbonate sediments: Jour. Sed. Petrology, v. 51, p. 961–985.

Epstein, S., Buchsbaum, R., Lowenstam, H.A., and Urey, H.C., 1953, Revised carbonate-water temperature scale: Geol. Soc. America Bull. v. 62, p. 417–426.

Evamy, B.D., 1969, The precipitational environment and correlation of some calcite cement deduced from artificial staining: Jour. Sed. Petrology, v. 39, p. 787–793.

Fairbridge, R.W., 1967, Phases of diagenesis and authigenesis, *in* G. Larsen and G.V. Chilingar (eds.), Diagenesis in Sediments: Amersterdam, Elsevier Scientific Publishing Co., p. 19–89.

FAIRCHILD, I.J., 1983, Chemical controls of cathodoluminescence of natural dolomites and calcites: New data and review: Sedimentology, v. 30, p. 579–583.

FEAZEL, C.T., AND SCHATZINGER, R.A., 1983, Prevention of carbonate cementation in petroleum reservoirs [abs.]: Am. Assoc. Petroleum Geologists Bull., v. 67, p. 459.

FISCHER, A.G., 1980, Gilbert-bedding rhythms and geochronology, *in* E.L. Yochelson (ed.), The Scientific Ideas of G.K. Gilbert: Jour. Soc. America Spec. Pub. 183, p. 93–104.

FOLK, R.L., AND ASSERETO, R., 1974, Giant aragonite rays and baroque white dolomite in tepee-fillings, Triassic of Lombardy, Italy [abs.]: Amer. Assoc. Petroleum Geologists, Abstracts with Programs, Annual Meeting, San Antonio, p. 34–35.

FRANK, J.R., CARPENTER, A.B., AND OGLESBY, T.W., 1982, Cathodoluminescence and composition of calcite cement in the Taum Sank Limestone (Upper Cambrian), Southeast Missouri: Jour. Sed. Petrology, v. 52, p. 631–638.

FRIEDMAN, G.M., 1959, Identification of carbonate minerals by staining methods: Jour. Sed. Petrology, v. 29, p. 87–97.

______, 1975, The making and unmaking of limestones or the downs and ups of porosity: Jour. Sed. Petrology, v. 45, p. 379–398.

FRUTH, L.S., JR., ORME, G.R., AND DONATH, F.A., 1966, Experimental compaction effects in carbonate sediments: Jour. Sed. Petrology, v. 36, p. 747–754.

GARRISON, R.E., AND KENNEDY, W.J., 1977, Origin of solution-seams and flaser structure in Upper Cretaceous chalks of southern England: Sed. Geology, v. 19, no. 2, p. 107–137.

GILBERT, G.K., 1900, Rhythms and geologic time: Amer. Assoc. Adv. Sci. Proc., v. 49, p. 1–19.

GILLETT, S.L., 1983, Major, through-going stylolites in the Lower Ordivician Goodwin Limestone, southern Nevada: Petrography with dating from paleomagnetism: Jour. Sed. Petrology, v. 53, p. 209–219.

GINSBURG, R.N., 1953, Beach rock in south Florida: Jour. Sed. Petrology, v. 23, p. 85–92.

______,1956, Environmental relationships of grain size and constituent particles in some south Florida carbonate sediments: Amer. Assoc. Petroleum Geologists Bull., v. 40, p. 2384–2427.

______,1957, Early diagenesis and lithification of shallow-water carbonate sediments in south Florida, *in* R.J. Leblanc and J.G. Breeding (eds.), Regional Aspects of Carbonate Deposition: Soc. Econ. Paleontologists and Mineralogists, Spec. Pub. 5, p. 80–100.

GROSS, M.G., 1964, Variations in the O^{18}/O^{16} and C^{13}/C^{12} ratios of diagenetically altered limestones in the Bermuda Islands, Jour. Geology, v. 72, p. 170–194.

GROVER, G., JR., AND READ, J.F., 1978, Fenestral and associated vadose diagenetic fabrics of tidal flat carbonates, Middle Ordovician New Market Limestone, Southwestern Virginia: Jour. Sed. Petrology, v. 48, p. 453–473.

______, 1983, Paleoaquifer and deep burial related cements defined by regional cathodoluminescent patterns, Middle Ordovician carbonates, Virginia: Amer. Assoc. Petroleum Geologists Bull., v. 67, p. 1275–1303.

HALLAM, A., 1964, Origin of the limestone-shale rhythm in the Blue Lias of England: A composite theory: Jour. Geology, v. 72, p. 157–169.

HALLEY, R.B., 1981, Evolution of carbonate porosity during burial—Bahamas, Florida and Gulf Coast: Holocene to Jurassic [abs.]: Amer. Assoc. Petroleum Geologists Bull., v. 65, p. 2466–2467.

______, AND HARRIS, P.M., 1979, Freshwater cementation of a 1,000-year-old oolite: Jour. Sed. Petrology, v. 49, p. 969–988.

______, AND EVANS, C.C., 1983, The Miami Limestone: A guide to selected outcrops and their interpretation: Miami Geol. Soc., Miami, Fla., 67 p.

______, AND SCHMOKER, J.W., 1983, High-porosity Cenozoic carbonate rocks of south Florida: Progressive loss of porosity with depth: Amer. Assoc. Petroleum Geologists Bull., v. 67, p. 191–200.

HAMILTON, E.L., 1959, Thickness and consolidation of deep-sea sediments: Geol. Soc. America Bull., v. 70, p. 1399–1424.

HARMS, J.C., AND CHOQUETTE, P.W., 1965, Geologic evaluation of a gamma-ray porosity device, *in* 6th Annual SPWLA Logging Symposium, Transactions: Soc. Prof. Well Log Analysts, v. 2, p. C1–C37.

HARPER, M.L., AND SHAW, B.B., 1974, Cretaceous-Tertiary carbonate reservoirs in the North Sea: Offshore North Sea Technology Conference, Stavanger, Norway, Paper G IV/4, 20 p.

HARRISON, R.S., 1975, Porosity in Pleistocene grainstones from Barbados: Some preliminary observations: Bull. Canadian Petroleum Geol., v. 23, p. 383–392.

HATHAWAY, J.C., AND ROBERTSON, E.C., 1961, Micro-texture of artificially consolidated aragonite mud: U.S. Geol. Survey Prof. Paper 424-G, p. 301–304.

HESSE, R., AND BUTT, A., 1976, Paleobathymetry of Cretaceous turbidite basins of the East Alps relative to the calcite compensation level: Jour. Geology, v. 84, no. 5, p. 505–533.

HECKEL, P.H., 1983, Diagenetic model for carbonate rocks in midcontinent Pennsylvanian eustatic cyclothems: Jour. Sed. Petrology, v. 53, p. 733–760.

HUDSON, J.D., 1975, Carbon isotopes and limestone cement: Geology, v. 3, p. 19–22.

KAHLE, C., 1966, Compaction in oolites: Compass, v. 44, p. 19–29.

KEITH, M.L., AND WEBER, J.N., 1964, Carbon and oxygen isotopic composition of selected limestones and fossils: Geochim. Cosmochim. Acta, v. 28, p. 1787–1816.

KLOSTERMAN, M.J., 1981, Applications of fluid inclusion techniques to burial diagenesis in carbonate rock sequences: Louisiana State University, Applied Carbonate Research Program, Technical Series, Contribution #7, 101 p.

LAND, L.S., 1980, The isotopic and trace element geochemistry of dolomite: The state of art, *in* D.H. Zenger, J.B. Dunham, and R.L. Ethington (eds.), Concepts and Models of Dolomitization: Soc. Econ. Paleontologists and Mineralogists, Spec. Pub. 28, p. 87–110.

LAWLER, J.P., AND CRAWFORD, M.L., 1983, Stretching of fluid inclusions resulting from a low-temperature microther-

mometric technique: Econ. Geology, v. 78, p. 527–529.
LLOYD, R.M., 1977, Porosity reduction by chemical compaction—stable isotope model [abs.]: Amer. Assoc. Petroleum Geologists Bull., v. 61, p. 809.
LOCKRIDGE, J.P., AND SCHOLLE, P.A., 1978, Niobrara gas in eastern Colorado and northwestern Kansas: Rocky Mtn. Assoc. Geologists 1978 Symposium, p. 35–49.
LOGAN, B.W., AND SEMENIUK, V., 1976, Dynamic metamorphism; processes and products in Devonian carbonate rocks, Canning Basin, Western Australia: Geol. Soc. Australia Spec. Pub. 16, 138 p.
LONGMAN, M.W., 1980, Carbonate diagenetic textures from near surface diagenetic environments: Amer. Assoc. Petroleum Geologists Bull., v. 64, p. 461–487.
LORENS, R.B., 1981, Sr, Cd, Mn, and Co distribution coefficients in calcite as a function of calcite precipitation rate: Geochim. Cosmochim. Acta, v. 45, p. 553–561.
LOUCKS, R.G., 1977, Porosity development and distribution in shoal-water carbonate complexes—subsurface Pearsall Formation (Lower Cretaceous) south Texas, *in* D.G. Bebout and R.G. Loucks (eds.), Cretaceous Carbonates of Texas and Mexico: Texas Bureau Econ. Geology, Report of Investigations 89, p. 97–126.
MANCINI, E.A., AND BENSON, D.J., 1981, Smackover carbonate petroleum geology in southwest Alabama: Oil and Gas Journal, Dec. 28, 1981, p. 266–275.
MANUS, R.W., AND COOGAN, A.H., 1974, Bulk volume reduction and pressure-solution derived cement: Jour. Sed. Petrology, v. 44, p. 466–471.
MERRIAM, C.W., 1964, Cambrian rocks of the Pioche mining district, Nevada: U.S. Geol. Survey Prof. Paper 469, 59 p.
MEYERS, W.J., 1974, Carbonate cement stratigraphy of the Lake Valley Formation (Mississippian) Sacramento Mountains, New Mexico: Jour. Sed. Petrology, v. 44, p. 873–861.
______, 1978, Carbonate cements: their regional distribution and interpretation in Mississippian limestones of southwestern New Mexico: Sedimentology, v. 25, p. 371–400.
______, 1980, Compaction in Mississippian skeletal limestones, southwestern New Mexico: Jour. Sed. Petrology, v. 50, p. 457–474.
______, AND HILL, B.E., 1983, Quantitative studies of compaction in Mississippian skeletal limestones, New Mexico: Jour. Sed. Petrology, v. 53, p. 231–242.
MIMRAN, Y., 1975, Fabric deformation induced in Cretaceous chalks by tectonic stresses: Tectonophysics, v. 26, p. 309–316.
______, 1977, Chalk deformation and large-scale migration of calcium carbonate: Sedimentology, v. 24, p. 333–360.
MITRA, S., AND BEARD, W.C., 1980, Theoretical models of porosity reduction by pressure solution for well-sorted sandstone: Jour. Sed. Petrology, v. 50, p. 1347–1360.
MOORE, C.H., JR., 1973, Intertidal carbonate cementation, Grand Cayman, West Indies, Jour. Sed. Petrology, v. 43, p. 591–602.
______, AND DRUCKMAN, Y., 1981, Burial diagenesis and porosity evolution, Upper Jurassic Smackover, Arkansas and Louisiana: Amer. Assoc. Petroleum Geologists Bull., v. 65, p. 597–28.
MORELOCK, J., AND BRYANT, W.R., 1971, Consolidation of marine sediments, *in* R. Rezak and H.J. Vernon (eds.), Contributions on the Geological and Geophysical Oceanography of the Gulf of Mexico: Texas A & M University Oceanog. Studies, v. 3, p. 181–202.
MOSSOP, G.D., 1972, Origin of the peripheral rim, Redwater Reef, Alberta, Bull. Canadian Petroleum Geol., v. 20, p. 238–280.
MUCCI, A., AND MORSE, J.W., 1983, The incorporation of Mg^{2+} and Sr^{2+} into calcite overgrowths: influences of growth rate and solution composition: Geochim. Cosmochim. Acta, v. 47, p. 217–233.
NELSON, R.A., 1981, Significance of fracture sets associated with stylolite zones: Amer. Assoc. Petroleum Geologists Bull., v. 65, p. 2417–2425.
NELSON, R.C., 1974, Fluid inclusion study of the Fernvale Limestone, northern Arkansas: Unpubl. Masters Thesis, University of Missouri-Columbia, Columbia, Missouri, 95 p.
NEUGEBAUER, J., 1973, The diagenetic problem of chalk—the role of pressure solution and pore fluid: Neues Jahrbuch für Geol. Paläontol. Abh., v. 143, p. 223–245.
______, 1974, Some aspects of cementation in chalk, *in* K.J. Hsü, and H.C. Jenkyns (eds.), Pelagic Sediments: On Land and Under the Sea: Internat. Assoc. Sedimentologists Spec. Pub. 1, p. 149–176.
OGLESBY, T.W., 1976, A model for the distribution of manganese, iron and magnesium in anthigenic calcite and dolomite cements in the upper Smackover Formation in eastern Mississippi: Unpubl. MS Thesis, University of Missouri-Columbia, Columbia, Missouri, 122 p.
OLDERSHAW, A.E., 1971, The significance of ferroan and non-ferroan calcite cements in the Halkin and Wenlock Limestones (Great Britain), *in* O. P. Bricker (ed.), Carbonate Cements: Johns Hopkins University Studies in Geology, no. 19, p. 225–232.
______, AND SCOFFIN, T.P., 1967, The source of ferroan and non-ferroan calcite cements in the Halkin and Wenlock Limestones: Geol. Jour., v. 5, p. 309–320.
OTTMANN, R.D., KEYES, P.L., AND ZIEGLER, M.A., 1976, Jay Field, Florida—A Jurassic stratigraphic trap, *in* Braunstein, Jules, ed., North American Oil and Gas Fields: Amer. Assoc. Petroleum Geologists Memoir 24, p. 276–286.
PARK, W.C., AND SCHOT, E.H., 1968a, Stylolitization in carbonate rocks, *in* G. Müller and G.M. Friedman (eds.), Recent Developments in Carbonate Sedimentology in Central Europe: New York, Springer–Verlag, p. 64–74.
______, AND ______, 1968b, Stylolites: their nature and origin: Jour. Sed. Petrology, v. 38, p. 175–191.
PERRY, E., AND HOWER, J., 1970, Burial diagenesis in Gulf Coast pelitic sediments: Clays and Clay Minerals, v. 18, p. 165–177.

Pierson, B.J., 1981, The control of cathodoluminescence in dolomite by iron and manganese: Sedimentology, v. 28, p. 601–610.

Poland, E.L., 1982, Stretching of fluid inclusions in flourite at confining pressures up to 1 kilobar: MS Thesis, University of California, Berkeley, California, 90 p.

Pray, L.C., 1960, Compaction in calcilutites [abs.]: Geol. Soc. America Bull., v. 71, p. 1946.

Purser, B.H., 1978, Early diagenesis and the preservation of porosity in Jurassic limestones: Jour. Petroleum Geol., v. 1, p. 83–94.

Radke, R.M., and Mathis, R.L., 1980, On the formation of saddle dolomite: Jour. Sed. Petrology, v. 50, p. 1149–1168.

Richards, A.F., and Hamilton, E., 1967, Investigations of deep-sea sediment cores, III. Consolidation, *in* A.F. Richards (ed.), Marine Geotechnique: Proc. Internat. Res. Conf. on Marine Geotechnique, Urbana, Illinois, Univ. of Illinois Press, p. 93–117.

Robertson, E.C., 1967, Laboratory consolidation of carbonate sediment, *in* A.F. Richards (ed.), Marine Geotechnique: Internat. Res. Conf. on Marine Geotechnique, Urbana, Illinois, Univ. Illinois Press, p. 118–127.

Robin, P.F., 1978, Pressure solution at grain-to-grain contacts: Geochim. Cosmochim. Acta, v. 42, p. 1383–1389.

Roedder, E., 1979, Fluid inclusion evidence on the environment of sedimentary diagenesis, a review, *in* P.A. Scholle, and P.R. Schluger (eds.), Aspects of Diagenesis: Soc. Econ. Paleontologists and Mineralogists Special Publication 25, p. 89–108.

Runnells, D.D., 1969, Diagenesis, chemical sediments and the mixing of natural waters: Jour. Sed. Petrology, v. 39, p. 1188–1201.

Schlanger, S.O., and Douglas, R.G., 1974, The pelagic ooze-chalk-limestone transition and its implication for marine stratigraphy, *in* K.J. Hsü and H.C. Jenkyns (eds.), Pelagic Sediments: On Land and Under the Sea: Internat. Assoc. Sedimentologists Spec. Pub. 1, p. 117–148.

Schmidt, V., and McDonald, D.A., 1979, The role of secondary porosity in the course of sandstone diagenesis, *in* P.A. Scholle, and P.R. Schluger (eds.), Aspects of Diagenesis: Soc. Econ. Paleontologists and Mineralogists Spec. Pub. 26, p. 175–207.

Schmoker, J.W., and Halley, R.B., 1982, Carbonate porosity versus depth: A predictable relation for south Florida: Amer. Assoc. Petroleum Geologists Bull., v. 66, p. 2561–2570.

Scholle, P.A., 1971, Diagenesis of deep-water carbonate turbidites, Upper Cretaceous Monte Antola Flysch, northern Appennines, Italy: Jour. Sed. Petrology, v. 41, p. 233–250.

______, 1977, Chalk diagenesis and its relation to petroleum exploration: Oil from chalk, a modern miracle?: Amer. Assoc. Petroleum Geologists Bull., v. 61, p. 982–1009.

______, and Arthur, M.A., 1980, Carbon isotope fluctuations in Cretaceous pelagic limestones: Potential stratigraphic and petroleum exploration tool: Amer. Assoc. Petroleum Geologists Bull., v. 64, p. 68–87.

______, ______, and Ekdale, A.A., 1983, Pelagic environment, *in* P.A. Scholle, D.G. Bebout, and C.H. Moore (eds.), Carbonate Depositional Environments: Amer. Assoc. Petroleum Geologists Memoir 33, p. 619–690.

Shinn, E.A., Halley, R.B., Hudson, J.H., and Lidz, B.H., 1977, Limestone compaction: An enigma: Geology, v. 5, p. 21–24.

______, and Robbin, D.M., 1983, Mechanical and chemical compaction in fine-grained shallow-water limestones: Jour. Sed. Petrology, v. 53, p. 595–618.

Stockdale, P.B., 1922, Stylolites: their nature and origin: Indiana Univ. Studies, v. 9, p. 1–97.

______, 1926, The stratigraphic significance of solution in rocks: Jour. Geology, v. 34, p. 399–414.

Stueber, A.M., and Pushkar, P., 1983, Application of strontium isotopes to origin of Smackover brines and diagenetic phases, southern Arkansas [abs.]: Amer. Assoc. Petroleum Geologists Bull., v. 67, p. 553–554.

Terzaghi, R.D., 1940, Compaction of lime mud as a cause of secondary structure: Jour. Sed. Petrology, v. 49, p. 437–462.

Trurnit, P., 1968, Analysis of pressure solution contacts and classification of pressure-solution phenomena, *in* G. Müller, and G.M. Friedman (eds.), Recent Developments in Carbonate Sedimentology in Central Europe: Berlin, Springer, p. 75–84.

Veizer, J., and Compston, W., 1974, $^{87}Sr/^{86}Sr$ composition of seawater during the Phanerozoic: Geochim. Cosmochim. Acta, v. 38, p. 1461–1484.

Vinopal, R.J., and Coogan, A.H., 1978, Effect of particle shape on the packing of carbonate sands and gravels: Jour. Sed. Petrology, v. 48, p. 7–24.

Wagner, P.D., and Matthews, R.K., 1982, Porosity preservation in the Upper Smackover (Jurassic) carbonate grainstone, Walker Creek field, Arkansas: Response of paleophreatic lenses to burial processes: Jour. Sed. Petrology, v. 52, no. 1, p. 3–18.

Wanless, H.R., 1979, Limestone response to stress: Pressure solution and dolomitization: Jour. Sed. Petrology, v. 49, p. 437–462.

Waples, D.W., 1980, Time and temperature in petroleum formation: Application of Lopatin's method to petroleum exploration: Amer. Assoc. Petroleum Geologists Bull., v. 64, p. 916–926.

Weller, J.M., 1959, Compaction of sediments: Amer. Assoc. Petroleum Geologists Bull., v. 43, p. 273–319.

Weyl, P.K., 1959, Pressure solution and the force of crystallization—a phenomenological theory: Jour. Geophys. Res., v. 64, p. 2001–2025.

Wong, P.K., and Oldershaw, A., 1981, Burial cementation in the Devonian, Kaybob Reef complex, Alberta, Canada: Jour. Sed. Petrology, v. 51, p. 507–520.

Wolfe, M.J., 1968, Lithification of a carbonate mud: Senonian chalk in northern Ireland: Sed. Geology, v. 2, p. 263–290.

Zankl, H., 1969, Structural and textural evidence of early lithification in fine-grained carbonate rocks: Sedimentology, v. 12, p. 241–256.

SECTION FOUR: INTRODUCTION

Examples of nonmarine cements representing either the beginning or end of the carbonate diagenetic cycle, are shown in this final part. The diverse nature of fresh-water environments, variations in cement mineralogy and morphology and their identifying features are discussed by Chafetz, Wilkinson and Love. Although volumetrically insignificant, such cements have to be identified and separated from neomorphic textures resulting from meteoric diagenesis. Folk, Chafetz and Tiezzi describe uniquely shaped and crystallized cements from hot-spring travertines in central Italy and offer an unexpected glimpse of the complexity of carbonate precipitation processes and the extreme diversity in cement morphology. The paper by Mimran illustrates cementation controlled by tectonic features in chalks from Israel. Faulted chalks are friable because of solution by meteoric waters drained through these faults. This calcium carbonate reprecipitates as thick veins and as calcite spar cement in the matrix of unfaulted chalk.

N.S., P.M.H.

MORPHOLOGY AND COMPOSITION OF NON-MARINE CARBONATE CEMENTS IN NEAR-SURFACE SETTINGS

HENRY S. CHAFETZ[1], BRUCE H. WILKINSON[2], AND KAREN M. LOVE[1]
[1]Department of Geosciences, University of Houston, Houston, Texas 77004; [2]Department of Geological Sciences, University of Michigan, Ann Arbor, Michigan 48109

ABSTRACT

Non-marine (essentially freshwater) carbonate cements form in a diverse assemblage of near-surface settings, including pedologic, lacustrine, spelean, and spring-fluvial environments, giving rise to a plethora of textures and structures. Many deposits exhibit both phreatic and vadose zone textures. Cement mineralogy and composition vary considerably from deposit to deposit, as well as within individual deposits, depending on water chemistry and environmental setting. Similarly, cement habits range from highly acicular to nearly equant. The wide variety of textures, mineralogies, and compositions exhibited by both cement and associated sediments suggests that freshwater carbonate precipitation may involve complex processes. A survey of our present knowledge indicates that such cements are most commonly composed of low-magnesium calcite as crystals with rhombohedral terminations. Among calcite cement features commonly forming in freshwater but rarely in marine settings are rhombohedrons, trigonal prisms, and crystals with thorn-shaped vacuoles. However, more significant than the presence of any individual feature is the great complexity of carbonate cements that are commonly displayed within a sample. This complexity provides the surest means of identifying a carbonate cement as non-marine in origin. Although variation in crystal habit may be influenced by either magnesium or total cation or anion concentrations in the precipitating fluid, the concentration of solutes does not appear to be the sole controlling factor. Growth rate, influenced by a variety of parameters, such as P_{CO_2}, may be the most important factor in predicting crystal habits in freshwater carbonate cements.

INTRODUCTION

The purpose of this investigation was to try to determine what, if any, forms of Ca-Mg carbonate cement are indicative of precipitation in the non-marine (generally freshwater) realm. This topic is both timely and significant in scope in that most limestones accumulate in shallow marine to intertidal areas (cf. Wilson, 1975) and, therefore, are frequently affected by meteoric diagenesis in vadose and phreatic settings. Because near-surface diagenesis under the influence of non-marine water may profoundly affect many shallow marine carbonate deposits, recognition of non-marine cement is significant; the present state of the art of carbonate petrology suggests a need for such a synthesis at this time.

We approached this task by primarily reviewing investigations of cements from strictly non-marine settings and have used studies that recognized the effects of freshwater on marine deposits in only a complementary fashion. A similar approach was recommended by Friedman and Kolesar (1971). The Ca-Mg carbonate cements that we have focused on are the most common ones (i.e., low-magnesium calcite, high-magnesium calcite, aragonite, and dolomite); this group of minerals will be referred to as the carbonate(s) or carbonate cement(s). Our primary concern was with the mineralogy, crystal morphology, and size, and not with larger scale features of gross cement forms such as stalactites, crusts, etc. In addition, we have focused primarily, though not exclusively, on recent deposits in order to maximize our ability to recognize varieties of cements that have unquestionably formed under non-marine conditions.

Carbonate cements form in a host of non-marine settings. In order to facilitate discussion, the non-marine environments have been grouped into the following categories: pedologic, lacustrine, spelean, and spring-fluvial. In many cases the assignment of a particular deposit to one category rather than another is quite arbitrary. This is a consequence of three factors: (1) individual deposits may form by a variety of processes (e.g., cementation within pores in caliche deposits that range from micropores to decimeter size cavities (small caves)), (2) very similar processes may be important in different settings; in such instances there is no real meaningful distinction as far as the processes of carbonate precipitation are concerned between various deposits from different environments, and (3) some deposits form in settings transitional between the four major groupings employed here (e.g., sublacustrine travertine) and, as such, do not readily fit into a particular category. Another problem we have encountered is the general paucity of petrologic detail concerning carbonate cements from non-marine environments.

PEDOLOGIC DEPOSITS

Under the rubric of pedologic deposits we include material that formed due to a variety of processes which most commonly result in the precipitation of salts due to evaporation in the near-surface setting (e.g., nari, caliche, calcrete). The most commonly precipitated carbonate within pedologic units, in general, is a micritic low-magnesium calcite. However, the mineralogy, crystal size, and morphology of individual grains which comprise these deposits vary greatly. The carbonate comprising the precipitated material within an individual pedologic deposit may be all low-magnesium calcite (James, 1972; Hay and Reeder, 1978; Semeniuk and Meagher, 1981), both low- and high-magnesium calcite (Read, 1974; Stalder, 1975), calcite and aragonite (Panos and Stelcl, 1968; Konishi and Sakai, 1972), calcite and dolomite (Arakel and McConchie, 1982), or low- and high-magnesium calcite, aragonite, and dolomite (Watts, 1980). In addition, a host of other authigenic precipitates, such as gypsum, opal, sepiolite, etc. may accompany the carbonates (Arakel and McConchie, 1982). Individual crystals range in size from micritic (Semeniuk and Meagher, 1981; Arakel, 1982) to bladed calcite crystals up to 2.5 cm long (Tandon and Narayan, 1981). Crystal morphology ranges from needle-fibrous (James, 1972) to bladed (Harrison, 1977) to equant and blocky (Knox, 1977), to list a few of the adjectives found in the literature, and all of these morphologies have been used to describe crystals of low-magnesium calcite. It is evident that the mineralogy, size, and morphology of authigenic carbonate in pedologic environments are highly variable and, in general, overlap with carbonate crystals originating in the submarine realm.

There are attributes of pedological carbonates, on the other hand, that have not been described or appear to be relatively rare in marine counterparts. These are well-developed rhombohedrons, trigonal prisms, needle-fibre crystals, flower spar, micrite bridges, meniscus and dripstone or gravitational cements. Low-magnesium calcite crystals with well-developed rhombohedral morphology range in size from 1–2 μm (Klappa, 1979) to 6–11 μm (Folk, 1971) to 15–40 μm (Chafetz and Butler, 1980; (Fig. 1 herein). In addition, Wright (1982, p. 20) has described "rhombic calcite crystals which may reach lengths up to 200 μm". . . from Lower Carboniferous calcrete deposits. Reports of calcite crystals with trigonal prism morphologies are rare in the literature, but have been recognized from pedologic units (Fig. 2 herein; Chafetz and Butler, 1980) as well as from several other non-marine settings (Binkley *et al.*, 1980; Schreiber *et al.*, 1981). Needle-fibre crystals can form as either a mesh of acicular, randomly oriented needles in void space (Fig. 3) or may tangentially coat particles (James, 1972); they are volumetrically very small and range in size from 1–5 μm in diameter and between 10–400 μm in length (James, 1972; Harrison, 1977). Needle-fibres may be the result of precipitation either in or along fungal hyphae (Harrison, 1977) and therefore, may not constitute a true cement.

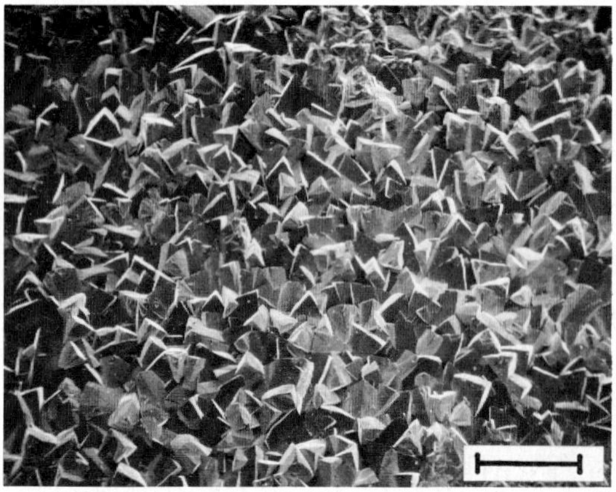

FIG. 1.—Scanning electron micrograph of well-developed rhombohedral low-magnesium calcite crystals which have grown on the surface of a thin recent caliche crust developed on the Edwards Formation of central Texas. Scale is 1.0 mm long.

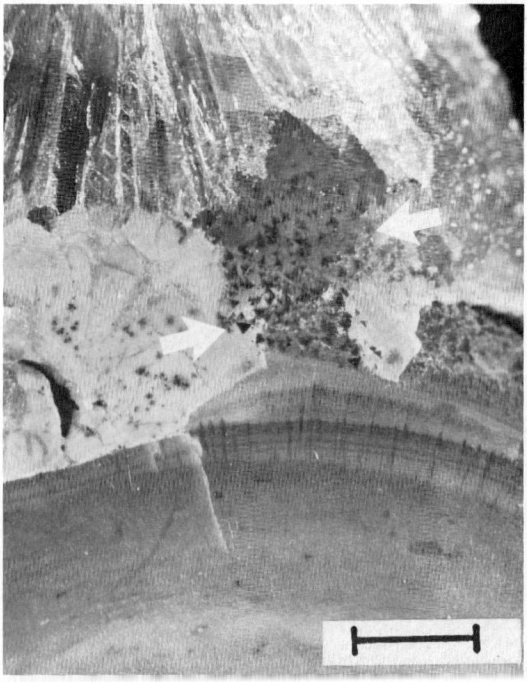

FIG. 2.—Etched slab of recent pisolitic caliche developed on the Edwards Formation of central Texas. Numerous spar crystals display a cross-section of their trigonal prism morphology (arrows). Scale is 0.5 cm long.

Flower spar, micrite bridges, meniscus cement and dripstone or gravitational textures are properties of aggregates of cement crystals rather than of individual crystals. Flower spar refers to radiating or fan-shaped patterns developed as bladed crystals grow outward from a common point on a substrate (James, 1972; Harrison, 1977). Micrite bridges (Knox, 1977) are localized deposits of bundles of rods or fibers that occur at grain contacts; fibers are approximately 10 μm in length and maximum length of the bridges is approximately 300 μm. Meniscus cements (Dunham, 1971) are similar to micrite bridges, but are a more commonly recognized form of cement in the vadose zone (Knox, 1977; Halley and Harris, 1979; Tandon and Narayan, 1981; Arakel and McConchie, 1982; and many others). Commonly associated with meniscus cements are dripstone, gravitational, or microstalactitic cement (Fig. 4 herein; Muller, 1971; Ward, 1975; Stalder, 1975; Chafetz and Butler, 1980; Tandon and Narayan, 1981; etc.). Compilation of the information briefly outlined above strongly reinforces the statement by Harrison (1977, p. 139) that "One of the most diagnostic criteria of caliche is its varied and often complex pattern of cementation, a reflection of the complex and variable diagenetic processes attendant on the near-surface environment. . . Crystal shape may be equant, bladed, radially bladed, fibrous, or needle-like; average crystal size can range from a few microns in diameter to lengths exceeding 100 μm. This tremendous variability is commonly visible in a single thin section of caliche . . . Multiple, petrographically distinguishable generations of cement within a single void are common in the caliche profiles".

LACUSTRINE DEPOSITS

Carbonate deposits that have formed in lacustrine environments display a range in mineralogies, crystal sizes, and morphologies similar to those described for pedologic deposits. Muller *et al.* (1972) summarized water chemistry and carbonate mineralogy data for 25 lakes scattered from Europe through the Middle East and into Africa. The data from that and other studies demonstrate that the common carbonate minerals (calcite, both low- and high-magnesium, aragonite, and dolomite) as well as rarer carbonates (huntite and magnesite), readily form within lake environments (Muller *et al.*, 1972; Hsu and Kelts, 1978; Muller and Wagner, 1978; Riding, 1979). In comparing carbonate minerals

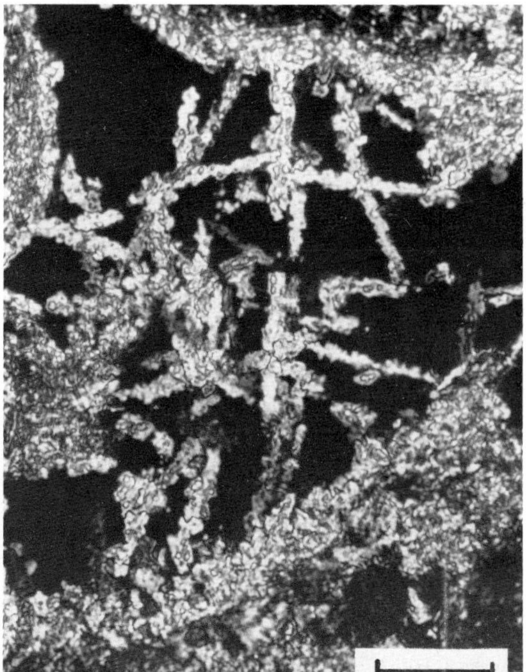

FIG. 3.—Fibrous whisker crystals of low-magnesium calcite precipitated within an intergranular pore in a calichied unit of the Pleistocene Key Largo Limestone, Florida. Scale is 100 μm long.

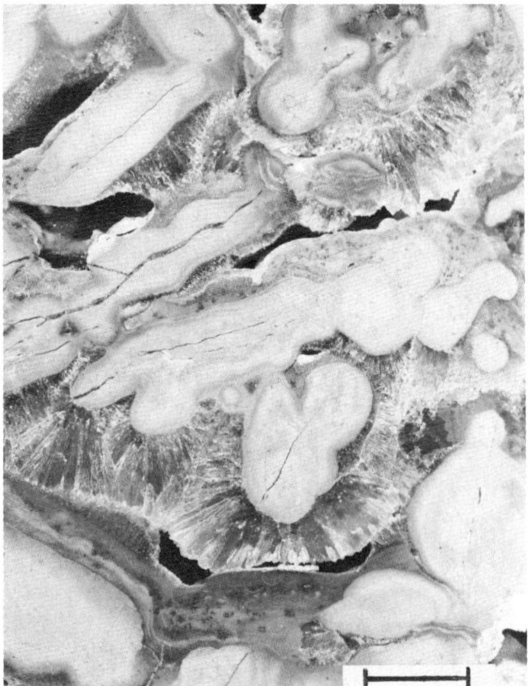

FIG. 4.—Etched slab of nodular caliche developed on the Edwards Formation of central Texas. Bladed low-magnesium calcite spar is best-developed in pendulous fashion with internal geopetal sediment resting on the tops of many of the nodules. Scale is 2 cm long.

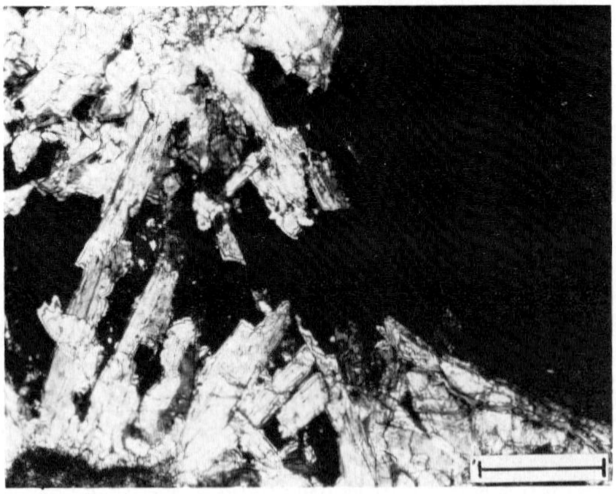

FIG. 5.—Meniscus cement in pisolitic gravel along the shore of Ore Lake, Michigan. These elongate crystals are exclusively low-magnesium calcite and were precipitated as a beachrock cement in a vadose setting. The pore at the right was larger, and contained a gas bubble whereas the smaller pore at the left was filled with fluid during cement precipitation. Scale is 0.2 mm long.

deposited in lacustrine environments with those in marine environments, Muller *et al.* (1972, p. 163) concluded that "the only difference in carbonate mineralogy between the different environments is that low-Mg calcite does not form under marine (or marine-derived) conditions". This they attributed to the relatively constant, high Mg/Ca ratios which occur in marine environments. Because it is now known that low-magnesium calcite may form in marine setting as well (e.g., Schlager and James, 1978; Rao, 1981) there is, in fact, no qualitative difference in the mineralogy of marine and lacustrine carbonates. The size and morphology of lacustrine carbonate crystals range from micrite (Boyer, 1981) to blocky subhedral spar, to elongate spikes and "tooth-like" crystals up to 200 μm in length (Riding, 1979). Lacustrine cements deposited subaqueously can display isopachous textures analogous to submarine cements (Riding, 1979), while those formed in vadose environments along lake margins can exhibit meniscus fabrics (Fig. 5 herein; Binkley *et al.*, 1980). In addition, both trigonal prisms (Fig. 6 herein; Binkley *et al.*, 1980) and well-developed rhombohedra (Fig. 7 herein; Boyer, 1981) occur in lacustrine deposits.

FIG. 6.—Elongate low-magnesium calcite crystal from Ore Lake, Michigan. These vadose cements (cf. fig. 5) are, in fact, aggregate crystals of trigonal prisms, each with a rhombohedral termination, which together form a larger trigonal prism such as the one shown here. Scale is 5 μm long.

FIG. 7.—Rhombohedral low-magnesium calcite spar lining a pore in a lacustrine grainstone unit from the mid-Tertiary Camp Davis Formation in northwestern Wyoming. Scale is 50 μm long.

Unlike pedologic carbonate cement, which largely forms as a pore-filling phase, lacustrine carbonate crystals may precipitate in three general settings: (1) as isolated crystals and as aggregate crystal encrustations on nannoplankton in warm epilimnic water during summer months (e.g., Hsu and Kelts, 1978), (2) as isolated crystals and as aggregate crystal encrustations on macrophytes in shallow benthic settings (e.g., Schottle and Muller, 1968), and (3) as pore-filling cement in vadose and phreatic settings below the sediment surface (e.g., Eggleston and Dean, 1976). Lacustrine settings are commonly sites of carbonate cement precipitation literally at the sediment-water interface. In these instances, cement may grow on the sediment surface either as equant (Figs. 8, 9) or as acicular crystals (Fig. 10). Although such benthic cements have been reported from some ancient marine sequences, they are fairly common in many lacustrine settings.

SPELEAN DEPOSITS

White (1976) listed 76 different minerals that have been identified as forming within caves due to chemical precipitation. Low-magnesium calcite is the most common, followed by gypsum and aragonite (White, 1976); dolomite is rare. There is a paucity of reports citing the presence of high-magnesium calcite speleothems (Rogers and Williams, 1982), however, this lack of recognition may be due in large part to a lack of interest in discriminating between calcite and low- and high-magnesium calcite by many of those who study caves and speleothem deposits. In addition to the other forms of carbonate, lublinite, a variety of calcite composed of *en echelon* stacked needles of calcite, has been identified in some cave deposits (Stoops, 1976; Bogli, 1980). It is not uncommon for some cave deposits, such as stalactites, to display alternating layers or rings of calcite and aragonite (Siegel, 1965); therefore mineralogy can vary not only from cave to cave and within a cave, but also within specific speleothems. One aspect of the carbonate comprising speleothems which appears to be almost universal is a characteristic yellow color, which primarily is due to included humic substances derived from the overlying soils (White, 1981).

Crystal size and morphology also display great variability within cave deposits (Hill, 1976) and probably exhibit the greatest range in size of any near-surface

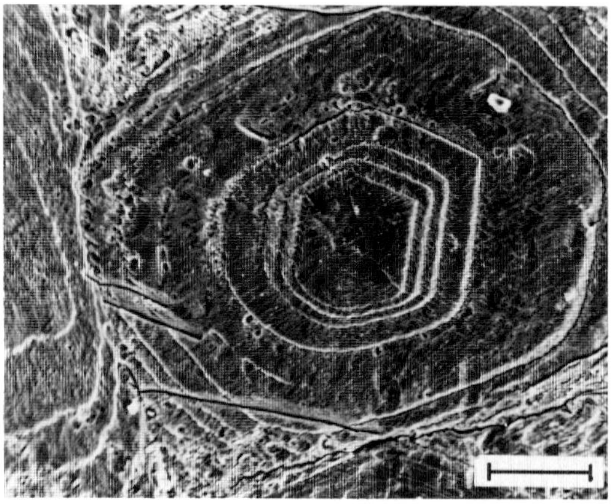

FIG. 9.—Section through a benthic cement crystal from Pyramid Lake, Nevada. Observe that early crystal faces were euhedral and hexagonal, whereas later faces became progressively more rounded. Scale is 30 μm long.

FIG. 8.—Nearly equant subhedral crystals of benthic low-magnesium calcite cement from the shallow shoreface of Pyramid Lake, Nevada. Scale is 10 μm long.

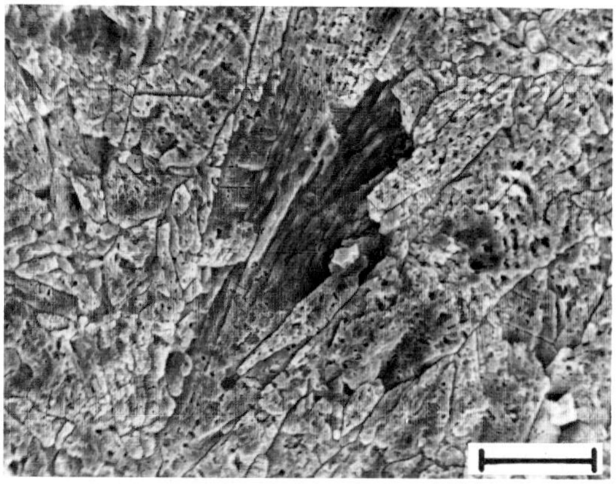

FIG. 10.—Elongate rays of benthic calcite cement which precipitated at the sediment-water interface in shallow sublacustrine portions of the mid-Tertiary Lake Camp Davis Formation, northwestern Wyoming. Scale is 30 μm long.

environment, ranging from micron-sized crystals to entire stalactites composed of a single crystal (White, 1976; Rogers and Williams, 1982). Portions of caves with poor air communication with the surface generally maintain constant temperatures, high humidity, and a P_{CO_2} in the air closer to that of the percolating solutions than normal atmospheric conditions, providing an excellent environment for the continuous deposition of carbonate and, therefore, the growth of large crystals (Fig. 11). Individual crystals range from acicular (Kendall and Broughton, 1978) to columnar crystals with rhombohedral terminations (Folk and Assereto, 1976) to scalenohedrons and rhombohedrons (Wells, 1971). Steep scalenohedrons are probably the most common crystal form of calcite in caves with the crystals ranging in size from a few microns to well-developed forms up to 15 cm in length (Bogli, 1980). Bogli (1980, p. 181) also has found that, in general, "rhombohedrons are quite rare". However, Wells (1971) and Rogers and Williams (1982) have found rhombohedrons to be a common form of calcite in the caves they have investigated. Wells (1971, p. 145) observed that when rhombohedrons and scalenohedrons occur in the same pool "the rhombohedral crystals are invariably the oldest . . . very frequently there is a change of habit and this invariably takes the form of scalenohedra growing on the faces of previously established rhombohedra or scalenohedra". He suggests that this common occurrence is due to a change in the chemical and/or physical environment. Another crystal form of calcite comprising speleothems is the trigonal prism. Kendall and Broughton (1978) mention its occurrence and, although not described as a trigonal prism in the texts, line drawings and photographs in Wells (1971, figs. 30C and 53) and in Rogers and Williams (1982, fig. 10) hint at its presence. In addition to the variations in mineralogy, crystal morphology, etc., calcite comprising speleothems can be either length-fast or length-slow (Folk and Assereto, 1976)(e.g., lublinite); however, this variation is not unique to non-marine cements (Chafetz, 1979). Other features recognized within speleothems are the common textural characteristics of cements formed in a vadose zone, e.g., gravitational and meniscus textures. In addition, gas- and/or liquid-filled vacuoles appear to be quite common as dense zones (Folk and Assereto, 1976; Kendall and Broughton, 1978) and/or as solitary thorn-shaped features (Fig. 12 herein; Kendall and Broughton, 1978). The complexity of cement crystal compositions and morphologies typical of pedologic carbonates also is characteristic of cements formed in the spelean environment. Cave carbonates display a great range in mineralogy, crystal size, and morphology and commonly these variations can be seen within a very small sample - a case in point is the 3cm thick piece of flowstone described by Folk and Assereto (1976).

SPRING—FLUVIAL DEPOSITS

There is a pronounced paucity of studies of recent cements from spring (travertine) and fluvial deposits. The few papers which we are aware of have found only low-magnesium calcite as the cementing phase, although we anticipate the finding of other carbonate phases. Nevertheless, the crystal size and morphology range widely within these few deposits. Crystals range from equant micrite to coarse prismatic crystals up to 300 μm wide and approximately 2.5 mm in length (Meredith, 1980). These crystals display the full range of morphologies cited for other environments, from acicular fibers to coarse prismatic crystals to equant grains (Braithwaite, 1979; Meredith, 1980). In addition, these crystals may display well-developed rhombohedral and scalenohedral forms. Trigonal prisms (Meredith, 1980; Schreiber *et al.*, 1981) and thorn-shaped inclusions are common within travertines. These deposits also show the features common in vadose cements, such as meniscus and gravitational textures (Meredith, 1980), as well as those in phreatic cements, such as isopachous rims of calcite rhombs (Geno and Chafetz, 1982). For a detailed description of some of the more exotic crystal forms that occur within travertines see Folk *et al.* (this volume).

FRESHWATER CARBONATE IN MARINE SEDIMENTS

A number of investigations have highlighted the effects of freshwater on recent marine sediments (e.g., Gavish and Friedman, 1969; Land, 1970; Schroeder, 1973; Steinen, 1974; Ward, 1975; Pingitore, 1976; Hattin and Dodd, 1978; Halley and Harris, 1979; Buchbinder and Friedman, 1980).

The stabilization of aragonite and high-magnesium calcite marine carbonates occurs through the dissolution of metastable phases and the precipitation of crystals of low-magnesium calcite. Such stabilization has commonly been perceived as taking place by one of two different diagenetic processes: (1) the complete dissolution of marine components and subsequent filling of either the resultant void or intergranular porosity by calcite, and (2) the dissolution of metastable components and the concomitant precipitation of calcite immediately adjacent to sites of aragonite dissolution (Bathurst, 1975). Calcite precipitated during the former process has frequently been termed "void-filling" calcite cement, while that formed during the latter process has commonly been referred to as "neomorphic calcite" (e.g., Folk, 1965; Bathurst, 1975). Although this subdivision implies that very different processes are operating, the two are commonly no more than portions of a continuum, defined on the basis of the size of the fluid-filled void that separates the dissolving and precipitating phases. If the volume of the fluid phase is sufficiently small that primary fabric elements such as solid (i.e., organic, argillaceous, or aragonite) inclusions are incorporated in the replacing calcite, then resultant fabrics are classically neomorphic. If, on the other hand, the volume of the fluid phase is large then void space will be filled with calcite cement, the fabrics of which are classically void-filling. Both fabrics, however, arise through the dissolution of marine carbonate and calcite precipitation. There is little difference, save scale, between the two.

The cements that have been associated with flushing of marine carbonate sediments by meteoric waters are essentially all low-magnesium calcites. These cements range from micron-sized crystals (Hattin and Dodd, 1978) to equant crystals up to several millimeters in diameter (Pingitore, 1976), while crystal morphology varies from fibrous 'whisker' crystals (Supko, 1971) or needle-fibers (Ward, 1975) to prisms to rhombohedrons (Fig. 13 herein; Steinen, 1974). Isopachous cement is common in the phreatic zone while meniscus (Fig. 14) and gravitational cements are common in the vadose zone (Land, 1970; Schroeder, 1973; Ward, 1975; Halley and Harris, 1979; Buchbinder and Friedman, 1980). Replacive freshwater calcite cements have been reported in many studies; several of these occurrences are as poikilotopic crystals. Gavish and Friedman (1969), Buchbinder and Friedman (1980) and Schneidermann *et al.* (1972) have interpreted such occurrences as neomorphic spar. Pingitore (1976) has described similar cements and interpreted them as precipitates which resulted from the recrystallization of a fine-grained (corallite) precursor. These form calcite mosaics which cross-cut original coral fabrics; single replacive crystals may extend outward from their substrates as void-filling cement (Fig. 15). Replacive crystal size is coarse; crystals several millimeters in diameter are not uncommon (Pingitore, 1976). In all three studies in which replacive poikilotopic cement was described, it developed in freshwater zones very early in the rocks' diagenetic history. Poikilotopic cements have not been described, to our knowledge, from the vadose zone and correlative with this, cementation, in general, is much more pervasive in the phreatic than in the vadose zone (Land, 1970; Pingitore, 1976). Marine sediments cemented under the influence of freshwater display the same complexity of cementation pattern recognized in the other depositional settings discussed above. "Within the same 1 or 2 mm of a thin section, some grains are rimmed with cement and others apparently lack cement; some pores are filled and some remain empty; some pores are rounded (meniscus cement) and others are lined with dog-tooth blades; some grains have microstalactitic druse and others are coated with radiating blocks of cement; etc." (Ward, 1975, p. 546).

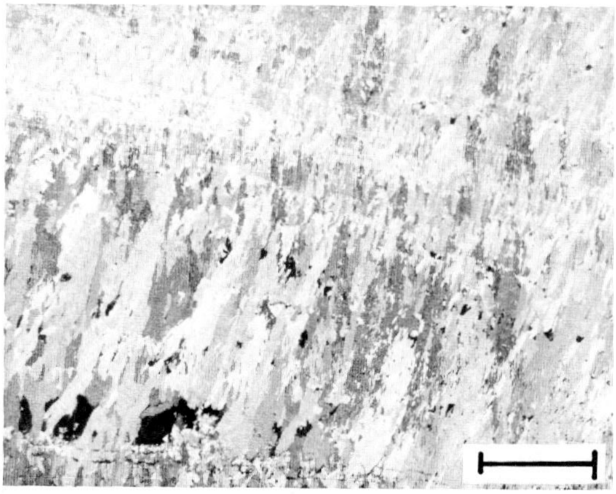

FIG. 11.—Photomicrograph of large columnar low-magnesium calcite crystals comprising a speleothem crust developed within the Edwards Formation of central Texas. Observe the feathery edges of the crystals and the fact that single crystals cross the banded structure of the crust. Scale is 5 cm long.

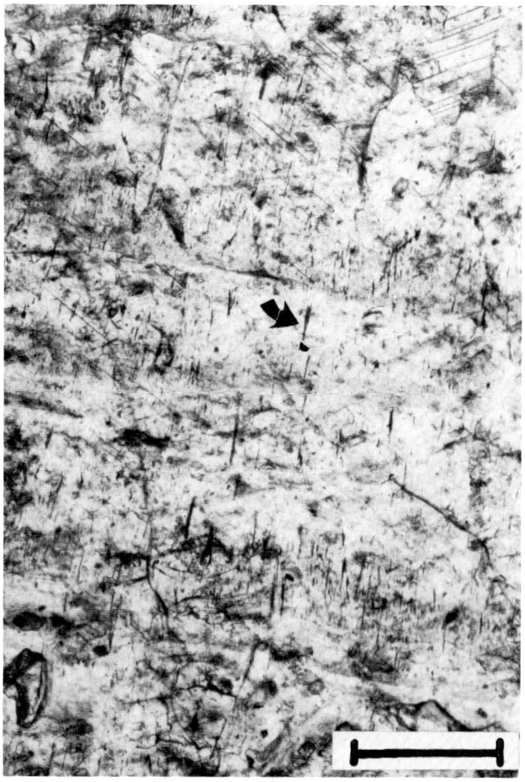

FIG. 12.—Photomicrograph of individual (arrow) and layers of thorn-shaped inclusions in banded speleothem crust developed within the Edwards Formation of central Texas. Scale is 0.2 mm long.

DISCUSSION AND SUMMARY

Based on these, and other studies, it is apparent that non-marine cements come in a wide range of mineralogies, crystal morphologies, and sizes, and therefore, exhibit a considerable degree of overlap with respect to the characteristics of marine cements. This should come as no surprise, because the waters from which these non-marine cements precipitate are highly variable, and may be compositionally analogous to waters in the marine environment. Nevertheless, particular morphological and mineralogical attributes of many non-marine cements are either rare or not reported from the shallow marine realm.

The most widely cited criterion for the recognition of freshwater cements, as compared with marine precipitates, is the dominance of low-magnesium calcite, especially displaying a well-developed rhombohedral form (Muller *et al.*, 1972; Folk, 1974; Badiozamani *et al.*, 1977). This criterion still appears to be valid in most cases, in spite of the fact that rhombohedral crystals of low-magnesium calcite can form in cool and/or deep marine settings (Schlager and James, 1978; Rao, 1981). The presence of needle-fibre or whisker crystal cements, although reported from several vadose settings, holds little promise for recognition in the rock record due to the size of the crystals and the likelihood that they either will be obliterated or completely enveloped and therefore hidden by later stages of cement precipitation. Similarly, lublinite holds little value due to its rarity in nature. The common presence of trigonal prisms of calcite within various non-marine settings suggests that this morphology may be useful for the distinction of marine from non-marine environments. However, A.C. Kendall (pers. comm., 1983) has found trigonal prisms that have formed under marine conditions. The large crystal size found comprising many speleothems, crystals on the order of many centimeters in length, may be useful for the recognition of spelean environments. However, large crystals can form in any environment of continuous precipitation within cavities. Numerous occurrences of coarse crystals have been reported as marine cement in reefs. Thorn-shaped liquid- and/or gas-filled

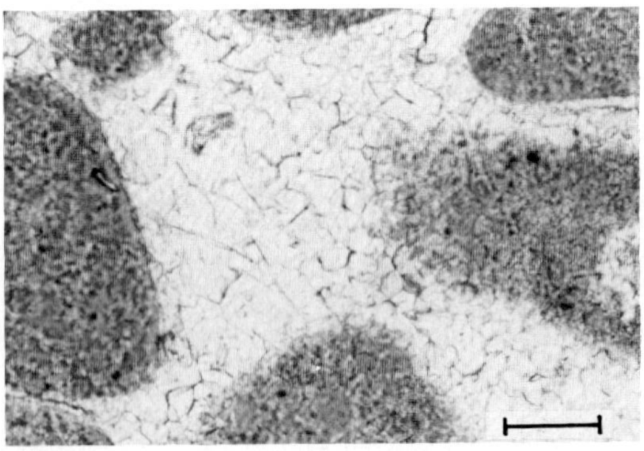

FIG. 13.—Photomicrograph of rhombohedral low-magnesium calcite spar, dark areas are oomoldic pores filled with blue-dyed epoxy. Sample from the Pleistocene Miami Oolite. Scale is 50 μm long.

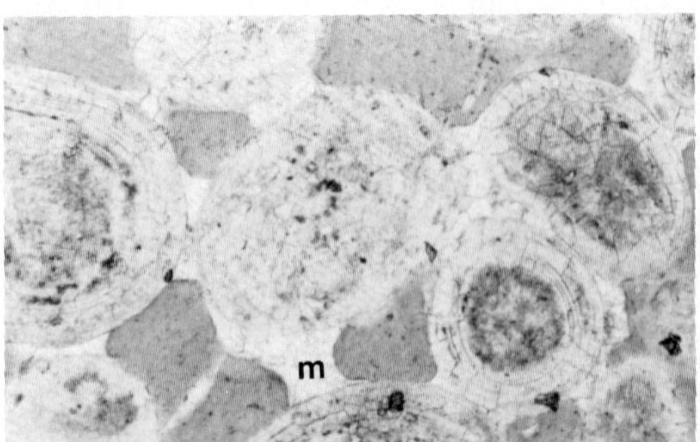

FIG. 11. Photomicrograph of recent oolitic material from the Bahamas displaying meniscus cement (m) at the sites of proximity of the allochems. Darker areas are blue-dyed epoxy-filled pore spaces; observe the rounding of the pore spaces due to the meniscus cement. Scale is 100 μm long.

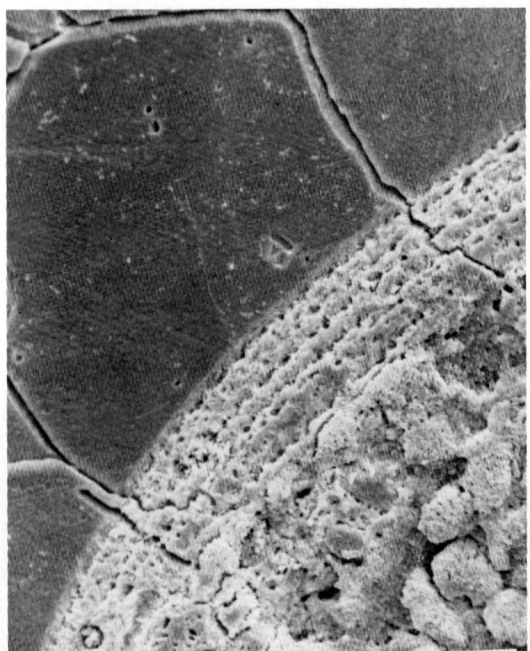

FIG. 15.—Partially replaced ooid from the Pleistocene Miami Oolite, Florida. This unit was subjected to meteoric fluids and is now infilled by equant low-magnesium calcite spar. Observe that the void-filling cement and the neomorphic calcite are in crystallographic registry such that common boundaries define both cement crystals as well as crystals of neomorphic spar. Scale is 10 μm long.

vacuoles that commonly occur within crystals comprising speleothem and travertine deposits indicate subaerial exposure. Although subaerial exposure does not of itself mean that the cements precipitated from freshwater, it is certainly a strong possibility. Other characteristics that are diagnostic of freshwater cementation are various textures that form in the vadose environment, such as meniscus and gravitational cements.

The nature of non-marine carbonate cements discussed above are, owing to the variety of settings in which they originate, little more than generalizations. Such cements are heterogeneous not only with respect to their mineralogy and composition, but also with respect to their habits and morphologies. This form variation is not solely a function of composition; very similar morphologies may develop in phases that are mineralogically and/or compositionally distinct while very different morphologies may characterize an individual cement phase.

From a compositional standpoint, the cement phase which is precipitated in any particular non-marine (or marine) setting is largely (but not entirely) determined by the composition of the fluid in which it forms. Muller *et al.* (1972) were perhaps the first to well-document the fact that the formation of primary and early diagenetic carbonate phases is largely dependent on the relative activity of Mg and Ca in environmental (or pore) waters, and suggested that phase variation was largely independent of the ionic strength of ambient fluids. Folk (1974) and many other workers (e.g., Longman, 1980) have generally accepted and expanded on this general premise; the composition and mineralogy of carbonate cement is largely determined by compositional parameters of precipitational fluids. Data on non-marine carbonate cements strongly support the general validity of this relationship.

From a morphologic standpoint, however, the relationships between fluid composition, cement composition, and cement crystal habit and morphology are almost completely unknown. Several authors have speculated on the nature of the relationships which may exist between the mineralogy and habit of carbonate cements and the Mg/Ca activities and/or the cation/anion activities of the surrounding fluids (Katz, 1973; Lippmann, 1973; Folk, 1974; Lahann, 1978). Despite the fact that, in general, marine water with high Mg/Ca and surface absorbed cation/anion ratios precipitate acicular high-magnesium calcite and acicular aragonite, whereas meteoric waters precipitate equant crystals of low-magnesium calcite, these generalizations which attempt to relate cement crystal morphology to bulk fluid compositions are most unsatisfactory.

Data on the form of non-marine freshwater cements strongly suggest that in many instances, highly variable morphologies may occur as compositionally identical cement crystals precipitated from nearly identical fluids. Non-marine low-magnesium calcite serves as an excellent case in point. In meteoric vadose and meteoric phreatic settings, low-magnesium calcite occurs in a great variety of forms ranging from whisker crystals (e.g., Longman, 1980) to acicular trigonal prisms (e.g., Binkley *et al.*, 1980) to equant rhombic spar (e.g., Halley and Harris, 1979). Within this spectrum, acicular cements appear to be restricted to vadose settings whereas cements displaying equant habits occur in both the meteoric phreatic and vadose zones. Vadose zones in this instance may include spelean, pedologic, and marginal lacustrine settings in addition to cementation in vadose pores developed in marine carbonate sediment during subaerial exposure.

Such morphologic variability within a single diagenetic system, a system which, at a scale perceived by most workers to be chemically homogeneous, requires that cement habits may be strongly determined by variables not directly related to the bulk composition of the fluid phase. This is not to suggest that there is no relationship between cement mineralogy and crystal habit. Dolomite invariably occurs as equant anhedral to euhedral crystals while most reported aragonite is strikingly acicular. Calcite, on the other hand, both as low-magnesium calcite and as high-magnesium calcite, is the dominant carbonate cement and exhibits morphological variations not related to either solid or fluid compositional variations.

Because morphological variations within such individual cement phases commonly are not related to variations in bulk fluid composition, kinetic factors must also play an important role in determining crystal habit. The rate of crystal growth, which may be influenced by a variety of parameters such as P_{CO_2}, rates of fluid flow, and temperature, must play a significant role in determining cement morphology. It is not surprising, therefore, that there is, at least qualitatively, a complete overlap in the morphologies of marine and non-marine cement crystals.

It is our belief that not one feature of freshwater carbonate cement is unique to the non-marine realm; however, combinations of characteristics can be used with confidence to indicate a freshwater origin. Certainly a low-magnesium calcite cement displaying meniscus and/or gravitational textures is almost definitely a product of the freshwater vadose zone. Also, the greater the complexity of the cements (the greater the variability of mineralogy, composition, crystal size, and morphology) within a specimen, the greater is the likelihood that the cement formed under non-marine conditions.

ACKNOWLEDGMENTS

We are grateful for the financial assistance from the Geology Foundation at the University of Houston and to Robert L. Folk (University of Texas), Walter E. Dean (U.S. Geological Survey), and Nahum Schneidermann and Paul M. Harris (Gulf Oil Corporation) for their valuable comments concerning this manuscript.

REFERENCES

Arakel, A.V., 1982, Genesis of calcrete in Quaternary soil profiles of the Hunt and Leeman lagoons, western Australia: Jour. Sed. Petrology, v. 52, p. 109–125.

______, and McConchie, D., 1982, Classification and genesis of calcrete and gypsite lithofacies in paleodrainage systems of inland Australia and their relationship to carnotite mineralization: Jour. Sed. Petrology, v. 52, p. 1149–1170.

Badiozamani, K., Mackenzie, F.T., and Thorstenson, D.C., 1977, Experimental carbonate cementation: Salinity, temperature and vadose-phreatic effects: Jour. Sed. Petrology, v. 47, p. 529–542.

Bathurst, R.G.C., 1975, Carbonate Sediments and Their Diagenesis (2nd ed.): Amsterdam, Elsevier, 658 p.

Binkley, K.L., Wilkinson, B.H., and Owen, R.M., 1980, Vadose beachrock cementation along a southeastern Michigan marl lake: Jour. Sed. Petrology, v. 50, p. 953–961.

Bogli, A., 1980, Karst Hydrology and Physical Speleology (translated by June C. Schmid): New York, Springer-Verlag, 284 p.

Boyer, B.W., 1981, Tertiary lacustrine sediments from Sentinel Butte, North Dakota and the sedimentary record of ectogenic meromixis: Jour. Sed. Petrology, v. 51, p. 429–440.

Braithwaite, C.J.R., 1979, Crystal textures of recent fluvial pisolites and laminated crystalline crusts in Dyfed, South Wales: Jour. Sed. Petrology, v. 49, p. 181–194.

Buchbinder, L.G., and Friedman, G.M., 1980, Vadose, phreatic, and marine diagenesis of Pleistocene-Holocene carbonates in a borehole: Mediterranean coast of Israel: Jour. Sed. Petrology, v. 50, p. 395–408.

Chafetz, H.S., 1979, Petrology of carbonate nodules from a Cambrian tidal inlet accumulation, central Texas: Jour. Sed. Petrology, v. 49, p. 215–222.

______, and Butler, J.C., 1980, Petrology of recent caliche pisolites, spherulites, and speleothem deposits from central Texas: Sedimentology, v. 27, p. 497–518.

Dunham, R.J., 1971, Meniscus cements, *in* O.P. Bricker (ed.), Carbonate Cements: Johns Hopkins Univ., Studies in Geology no. 19, p. 297–300.

Eggleston, J.R., and Dean, W.E., 1976, Freshwater stromatolitic bioherms in Green Lake, New York, *in* M.R. Walter (ed.), Stromatolites: Amsterdam, Elsevier, p. 479–488.

Folk, R.L., 1965, Some aspects of recrystallization in ancient limestones, *in* L.C. Pray, and R.C. Murray (eds.), Dolomitization and Limestone Diagenesis: Soc. Econ. Paleontologists and Mineralogists, Spec. Pub. 13, p. 14–48.

______, 1971, Caliche nodule composed of calcite rhombs, *in* O.P. Bricker (ed.), Carbonate Cements: Johns Hopkins Univ., Studies in Geology, no. 19, p. 167–168.

______, 1974, The natural history of crystalline calcium carbonate: Effects of magnesium content and salinity: Jour. Sed. Petrology, v. 44, p. 40–53.

______, and Assereto, R., 1976, Comparative fabrics of length-slow and length-fast calcite and calcitized aragonite in a Holocene speleothem, Carlsbad Caverns, New Mexico: Jour. Sed. Petrology, v. 46, p. 486–496.

Friedman, G.M., and Kolesar, P.T., 1971, Fresh-water carbonate cements, *in* O.P. Bricker (ed.), Carbonate Cements: Johns Hopkins Univ., Studies in Geology no. 19, p. 122–126.

Gavish, E., and Friedman, G.M., 1969, Progressive diagenesis in Quaternary to Late Tertiary carbonate sediments: Sequence and time scale: Jour. Sed. Petrology, v. 39, p. 980–1006.

Geno, K.R., and Chafetz, H.S., 1982, Petrology of Quaternary fluvial low-magnesian calcite coated grains from central Texas: Jour. Sed. Petrology, v. 52, p. 833–842.

Halley, R.B., and Harris, P.M., 1979, Fresh-water cementation of a 1,000-year-old oolite: Jour. Sed. Petrology, v. 49, p. 969–988.

Harrison, R.S., 1977, Caliche profiles: Indicators of near-surface subaerial diagenesis, Barbados, West Indies: Bull. Canadian Petroleum Geol., v. 25, p. 123–173.

Hattin, D.E., and Dodd, J.R., 1978, Holocene cementation of carbonate sediments in the Florida Keys: Jour. Sed. Petrology, v. 48, p. 307–312.

Hay, R.L., and Reeder, R.J., 1978, Calcretes of Olduvai Gorge and the Ndolanya Beds of northern Tanzania: Sedimentology, v. 25, p. 649–673.

Hill, C.A., 1976, Cave minerals: Huntsville, Nat. Speleological Soc., 137 p.

Hsu, K.J., and Kelts, K., 1978, Late Neogene chemical sedimentation in the Black Sea, *in* A. Matter and M.E. Tucker (eds.), Modern and Ancient Lake Sediments, Int. Assoc. Sedimentologists, Spec. Pub. 2, p. 129–145.

James, N.P., 1972, Holocene and Pleistocene calcareous crust (caliche) profiles: Criteria for subaerial exposure: Jour. Sed. Petrology, v. 42, p. 817–836.

Katz, A., 1973, The interaction of magnesium with calcite during crystal growth at 25–90°C and one atmosphere: Geochim. Cosmochim. Acta, v. 37, p. 1563–1586.

Kendall, A.C., and Broughton, P.L., 1978, Origin of fabrics in speleothems composed of columnar calcite crystals: Jour. Sed. Petrology, v. 48, p. 519–538.

Klappa, C.F., 1979, Calcified filaments in Quaternary calcretes: Organomineral interactions in the subaerial vadose environment: Jour. Sed. Petrology, v. 49, p. 955–968.

Knox, G.J., 1977, Caliche profile formation, Saldanha Bay (South Africa): Sedimentology, v. 24, p. 657–674.

Konishi, K., and Sakai, H., 1972, Fibrous aragonite of fresh-water origin in sealed Pliocene *Glycymeris yessoensis*: Japanese Jour. Geol. Geogr., v. 42, p. 19–30.

Lahann, R.W., 1978, A chemical model for calcite crystal growth and morphology control: Jour. Sed. Petrology, v. 48, p. 337–344.

LAND, L.S., 1970, Phreatic versus vadose meteoric diagenesis of limestones: Evidence from a fossil water table: Sedimentology, v. 14, p. 175–185.

LIPPMANN, F., 1973, Sedimentary carbonate minerals: New York, Springer-Verlag, 228 p.

LONGMAN, M.W., 1980, Carbonate diagenetic textures from nearsurface diagenetic environments: Amer. Assoc. Petroleum Geol. Bull., v. 64, p. 461–487.

MEREDITH, J.C., 1980, Diagenesis of Holocene-Pleistocene (?) travertine deposits: Fritz Creek, Clark County and Fall Creek, Bonneville County, Idaho: Unpub. Masters thesis, Univ. Houston, 263 p.

MULLER, G., 1971, "Gravitational" cement: An indicator for the vadose zone of the subaerial diagenetic environment, *in* O.P. Bricker (ed.), Carbonate Cements: Johns Hopkins Univ., Studies in Geology no. 19, p. 301–302.

______, IRION, G., AND FORSTNER, U., 1972, Formation and diagenesis of inorganic Ca-Mg carbonates in the lacustrine environment: Naturwissenschaften, v. 59, p. 158–164.

______, AND WAGNER, F., 1978, Holocene carbonate evolution in Lake Balaton (Hungary): A response to climate and impact of man, *in* A. Matter and M.E. Tucker (eds.), Modern and Ancient Lake Sediments, Int. Assoc. Sedimentologists Spec. Pub. 2, p. 57–81.

PANOS, V., AND STELCL, O., 1968, Physiographic and geologic control in the development of Cuban mogotes: Z. Geomorph., v. 12, p. 117–164.

PINGITORE, N.E., 1976, Vadose and phreatic diagenesis, processes, products and their recognition in corals: Jour. Sed. Petrology, v. 46, p. 985–1006.

RAO, C.P., 1981, Criteria for recognition of cold-water carbonate sedimentation: Berriedale Limestone (Lower Permian), Tasmania, Australia: Jour. Sed. Petrology, v. 51, p. 491–506.

READ, J.F., 1974, Calcrete deposits and Quaternary sediments, Edel Province, Shark Bay, Western Australia, *in* Evolution and Diagenesis of Quaternary Carbonate Sequences, Shark Bay, Western Australia: Amer. Assoc. Petroleum Geologists, Mem. 22, p. 250–282.

RIDING, R., 1979, Origin and diagenesis of lacustrine algal bioherms at the margin of the Ries crater, Upper Miocene, southern Germany: Sedimentology, v. 26, p. 645–680.

ROGERS, B.W., AND WILLIAMS, K.M., 1982, Mineralogy of Lilburn Cave, Kings Canyon National Park, California: Nat. Spel. Soc. Bull., v. 44, p. 23–31.

SCHLAGER, W., AND JAMES, N.P., 1978, Low-magnesian calcite limestones forming at the deep-sea floor, Tongue of the Ocean, Bahamas: Sedimentology, v. 25, p. 675–702.

SCHNEIDERMANN, N., SANDBERG, P.A., AND WUNDER, S.J., 1972, Recognition of early cementation of aragonitic skeletal carbonates: Nat. Phys. Sci., v. 240, p. 88–89.

SCHOTTLE, M., AND MULLER, G., 1968, Recent carbonate sedimentation in the Gnadensee (Lake Constance), Germany, *in* G. Muller and G.M. Friedman (eds.), Recent Developments in Carbonate Sedimentology in Central Europe: New York, Springer-Verlag, p. 148–156.

SCHREIBER, B.C., SMITH, D., AND SCHREIBER, E., 1981, Spring peas from New York State: Nucleation and growth of fresh water hollow ooliths and pisoliths: Jour. Sed. Petrology, v. 51, p. 1341–1346.

SCHROEDER, J.H., 1973, Submarine and vadose cements in Pleistocene Bermuda reef rock: Sed. Geol., v. 10, p. 179–204.

SEMENIUK, V., AND MEAGHER, T.D., 1981, Calcrete in Quaternary coastal dunes in southwestern Australia: A capillary-rise phenomenon associated with plants: Jour. Sed. Petrology, v. 51, p. 47–68.

SIEGEL, F.R., 1965, Aspects of calcium carbonate deposition in Great Onyx Cave, Kentucky: Sedimentology, v. 4, p. 285–299.

STALDER, P.J., 1975, Cementation of Pliocene-Quaternary fluviatile clastic deposits in and along the Oman Mountains: Geol. en Mijnbouw, v. 54, p. 148–156.

STEINEN, R.P., 1974, Phreatic and vadose diagenetic modification of Pleistocene limestone: Petrographic observations from the subsurface of Barbados, West Indies: Amer. Assoc. Petroleum Geol. Bull., v. 58, p. 1008–1024.

STOOPS, G.J., 1976, On the nature of "lublinite" from Hollanta (Turkey): Amer. Mineralogist, v. 61, p. 172.

SUPKO, P.R., 1971, "Whisker" crystal cement in a Bahamian rock, *in* O.P. Bricker (ed.), Carbonate Cements: Johns Hopkins Univ. Studies in Geology no. 19, p. 143–146.

TANDON, S.K., AND NARAYAN, D., 1981, Calcrete conglomerate, case-hardened conglomerate and cornstone - a comparative account of pedogenic and non-pedogenic carbonates from the continental Siwalik Group, Punjab, India: Sedimentology, v. 28, p. 353–367.

WARD, W.C., 1975, Petrology and diagenesis of carbonate eolianites of the northeastern Yucatan Peninsula, Mexico, *in* K.F. Wantland and W.C. Pusey (eds.), Belize Shelf-Carbonate Sediments, Clastic Sediments, Ecology: Amer. Assoc. Petroleum Geol. Studies in Geology no. 2, p. 500–571.

WATTS, N.L., 1980, Quaternary pedogenic calcretes from the Kalahari (southern Africa): Mineralogy, genesis and diagenesis: Sedimentology, v. 27, p. 661–686.

WELLS, A.W., 1971, Cave calcite: Studies in Speleology, v. 2, p. 129–148.

WHITE, W.B., 1976, Cave minerals and speleothems, *in* T.D. Ford and C.H.D. Cullingford (eds.), The Sciences of Speleology: London, Academic Press, p. 267–327.

______, 1981, Reflectance spectra and color in speleothems: Nat. Spel. Soc. Bull., v. 43, p. 20–26.

WILSON, J.L., 1975, Carbonate Facies in Geologic History: New York, Springer-Verlag, 471 p.

WRIGHT, V.P., 1982, Calcrete palaeosols from the Lower Carboniferous Llanelly Formation, South Wales: Sed. Geol., v. 33, p. 1–33.

BIZARRE FORMS OF DEPOSITIONAL AND DIAGENETIC CALCITE IN HOT-SPRING TRAVERTINES, CENTRAL ITALY

ROBERT L. FOLK[1], HENRY S. CHAFETZ[2], AND PAMELA A. TIEZZI[1]
[1]Department of Geological Sciences, University of Texas, Austin, Texas 78712, and [2]Department of Geosciences, University of Houston, Houston, Texas 77004

ABSTRACT

Travertines of central Italy have been deposited from warm, fresh-water springs very rich in H_2S. Probably as a result of this, the carbonate sediments are largely constructed by bacteria instead of algae. Even the chemically-precipitated forms of calcite show a variety of bizarre characteristics, some of them probably due to the interaction of S with the calcite crystal lattice. Among these are thick crusts of radial fibrous calcite crystals up to a meter long, that show probable daily lamination due to the diurnal activity of photosynthetic bacteria and have depositional rates of as much as one meter per year. Fine ray-crystal fans only one mm or so thick are made of helically-twisting radial ribbons of calcite and also show daily growth bands. Bacterial clumps are enclosed within an elliptical ball of calcite, and the bacteria plate out upon crystal planes; the ball is then surrounded by a tattered hollow carapace of very thin-walled calcite forming a single crystal outline. Fibrous nests of square-ended calcite rods with inclined extinction ("lublinite") occur among the bacterial calcite clumps. Pore-filling calcite cement in the rocks is also weird, sometimes showing forms like superimposed, curving gothic arches, or multiple spikes. Some calcite crystals also show features resembling screw-dislocations. Palisade calcite cement shows a nearly basal parting.

INTRODUCTION

Travertine from Central Italy has been used as prime building and ornamental stone since the late second century B.C. (Blake, 1947, 1959), included in such monuments as the "Temple of Sibylla" (at Tivoli itself), the Coliseum, colonnade around Piazza San Pietro, the famous fountains of Roma, as well as the modern University, apartment houses, and curbstones. During the first century B.C. its use became standard for Roman bridges and other heavy structures and at that time Vitruvius wrote that it was almost an ideal building stone because of its ability to bear heavy loads and resist weathering, yet it could be cut like wood with a toothed saw and the porosity facilitated binding for coats of stucco (Briggs, 1956, p. 27). Most of the quarries were near Tivoli, and in fact the name "travertine" comes from Latin "Lapis Tiburtinus" (stone of Tibur), the original name for Tivoli in use prior to the eighth century (Hare, 1884). Today it is probably the most widely exported and commercially valuable building stone in the world, being shipped worldwide from dozens of very active quarries.

The romantic charm of the grottos and waterfalls in the Tivoli region has attracted geologists for almost two centuries, having been discussed by such famous early giants as Lyell (1829), who cited earlier studies by von Buch and Brongniart. d'Halloy (1812) concluded correctly that the plain of Tivoli had once been a lake, subsequently filled by deposition of travertine from hot springs except for a few tiny surviving ponds such as the Acquae Albulae north of Bagni.

Our study (Chafetz and Folk, 1984) has concentrated on the many quarries around Bagni di Tivoli (30 km E. of Roma), but we have also examined quarries around Rapolano Terme (30 km E. of Siena, in Toscana).

At Tivoli, the material was deposited in a series of ponds, swamps, and lakes about thermal springs, some of which are active today. The travertine deposits, up to 85 m. thick, lie upon Pleistocene volcanic ash and probably date from the last 500,000 years (see Manfra *et al.* 1974, 1976). The warm springs contain abundant dissolved $Ca(HCO_3)_2$ derived from adjoining Mesozoic carbonates which form the fault-bounded Sabine mountains bordering the plain between Tivoli and Roma. Springs are also very rich in H_2S, and native sulphur is precipitated where water is flowing rapidly and becoming oxygenated (approximately $2H_2S + O_2 \rightarrow 2H_2O + 2S\downarrow$). The hot spring orifices at Acquae Albulae emit so much H_2S that dead birds litter the ground.

Sections of travertine are well exposed in quarry walls. These exhibit an upper few meters of punky "testina" overlying heavily cemented material with a consistency of well-lithified limestone, and capable of taking on a high polish. Gentle angular unconformities and some large solution cavities are also seen (Chafetz and Folk, 1984).

At Rapolano, the travertines cover an alluvial plain and are also draped over Cretaceous carbonate rocks and siliciclastic turbidites (Losacco, 1951). Fan-shaped travertine deposits resulted when springs emerged on hill slopes. Spring waters at Rapolano have a Mg/Ca ratio between 1:3 and 1:6, and contain 0.3–1.4 gm/L of SO_4^{2-} (Cipriani *et al.*, 1972).

SUMMARY OF TRAVERTINE PETROGRAPHY

The travertines consist entirely of low-magnesian calcite. Bulk analyses by Cortesi and Leoni (1955) showed that the rock contains 0.2–0.4 wt. % MgO and 0.5–1.2% SO_4^{2-}. We have detected nothing except calcite by x-ray diffraction methods. An occasional rhomb of calcitized dolomite occurs in a few of the samples, but we have found no gypsum to account for the SO_4^{2-}. The elementary building block of the bulk of the travertine is a bacterial clump about 10–40 μm in diameter, encased in a single calcite crystal only slightly larger than the clump (Folk and Chafetz, 1980, 1983; Chafetz and Folk, 1984). The bacteria occur as tiny elliptical dots about 1 μm long, and as threads or curving filaments ½ μm wide and up to 20 μm long. They make a dense tangle forming a ball that is white in reflected light, but brown to nearly opaque in transmitted light; they are responsible for the brilliant whiteness of the stone. In the Italian travertines, the bacteria have been oxidized and are visible only as cavities, but in Idaho travertines with similar characteristic micro-structures, bacterial bodies can still be seen with SEM on etched surfaces (Chafetz and Meredith, 1983; Chafetz and Folk, 1984).

Bacterial colonies assume a variety of shapes, including branching shrubs up to several cm high (Chafetz and Folk, 1984), laminated mats resembling blue-green algal mats, and "foam rock" made of bacterially-coated and rapidly lithified bubbles. All of these materials can be reworked to form silt-size mechanical sediment, or sand-to-gravel size intraclasts made of bacterial "rock". Bacteria also form layers of bumpy pisoids that often grade into bacterial shrubs (Folk and Chafetz, 1983). These biologic rock types eventually become cemented by inorganic sparry calcite, but many unfilled holes remain (responsible for the rows of cavities parallel with the bedding seen in building stone).

As will be demonstrated, there is gradation between the predominantly organic (bacterial) sediments with biologically-controlled form—which, however, became cemented by inorganic sparry calcite; and the predominantly inorganic precipitates which contain adventitious bacteria but whose radial crystal form is not controlled by them. On a gross basis, the travertines show three main lithologic end-members, which match with three main types of bedding (Chafetz and Folk, 1984).

1. Relatively horizontal, thin-bedded deposits (individual beds approximately 3 cm thick). These are made largely of bacterial shrubs, with some bacterial mats, bacterial silt and foam-rock; they formed in very shallow, level ponds in quiet water.
2. Areas of very irregular bedding, representing terraces and travertine dams sometimes meters in relief with vertical to overhanging slopes (Fig. 1A, B). Small depressions in terraces are frequently filled with inorganic pisoids. Steep terrace slopes and funnel-shaped sides of spring mouths are made mainly of the inorganic ray-crystals, formed in these highly agitated environments, where water was rapidly losing CO_2. Typically, as soon as horizontal beds begin to dip down a terrace scarp, the ray-crystals take the place of shrubs as the main rock component. Spectacular development of these can be seen in the quarries at Rapolano Serre, and at the modern active springs at San Giovanni Battista Terme, 3 km W. of Rapolano.
3. Rather steeply and uniformly sloping (up to 15°) broad areas with a fan-like conical form, representing sloping terraces at the foot of scarps, above stream valleys, or huge steep mounds built out from hillside springs. Travertines here are composed mainly of clastic bacterial sediment, with few *in situ* shrubs or ray-crystals.

VARIETIES OF DEPOSITIONAL AND DIAGENETIC CALCITE

The present paper focuses upon the mainly inorganic calcite precipitates, and the following types will be discussed: (1) large ray-crystal crusts, generally a few cm thick; (2) fine ray-crystal fans, a few mm thick; (3) cements developed around the bacterial clumps; (4) calcite rafts; (5) odd forms of rhombic calcite cement; and (6) coarse palisade calcite as a final cement. Also there are inorganic pisoids found in pockets and spring orifices, but these are only briefly mentioned, as they have been discussed previously (Folk and Chafetz, 1983).

The ray-crystals are primary material, crystallized directly from solution and forming a hard crust (often called "sinter") where waters are agitated and precipitation of calcite is rapid; they incorporate varying amounts of microscopic bacterial structures, though the bacteria do not control their major, radiating crystalline fabric. Inorganic pisoids are formed very rapidly in highly agitated water, and show no evidence of bacteria. Cements developed around the bacterial clumps seem to have crystallized contemporaneously with each colony, e.g. on a time scale of a few minutes, and in contact with less agitated water. Calcite rafts grew more slowly on the surfaces of standing water. Most pore spaces and vugs are partially filled with curiously formed early diagenetic crystals that are variants on the basic rhombohedron. Finally, palisade calcite cement is the last pore-filling in large cavities.

Large Ray-Crystal Crusts

Ray-crystal crusts (Fig. 1) are commonly 2 to 8 cm thick, but on sawn blocks we have measured some as thick as 70 cm, and in ancient travertine "dams" on inaccessible quarry faces we have seen some that may be as thick as one or two meters (e.g., Rapolano Serre). Apparently these are the largest calcite crystals yet known from any non-spelean stratigraphic unit, though Assereto and Folk (1980) describe aragonite ray-crystals as long as 20 cm from Triassic paleo-speleothems of Italy (Calcare Rosso). Risacher and Eugster (1979, p. 264) describe crusts of radial calcite crystals from hot spring deposits that are similar to our

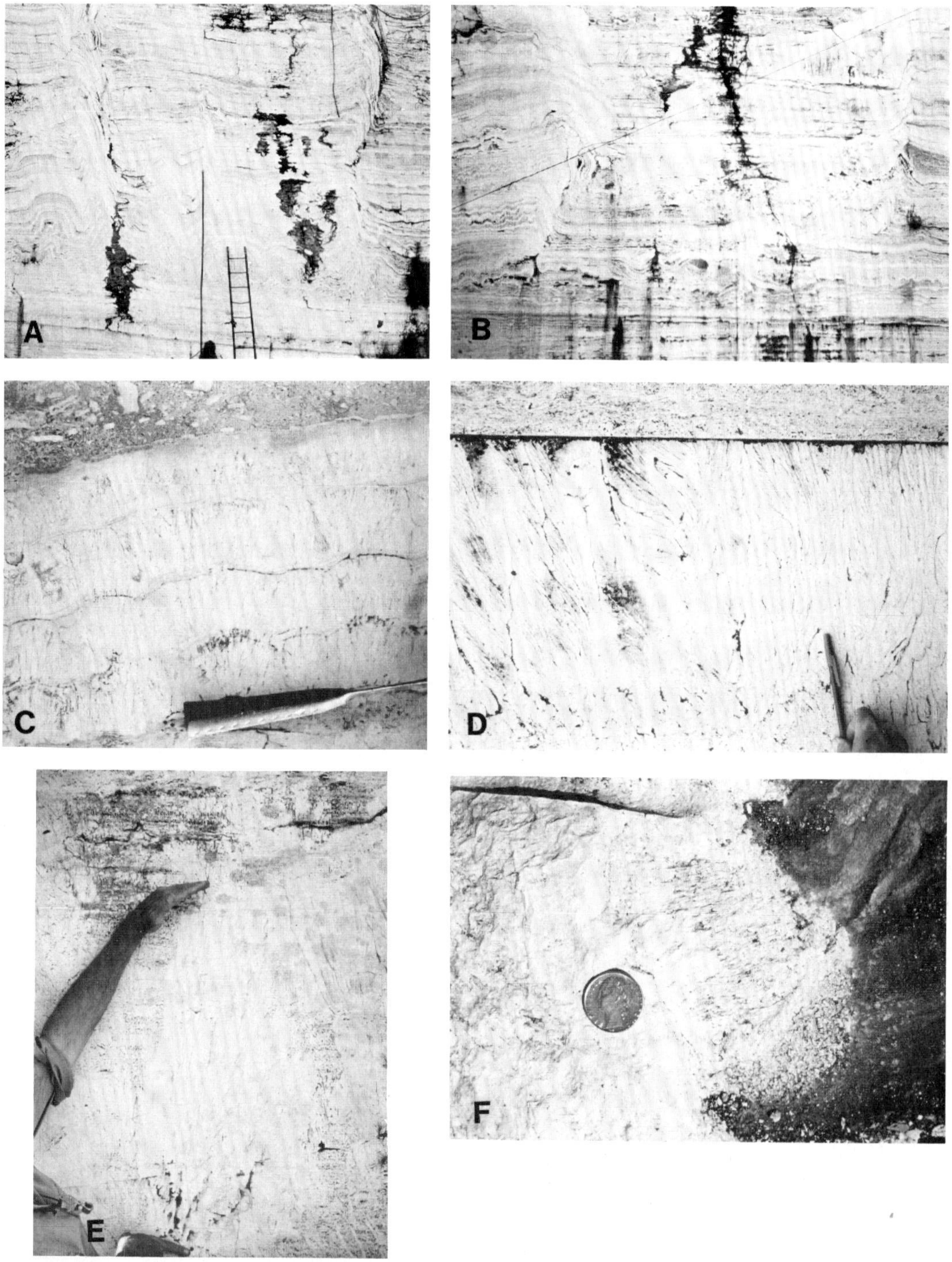

FIG. 1.—*A*, Quarry wall near Rapolano (ladder for scale). Observe terraces with steeply dipping slopes, indicating water flow toward the right. *B*, Same locality, overhanging rims of ancient terrace mounds; large ray-crystals occur at the curving lips. *C*, Four typical ray-crystal crusts, each one several cm thick. Bundles of crystals diverge slightly upward. *D*, Large, feathery ray crystals in a slab built into the wall of the railroad station at Ferentino. Crystals tend to branch toward the left of the main stem. *E*, Quarry wall at Rapolano Serre, large fan of ray crystals about 70 cm long. Nests of pisolites occur between large bundles of ray crystals. *F*, View down on the bedding plane of a quarried block. The ray crystals "outcrop" as parallel ridges (to the right of the coin), indicating water flow parallel to their trend.

crystals, but theirs are no more than 1 mm thick. Generally the travertine crusts are made of crystals that are almost parallel, but they can range to narrow fans with an axial divergence at about 10–30° (Fig. 1C). Seen from above, crusts reveal grossly mammilated surfaces, with spherical convexities as in the top surface of a head of broccoli. The convexities had a relief of at most about 1 or 2 cm above the "normal" bedding planes. Tops of some weathered bedding surfaces reveal subparallel trends of linear protruding crystals (Fig. 1F), going for many cm across the tops of the crusts, as if the crystals really were large plates stuck on edge in parallel fashion. In a few rare outcrops it can be observed that these lineated crystals radiate out from fissure-shaped orifices. Crystals similar in shape but much smaller have been observed by Tiezzi and Folk at Yellowstone (unpubl.); though the *c* axes of the fibers are radial to the encrusted surface, the elongate crystals have a lateral "wing". B.C. Schreiber (pers. comm., 1982) observed gypsum crystals that grew toward the source of nourishing fluids; thus we conclude that the ray-crystals in the travertine crusts tended also to grow parallel to the flow direction of fluids, as shown also in ore deposits by Newhouse (1941).

In some remarkable examples (e.g., the slabs covering the old railroad station at Ferentino near Frosinone, Lazio) large supercrystals have a feathery morphology (Fig. 1D); the supercrystals curve gently (like frost crystallized on a window pane) and are 35–50 cm long with pinnate branches occurring off the main stem, the whole structure being 1–3 cm wide. The branches grow larger more on one side than the other of the main "trunk", and this asymmetry is generally consistent within any one quarry block.

Most ray-crystal crusts, especially the larger ones, show exquisite crosswise laminations particularly when accentuated by weathering or black dirt (Fig. 2) as in many curbstones of Roma. The laminae are extraordinarily uniform and range from a spacing of 3 mm down to about 0.3 mm. They curve across the radial or feathery crystals and represent growth zones. Similar 1-2 mm laminae are pictured by Bargar (1978, Fig. 20) from Mammoth Hot Springs, Wyo., and Tiezzi and Folk (unpubl.) also found them in shrub-like forms in Yellowstone travertines.

Ray-crystal crusts are found not only at the base of quarries where the travertine is highly indurated and virtually non-porous, but also at the top ("testina") where the material is as crumbly as stale white bread. Thus the rays are not a late diagenetic feature, but developed as "primary" material contemporaneously with sedimentation. Italian quarry operators do not like to encounter ray-crystals in sawn slabs, because of their sometimes punky character. To them it is known as "gesso" (Italian for gypsum), but this is a misnomer as the material is in fact all low-Mg calcite.

Microscopic appearance.—Most ray-crystals show (1) bacterial bands representing cyclic accumulations as growth proceeded (Fig. 2E) transected by (2) radial feathery calcite crystals. The crystals (Fig. 2F) tend to grow like a sub-parallel, outward branching array of needles (typically .05 × 2 mm) similar to the "cedar-tree needles" of calcite described by Kitano (1963) from thermal springs in Japan. Sometimes the "cedar needles" are individual and distinct, but in other examples they merge to form a composite crystal with continuously sweeping undulose extinction. In many specimens the "cedar needles" branch off only one side of the trunk crystal, presumably facing the direction from which the nourishing solutions came. Crystals are length-fast with diverging radial c-axes. Other ray-crystal crusts appear similar in the field and hand specimen, but under crossed nicols no evidence of crystal continuity is visible; it is presumably present, but is too subtle to be observed. SEM observations show the "cedar tree needles" to be very poorly built and ragged (Fig. 2G) expanding outward at the tips like a flower, and full of cavities. They somewhat resemble the "applelite" crystallites described by Ash (1975) from an Indiana cave and thought by him to be influenced by bacterial or algal microflora.

Bacteria have apparently played no role in controlling the overall smoothly radial morphology of our ray-crystals, and have been merely incorporated as the crystals grew. Crystals would have assumed the same regular sub-botryoidal radial form even if there had been no bacteria. Thus they differ from the shrubby growths, whose delicately branching gross morphology is strongly controlled by bacterial colonies (Folk and Chafetz, 1980; Chafetz and Folk, 1984).

Growth bands in the large ray-crystals consist of couplets (Fig. 2E); a quasi-horizontal lamina 0.1 mm thick, made of tiny bacterial clumps 5–10 μm in diameter, succeeded by a forest-like layer of minute bac-

FIG. 2.—*A*, Quarried block at Bagni di Tivoli, showing fine laminations that become undulose toward the top. *B*, Daily laminations on a fractured block; lichens have brought out the delicate structure. Coin is 24 mm. *C*, Laminae with a spacing of over 2 mm. (Coin = 24 mm) *D*, Very fine and regular laminae with a spacing of about 0.5 mm. *E*, Views of laminae in thin section. The left view in reflected, the right view in transmitted light. We interpret the small shrubs as having been formed by photosynthetic bacteria during the daylight hours, while the nearly horizontal layers are the production of the night. The single layer of very dense bacterial colonies between the night and the next day represents the dawn layer. About five days' accumulation is shown here. *F*, Some of the large ray crystal crusts are made of length-fast "cedar-needle" crystals, shown here. Crossed nicols. *G*, SEM view of the terminations of the "cedar-needle" crystals. They are made of sub-units and tend to show a splayed pattern, with *c* axis parallel to the long dimension of the crystal-aggregate.

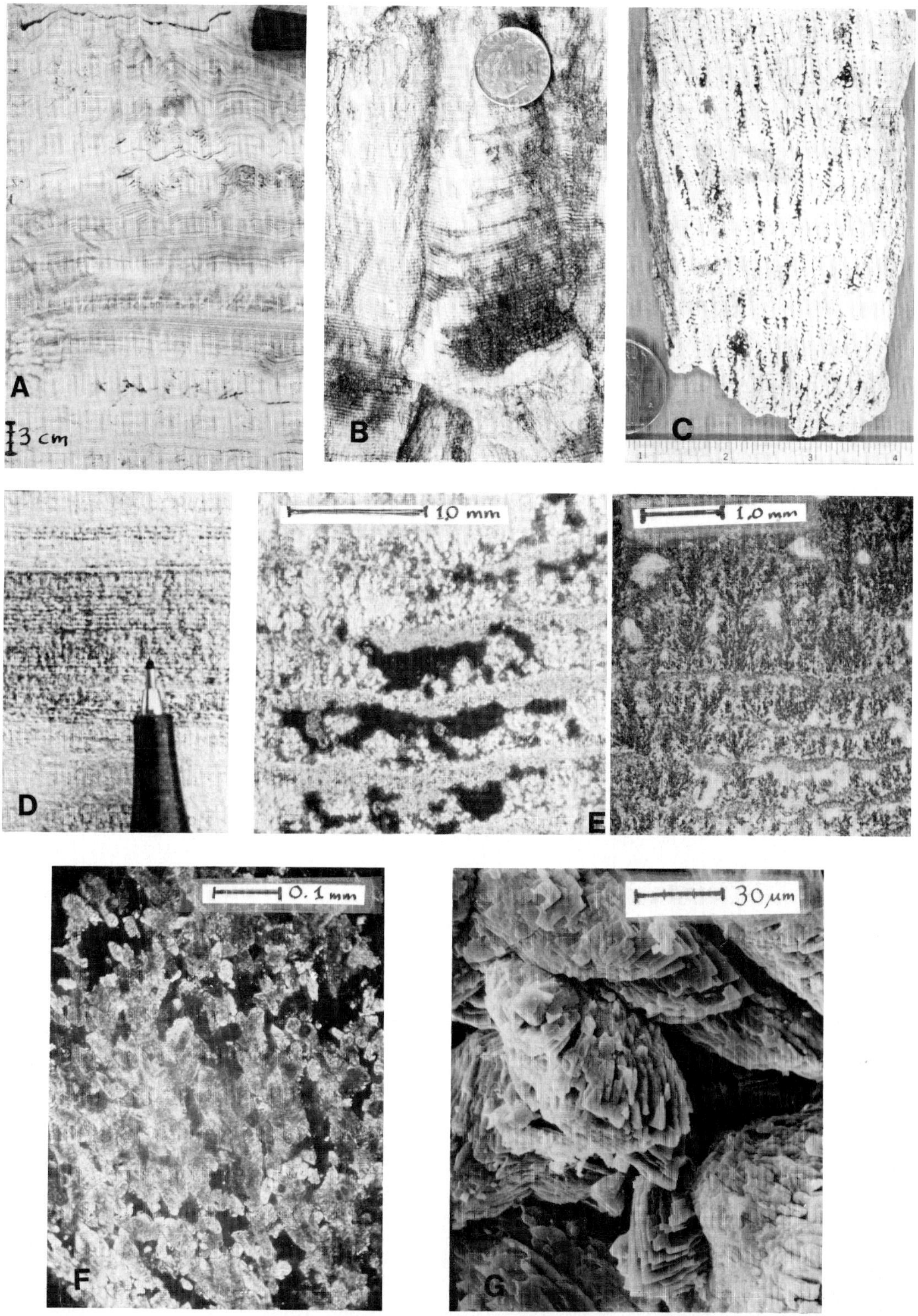
A
3 cm
B
C
1
2
3
4
D
1,0 mm
E
1,0 mm
0.1 mm
F
30 μm
G

terial shrubs about 0.4 to 2 mm high, each shrub made up of 10–25 μm bacterial clumps. Shrubs branch uniformly upward, and are similar in every way except size to the cm-scale bacterial shrubs (Chafetz and Folk, 1984). Layer thicknesses in any one specimen are exceedingly uniform, but in different specimens layers vary from 0.3–3 mm thick (see Table 1). The quasi-horizontal laminae must have had considerable strength, perhaps like that of the skin on heated milk, because they stretch across the tops of bacterial shrubs with only slight sagging. At the top of many lamina, there is a single layer of denser bacterial clumps, slightly larger (15–20 μm), and more intense white in reflected light (Fig. 2E). Presumably, because of the upward-branching nature of the micro-shrubs (Fig. 2E), the bacteria that formed them are influenced by light and are photosynthetic sulfur-oxidizers (Chafetz and Folk, 1984). Thus the bacterial shrub layers are probably the product of the day, and the thin quasi-horizontal sediment laminae are the deposits laid down at night by non-photosynthetic bacteria. The dense layer of bacterial clumps at the top of the night layer probably represents the dawning of each day, when the bacteria first commence photosynthesis.

Further evidence on the nature of the banding and deposition rates.—Such amazing uniformity in banding implies some very cyclic phenomenon, alternating between: (1) rapidly branching growth of bacterial colonies, with inorganic crystallization only enough to coat the shrubs; and (2) a relative resting period for the bacteria, during which time they developed no branching habit.

One can think of daily, monthly, or annual cycles to explain the extreme regularity of the laminae (Table 2). Deciding which is not easy, but monthly is *prima facie* not very likely in this non-tidal situation. Here the large cm-scale bacterial shrubs can offer some instructive aid. In some localities (e.g., Guidonia) the shrubs show very rhythmic banding on the order of 2–4 cm, with a quasi-horizontal sediment band a few mm thick, succeeded by a layer of branching bacterial shrubs that makes up the bulk of each layer. Again we faced the problem: daily or yearly cycles? Luckily we found a quarried block whose corner had been subjected to meteoric water etching, and this block showed in addition to the gross 3 cm layering a very minute banding on the order of .01–.05 mm within the shrubs; we subsequently found this microlayering on some travertine blocks that had been exposed as building stones for several centuries in Roma (e.g., the twin Churches on the S. side of the Piazza del Popolo and on top of the Spanish Steps). It thus seems logical to assume that the fine-scale banding (approximately 100–200 visible bands per large layer) is a daily phenomenon, and the larger-scale banding is annual with winter deposition being represented by the thinner horizontal layer. The shortfall from 365 days is due to invisibility or absence of growth laminae during the winter season. In a very few blocks we were able to trace the fine-scale lamination in the shrub layers into the fine-scale lamination of the ray-crystals. So we conclude that in both lithic types the fine-scale lamination (10 microns to a few mm) is in fact daily, and the coarse banding in the shrub layers (few cm) is annual.

In 1983 we made a fortuitous visit with V. Pursell (thesis in progress) to Opal Terrace, Mammoth Hot Spring, Yellowstone, and there found further evidence for the daily origin of travertine laminae. A tourist boardwalk intersects the edge of Opal Terrace, and a cut 1.5 meter high along the walkway showed that in the 1940's a vertical wall of burlap sandbags had been erected to stop encroachment of the travertine terrace. The travertine built up to the sandbag wall and partially covered it. Parts of the travertine are now exposed, and locally show lamination on a scale of 0.6–0.8 mm, like the lamination in the Italian ray-crystal crusts. If the lamination were yearly, this would calculate out to over 2000 years for the travertine to build up against the sandbag wall, an obvious absurdity. If the lamination is daily, it would require about 10 years, which fits well with the known history.

If the ray-crystal bands in the Italian travertines are indeed daily, this calculates out to an annual growth rate of between 10 cm and a meter per year, surely one of the fastest precipitational growth rates known for any kind of rock. However, there is data from other localities to show that this is not an unreasonable figure; Emig (1917, p. 37, 40) describes algal travertine in Oklahoma rapids as growing up to 7–13 cm in a summer, and (p. 58) cites Italian hot springs with travertine growth rates of 15 cm/yr, and Sicilian springs depositing 30 cm in only four months. By observing layers of deciduous leaves at regular intervals in hot spring deposits of Japan, Kitano (1963) also showed deposition rates of 20–100 cm/yr. Algal mats can also grow 1 mm per day (Golubic and Focke, 1978;

TABLE 1.—THICKNESS DATA FOR THREE LAMINATED SPECIMENS

$\bar{x}$	σ	[1]CV $\left(\frac{\sigma}{\bar{x}}\right)$
0.56 mm	.08 mm	14%
1.4 mm	.25 mm	18%
2.1 mm	.17 mm	8%

[1]CV is the coefficient of variation. The average CV of 13% signifies that two-thirds of all measured laminae lie within about ± 1/8 of the mean thickness value.

TABLE 2.—CHARACTER OF BANDING AND LAMINATION IN THE SEVERAL TYPES OF LITHOLOGIES IN ITALIAN TRAVERTINES, AND LAMINATION IN A MODERN STREAM CARBONATE FROM A TEXAS CREEK.

	BACTERIAL		CHEMICAL	
	SHRUBS	LARGE RAY-CRYSTAL CRUSTS	FINE FANS	TEXAS CREEK
GROSS LAYERS (ANNUAL)	10–50 mm thick Summer shrub growth Winter quiescence—flat layers	X	1–2 mm thick Summer—rapid ppt, thicker laminae Winter—slow deposition, thinner laminae	X
FINE LAMINATION (DAILY)	.01–.05 mm	0.4–3 mm Daytime microshrubs; Nighttime thin blanket	1–4 μm Daytime ppt; Nighttime halt	3–6 μm Daytime ppt; Nighttime halt

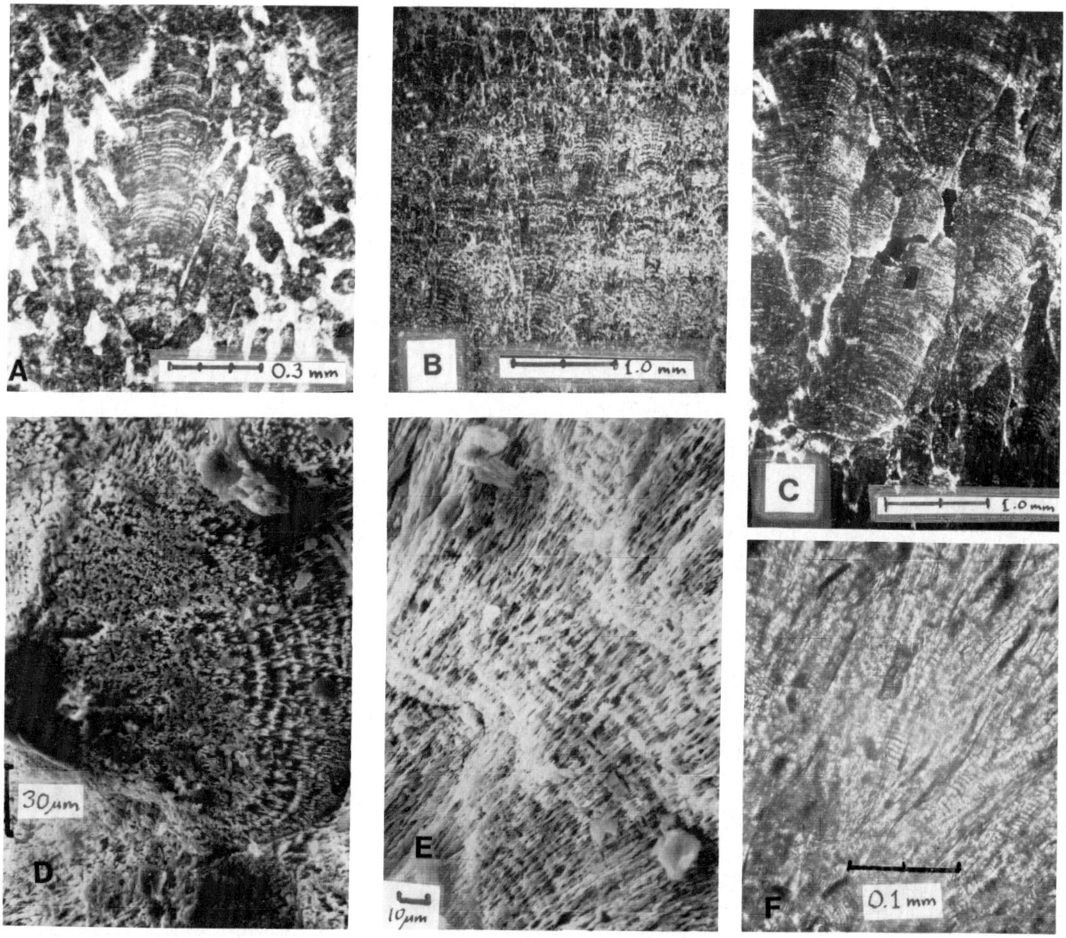

FIG. 3.—*A*, Glove-shaped fan with daily laminations. Dark bands are very microporous. Ordinary light. *B*, Layer of scalloped fans, showing wider spacing of bands toward the middle (interpreted as summer deposition). Bands near the top and bottom of the photo become so thin as to be indistinguishable. Ordinary light. *C*, Group of fans showing smoothly undulose extinction in crossed nicols. *D*, SEM view showing a very porous fan. *E*, Fan with daily pore-rich bands. *F*, Modern stream travertine from central Texas, showing clear calcite with micron-wide gaps probably representing nightly halts in deposition. Compare with Fig. A and B on this figure. Crossed nicols.

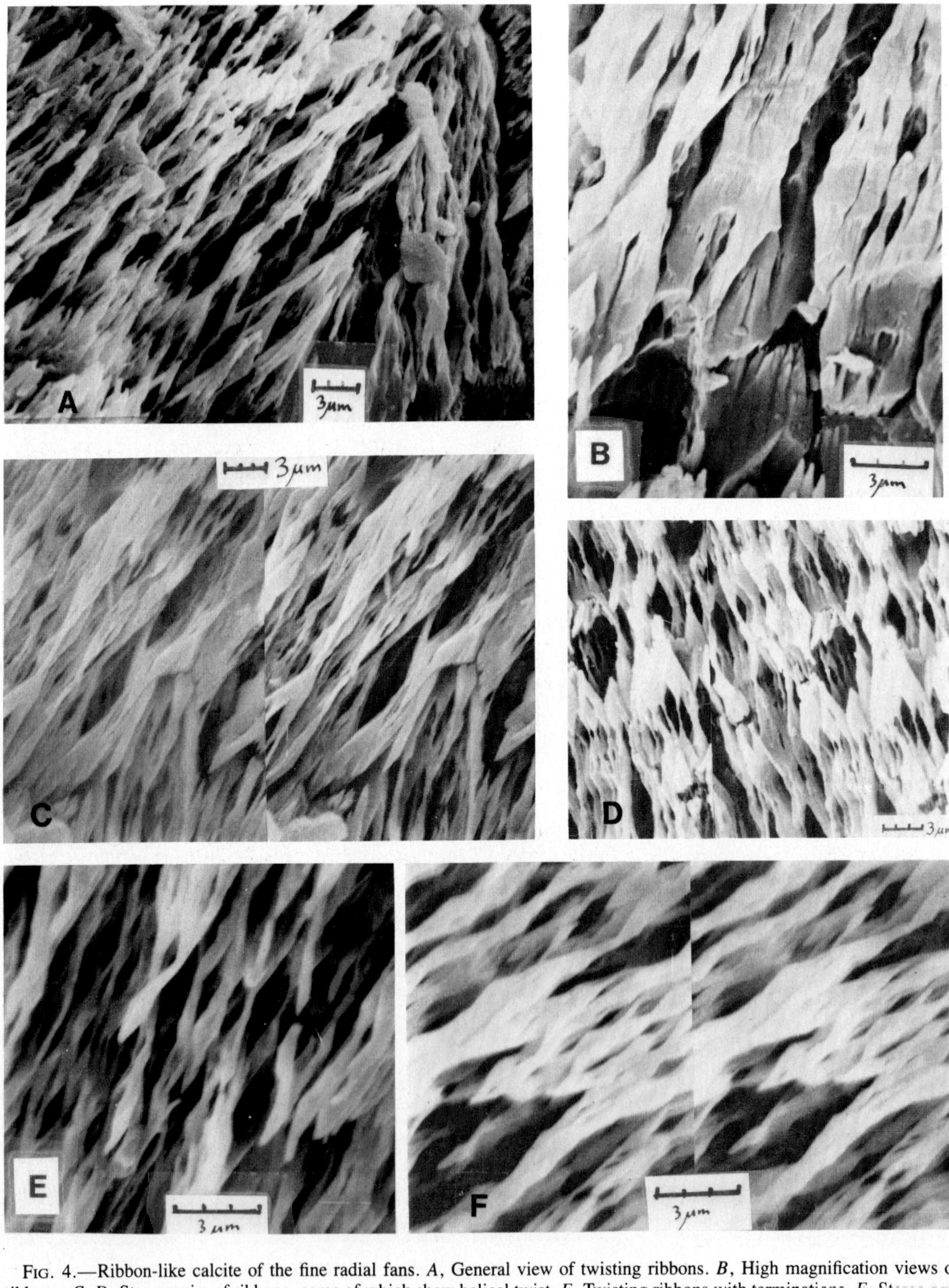

FIG. 4.—Ribbon-like calcite of the fine radial fans. *A*, General view of twisting ribbons. *B*, High magnification views of ribbons. *C*, *D*, Stereo-pairs of ribbons, some of which show helical twist. *E*, Twisting ribbons with terminations. *F*, Stereo pair of twisting calcite ribbons.

Gebelein, 1969). Bargar (1978) shows growth rates in Mammoth Hot Springs (Yellowstone) of 20 cm/yr. In Yellowstone, Tiezzi and Folk (unpubl.) observed bacterial shrubs with lamination of .02–.2 mm, presumably also daily growth increments.

We can see from field relationships that the most rapid growth rates (e.g. widest spaced daily laminae) occur in ray-crystals around spring orifices or on travertine dams, where thermal waters, highly charged with $Ca(HCO_3)_2$, lose their CO_2 rapidly with the sudden drop in pressure and extreme agitation. This is apparently also a fertile area for bacterial colonization.

Fine Ray-Crystal Fans

These fans form horizontal layers only 1–3 mm thick (Fig. 3) and differ in structure from the thicker ray-crystal crusts because the fine ones are made of very obvious radial calcite crystals in both thin section and SEM; the cross-banding is just as regular but is on a scale of a few microns instead of mm; and the bands do not show evidence that they contained any bacteria. Despite their similar radiating nature, there seems to be very little transition between the two types of radial-crystal fabrics.

Crystals form radial fans spanning about 10–45 degrees, and are made of length-fast calcite crystals that show generally continuous sweeping extinction (Fig. 3C).

In overall shape the fine ray-crystal fans are much less regular than the large ones, often resembling miniature glove-shaped stromatolitic colonies with digitate form always branching upward (Fig. 3A).

The fine ray-crystals are also characterized by minute, very evenly spaced crosswise growth bands, with a spacing of a few microns. Most of them are scalloped, convex-up (Fig. 3). Laminae consist of alternating layers of clear calcite, and fine brown laminae (white in reflected light); brown layers are riddled with cavities (Fig. 3D, E). Similar fine ?daily laminae have been illustrated from Paleozoic rocks by Termier and Termier (1975, Fig. 5).

The fans that show the best-developed crosswise growth bands in the SEM are found, at high magnification, to be extraordinarily bizarre (Fig. 4). They are made of thin radial-parallel ribbons of calcite that are hundreds of microns long, but in detail each ribbon is made of an alternation of (1) flat diamond-shaped plates and (2) thin connecting strands. The plates are about 1–2 μm wide and 0.1 μm thick, and the strands are also about 0.1 μm thick. Neighboring calcite ribbons have a congruent pattern of diamond-shaped plates and strands, producing the cross-wise growth bands. The whole array looks like an open lattice-work fence in that it is very largely empty space; the bands that appear brown in the light microscope are the rows of diamond-shaped holes left between the diamond-shaped plates.

SEM stereopairs show that some of these calcite ribbons occur as parallel-oriented arrays like a set of picket fences, and show no obvious twisting. But many of these calcite ribbons have a helical twist (Fig. 4) and a whole series of ribbons seems to twist in unison (resembling the helical quartz fibers in zebraic chalcedony (Milliken, 1979)).

The *c* axes of all the ribbons are parallel with their length (and growth direction); the flat sides of the diamond-shaped plates must then be prism faces (two of which are greatly developed, while the other four are suppressed), and the edges of the diamond appear to be truncated by $40\bar{4}1$ faces. Some overall crystallographic influence must have made the diamond-shaped plates also become oriented parallel with each other.

In summary, the calcite crystals showed extraordinary growth in the *c* direction, but only one of the *a* axes grew while the other two were suppressed. Here we see a common thread uniting much of the travertine crystallography—these helical ribbons on a 1 micron scale, the lateral "wings" on calcite needles seen both in Yellowstone and Italy (ten-micron scale) and the cm-scale edgewise-plate structure of the large ray crystals. In the latter case the plates apparently grew parallel with the direction of water flow. All of these forms show major growth in the *c* axis direction, perpendicular to the encrusted surface, and selective lateral growth on only one of the three *a* axes, with the lateral growth recurring in the same direction on many adjoining crystals.

Some of the fine ray-crystal fans show much less porosity in the SEM (Fig. 5A, B, C), instead looking like a solid mass of tightly-fitting bundles (fasces) of prismatic calcite crystals 2–10 μm wide, with only occasional cross-wise gaps formed by upwardly- and downwardly-projecting steep rhombic terminations ($20\bar{2}1$) off the bundled calcite prisms (Fig. 5C). Some of these bundles also show curvature or a braided, winding morphology (Fig. 5A, B) and they probably represent helical ribbons that have had their intervening spaces filled in by early diagenetic calcite.

The cavity-rich bands are interpreted as daily laminations, by the arguments given previously. Fans are a millimeter or two thick, and the central part shows wider spacing of cavity bands ($\simeq$10 μm)—(?summer)—and the spacing decreases to invisibility nearing the top and bottom of the fan —(?winter) - (Fig. 3B). If the bands are indeed daily, then the fans accomplished the bulk of their growth in one or two summer months.

Some reinforcement of this idea of daily lamination comes from examination of stream travertine now being deposited on a branch of Bee Creek, Travis

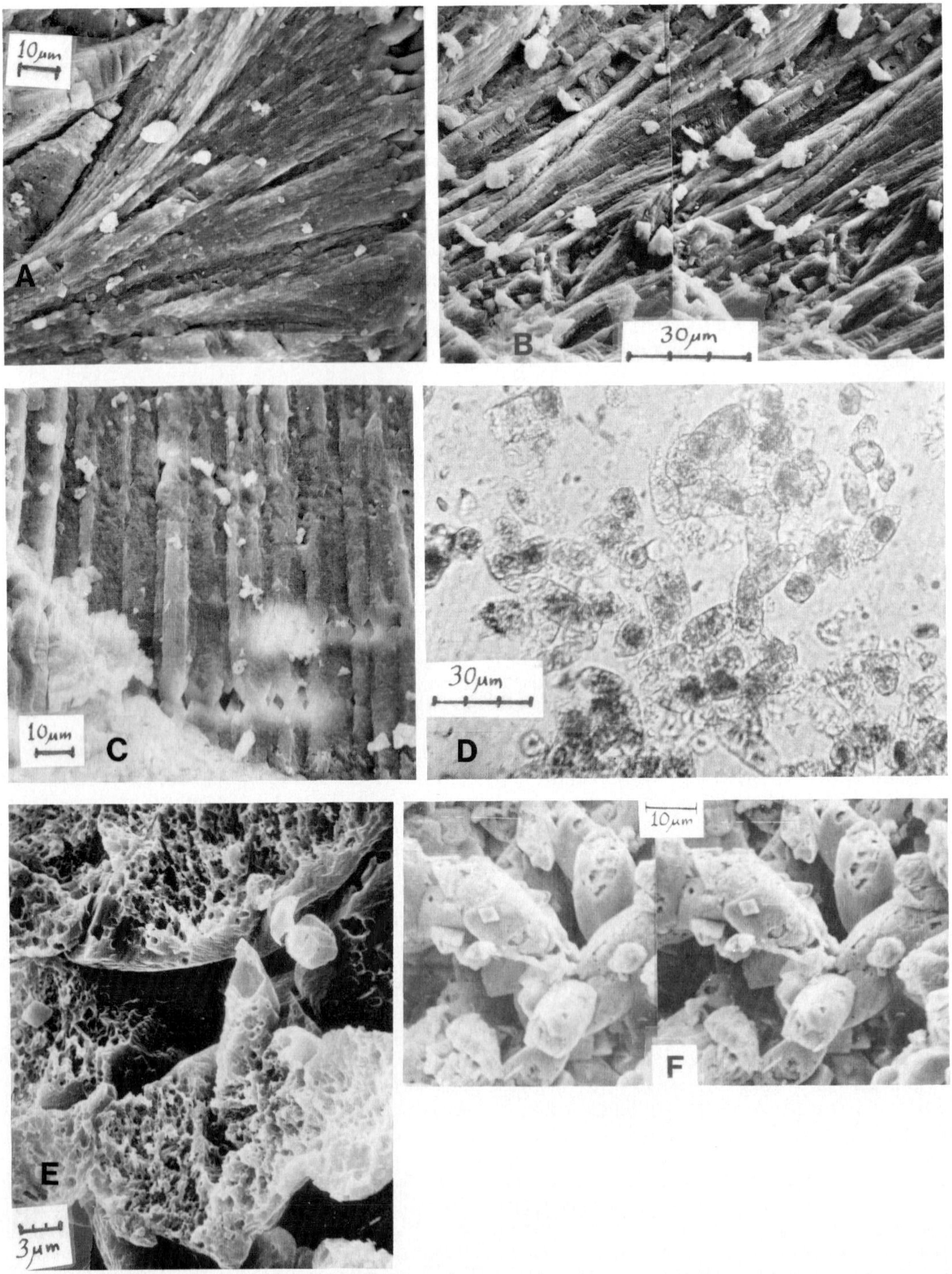

FIG. 5.—*A*, SEM view of fine ray-crystal fan showing pronounced curvature of the individual crystallites. *B*, SEM stereo pair; note crystals winding over each other. *C*, SEM of ray crystal fan that is relatively non-porous and tightly-bundled (fasces). Observe short row of diamond-shaped pores. *D*, Calcite crystals of bacterial mat. Nearly circular dark centers are the bacterial clumps; the clear outer rim is a single calcite crystal. Ordinary light. *E*, SEM of sample very slightly etched in very dilute HCl. Crystals are riddled with bacterial holes. *F*, SEM stereo-pair showing "wormy pecan" calcite crystals of the bacterial shrubs.

County, central Texas. Folk built a small dam of rocks across the stream, and it took five years for a travertine crust approximately 3 mm thick to accumulate. Microscopically, this travertine shows bands made up of a gap 1 μm thick, followed by a calcite band 3–6 μm thick (Fig. 3F). There are no bacteria microscopically evident in this low-sulfur fresh-water creek. Thickness calculations indicate about 100–200 small bands per year, which probably represents daily growth during those parts of the year when the stream is flowing well (it is normally dry in summer). Presumably precipitation of a band of calcite occurs during the heat of the day. However, the reason for the gap—presumably formed at night—is not known. We can postulate that it may have been a thin layer of mucus-like organic matter that oxidized and disappeared, similar to the continuous humate membranes described from oolites by Suess and Fütterer (1972).

Calcite of Bacterial Mats and Shrubs

Gently undulose bacterial mats and fern-like bacterial shrubs several cm tall are a common feature of the Italian travertine (Chafetz and Folk, 1984). These are made of calcite crystals that are full of cavities and resemble old pecans that have had their centers eaten out by worms (Honda and Folk, 1982). The crystals are prolate ellipsoids 10–30 μm long and about ½ to ⅓ as wide. They vary from rather smoothly rounded forms into those with indistinct crystal faces, with hexagonal prisms grading into steep rhombs or scalenohedrons.

Under the optical microscope these calcite crystals are single crystallographic units (Fig. 5D) but consist of two conjoined parts: a central, smoothly elliptical core that is dark, almost opaque (white in reflected light) because it is richly crowded with bacterial cavities (Fig. 5E) separated by a largely empty moat from an outer carapace crystal that is clear and shows reasonably good crystal form. The whole mass of crystals shows loose packing, like an uncompacted heap of pecans (Figs. 5F, 6A).

The core has closely-spaced bands of cavities (from <0.2 to 1 μm wide) representing former bacteria, that are plated out along the basal planes of the calcite crystal (Fig. 6A). These bacteria grew *pari passu* with the calcite, as basal sheets of the crystal formed. Bradley (1963) mentions similar pecan-shaped calcite crystals 20 μm long in an H_2S-rich mud at the bottom of a New York lake; they were black with occluded bacteria that also had plated out on crystallographic planes as the crystals grew.

After the bacteria-rich core attained its smoothly elliptical shape and stopped growing, something formed a "moat" between the core and the carapace crystal (Fig. 6A, B). Perhaps the bacteria secreted a mucus that temporarily coated the core. Then the core grew an optically-continuous carapace, and many of these are made of extraordinarily thin walls of calcite .05–.1 μm thick, so that the whole crystal resembles an ear of corn with a tattered shuck enclosing it. The carapace crystals often have large holes in them, but we cannot tell whether these were original gaps left as the crystal grew, or are slightly later micro-solution features. In the specimens there is no overt evidence of later solution such as larger (mm-scale) cavities. As 0.1 μm also seems to be about the limiting thickness of calcite strands in the fine helical fans, this may be a minimal size for calcite to exist in this environment, i.e. all smaller crystal lattices are quickly dissolved.

One specimen of bacterial mat contained a layer with nests of minute calcite rods (Fig. 6C). These measure 3–20 μm long and 0.2–0.5 μm wide, have squared-off ends and appear to have either four or six sides in cross section[1]. They are oriented like a pile of matchsticks, with some of them in sub-parallel bunches. Optical data shows that they are crystals of calcite, not aragonite; they do not extinguish parallel to the rod direction as would be normal for calcite, but instead extinguish at angles of 30 to 60° like the ones described by Moore and Bukry (1968) from "moonmilk" in a cave. Thus they apparently are dendrite-like needles of calcite that have grown along the intersection of two rhomb faces, similar to that variety of calcite once known as "lublinite" (see Assereto and Folk, 1980, p. 388).

Most calcite rods are single crystal units with smooth edges like a length of board (Fig. 6D) but others are made up of overlapping imbricated rhombs like the "whisker" crystals of Supko (1971) or the vadose needle-fiber calcite depicted by Ward (1974, 1975) and from a cave by Stoopes (1976). James (1972), Knox (1977), Harrison (1977) and Klappa (1979) have shown similar needles in caliche soils. Many of these authors have suggested that the needle crystals nucleated upon fungal filaments, but in ours the crystals lie free in the cavities and are squared-off at both ends; had they been nucleated about fungal filaments we would have expected them to bridge the cavities. We feel that our needle crystals formed during evaporation as the water level dropped and the bacterial-mat pond dried out.

Calcite Rafts

Paper-thin sheets of calcite often form on the surface of bodies of still water, where CO_2 degassing takes place. They are well known from caves (Baker and Frostick, 1951; Black, 1953), algal mounds (Riding, 1979), surficial carbonates (Chafetz and Butler, 1980), and hot spring pools (Weed, 1887, p. 641; Kitano, 1963) and have recently been described from peritidal tepee structures by Chambers *et al.* (1982, p. 1135) under the description "floe" carbonate. Agitation causes the rafts to break up and sink (Weed, 1887)

[1]In 1983 Tiezzi found in a quarry at Tivoli small cavities containing clear needles of calcite measuring 0.5 × 10 mm, with rhombic cross sections. These are apparently a super large-scale development of the ones seen in SEM.

forming an aggregate of flakes on the bottom. Similar rafts are present in Italian travertines, forming sheet-like fragments 0.2–20 mm long. The fragments have a dark micrite film in the center, and are lined by calcite crystals on both surfaces. The public men's room in Siena (Toscana) has walls which show beautiful calcite rafts with exquisite tiny bacterial shrubs only a few mm tall, encrusting only the top surfaces of the rafts and perpendicular to them (Fig. 7, A, B)—this is important in showing that the bacteria were photosynthetic and the rafts must have been horizontal when exposed to light, then were subsequently broken and tilted. Rafts could probably also form without the aid of photosynthetic bacteria in any enclosed system of cavities where water stands and degassing of CO_2 occurs at the surface, and where they would be sheltered from any wave action.

Rhombic Calcite Cement and Its Bizarre Variants

Travertine has an abundance of pores ranging from micron-sized ones between elemental bacterial clumps and the surrounding "wormy pecan" crystals, larger ones between bacterial shrubs, unfilled holes a few mm in diameter in "foam rock", to decimeter-long horizontal pores so characteristic of building stone. Almost all of these represent primary and original unfilled pore

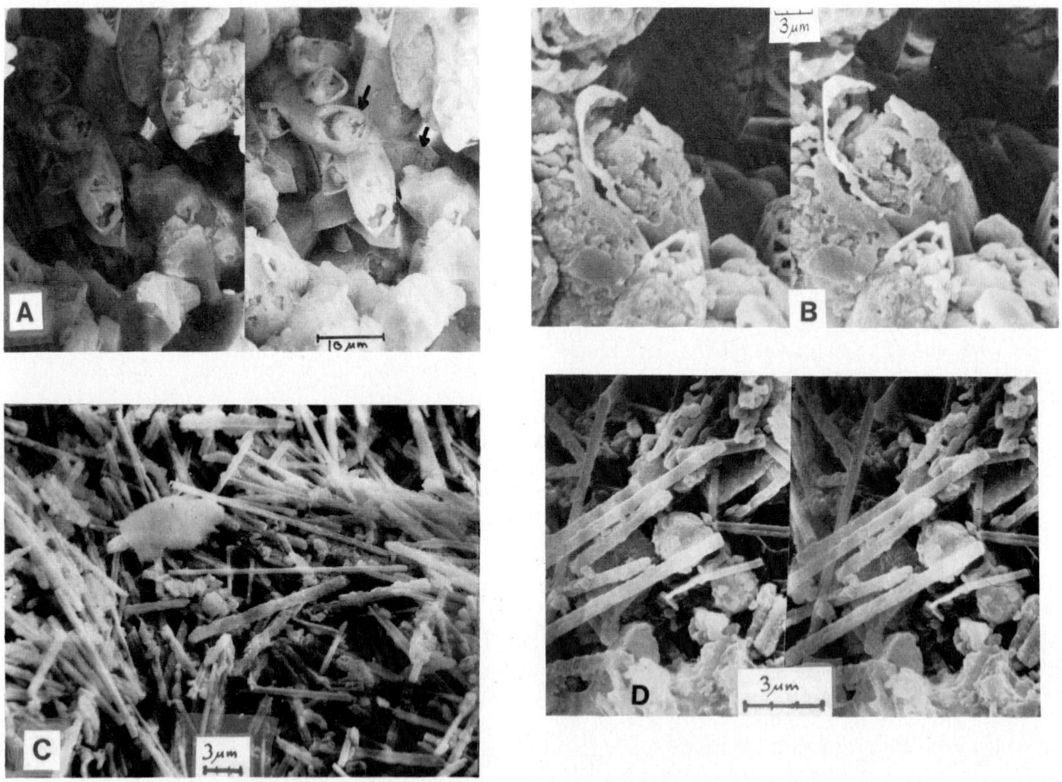

FIG. 6.—*A*, SEM stereo-pair showing central bacterial-rich core crystals that have transverse banding (arrows) due to denser concentrations of bacteria. These are surrounded by moth-eaten carapace crystals of very thin calcite walls. *B*, SEM stereo-pair showing fragile and tattered character of the carapace crystals. *C*, SEM of nest of needles in bacterial mat. *D*, SEM stereo-pair showing needles of diagonally-extinguishing calcite ("lublinite").

FIG. 7.—Rafts and Gothic-Arch calcite *A, B*, Slabs on the wall of the public men's room, Siena. Calcite rafts grew horizontally at the surface of standing pools of water and were encrusted on the topsides by tiny bacterial shrubs, probably photosynthetic. In A, the rafts were broken and the shrubs tilted along with them (arrow). In B, the rafts are substantially *in situ*. *C*, Gothic-arch calcite with each crystal made of overlapping scales. Ordinary light. *D*, End-on view of gothic-arch calcite to show the outriding slabs in optical continuity with the triangular mother crystal. Crossed nicols (right) and ordinary light (left). *E*, SEM stereo-pair of gothic-arch calcite, showing curvature of crystals. *F*, Overall view of gothic-arch calcite with the characteristic slabs on the rhomb faces. SEM view. *G*, SEM close-up of the gap between the slabs. (arrow)

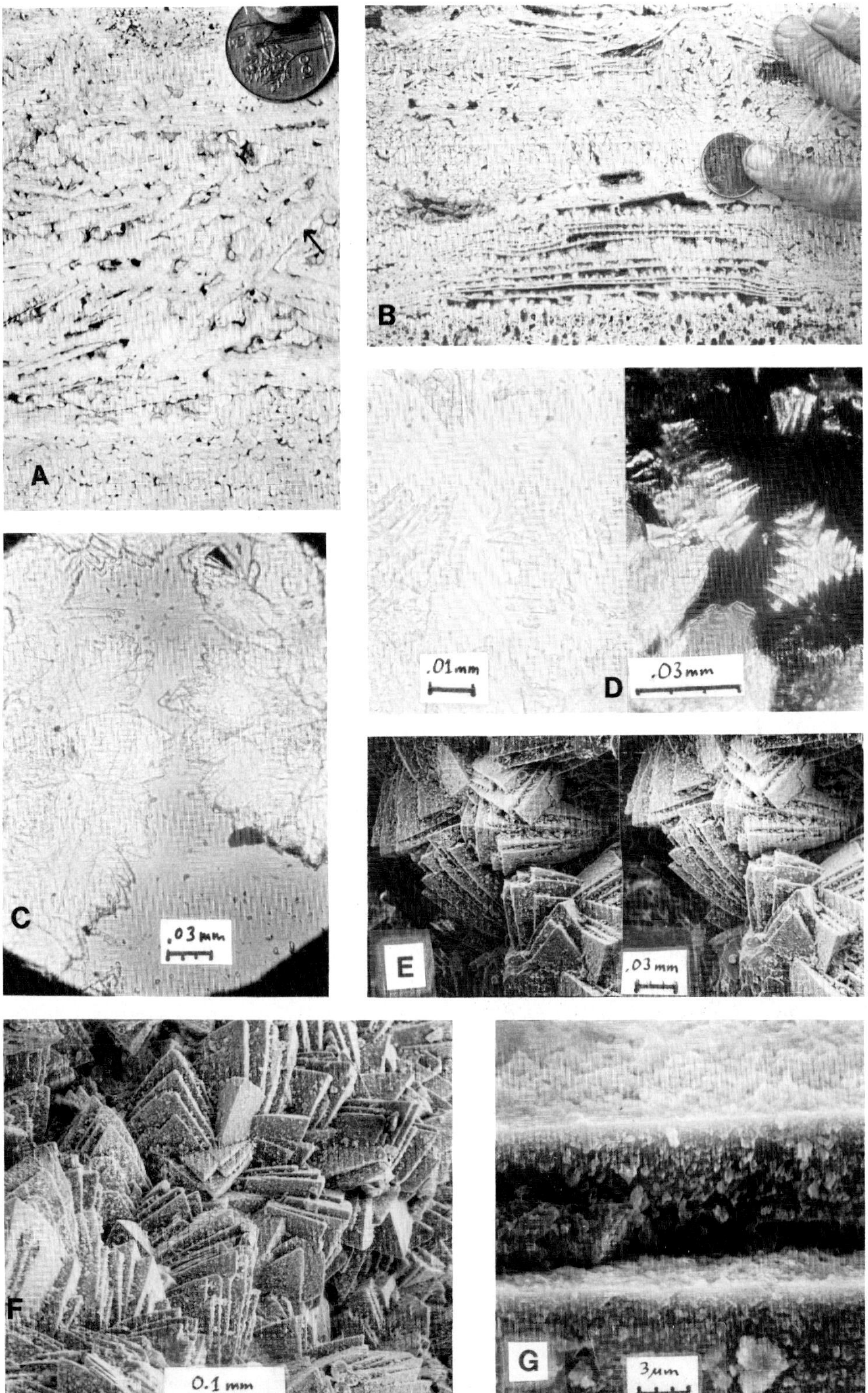
A
B
C
.03 mm
.01 mm
D
.03 mm
E
.03 mm
F
0.1 mm
G
3 μm

space, though locally in quarries some major solution cavities and solution-enlarged fractures do occur.

Pores larger than a few tenths of a millimeter are almost universally lined with clear, equant euhedral crystals of sparry calcite that are typically finely crystalline (.02–.06 mm) and have the steeper rhomb terminations of $20\bar{2}1$ and less commonly $40\bar{4}1$. They are steep rhombs, not scalenohedrons, because viewed down the *c* axis they give triangular (not 6-sided) cross sections. The unit rhomb $(10\bar{1}1)$ and flat rhomb $(10\bar{1}2)$ are very seldom observed, and we have not observed any basal pinacoid faces (0001). Very little is yet known about the significance or cause of the various crystal forms of calcite, though Heddle (1901, p.126) noted that in pure $CaCO_3$ solutions the elemental rhombs $(10\bar{1}1)$ formed, and as impurities were added to the solutions in the laboratory, various more complex faces were developed. Vater (1899) also experimentally determined that rhomb faces became steeper upon addition of SO_4^{2-} to the system, and Bischoff and Fyfe (1968) showed that poorly-developed scalenohedra developed in laboratory solutions containing SO_4^{2-}.

In many of the travertines, the basic rhombohedron of sparry calcite is modified into aberrant varieties that create spectacular SEM views. There are no prior names to describe these, so for brevity and descriptive facility we shall refer to them as "gothic-arch", "spiky", and "edge-guttered" calcite.

Gothic-arch calcite.—These crystals are typically fine crystalline (.02–.04 mm), and of more or less equant shape. In the optical microscope they appear as steep-rhombic crystals having a series of triangular flat-iron-like slabs added to the rhomb faces (Fig. 7C–G); the "mother" crystal and the slabs may have slight convex curvature and sometimes have a slight sweeping extinction. The slabs are separated from the central crystal by a gap a fraction of a micron wide

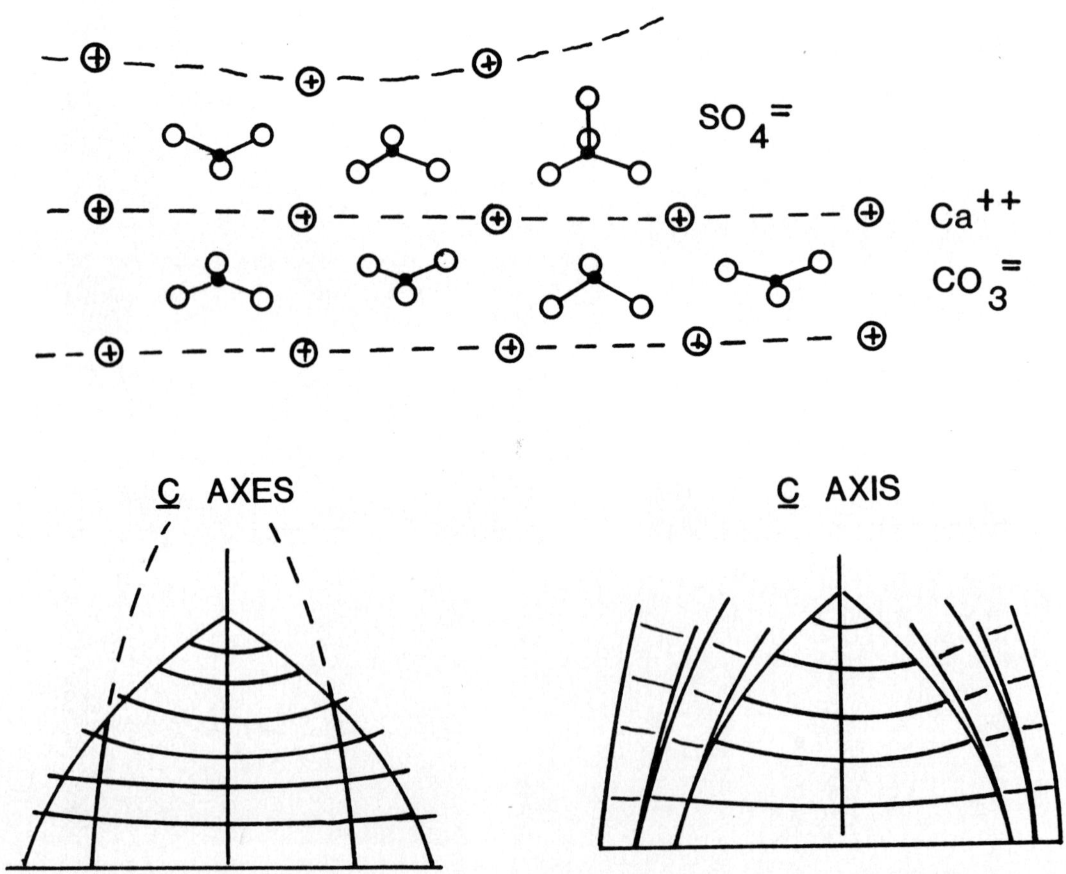

FIG. 8.—Proposed mechanism to explain crystal curvature in gothic-arch calcite. The SO_4^{2-} group is too big to fit into the carbonate sheet, so forces the Ca layer to be bent up. Thus, as the crystal builds out into a cavity, the carbonate sheets of the crystal lattice will become concave outward. At a certain stage (lower right) the distortion becomes so great that the crystal cannot tolerate it, and a new slab starts forming, slightly detached from the mother crystal.

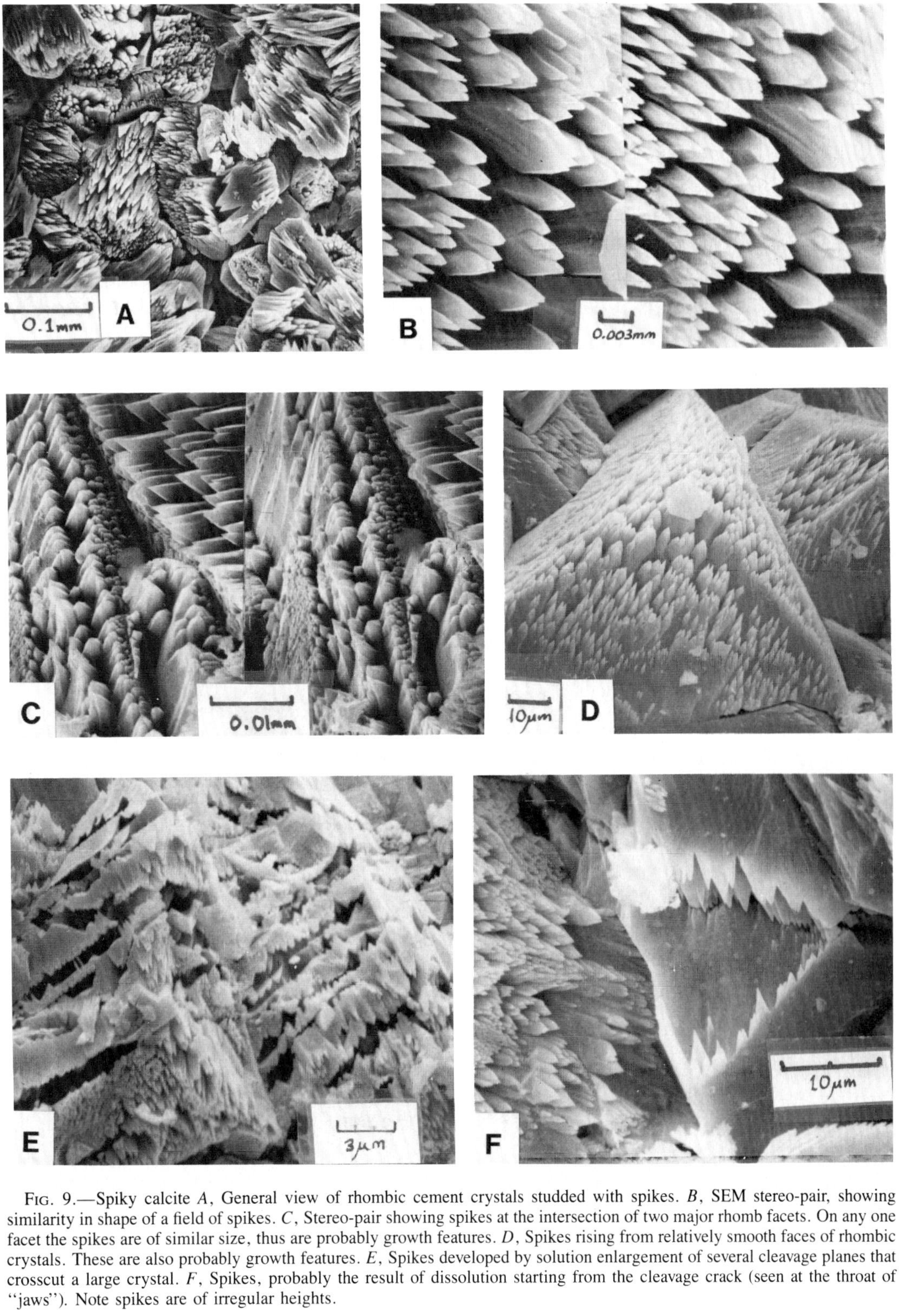

FIG. 9.—Spiky calcite *A*, General view of rhombic cement crystals studded with spikes. *B*, SEM stereo-pair, showing similarity in shape of a field of spikes. *C*, Stereo-pair showing spikes at the intersection of two major rhomb facets. On any one facet the spikes are of similar size, thus are probably growth features. *D*, Spikes rising from relatively smooth faces of rhombic crystals. These are also probably growth features. *E*, Spikes developed by solution enlargement of several cleavage planes that crosscut a large crystal. *F*, Spikes, probably the result of dissolution starting from the cleavage crack (seen at the throat of "jaws"). Note spikes are of irregular heights.

(Fig. 7G). In end-on view, the mother crystals appear as triangles, with the accompanying slabs stacked beside each rhomb face (Fig. 7D).

In the SEM, the crystals appear as gothic arches, i.e. curving triangular faces that converge to a point, with successive slabs of similar gothic-arch shape attached to each rhomb face (Fig. 7E, F). Again, gaps between the slabs and the mother crystal are clearly visible and about 0.2 μm wide. Some slabs fatten as they approach the *c*-axis point of the mother crystal. Among the 2544 crystal forms of calcite pictured in vol. 2 of Goldschmidt's (1913) Atlas, only figure 1403 approaches these curious shapes.

At this point we can only propose a rather speculative hypothesis to explain the gothic-arch form and semi-detached flatiron shells. We propose that, in this S-rich environment, an occasional SO_4^{2-} group takes the place of the planar CO_3^{2-} group in the sheet-structured calcite lattice (Fig. 8). The tetrahedral SO_4^{2-} group with four oxygens is too large, so that the following carbonate sheets tend to be bent upwards. If this happens at regular intervals, the crystal will have basal CO_3^{2-} sheets that are concave outward into the cavity (Fig. 8). Furthermore, because the sheets increase their separation as they grow laterally, a point will be reached beyond which the crystal can grow no further because distortion has become too great. At this point, crystal growth halts, and a new slab with proper CO_3^{2-}-sheet spacing begins to form. As it grows, it becomes distorted in the same manner by small amounts of SO_4^{2-} until it is forced to halt and another new slab forms. This would explain why the sheets appear to be similar in thickness on all three sides of the rhomb. Sulfate has not yet been detected by Energy Dispersive X-Ray Analysis, but perhaps only a few ppm could account for the curvature. Cortesi and Leoni (1955) report that bulk travertine samples from Bagni run 0.5–1.2 wt. % SO_4^{2-} and much of this may be incorporated into the carbonate lattice. Experiments of Bischoff and Fyfe (1968) indicated that SO_4 was able to be absorbed by the calcite lattice.

Spiky calcite.—This extraordinary type of calcite provides some of the most astounding views ever seen by the SEM (Chafetz, 1983). Crystals that at first appear to be ordinary rhombic crystals of sparry calcite are found upon increasing magnification to be covered with a forest of spikes like a fakir's bed (Fig. 9). Spikes all point in the direction of the *c* axis; viewed end-on or in stereo (Fig. 9C) some have hexagonal cross-sections, some are triangular and some are thin lenticular blades like spear points. There is considerable variation in size but most spikes protrude about 1–10 μm.

Occurrences range from smooth crystal faces having only a few spikes to rhomb faces that consist entirely of spikes all joined at the base like sawteeth.

Deciding whether the spikes are growth features or dissolution features is not easy. In the light microscope, the crystals lining the pores appear to be euhedra and do not show any evidence of a later solution episode. In sandstones, other minerals such as feldspars, amphiboles, and pyroxenes develop spectacular spikes upon dissolution, but in them the petrographic evidence of alteration is generally obvious and the spikes represent deep solution along cleavage planes. The comprehensive series of experiments carried out by Honess (1918, 1927) on calcite etching produced only pits upon flat crystal faces, not spikes. Goldschmidt and Wright (1904) produced some etch hillocks by attacking calcite with phosphoric acid, and judging from their figure (Taf. XXX, Fig. 11) these formed spikes oriented parallel to the *c* axis.

At this point we feel that the spikes may originate both by growth and by dissolution. The isolated spikes that rise to the same altitude from a planar crystal face (like a few nails driven through a board, Fig. 9D) are probably growth features (such as the hillocks depicted by Elwell and Scheel, 1975, p. 188). Moreover closely-spaced spikes that touch at their bases, if they rise to the same elevation, are suggestive of a growth feature. On the other hand, we have a few examples wherein the spar crystals have a crosscutting cleavage plane (Fig. 9E, F) and this enlarged cleavage opening is also lined with spikes apparently as a result of solutions attacking the host crystal. These solution spikes appear to be less uniform in shape, and their heights are dissimilar. As calcite dissolves most easily perpendicular to the basal plane, solution spikes should result paralleling the *c* axis direction.

We suspect that the true answer is a composite. Some chemical imbalance or lattice defect (perhaps a spiral dislocation center) that was spaced rather regularly along the calcite crystal lattice caused more rapid growth of the spikes in the *c* axis direction. First, spikes formed at regular intervals along the basal carbonate sheet, and later the rest of the crystal "caught up" and buried the spikes with normal calcite. Still later, when solution occurred, the phenomenon took place in reverse with selective solution now "exhuming" the original spikes buried within a normal-appearing crystal. Dissolution probably occurred very shortly after growth—perhaps during the next night or next cool season. Energy-Dispersive X-Ray Analysis has not revealed any composition differences between the spike tips and the ordinary crystal faces of the calcite from which they protrude, but again a few ppm of Mg or SO_4^{2-} might have profound crystallographic effects but not show up on analysis. A SO_4^{2-}-group, too large for its position in the CO_3^{2-} sheet (Fig. 8), might "kick-up" the next sheet and initiate a spiral dislocation to stimulate local growth in the same manner as that postulated by Frank (1952) for foreign particles settling on the crystal surface. Hendricks (1955), Elwell and Scheel (1975) and Sunagawa and Bennema (1979) show examples of spiral defects or screw dis-

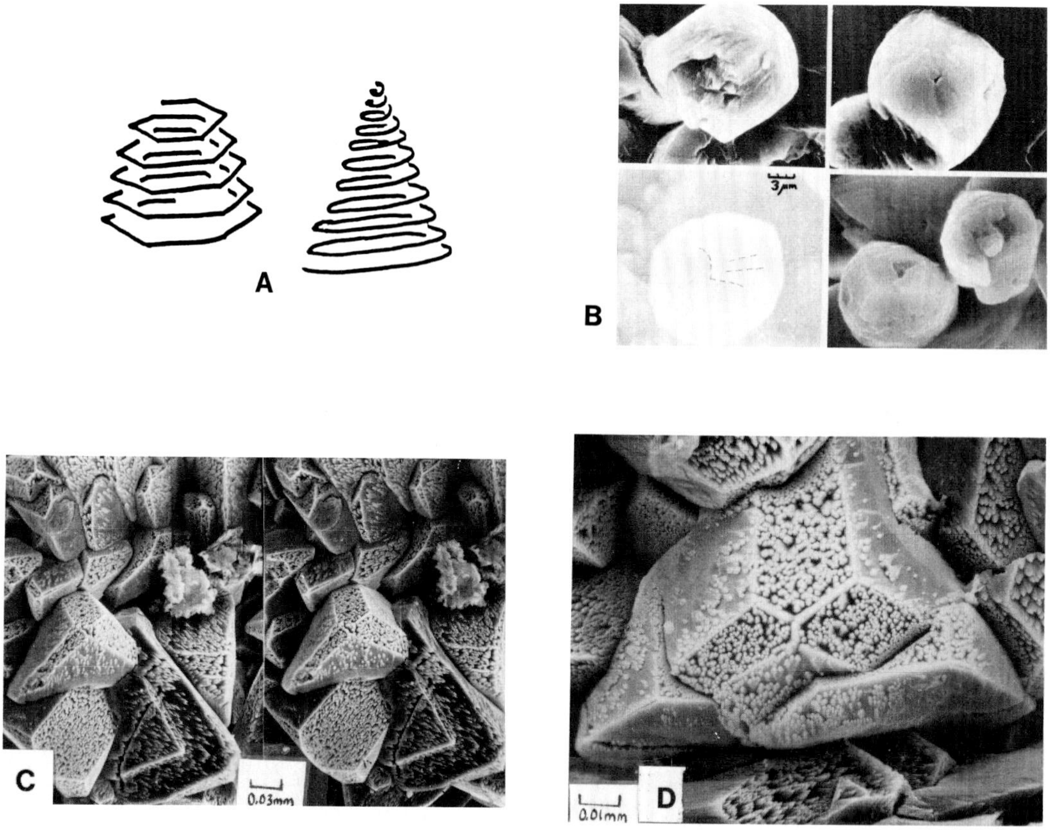

FIG. 10.—Probable screw-dislocations and edge-guttered calcite *A*, Conical or conical-hexagonal form of the calcite spikes of Fig. 9 may be caused by a screw dislocation (due to the too-large SO_4^{2-} group). It is proposed that the sheets of the calcite lattice wind helically about the *c* axis. *B*, Four of the many probable screw-dislocation crystals that occur abundantly in one of our specimens. All taken looking down the *c* axis which is marked by a centrally-located pit; sheets of calcite then grow in a helical manner around this axis. Lower left figure had to be retouched because the SEM filament blew before we could get a better photograph of the crystal. *C*, SEM stereo-pair showing a field of edge-guttered calcite cement crystals lining a pore. *D*, Close-up of C, looking down the *c* axis of a crystal. Note abundance of spikes (?dissolution) on three rhomb faces, and much smoother steep-rhomb faces.

locations, all developed as gentle swells on essentially flat crystal faces. Perhaps our spiral dislocations develop into steep conical spikes (exaggerated hillocks, as in Fig. 10A), rather than gentle domes. Whisker crystals of NaCl form because of screw dislocations propagated along the fiber, and these also (Eshelby, 1953), give rise to a slight winding of the crystals known as the "Eshelby twist" (Uebach and Bradaczek, 1982). Our steep conical spikes and helically-twisted ribbons of calcite in the fans may form by a similar mechanism.

One of our samples of large-ray crystal crusts is made up of pecan-shaped calcite crystals that when viewed in SEM end-on (down the *c* axis) show what appears to be many examples of screw-dislocations (Fig. 10A, B). These layers appear to be about 0.1 μm thick (thus a step including many unit cells) and are more properly termed "macrospirals" (Elwell and Scheel, 1975, p. 180). Such macrospirals can be made of layers many tenths of microns thick, and can have etch-pits at the center of the spirals (Amelinckx *et al.*, 1957) like those shown in our calcite crystals.

Edge-guttered crystals.—These crystals consist of two sets of rhomb faces (Fig. 10C, D). One steeper rhomb set (f, $20\bar{2}1$) forms three major relatively smooth facets as viewed down the *c* axis. The other rhomb set is the unit negative rhomb (r, $01\bar{1}1$) which developed along the edges of the first rhomb set, and

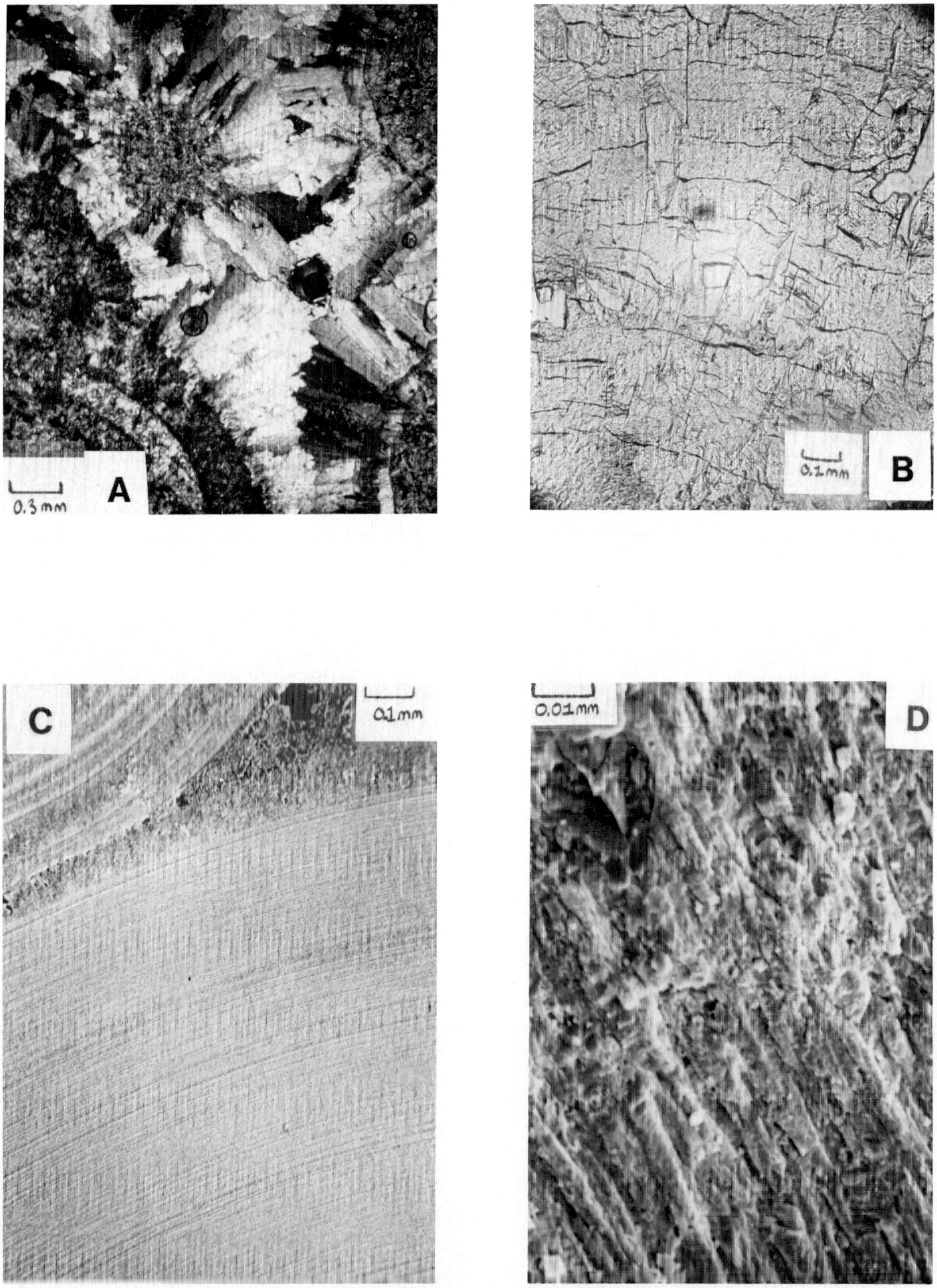

FIG. 11.—Palisade calcite and inorganic pisolites. *A*, Palisade calcite filling a pore and encrusting *Chara* stems; crossed nicols. *B*, Close-up to show near-basal parting (running almost horizontally across the photo). Center of the encrusted surface is below the edge of the photo. Ordinary light. *C*, SEM view of inorganic pisolite showing very delicate banding. *D*, SEM enlargement of C. The pisolite is made of tightly-bonded calcite crystals (like fasces) that are radial in orientation.

these faces are extensively corroded into spikes and hillocks like a miniature version of Chinese tower karst. The overall crystal shape resembles Figure 231 in Dana and Ford (1932). Thin surviving walls divide the facets from each other. It is well established that the positive and negative rhomb faces in calcite show differential solubility (e.g. Dana and Ford, 1932, p. 122) as well as there being major solubility differences between rhombs and scalenohedra of different indices (Heddle, 1901, p. 116; Honess, 1918). Goldschmidt's (1913) Atlas shows several examples of similar crystals with one set of rhombs dull and the other set normal (Figs. 1209, 1959, 2166, 2247).

Palisade Calcite Cement

These crystals tend to line the largest cavities, particularly those found surrounding colonies of the freshwater alga, *Chara*. These drusy cavities can be several cm in diameter and are a delight to find in the field. Crystals are typically .3 to 2 mm long and .1–.2 mm wide, length-fast, and occur in subparallel to radial palisades about *Chara* stems and as cavity linings (Fig. 11A, B). In the binocular microscope their terminations appear similar to the "gothic arch" calcite in that they are made of offlapping parallel slabs of rhomb faces, but they show no evidence of curvature. The crystals are terminated mainly by moderately steep rhomb faces, (f = $20\bar{2}1$), with some steep rhombs (M, $40\bar{4}1$). All cross-sections of terminations are triangular, so that scalenohedral faces are lacking. Furthermore, we have never observed the basal pinacoid face. The unusual feature of the palisade calcite is that it shows what at first appears to be a rather ragged basal parting that often runs continuously across several crystals (Fig. 11B). What this probably represents is a rough parting that oscillated about the flat-rhomb ($10\bar{1}2$) plane (which is a common parting direction for calcite) combined with the very weak parting on the basal plane, (0001) (Palache *et al.*, 1951, v. II, p. 150).

Palisade calcite was the last to form, and because it encrusts all sides of the cavities equally, was probably formed in the phreatic realm. Its relatively conventional crystal habit indicates it was probably formed in water approaching normal fresh water. Continuing our sulfate bias, perhaps at certain time intervals enough SO_4^{2-} accumulated in the water to plate across growing crystals as a sort of time line, interrupting growth and forming a surface of weakness now expressed as the parting.

Calcite "Fasces" of the Inorganic Pisoids

Travertine pisoids are of two types: (1) irregular bumpy ones made out of radiating bacterial shrubs, and (2) extraordinarily smooth ones made of extremely fine and regular concentric bands, formed inorganically under high agitation within hot spring mouths (Fig. 11 C). As these two types have been discussed in detail by Folk and Chafetz (1983), only a few points will be mentioned here. Many of the inorganic pisoids show a prominent extinction cross, and these are made of radial length-fast calcite crystals (Fig. 11D) that are tightly bundled together like the fasces of the ancient Roman Empire (signifying "in unity there is strength"). The crystals must have grown by rapid precipitation of calcite as radial crystals at the smooth, advancing surface of the pisoid and are not mechanically-accreted or diagenetic features. The radial crystals are 2–3 μm wide and resemble almost exactly the radially-bundled crystals in the more solid-appearing fine fans (see Fig. 5, C). Thus the radial orientation of calcite here has nothing to do with the physical energy (agitation) of the environment, but depends instead upon an inherent growth bias of calcite that normally crystallizes with radial *c* axes.

CONCLUSIONS

The travertines of central Italy show an extraordinary variety of strange carbonate rock types: bacterial "bioherms" and mats, foam-rock, ray-crystal crusts, pisoids, etc. In addition to this they have near-record precipitational rates of up to one meter per year, and show annual and daily lamination with the first known "dawn layer". Furthermore, we have presented a veritable "zoo" of strange types of primary and diagenetic calcite: "cedar-needles" that aggregate into ray-like supercrystals up to a meter long, "wormy pecans" centered by bacteria-rich nuclei, fine fans made up of helically-twisted ribbons of calcite, curved gothic-arch calcite crystals with flatiron-like slabs, crystals whose edges are guttered or whose faces are studded with spikes, and palisade crystals with quasi-basal parting. The characteristics that make this type of limestone differ from "normal" fresh water carbonates are the high initial H_2S content of the water, the warm temperatures, and the abundance of bacteria. We have suggested that some of the strange crystallography is due to incorporation of SO_4^{2-} into the carbonate lattice in very minute quantities, something that it is beyond present experimental capacity to verify. We hope to extend this study to other travertines of sulfur-associated carbonates to see if some consistent pattern emerges. Until it does, we continue to be amazed and intrigued by each new sample that goes into the SEM.

ACKNOWLEDGMENTS

It is a pleasure to acknowledge the assistance of Patrizia Reale Massano for providing logistic support through the University of Roma; and Hiromi Honda for some of the initial SEM work. Larry Mack, Jim Wittke and Matt Coldwell (University of Texas) nursed along our cranky SEM in Austin. Richard Morales ground the difficult thin sections, and David Stephens did the conversion of the final photo prints. Our thanks go also to Lynton Land and Phil Sandberg for their careful editing and enlightened skepticism. The research was supported by the Geology Foundation and Carleton Professorship fund of the University of Texas at Austin and the Geology Foundation of the University of Houston after repeated rejections by other funding sources. The Owen-Coates fund defrayed some printing costs.

REFERENCES

AMELINCKX, S., BONTINCK, W., AND DEKEYSER, W., 1957, Helical dislocations and spiral etch-pits: Philosophical Mag., v. 2, p. 1264–1270.

ASH, D.W., 1975, Appleite: A new calcite structure from Apple Cave, Orange County, Indiana: Nat. Speleological Soc. Bull., v. 37, no. 1, p. 35–39.

ASSERETO, R., AND FOLK, R.L., 1980, Diagenetic fabrics of aragonite, calcite and dolomite in an ancient peritidal-spelean environment: Triassic Calcare Rosso, Lombardia, Italy: Jour. Sed. Petrology, v. 50, p. 371–394.

BAKER, G., AND FROSTICK, A.C., 1951, Pisoliths, ooliths, and calcareous growths in limestone caves at Port Campbell, Victoria, Australia: Jour. Sed. Petrology, v. 21, p. 85–104.

BARGAR, K.E., 1978, Geology and Thermal History of Mammoth Hot Springs, Yellowstone National Park, Wyoming: U. S. Geol. Survey Bull. 1444, 55 p.

BISCHOFF, J.L., AND FYFE, W.S., 1968, Catalysis, inhibition, and the calcite-aragonite problem. 1. The aragonite-calcite transformation. Amer. Jour. Sci., v. 266, p. 65–79.

BLACK, D.M., 1953, Aragonite rafts in Carlsbad Caverns, New Mexico: Science, v. 117, p. 84–85.

BLAKE, M.E., 1947, Ancient Roman Construction in Italy: Carnegie Inst. Washington Pub. 570.

______, 1959, Ancient Roman Construction in Italy: Carnegie Inst. Washington Pub. 616.

BRADLEY, W.H., 1963, Unmineralized fossil bacteria: Science, v. 141, p. 919–921.

BRIGGS, M.S., 1956, Building and construction, *in* Charles Singer *et al.* (eds.), A History of Technology, v. II: Oxford, Clarendon Press, p. 397–448.

CHAFETZ, H.S., 1983, Sedimentology photo, sparry calcite cement, Tivoli, Italy: Jour. Sed. Petrology, v. 53, p. 274.

______, AND BUTLER, J.C., 1980, Petrology of Recent caliche pisolites, spherulites and speleothem deposits from central Texas: Sedimentology, v. 27, p. 497–510.

______, AND FOLK, R.L., 1984, Travertines: Depositional morphology and the bacterially-constructed constituents: Jour. Sed. Petrology, v. 54, p. 289–316.

______, AND MEREDITH, J.C., 1983, Recent travertine pisolites (pisoids) from southeastern Idaho, U.S.A., *in* T.M. Peryt (ed.), Coated Grains: New York, Springer-Verlag, p. 450–455.

CHAMBERS, L.A., BURNE, R.V., AND FERGUSON, J., 1982, Lithification of peritidal carbonates by continental brines at Fisherman Bay, South Australia, to form a megapolygon/spelean limestone association: Jour. Sed. Petrology, v. 52, p. 1127–1145.

CIPRIANI, N., ERCOLI, A., MALESANI, P., AND VANNUCCI, S., 1972, I travertini di Rapolano Terme (Siena): Mem. Soc. Geol. Ital., v. 11, p. 31–46.

CORTESI, C., AND LEONI, M., 1955, Studio sedimentologico e geochimico del travertino di un sodaggio a Bagne di Tivoli: Periodico di Mineralogia, v. 27, p. 407–458.

DANA, E.S., AND FORD, W.E., 1932, A Textbook of Mineralogy (4th ed.): New York, John Wiley, 851 p.

D'HALLOY, J.J., 1812, Sur l'existence de calcaire d'eau douce dans les departements de Rome et de l'Ombrone, et dans le Royaume de Wurtemberg: Jour. des Mines (Paris), v. 32, p. 401–410.

ELWELL, D., AND SCHEEL, M.J., 1975, Crystal Growth from High-Temperature Solutions: New York, Academic Press, 634 p.

EMIG, W.H., 1917, The Travertine Deposits of the Arbuckle Mountains, Oklahoma: Oklahoma Geol. Surv. Bull. 29, 76 p.

ESHELBY, J.D., 1953, Screw dislocations in thin rods: Jour. Applied Physics, v. 24, p. 176–179.

FOLK, R.L., AND CHAFETZ, H.S., 1980, Quaternary travertine of Tivoli (Roma) Italy: Bacterially constructed carbonate rock [abs.]: Geol. Soc. America Abs. Prog., v. 12, p. 428.

______, AND ______, 1983, Pisolites (pisoids) in Quaternary travertines of Tivoli, Italy, *in* T.M. Peryt (ed.), Coated Grains: New York, Springer Verlag, p. 474–487.

FRANK, F.C., 1952, Crystal growth and dislocations: Advances in Physics (Philosophical Mag. Supplement), v. 1, p. 91–109.

GEBELEIN, C.D., 1969, Distribution, morphology, and accretion rate of Recent subtidal algal stromatolites, Bermuda: Jour. Sed. Petrology, v. 39, p. 49–69.

GOLDSCHMIDT, V., 1913, Atlas der Krystallformen: Heidelburg, Carl Winters Univ., 9 volumes.

______, AND WRIGHT, F.E., 1904, Ueber Losungskorper und Losungsgeschwindigkeiten von Calcit.: N. Jb. Geol. Paläontol. Beilegeband, v. 18, p. 335–376.

GOLUBIC, S., AND FOCKE, J.W., 1978, *Phormidium hendersonii* Howe: Identity and significance of a modern stromatolite-building microorganism: Jour. Sed. Petrology, v. 48, p. 751–764.

HARE, A.J.C., 1884, Cities of Central Italy, v. II: London, Smith, Elder, & Co., 180 p.

HARRISON, R.S., 1977, Caliche profiles: Indications of near-surface subaerial diagenesis, Barbados, West Indies: Bull. Canadian Petroleum Geol., v. 25, p. 123–173.

HEDDLE, M.F., 1901, The Mineralogy of Scotland: Edinburgh, David Douglas, 2 volumes, 247 p.

HENDRICKS, S.B., 1955, Screw dislocations and charge balance as factors of crystal growth: Am. Mineralogist, v. 40, p. 139–146.

HONDA, H., AND FOLK, R.L., 1982, Sedimentology photo: Rotten calcite, travertines, Tivoli: Jour. Sed. Petrology, v. 52, p. 1294.

HONESS, A.P., 1918, On etching figures of the dihexagonal alternating type: Amer. Jour. Sci., v. 45, p. 201–221.

______, 1927, The Nature, Origin and Interpretation of the Etch Figures on Crystals: New York, John Wiley, 171 p.

JAMES, N.P., 1972, Holocene and Pleistocene calcareous crust (caliche) profiles: Criteria for subaerial exposure: Jour. Sed. Petrology, v. 42, p. 817–838.

KITANO, Y., 1963, Geochemistry of calcareous deposits found in hot springs: Nagoya Univ. Jour. Earth Sci., v. 11, p. 68–100.

KLAPPA, C.F., 1979, Calcified filaments in Quaternary calcretes: Organo-mineral interactions in the subaerial vadose environment: Jour. Sed. Petrology, v. 49, p. 955–968.

KNOX, G.J., 1977, Caliche profile formation, Saldanha Bay (South Africa): Sedimentology, v. 24, p. 657–674.

LOSACCO, U., 1951, La struttura del Territorio di Rapolano e Lucignano (Siena e Arezzo): Boll. Soc. Geol. Ital., v. 70, p. 402–434.

LYELL, C., 1829, On a recent formation of freshwater limestone in Forfarshire and on some recent deposits of freshwater marl and an appendix on the gyrogonite or seed-vessel of the *Chara*: Geol. Soc. London Trans. (ser. 2), v. 2, p. 73–96.

MANFRA, L., MASI, U., AND TURI, B., 1974, Effetti isotopici nella diagenesi dei travertini: Geol. Romana, v. 13, p. 147–156.

______, ______, AND ______, 1976, La composizione isotopica dei travertini del Lazio: Geol. Romana, v. 15, p. 127–174.

MILLIKEN, K.L., 1979, The silicified evaporite syndrome—Two aspects of silicification history of former evaporite nodules from southern Kentucky and northern Tennessee: Jour. Sed. Petrology, v. 49, p. 245–256.

MOORE, G.W., AND BUKRY, D., 1968, Cover photograph, Moonmilk from Texas: Nat. Speleological Soc. News, v. 26, p. 121–122.

NEWHOUSE, W.H., 1941, The direction of flow of mineralizing solutions: Econ. Geol., v. 36, p. 612–629.

PALACHE, C., BERMAN, H., AND FRONDEL, C., 1951, Dana's System of Mineralogy (7th ed.): New York, John Wiley, 3 volumes.

RIDING, R., 1979, Origin and diagnesis of lacustrine algal bioherms at the margin of Ries Crater, upper Miocene, S. Germany: Sedimentology, v. 26, p. 645–680.

RISACHER, F., AND EUGSTER, H.P., 1979, Holocene pisoliths and encrustations associated with spring-fed surface pools, Pastos Grandes, Bolivia: Sedimentology, v. 26, p. 253–270.

STOOPES, G.J., 1976, On the nature of "Lublinite" from Hollanta (Turkey): Amer. Mineralogist, v. 61, p. 172.

SUESS, E., AND FÜTTERER, D., 1972, Aragonitic ooids: Experimental precipitation from seawater in the presence of humic acids: Sedimentology, v. 19, p. 129–139.

SUNAGAWA, I., AND BENNEMA, P., 1979, Modes of vibrations in step trains: Rhythmical bunching: Jour. Crystal Growth, v. 46, p. 451–457.

SUPKO, P.R., 1971, "Whisker" crystal cement in a Bahamian rock, *in* O.P. Bricker (ed.), Carbonate Cements: Johns Hopkins Univ. Studies in Geol., v. 19, p. 143–146.

TERMIER, H., AND TERMIER, G., 1975, Sedimentary behavior and skeletal textures available in growth cycle analysis, *in* G.D. Rosenburg and S.K. Runcorn (eds.), Growth Rhythms and the History of the Earth's Rotation: New York, John Wiley, p. 89–901.

UEBACH, W., AND BRADACZEK, H., 1982, Unwinding of the Eshelby Twist in sodium chloride whiskers: Jour. Crystal Growth, v. 58, p. 203–207.

VATER, H., 1899, Ueber den Einfluss der Losungsgenossen auf die Krystallisation des Calciumcarbonates. VII. Der Einfluss des Calciumsulfates, Kaliumsulfates und Natriumsulfates: Zeitschr. Kristallographie, Kristallogeometrie, Kristallophysik, Krystallochemie, v. 30, p. 485–508.

WARD, W.C., 1974, Collector's guide to carbonate cement types, northeastern Yucatan Peninsula, *in* A.E. Weidie (ed.), Field Seminar in Yucatan: New Orleans Geol. Soc. Guidebook for Field Trip 2, Miami Geol. Soc. America Meeting, p. 176–198.

______, 1975, Petrology and diagenesis of carbonate eolianites of northeastern Yucatan Peninsula, Mexico: Amer. Assoc. Petroleum Geologists Stud. Geol., v. 2, p. 500–571.

WEED, W.H., 1887, The formation of travertine or siliceous sinter by the vegetation of hot springs: U. S. Geol. Survey Ann. Rep. 9, p. 613–676.

TECTONICALLY CONTROLLED FRESHWATER CARBONATE CEMENTATION IN CHALK

YA'ACOV MIMRAN
30, Malkhei Yisrael Street, Jerusalem, Israel 95501

ABSTRACT

A one kilometer long roadcut exposing Santonian chalk along the western monocline of the Fari'a anticlinorium, central Israel, was examined in the vicinity of a main normal fault in order to study possible effects of structural control over cementation. The fault is replaced in the chalk by a series of secondary faults. Where cut by secondary faults the country rock consists of soft friable chalk and the fault surfaces are associated with thick calcite veins. Where unaffected by that fault system the chalk is hard and shattered. Pure calcite from the veins was found to be characteristic of meteoric water ($\delta^{13}C = -5‰$ PDB; $\delta^{18}O = -9‰$ PDB; $Sr^{2+} = 300$). The soft chalk was found to be characteristic of sea water in respect of the above parameters, with relatively high insoluble residue, whereas the hard chalk was found to have intermediate values with relatively low insoluble residue.

The soft chalk underwent flushing and dissolution of calcium carbonate by meteoric water drained efficiently by the secondary faults. The dissolved calcium carbonate crystallized partially as thick veins and partially as sparry calcite cement in the matrix of the nonfaulted chalk thus hardening it. A further deformation of the area induced twinning of calcite along with brittle deformation in the veins and intense fracturing and some twinning of sparry calcite crystals in the hard chalk.

INTRODUCTION

Numerous examples of both modern and ancient freshwater calcite cements have previously been described from various environments by Longman (1980). However, the majority of these examples deal with shallow water limestones of diverse original mineralogy. Chalks are essentially pure nannomicrites rich in planktonic foraminifera, and often act as simple and stable chemical systems. Because chalks have uniform facies with wide distribution and normally are exposed to changes in water chemistry only during later diagenesis, accompanying broad regional uplift or deep burial, they show patterns of diagenesis which are far simpler and more regional than those of most other carbonate rocks. One of the dominating processes in chalk diagenesis is burial cementation which takes place by dissolution of certain calcite grains. The dissolved carbonate goes into solution and is reprecipitated locally in a suitable micro-environment, to form indurated hard chalk (Bathurst, 1975, chapter 8). Most of the literature concerns nontectonically induced diagenetic processes (e.g. Neugebauer, 1973; Matter, 1974; Scholle, 1974; Kennedy and Garrison, 1975).

Examples of cement precipitation controlled by tectonic factors are mostly of more local nature, and are more scarce in the literature. White (1921, p. 155) noticed such phenomena in the Isle of Wight and Arkell (1947, p. 195) suggested that the local faulting, staining and hardening of some of the Dorset chalk was caused by "tectonic movements". Mimran (1977) described such an example from Upper Cretaceous folded chalks in Southern England and Diemu-Tshiband (1980) in Miocene limestones of Southern France.

The measurements of minor elements (e.g. strontium and sodium) as well as that of oxygen and carbon isotopic values lead to a better quantitative understanding of the processes involved in diagenesis. Brand and Veizer (1980, 1981) present just two recent examples of geochemical and stable isotope studies of diagenetic processes and their environments.

The present study deals with the effect of Neogene normal faulting on cement precipitation in Santonian chalk. Large fractions of calcite cement have been detected by both petrographic and geochemical methods, and found to correlate with secondary permeability dominated by fracturing.

Geological Setting

The chalk studied herein belongs to the Santonian Menuha Formation. The chalk is soft and consists of almost pure calcium carbonate. It is a nannomicrite with mainly planktonic foraminifera, calcispheres, mollusc fragments, ostracoda and sponge spicules. The chalk is assumed to have been deposited in a lower shelf environment, at normal marine conditions (Flexer, 1971). The outcrops examined in the present study are located on the western flank of the Fari'a anticlinorium which developed intermittently since the early Senonian. The anticlinorium is cut by a series of transversal normal faults which form NW-WNW trending horsts and grabens. The age of faulting is Upper Neogene - Pleistocene as proven by the mainly clastic sediments filling the grabens. The faults reach their maximum throws along the axial zone of the anticlinorium, where two of them exceed 1000 m. The displacements are less towards both flanks, being

100–400 m in the Upper Cenomanian- Turonian hard dolomites and limestones. A detailed photogeologic and field examination of one of the above faults showed that when cutting through the Senonian chalk the fault reduces its displacement sharply from about 300 m to only 50 m (Fig. 1). This observation aroused the interest of the mechanism of absorption of faulting by soft chalk, and in the diagenetic processes associated with such absorption.

Methods of Study

Six outcrops of the Menuha chalk of approximately the same stratigraphic level were studied and sampled along the roadcut on the highway Nablus-Tubas (Fig. 1). Twenty specimens were taken from each outcrop and their compressional strength was determined in the field by a hydraulic point-load device. Four specimens from each outcrop were selected for the following further analyses:

(1) Density was determined by weighing the specimens in air and in water after drying: (2) Specimens of both bulk chalk and calcite veins were ground and dissolved in diluted hydrochloric acid for determining the insoluble residue in weight percentage and the concentration of strontium by atomic absorption; (3) Parts of the above ground specimens were reacted with 95% phosphoric acid at 25°C, and the CO_2 generated was analysed with an Atlas M86 mass spectrometer. The results are expressed in the conventional 'δ' per mil notation relative to the PDB standard; and (4) Scanning electron microscopy detailed the fabric of the chalks and calcite veins. The best results were achieved from polished specimens which were etched for 5 seconds in hydrochloric acid.

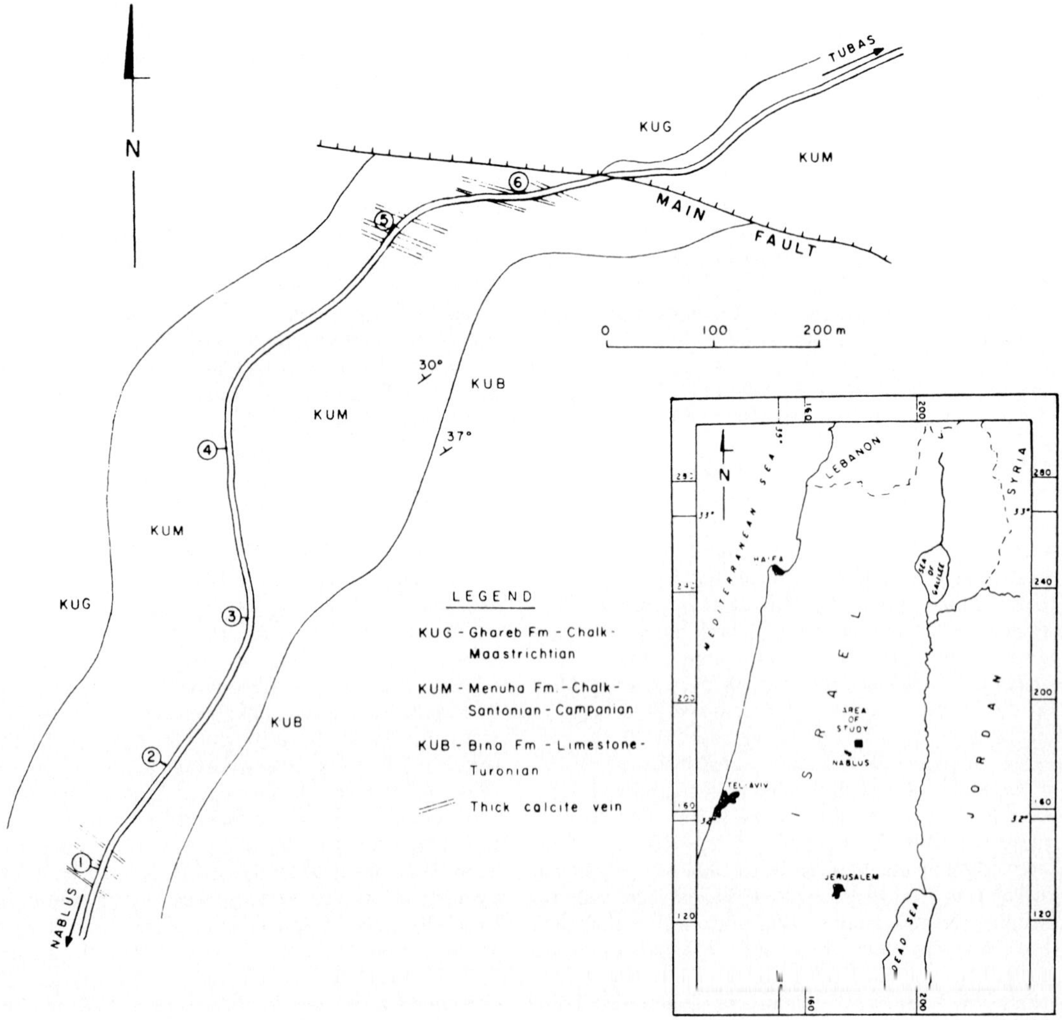

FIG. 1.—Location map and geologic map of the area of study. Outcrop areas are numbered.

RESULTS

Petrofabric

A brief examination of the chalk exposed along the roadcut (Fig. 1) revealed that two types of chalk can be distinguished, i.e. mechanically strong chalks, and soft weak chalks. This distinction was found later to have expression also in chemical, isotopic and petrographic data.

Outcrops 1, 5 and 6 (located on Fig. 1) consist of soft chalks crosscut by a system of veins associated with normal faults displacing up to 2 m and trending E-W to WNW. This trend is that of the main transversal fault which passes from the dolomites and limestones of Cenomanian-Turonian age into the soft Senonian chalks (Fig. 1). The veins in the above system are 2–20 cm thick and consist of coarse clear calcite which is occasionally mixed with chalk fragments (Figs. 2,3). The vein illustrated in Figure 3 also contain fragments of coarsely crystalline secondary calcite which were deformed during the shear along the fault. It is likely therefore that the calcite veins developed syntectonically.

The hard chalks exposed in outcrops 2, 3 and 4 (located on Fig. 1) are fractured very intensely (Fig. 4). A closer inspection reveals that 20–40 cm wide shear zones exist where the chalk is especially shattered (Fig. 5). Only scarce and very thin calcite veins crosscut the hard chalk.

The density of the soft chalk ranges usually from 1.50–1.80 g/cm^3 and that of the hard chalk from 1.90–2.30 g/cm^3. Mechanical point-load tests on more than a hundred irregular specimens in the field gave an average compressive strength of 8 kg/cm^2 for the soft chalk and of 33 kg/cm^2 for the hard chalk.

When surfaces of the different types of chalk are viewed with the scanning electron microscope, important details of the nature of the matrix and of the mutual relations between the matrix and the above physical properties come to light. The soft chalk consists of fairly well rounded calcite crystals (Fig. 6). Many coccoliths were found to consist of individual scales which underwent rounding and touch each other only slightly. The matrix of the hard chalk is extremely different.

FIG. 2.—WNW trending, thick calcite vein cutting through soft chalk. Outcrop 1.

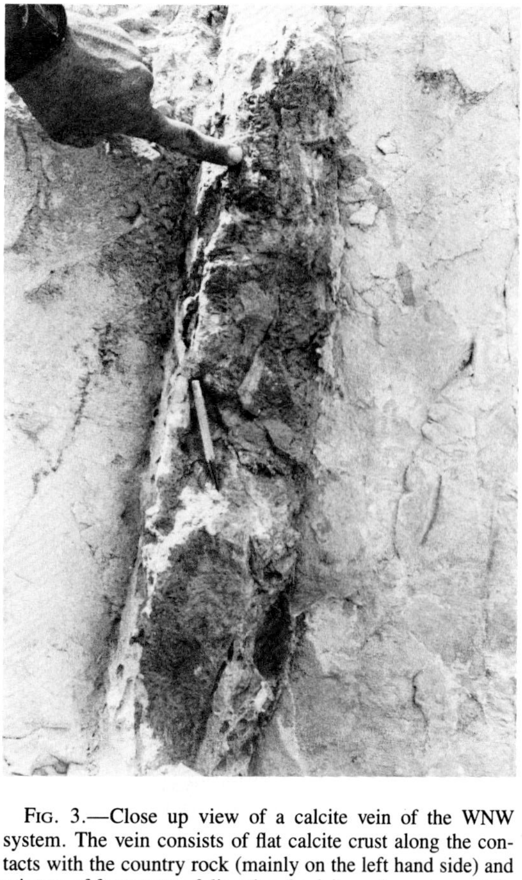

FIG. 3.—Close up view of a calcite vein of the WNW system. The vein consists of flat calcite crust along the contacts with the country rock (mainly on the left hand side) and mixture of fragments of disordered calcite and country rock in the middle. The country rock is hardly fractured. Outcrop 1.

Figures 8, 9 and 11 illustrate overgrown coccoliths in a very dense fabric. Occasionally coccoliths are well obscured and difficult to distinguish due to the intense overgrowth. The calcite crystals constituting the hard chalk are much more angular than those in the soft chalks. (Attention must be paid to the fact that the crystals tend to loose their high angularity during etching). Coarse calcite crystals were accommodated in veins in the soft chalk. As these veins started crystallizing syntectonically (as indicated above by field observations, Fig. 3), twinning was induced by further deformation (Fig. 7). Individual calcite crystals which appear frequently within the matrix of hard chalk have also been twinned. Figures 10 and 11 illustrate blocky subhedral calcite crystals grown both as cement and intrabiotic fills of calcispheres and coccoliths. Similar calcite crystals were found to fill foram tests in places.

Geochemistry

A lower shelf environment of deposition like that which prevailed during the accumulation of the Santonian chalk in the area of study is likely to produce fairly homogeneous rocks over short lateral distances. Hence significant variations in various geochemical parameters may serve as important clues to the establishment of a diagenetic model.

The average acid-insoluble residue content in the soft chalk is higher than that in the hard chalks (6% and 3–4% respectively, Fig. 12). If we attribute this difference to relative enrichment due to loss of calcium carbonate by the soft chalk, an order of 20–30% of such a loss must be considered (compare Mimran, 1977).

Strontium concentration is also significantly different in the soft and hard chalks. While ranging from 1100–1400 ppm with an average of 1200 ppm in the soft chalk, it averages as low as 900 ppm with a total range of 700–1000 ppm in the hard chalk (Fig. 12). The strontium content of the soft chalk is a little high compared with that reported by Schneidermann (1970) from the same stratigraphic unit (Menuha Fm.) elsewhere in Israel. Figure 14 illustrates that the increase in density correlates very well with the reduction in strontium concentration. The three lowest data points in Figure 14 indicate three analysed calcite veins. The significance of strontium concentration to diagenetic solutions of carbonate rocks can be estimated by as-

FIG. 4.—Fractured hard chalk with very scarce thin calcite veins. Outcrop 3.

FIG. 5.—Intensely fractured shear zone in hard chalk. The shear zone is 20–40 cm thick and runs through the middle portion of the photograph. Outcrop 5.

FIG. 6.—Characteristic loose fabric of soft chalk in SEM. Notice the roundness of crystals and the partly dissolved coccoliths. Outcrop 1, fracture surface (not etched).

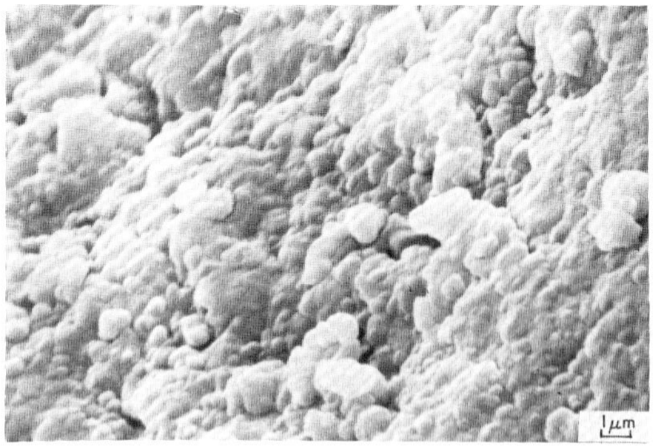

FIG. 9.—Close up SEM view of the matrix of the sample in Figure 8. Notice the particularly dense texture and the coccoliths which are partly "obscured" by overgrowths.

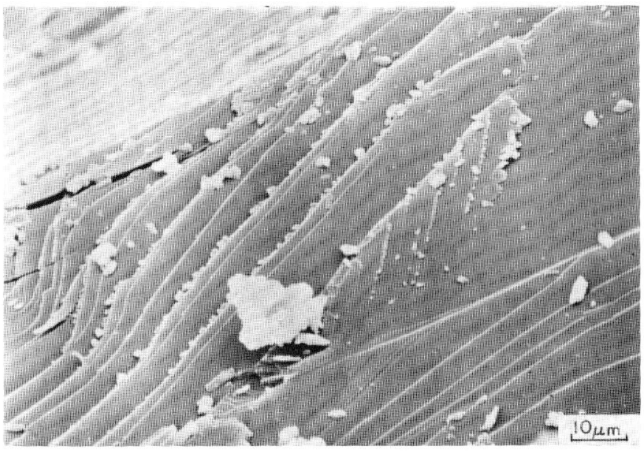

FIG. 7.—Twinned calcite crystal in calcite vein indicating either deformation following primary stage of veining or syntectonic calcite growth in veins. SEM of sample from outcrop 6, fracture surface.

FIG. 10.—Coarse sparry calcite cementing hard chalk. Notice the intrabiotic calcite filling calcispheres. SEM of sample from outcrop 3, fracture surface.

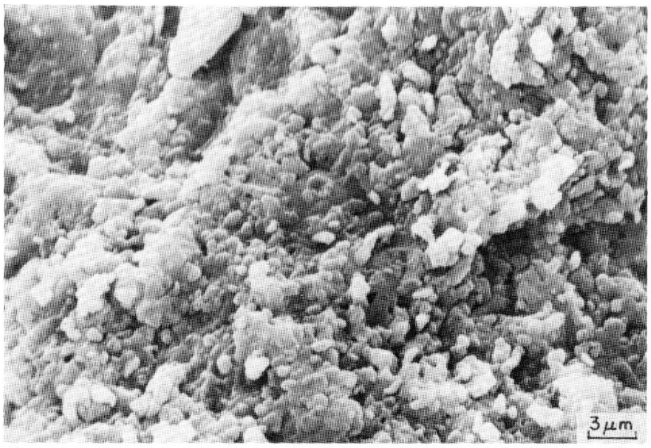

FIG. 8.—Dense fabric of hard chalk in SEM showing abundant overgrown coccoliths and matrix crystals. Outcrop 3, fracture surface.

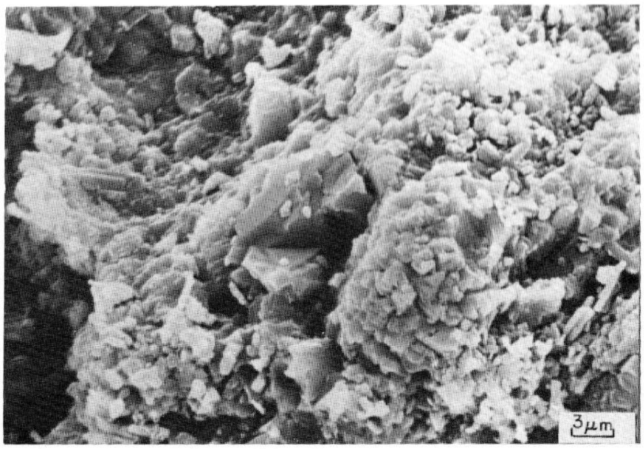

FIG. 11.—SEM of subhedral calcite both filling coccolith cavities and in the matrix of hard chalk. Outcrop 3, fracture surface.

suming a distribution coefficient of that element. Kinsman (1969) and Katz *et al.* (1972) calculated different distribution coefficients, however, whatever coefficient we adopt we may conclude that the strontium content in the calcite veins is indicative of meteoric water.

The carbon and oxygen stable isotope compositions of the chalks are illustrated in Figures 12 and 13. Al-

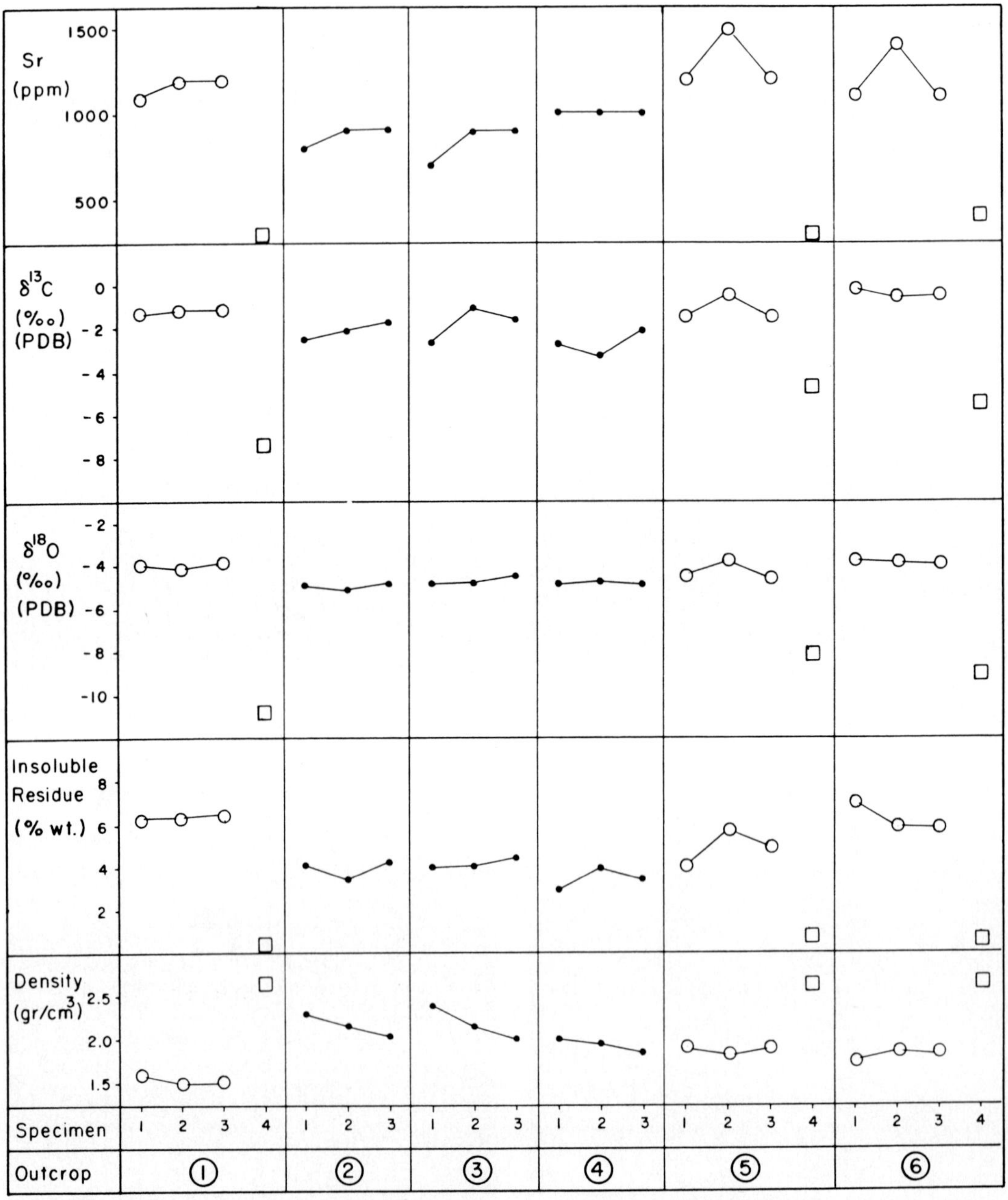

FIG. 12.—Density, acid insoluble residue, stable isotope composition and strontium contents in the analysed specimens. Solid circles designate bulk hard chalk specimens (outcrops 2–4), open circles—bulk soft chalk specimens (outcrops 1, 5, and 6), and open squares—calcite veins.

though the carbon isotope composition is more variable within outcrops, both parameters are significantly lower in the hard chalk than in the soft chalk as proven by a statistical t test. The test was run on the data of both stable isotopes and the results show that the difference in the isotope composition between the soft and hard chalk is highly significant ($\alpha < 0.005$). Needless to mention that the above isotopic compositions are indicative of the environment of crystallization of the

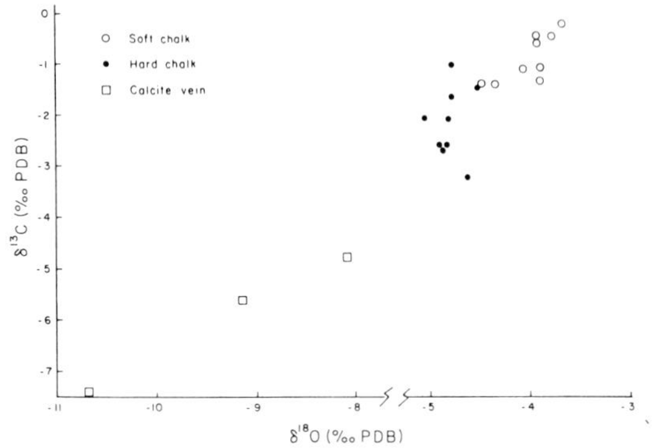

FIG. 13.—Carbon vs. oxygen stable isotope composition of all bulk chalks and calcite veins analysed.

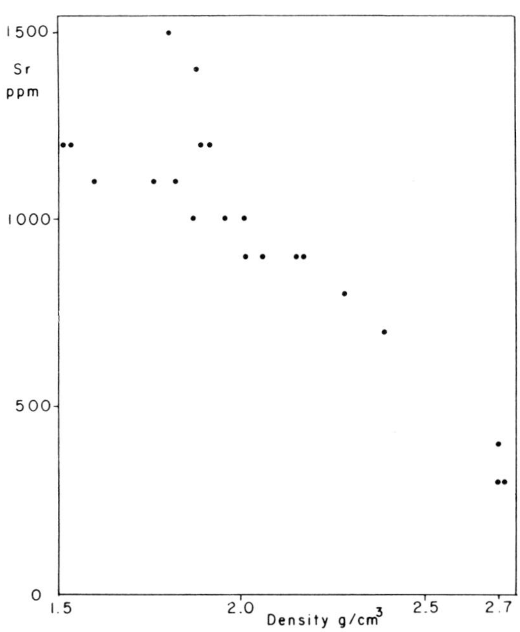

FIG. 14.—Strontium concentration vs. density in all analysed specimens illustrated in Figure 12. The three data points in the lower right represent the three calcite veins. The clear negative correlation indicates an increase in fraction of meteoric water calcite with the increase in density.

calcite constituting the rocks concerned. The isotopic composition of the three analysed calcite veins seems to be indicative of freshwater environment (e.g. Craig, 1965). It is possible to calculate the temperature of precipitation of the vein and cement calcite by assuming equilibrium between the solution and the precipitating calcite using the $\delta^{18}O$ values. The relation between the $\delta^{18}O$ of the calcite and the $\delta^{18}O$ in the water from which the calcite was precipitated, is given by Craig (1965):

$$T = 16.9 - 4.2\,(\Delta\delta) + 0.13\,(\Delta\delta)$$

$$\Delta\delta = \delta^{18}O \text{ (carbonate vs. PDB)} - \delta^{18}O \text{ (water vs. SMOW)}$$

Employing this equation and using a $\delta^{18}O$ of $-4‰$ vs. SMOW for modern fresh water in the discussed area a temperature of 41°C is obtained. This should be regarded as a minimum temperature since the calculation entailed the assumption that the diagenetic water was fresh. A normal geothermal gradient is not sufficient to account for such temperature, since an overburden of the order of 700 m is required. From what is known about the burial history of the area, the overburden could not have exceeded 300–400 m. Two other models can be suggested to account for the high temperature of the diagenetic environment:

1) Penetration of meteoric water from the elevated Eocene recharge area, through fractured Senonian chalk, into the deeper Cenomanian aquifer, thus pushing up deep ground water (heated due to normal geothermal gradient) through the major fault.
2) Heating of the host rock where precipitation of secondary calcite took place by concurrent magmatic intrusion.

On the basis of the presented geochemical data it is impossible to prefer either of the models. However it should be noted that Ilani and Mimran (1982) reported a geochemical anomaly (high concentrations of Cu, Mn, Ni and Co) on the same fault. The anomaly is located 9km to the southeast and is attributed to an effect of a magmatic intrusion.

DISCUSSION

The petrophysical and geochemical parameters serve as good indicators in the distinction between the hard and soft chalks. The purpose of this section is to integrate the various parameters into a reasonable model of diagenesis. The only assumption required for the construction of such a model is that the original Santonian chalk, which was deposited on a lower shelf environment in the area of study, had not differed significantly over short distances.

The NW-WNW normal transversal faults which cut the Fari'a anticlinorium (see above), were replaced by secondary normal faults of the same trend in the chalky Santonian Menuha Formation on the western flank. The main fault of interest continues into the chalky lithology but its displacement decreases sharply from an order of 300–400 m to about 50 m (Fig. 1). The calcite veins indicated on Figure 1 are in fact secondary normal faults. The association of deformed and non-deformed coarse calcite in the veins (Fig. 3) indicate either a syntectonic development of the vein or at least two phases of precipitation, one preceeding and one following the main shear along the vein surface. The existence of twinned sparry calcite cement crystals in the hard chalk, also confirms a further deformation following the initial crystallization. The soft chalk, which is cut by the secondary faults accommodating the veins, lacks such sparry cement. It is suggested, therefore, that the secondary faults played an important role in the chalk diagenesis. The surfaces served as highly permeable hydraulic conduits which enabled efficient circulation of groundwater through the rock mass.

The petrofabric properties of chalk thus enable an absorption of faulting by both external and internal mechanisms of deformation. The external deformation mechanism is the substitution of the major fault by minor secondary faults, and the internal mechanism is by dissolution and redistribution of calcite controlled by faulting, and further shearing within blocks between adjacent secondary faults.

The oxygen and carbon stable isotope composition, as well as the strontium concentration of the pure secondary calcite from the veins suggest a crystallization in warm meteoric water. The same three parameters in the bulk hard chalk specimens indicate an impregnation of the same kind of calcite into the chalk (Figs. 8–11). The fraction of the impregnated calcite correlates very well with both the increase in the bulk rock density and with the relative depletion in the total acid insoluble residue (Fig. 12, 14). The process of intense flushing of meteoric water through the soft chalk was responsible for an extensive dissolution and removal of calcium carbonate from the rock, thus leaving a loose fabric consisting of fairly well rounded calcite crystals (Fig. 6). The relative enrichment in insoluble residue (Fig. 12) indicates up to 30% loss of calcium carbonate in extreme cases. Part of the calcium carbonate removed was precipitated along the secondary fault surfaces whereas the rest was precipitated as sparry cement in the adjacent nonfaulted chalk. The impregnated chalk became harder and when further shear occurred, it took place by brittle deformation, and hence intense fracturing developed (Figs 4, 5), associated with minor twinning in the sparry calcite cement.

SUMMARY AND CONCLUSIONS

The effect of faulting on chalk diagenesis has been described in an area in central Israel. A WNW trending main fault is replaced by a series of secondary faults associated with thick calcite veins. The vein material is characteristic of meteoric water calcite. Two types of chalk are distinguished within a distance of a few hundred meters from the main fault: (a) soft, porous chalk relatively rich in insoluble residue and with carbon and oxygen isotope composition and strontium concentration characteristic of sea water, and (b) hard dense chalk relatively depleted in insoluble residue and with carbon and oxygen isotope composition and strontium concentration approaching those of meteoric water calcite. The thick calcite veins are restricted to the soft chalk.

It is concluded that the secondary faults replacing the main fault account for the variability in chalk diagenesis over short distances. Meteoric water flushed through the faulted chalk and dissolved calcium carbonate from the matrix on their way. The dissolved material precipitated either in veins within the fault surfaces or as sparry calcite within the fabric of the adjacent nonfaulted chalk thus hardening it. Further displacement along the secondary faults induced both plastic (calcite twinning) and brittle deformation along the veins in the soft chalk in association with calcite twinning in the sparry calcite and intense fracturing in the hard chalk.

ACKNOWLEDGMENTS

Thanks are due to N. Schneidermann, P.M. Harris and W.D. Wiggins for the review of the manuscript. Thanks are also due to M. Magaritz for the stimulating discussions and for the stable isotope analyses. N. Dalman is acknowledged for the chemical analyses. Additionally, my thanks go to L. Michaelesco, B. Lang, and M. Arnon for their assistance in the field work and in the mechanic analyses. Special thanks go also to A. Peer for the drafting work, to M. Dvorchak and Y. Levi for the operation of the SEM and the photographic work, and to R. Gordon and V. Arie for typing the manuscript.

REFERENCES

ARKELL, W.J., 1947, The geology of the country around Weymouth, Swanage, Corfe and Lulworth: Mem. Geol. Surv. U.K. 341, 342, 343.

BATHURST, R.G.C., 1975, Carbonate Sediments and Their Diagenesis (2nd ed.): Elsevier, Amsterdam, 658 p.

BRAND, U., AND VEIZER, J., 1980, Chemical diagenesis of a multicomponent carbonate system—1: Trace elements: Jour. Sed. Petrology, v. 50, p. 1219–1236.

______, AND ______, 1981, Chemical diagenesis of a multicomponent carbonate system—2: Stable isotopes: Jour. Sed. Petrology, v. 51, p. 987–997.

Craig, H., 1965, The measurement of oxygen isotope paleotemperatures, *in* T.E. Consiglio (ed.), Stable Isotopes in Oceanographic Studies and Paleotemperatures: Pisa, Italy, Nazionale Richerche, p. 3–24.

Diemu-Tshiband, K., 1980, Relationships between tectonic stress, pressure solution and calcite cementation in the Miocene limestones of southern France: Internat. Assoc. Sedimentologists 1st Europ. Mtg., Bochum (abstract).

Flexer, A., 1971, Late Cretaceous paleogeography of northern Israel and its significance for the Levant geology: Paleogeography, Palaeoclimatol., Palaeoecol., v. 10, p. 293–316.

Ilani, S., and Mimran, Y., 1982, E-W trending mega-lineaments crossing Samaria: Israel Geol. Surv. Current Res. pp. 43–45.

Katz, A., Sass, E., Starinsky, A., and Holland, H.D., 1972, Strontium behaviour in the aragonite - calcite transformation: an experimental study at 40-98°C: Geochim. Cosmochim. Acta, v. 36, p. 481–496.

Kennedy, W.J., and Garrison, R.E., 1975, Morphology and genesis of nodular chalks and hardgrounds in the Upper Cretaceous of Southern England: Sedimentology, v. 22, p. 311–386.

Kinsman, D.J.J., 1969, Interpretation of Sr^{+2} concentrations in carbonate minerals and rocks: Jour. Sed. Petrology, v. 39, p. 486–508.

Longman, M.W., 1980, Carbonate diagenetic textures from nearsurface diagenetic environments: Amer. Assoc. Petroleum Geologists Bull. v. 64, p. 461–487.

Matter, A., 1974, Burial diagenesis of pelitic and carbonate deep sea sediments from the Arabian Sea, *in* Initial Reports of the Deep Sea Drilling Project, V XXIII (Whitmarsh, R.B., Weser, D.E., Ross, D.A. *et al.*), Washington, U.S. Government Printing Office, p. 421–469.

Mimran, Y., 1977, Chalk deformation and large scale migration of calcium carbonate: Sedimentology, v. 24, p. 333–360.

Neugebauer, J., 1973, The diagenetic problem of chalk—The role of pressure solution and pore fluid: Neues Jb. Geol. Paläont. Abh. v. 143, p. 223–245.

Schneidermann, N., 1970, Genesis of some Cretaceous carbonates in Israel: Israel Jour. Earth Sci. v. 19, p. 97–115.

Scholle, P.A., 1974, Diagenesis of Upper Cretaceous chalks from England, Northern Ireland and the North Sea, *in* K.J. Hsü and H.C. Jenkyns (eds.), Pelagic Sediments on Land and Under the Sea: Int. Assoc. Sedimentologists Spec. Pub., v. 1, p. 177–210.

White, H.J.O., 1921, A short account of the geology of the Isle of Wight: Mem. Geol. Surv. U.K. 330, 331, 334 & 335.